ANNUAL REVIEW OF ENTOMOLOGY

ANNUAL REVIEW OF ENTOMOLOGY

RAY F. SMITH, *Editor*

University of California

THOMAS E. MITTLER, *Editor*

University of California

CARROLL N. SMITH, *Editor*

U.S. Department of Agriculture, Ret.

Volume 20

1975

ANNUAL REVIEWS INC. 4139 EL CAMINO WAY PALO ALTO, CALIFORNIA 94306

ANNUAL REVIEWS INC.
Palo Alto, California, USA

International Standard Book Number: 0-8243-0120-X
Library of Congress Catalog Card Number: A56-5750

REPRINTS

The conspicuous number aligned in the margin with the title of each article in this volume is a key for use in ordering reprints. Available reprints are priced at the uniform rate of $1 each postpaid. Effective January 1, 1975, the minimum acceptable reprint order is 10 reprints and/or $10, prepaid. A quantity discount is available.

PRINTED AND BOUND IN THE UNITED STATES OF AMERICA

PREFACE

After two decades of uninterrupted publication of Annual Review of Entomology, it would seem appropriate to review the charge of the original Editorial Committee. In the preface of Volume 1 the objectives were defined thus: "To publish authoritative and concise treatments of definitive subjects of current interest. . . . These reviews will be solicited judiciously from the leaders in the fields concerned. It is our hope that each review will present a critical analysis of recent literature and, insofar as is feasible, an appraisal of the present status of the subject."

During this period, 440 internationally recognized authorities have consistently met these objectives by providing us with 367 articles reviewing the latest developments in entomology. In addition, 25 authors joined in writing a comprehensive volume on the history of entomology, which was published in 1973. The members of the Editorial Committee wish to express their sincere appreciation to all these contributors. Moreover, the unique partnership that was forged between the Entomological Society of America and Annual Reviews Inc. has continued to function and prosper. The financial status of the Review is sound, thanks in large measure to the support which individual entomologists and libraries have given the publication through their continuing subscriptions.

The emergence of Annual Review of Entomology coincided with an especially dynamic period in the history of entomology. For example, the development of the concept of integrated pest management in the broadest context has provided many new opportunities and challenges for the professional entomologist. A review of the tables of contents of the past 20 volumes clearly indicates that authors have identified and critically analyzed the significant developments associated with this exciting scientific discipline.

Finally, the high technical quality of the Review maintained over the past two decades owes much to the painstaking efforts of our Assistant Editors and to the meticulous care and skill of the compositors and printers.

THE EDITORIAL COMMITTEE

CONTENTS

Biology and Control of Imported Fire Ants, *C. S. Lofgren, W. A. Banks, and B. M. Glancey*	1
Biological Control of Aquatic Weeds, *Lloyd A. Andres and Fred D. Bennett*	31
The Biology and Ecology of Armored Scales, *John W. Beardsley Jr. and Roberto H. Gonzalez*	47
Physiology of Tree Resistance to Insects, *James W. Hanover*	75
The Status of Viruses Pathogenic for Insects and Mites, *W. A. L. David*	97
Federal and State Pesticide Regulations and Legislation, *Errett Deck*	119
Neurosecretion and the Control of Visceral Organs in Insects, *Thomas A. Miller*	133
Neuromuscular Pharmacology of Insects, *T. J. McDonald*	151
Inherited Sterility in Lepidoptera, *David T. North*	167
Evolution and Classification of the Lepidoptera, *I. F. B. Common*	183
Recent Developments in Insect Steroid Metabolism, *J. A. Svoboda, J. N. Kaplanis, W. E. Robbins, and M. J. Thompson*	205
Recent Research Advances on the European Corn Borer in North America, *T. A. Brindley, A. N. Sparks, W. B. Showers, and W. D. Guthrie*	221
Recent Advances in Our Knowledge of the Order Siphonaptera, *Miriam Rothschild*	241
Adaptations of Arthropoda to Arid Environments, *J. L. Cloudsley-Thompson*	261
Responses of Arthropod Natural Enemies to Insecticides, *B. A. Croft and A. W. A. Brown*	285
Plant Resistance to Insects Attacking Cereals, *R. L. Gallun, K. J. Starks, and W. D. Guthrie*	337
Brain Structure and Behavior in Insects, *P. E. Howse*	359
Structure of Cuticular Mechanoreceptors of Arthropods, *Susan B. McIver*	381
The Brazilian Bee Problem, *Charles D. Michener*	399
Insect Growth Regulators with Juvenile Hormone Activity, *G. B. Staal*	417
Genetical Methods of Pest Control, *M. J. Whitten and G. G. Foster*	461
Author Index	477
Subject Index	493
Cumulative Index of Contributing Authors, Volumes 11 to 20	507
Cumulative Index of Chapter Titles, Volumes 11 to 20	509

ANNUAL REVIEWS INC. is a nonprofit corporation established to promote the advancement of the sciences. Beginning in 1932 with the *Annual Review of Biochemistry,* the Company has pursued as its principal function the publication of high quality, reasonably priced Annual Review volumes. The volumes are organized by Editors and Editorial Committees who invite qualified authors to contribute critical articles reviewing significant developments within each major discipline.

Annual Reviews Inc. is administered by a Board of Directors whose members serve without compensation.

Annual Reviews are published in the following sciences: Anthropology, Astronomy and Astrophysics, Biochemistry, Biophysics and Bioengineering, Earth and Planetary Sciences, Ecology and Systematics, Entomology, Fluid Mechanics, Genetics, Materials Science, Medicine, Microbiology, Nuclear Science, Pharmacology, Physical Chemistry, Physiology, Phytopathology, Plant Physiology, Psychology, and Sociology (to begin publication in 1975). In addition, two special volumes have been published by Annual Reviews Inc.: *History of Entomology* (1973) and *The Excitement and Fascination of Science* (1965).

BIOLOGY AND CONTROL OF IMPORTED FIRE ANTS[1]

❖6078

C. S. Lofgren, W. A. Banks, and B. M. Glancey
Insects Affecting Man Research Laboratory, Agricultural Research Service, U.S. Department of Agriculture, Gainesville, Florida 32604 and Imported Fire Ant Research Laboratory, Agricultural Research Service, U.S. Department of Agriculture, Gulfport, Mississippi 39501

Two species of imported fire ants, *Solenopsis invicta* and *Solenopsis richteri,* were introduced into the United States at Mobile, Alabama, about 35 and 56 years ago, respectively. *S. richteri* is now found in a relatively small area in northeastern Mississippi and northwestern Alabama, while *S. invicta* has spread and occupied a much greater area, since it is found in nine states from the Carolinas to Texas. The warm, wet weather of the South is ideal for the imported fire ants, and their colonies have flourished in the prime grazing and crop land, along roadsides, and in parks and lawns. The most extensive and rapid spread of the imported fire ants occurred during the 1940s and 1950s. Eventually, the mound-building and stinging habits of the ants caused farmers in the infested areas (mounds sometimes numbered 125 to 150/ha) to demand relief. Thus, in late 1957, the U.S. Congress authorized a cooperative federal-state eradication or control program. Since that time, a continuing, but at times limited, program of research has investigated the biology, zoogeography, and taxonomy of the fire ants in the United States and in their homeland in South America. Also, considerable work has been devoted to development and improvement of methods of chemical control. More recently, studies have begun on biological control. Taxonomy of the ants has become more critical with the initiation of this work since many biological control agents are species specific. Our purpose here is to review this research, although because of space limitations we cannot fully review much of the recent work concerning the chemical toxicology and persistence of mirex, the chemical currently used for control of imported fire ants.

[1]Mention of a pesticide or commercial or proprietary product in this paper does not constitute a recommendation or an endorsement of this product by the USDA.

TAXONOMY OF FIRE ANTS

The taxonomy of the many species of fire ants on the South American continent remains largely unresolved. Wilson (117) recognized quite early that there might be two forms of the imported fire ant in the United States, but the characterizations were made on color only and no taxonomic separations were proposed. Since he recognized that the red form (or light phase) ants were more vigorous and were spreading more rapidly than the original black phase, he postulated that the red form was an offshoot that had arisen from the original population after its invasion. A year later, Wilson (118) changed his opinion and postulated that the light phase ants arose as the result of hybridization between forms that he treated as *Solenopsis saevissima richteri* and *Solenopsis saevissima saevissima* in a postulated blend zone in South America, and that there had been a second importation of one of these hybrid forms. The double invasion hypothesis was largely ignored by subsequent authors until Buren (25) in his taxonomic revision separated six species of the *S. saevissima* complex, including the two species that have invaded the United States. The first of these two species, the species that was detected in Alabama by Creighton (33), was restricted and separated as *S. richteri* Forel; it is now known as the black imported fire ant. The second and more important of the two importations was named and described as S. *invicta* Buren and given the common name, red imported fire ant. In this taxonomic revision, Buren demonstrated that (*a*) there are two species of imported fire ants in the United States, and that they may be distinguished by the morphology of the head, thorax, and postpetiole, as well as by color; (*b*) that the two species have remained phenotypically constant since the time of importation and moreover are unchanged from parental populations in South America; (*c*) that hybridization appears rare; and (*d*) that the two homeland ranges in South America are geographically distinct.

Cupp et al (36) have suggested that species discreteness in *Solenopsis* is not complete. They worked with purported field-collected hybrids of *S. richteri* and *S. invicta.* However, their distinctions were based only on color and no morphological characters were given. Nevertheless, these investigators were able to force-mate the two species in the laboratory and obtain known hybrids. Their studies need to be continued and broadened to include very careful morphological examination and comparison of the laboratory-produced hybrids with specimens from field populations to determine the extent of natural hybridization. However, determinations of hybridization by color appear questionable. Buren, in published (25) and unpublished work, has found that the color of *S. invicta* is variable, and ranges from light reddish brown to strongly dark brown; also the light-colored spot on the gaster may be present or absent, even in nestmates. In the absence of correlative morphological characters, the color differences appear to be normal intraspecific variation or, at most, localized color strains. If extensive hybridization were occurring, species discreteness would eventually break down, and there is no evidence that this happened to *S. richteri* or *S. invicta* in North America. Indeed, specimens of *S. richteri* (ca 150 nest collections examined by Buren) taken in northeastern Mississippi and northwestern Alabama from 1968 to 1972 appear morphologically identical to specimens of *S. richteri* taken at Mobile in 1927–1928 by W. S. Creighton.

The present authors feel that hybridization is not occurring to any noteworthy extent in *Solenopsis* and that the limited amount that may be occurring is having no measurable effect on the phenotypes of the species. Bigelow (14) and Wing (125) give possible reasons why a limited amount of hybridization may not measurably affect a population gene pool even when the hybrids are fully fertile.

ZOOGEOGRAPHY

Distribution in South America

The postulated distribution of *S. invicta* and *S. richteri* in South America (26) is shown in Figure 1. Further surveys will be necessary to fully delimit the actual distribution of the two species. The work to date indicates that the distribution of *S. richteri* is much more restricted than Wilson (118) believed. The homeland range of *S. richteri* was shown to be in southernmost Brazil (Rio Grande do Sul), Uruguay, and Argentina. The state of Mato Grosso in Brazil, and specifically the Pantanal (the large flood plain of the head waters of the Paraguay River and its fringes), was shown to be the homeland of *S. invicta* (5, 26). Buren et al (26) also showed that the species extended northward along the Guapore River to Porto Velho, Rondonia, and to the south along the Paraguay River at least as far as the province of Chaco, Argentina. The regions near the Paraguay River in Paraguay and those in easternmost Bolivia into which the Pantanal extends have not yet been sampled, but they are probably a part of the range of *S. invicta.* The species has not been taken, however, in the Campo Cerrado areas to the east of the Pantanal or in any part of the state of Sao Paulo, or in any of the upland areas of Bolivia, though other species of *Solenopsis* are indigenous. Buren et al (26) speculate as to why the range of *S. invicta* in South America is as circumscribed as it appears to be.

Importation and Spread in the United States

Since there are two species of imported fire ants rather than one, it follows that there were two importations rather than one. There are many references to the supposed 1918 date of introduction, giving a period of 56 years for "the imported fire ant" to have reached its present distribution. Now the thinking on this subject must obviously be changed. Only *S. richteri* was imported as early as 1918, and this species has been only moderately successful in its spread. *S. invicta,* however, may have been in the United States only 30 to 35 years. It has achieved its phenomenal spread because of dispersal in mating flights and because of transport by man (34, 75), particularly on nursery stock in the 1940s and early 1950s. The matter has been further confounded by the fact that importation of both *S. richteri* and *S. invicta* apparently occurred in the Mobile, Alabama area. The reason why will probably never be fully resolved. Buren et al (26) suggest that *S. invicta* was successful because *S. richteri* had preconditioned the Mobile area. Perhaps *S. richteri* exerted considerable competitive pressure against the Argentine ant, *Iridomyrmex humilis,* and against native ants that were present without fully occupying the available

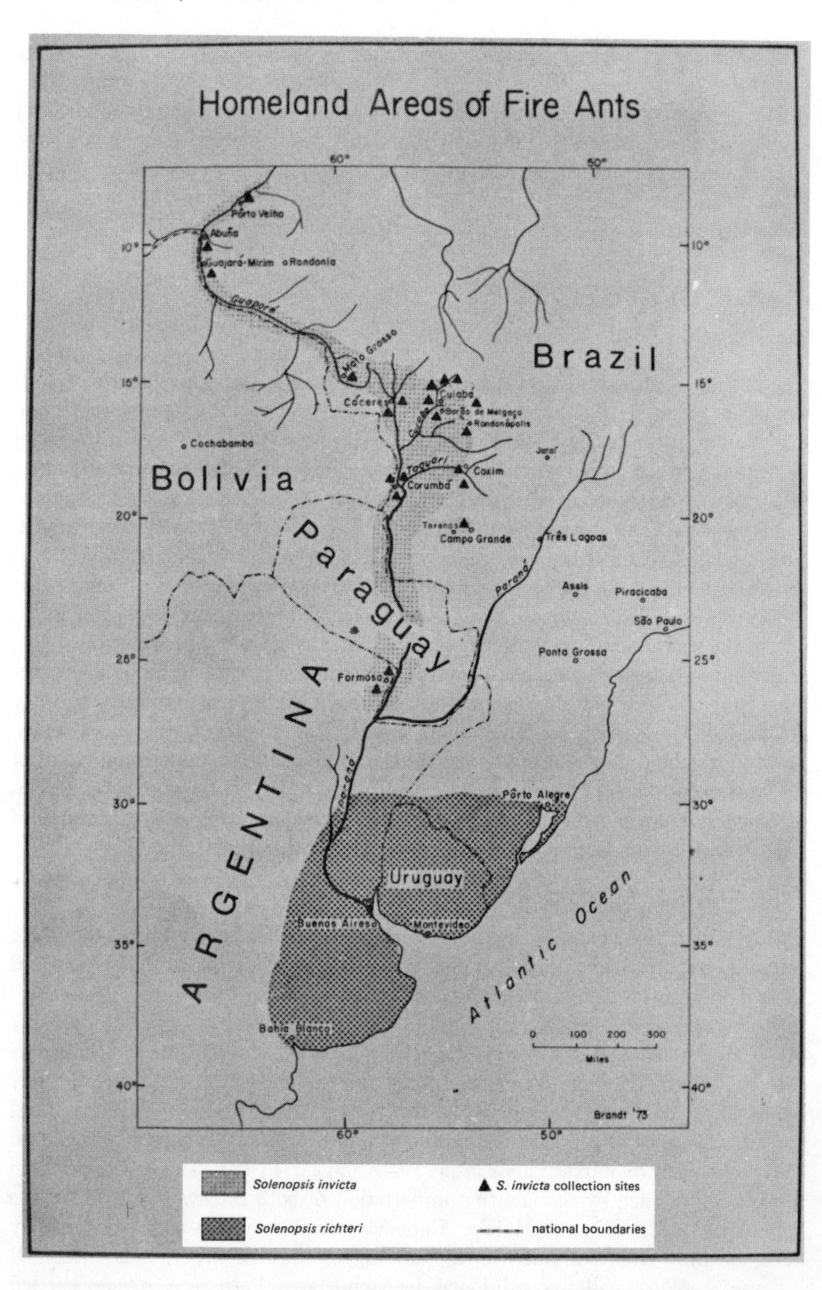

Figure 1 Distribution of *Solenopsis invicta* and *S. richteri* in their homeland in South America [used by permission from Buren et al 1974 (26)].

ecological niche. Massive populations of *I. humilis* existed at New Orleans and in other parts of the Southeast between 1933 and 1945 and may have prevented *S. invicta* from gaining an initial foothold at New Orleans.

Lennartz (57) established, by personal communication with W. S. Creighton and from other data, that *S. invicta* was probably imported into the Mobile area between 1933 and 1945. She was not able to definitely associate the importation with any specific cargo. The Brazilian coffee that was imported in the 1930s and 1940s came principally from the states of Sao Paulo and Paranà where *S. invicta* has never been found. The few recorded shipments of mahogany to Mobile during the 1933–1945 period were entirely from Central America. No records show that potted plants from Brazil entered during the critical period, and other imports, rubber, Brazil nuts, manganese ore, Quebracho wood, and skins and hides, were not likely transport media for various reasons. Shipping records show that while much grain was shipped from Argentina during the drought years in the United States (the late 1930s), the amounts shipped to Galveston and New Orleans greatly exceeded the amount shipped to Mobile. It seems likely that the exact mode of entry of the ants into the United States will never be determined.

The ants have been spectacularly successful in establishing themselves in the southeastern United States (*S. invicta* more so than *S. richteri*) since they now occupy more than 52 million ha of land (Figure 2). With the exception of some resistance which the native fire ant, *Solenopsis geminata,* appears to be exerting in parts of Florida, the ultimate distribution of the imported fire ants in North America may be dependent on abiotic factors (26). Winter severity may limit the northward progression since *S. invicta* lacks the ability to hibernate and winter kill probably is roughly proportional to the depth at which the soil becomes frozen. The deserts of western Texas should halt the natural spread of the species to the west. If inadvertently transported by man's agency, we feel that the species could establish in California and be successful in watered lawns or irrigated areas of the Southwest. *Solenopsis xyloni* and *I. humilis* are successful in these areas and could be displaced by *S. invicta.*

BIOLOGY

Colony Founding, Growth, and Seasonal Life Cycle

COLONY FOUNDING The founding of colonies by imported fire ants is typical of most nonparasitic myrmicine ants and begins when new queens alight on the ground after their nuptial flight. Typically, a colony is started by a single queen; however, in heavily infested areas more than a dozen dealated queens often group together under debris such as paper, cans, and pieces of wood. Markin et al (72) reported that more than half of the new nests of *S. invicta* in a field near Gulfport, Mississippi, were each started by one queen, but the remaining nests contained 2–5 queens. The queen usually breaks off her wings and begins excavation of her burrow within 4 hr after the mating flight. The burrow may be made in open areas or under solid objects. Within 6–7 hr, 90% of the queens may have completed their burrows and

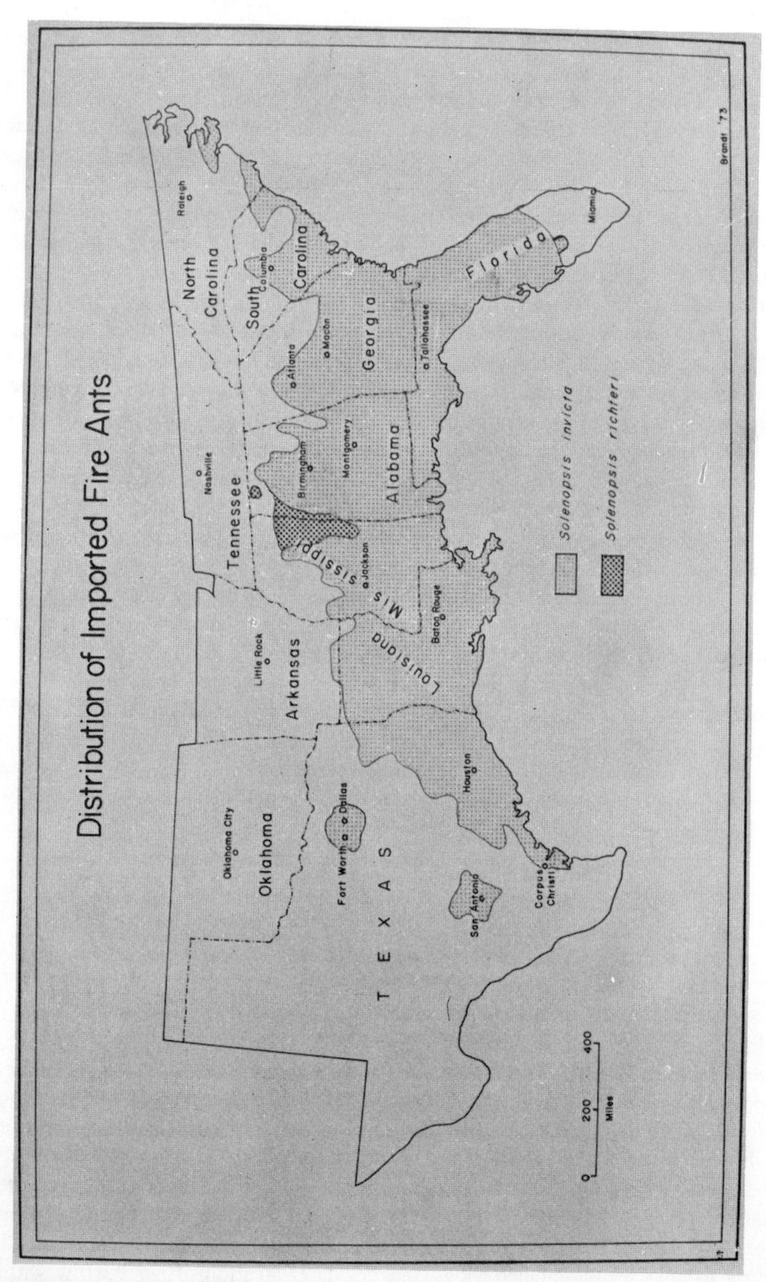

Figure 2 Distribution of *Solenopsis invicta* and *S. richteri* in the United States [used by permission from Buren et al 1974 (26)].

plugged the entrances. The burrow of *S. invicta* consists of a vertical tunnel 3–12 cm deep with a small cell located about 1 cm from the upper end and one, or rarely several, small cells at the bottom (72). The cells are ca 7–10 mm wide and 3–12 mm deep. Green (48) describes the burrow of *S. richteri* as running vertically for about 12 mm and then turning parallel to the soil surface for about 5 cm; a cell is located at the end. Both species of fire ants normally lay their first eggs within 24–48 hr after completing the burrow.

In laboratory studies, *S. invicta* queens laid 15 to 20 eggs within 2–3 days and had produced a total of 20–125 eggs by the time the first larvae emerged (72, 103). The queen constantly attends the eggs and thus probably prevents their decimation by fungi or bacteria (86). Glancey et al (42) reported that 26–58% of the first eggs are trophic eggs which serve as food for the first- and second-stage larvae. When the fertile eggs hatch (within 4 to 7 days), the larvae rotate their heads until they locate a trophic egg which they begin to eat. However, O'Neal & Markin (86) claim that both trophic and viable eggs are eaten indiscriminately by the larvae and that queens were observed feeding both types of eggs to second-instar larvae. Such a habit by the queen would seem to be self-defeating since it would decrease the number of minim workers that could care for her. Further observations are needed to clarify this point. Late second-, third-, and fourth-instar larvae receive only regurgitated food which is derived from the dissolution of the wing muscles of the queen (86). Total development time for the 4 larval instars ranged from 6 to 10 days at optimum temperatures in the laboratory (29.5 to 32°C); pupal development required 7–8 days. The entire developmental time from egg deposition to emergence of the minim worker at these temperatures was 20–21 days. Field tests conducted from May through July showed that 22–33 days were required for minim workers to develop when the queens were held in vials or cages buried in the soil. Observations on the naturally occurring nests indicated that 24–30 days were needed to complete development (72).

The success of the colony founding process is contingent on many factors which are ill-defined but they must include the physical properties of the soil, the climate, the vegetation, the availability of food, inter- and intraspecific competition for food and living space, and the presence of pathogens, parasites, and predators. Green (47) indicated that soil moisture (from frequent rain showers) is an extremely important factor in colony founding by *S. richteri.* His subsequent discovery (48) that the burrow of *S. richteri* is in the top 2.5 cm of the soil would make soil moisture a more critical factor for this species than for *S. invicta* which builds a deeper burrow. Queens from nuptial flights that occur between April and August have the best chance for successful colony establishment because climatic conditions are most favorable and soil temperatures are optimum (25–30°C) during this period. In addition, there is usually sufficient time for the queen to produce several thousand workers that can excavate a nest of sufficient depth to provide protection from winter temperatures (66, 74).

Hundreds of small colonies (in some cases over 2500/ha) are commonly found in newly infested land (without mature colonies) in the fall (48, 74), but the majority do not survive the winter. Therefore, eventually, the number stabilizes at some level that is probably governed by the availability of food and the presence of suitable nest sites. In many instances workers from one small colony may abandon their queen and amalgamate with another colony (74, 90). This probably accounts for some of the decline in the number of colonies. Amalgamation might also result in very rapid development of the colony due to the increased complement of workers. In the laboratory we have noted substantial differences in the rate of development of colonies established by different queens despite similar food availability and identical environmental conditions. If the same differences occur in the field, then slowly developing colonies obviously could not compete for food and territory, and they would be absorbed by other colonies or die.

The effect of predators on colony founding is discussed in the section on biological control of imported fire ants.

COLONY GROWTH AND SEASONAL LIFE CYCLE The growth of the colonies and the seasonal life cycle of the red imported fire ant were discussed in detail by Markin & Dillier (73) and Markin et al (74). They reported that the first workers (10–20 minims) appeared about 30 days after the queen laid her first eggs. By 60 days, the number of workers (minims and a few minor workers) increased to about 85 and in 90 days to 225. At 5 months, colonies averaged over 1,000 minor workers and a few major workers (minims had disappeared); after 7 months, there were 6,500 to 14,000 workers with about 3% major workers; after 1 year colonies averaged 11,000 workers, after 1½ years, 30,000, and after 2½ to 3 years contained from 52 to 69,000 workers. The colonies were generally considered mature after 3 years when some contained as many as 230,000 workers. (The number of workers generally declined during the winter since brood production essentially ceased.)

The estimates of populations in these cited studies (73, 74) were made by driving a large cylinder into the soil around the mound and slowly flooding the mound to force the ants out onto the surface of the water. The ants were then taken to the laboratory where the numbers were determined. The investigators (74) estimated, but did not offer supporting evidence, that they collected 90% of the ants within a colony. Observations in the laboratory suggest that much more than 10% of the ants are away from the nest foraging at any given time, particularly when the weather is favorable. Thus, we believe they have grossly underestimated worker numbers, especially in the older colonies.

Although mature colonies have a seasonal pattern of brood production, considerable variation occurs between colonies in the southern and northern areas of the range of *S. invicta.* For example, brood production essentially ceases in the winter in areas north of 30° latitude, but in central and southern Florida, brood has been noted in some mounds throughout the year. In general, the seasonal reproductive cycle begins in March. The first eggs that are laid give rise to both worker and sexual castes even though the sexuals may appear to predominate because of their size. Even at the time of maximum abundance of sexual larvae and pupae in May, they comprise only about 8.7% of the total colony biomass as compared with about 2.5% of the biomass in September. The highest percentage of adult sexuals (9.6% of the biomass) is found in June and there is a steady decline subsequently, except that Green (46) reported *S. richteri* to have a second period of high sexual brood production in the fall.

The data of Markin & Dillier (73) show that the biomass of females is greatest in August and September rather than in June when the biomass of males is the greatest. They estimated that over an entire year 76.3% of the colony is comprised of worker forms. In the spring the numbers of worker larvae and pupae peak in May (31% of biomass) and the percentage then declines during the summer to about 20%. In September, it increases again to 30% and finally reaches a maximum peak for the year (34% of the biomass) in October. Thereafter the worker brood drops sharply: the lowest numbers occur during January, February, and March.

Markin et al (74) found that the first reproductive forms (males) appeared in the mounds at Gulfport, Mississippi at 7 months. Rhoades & Davis (90) found reproductives in 5–6-month-old colonies in north Florida (no indication was given as to the sex). We have found that field colonies in Florida contain reproductives of both sexes when the colony is 5 months old, though this is not the norm. Markin et al (74) have further indicated that 1 to 1½-yr-old colonies produced very few sexual forms; however, our studies in both Mississippi and Florida show that colonies of this age have definitely produced sexuals that participated in mating flights. (In one field near Tampa, Florida, 46% of the colonies contained sexuals when they were 1 yr old.) Also, in the laboratory our studies and those of Wilson (120) showed that 1-yr-old colonies produced numerous sexuals.

Markin et al (74) reported that when the nest is opened by the first minim workers, it consists of nothing more than the original burrow dug by the queen. The workers immediately begin deepening the tunnel: By 60 days, several distinct chambers are apparent. By 90 days, a soil surface mound (3–7 cm diam and 5–7 cm high) containing numerous chambers is evident, and horizontal tunnels radiate out from the mound. From 5 to 15 additional vertical tunnels branch at various depths from the main vertical tunnel. By 5 months the surface mound is still larger and has taken on the spongelike appearance of older mounds. After 2½ to 3 years, the mound has taken on a typical dome-like shape and the galleries and tumulus occupy a volume of up to 40 liters. The vertical tunnels extend a meter or more into the ground to the water table.

The size of the aboveground mounds may vary greatly with different soil types, soil moisture, and vegetation. Mounds in sandy areas tend to be flat and rather broad, while mounds in clay soils may be ½ to 1 m high and 1 m wide at the base. Also, recent evidence indicates that large colonies may utilize several mounds. Some colonies maintain the same mound for many years, and others may move their nest frequently for distances of from a meter to over 30 m. The mounds allow the colonies to maintain brood at the optimum humidity and temperature since it may be moved up or down as the need arises. In hot summer weather, brood is usually found deep within the underground tunnels while in the spring or fall, brood is kept on the warm sunny side of the mound near the surface.

Reproduction

NUMBER OF QUEENS PER COLONY Until recently, it was generally accepted that mature fire ant colonies contained only one queen; however, recent information has shown that there may be more than one fertilized queen in many colonies. Glancey et al (41) collected 20 inseminated queens from one colony and two from another near Jackson, Mississippi. In a study near Gulfport, Mississippi, they found that 2 of the 35 colonies they examined had two fertile queens each. (Fertility was determined by holding the queens in the laboratory with workers to determine whether they had laid fertilized eggs, that is, whether worker brood developed.) Similarly, Markin et al (74) found that at least 5% of the colonies in southern Mississippi that were over 2 yr old contained more than one fertile queen.

Still more recently, Glancey et al (submitted for publication) found an even more extreme instance of polygyny in *S. invicta* colonies. About 30 mounds of *S. invicta* were found on a relatively narrow ditch bank at a garbage dump near the Mississippi-Alabama boundary. A portion of the tumulus of a mound (with the ants) was scattered on the asphalt highway; then the queens could be easily collected by observing the clustering of many workers about them. Ten mounds contained from as few as 7 to as many as 677 dealated queens. The presence of sperm in the spermathecae of 521 queens indicated that they had mated. Other queens that were placed individually in containers with workers oviposited and produced workers. Eventually, over 3000 fertile queens were collected from these mounds. Further studies are needed to determine the prevalency of this type of polygyny and to determine its significance in the survival, spread, and economic importance of imported fire ants.

MATING BEHAVIOR Although nuptial flights of imported fire ants have been noted in every month, peak activity occurs from late May through August subsequent to the periods of highest production of sexual brood (75). Glancey (unpublished data) has found that about 7–10 days are required for the males and queens to reach sexual maturity after they eclose from their pupal stage.

Little is known of the factors that trigger mating flight activity in a colony. Flights may be localized with only a few colonies participating or they may be very generalized and may include many colonies in several states (75). However, the flights will generally occur within 1 to 2 days after a rain, especially if the rain has been preceded by a period of dry weather. Also, Rhoades & Davis (90) found that the relative humidity during flights was always over 80%; observed no flights of *S. invicta* on days when soil temperature at a depth of 4 inches was below 18°C; and found that all flights occurred when the ambient air temperatures were between 24 and 32°C. In contrast, Green (48), in Mississippi, noted flights of *S. richteri* when the air temperature was 21°C, and Markin et al (75) reported flights of *S. invicta* in the winter in Louisiana when the temperature ranged from 20 to 26°C. No correlation of flight activity with cloudiness or barometric pressure has been found. In most cases the wind velocity at the time of flight was less than 5 mph; however, a few flights were observed when some wind gusts exceeded 15 mph (75).

The activity of the workers gives an indication of impending mating flights from mounds. About 30 min to 1 hr before a flight, large numbers mill excitedly over the surface of the mound and the surrounding vegetation and soil surface. During this time they open exit holes, 6 to 12 mm in diameter, in the surface of the mound. The males take flight about 30 min to 1 hr before the females. Flights usually occur from about mid-morning to late afternoon.

Markin et al (75) studied the aerial distribution of alates by collecting them in nets fixed to an aircraft. They found that males concentrated at an altitude of 90 to 150 m and remained airborne for several hours. Female and male alates were collected at a maximum altitude of 150 and 300 m, respectively. Since 98% of the alates captured were males, the females must ascend into the male swarm, become mated quickly, and then descend rapidly to the gound. It is apparent that the imported fire ants do not produce the prominent mating swarms typical of some other species of ants; however, their layering type of swarm is evidently efficient in assuring mating. Markin et al (75) reported that 95% of the queens picked up from the soil after a flight were inseminated though they calculated that there is probably no more than 1 male per 90 m^3 of air during a normal flight. Logically then, one or more pheromones must be utilized during the mating flight, to initiate activity and to assure that contact is made between the males and females in the air.

Markin et al (75) felt that the queens selected the sites in which they landed after a mating flight since they found more queens and new mounds on recently cultivated fields or dirt roads than they did on heavily vegetated areas. They also interpreted their finding that more male alates were captured over pasture than over adjacent swamp as further evidence of habitat selection. The queens may indeed select the sites on which they land; however, the findings by Markin et al (75) do not necessarily lead to this conclusion. The ease with which queens can secret themselves would make it much more difficult to find them on vegetated areas than on cleared areas, even though about the same number might be present. It would appear logical that more mounds would be found on cultivated fields or dirt roads since the soil is usually soft and the queens can quickly excavate a burrow and escape predation and dehydration to a greater extent than on some other sites. The heavier concentrations of males over the pasture rather than over the swamp probably can be attributed to the fact that the mounds producing the flights were numerous in the pasture and very sparse or absent in the swamp. The alates simply congregated over the area from which they arose. Further study is needed on habitat selection by the queens.

Considerable data have been accumulated in recent years on the numbers of alates produced by individual colonies though much of it is not yet published. The most extensive research has been done in north Florida. There, Morrill (80) found that in June the number of alates that left a mound for a flight averaged 690 and that the average number of alates produced per hectare per year from mounds located in four habitats (pasture, road ditch, pine woods, pipeline) was about 462,000. During vigorous flights up to 100 alates per minute were observed leaving mounds. Although 90% of the flights observed from any particular mound were composed predominately ($> 74\%$) of one sex, enough mounds containing each sex produced flights so that total numbers of each sex were about equal, and the sex ratio for the entire area was about 1:1. Colonies were found that produced predominately one sex at one time of the year and the other sex at a different time, but no colonies produced only one sex throughout the year. Total alate production per acre did not necessarily correlate with mound density, e.g. a pasture with 50 mounds/ha produced more alates than did a ditch with 125 mounds/ha.

Studies in Arkansas (93) gave results similar to those obtained in Florida. The largest flight from one mound produced 3168 alates within a 24-hr period, and the largest number collected during the study period (152 days) from one mound was 7600. Flights of *S. invicta* occurred from 1 to 3 PM; flights of the southern fire ant, *S. xyloni*, occurred primarily from 3:30 to 5:30 PM.

Food, Food Gathering, and Exchange

FOOD Fire ants have been described as voracious feeders on crops and domestic and wild animals. These descriptions have been based, to a great extent, on hearsay instead of scientific evidence. The ants are highly omnivorous and opportunistic so they may feed upon whatever plant or animal material they encounter. Several scientific reports (48, 51, 122, 124) have established that the primary diet of fire ants is insects, spiders, myriapods, earthworms, and other small invertebrates. Although some feeding on plants does occur, it seems to be the exception rather than the rule, and it usually involves eating the germ plasm of newly germinated seeds or girdling the stems of small seedlings. The damage occurs most often when a previously uncultivated area is brought under cultivation and the ants are thereby deprived of their normal food sources. Vinson (108) observed the red imported fire ant feeding on *Paspalum* seeds but found on close examination that only seeds infested with ergot were collected. He concluded that the fungus was reacting with the seed to produce certain carbohydrates desired by the ants.

Efforts to develop effective toxic baits for control of imported fire ants led to food acceptance studies to find materials attractive enough to the ants to elicit collection of a bait applied area wide in the field. Hays & Arant (50) and Lofgren et al (62, 64) evaluated a wide variety of food materials and found that fats and oils were consistently the most acceptable foods. An evaluation of the food preferences and nutritional needs of the ants conducted at Mississippi State University (91, 92, 106, 107, 110) showed that practically any insect was acceptable as food for the ants, but that some were more acceptable than others. Soybean oil was distributed to all castes in the colony to a greater extent than were protein or carbohydrate, and two natural components of the soybean oil (linoleic and linolenic acid) proved to be phagostimulants for imported fire ants. Two amino acids, leucine and valine, were well accepted by the ants; however, aqueous solutions of various electrolytes did not induce any strong feeding response. Lipase was the most widely distributed digestive enzyme present in the workers.

FOOD GATHERING Food for the fire ant colony is collected by the foraging workers who leave the mound through foraging tunnels that radiate from the mound at a depth of 6 to 12 mm under the soil surface. Green (48) reports that a mound usually will have 5 or 6 such tunnels extending from the mound in somewhat fixed positions though the extremities are temporary and change according to need. These tunnels branch repeatedly over the foraging area, and extend as far as 15–25 m from the mound and have openings to the soil surface at irregular intervals. The worker ant exits to the soil surface through one of these openings and forages at random until a source of food is located. Once food is located, the forager returns to the tunnel laying a trail with the trail pheromone. The additional workers recruited by the original forager then establish a continuous stream of ants between the food source and the nest.

The size of the area foraged by a colony depends upon the food requirements of the colony and the availability of food in the area. Wilson et al (123) reported that the area foraged by a given mound is discrete, and little or no interrelationship or food exchange exists between adjacent mounds. However, more recent evidence indicates that though a given mound may be an entity, there is often communication between many adjacent mounds in an area. In tests in central Florida (W. A. Banks et al, unpublished data) a dyed food that had been buried within a selected mound appeared in the ants in 22 other mounds within 23 m of the baited mound. Moreover at only two of the ten test sites did the dye fail to appear in at least one other mound. The exact mechanism of food transfer is unknown. The same test repeated in Texas (J. W. Summerlin, personal communication) had similar results. Also, when ants were sprayed with fluorescent or aluminum paint at one mound some were collected later from a different mound—an indication of a possible exchange of workers between adjacent mounds (B. M. Glancey, unpublished data). Likewise, Morrill (80) reported that after alate ants taking flight from a mound were captured, marked with paint, and returned to the mound, he later captured these same alates taking flight from a different mound. Mounds of the imported fire ants, therefore, are not necessarily distinct entities, and there may be food exchange and exchange of workers and alates between mounds.

The foraging of the ants is regulated to a great extent by the prevailing soil temperatures. Markin (personal communication) found that no significant foraging occurred when soil temperature at the 5 cm depth was below 15°C or above 37°C. Maximum foraging occurred between 21 and 35°C. Our studies (unpublished) confirm these observations, though we found that limited foraging occurred between 10 and 15°C if the day was sunny (not if the sky was overcast).

FOOD EXCHANGE The return of the foraging worker to the nest with food begins a fairly complex system of distribution of the food through the colony biomass. Wilson (121) indicates that this exchange not only provides nourishment but plays an important role in the social organization. Food exchange occurs between all castes in a colony. Dye incorporated into food offered to worker ants shows up quickly in the larvae, and more than 90% of the individuals in the colony have received the dye within 48–72 hr. G. P. Markin (personal communication) found that an insecticide administered in peanut butter to the workers appeared in highest concentration first in the minor workers and then in the small larvae. During the seven days of the test, the highest concentrations were found in the large larvae with lesser amounts being found in the small larvae, minor workers, major workers, and alate females in that order. Vinson (106) found that oil, carbohydrate, and protein were distributed throughout a colony after being taken in by the foraging workers; the larvae received a greater proportion of all three foods than any other caste. Our own unpublished studies have shown that a ^{32}P-labeled diet high in lipid and protein fed to 12 colonies reached 99% of the minor workers, 95% of the medium workers, 93% of the major workers, 80% of the larvae, and 10 of 12 mother queens within 9 hr. However, levels of radioactivity were highest in the larvae so individual larvae received more of the labeled food though the percentage of larvae fed was lower than for the other castes.

O'Neal & Markin (86) showed that both oral and proctodeal food exchange occurs from larvae to workers. Our own unpublished studies confirm the transfer of material from larvae to worker; however, we did not establish the mode of exchange.

Glancey et al (45) found evidence that the larger major workers in the colony retain oil in their crops for many months. Thus these workers may be a replete caste such as occurs in the genera *Myrmecocystus* and *Prenolepis.* Such repletes could provide food for the colony in times of stress or food shortage.

Chemical Secretions of Fire Ants

VENOM

Chemistry The chemistry of the venoms of several species of *Solenopsis* was carefully studied under the direction of M. S. Blum at the University of Georgia (19, 20, 69, 70). These investigators isolated, identified, and synthesized the following five major alkaloidal compounds of the venom of *S. invicta:* solenopsin A (*trans*-2-methyl-6-*n*-undecylpiperidine); solenopsin B (*trans*-2-methyl-6-*n*-tridecylpiperidine); solenopsin C (*trans*-2-methyl-6-*n*-pentadecylpiperidine); dehydrosolenopsin B [*trans*-2-methyl-6(*cis*-4-tridecenyl) piperidine]; and dehydrosolenopsin C [*trans*-2-methyl-6(*cis*-6-pentadecenyl) piperidine]. However, they found no *trans* C_{15} or *trans* $C_{15:1}$ alkaloids in *S. richteri.*[2] Thus this species lacks two of the major constituents of the venom of *S. invicta.* The venom of workers of *S. xyloni* contains *cis*-2-methyl-6-*n*-undecylpiperidine (*cis* C_{11}) as the main alkaloidal constituent, and *trans*-2-methyl-6-*n*-undecylpiperidine (*trans* C_{11}) is present in about one fourth the amount of the *cis* C_{11} isomer. Significantly, neither the *trans* C_{13} nor the *trans* C_{15} alkaloids are present in the venom of *S. xyloni.* Also, the venom of workers of *S. geminata* lacks the C_{13} and C_{15} alkaloids, and only the C_{11} dialkylpiperidines are present. However, there is about 1½ times more *cis* C_{11} than *trans* C_{11} so the venom of *S. geminata* differs from the venom of *S. xyloni* which contains 4 times more *cis* C_{11} than *trans* C_{11}.

The same investigators compared the ratio of *cis* C_{11} to *trans* C_{11} in queens and workers. The workers of *S. xyloni* and *S. geminata* contain much more *cis* C_{11} than

[2]The abbreviated terms C_{11}, C_{13}, C_{15}, and $C_{15:1}$ are used to designate the venom components. They indicate the carbon length and the presence or absence of unsaturation in the 6-alkyl substituent on the piperidine ring but do not signify the total number of C-atoms in the molecule.

trans C_{11}. The workers of *S. richteri* and *S. invicta* contain much less *cis* C_{11} than *trans* C_{11}. However, the queens of all four species always contain more *cis* C_{11} than *trans* C_{11}. The venom of workers of *S. invicta* contains primarily *trans* C_{13} and C_{15} dialkylpiperidines; the venom of the queens lacks these compounds. Similarly, the venom of *S. richteri* workers is dominated by C_{13} alkaloids; the venom of the queen lacks these compounds.

Since the difference in the type of alkaloids in the workers and their ratios to each other proved to be consistent for the species studied, they can serve as fingerprints that may be and have been used to substantiate the validity of the taxonomic classification of the species.

The venom of another, probably undescribed, species of *Solenopsis* from Brazil was found (71) to contain only the *cis* and *trans* isomers of a compound that was previously undiscovered in fire ants, *trans*-2-methyl-6-*n*-nonylpiperidine.

Evolutionary implications Brand et al (21) constructed, and MacConnell et al (71) expanded upon, a hypothetical model for the evolution of the fire ants based upon the venoms of the different species. It would seem unwise to speculate further on this hypothesis until the taxonomy of the fire ants is much further advanced and the venom chemistry of more species is known.

Toxicology The venom of the fire ants is used to immobilize or kill prey for food and has been shown to possess insecticidal, bactericidal, and fungicidal activity (16, 55). It is also an extremely effective defensive weapon for preventing large animals from disturbing their mounds. In the typical stinging of man, the worker ant bites or attaches to the skin with the mandibles and then lowers the tip of the abdomen to insert the sting. Several separate stings may be produced by a single ant if it is not removed. The morphology and histology of the sting apparatus, venom gland, and poison sac was described by Callahan et al (27). The immediate reaction to the injected venom is an intensive burning sensation that explains the common name of the ants. When the sensation subsides after a few minutes, a wheal appears at the site of venon injection. A few hours later, a superficial vesicle containing clear fluid appears. By 24 hr, the fluid becomes purulent as a result of the necrotizing properties of the venom. The pustule is sterile and may persist as long as a week if it is not broken. During this time the pus is absorbed leaving a crust and, in many instances, scar tissue. Occasionally an individual may have systemic reactions including nausea, vomiting, dizziness, perspiration, cyanosis, asthma, and other symptoms typical of severe allergic reactions. In severe cases, if medical assistance is not received, the individual may die (102).

PHEROMONES Pheromones play a highly important role in the organization and coordination of activity in ant colonies. In fact, Blum (15) stated that the social cohesiveness of an insect colony appears to be maintained primarily through chemical communication. Wilson (120) recognized at least five distinct areas of chemical communication among workers of the imported fire ants: nest odor, body surface attractants, food exchange, trail following, and alarm. Various investigators have demonstrated communication among the ants by a trail-following pheromone and by surface contact pheromones produced by the queen and the brood to elicit care by the workers.

Trail-following pheromone More study has been devoted to the trail-following pheromone than to any other pheromone produced by fire ants. Wilson (119) reported that the trail pheromone is produced in the Dufours gland of the worker ant and is extruded through the stinger. Subsequently (120, 121) he discussed in detail the function of the trail pheromone in recruiting and orientating the workers during mass foraging and reported (120) that it was highly species specific. More recent work (M. R. Barlin and M. S. Blum, personal communication; 56) has shown that each species is generally more responsive to the pheromone of its own species, but specificity is not as strong as Wilson (120) indicated (see Table 1). Indeed, some of the differences that are noted probably result from differences in bioassay techniques and in the criteria for determining a positive response. For instance, Wilson (120) used only extracts of Dufours gland; Barlin and Blum (personal communication) used extracts of Dufours glands and natural trails; and Jouvenaz et al (56) used partially purified whole ant extracts and naturally laid trails. Moreover, the locations from which Wilson (120) obtained his specimens of *S. geminata* (Costa Rica) and *S. xyloni* (California) present the possibility that these tests were not made with the same species used by Barlin, Blum, and Jouvenaz since there is still some question as to the taxonomy of these species.

Studies are currently underway to identify the chemical nature of the trail pheromones of both the red and the black imported fire ants. Walsh et al (111) reported

Table 1 Response of fire ant workers to trail pheromone of their own and other species as reported by three authors[a]

Source species	Author[b]	Response of indicated test species[c] *Solenopsis geminata*	*Solenopsis xyloni*	*Solenopsis invicta*	*Solenopsis richteri*
Solenopsis geminata	W	4	1	0	NT[d]
	B	4	3	0	1
	J	4	3	1	1
Solenopsis xyloni	W	3	4	4	NT
	B	3	4	0	3
	J	3	4	1	1
Solenopsis invicta	W	0	2	4	NT
	B	0	2	4	3
	J	4	1	4	4
Solenopsis richteri	W	NT	NT	NT	NT
	B	0	1	2	4
	J	2	1	4	4

[a]The numbers shown are relative values based on data of these authors.
[b]W = Wilson, B = Barlin & Blum, J = Jouvenaz.
[c]Intensity of response 4 = highest to 1 = lowest. 0 = no response.
[d]NT = not tested.

that they had purified, but not identified, a trail pheromone (apparently of *S. invicta*). Elucidation of the nature of the trail pheromone is proving to be somewhat difficult. Present evidence suggests (M. S. Blum, personal communication) that the pheromone is a sesquiterpene type $C_{16}H_{26}$ hydrocarbon. Bioassays of known sesquiterpenes, however, have not elicited any response from the ants at the concentrations tested. Bioassays with purified fractions from GLC chromatographic separation of whole ant extracts show excellent activity over at least 10^6-fold range of dilutions.

Queen-tending pheromone A fertile queen confined to a spot on a square of blotter paper releases an attractive substance that makes the spot more attractive to worker ants than other parts of the paper (54). The attractive substance is soluble in organic solvents and remains attractive for 72 hr when impregnated onto absorbent paper. B. M. Glancey (unpublished data) applied hexane extracts of *S. invicta* queens to queens of other ant species (*S. richteri, S. geminata, Camponotus pennsylvanicus*). The foreign queens were then temporarily treated as *S. invicta* queens when they were placed with *S. invicta* workers; untreated queens of the other species were immediately killed by the *S. invicta* workers. Neither the chemical nature nor the site of production of the queen pheromone is yet known.

Brood pheromone Glancey et al (44) suggested the presence of a brood pheromone in fire ants when they found that inanimate objects coated with a hexane slurry of fire ant larvae caused the objects to be treated as larvae. An in-depth behavioral study (112) with the brood pheromone revealed that it occurs uniformly over the body surface of the larvae and is apparently nonvolatile, requiring contact by the workers to be effective.

ARTHROPODS ASSOCIATED WITH IMPORTED FIRE ANT COLONIES

Many arthropods have been collected from nests of imported fire ants in the United States and South America (30, 97), however, only three of the species commonly found in nests in the United States can be classified as inquilines: a scarabaeid beetle, *Myrmecaphodius excavaticollis;* a staphylinid beetle, *Myrmecosaurus* sp.; and an undescribed thysanuran. The scarabaeid beetle was originally described from Argentina (127); the staphylinid belongs to a genus (not previously reported from the United States) associated with ants in South America. Thus, the two beetle species may have been introduced into the United States with the original introduction of the imported fire ant. Also *Gymnolaelops shealsi* (53), a genus of mites normally associated with ants, has been collected from imported fire ants in Mississippi.

ECONOMIC IMPORTANCE OF IMPORTED FIRE ANTS

The literature fails to give much valid data concerning the economic importance of the imported fire ants. Assessments have too often been based on hearsay and personal emotions rather than scientific fact. Thus, the ants are variously viewed by both scientists and laymen as serious economic and public health pests, as minor nuisance pests, as totally innocuous, or even as beneficial. Obviously, their real importance lies somewhere between the extremes.

The greatest difficulty in evaluating the importance of fire ants probably derives from the fact that they affect a broad spectrum of "things" but are not major pests on any one thing. However, when all areas in which fire ants are pestiferous are summed, the ants must be considered an important economic pest in the southern United States. The majority of entomologists who are conversant with imported fire ants, agree that they are primarily nuisance pests. The fact, however, does not justify claims that they are not important pests to man and his possessions. For example, the state of Florida and the various mosquito control districts within the state spend more than $10 million annually to control mosquitoes whose primary importance is that they bite and annoy man (J. A. Mulrennan, personal communication).

Some of the more important problems created by imported fire ants include feeding on plants (particularly seedlings or germinating seeds and okra flowers), stinging of livestock, damage to farm machinery that strike mounds, loss of hay and grazing area, refusal of workers to enter heavily infested fields to cultivate or harvest crops, and hazards to human health from stings that may cause systemic reactions or complications from secondary infections (1, 32, 48, 83, 103). Also, reports have been received and documented of telephone cable and lighting cable for airport runways being gnawed by the ants to the extent that water entered and created short circuits; the ants have been observed tending aphids and mealy bugs on plants (a habit that could result in increased plant disease transmission on some crops), and girdling young citrus trees. In addition there are documented cases of destruction of the nest and young of ground nesting birds and other animals, killing of small pigs, and predation of beneficial arthropods.

On the other hand, predation by imported fire ants has in some cases played an important role in reducing damage by pest insects, e.g. that done to sugarcane by sugarcane borers *Diatraea saccharalis* (89). They are, however, nondiscriminatory in their predation and attack beneficial as well as harmful arthropods. Thus, elimination of fire ants in some areas has resulted in an upsurge of harmful pests, largely because the ant had previously eliminated other insects that would have preyed upon these pests. Biological control authorities agree that the more diverse the predator population, the more likely that effective pest management can be attained. It is therefore regrettable that in areas where imported fire ants abound, they have become the dominant and in some areas the sole predator. For example, Roe (93) showed that *S. invicta* competitively displaced *S. xyloni* and severely reduced populations of some other ant species in Arkansas. Whitcomb et al (115) found that *S. invicta* sharply reduced the diversity of ant species of Florida soybean fields.

Reductions of populations of lone star ticks, *Amblyomma americanum,* and Gulf Coast ticks, *Amblyomma maculatum,* in Louisiana were attributed to imported fire ants. Supporting evidence is rather meager, however. Harris & Burns (49) released ticks in fields infested with fire ants and in fields that had been cleared of ants by using mirex bait. About 75% of the ticks released in the mirex-treated plot were recovered after 5–6 months while none were recovered from the ant-infested plots. It is likely that any arthropod released into an ant-infested field would be preyed upon by the ants, although possibly not being preyed upon in a natural situation. Further work is needed to determine if the ants are indeed reducing natural populations of the ticks.

An estimate of the problems and losses caused by imported fire ants was made in 1973 when a private research agency (Chilton Research Services, Radner, Pennsylvania) was commissioned by the Allied Chemical Corporation to conduct a survey of farmers and county agricultural agents in the infested states (F. L. Bailey, personal communication). The individuals surveyed were selected randomly and requested by telephone or personally to respond to a questionnaire designed to provide their estimate of problems and losses caused by imported fire ants and to determine the attitude of the respondent toward control of the ants. From 7 to 23% of 400 farmers indicated that fire ants had caused them to lose hay, had damaged equipment, or had caused them to spend money for veterinary or medical services. Calves were cited as the farm animal most often harmed. Also, 50% had spent money for pesticides to control the ants. In the 261 generally infested counties surveyed (400 counties are known to be infested), economic losses in 1972 were estimated to be $48 million. Net decrease in land value was estimated at $500 million. While the validity of such a survey might be questioned, since documentation of actual losses was not required, the fact that 50% of the farmers spent money on pesticides indicates that they consider imported fire ants to be important economic pests. Further documentation of actual losses due to the ants is needed before we can fully assess the impact of the ants on the economy of the affected states.

Recent medical studies indicate that the imported fire ant is assuming greater importance as a public health hazard. Triplett (102) and Brown (23) found that allergic symptoms exhibited by 20 patients stung by imported fire ants ranged from urticaria and angioedema through respiratory problems, nausea, and vomiting to anaphylatic shock and collapse. In addition, in 1971, Triplett, who is at the Mississippi Allergy Clinic, Jackson, Mississippi, surveyed physicians in selected areas of Mississippi, Alabama, and Georgia to determine the extent of medical problems associated with the fire ant. Of the 2,485 physicians contacted in this survey (as yet unpublished), 1,336 responded, and 901 reported treating patients for fire ant stings or complications arising from stings. These physicians had treated totals of 9,224 and 11,937 patients for fire ant stings in 1969 and 1970, respectively, and had seen 12,438 patients for stings in the first 7–9 months of 1971. Average medical costs for each patient treated in 1971 was $28.32. The physicians indicated that the incidence of complications from stings appeared to be increasing. We know of at least two deaths in southern Mississippi since 1971 that were caused by complications arising from fire ant stings. The available evidence suggests that the medical importance of the imported fire ants may be far greater than heretofore suspected.

CONTROL

Efforts to control imported fire ants in the southeastern United States have stirred a controversy that has raged for more than a decade (24, 28, 31, 38, 100, 105). In a previous section of this paper we discussed the conflicting views regarding the status of the ant as a pest. Irrespective of these views, most landowners in generally infested areas want to be rid of the ants, and almost every conceivable method of control has been utilized by some property owners. These include the use of every available insecticide, burning the nest with a flammable such as gasoline, and physical destruction of the nests by digging or plowing.

Excellent reviews exist of organized control programs against imported fire ants with residual insecticides (heptachlor and dieldrin) and with mirex bait (6, 13). We will therefore restrict our discussion on control to the development and use of mirex bait for this burpose.

Development of Toxic Baits

Studies conducted by Auburn University in the late 1950s (50) determined that a bait consisting of 0.125% Allied Chemical GC-1189 (dodecachlorooctahydro-1,3,4-metheno-2*H*-cyclobuta[*cd*]pentalene-2-one) (Kepone®) in peanut butter gave excellent control of imported fire ants. Concurrent studies by scientists at a US Department of Agriculture Methods Development Laboratory, established at Gulfport, Mississippi, in 1957, revealed that mirex, an analog of Kepone, was as effective as Kepone as a bait toxicant and had a more favorable mammalian toxicity (65). The scientists subsequently (60, 61, 64, 99) developed a toxic bait formulated with once-refined soybean oil as the food, corncob grits as the carrier, and mirex as the toxicant, that gave excellent control of the ants. In 1962 it replaced heptachlor as the standard control agent. Initially, the bait was applied at a rate of 11.2 kg/ha of a 0.075% formulation; however, further studies (12, 60) showed that an increase in the content of mirex in the bait would permit reductions in the bulk rate of application. Thus, in rapid sequence, application rates were reduced to 5.6 kg/ha of 0.15% bait, then to 2.8 kg/ha of 0.3% bait and finally to the current rate of 1.4 kg/ha of 0.3% bait. The amount of toxicant applied remained constant at 8.4 g/ha until the last rate change when a 50% reduction, to 4.2 g/ha, was achieved.

Field studies showed that wherever the bait was properly applied, excellent control of the ants was obtained. The average control obtained on 63 plots of ca 0.4 ha was 98% (12, 60, 61, 99); the average control obtained on large areas (259–405,000 ha treated by aircraft) was 96% (8, 10, 11, 60).

Research, however, has continued in an attempt to improve the performance and extend the field life of the bait. Markin & Hill (78) demonstrated that the technique of encapsulating insect pathogens and insecticides (39, 88) could be used successfully with the soybean oil-mirex fire ant bait. The microcapsules (200–800 μ diameter) were composed of ca 85% soybean oil, 2% mirex, and 13% capsular wall material and were as effective at a rate of 247 g/ha as 1.4 kg/ha of the 0.3% standard corncob grit bait. Also, the effective field life of the capsular bait was greater than 30 days vs 3–4 days for the corncob grit bait. The encapsulated bait, although superior to the standard, has not gained wide usage because of high production costs and lack of available equipment for application. (The fragility of the capsules prohibits the use of standard equipment.)

Studies revealed that coating the bait with acrylic latex allowed more oil to be added to the bait and increased the availability of oil and mirex to the ants (9, 11). In subsequent studies the method of formulation was changed (10) so that one coat of neat soybean oil was applied to the corncob grits and a second coat of oil containing the mirex was applied 24 hr later. The latex coating was then applied. This procedure concentrates the mirex on or near the surface of the corncob grit where it can be more easily removed by the ants. Field tests in Georgia in 1971 (10) showed that 1.12 kg/ha of this formulation containing 0.1% mirex (1.12 g/ha active ingredient) was as effective as the 0.3% standard bait at 1.4 kg/ha (4.2 g/ha AI). Some more recent tests (W. A. Banks et al, unpublished data) indicate that the latex coating may be omitted from the formulation without serious decline in effectiveness. However, difficulties in formulation, application, and storage life have slowed the acceptance of the coated baits as control agents.

In 1967, a special committee established by the National Academy of Sciences (79) concluded from information available at that time that eradication of the fire ant was biologically and technically impossible, and inadvisable were it possible. However, a series of eradication trials conducted subsequent to that report indicated that eradication is technically feasible (8). Lofgren & Weidhaas (66) in a theoretical appraisal of eradication calculated that from 3 to 9 applications of mirex bait could be required, dependent upon the effectiveness of a given application and the effective rate of increase of the species. We used only 3 applications in the eradication trials (8) and the results obtained indicated that eradication could be accomplished with 3 or 4 sequential applications that were properly timed if contiguous areas were progressively treated.

About this time, however, several laboratory studies showed that mirex had undesirable effects on nontarget animals (17, 40, 67, 68, 81, 104, 113), so questions arose concerning the environmental safety of the bait. Subsequently, data became available that showed that residues of mirex were appearing in nontarget organisms after large area applications for fire ant control (7, 18, 29, 76, 82). As a result, the Environmental Protection Agency in 1971 issued notice of cancellation of the registration of products containing mirex. The registration was ultimately reinstated but with severe restrictions on the application of the bait in estuarine and prime wildlife habitats (94). Therefore, since 1971, the federal-state program has been aimed at providing relief in areas infested with large populations of the ants.

TECHNIQUES AND EQUIPMENT FOR APPLICATION OF CHEMICALS Chemical control of fire ants before 1957 was achieved by individual mound treatment or by broadcast treatments on limited areas with ground equipment. Eden & Arant (37) demonstrated that individual mound treatment or limited area treatment gave good control with residual insecticides if the materials were properly applied. However, mound treatments with mirex bait at rates as high as 14 g/mound, gave poor control in two series of tests (C. S. Lofgren et al, unpublished data).

In any case the large acreages involved in the cooperative federal-state program made treatment of individual mounds impractical and ground equipment was also impractical except for limited areas. Therefore, aircraft have been used almost exclusively in this program. Initially, single engine planes were used. However, by 1959–1960, some multiengine aircraft were in use. By the late 1960s, numerous multiengine planes, mostly converted World War II bombers, had been equipped for application of mirex bait (2). Dispersal equipment has been continually improved so most large applicator planes now have auger or chain-fed systems mounted within the wings of the aircraft. However, the development of the microencapsulated bait has now created a need for a new dispersal system, and the Aircraft Operations Section of the Animal and Plant Health Inspection Service, US Department of Agriculture at Beltsville, Md., has developed the prototype for a system to handle this bait (77).

The application aircraft were guided by helium-filled balloons (Kytoons®) until about 1966 when electronic guidance was introduced. The guidance systems, which consist of a master radio transmitter and two slave stations arranged in a triangular pattern within the area to be treated, produce electronic signals in a hyperbolic grid over the area. Numbers are assigned to the lines of the grid, and computerized receiving equipment in each aircraft is preset to home on a given numbered coordinate. Deviation of the aircraft from the prescribed flight path is indicated by an indicator needle in the cockpit and on a chart recorder. The operation and the use of the guidance systems in the fire ant program have been discussed in a number of articles (3, 43, 52, 87).

Standard operational procedures until 1970 required that the aircraft disperse the bait at an altitude of ca 23–46 m. However, a series of studies (Glancey et al unpublished data) showed that dispersal of the bait from an aircraft flying at an altitude of 213.5 m produced a wider swath and more uniform distribution on the ground. When wind speed and direction at dispersal altitude were determined with pilot balloons and a theodolite, the increased altitude coupled with crosswinds, increased the swath width 2 to 3 times the normal effective swath width of the aircraft. Since the increased swath width greatly increased the coverage of an aircraft per unit of flying time, costs of application were substantially reduced because the bid price for single spacing was reduced 25% and 50% for acreage flown at double or triple spacing, respectively.

Future Prospects for Control

CHEMICAL The outlook for alternative chemicals for control of imported fire ants is not promising. Screening programs to evaluate chemicals as residual soil surface treatments or as bait toxicants have been conducted since 1958 by the USDA and its cooperators. More than 150 chemicals were evaluated as residual soil surface treatments (59; W. A. Banks et al, unpublished data), but the only chemicals that were very effective in the field were cyclodiene type compounds. Since the use of these compounds has been very restricted since the early 1960s, only a limited amount of work has been done with residual compounds since 1963.

The bait screening tests have evaluated more than 2050 chemicals (58, 63, 126; R. Levy et al, unpublished data) from all the different classes of insecticides available. Stringer et al (99) determined that an effective bait toxicant must (*a*) exhibit delayed toxicity (less than 15% mortality after 24 hr and more than 90% mortality after 14 days) over at least a tentold and preferably above a hundredfold range of doses; (*b*) be readily transferred from one ant to another and result in mortality of the recipient; and (*c*) not be repellent to ants. Only 23 chemicals of the more than 2050 tested have met these requirements; 3 of the 23 were not available for field testing, and field tests with 11 showed that mirex and closely related compounds are the only ones that consistently give control effective enough to suppress populations (i.e. 90% or higher kill). The other 9 compounds were only recently tested, and sufficient quantities are not yet available for field tests; however, they are all organophosphorus chemicals, and all previous field tests with such materials have resulted in very poor control. The laboratory evaluation of additional chemicals is continuing.

Recent laboratory tests with several juvenile hormone analogs (35, 101, 109) indicate that these compounds may have some potential as control agents for fire ants. Methoprene and hydroprene caused anomalous prepupae, pupae, and adults and prevented larval metamorphosis. Some other compounds were found to produce similar effects (109). In addition, temporary sterility was induced by methoprene and hydroprene (85, 101). Field tests with these compounds have caused some deformities and some worker ant mortality, but have not effectively reduced infestations of the fire ants.

BIOLOGICAL

Pathogens Surveys of fire ant populations in the areas infested with *S. invicta* and *S. richteri* in the United States revealed few indigenous pathogens (B. A. Federici, personal communication; 22). A species of *Pseudomonas* bacteria was found in dead ants taken from mounds in southwestern Mississippi. Subsequent investigations in the laboratory showed the bacterium to be highly toxic to fire ants, since it killed within 5 days 100% of the larvae fed on a glucose solution containing vegetative cells. Also, a glucose-bacteria suspension placed on corncob grits killed 50% of the worker ants allowed to feed on the grits. Three fungi were found to be pathogenic to both species of fire ants in the laboratory. *Beauveria bassiana, Metarrhizium anisopliae,* and *Aspergillus flavus* caused 80%, 40%, and only occasional mortality, respectively, to fire ants within 5 to 10 days after exposure (22, 95). *B. bassiana* and *A. flavus* have been recorded previously from *Atta texana* in Louisiana, and *M. anisopliae* has been recorded from *Solenopsis richteri* in Uruguay (98). Allen (4) indicates that *B. bassiana* and *M. anisopliae* appear to be important pathogens of ants in South America. There is no evidence, however, that any of these pathogens are exerting any appreciable control of any field population of imported fire ants in the United States.

The most promising development from the standpoint of possible biological control of fire ants is the finding by Allen & Buren (4) that microsporidian pathogens occur in *S. invicta, S. saevissima,* and other species of the *S. saevissima* complex in Brazil. These are the first protozoan pathogens to be reported in Formicidae. In some colonies 50 to 95% of the adults appear to be infected. Workers with a heavy infection may have the gaster packed with spores and the abdominal organs greatly reduced. How the disease is transmitted, whether each microsporidian is host specific, the length and course of the infection process, and whether these pathogens could be effectively utilized as biological control agents are all subjects of recently initiated research.

Predators and parasites The role that predators and parasites play in regulating populations of imported fire ants in the United States has not been studied in depth. Whitcomb et al (114) reported that the queens of *S. invicta* are attacked and killed by a variety of predators from the time they take flight until successful colony establishment is achieved. O'Neal (84) recorded predation on larvae of *S. richteri* by *Neivamyrmex opacithorax.* Also, the thief ant, *Solenopsis molesta,* was found living in the nests of both *S. invicta* and *S. richteri* and preying on the eggs and young larvae. Another ant, *Paratrechina melanderia arenivaga,* was found preying on the eggs of *S. invicta* in nests in southern Mississippi.

A nematode, as yet unidentified, has been found in *S. richteri* in Oktibbeha County, Mississippi, and has been established as being pathogenic to the ants (B. R. Norment, personal communication).

No evidence exists that any of the foraging parasites or predators are exerting sufficient restraints to reduce populations of fire ants in any area to even tolerable levels, nor does it appear likely that they will be able to do so in the future.

Studies have shown that the endoparasite, *Neoaplectana dutkyi,* (DD-136) is effective against the fire ants in the laboratory; however, field tests have not been extremely encouraging to date (Norment, personal communication).

In the homeland of the fire ants in Brazil, the states of Mato Grosso and Sao Paulo, *Solenopsis* species are beset by a large number of parasites. R. N. Williams and W. H. Whitcomb (submitted for publication) list or discuss 14 species of *Pseudacteon* and one species of *Apodicrania* (Phoridae), two species of *Orasema* (Eucharitidae), and three species of parasitic *Solenopsis. Pseudacteon* has been observed attacking fire ants (116) but as yet no one has observed the immature forms of these flies in the ants or determined their potential effect on the colonies. Nematodes have been observed in the gasters of *S. invicta* from Rondonopolis, Mata Grosso (G. E. Allen, personal communication). The possible suppression of *Solenopsis* by competitive species of ants has not been studied but has been speculated upon by Buren et al (26). The range of *S. invicta* in South America well may be restricted by many strong biotic and abiotic factors.

Silveira-Guido et al (96, 97) made an extensive study of the social parasite, *Solenopsis daguerri* (formerly *Labauchena daguerri*), and its effects on fire ant colonies. The parasite is confined to *S. richteri,* the imported fire ant of lesser importance in the United States, and it infests only a small percentage of the colonies. Present evidence suggests that it is not exerting any effective control on fire ants in South America, nor does it appear likely that it would do so on *S. richteri* in the United States.

Pheromones In a previous section of this paper we discussed the role of pheromones in the colony organization of fire ants. Research aimed at the elucidation of the chemical nature of the trail pheromone of both United States species of imported fire ants is well advanced. The use of this pheromone as a confusion lure, as suggested by Brown (24), may have application in future control of the fire ants. Elucidation of the chemical natures of the brood-tending (44) and queen pheromones (54) may provide effective lures and means of making baits more species specific.

Sterilization or genetic manipulation Neither sterilization nor genetic manipulation appear practical for the control of imported fire ants.

Chemosterilants cannot be used in the field because there are no compounds that are species specific and the effects of a nonspecific chemosterilant on nontarget species could be undesirable. In addition, the sterile male technique that has been used so successfully with some insects is totally impractical. For example, no methods are available for rearing large numbers of males in the laboratory and field studies (80) have shown that nuptial flights can occur on at least 138 days in north Florida producing an average of 239,590 females/ha (an average of ca 1736 females per ha/flight) (114). Then, to overflood with a 9:1 ratio of sterile males to females, we would have to release 15,624 males per ha/flight. If we assume that the releases would have to cover a 259 km^2 area, a total of 404.9 million males would be required per release. Nuptial flights can occur on any day in the year in any part of the infested area (75), however, those flights that occur during the cooler months may not result in successful colony founding. If we assume that colonies can become established only from flights that occur between April 1 and August 15, then colonies could result from flights on 137 days. Releases would have to be made every day during this period over the entire 259 km^2 since there are no reliable criteria for predicting time or location of a nuptial flight. Releases over this period on the 259 km^2 would require more than 55 billion males. These calculations are based upon the average number of females per hectare per flight. Roe (93) has reported a maximun of 3118 females flying from a single nest of *S. invicta.* Actual releases would have to be made on the basis of maximum numbers expected so the calculations are low by a factor of four times or more. It quickly becomes evident that the logistics of rearing sufficient males is overwhelming. In addition, sterile males would have no effect upon the parent colonies, which would continue to produce more potential queens unless an insecticide or some other control was used to eliminate the colony.

At present, our knowledge of the genetics of fire ants is insufficient to allow us to evaluate fully the prospects of genetic manipulation. Techniques for mass forced mating and rearing large numbers of genetically aberrant sexual forms are not available and the chances of developing such techniques appear remote. As with the sterile male approach, even if genetic aberrants could be introduced, the existing population would continue to flourish essentially unaffected by the aberrants.

CONCLUDING REMARKS

Probably no insects have created so much controversy as the imported fire ants. Despite the many hours of research and the many millions of dollars spent during the past two decades, the ants continue to annoy man and animals and to perplex those who seek new and better methods of controlling them. Eradication of these ants from the United States would be most desirable; however, the prospects of this being achieved appear remote, even though studies have shown that it could probably be accomplished with mirex bait. We must, therefore, find new methods for alleviating the many problems caused by the fire ants in the southeastern United States.

ACKNOWLEDGMENTS

The authors acknowledge with gratitude the assistance of Dr. W. F. Buren, Department of Entomology and Nematology, University of Florida, in preparation of the sections on the taxonomy and zoogeography, and of D. P. Wojcik, Agricultural Research Service, in providing many of the references.

Literature Cited

1. Adkins, H. G. 1970. The imported fire ant in the southern United States. *Assoc. Am. Geography* 60:578–92
2. 1969. Fire on fire ants. *Aerial Applicator* Apr. 1969:5, 8
3. 1971. Electronic tract guidance in the USA. *Agr. Aviat.* 13:29–30
4. Allen, G. E., Buren, W. F. 1974. Microsporidian and fungal diseases of *Solenopsis invicta* Buren, in Brazil. *J. NY Entomol. Soc.* 82:In press
5. Allen, G. E., Buren, W. F., Williams, R. N., de Menezes, M., Whitcomb, W. H. 1974. The red imported fire ant, *Solenopsis invicta:* Distribution and habitat in Mato Grosso, Brazil. *Ann. Entomol. Soc. Am.* 67:43–46
6. Alley, E. G. 1973. The use of mirex in control of the imported fire ant. *J. Environ. Qual.* 2:52–61
7. Baetcke, K. P., Cain, J. D., Poe, W. E. 1972. Residues in fish, wildlife, and estuaries. *Pestic. Monit. J.* 6:14–22
8. Banks, W. A. et al 1973. Imported fire ants: Eradication trials with mirex bait. *J. Econ. Entomol.* 66:785–89
9. Banks, W. A., Jouvenaz, D. P., Lofgren, C. S., Hicks, D. M. 1973. Evaluation of coatings of corncob grit-soybean oil bait used to control imported fire ants. *J. Econ. Entomol.* 66:241–44
10. Banks, W. A., Lofgren, C. S., Jouvenaz, D. P., Wojcik, D. P., Summerlin, J. W. 1973. An improved mirex bait formulation for control of imported fire ants. *Environ. Entomol.* 2:182–85.
11. Banks, W. A., Markin, G. P., Summerlin, J. W., Lofgren, C. S. 1972. Four mirex bait formulations for control of the red imported fire ant. *J. Econ. Entomol.* 65:1468–70
12. Banks, W. A., Stringer, C. E., Pierce, N. W. 1971. Effect of toxicant concentration and rate of application of mirex bait on control of the imported fire ant, *Solenopsis saevissima richteri. J. Ga. Entomol. Soc.* 6:205–7
13. Bellinger, F., Dyer, R. E., King, R., Platt, R. B. 1965. A review of the problem of the imported fire ants. *Ga. Acad. Sci. Bull.* 23(1):1–22
14. Bigelow, R. S. 1965. Hybrid zones and reproductive isolation. *Evolution* 19: 449–58
15. Blum, M. S. 1970. The chemical basis of insect sociality. In *Chemicals Controlling Insect Behavior,* ed. M. Beroza, 61–94. New York: Academic. 170 pp.
16. Blum, M. S., Walker, J. R., Callahan, P. S., Novak, A. F. 1958. Chemical, insecticidal and antibiotic properties of fire ant venom. *Science* 128(3319): 306–7
17. Bookhout, C. G., Wilson, A. J. Jr., Duke, T. W., Lowe, J. I. 1972. Effects of mirex on the larval development of two crabs. *Water, Soil, Air Pollut.* 1:165–80
18. Borthwick, P. W. et al 1973. Residues in fish, wildlife and estuaries. *Pestic. Monit. J.* 7:6–26
19. Brand, J. M., Blum, M. S., Barlin, M. R. 1973. Fire ant venoms: Intraspecific and interspecific variation among castes and individuals. *Toxicon* 11:325–31
20. Brand, J. M., Blum, M. S., Fales, H. M., MacConnell, J. G. 1972. Fire ant ven-

oms: Comparative analyses of alkaloidal components. *Toxicon* 10:259–71
21. Brand, J. M., Blum, M. S., Ross, H. H. 1973. Biochemical evolution in fire ant venoms. *Insect Biochem.* 3:45–51
22. Broome, J. R. 1973. *Microbial control of the imported fire ant, Solenopsis richteri Forel.* PhD thesis. Mississippi State Univ., State College, Miss.
23. Brown, L. L. 1972. Fire ant allergy. *S. Med. J.* 65:273–77
24. Brown, W. L. Jr. 1961. Mass insect control programs: Four case histories. *Psyche* 68:75–109
25. Buren, W. F. 1972. Revisionary studies on the taxonomy of the imported fire ants. *J. Ga. Entomol. Soc.* 7:1–27
26. Buren, W. F., Allen, G. E., Whitcomb, W. H., Lennartz, F. E., Williams, R. N. 1974. Zoogeography of the imported fire ants. *J. NY Entomol. Soc.* 82:In press
27. Callahan, P. S., Blum, M. S., Walker, J. R. 1959. Morphology and histology of the poison glands and sting of the imported fire ant (*Solenopsis saevissima* var. *richteri* Forel). *Ann. Entomol. Soc. Am.* 52:573–90
28. Carson, R. 1962. *Silent Spring.* Greenwich, Conn.: Fawcett. 304 pp.
29. Collins, H. L., Davis, J. R., Markin, G. P. 1973. Residues of mirex in channel catfish and other aquatic organisms. *Bull. Environ. Contam. Toxicol.* 10: 73–77
30. Collins, H. L., Markin, G. P. 1971. Inquilines and other arthropods collected from nests of the imported fire ant, *Solenopsis saevissima richteri. Ann. Entomol. Soc. Am.* 64:1376–80
31. Coon, D. W., Fleet, R. R. 1970. The ant war. *Environment* 12:28–38
32. Crance, J. H. 1965. Fish kills in Alabama ponds after swarms of the imported fire ant. *Progr. Fish Cult.* 27: 91–94
33. Creighton, W. S. 1930. The new world species of the genus *Solenopsis. Proc. Am. Acad. Arts Sci.* 66(2):87–89
34. Culpepper, G. H. 1953. Status of the imported fire ant in the southern states in July 1953. *US Dep. Agr. Agr. Res. Serv.* #-867. 8 pp.
35. Cupp, E. W., O'Neal, J. 1973. The morphogenetic effect of two juvenile hormone analogues on larvae of imported fire ants. *Environ. Entomol.* 2:191–94
36. Cupp, E. W., O'Neal, J., Kearney, G., Markin, G. P. 1973. Forced copulation of imported fire ant reproductives. *Ann. Entomol. Soc. Am.* 66:743–45
37. Eden, W. G., Arant, F. S. 1950. Control of the imported fire ant in Alabama. *J. Econ. Entomol.* 42:976–79
38. Ferguson, D. E. 1970. Fire ant: Whose pest? *Science* 169:630
39. Fogle, M. V. 1968. Microencapsulation promises big potential in agricultural field. *Croplife* Feb. 1968:22, 27, 28, 30
40. Gaines, T. B., Kimbrough, R. D. 1970. Oral toxicity of mirex in adult and suckling rats. *Arch. Environ. Health* 21:7–14
41. Glancey, B. M., Craig, C. H., Stringer, C. E., Bishop, P. M. 1973. Multiple fertile queens in colonies of the imported fire ant, *Solenopsis invicta. J. Ga. Entomol. Soc.* 8:237–38
42. Glancey, B. M., Stringer, C. E., Bishop, P. M. 1973. Trophic egg production in the imported fire ant, *Solenopsis invicta. J. Ga. Entomol. Soc.* 8:217–20
43. Glancey, B. M., Stringer, C. E. Jr., Bishop, P. M., Craig, C. H., Martin, B. B. 1973. Evaluation of an electronic guidance system for aircraft for bait application in the imported fire ant program *Agr. Aviat.* 15:38–40
44. Glancey, B. M., Stringer, C. E., Craig, C. H., Bishop, P. M., Martin, B. B. 1970. Pheromone may induce brood tending in the fire ant, *Solenopsis saevissima. Nature* 226(5248):863–64
45. Glancey, B. M., Stringer, C. E. Jr., Craig, C. H., Bishop, P. M., Martin, B. B. 1973. Evidence of a replete caste in the fire ant, *Solenopsis invicta. Ann. Entomol. Soc. Am.* 66:233–34
46. Green, H. B. 1952. Biology and control of the imported fire ant in Mississippi. *J. Econ. Entomol.* 45:593–97
47. Green, H. B. 1962. On the biology of the imported fire ant. *J. Econ. Entomol.* 55:1003–4
48. Green, H. B. 1967. The imported fire ant in Mississippi. *Miss. State Univ. Exp. Sta. Bull.* 737, 23 pp.
49. Harris, W. G., Burns, E. C. 1972. Predation on the lone star tick by the imported fire ant. *Environ. Entomol.* 1: 362–65
50. Hays, S. B., Arant, F. S. 1960. Insecticidal baits for control of the imported fire ant, *Solenopsis saevissima richteri. J. Econ. Entomol.* 53:188–91
51. Hays, S. B., Hays, K. L. 1959. Food habits of *Solenopsis saevissima richteri* Forel. *J. Econ. Entomol.* 52:455–57
52. Henderson, D. K. 1966. Decca system looks good for large scale ag operations. *Am. Aviat.* 30(3):30–34
53. Hunter, P. E., Costa, N. 1971. Description of *Gymnolaelops shealsi,* n. sp., as-

sociated with the imported fire ants. *J. Ga. Entomol. Soc.* 6:51–53

54. Jouvenaz, D. P., Banks, W. A., Lofgren, C. S. 1974. Fire ants: Attraction of workers to secretions of queens. *Ann. Entomol. Soc. Am.* 67:In press
55. Jouvenaz, D. P., Blum, M. S., MacConnell, J. G. 1972. Anti-bacterial activity of venom alkaloids from the imported fire ant, *Solenopsis invicta* Buren. *Antimicrob. Ag. Chemother.* 2:291–93
56. Jouvenaz, D. P., Lofgren, C. S., Carlson, D. A., Banks, W. A. 1974. Studies on the trail following pheromone of four species of fire ants, *Solenopsis.* Pap. presented at *Ann. Meet. Southeast. Br. Entomol. Soc. Am., 48th, Memphis*
57. Lennartz, F. E. 1973. *Modes of dispersal of Solenopsis invicta from Brazil into the continental United States—A study in spatial diffusion.* MS thesis, Univ. of Florida, Gainesville. 242 pp.
58. Levy, R., Chiu, Y. J., Banks, W. A. 1973. Laboratory evaluation of candidate bait toxicants against the imported fire ant, *Solenopsis invicta. Fla. Entomol.* 56:141–46
59. Lofgren, C. S., Banks, W. A., Stringer, C. E. 1964. Toxicity of various insecticides to the imported fire ant. *US Dep. Agr. Agr. Res. Serv.* 81–11. 15 pp.
60. Lofgren, C. S., Bartlett, F. J., Stringer, C. E. Jr., Banks, W. A. 1964. Imported fire ant toxic bait studies: Further tests with granulated mirex-soybean oil bait. *J. Econ. Entomol.* 57:695–98
61. Lofgren, C. S., Bartlett, F. J., Stringer, C. E. 1963. Imported fire ant toxic bait studies: Evaluation of carriers for oil baits. *J. Econ. Entomol.* 56:62–66
62. Lofgren, C. S., Bartlett, F. J., Stringer, C. E. 1964. The acceptability of some fats and oils as food to imported fire ants. *J. Econ. Entomol.* 57:601–2
63. Lofgren, C. S., Stringer, C. E., Banks, W. A., Bishop, P. M. 1967. Laboratory tests with candidate bait toxicants against the imported fire ant. *US Dep. Agr. Agr. Res. Serv.* 81–14. 25 pp.
64. Lofgren, C. S. Stringer, C. E., Bartlett, F. J. 1961. Imported fire ant toxic bait studies: The evaluation of various food materials. *J. Econ. Entomol.* 54:1096–1100
65. Lofgren, C. S., Stringer, C. E., Bartlett, F. J. 1962. Imported fire ant toxic bait studies: GC–1283, a promising toxicant. *J. Econ. Entomol.* 55:405–7
66. Lofgren, C. S., Weidhaas, D. E. 1972. On the eradication of imported fire ants: A theoretical appraisal. *Bull. Entomol. Soc. Am.* 18:17–20
67. Lowe, J. I., Parrish, P. R., Wilson, A. J. Jr., Wilson, P. D., Duke, T. W. 1971. Effects of mirex on selected estuarine organisms. *Trans. Conf. N. Am. Wildl. Natur. Resources, 36th,* 171–86
68. Ludke, J. L., Finley, M. T., Lusk, L. 1971. Toxicity of mirex to crayfish, *Procambarus blandingi. Bull. Environ. Contam. & Toxicol.* 6:89–96
69. MacConnell, J. G., Blum, M. S., Fales, H. M. 1970. Alkaloid from fire ant venom: Identification and synthesis. *Science* 168:840–41
70. MacConnell, J. G., Blum, M. S., Fales, H. M. 1971. The chemistry of fire ant venom. *Tetrahedron* 26:129–39
71. MacConnell, J. G., Williams, R. N., Brand, J. M., Blum, M. S. 1974. New alkaloids in the venoms of fire ants. *Ann. Entomol. Soc. Am.* 67:134–35
72. Markin, G. P., Collins, H. L., Dillier, J. H. 1972. Colony founding by queens of the red imported fire ant, *Solenopsis invicta. Ann. Entomol. Soc. Am.* 65:1053–58
73. Markin, G. P., Dillier, J. H. 1971. The seasonal life cycle of the imported fire ant, *Solenopsis saevissima richteri,* on the Gulf Coast of Mississippi. *Ann. Entomol. Soc. Am.* 64:562–65
74. Markin, G. P., Dillier, J. H., Collins, H. L. 1973. Growth and development of colonies of the red imported fire ant, *Solenopsis invicta. Ann. Entomol. Soc. Am.* 66:803–8
75. Markin, G. P., Dillier, J. H., Hill, S. O., Blum, M. S., Hermann, H. R. 1971. Nuptial flight and flight ranges of the imported fire ant, *Solenopsis saevissima richteri. J. Ga. Entomol. Soc.* 6:145–56
76. Markin, G. P. et al. 1972. The insecticide mirex and techniques for its monitoring. *US Dep. Agr. APHIS* 81–3. 19 pp.
77. Markin, G. P., Henderson, J. A., Collins, H. L. 1972. Aerial application of microencapsulated insecticide. *Agr. Aviat.* 14:70–75
78. Markin, G. P., Hill, S. O. 1971. Microencapsulated oil bait for control of the imported fire ant. *J. Econ. Entomol.* 64:193–96
79. Mills, H. B. 1967. *Chairman, Report of committee on imported fire ant, National Academy of Sciences, to administrator, Agr. Res. Serv., US Dep. Agr.* 15 pp.
80. Morrill, W. L. 1974. Production and flight of alate red imported fire ants. *Environ. Entomol.* 3:265–71
81. Naber, E. C., Ware, G. W. 1965. Effect of kepone and mirex on reproductive

performance in the laying hen. *Poultry Sci.* 44:875–80

82. Oberheu, J. C. 1972. The occurrence of mirex in starlings collected in seven southeastern states. *Pestic. Monit J.* 6:41–42
83. Oliver, A. D. 1960. Infestation of the imported fire ant, *Solenopsis saevissima* v. *richteri*, at Fort Benning, Ga. *J. Econ. Entomol.* 53:646–48
84. O'Neal, J. 1974. Predatory behavior exhibited by three species of ants on the imported fire ants, *Solenopsis invicta* Buren and *S. richteri* Forel. *Ann. Entomol. Soc. Am.* 67:140
85. O'Neal, J., Mangum, C. 1974. The sterility effects of juvenile hormone analogues on queens of the imported fire ants, *Solenopsis invicta* Buren. Pap. presented *Ann. Meet. Southeast. Br. Entomol. Soc. Am., 48th, Memphis*
86. O'Neal, J., Markin, G. P. 1973. Brood nutrition and parental relationships of the imported fire ant, *Solenopsis invicta. J. Ga. Entomol. Soc.* 8:294–303
87. Paget-Clarke, C. D. 1971. Electronic guidance system used in fire ant eradication programmes. *Agr. Aviat.* 13:89–90
88. Raun, E. S., Jackson, R. D. 1966. Encapsulation as a technique for formulating microbial and chemical insecticides. *J. Econ. Entomol.* 59:620–22
89. Reagan, T. E., Coburn, G., Hensley, S. D. 1972. Effects of mirex on the arthropod fauna of a Louisiana sugarcane field. *Environ. Entomol.* 1:588–91
90. Rhoades, W. C., Davis, D. R. 1967. Effects of meterological factors on the biology and control of the imported fire ant. *J. Econ. Entomol.* 60:554–58
91. Ricks, B. L., Vinson, S. B. 1970. Feeding acceptability of certain insects and various water-soluble compounds to two varieties of the imported fire ant. *J. Econ. Entomol.* 63:145–48
92. Ricks, B. L., Vinson, S. B. 1972. Digestive enzymes of the imported fire ant, *Solenopsis richteri. Entomol. Exp. Appl.* 15:319–34
93. Roe, R. A. II. 1973. *A biological study of Solenopsis invicta Buren, the red imported fire ant, in Arkansas with notes on related species.* MS thesis, Univ. of Arkansas, Fayetteville. 135 pp.
94. Ruckelshaus, W. D. 1972. Products containing mirex; determination and order. *Fed Register* 37(130):299–300
95. Sikorowski, P. P., Norment, B. R., Broome, J. R. 1973. Fungi pathogenic to imported fire ants. *Miss. Agr. Forest. Exp. Sta. Inform. Sheet 1212*
96. Silveira-Guido, A., Bruhn, J. C., Crisci, C., San Martin, Y. P. 1968. *Labauchena daguerri* Santschi, como parasito social de la hormiga *Solenopsis saevissima richteri* Forel. *Agron. Trop.* 18(2): 207–9
97. Silveiro-Guido, A., Carbonell, J., Crisci, C. 1973. Animals associated with the *Solenopsis* (fire ants) complex, with special reference to *Labauchena daguerri. Proc. Tall Timbers Conf. Ecol. Anim. Contr. Habitat Manag.* 4:41–52
98. Steinhaus, E. A., Marshall, G. A. 1967. Previously unreported accessions for diagnosis and new records. *J. Invertebr. Pathol.* 9:436–38
99. Stringer, C. E. Jr., Lofgren, C. S. Bartlett, F. J. 1964. Imported fire ant toxic bait studies: Evaluation of toxicants. *J. Econ. Entomol.* 57:941–45
100. 1970. Fighting the fire ant. *Time* Nov. 2, 1970:40
101. Triosi, S. J., Riddiford, L. M. 1974. Juvenile hormone effects on metamorphosis and reproduction of the fire ant, *Solenopsis invicta. Environ. Entomol.* 3:112–16
102. Triplett, R. F. 1973. Sensitivity to the imported fire ant: Successful treatment with immunotherapy. *S. Med. J.* 66:477–80
103. U.S. Department of Agriculture. 1958. Observations on the biology of the imported fire ant. *US Dep. Agr. Agr. Res. Serv. 33–49.* 21 pp.
104. VanValin, C. C., Andrews, E. K., Eller, L. L. 1968. Some effects of mirex on two warm-water fishes. *Trans. Am. Fish. Cult.* 97:185–96
105. Vickers, L. 1973. Fire ant control could trigger fish problems. *Nat. Fisherman* April 1973:10–11C
106. Vinson, S. B. 1968. The distribution of an oil, carbohydrate, and protein food source to members of the imported fire ant colony. *J. Econ. Entomol.* 61: 712–14
107. Vinson, S. B. 1970. Gustatory response by the imported fire ant to various electrolytes. *Ann. Entomol. Soc. Am.* 63: 932–35
108. Vinson, S. B. 1972. Imported fire ant feeding on *Paspalum* seed. *Ann. Entomol. Soc. Am.* 65:988
109. Vinson, S. B., Robeau, R., Dzuik, L. 1974. Bioassay and activity of several juvenile hormone analogues on the imported fire ant, *Solenopsis invicta. Environ. Entomol.* 3:In press
110. Vinson, S. B., Thompson, J. L., Green, H. B. 1967. Phagostimulants for the imported fire ant, *Solenopsis saevissima*

var. *richteri, J. Insect Physiol.* 13: 1729–36
111. Walsh, C. T., Law, J. H., Wilson, E. O. 1965. Purification of the fire ant trail substance. *Nature* 207:320–21
112. Walsh, J. P., Tschinkel, W. R. 1974. Brood recognition by contact pheromone in the imported fire ant, *Solenopsis invicta. Anim. Behav.* 22:In press
113. Ware, G. W., Good, E. E. 1967. Effects of insecticides on the reproduction in the laboratory mouse. II. Mirex, telodrin and DDT. *Toxicol. Appl. Pharmacol.* 10:54–61
114. Whitcomb, W. H., Bhatkar, A., Nickerson, J. C. 1973. Predators of *Solenopsis invicta* queens prior to colony establishment. *Environ. Entomol.* 2:1101–3
115. Whitcomb, W. H., Denmark, H. A., Bhatkar, A. P., Greene, G. L. 1972. Preliminary studies on the ants of Florida soybean fields. *Fla. Entomol.* 55:129–42
116. Williams, R. N., Panaia, J. R., Gallo, D., Whitcomb, W. H. 1973. Fire ants attacked by phorid flies. *Fla. Entomol.* 56:259–62
117. Wilson, E. O. 1951. Variation and adaptation in the imported fire ant. *Evolution* 5(1):68–79
118. Wilson, E. O. 1952. O complexo *Solenopsis saevissima* na America do Sul. *Inst. Oswaldo Cruz Mem.* 50:49–68
119. Wilson, E. O. 1959. Source and possible nature of the odor trail of fire ants. *Science* 129(3349):643–44
120. Wilson, E. O. 1962. Chemical communication among workers of the fire ant, *Solenopsis saevissima* (Fr. Smith). 1. The organization of mass-foraging. 2. An informational analysis of the odour trail. 3. The experimental induction of social responses. *Anim. Behav.* 10: 134–64
121. Wilson, E. O. 1971. *The Insect Societies.* Cambridge, Mass.: Belknap Press of Harvard Univ. Press. 548 pp.
122. Wilson, E. O., Eads, J. H. 1949. A report on the imported fire ant, *Solenopsis saevissima* var. *richteri,* in Alabama. *Spec. Rep. Ala. Dep. Conserv. Mimeo.* 54 pp.
123. Wilson, N. L., Dillier, J. H., Markin, G. P. 1971. Foraging territories of imported fire ants. *Ann. Entomol. Soc. Am.* 64:660–65
124. Wilson, N. L., Oliver, A. D. 1969. Food habits of the imported fire ant in pasture and pine forest areas in southeastern Louisiana. *J. Econ. Entomol.* 62: 1268–71
125. Wing, M. W. 1968. Taxonomic revision of the Nearctic genus *Acanthamyops. Memoir 405, Cornell Univ. Agr. Exp. Sta.* 173 pp.
126. Wojcik, D. P., Banks, W. A., Plumley, J. K., Lofgren, C. S. 1973. Red imported fire ant: Laboratory tests with additional candidate bait toxicants. *J. Econ. Entomol.* 66:550–51
127. Woodruff, R. E. 1973. The scarab beetles of Florida. Arthropods of Florida and neighboring lands. *Fla. Dep. Agr. Conserv. Serv., Div. Plant Ind.* Vol. 8. 220 pp.

BIOLOGICAL CONTROL OF AQUATIC WEEDS

◆6079

Lloyd A. Andres and Fred D. Bennett

Biological Control of Weeds Laboratory, U.S. Department of Agriculture, Albany, California 94706 and Commonwealth Institute of Biological Control, Trinidad, West Indies

Although insects had been used to control terrestrial weeds on numerous occasions (36, 39, 40, 75), the introduction of the chrysomelid beetle, *Agasicles hygrophila,* into the southeastern United States in 1964 for the control of alligatorweed (*Alternanthera philoxeroides*) marked their first use as aquatic weed control agents (35, 77). The relative success of that project served to reduce skepticism as to whether the insects associated with aquatic plants were sufficiently monophagous to permit their use for biological control (75) and has focused increased attention on this area.

Although much has been written about the biological control of terrestrial weeds, the literature relating to the biological control of aquatic weeds is still limited but increasing rapidly. An excellent starting point on the subject of water weeds and their control is the bibliography by Little (45). Blackburn et al (14) prepared a general review of the kinds of natural enemies used to control aquatic plants, including the use of polyphagous organisms in the general reduction of aquatic vegetation. The use of insects for aquatic weed control was discussed by Bennett (9–11).

As with terrestrial plant communities, the elements which limit and regulate the abundance of aquatic plant populations are primarily climatic, edaphic (hydrologic in this case), and biotic. The successful manipulation of the latter to control aquatic weeds depends on: (*a*) the particular weed problem, e.g. the weed species involved, the habitat, the degree of control desired; (*b*) the availability of suitable natural enemies, e.g. host-specific or polyphagous insects, herbivorous fish, plant pathogens; and (*c*) the interactions between *a* and *b.* This review examines these three elements, particularly with respect to the use of insects as natural enemies.

AQUATIC WEEDS

The Problem

Plants form an important element of the aquatic environment, manufacturing and providing food and shelter for insects, fish, and various forms of wildlife. Frequently these plants become over abundant and create problems by obstructing water flow and navigation, by limiting recreational use of the water, by increasing water loss through evapotranspiration, and by contributing to health hazards. Holm et al (37) estimated that aquatic ditchbank weeds cause an annual loss of 1,272,480 acre-feet of water at a cost of $39¼ million in 17 western states.

In North America over 170 species of aquatic plants are classified as weeds, of which 40 to 50 species are considered of major importance (65, 72). For purposes of this review the plants are grouped into two broad categories: (*a*) submersed weeds (i.e. those with most or all of their vegetative structures underwater, e.g. *Hydrilla* and *Najas*), and (*b*) emersed or floating weeds (i.e. those rooted underwater with the major portion of their vegetation growing above it, e.g. *Alternanthera;* or those that float freely on the water surface, e.g., *Eichhornia* and *Salvinia*).

The importance of each category as weeds will vary. Holm et al (37) considered the submersed weeds to be the more serious since they drastically reduce water flow and recreational value, are quick to invade new areas, and are difficult to control. Sculthorpe (60), on the other hand, considered submersed weeds to be the least troublesome, although perhaps of local importance, rating the stoloniferous free-floating weeds the more serious problem worldwide [e.g. *Eichhornia crassipes, Salvinia molesta* (=*auriculata*), and *Pistia stratiotes*].

Most bodies of water will support at least some type of flora—the plant species and their abundance depending upon their inherent characteristics (such as temperature-tolerance, growth cycle, and adaptation for rapid dispersal), and factors of the habitat (such as topography, substrate, and water quality) (34, 38, 42). Eutrophication has led to increasing weed problems in reservoirs (18), and was cited by Little (45) as one of the chief causes of recent aquatic weed proliferations in advanced countries. An excess of nutrients will affect each plant species differently, depending on its rate of nutrient uptake and the concentrations needed for optimal growth. Algae often increase rapidly following the addition of nutrients to water, followed by the vascular species (52). Shifting nutrient levels have been cited as a cause of plant succession in the man-made Kariba and Volta lakes of Africa (26).

Control

In seeking the solution to aquatic weed problems, one frequently must work with several unrelated plant species, some of which are indigenous, and beneficial at one level of abundance, while at different levels they are detrimental. Sculthorpe (60) listed four items to consider in solving this dilemma: 1. Determine how much aquatic vegetation is needed to provide food and shelter for fish and wildlife, edaphic stability, and other benefits essential for the proper development of the aquatic habitat [e.g. plants remove excess nutrients from the water, and beds of emergent plants trap sediments and aid in building up rich soil along the shores at time of water drawdown, and also improve fish production (26)]. 2. Determine the minimum amount of vegetation that will retard water flow, deplete dissolved oxygen supplies, or otherwise interfere with multipurpose water use. In irrigation and drainage ditches, and areas used for swimming or water-skiing, at least 80% of the water should be weed-free; where fishing, hunting and water storage are the main consideration, 30 to 50% control is usually adequate; in water held for drinking purposes, any aquatic growth is considered undesirable (F. L. Timmons, personal communication). 3. Determine the plant abundance needed to achieve a practical balance between points 1 and 2. 4. Decide how to maintain the vegetation at the desired level at a minimum cost. It is probably safe to say that for most weed problems, adequate information is lacking on all four points. At first glance an acceptable balance between vegetation and open water would seem to be a suitable goal; however, attention should also be given to the characteristics and values of each plant species involved. Hence, each weed problem is unique and demands unique solutions.

The several control options available to the aquatic weed scientist were listed and briefly discussed by Little (45) and Holm et al (37). Aquatic weeds are commonly removed by herbicides and mechanical equipment, but, in addition to other drawbacks, their effects are of limited duration. Other options include the harvesting of aquatic weeds, either mechanically for processing into protein or other food supplements, plant mulch, etc (17, 57), or biologically by conversion into protein by herbivorous fish (41). Regarding the latter, questions of host preference and impact on other fauna and the environment need careful attention before introducing new fish into an area (51, 64, 67).

Biological control with insects and other organisms presents still another alternative depending on the degree and selectivity of control needed. Although the majority of aquatic weeds are indigenous species, the plants attracting major attention are exotics (e.g. *E. crassipes, Myriophyllum spicatum, S. molesta*). These alien weeds frequently dominate large areas as relatively homogeneous populations. Certain terrestrial weeds similarly distributed have proven well suited for biological control with selective insect enemies (30, 39).

The mixed blessings provided by most aquatic plants may restrict the use of free-ranging natural control organisms, since their impact cannot be predicted nor limited to specific weedy sites. For example, Ensminger (25) states that alligatorweed is considered important as food for cattle, nutria, and white-tailed deer in the coastal marshes of Louisiana, and that it protects the stream banks from erosion during flooding. He further notes that the introduced *Agasicles* beetle has already damaged the plant in some of the marsh areas where it is grazed by cattle. Such problems of conflicting interests in aquatic weeds will be of increasing importance.

BIOLOGICAL CONTROL AGENTS

Any plant-feeding organism or parasite may be used to control aquatic weeds, providing it does not harm plants of value or create undesirable imbalances in the plant community. Some of the natural enemies that have been considered for the control of submersed, emersed, and floating weeds are discussed below.

Herbivorous Fish

The white amur, *Ctenopharyngodon idella,* a fish native to the Amur River areas of China and the USSR, has been cultured on aquatic plants in ponds; employed for weed control in China, Hungary, and Japan; and is felt to have promise in other regions (3). It feeds on a range of submersed, emersed, and floating plants, but prefers plants having soft tissue. Its rate of growth and development vary with the food source (3, 67). The white amur displays good high and low temperature tolerance, and is not known to reproduce naturally outside of its native waters (14). Other herbivorous fish have been considered for control of aquatic weeds (14).

Snails

The snail, *Marisa cornuarietis,* feeds on a number of aquatic plants and was considered to have weed control potential (61). However, its usefulness is limited by its ability to feed on young rice plants and poor tolerance to water temperatures below 10°C. On the other hand, its ability to destroy the breeding sites of the snail vector of bilharzia may allow its introduction into areas where rice culture is not important (15). *Pomacea australis,* a South American snail, is also being considered for weed control (14).

Plant Pathogens

The role of pathogens in aquatic weed control was reviewed by Zettler & Freeman (79). A number of pathogens have been identified, but not yet employed in biological control.

Miscellaneous Agents

The manatee, *Trichechus manatus,* has been used against aquatic weeds in South America (1) and tested in the United States. It breeds slowly and is highly sensitive to disturbance (14). The crayfish, *Orconectes causeyi,* was credited with controlling submersed weeds in a trout lake (23). An example of biological control by a plant competitor is the use of *Eleocharis* spp. plants to blanket the bottoms of canals in California, to exclude noxious bottom-rooted plants (76). Floating plants and plants growing along the banks of waterways can also shade out aquatic weeds (67).

Each of the above types of organisms has its advantages and limitations. The numbers of polyphagous grazers per unit area must be regulated to minimize the effects of host preference and to obtain optimal control action.

Insects

The co-evolution of plants and phytophagous insects has led to a high degree of host specialization among the latter. The remainder of this review will deal with the availability, host-feeding range, and interaction between insects and their aquatic plant hosts.

AVAILABILITY OF INSECTS Host records from the literature and surveys of living insects associated with an aquatic plant throughout its range are used to detect insects of potential biological control value. The growth habit of the plant will influence the number of insects recorded. McGaha (50), working with the insects on several emersed plant species, and Berg (13), in a study of the insects associated with several *Potamogeton* spp. having submersed, floating, and emersed parts, noted a greater variety of insects on the emersed as compared with the submersed plant structures. Surveys of insects on emersed, floating, and submersed weeds tend to bear this out. The thoroughness and timing of surveys, as well as the history of the area surveyed, will affect the number of insects collected. The 30-plus phytophagous species collected from waterhyacinth (12; A. Silveira-Guido, unpublished report) and the 40 to 53 species from alligatorweed (69, 70) in their native ranges compare favorably with the number of insect species collected from certain terrestrial European thistles (80). The numbers of insect species collected from submersed plants tend to run somewhat lower: 11 species were collected from watermilfoil (*Myriophyllum* spp.) in Pakistan (29), while 15 species were taken from watermilfoil in Yugoslavia (44); 8 species of insects were reported from *Hydrilla verticillata* (5). The greater number of phytophagous insects associated with emersed or floating plants than submersed plants suggests a greater availability of species for use against weeds of the former type. However, generalizations of this nature are questionable, since the capacity of an insect to regulate the abundance of its host plant must be judged on a plant-by-plant, insect-by-insect basis.

HOST SPECIFICITY When introducing an exotic phytophagous insect to control a weed, great attention is given to its host specificity and whether it will damage plants of value. The several criteria used to judge the host specificity of weed-feeding insects, such as insect biologies and host-determined adaptations, plants attacked by related insects, host range studies, and causal analysis of host specificity (81), have also been considered in studies with insects attacking aquatic weeds.

Frequently an insect will have certain physiological, behavioral, morphological and/or other adaptations to particular plant structures, species, or habitats, which serve to limit its host range. The *Agasicles hygrophila* larvae require an adequately inflated alligatorweed stem as a pupation site (46). The larvae of the weevil, *Neochetina eichhorniae,* which mine the petioles of waterhyacinth, need roots of the type possessed by waterhyacinth for successful pupation (56). The stem size and texture required for successful oviposition by *Cornops aquaticum* were apparently sufficient to limit reproduction to waterhyacinth, although the developing eggs and nymphs could be reared to maturity on plants in related families in the laboratory (7, 63).

Insects attacking submersed plant parts frequently have adaptations which enable them to utilize oxygen from the water or the host plant (66). For example, *Parapoynx stratiotata* larvae, which feed on the submersed leaves of Eurasian watermilfoil (*Myriophyllum spicatum*), take oxygen directly from the water through gills, later cutting a hole in the plant stalk to obtain oxygen from the plant during pupation (43). Adaptations of this nature ensure that emersed plant tissues will not be attacked, but generally do not limit the insect to specific host plants (43). Likewise, the need of the grasshopper *Paulinia acuminata* to place her eggs on the underwater surface of the host plant restricts this species to the aquatic environment, but not to a specific host plant (6).

The host range of insects closely related to the weed-feeding candidate biological control agent will often give a clue to the evolution of the latter's host relationship.

The fact that the *Agasicles* spp. beetles are all closely linked to the aquatic amaranths of the genus *Alternanthera* substantiated the conclusion that *A. hygrophila* would not attack plants other than the weedy *Alternanthera philoxeroides* (G. B. Vogt, J. U. McGuire, and A. D. Cushman, submitted for publication). The lack of host information for many aquatic and semi-aquatic insects limits the value of this evaluation at the present time.

Field observations and laboratory feeding, oviposition, and developmental studies are used to assess the stability of an insect-host plant relationship. Host specificity studies on *Agasicles hygrophila* (D. M. Maddox, unpublished data), *Amynothrips andersoni* (47), and *Vogtia malloi* (D. M. Maddox and R. D. Hennessey, unpublished data) from alligatorweed; on *N. eichhorniae* (56), *Neochetina bruchi* (B. D. Perkins and D. M. Maddox, submitted for publication; C. J. DeLoach, submitted for publication), and *Epipagis albiguttalis* (58) from waterhyacinth; and on *Samea multiplicalis, Cyrtobagous singularis,* and *P. acuminata* (6) from *Salvinia* spp. indicated sufficient host specificity to allow these insects to be safely introduced into areas outside their native range. Although limited studies on insects attacking submersal plants [e.g. *P. stratiotata* on *M. spicatum* (43); the ephydrid, *Hydrellia* sp. on *H. verticillata* (5)] indicate a high degree of host preference, there is as yet insufficient evidence to support their clearance for release to other areas.

Zwölfer & Harris (81) note that an insect's host range can be precisely determined if we know the specific visual, chemical, and tactile stimuli used by the insect in the finding and acceptance of its host plant. If these stimuli in toto are characteristic only of the host plant, other plants will not be attacked (32). For example, a flavone feeding stimulant extracted from alligatorweed is considered to be partially responsible for the feeding of the alligatorweed flea beetle, *A. hygrophila* (78). It is quite likely that plants lacking this factor will not be attacked. Unfortunately no other studies have been carried out on the causal analysis aspect of the insect-aquatic weed relationship.

THE USE OF INSECTS AGAINST AQUATIC WEEDS Insects have already proven of value against an emersed plant [alligatorweed (22, 48)] and several species have been introduced to control floating weeds [e.g. *Salvinia* (9); waterhyacinth (10, 54)]. However, the value of their use against floating and submersed weeds remains untested. Certainly insects damage submersed plants (13, 43) and play a role in determining the composition of the submersed plant community (24), but whether those insects attacking submersed plants are sufficiently host specific to meet the current criteria for safe introduction remains in doubt. Bennett (9) suggested that clearance might be safely granted if, 1. field observations or other criteria indicate that the insect is restricted to the aquatic environment, 2. the insect does not damage plants of positive economic value, and 3. it will not cause irreparable harm to the ecosystem. The paucity of economically important plants in the aquatic habitat suggests that insects with a somewhat broader host range might be used safely against submersed weeds; however, increased understanding of the aquatic plant community in relation to the productivity of aquatic ecosystems is needed.

INTERACTION: PLANTS, INSECTS, AND ENVIRONMENT

There are three elements necessary for the successful outcome of a biological control project: 1. establishment or presence of natural enemies, 2. buildup of the natural enemy population, and 3. control of the target weed. Completion of each step depends upon a balanced interaction between the weed, the environment, and the weed-feeding insect. Establishment and buildup depend primarily on the acceptability of the host to the insect (i.e. plant quality), the climate, and the absence of indigenous natural enemies. For establishment to be successful, the reproduction of the insect should not be precluded by environmental extremes. The rate and level of population buildup depends on how closely environmental conditions approximate those required for optimum insect development. Control depends on the insect's ability to damage the plant (type of attack) and timing of the damage to some critical point in the plant's developmental cycle. Other environmental pressures on the plant, e.g. plant competition, can affect the outcome of control. It is difficult to weigh the importance of each factor in terms of a particular biological control project without detailed study and experimentation. Although some work of this nature has been done with alligatorweed, information in these areas is very limited as regards the biological control of other aquatic weeds.

Establishment and Buildup of Natural Enemies

PLANT QUALITY There is considerable field evidence suggesting that site to site variations in the quality of alligatorweed influenced the establishment and buildup of *Agasicles.* Beetles released on the Savannah National Wildlife Refuge in March 1964 caused sporadic heavy damage to alligatorweed and spread to other areas of the refuge, but persisted at low levels only (35). An unexplained leaf drop in the summer of 1964 followed by a decline in the condition of the plants, apparently the result of mineral deficiencies and water level management practices, were credited with halting the beetle buildup (22). High beetle populations in a nearby canal in Savannah, at another release site near Charleston, S. Carolina, and on the Santee River, S. Carolina also point towards plant quality rather than temperature as the limiting factor in the wildlife refuge (74).

Alligatorweed with spindly top-growth or fibrous stems with hardened nodes and shortened internodes is avoided by adult beetles (49). Bottom-rooted plants, frequently noted to have spindly stems, were often unattractive to the beetle, thus limiting buildup at sites in the Florida Panhandle (22). Laboratory plants with induced phosphorus, calcium, and magnesium deficiencies exhibited growth characteristics similar to the unthrifty field plants and were less attractive to the beetles, resulting in reduced fecundity (D. M. Maddox and M. Rhyne, unpublished manuscript).

CLIMATE Winter temperatures are thought to be the most important factor influencing the establishment and effectiveness of nondiapausing *Agasicles* over the range of alligatorweed in the United States (74). The zone of beetle effectiveness appears to be limited to areas with average January temperatures of 49–60°F with a frost-free period of 240 to 250 days. A gradual northward extension of the beetle's range is apparently the result of adaptation to cold (22).

A midsummer slowdown of *Agasicles* buildup in Florida, Texas, and Louisiana is apparently the result of high temperatures (22, 74). On one occasion a drenching rain washed the beetles from the plant, stopping *Agasicles* buildup without particularly harming the plant (49).

The low night temperatures due to high altitude were considered an important factor in the nonestablishment of *Paulinia acuminata* on *Salvinia molesta* in Lake Naivasha, Kenya (9).

NATURAL ENEMIES

General predators There has been relatively little study on the impact of general predators on aquatic phytophagous insects other than general records on the several predators associated with *Agasicles* (46, 48), *Paulinia acuminata* (11), *Cornops aquaticum* (63), and *C. longicorne* (12). Bennett (11) speculates that predation by invertebrate predators, fish, and other aquatic organisms on those insect stages not passed within the host plant may be of importance.

Parasites Indigenous parasites in the United States have been recorded attacking the introduced *Agasicles* (48) and *Vogtia malloi* (19), but the data are insufficient to assess their impact on the buildup of the host. A higher relative abundance of entomophagous to phytophagous insects collected from emersed vs submersed hydrophytes (13) suggests that parasites may play only a minor role in regulating weed-feeding insects introduced to control submersed weeds. Bennett (11) noted the absence of parasites attacking several of the phytophagous insects collected from *Salvinia* and *Eichhornia* in South America, but states that this does not extend to the lepidopterous phytophages on these plants.

Pathogens The fact that *Acigona infusella* and *Epipagis albiguttalis* on *Eichhornia* are reported to be infected by microsporidians in their native South America led Bennett (11) to suggest that pathogens may be a more important factor in limiting the populations of insects for aquatic weed control than for terrestrial weeds.

Certainly further study is needed to clarify the role of natural enemies in limiting the phytophages of aquatic plants. The recent discovery of a microsporidian and nematode parasitic on *Neochetina eichhorniae* and *Neochetina bruchi* in South America (C. J. DeLoach, personal communication) partially answers the concern of Bennett (11) that if natural enemies are not holding aquatic weed-feeding insects in check in their native areas, how can we expect higher populations in areas where they may be released to control weeds? In the case of *N. eichhorniae,* apparently, there are natural enemies holding this insect in check in its native area. Populations of *N. eichhorniae* are estimated to be reaching higher levels where released in Florida than in South America (B. D. Perkins, personal communication).

OTHER FACTORS Changing water levels and silting were blamed for the elimination of *Agasicles* from alligatorweed in California (27). Fluctuating water levels were also felt to hamper the buildup of *Bagous* sp., weevils feeding on milfoil, in Pakistan (4). Insect attack on submersed plants is limited by silting (discourages oviposition and larval entry), pollution (oil and scum entrap insects), and water disturbance (boat traffic and wave action dislodge insects) (50). The density of insects on *Potamogeton* spp. was higher in sheltered bodies of water where wave and wind action was minimal (13). Certainly other factors of the aquatic environment, e.g. water flow and insecticide runoff from nearby cropland, can influence the impact of insects in controlling aquatic weeds.

Control

TYPE OF ATTACK The relationship between type of insect attack and successful weed control is little understood. Defoliation by *Agasicles* larvae and adults and the cutting of stem vascular tissues by tunneling *V. malloi* larvae have proven effective in the control of alligatorweed (20, 22). Feeding by *Cyrtobagous singularis* adults on the developing buds of *Salvinia* and the invasion of the wound by pathogenic organisms prompted Bennett (9) to speculate that pathogens may be of greater importance in aquatic weed control than in the biological control of terrestrial weeds. Also, feeding damage by the weevil *N. eichhorniae* permits pathogens to enter waterhyacinth—the most virulent being the leafspot, *Acremonium zonatum* (21). Whether pathogens will have increased value against aquatic weeds remains to be proven.

TIME OF ATTACK The timing of the weed insect's attack can be equally as important as the type of damage it may cause. The removal of new alligatorweed growth by *Agasicles* prior to and at a time when the plant's energy reserves were at a low level, contributed to the rapid destruction of the plant at Jacksonville, Florida (2). On the other hand, the webworms, *Herpetogramma bipunctalis* and *Hymenia recurvalis* (=*fascialis*), which can also cause heavy defoliation (77), do so late in the season at a time when the energy reserves are high, with little apparent consequence. To be effective, the *Agasicles* overwintering population should be of sufficient size to permit the beetles to reach effective control levels by June (22). Site to site variation in beetle population size and the growth pattern of alligatorweed will cause variations in control attained (74). The buildup of *Vogtia malloi* throughout the season should offset the *Agasicles* midsummer population dip and improve control where both occur (19).

OTHER FACTORS The action of other elements can often supplement or nullify the impact of the controlling insect. Competition from smartweed (*Polygonum* sp.) and water primrose (*Jussiaea* sp.) in conjunction with *Agasicles* damage, reduced the stand of alligatorweed by as much as 90% near Charleston, S. Carolina (74). Wind action forced waterhyacinth plants onto *Agasicles*-weakened alligatorweed mats at several sites, hastening its control. Weldon et al (74) also claim that low dosages of herbicides (e.g. 2,4-D) speed and enhance the control of alligatorweed by *Agasicles.*

Evaluation Studies

Studies correlating weed control with the establishment and increase of an insect population will be of greatest value if they discern the key factors involved. Brown (19) used a system of canonical correlations in conjunction with multivariate analysis in studying the relationship between alligatorweed stem length, height, internode length and diameter, number of stems, insect abundance, and several environmental factors, which enabled him to speculate on the importance of each in relation to control exerted by *Vogtia* and *Agasicles.* Preintroduction surveys of the indigenous natural enemies associated with the problem weed are of value in determining which insects to select for introduction (e.g. waterhyacinth surveys) (28, 68).

PROJECTS

Alligatorweed (Alternanthera philoxeroides)

This native South American plant was introduced into North America in the 1890s (73). It grows primarily as an emersed plant, rooted along the water's edge or on the bottom in shallow areas; its buoyant hollow stems form floating mats on the water surface. When extensive, these mats block waterways. Maddox et al (48) noted that, apart from waterhyacinth, few aquatic plants can successfully compete with alligatorweed over much of its range in the southeastern United States. Its ability to regenerate from any stem node limits attempts at mechanical control. Poor translocation, among other characteristics, reduces the effective use of herbicides.

Between 40 and 53 species of insects were recorded associated with the plant in South America, of which 5 were earmarked as major biotic suppressants (69, 70). Of these, *Agasicles hygrophila,* a leaf-feeding chrysomelid; *Vogtia malloi,* a stem-tunneling phycitid; and *Amynothrips andersonii* were cleared for release in the United States after appropriate host specificity tests. All 3 species are multivoltine and nondiapausing (48).

Agasicles is established in South Carolina, Florida, Mississippi, Georgia, Texas, Alabama, and Louisiana and perhaps North Carolina and Arkansas (22). Effective control appears to be correlated with the size of the overwintering populations and the rapidity of the beetle's spring population buildup. *Amynothrips,* while established at sites in South Carolina, Georgia, and Florida, has remained localized and has had little or no impact on alligatorweed (22).

Vogtia, though only recently released, is already established in Florida, Georgia, and South Carolina and has effected considerable control at one site in Florida (20). It has survived in relatively cold sites and seems unaffected by high summer temperatures in southern Florida (19).

Waterhyacinth (Eichhornia crassipes)

This plant is also native to South America, but is now widely distributed over much of the tropical and subtropical areas of the world (60). Its floating habit enables it to cover rivers and lakes, increasing water loss through evapotranspiration, and interfering with navigation, fishing, and other water uses (37).

Over 30 phytophagous insect species were reported from this plant in its native Uruguay (A. Silveira-Guido, unpublished report), Trinidad, Surinam, Guyana, and Brazil (12). Several of these insects have high potential for control (53). A number of phytophages have been recorded from this plant in the United States (28), India (V. P. Rao, unpublished report), and several West Indian Islands and British Honduras (8), where it is naturalized. Few of these are considered to have control value. One exception might be the indigenous lepidopteran, *Bellura densa,* which inflicts considerable damage to waterhyacinth in the southern United States but is limited by parasites (68). It apparently transferred to waterhyacinth from the related *Pontederia lanceolata.*

Among the insect enemies found in South America, the newly described weevil, *Neochetina eichhorniae* (71), was cleared and released in the United States where initial field observations indicate it is affecting the vigor and growth rate of its host plant (54, 56; B. D. Perkins, personal communication). It was shipped from Trinidad and Argentina to India for further studies under quarantine (58). *Neochetina bruchi,* another petiole-mining weevil, was tested by C. J. DeLoach (unpublished observation) and B. D. Perkins and D. M. Maddox (unpublished observation), and found to be equally as host specific as *N. eichhorniae.* Preliminary host specificity studies on *Acigona infusella* were reported by Silveira-Guido (62) and Bennett (8), and continuing observations on this and another petiole-mining pyraustine, *Epipagis albiguttalis,* are underway in Trinidad (8), India (58), and Argentina (C. J. DeLoach, personal communication). While host specificity tests with *E. albiguttalis* in Trindad were not conclusive, due to a high incidence of disease-induced mortality in the control, a clean culture was developed in India by the use of antibiotics, its host oligophagy proved, and material shipped to Zambia for release on the Kafue River and to Rhodesia (10). The oribatid mite, *Orthogalumna terebrantis,* indigenous to South America and already naturalized in the southern United States, is highly specific to *E. crassipes.* The nymphs tunnel and feed in the leaf tissue (55; H. Cordo and C. J. DeLoach, unpublished observation). This mite has been introduced and established in Zambia (10). Host specificity reports on the grasshoppers, *Cornops aquaticum* (63) and *Cornops longicorne* (8), indicate that further testing will be required.

Eurasian Watermilfoil (Myriophyllum spicatum)

Although this submersed weed was present in the United States prior to 1900, it has developed into a major weed problem in the Chesapeake Bay and the reservoirs of the TVA only since the late 1950s. It is currently spreading in other states (16). It is deleterious to fishing, commercial shellfish beds, and other water uses (37). The plant is found throughout Europe and Asia. Eleven species of insects were reported from milfoil in Pakistan (29). Several species showed a relatively high degree of host specificity, but bred only on plants or plant parts outside of the water (4). Of 15 insects reported in Yugoslavia, only a pyraustine moth, *Parapoynx stratiotata,* and the weevil *Litodactylis leucogaster* demonstrate sufficient specificity and control potential to be of possible value (43).

Salvinia (Salvinia molesta)

This floating aquatic fern has become a serious pest in parts of Asia and Africa where it is naturalized from the neotropics. Dense mats of salvinia interfere with navigation, reduce dissolved oxygen to levels inadequate for fish, pose a threat to hydroelectric turbines and have enhanced the spread of the snail vector of bilharzia (6).

Following the establishment of *Salvinia* on Lake Kariba, Africa, surveys of Trindad, Guyana, and parts of northern Brazil in 1961–1963 disclosed several insects on *Salvinia* spp. Host specificity and biological studies of three of these were carried out in 1964–1965 (6). The leaf- and stem-mining weevil, *Cyrtobagous singularis,* the leaf-feeding grasshopper, *Paulinia acuminata,* and the leaf-feeding pyralid moth, *Samea multiplicalis,* have been released in areas of Africa. Strains of *Paulinia* from both Uruguay and Trindad are well established on Lake Kariba (9); this insect has spread 30 to 50 miles from the release sites and populations of 25 to 50 grasshoppers per square meter have been found in suitable habitats (D. J. W. Rose, personal communication).

Miscellaneous Projects

Sankaran & Rao (59) listed a number of insects from 24 species of aquatic plants surveyed throughout India. The weevil, *Nanophyes* sp. nr. *nigritulus,* is considered promising against *Ludwigia adscendens,* and the noctuid *Namangana pectinicornis,* against *Pistia stratiotes.* Other insects listed either required further study or were considered to be of no value for biological control. In Pakistan several insects attacking *Hydrilla verticillata* have been under study; of these, *Hydrellia* sp., *Bagous* sp., and *Nymphula diminualis* appear to be promising candidates for biological control (5).

CONCLUDING REMARKS

The principles underlying the biological control of aquatic and terrestrial weeds are essentially the same, although new considerations are sure to arise as information on aquatic weeds and their associated insects accumulates. Perhaps, as Bennett (9) suggests, it will become possible to selectively relax the stringent host-specificity requirements currently enforced for all insects introduced for weed control, if it can be demonstrated that those attacking aquatic weeds are closely tied to the water environment. The role of pathogens in relation to aquatic plants and insects certainly needs elucidation before speculation on their impact can be made. The high incidence of microsporidian diseases among the insects attacking waterhyacinth suggests that every precaution should be taken to assure that all insects under consideration for biological control are free of pathogens. The recent discovery that Benomyl is an antimicrosporidial agent (33) may facilitate a solution as regards waterhyacinth.

Insect enemies have proven of considerable value against exotic range and pasture weeds, especially where the selective removal of preferred forage species by grazing animals has favored weed encroachment. In the aquatic environment, on the other hand, it is the use of water that is of primary interest to man, and not the plants. Although low levels of plant growth are often encouraged to maximize water usage, only limited interference on their part will be tolerated. Thus on rangelands where the growth of desirable forage plants is encouraged, plant competition assists the biological control agents suppressing weeds. In aquatic areas, assistance from competing plants may be minimal from the biological control standpoint, and of value only where relatively high levels of plant growth are tolerated (i.e. fishing and water storage areas).

Invasion by a well-adapted exotic or proliferation of indigenous plant species following alteration of the habitat (e.g. influx of nutrients) are two major types of aquatic weed problems. If the weed problem concerns an exotic invader, biological control with host-specific insect enemies may prove useful. If the problem concerns general vegetation increase due to habitat alteration, then selective control with host-specific insects will merely lead to the replacement of one plant species by another. In the latter instances polyphagous feeders (e.g. herbivorous fish, snails, manatee), may prove of value. However, even these exhibit host-feeding preferences which could lead to a buildup of the nonpreferred plants, and weed problems similar to those encountered on overgrazed terrestrial areas. Van Zon (67) noted that in Holland there are a few plants not eaten by the grass carp that could become weedy. Periodic chemical treatment of these species, e.g. *Utricularia vulgaris, Ranunculus* spp., may help with this problem.

How to evaluate the control potential of an insect, prior to introduction against a weed, has perplexed biological control workers for years. Harris (31) suggests a quantitative rating system to be substituted for the current intuitive analysis method. In those instances where even minor insect feeding may allow the entry of pathogens into an aquatic plant, a meaningful rating can be developed only in cooperation with plant pathologists as well as other plant scientists and entomologists.

The instances of attempted biological control of aquatic weeds with insects are still too few to do more than speculate on the future of this method for aquatic weed control. The success obtained in the biological control of alligatorweed and the initial success in obtaining establishment of *Paulinia* against salvinia on Lake Kariba and *Neochetina eichhorniae* on waterhyacinth in Florida suggests that this approach may have a bright future against certain aquatic weed problems.

Literature Cited

1. Allsopp, W. H. L. 1969. Aquatic weed control by manatees—its prospects and problems. In *Man-Made Lakes, The Accra Symposium,* ed. L. E. Obeng, 344–51. Accra: Ghana Univ. Press.
2. Andres, L. A., Davis, C. J. 1974. The biological control of weeds with insects in the United States. *Proc. Int. Symp. Biol. Contr. Weeds, 2nd, Rome, 1971*, 11-28
3. Avault, J. W. 1965. Preliminary studies with grass carp for aquatic weed control. *Progr. Fish Cult.* 27:207–9
4. Baloch, G. M., Khan, A. G., Rahim, A., Ghani, M. A. 1972. Phenology, biology and host-specificity of some stenophagous insects attacking *Myriophyllum* spp. in Pakistan. *Hyacinth Contr. J.* 10:13–16

5. Baloch, G. M., Sana-Ullah. 1974. Insects and other organisms associated with *Hydrilla verticillata* (L.f.) L.C. in Pakistan. *Proc. Int. Symp. Biol. Contr. Weeds, 3rd, Montpellier, 1973*
6. Bennett, F. D. 1966. Investigations on the insects attacking aquatic ferns, *Salvinia* spp., in Trinidad and northern South America. *Proc. S. Weed Conf.* 19:497–504
7. Bennett, F. D. 1968. Insects and mites as potential controlling agents of waterhyacinth, *Eichhornia crassipes. Proc. Brit. Weed Contr. Conf., 9th, Brighton, England, 1968* 2:832–35
8. Bennett, F. D. 1970. Insects attacking waterhyacinth in the West Indies, British Honduras and the U.S.A. *Hyacinth Contr. J.* 8:10–13
9. Bennett, F. D. 1972. *Current Status of the Biological Control of Salvinia in Africa.* Presented at Int. Congr. Entomol., Canberra
10. Bennett, F. D. 1974. Biological control. In *Aquatic Vegetation and Its Use and Control,* ed. D. S. Mitchell, 99–106. Paris: UNESCO
11. Bennett, F. D. 1974. Some aspects of the biological control of aquatic weeds. *Proc. Int. Symp. Biol. Contr. Weeds, 2nd, Rome, 1971,* 63–73
12. Bennett, F. D., Zwölfer, H. 1968. Exploration for natural enemies of the waterhyacinth in northern South America and Trinidad. *Hyacinth Contr. J.* 7:44–52
13. Berg, C. O. 1949. Limnological relations of insects to plants of the genus *Potamogeton. Trans. Am. Microsc. Soc.* 68:279–91
14. Blackburn, R. D., Sutton, D. L., Taylor, T. M. 1971. Biological control of aquatic weeds. *J. Irrig. Drain. Div. Am. Soc. Civil Eng.,* 97(IR3):421–32, Proc. Paper 8362
15. Blackburn, R. D., Taylor, T. M., Sutton, D. L. 1971. Temperature tolerance and necessary stocking rates of the *Marisa cornuarietis* for aquatic weed control. *Proc. Eur. Weed Res. Counc., 3rd Int. Symp. Aquatic Weeds, Oxford,* 79–86
16. Blackburn, R. D., Weldon, L. W. 1967. New underwater menace, Eurasian watermilfoil. *Weeds, Trees and Turf,* November: 10–13
17. Boyd, C. E. 1968. Fresh-water plants: a potential source of protein. *Econ. Bot.* 22:359–68
18. Boyd, C. E. 1971. The limnological role of aquatic macrophytes and their relationship to reservoir management. *Reservoir Fisheries and Limnology, Special Publ. No. 8, Am. Fish. Soc.,* 153–66
19. Brown, J. L. 1973. *Vogtia malloi, a newly introduced pyralid for the control of alligatorweed in the United States.* PhD thesis. Univ. of Florida, Gainesville. 71 pp.
20. Brown, J. L., Spencer, N. R. 1973. *Vogtia malloi,* a newly introduced phycitid to control alligatorweed. *Environ. Entomol.* 2:521–23
21. Charudattan, R., Perkins, B. D. 1974. *Fungi Associated with Insect Damaged Waterhyacinth in Florida and Possible Effects on Plant Host Population.* Paper presented at meeting of Hyacinth Control Society, Winterhaven, Florida
22. Coulson, J. R. 1973. *Biological Control of Alligatorweed 1959–1972.* Presented at U.S. Army Corps of Engineers Aquatic Plant Control Program, Vicksburg, Miss., October, 1973
23. Dean, J. L. 1969. Biology of the crayfish, *Orconectes causeyi* and its use for control of aquatic weeds in trout lakes. *Tech. Paper No. 24. Bur. Sport Fish. Wild. US Dep. Interior.* 15 pp.
24. Deonier, D. L. 1971. A systematic and ecological study of nearctic *Hydrellia. Smithson. Contrib. Zool.* No. 68. 147 pp.
25. Ensminger, A. 1973. Flea beetle project detrimental. *Catalyst for Environmental Quality* 3(3):4, 6. Letter to editor
26. Gaudet, J. J. 1973. *Report on the Development and Control of Aquatic Weeds in Volta Lake.* Dep. Bot., Makerere Univ., Kampala, Uganda. 18 pp. (mimeo)
27. Goeden, R. D., Ricker, D. W. 1971. Imported alligatorweed insect enemies precluded from establishment in California. *J. Econ. Entomol.* 64:329–30
28. Gordon, R. D., Coulson, J. R. 1971. Report on field observations of arthropods on waterhyacinth in Florida, Louisiana, and Texas, July, 1969. Appendix F, 29 pp. In Gangstad, E. O., Raynes, J. J., Mobley, G. S. Jr., Zeiger, C. F. 1971. *Technical Report of the Potential Growth of Aquatic Plants of the Cross-Florida Barge Canal, Review of the Aquatic Plant Control Research Program and Summary of the Research Area Development Operations in Florida.* U.S. Army Corps of Engineers, Serial No. 34, 191 pp.
29. Habib-ur-Rehman, Mustaque, M., Baloch, G. M., Ghani, M. A. 1969. Preliminary observations on the biological control of watermilfoils (*Myriophyllum* spp.). *Commonw. Inst. Biol. Contr. Tech. Bull.* 11:165–71

30. Harris, P. 1971. Current approaches to biological control of weeds. In *Biological Control Programmes Against Insects and Weeds in Canada,* pp. 67–76. *Commonw. Inst. Biol. Contr. Tech. Commun.* No. 4. 266 pp.
31. Harris, P. 1973. The selection of effective agents for the biological control of weeds. *Can. Entomol.* 105:1495–1503
32. Harris, P., Zwölfer, H. 1968. Screening of phytophagous insects for biological control of weeds. *Can. Entomol.* 100:295–303
33. Hsiao, T. H., Hsiao, C. 1973. Benomyl: a novel drug for controlling a microsporidian disease of the alfalfa weevil. *J. Invertebr. Pathol.* 22:303–4
34. Haslam, S. M. 1971. Physical factors and some river weeds. *Proc. European Weed Res. Counc., 3rd Int. Symp. on Aquatic Weeds, Oxford,* 29–39
35. Hawkes, R. B., Andres, L. A., Anderson, W. H. 1967. Release and progress of an introduced flea beetle, *Agasicles* n. sp., to control alligatorweed. *J. Econ. Entomol.* 60:1476–77
36. Holloway, J. K. 1964. Projects in biological control of weeds. In *Biological Control of Insect Pests and Weeds,* ed. P. DeBach, 650–70. New York: Reinhold
37. Holm, J. G., Weldon, L. W., Blackburn, R. D. 1969. Aquatic weeds. *Science* 166:699–709
38. Hoogers, B. J., Van der Weij, H. G. 1971. The development cycle of some aquatic plants in the Netherlands. *Proc. Eur. Weed Res. Counc., 3rd Int. Symp. on Aquatic Weeds, Oxford,* 3–18
39. Huffaker, C. B. 1959. Biological control of weeds with insects. *Ann. Rev. Entomol.* 4:251–76
40. Huffaker, C. B. 1964. Fundamentals of Biological Weed Control. In *Biological Control of Insect Pests and Weeds,* ed. P. DeBach, 631–49. New York: Reinhold
41. Krupauer, V. 1971. The use of herbivorous fish for ameliorative purposes in Central and Eastern Europe. *Proc. Eur. Weed Res. Counc., 3rd Int. Symp. on Aquatic Weeds, Oxford,* 95–103
42. Ladle, M., Casey, H. 1971. Growth and nutrient relationship of *Ranunculus penicillatus* var. *calcareus* in a small chalk stream. *Proc. Eur. Weed Res. Counc., 3rd Int. Symp. on Aquatic Weeds, Oxford,* 53–64
43. Lekic, M. 1970. Ecology of the aquatic insect species, *Parapoynx stratiotata* L. *J. Sci. Agr. Res.* 23:49–63
44. Lekic, M., Mihajlovic, L. 1970. Entomofauna of *Myriophyllum spicatum* L., an aquatic weed on Yugoslav territory. *J. Sci. Agri. Res.* 23:59–74
45. Little, E. C. S. 1968. The control of water weeds. *Weed Res.* 8:79–105
46. Maddox, D. M. 1968. Bionomics of an alligatorweed flea beetle, *Agasicles* sp., in Argentina. *Ann. Entomol. Soc. Am.* 61:1299–1305
47. Maddox, D. M. 1973. *Amynothrips andersoni,* a thrips for the biological control of alligatorweed. 1. Host specificity studies. *Environ. Entomol.* 2:30–37
48. Maddox, D. M., Andres, L. A., Hennessey, R. D., Blackburn, R. D., Spencer, N. R. 1971. Insects to control alligatorweed, an invader of aquatic ecosystems in the United States. *Bioscience* 21:985–91
49. Maddox, D. M., Hambric, R. N. 1970. Use of alligatorweed flea beetle in Texas: an exercise in environmental biology. *Proc. S. Weed Sci. Soc.* 23: 283–86
50. McGaha, Y. J. 1952. The limnological relations of insects to certain aquatic flowering plants. *Trans. Am. Microsc. Soc.* 71:355–81
51. Michewicz, J. E., Sutton, D. L., Blackburn, R. D. 1972. Water quality of small enclosures stocked with white amur. *Hyacinth Contr. J.* 10:22–25
52. Mulligan, H. F., Baranowski, A. 1969. Growth of phytoplankton and vascular aquatic plants at different nutrient levels. *Verh. Int. Verein. Limnol.* 17: 802–10
53. Perkins, B. D. 1973. Potential for waterhyacinth management with biological agents. *Proc. Tall Timbers Conf. Ecol. Anim. Contr. Habitat Manage.* 4:53–56
54. Perkins, B. D. 1973. Release in the United States of *Neochetina eichhorniae* Warner, an enemy of waterhyacinth. *Proc. Ann. Meet. S. Weed Sci. Soc., 26th* 368 (Abstr.)
55. Perkins, B. D. 1974. Preliminary studies on a strain of the waterhyacinth mite. *Proc. Int. Symp. Biol. Contr. Weeds, 2nd, Rome, 1971,* 179–84
56. Perkins, B. D. 1974. Biocontrol of water hyacinth. *Aquatic Plant Control Program Tech. Rep. No. 6, Biological Control of Waterhyacinth with Insect Enemies, U.S. Army Engineers Waterways Exp. Sta., Vicksburg, Miss.* pp. E3–E17
57. Pirie, N. W. 1970. Weeds are not all bad. *Ceres* 3:31–39
58. Rao, V. P., Sankaran, T., Ramaseshiah, G. 1972. Studies on four species of natu-

ral enemies for biological control of *Eichhornia crassipes. Commonw. Inst. Biol. Contr.* 10 pp. (cyclostyled)
59. Sankaran, T., Rao, V. P. 1972. An annotated list of insects attacking some terrestrial and aquatic weeds in India with records of some parasites of the phytophagous insects. *Commonw. Inst. Biol. Contr. Tech. Bull.* 15:131–57
60. Sculthorpe, C. D. 1967. *The Biology of Aquatic Vascular Plants.* New York: St. Martins Press. 610 pp.
61. Seaman, D. E., Porterfield, W. A. 1964. Control of aquatic weeds by the snail *Marisa cornuarietis. Weeds* 12:87–92
62. Silveira-Guido, A. 1971. Datos preliminares de biologia y especificidad de *Acigona ignitalis* Hamps. sobre el hospedero *Eichhornia crassipes* (Mart.) Solms-Laubach. *Rev. Soc. Entomol. Argent.* 33:137–45
63. Silveira-Guido, A., Perkins, B. D. 1974. Biology and host specificity of *Cornops aquaticum* (Bruner), a potential biological control agent for waterhyacinth. *Environ. Entomol.* In press
64. Stott, B., Cross, D. G., Iszard, R. E., Robson, T. O. 1971. Recent work on grass carp in the United Kingdom from the standpoint of its economics in controlling submerged aquatic plants. *Proc. Eur. Weed Res. Counc., 3rd Int. Symp. on Aquatic Weeds, Oxford,* 105–16
65. Timmons, F. L. 1970. Control of aquatic weeds. *FAO Int. Conf. Weed Control, Davis, California,* 357–73
66. Usinger, R. L., Ed. 1956. *Aquatic Insects of California.* Berkeley, Calif.: Univ. California Press. 508 pp.
67. Van Zon, J. C. J. 1974. Studies on the biological control of aquatic weeds in The Netherlands. *Proc. Int. Symp. on Biol. Contr. Weeds, 3rd, Montpellier, 1973*
68. Vogel, E., Oliver, A. D. 1969. Evaluation of *Arzama densa* as an aid in the control of waterhyacinth in Louisiana. *J. Econ. Entomol.* 62:142–45
69. Vogt, G. B. 1960. Exploration for natural enemies of alligatorweed and related plants in South America. *USDA, Agr. Res. Serv. Spec. Rep. PI-4,* 58 pp. (mimeo)
70. Vogt, G. B. 1973. Exploration for natural enemies of alligatorweed and related plants in South America. *Aquatic Plant Control Program Tech. Rep. No. 3, Biological Control of Alligatorweed, U.S. Army Engineers Waterways Exp. Sta., Vicksburg, Miss.* pp. B1–B66
71. Warner, R. E. 1970. *Neochetina eichhorniae,* a new species of weevil from waterhyacinth, and biological notes on it and *N. bruchi. Proc. Entomol. Soc. Wash.* 72:487–96
72. Weed Science Society of America. 1971. Composite List of Weeds. Report of Subcommittee on Standardization of Common and Botanical Names of Weeds. *Weed Sci.* 19:435–76
73. Weldon, L. W. 1960. A summary review of investigations on alligatorweed and its control. *USDA Agr. Res. Serv. Crops Res. Div. CR 33-60,* 41 pp.
74. Weldon, L. W., Blackburn, R. D., Durden, W. C. 1973. Evaluation of *Agasicles* n. sp. for biological control of alligatorweed. *Aquatic Plant Control Program Tech. Rep. No. 3, Biological Control of Alligatorweed, U.S. Army Engineers Waterways Exp. Sta., Vicksburg, Miss.* pp. D1-D54
75. Wilson, F. 1964. The biological control of weeds. *Ann. Rev. Entomol.* 9:225–44
76. Yeo, R. R., Fisher, T. W. 1970. Progress and potential for biological weed control with fish, pathogens, competitive plants and snails. *FAO Int. Conf. on Weed Control, Davis, California,* 450–63
77. Zeiger, C. F. 1967. Biological control of alligatorweed with *Agasicles* n. sp. in Florida. *Hyacinth Contr. J.* 6:31–34
78. Zielske, A. G., Simons, J. N., Silverstein, R. M. 1972. A flavone feeding stimulant in alligatorweed. *Phytochemistry* 11:393–96
79. Zettler, F. W., Freeman, T. E. 1972. Plant pathogens as biocontrols of aquatic weeds. *Ann. Rev. Phytopathol.* 10:455–70
80. Zwölfer, H. 1965. Preliminary list of phytophagous insects attacking wild cynareae (Compositae) species in Europe. *Commonw. Inst. Biol. Contr. Tech. Bull.* No. 6:81–154
81. Zwölfer, H., Harris, P. 1971. Host specificity determination of insects for biological control of weeds. *Ann. Rev. Entomol.* 16:159–78

THE BIOLOGY AND ECOLOGY OF ARMORED SCALES [1,2]

❖6080

John W. Beardsley Jr. and Roberto H. Gonzalez
Department of Entomology, University of Hawaii, Honolulu, Hawaii 96822 and Plant Production and Protection Division, Food and Agriculture Organization, Rome, Italy

The armored scales (Family Diaspididae) constitute one of the most successful groups of plant-parasitic arthropods and include some of the most damaging and refractory pests of perennial crops and ornamentals. The Diaspididae is the largest and most specialized of the dozen or so currently recognized families which compose the superfamily Coccoidea. A recent world catalog (19) lists 338 valid genera and approximately 1700 species of armored scales. Although the diaspidids have been more intensively studied than any other group of coccids, probably no more than half of the existing forms have been recognized and named.

Armored scales occur virtually everywhere perennial vascular plants are found, although a few of the most isolated oceanic islands (e.g. the Hawaiian group) apparently have no endemic representatives and are populated entirely by recent adventives. In general, the greatest numbers and diversity of genera and species occur in the tropics, subtropics, and warmer portions of the temperate zones.

With the exclusion of the so-called palm scales (*Phoenicococcus, Halimococcus,* and their allies) which most coccid taxonomists now place elsewhere (19, 26, 99), the armored scale insects are a biologically and morphologically distinct and homogenous group. The definition of subfamily and tribal categories within the Diaspididae is a matter of some disagreement among taxonomists, although most recognize at least four major subdivisions: the aspidiotines, diaspidines, parlatorines, and odonaspidines. Borchsenius (19) treated these as subfamilies, whereas earlier workers (8) gave them tribal status within the nominate subfamily of a more inclusive family Diaspididae. Recently Takagi (99) has proposed dividing the Diaspididae into seven tribes (the larger number derived by splitting Balachowsky's Diaspidini and Parlatorini), without designating subfamilies. Takagi's scheme, which is based

[1]Published with the approval of the Director of the Hawaii Agricultural Experiment Station as Journal Series No. 1714.

[2]The survey of literature pertaining to this review was completed in December 1972. The authors express their appreciation to Dr. P. DeBach, University of California, Riverside, for making available IBP funds to assist them in the library search.

largely on the comparative morphology of first instar nymphs, appears to have considerable merit but has not yet achieved general acceptance.

A relatively small proportion of the known species of armored scales are agricultural or horticultural pests of consequence. Schmutterer, Kloft & Lüdicke (93) listed about 135 species in their treatment of diaspidid pests. Perhaps a quarter of these can be considered as pests of major importance. Published biological and ecological information on diaspidids has been derived largely from studies on the major pest species listed in Table 1.

HOST RELATIONSHIPS

Host Specificity

Polyphagy is common among diaspidids. San Jose scale, for example, is known to infest representatives of 34 different plant families (114) while the oleander scale has been recorded from hundreds of host species in more than 100 plant families (16). On the other hand, many armored scales have either limited oligophagous or monophagous host ranges. Species of the genus *Kuwanaspis,* for example, are confined to bamboos. Closely related species sometimes exhibit very different degrees of host specificity. For example, *Carulaspis carueli* is known to attack several genera of conifer hosts, but *Carulaspis juniperi* is restricted exclusively to *Juniperus communis.* A third related species, *Carulaspis visci,* lives exclusively on the leaves of *Viscum album* (a mistletoe semiparasitic on *Pinus sylvestris*), and does not migrate to *Juniperus* even if the foliage of the two conifers are in contact (53).

With respect to varietal susceptibility of host plants, very little information is known. Agarwal & Sharma (1) studied the occurrence and abundance of *Melanaspis glomerata* on different varieties and ecotypes of sugar cane and found a direct relationship between a heavy attack and a high stomatal density of the stem. Five apple varieties studied in New Zealand exhibited no difference in susceptibility to attack of San Jose scale (91). Similarly, sour cherry hybrids tested for resistance to this scale in Hungary failed to show differences in susceptibility (62).

Host Parts Attacked

Most armored scales are eurymerous, that is, they may feed on various parts (organs) of the host plant without apparent problems (6, 115). Stenomerous species, restricted to only one plant organ, are less numerous and usually found on grasses or other herbaceous plants. Practically all plant organs, including those with a thick epidermal layer, are suitable for scale feeding and reproduction. Subterranean parts (roots, tubers, and rhizomes) are not generally attacked unless they are exposed, because armored scales are not adapted to the hypogeal environment.

When two or more species with similar feeding and settlement requirements coexist, competition between the species may affect their relative abundance and distribution on the host. Competitive displacement may occur whereby one species of scale completely displaces the other. In other instances, two species coexisting in the same host may reach a certain degree of equilibrium. Gerson (50) reported

Table 1 Principal armored scale pests of the world

	Common Name
Tribe Aspidiotini	
Aonidiella aurantii (Maskell)	California red scale[a]
Aonidiella citrina (Coquillett)	yellow scale[a]
Aspidiotus destructor Signoret	coconut scale[a]
Aspidiotus nerii Bouché [= *A. hederae* Vallot]	oleander scale[a]
Chrysomphalus dictyospermi (Morgan)	dictyospermum scale[a]
Chrysomphalus ficus Ashmead [= *C. aonidum* (L.) of authors]	Florida red scale[a]
Hemiberlesia lataniae (Signoret)	latania scale
Hemiberlesia rapax (Comstock)	greedy scale[a]
Pseudaonidia duplex (Cockerell)	camphor scale[a]
Quadraspidiotus forbesi (Johnson)	Forbes scale[a]
Quadraspidiotus juglansregiae (Comstock)	walnut scale[a]
Quadraspidiotus ostreaeformis (Curtis)	European fruit scale[a]
Quadraspidiotus perniciosus (Comstock)	San Jose scale[a]
Quadraspidiotus pyri (Lichtenstein)	false San Jose scale
Selenaspidus articulatus (Morgan)	rufous scale
Tribe Parlatorini	
Parlatoria blanchardi (Targioni-Tozzetti)	parlatoria date scale[a]
Parlatoria oleae (Colvee)	olive scale[a]
Parlatoria pergandii Comstock	chaff scale[a]
Parlatoria proteus (Curtis)	proteus scale
Parlatoria theae Cockerell	parlatoria tea scale
Parlatoria ziziphi (Lucas)	black parlatoria scale
Tribe Diaspidini	
Aulacaspis rosae (Bouché)	rose scale[a]
Aulacaspis tegalensis (Zehntner)	sugarcane scale
Carulaspis carueli (Targioni-Tozzetti) [= *C. minima* (T.-T.) of authors]	minute juniper scale
Carulaspis juniperii (Bouché)	juniper scale[a]
Diaspis bromeliae (Kerner)	pineapple scale[a]
Diaspis boisduvalii Signoret	Boisduval scale[a]
Epidiaspis leperii (Signoret)	Italian pear scale[a]
Fiorinia fioriniae (Targioni-Tozzetti)	fiorinia scale
Fiorinia theae Green	tea scale[a]
Howardia biclavis (Comstock)	mining scale[a]
Ischnaspis longirostris (Signoret)	black thread scale[a]
Lepidosaphes beckii (Newman)	purple scale[a]
Lepidosaphes ficus (Signoret) [= *L. conchiformis* (Gmelon) of authors]	fig scale[a]
Lepidosaphes gloverii (Packard)	Glover scale[a]
Lepidosaphes ulmi (L.)	oystershell scale[a]
Phenacaspis pinifoliae (Fitch)	pine needle scale[a]
Pinnaspis aspidistrae (Signoret)	fern scale[a]
Pinnaspis strachani (Cooley) [= *P. minor* (Maskell) of authors]	lesser snow scale
Pseudaulacaspis pentagona (Targioni-Tozzetti)	white peach scale[a]
Unaspis citri (Comstock)	citrus snow scale
Unaspis euonymi (Comstock)	euonymus scale[a]
Unaspis yanonensis (Kuwana)	arrowhead scale

[a]Entomological Society of America approved common name.

Parlatoria pergandii to be more numerous in the summer, while *Parlatoria cinerea,* coexisting on the same citrus host, was more abundant in the winter season. A different type of accommodation is found when two species of *Aonidiella* infest citrus simultaneously: *Aonidiella aurantii* occupies all the aerial portion of the tree including trunk and branches, while *Aonidiella citrina* is restricted to the leaves and fruits (46).

McLaren (74) found that in certain areas of Victoria, Australia, red scale remains markedly more abundant on citrus than yellow scale. The intrinsic rate of natural increase (r_m) of red scale was greater at low temperatures than that of yellow scale, although the r_m of the latter was greater during the warmer summer months. The threshold of population development was 15°C for red scale and 18°C for yellow scale, which indicated that during an average year red scale would show a positive population growth rate for approximately ten months compared to six months for yellow scale. McLaren concluded that greater survival and growth rates during the colder months were largely responsible for the preponderance of red scale in these areas.

In some eurymerous species the location of the insect on the host plant may affect its morphological characteristics. Such effects sometimes are so pronounced that individual scales from different parts of the same host formerly were considered to be distinct species and sometimes were placed in different genera. Takahashi (102) recognized that certain Japanese species which had been placed in the genera *Phenacaspis* and *Chionaspis* were, in fact, site-determined morphological forms of the same species. In these species leaf-infesting individuals developed pygideal characters typical of the genus *Phenacaspis,* while bark-inhabiting individuals had the typical *Chionaspis* pygidium. Takagi & Kawai (100) reviewed the Japanese and North American species of the *Chionaspis* group and demonstrated the existence of forms intermediate between the *Chionaspis sylvatica* and *Phenacaspis nyssae* morphs of the species which is now known as *Chionaspis nyssae.* Similarly, *Aspidiotus ancylus* develops in typical form on the twigs of elm and sugar maple, but distinctly different morphological forms of this species develop on the leaves of these hosts. These differences are consistent, and the leaf-infesting forms formerly were considered separate species: *Aspidiotus howardi* on elm leaves and *Aspidiotus comstocki* on sugar maple leaves (97).

In the fig scale, *Lepidosaphes ficus,* the overwintering females on twig wood differed so greatly in size and pygideal characteristics from the summer form found primarily on the leaves that the latter was considered to be a separate species, until biological studies proved otherwise (66, 96). Danzig (34) has also demonstrated the existence of site-determined morphs, which formerly were considered distinct species, in several Russian diaspidids.

Stylet Renewal

A problem of particular interest in the biology of the armored scale insects has been the study of the processes involved in their feeding habits. For an adequate understanding of the mechanisms involved in the stylet penetration into the plant tissue, it is necessary to examine the various processes related to the renewal of the stylet during the molt.

A diaspidid embryo, shortly before birth, shows the long maxillary and mandibular stylets separately coiled on each side of the head. During birth, the separate stylets travel down from the head, their points converging to meet at the base of the labrum where the stylets coalesce and issue as a single functional organ, the stylet fascicle. As it is extruded from the head, the stylet fascicle is stored as an elongated loop within the crumena, which is a deep membranous, cuticular invagination opening at the base of the labium (83).

The process of renewal of the stylets has been studied in detail by Heriot (56). As soon as the larva has succeeded in introducing the stylet into the plant tissue, the hypodermal cells of the base of the rostrum produce a small fold of tissue that extends like the finger of a glove on each side of the rostrum. The tip of the new stylet, which eventually will serve the second instar, begins to form at the distal end of this fold. As the new stylet is laid down from the tip, the mass of hypodermal cells lengthens out to form, subsequently, a coil in which the core of the new stylet is gradually built up. On completion of the requisite length, the creative hypodermal cells come together at the base of the rostrum to repeat the process for the subsequent instar.

Inasmuch as the stylets are simple, hollow, cuticular structures, they bear a close resemblance to other such cuticular processes. However, once they are inserted into the plant tissue, no withdrawal takes place. Consequently the stylets are never found with other cuticular processes attached to the molted exuviae (113). The old stylets are broken off during molting. Sections through the bark of twigs bearing adult female scales sometimes show the cast-off stylets of the previous two instars in their relative insertion position. The stylets of each instar can be distinguished by their relative lengths and diameters. For example, the new stylets acquired by the second instar of *Lepidosaphes ulmi* are twice the diameter and twice as long as those of the first instar (56).

Penetration of Plant Tissue

Several early workers studied feeding mechanisms of the diaspidid insects (27, 108, 118). Stylet penetration, the role of the saliva during feeding, and the reaction of the host plant, were the main subjects of interest to these investigators. The means by which insects as small as scale insects are able to penetrate even woody plant tissue with elongate hair-like feeding stylets was long considered a major entomological mystery. The explanation of Weber (117), summarized by Snodgrass (95), seems largely adequate, although Pesson (83) has questioned certain details of Weber's explanation as these apply to the Coccoidea and has suggested some modifications. It appears that some of the finer details of the process remain to be worked out.

The four stylets which comprise the stylet fascicle are intimately associated throughout their lengths so that they function as a unit. A cross section of the stylet fascicle reveals that the two mandibular stylets form the outer, lateral elements of the fascicle, enclosing the interlocking maxillary stylets between them. The inner faces of the maxillary stylets are so formed that their juxtaposition creates two tubular canals which extend the entire length of the fascicle: a larger dorsal food channel and a smaller ventral salivary channel.

Apparently the stylets, the crumena, and the labium each play a part in the mechanics of penetration, and a repetitive sequence of events drives the stylet tips, step by step, into the host tissue. In penetration, the group of protractor muscles at the bases of each of the stylets contract sequentially. First one mandibular stylet, then the other, followed by the maxillary stylets (either together or individually), are advanced equally a short distance. The individually protracted stylets slide along the contiguous surface or surfaces of the others, but the stylet fascicle remains intact, held so by interlocking grooves and ridges in the case of the maxillary stylets, and probably also by the crumena. Pesson (83) has shown that because of its flexible nature the crumenal sac is stretched tightly around the loop of the stylet fascicle in such a way that both efferent and afferent legs of the loop are essentially individually surrounded by the adherent crumenal wall. The crumena, therefore, appears to hold the stylet fascicle together and prevents separation of the individual stylets during protraction. The elastic labial gutter may also play a similar role. The net effect therefore is to transmit the force of protraction to the tip of the protracted stylet without deforming the fascicle loop. When all four stylets have been protracted a specialized clamp-like portion of the labial gutter is closed to maintain the position of the apical portion of the fascicle while the crumenal loop is being shortened. The stylet retractor muscles are then simultaneously contracted and the protractor muscles are relaxed so that the stylet bases are returned to their original positions ready for the next series of protractions. It is the simultaneous action of the stylet retractor muscles, possibly in concert with the labial clamp mechanism, which serves to shorten the crumenal loop while the stylet tips remain in their new position in the host tissue.

Theodoro (108) found that the path of armored scale stylets in leaf tissues traveled parallel to the surface. When an obstacle such as a vascular bundle was encountered, the stylet bundle was withdrawn a little and pushed forward again in a different direction. Heriot (56) suggested that the elm scale inserts its stylets into apple twigs without directing them toward a specific tissue. Sections through the annual growth of scale-infested twigs showed stylets passing straight through the cortex, being turned aside by bundles of pericycle fibers, continuing their course through the phloem, crossing the cambium, and invading immature xylem. He suggested that tissue penetration is very slow in the elm scale as compared to the wooly apple aphid, and that the scales feed on the contents of any cells which the tips of the stylets invade.

Feeding Effects

The feeding sites of armored scales often are associated with depressions, discolorations, and other distortions of host tissues, such as leaf crinkling, etc. Splitting of bark, defoliation, dieback of twig terminals, and sometimes the eventual death of the host may follow heavy infestation. For example, heavy concentrations of *Odonaspis ruthae* were reported to cause the death of Bermuda grass in Arkansas, while low populations weakened the grass and made it susceptible to pathogenic fungi and winter injury (32). Heavy infestations of two accidentally introduced scales, *Lepidosaphes newsteadi* and *Carulaspis juniperi* destroyed many of the native "cedars" (*Juniperus bermudiana*) of Bermuda (116).

Leaf chlorosis and other localized toxic effects are commonly associated with armored scale infestations. Infestation by *Aspidiotus destructor,* for example, normally is accompanied by pronounced yellowing of affected foliage on coconut and other palms (10). Feeding on citrus leaves by the purple scale produces chlorotic spots around the insects. On the fruit, however, the tissue underneath the scale remains green while the rest of the rind approaches maturity. Likewise, when light-colored fig varieties infested with fig scale are ripening, the portions beneath and immediately surrounding the scales remain dark green as the rest of the fig turns light green and yellowish (96). Green olives infested by olive scale show purple spots which later become more pronounced and turn straw colored as the fruits mature (73). With *Melanaspis bromeliae* on pineapple leaves, Carter (29) reported that the tissue directly beneath the scale was darker green than normal leaf color, forming a dark green spot which was surrounded by a much larger chlorotic ring. Development of the dark central spot began shortly after settling of the crawler on the leaf. The chlorotic ring developed later and eventually became depressed leaving the green center as an elevated island. Carter suggested that such bizonate symptoms are due to the actions of two or more components of the saliva.

The destructive effect produced by the San Jose scale in woody tissue has been investigated by Enser (42). Sections of wood cut through the points of attachment of the insect demonstrate that the stylets penetrate the cortex into cambium. Growth of the pierced cambium ceases or is impaired. If impaired, the associated xylem and phloem cells differ in number, size, and arrangement compared to those in normal tissue. Gentile & Summers (49) indicate that the feeding of San Jose scale on peach twigs produces a characteristic halo-like discoloration which appears 24 hr after the crawlers settle on tender wood. The discolored area increases in diameter with the age of the nymph, the cortical tissue swells with accumulating sap, the bark often cracks, and finally a gradual dessication of the cortical tissue follows.

The formation of plant galls seems to be a relatively uncommon phenomenon associated with feeding by armored scales. The Australian *Maskellia globosa* is unique among diaspidids in that it produces no external scale and develops completely enclosed within globular galls on young twigs of *Eucalyptus* (47). The formation of galls on mistletoe leaves resulting from the feeding action of *Carulaspis visci* has been described by Goidanich (53). Atrophied cells are found in the vicinity of the stylets while hypertrophied tissue develops in the more distal part of the mesophyll. As a consequence, the scale becomes practically embedded in the leaf tissue. In this respect the damage produced is similar to that of the pit scale, *Asterolecanium variolosum,* on oak twig, which is also due to the enzymatic activity of the saliva (82). Information on the nature of enzymes occurring in the saliva of diaspidids is scanty. The invertase activity is greater than that of amylase in both Florida red and California red scales, and salivary composition varies according to the feeding activity and the temperature of development of the insects (60).

THE DIASPIDID DIGESTIVE SYSTEM

Armored scale insects do not excrete honeydew and lack the filter chamber type of digestive system found in most other Coccoidea. In diaspidids the esophagus opens into a large pouch-like structure termed the stomach or first ventriculus (30, 80). This is followed by a second, smaller pouch, termed the Malpighian bulb, into which the two large Malpighian tubules discharge. From the bulb the hind gut empties into an elongate rectum. The stomach is connected to the Malpighian bulb and rectum by several fine ligaments. An unusual anatomical feature of the armored scale digestive system is the apparent lack of continuity between the stomach and the hind intestine. Berlese (13) first observed that the stomach was entirely disconnected from the Malpighian bulb and rectum and correctly described both the bulb and the pair of Malpighian tubes as independent structures, separated from the stomach sac. Childs (30) and Nel (80) disputed these findings and contended that the stomach was indeed connected to the rectum, but later investigators confirmed the validity of Berlese's results (5, 7, 14, 39, 83).

Berlese (13) postulated that the products of digestion diffused from the stomach into the hemolymph and that wastes passed from the hemolymph into the Malpighian tubules and thence into the rectum. Pesson (83) suggested that the Malpighian bulb and rectum possibly have no digestive functions, and that undigested food residues are stored permanently as a fecal meconium in the hind portion of the stomach. However, histological studies in the Malpighian tubes indicate that these structures take an active part in the excretory processes (5, 39). It has been shown that the material excreted through the anus is used in the construction of the scale covering (9, 67, 101).

LIFE HISTORY

In the Diaspididae, as in all Coccoidea, the females are neotenic; except for precocious sexual maturity, the adult female is morphologically a larva. Postembryonic development in armored scales consists of three instars in females and five instars in males (88). *Pracocaspis diversa* apparently is an exception; adult females are said to develop directly from the first instar without any evident second stage (45). Generally, the sexes are indistinguishable morphologically in the first instar, although a few species (e.g. *Pseudaulacaspis pentagona*) exhibit sex-linked color differences during the crawler phase.[3] Sexual dimorphism becomes manifest in the second instar. In that stage, differences in the number and distribution of secretory pores, the form of the pygidium, and the development of structures on the pygideal margin are discernible in species in which the sexes have been closely compared (18). According to Van Dinther (113) second instar males of *Chionaspis salicis* can be distinguished from females by the number of abdominal segments discernible; six in females, seven in males.

[3]Since this manuscript was completed, Stoetzel and Davidson (1974. *Ann. Entomol. Soc. Am.* 67:138–40) have demonstrated chaetotaxal sexual dimorphism in all stages, including the first instar, in representatives of several genera of Aspidiotini.

Toward the end of the second stadium, males become more elongated and show the beginning of occular pigmentation spots and appendage histoblasts. The third and fourth instars in the male developmental cycle are the nonfeeding prepupal and pupal stages. The pygidium is no longer evident after the second molt, and appendage buds are enlarged. The pupal stage possesses well-defined rudimentary appendages, genital style, eye spots, and the beginnings of the imaginal thoracic sclerites (80). Bodenheimer & Harpez (17) pointed out that the metamorphosis of male Coccoidea is essentially similar to the complete metamorphosis which is characteristic of the holometabolous insect orders.

The life histories of many of the economically important diaspidids have been worked out in detail. In a particular species, the rate of development and the number of generations per year may vary substantially in different regions. Climatic conditions, particularly temperature, humidity, and rainfall, appear to be the principal controlling factors. California red scale, for example, may have anywhere from two to as many as five or six generations per year, the number being greatest in areas of relatively uniform warm, dry conditions, such as Queensland, Australia (16). In California this scale averages about two generations per year in cool coastal areas and slightly more than three generations per year in interior areas with hot, dry summers (40). With the camphor scale, Cressman et al (31) found that rate of development was closely correlated with mean temperature. Under insectary conditions they obtained indices of correlation between length of stadia and mean temperature of 0.98 for the first stadium and 0.96 for the second. These authors used average daily mean temperature as a means of predicting the date of appearance of various stages of this scale in the field.

In areas which are subject to severe winters, only scales of a particular age class may be able to survive. For example, in Quebec the oystershell scale overwinters only in the egg stage; eggs laid in September hatch the following May or June (92). In central Europe the San Jose scale overwinters exclusively as first instar nymphs in the black cap phase of development (at this stage the first instar scale has been completed but the first ecdysis has not yet taken place). This stage is extremely cold-resistant, and all other developmental stages are killed (64). In southern Europe all stages of the scale may overwinter. Melis (75) reported that in areas characterized by mild winters, as in southern Europe, true hibernation does not occur and although development may be retarded by cool weather, it does not cease. Gentile & Summers (49) found that in California's interior valley some scales overwintered as mature females, but many remained as first instar nymphs in the black-cap phase until January. They concluded that the spring brood of new adults originated principally from overwintered first nymphs. Several authors have pointed out that not only low temperatures, but also dormancy of deciduous hosts, may affect the winter development of species such as San Jose scale.

Summer diapause also occurs in certain armored scales. In San Jose scale, for example, it has been shown that a portion of the two overlapping summer generations remains as unmolted first instar nymphs until the end of summer. In California this condition persisted for 40 to 45 days. The factors responsible for the onset of this diapause are not known (49). Summer diapause of adult females has been reported to occur due to detrimental climatic conditions. Purple scale on citrus is severely affected in the summer by the warm winds blowing in the Mediterranean area, and part of the females, generally those produced by the latest born individuals from the spring generation, enter diapause. These females resume activity in the fall, lay eggs, and produce young which coincide in their appearance with those normally resulting from the third annual generation of the scale (11). Winter adult diapause occurs mostly in univoltine diaspidids, the females entering into diapause before the end of the summer and recommencing activity in the next spring.

From the standpoint of controlling diaspine pests, forecasting the field appearance of crawlers acquires a great significance. Some methods developed require laboratory observations on the maturation of ova. The time of appearance of the arrowhead scale, *Unaspis yanonensis*, can be predicted about two months ahead in the first generation and one month prior to the second generation by observing the blastoderm formation and yolk formation, respectively (103).

REPRODUCTION

Reproduction in most armored scales is bisexual, although several widespread species occur as both bisexual and parthenogenetic races (e.g. *Aspidiotus nerii, Hemiberlesia lataniae,* and *Lepidosaphes ulmi*), and some others are known only in the parthenogenetic state (e.g. *Odonaspis ruthae*) (23). Despite some older published reports to the contrary, the occurrence of facultative parthenogenesis in the Diaspididae has not been confirmed.

Species having both uniparental and biparental races perhaps would be better treated as sibling species pairs. DeBach & Fisher (36) tested crawlers of the two such races of oleander scale and found distinctly different responses to temperature. The optimum temperature for the biparental form was 65°F, whereas 75°F appeared optimum for the uniparental form. They concluded that the differential responses of the two forms plus their obvious genetic isolation were sufficient to indicate that sibling species were involved. In Europe, the biparental and uniparental races of *L. ulmi* infest different groups of hosts. Both forms apparently have wide host ranges, but the uniparental race is reported to be the only one which occurs on fruit trees such as apple (109). Danzig (33) noted several biological differences between the two races of this species. The biparental race was the more vigorous, had a wider host range, higher fertility, and a greater resistance to adverse climatic conditions. According to Ferris (44) only the uniparental form of *L. ulmi* occurs in North America.

Adult Male Behavior

Although wingless adult males sometimes occur in a few species (e.g. *Chionaspis salicis*) (51), most armored scale males are winged and are capable of flight. However, they are tiny, fragile, and short-lived. Lacking functional mouthparts they cannot feed and the longevity of this stage generally is limited to a few hours. Development of the two sexes normally is closely correlated so that, in a given brood, adult (fifth instar) males emerge at about the same time that adult (third instar) females attain sexual maturity.

The location of receptive females by flying males probably is largely in response to female sex pheromones, although, to date, pheromones have been demonstrated in only two species, the California red scale (106) and the yellow scale (79). By exposing various dissected portions of virgin females to males, Moreno (78) showed that the probable site of pheromone origin in these scales is a pair of large internal pygideal glands, described by Nel (80), which discharge through a fine duct leading to the anal opening.

Few careful observations on the behavior of adult male scales have yet been published. Tashiro & Beavers (105) showed that longevity in adult males of California red scale was influenced neither by the presence nor the absence of virgin females. Under laboratory conditions male mortality commenced 2 hr after emergence, 50% died within 6–7 hr, and all were dead in 14 hr. Although Bodenheimer (16) believed that males of this scale were nocturnal, Tashiro & Beavers (105) found that males held in complete darkness failed to emerge. Males normally emerged during the late afternoon, apparently under influence of decreasing light intensity, at illumination levels of 350–525 fc. Emergence ceased when illumination fell below 1 fc. Moreno et al (79) suggested that males of yellow scale and California red scale must be exposed to a definite quantum of light (about 8 hr under laboratory illumination) before they are stimulated to emerge.

In dispersal tests, marked males of California red scale were recovered up to 189 m downwind and up to 92 m upwind from release points on traps baited with female pheromone (90). However, males were unable to fly upwind when wind velocities exceeded 1 mph. During colder fall and spring months maximum flight activity occurred just before sunset, and during summer it peaked between sunset and dark. A temperature of 26.7°C and relative humidity (r.h.) of 75 ± 5% were optimum for male flight under laboratory conditions. In tests with colored cards, significantly greater numbers of marked males were attracted to yellow, but when female pheromone was applied to cards all colors were equally attractive (89, 90). These results suggest that dispersing males may orient toward foliage (color) but that mate-finding is dependent ultimately upon female pheromones.

Mating Behavior

In California red scale (107) the pygidium extends behind the lateral lobes of the thorax in virgin females and can be extended to the edge of the scale covering or, occasionally, slightly beyond. Within 24 hr after mating, irreversible retraction of the pygidium commences, and, after three days, the typical reniform body shape, with lateral thoracic lobes extending posterior to the pygidium, is attained. After being inseminated females become plumper, more turgid, and attached to the scale covering. Also, rotation of the female within the scale cover ceases. During mating males orient themselves on the edge of the scale cover, facing toward its center, and insert the aedeagus under the edge by thrusting it downward and forward. In laboratory tests individual males were able to inseminate up to 30 females and the mean number inseminated was 11.9 per male. Males were able to copulate as soon as they emerged from their scales and there was no apparent difference in mating potential between males allowed to mate immediately after emergence and those held for up to 4 hr prior to mating. Abundance of females also had no appreciable effect on male mating capacity. Female California red scales mated with as many as eight different males, but multiple matings did not influence their reproductive potential. Virgin females became attractive to males from the time the gray margin of the scale began to be formed (about 23 days after settling) and, if unmated, remained attractive up to a maximum of 107 days old. They were most attractive during the first two weeks after formation of the gray margin began. Within 24 hr after insemination females ceased to be attractive. Males placed with attractive females were capable of mating at any time of day but greatest activity occurred during the period of normal male emergence during late afternoon. The level of illumination apparently had no influence on mating. McLaren (74) reported that delayed mating of overwintering virgin females of this scale caused an initial high spring natality and suggested this was due to continued ovulation in unfertilized females which resulted in a large initial output of crawlers once mating occurred.

Virgin females which are prevented from mating for long periods may exhibit unusual behavior. Ezzat (43) reported that unmated olive scale females produced abnormal elongate scales. Later they exerted their pygidia from beneath the edge of the scale, and ultimately vacated the scale covering entirely. Extension of the pygidium beyond the scale covering was also observed in unmated females of San Jose scale (49) and white peach scale (112). Virgin females of the latter species eventually vacated their scales and produced an anomalous external mass of secretory filaments. The habit of exerting the pygidium may serve to increase the attractiveness of female scales through the discharge of sex pheromones directly into the atmosphere, but vacating the scale entirely does not appear to have any obvious survival value and is difficult to explain.

Sex Ratio and Sex Determination

Published data on sex ratios in biparental armored scales indicate that within a given species considerable variation frequently occurs. Sex ratios often vary with season; for example, widely different ratios have been reported for California red scale at different times of year (63, 80), although when calculated for the entire year the ratio may approach unity (16, 63). Also, individual females within a given population may yield progeny of widely varying sex ratios in some species (e.g. *Aspidiotus simulans*) (25).

The factors which control sex determination, and therefore govern sex ratios in armored scales, are as yet poorly understood. In most armored scales adult males are haploid, yet in all bisexual forms which have been carefully studied unmated females produce no progeny of either sex, thus ruling out the usual type of haploid parthenogenesis. In the white peach scale (24) and in numerous other species (23), males develop from embryos in which the paternal chromosome set is eliminated during early embryogeny, and males continue development as haploids. Other chromosome systems occur in a few species (22, 23) but in every case there is no clear-cut genetic determination of sex. A general explanation for widely varying sex ratios which are encountered among those Coccoidea which lack sex chromosomes (such as the Diaspididae) was offered by Hughes-Schraeder (58) who suggested that sex may be determined by rather weak genetic factors which are easily overridden by environmental influences. The identification of these environmental influences and how they affect sex determination appears to be a promising area for future research.

In the white peach scale the deposition of eggs by an individual female is sexually dichronistic (12, 24). The insect first produces a series of coral-colored eggs containing female embryos, followed by a series of pinkish-white, male-containing eggs. Furthermore, aging of females prior to mating resulted in a marked increase in the proportion of male progeny produced. In this species it appears that the sex of progeny is predetermined by ovarian conditions which change with aging. Therefore, a scarcity of males which could result in the delayed mating of many females, might be expected to cause a relative abundance of males in the generation following. Sexual dichronisms of this nature have not been reported in other armored scales, but may have been overlooked in forms which lack the easily observed sexual color differences characteristic of white peach scale embryos. Direct effects by external environmental factors on sex ratio in progeny of armored scales have not been well documented, although they probably occur. In the citrus mealybug, which also lacks a clear-cut genetic mechanism of sex determination, relative humidity had a minor influence on sex ratio of progeny of females which were aged prior to mating (61).

DISPERSAL

There are two principal ways by which scale insects spread: by passive transport on infested plant material and as unsettled first instar nymphs (crawlers). Spreading by crawlers generally has been assumed to be limited to relatively short distances, although a recent paper (54) places doubt on the universal validity of this assumption. Long-range dispersal, at least over distances of several hundred kilometers or more, probably almost always results from transport of infested host material, principally in the form of viable propagative stock.

Dispersal by Infested Plant Material

The majority of economic problems involving armored scales result from spread of pest species into new geographical areas, and relatively few species have become serious pests in areas where they are indigenous. Most of the major diaspidid pests have achieved wide geographical distribution through the activities of man. The spread of pest species resulting from transport of infested plant material has been well documented. The San Jose scale, for example, apparently is indigenous to northern China (70). This pest, which spread chiefly on infested nursery stock, now occurs in deciduous fruit-growing areas throughout temperate and subtropical regions of the world (3). Similarly, the California red scale, which probably originated in Southeast Asia, now occurs virtually everywhere citrus is grown commercially (4). The promulgation and enforcement of plant quarantine and nursery inspection regulations has materially slowed the spread of many pest scales. However, despite quarantines, some species have continued to expand their ranges. The olive scale, for example, was discovered in California near Fresno in 1934 and soon became a serious pest (73). The sugar cane scale, *Aulacaspis tegalensis,* which probably is indigenous to Indonesia, recently has become a serious pest in East Africa where it was unknown until 1946 (2). The coconut scale, *Aspidiotus destructor,* appeared in the Hawaiian Islands for the first time in 1968 (10). These examples, and others, suggest that many pest scales are destined to extend their ranges until they occupy virtually all regions where suitable host plants and favorable climate occur. A single gravid female may be all that is necessary to initiate an infestation. The small size and sessile habits of armored scales make incipient infestations difficult to detect, and usually by the time a scale pest is discovered in a new region it is already too widespread to warrant an eradication attempt; or, as in the case of the olive scale in California, its potential seriousness is not sufficiently apparent to generate the support necessary for an effective eradication campaign (73). That eradication may at times be feasible was demonstrated in the southwestern US with the parlatoria date scale (20). However, the heroic measures which were necessary to eliminate this relatively isolated infestation of a species confined largely to a single host are not often likely to be applied.

Infested plant material can be responsible for local as well as long-range dispersal. For example, Williams (118) has shown that the sugar cane scale may be spread into new fields by planting scale-infested setts. The local spread of pest scales on infested nursery stock has been documented repeatedly.

The probability of armored scales being spread by means of nonpropagative plant material such as fruits, edible tubers, cut flowers, and the like, appears to be low, as establishment requires the chance placement of infested material in close proximity to suitable growing hosts. Schweig & Grunberg (94) found that infested fruit was of little or no importance in spreading Florida red scale in citrus orchards in Palestine, and regarded infested seedlings as the principal means of spread to new areas. Infested fruit piled under clean trees did not result in infestation except where branches were low enough to sweep the ground. These authors believed fallen infested leaves blown by the wind were of greater importance in spreading the scale within orchards than was infested fruit. Similarly, Melis (75, 76), after testing the ability of crawlers to cross exposed soil, concluded that infested fruits were of no importance in spreading San Jose scale unless fruit was placed in direct contact with susceptible hosts.

Crawler Behavior

The unfed first instar or crawler is the primary dispersal phase in the life cycle of all Coccoidea. Although many soft scales, mealybugs, and the like are able to move from one feeding site to another at various times during their developmental cycles, armored scales are incapable of further wandering once they have settled and commenced feeding. Dispersal of armored scale crawlers is accomplished mainly by active wandering and by wind. Occasionally other agencies such as birds, insects, and other animals, including man, may serve as accidental carriers. A report (69) that San Jose scale crawlers fasten themselves to legs and other parts of ants, and are transported in a phoretic manner, needs to be confirmed.

The free-living crawler is possibly the most vulnerable stage in the developmental cycle of most armored scales. Mortality, particularly that due to abiotic factors, must be very high for most species, although data dealing with crawler mortality are few. Oda (81) calculated 89% mortality in crawlers of the white peach scale, while mortality during the period between settling and the onset of oviposition was about 30% of the survivors. The principal hazards which crawlers face include extremes of temperature, low humidity, rain, and lack of suitable settling sites (54, 57). Wind is both a mortality agent and a dispersal agent. Large numbers of crawlers may be dislodged by wind, but a few eventually may be deposited on suitable hosts which they otherwise would be unable to colonize. There are virtually no data available on the viability of wind-borne crawlers, a parameter which has important bearing on the problem of long-range dispersal.

Newly hatched crawlers generally remain sequestered for a time beneath the maternal scale, although this period may be very brief. Jones (63) reported that in California red scale the period between the birth of crawlers and their emergence from beneath the female scale, varied from half an hour to more than two days and was dependent upon external environmental conditions. Delayed emergence coincided with unfavorable climatic conditions such as low temperature or very high relative humidity. Gentile & Summers (49) found that San Jose scale crawlers began to migrate between 2 min and 24 hr after birth, but those which remained beneath the mother for a long period often died without settling. Bliss et al (15) state that crawlers of the camphor scale seldom emerged at night and tended to emerge earlier on warm days than on cool. When the mean temperature between 9 AM and noon was 25°C, 80% of the daily emergence occurred during that time, but when temperature during this period was 31°C, so many crawlers emerged before 9 AM that the portion emerging later in the morning was reduced to 40% of the total.

The dispersal and settling behavior of crawlers following emergence appears to be controlled primarily by three types of factors: 1. innate behavior patterns which initiate wandering and settling; 2. availability of acceptable settling sites; and 3. ambient environmental conditions such as illumination, temperature, humidity, wind velocity, etc. Numerous workers have attempted to measure behavioral parameters such as speed of crawler movement, distance traveled, duration and orientation; and to assess the influences of various environmental factors on these. Much of such work has been based upon laboratory experiments under artificial conditions and the results are sometimes difficult to relate to field situations. The literature is also replete with conjectural explanations for field-observed behavioral phenomena.

The duration of wandering differs from species to species and is influenced strongly by such factors as temperature and humidity, nature of the substrate, and sex of the crawlers. Jones (63) reported that the wandering period of California red scale crawlers in Southern Rhodesia lasted from one half to three days, but that the majority usually settled within one day. In one experiment 91.8% of the crawlers which eventually settled did so during the first day, 8% the second day, and 0.2% in the third day. Bodenheimer (16) suggested that crawlers of this scale which are still wandering after the second day have little chance of settling successfully. Gentile & Summers (49) observed that most San Jose scale crawlers settled within 1–4 hr after leaving the mother.

The speed and distances which crawlers travel are highly dependent upon the nature of the substrates over which they move. In laboratory tests on relatively smooth surfaces such as frosted glass or smooth paper, crawlers have been clocked at rates ranging from a few millimeters to a few centimeters per minute (15, 16, 65, 71, 84). Quayle (85) calculated the hypothetical maximum distance which purple scale and red scale crawlers might traverse as 444 feet, assuming a maximum active life of 4 days, an optimum temperature of about 90°F, and a constant speed of about 23 mm per minute. It seems likely that the maximum distance which could be covered by a walking crawler in nature would be far less than this value.

Various authors have suggested that the crawler wandering generally serves to disperse young scales away from the mother onto new growth of the same host (12, 21). Except for wind transport, movement between plants seldom occurs unless the crowns of such plants are in contact (16). There is considerable evidence that diaspidid crawlers can move across sand or bare soil only very short distances and with great difficulty. Mathis (71), for example, reported that Florida red scale crawlers required a minimum of 2 hr to cross a 6 inch circle of sandy soil. Melis (76) found that San Jose crawlers were unable to migrate more than about 1.5 inches over soil. Quayle (85) obtained similar results with California red scale and purple scale crawlers crossing sand, but red scale crawlers were able to travel up to 2.3 feet across a compacted irrigation furrow.

Temperature influences both the initiation and rate of crawling, as well as crawler survival, and probably also affects direction of movement and selection of settling sites. The threshold for crawler activity appears to be between 13 and 20°C in most species studied (48, 71). Bodenheimer (16) states that in California red scale crawler activity occurs between 14 and 39°C, with 25–32°C being the optimum range. He suggested that temperature preferences of settling crawlers may determine distribution of scales within the crowns of trees, and cites California red scale which settles more densely in the shaded inner part of citrus in the hot Jordan Valley than it does in the cooler coastal plain. Florida red scale tends to settle in shaded parts of trees during hot summer months and on sunny parts during cooler periods of the year (94). These findings suggest possible directional responses to temperature gradients on the part of scale crawlers. However, such responses might just as well be attributed to humidity gradients or be due to combinations of temperature and humidity stimuli. Although desiccation would seem to be an important factor in crawler mortality, there have been few studies designed to assess the influence of humidity upon crawler behavior and survival. Greathead (54), experimenting with *Aulacaspis tegalensis* crawlers, showed that both high temperature and low humidity increased crawler mortality; but even under the most unfavorable conditions which he tested (30°C, 30% r.h.), 50% of crawlers survived for 8 hr. He found also that very high humidity inhibited crawler activity in this scale. DeBach & Fisher (36) found that mortality of oleander scale crawlers due to raising the temperature from 65 to 80°F was much greater than that attributable to lowering relative humidity from 80 to

40%. Avoidance of direct sunlight, reported for crawlers of several species (16, 49), may well be a response to high temperature rather than to light.

Light appears to be a major factor influencing the direction of wandering in scale crawlers. Bennassy (11) concluded that, after the temperature threshold had been reached, phototaxis played the dominant role in dispersal of white peach scale crawlers. The two sexes of crawlers of this species responded differently to light, but in total darkness both sexes tended to settle in the immediate vicinity of the mother. Hulley (59) reported a definite positive orientation toward a directional light source in purple scale crawlers, although the accuracy of orientation was not great. In his experiments an increasing percentage of crawlers responded as light intensity was increased from 0 to about 42 m-candles, when 100% response was obtained. He found no clearcut threshold response to light, and in nondirectional light, crawler movement appeared to be random. Experiments with camphor scale crawler (15) also showed a definite, but somewhat irregular movement toward light. *Aulacaspis tegalensis* crawlers also proved to be positively phototactic, moving upwards toward light during the morning. However, in darkness they moved downward (54).

The nature of the substrate over which crawlers move not only affects the rate and direction of movement, but also has a profound influence on the duration of wandering. In some species it appears that settling behavior may be initiated as soon as a suitable site is encountered, while in others a definite period of wandering appears to be necessary before settling can be induced (11). In scales which utilize crevices, other surface irregularities or contact surfaces, or which normally occur in closely packed aggregations, settling may be induced largely by thigmotaxis (11, 28).

It has been often noted that heavy scale infestations frequently occur in the presence of deposits of dust or other particulate matter on host foliage. The most commonly accepted explanation for this phenomenon is that such particles interfere with the activities of natural enemies (35). Dust particles also affect crawler behavior and, therefore, scale distribution and density. Hulley (59) reported that the average wandering time of purple scale crawlers on dusty orange leaves was 63.8 ± 3.9 min at 28°C, and 91.0 ± 8.8 min on clean leaves. In species with thigmotactically induced settling behavior, such as purple scale, dust patches on foliage may result in crawlers settling where they otherwise would not, and the shortening of wandering time also may result in the settling of crawlers which otherwise might be dislodged by wind (59).

Geotactic responses of scale crawlers seldom have been investigated. In purple scale, no clearcut response could be demonstrated (59). However, McLaren (74) states that the majority of California red scale crawlers settled on upper surfaces of peripheral citrus leaves because of a negative geotaxis. Chemotactic phenomena in scale crawlers also have received little attention. Bodenheimer (16) concluded that citrus odor had a positive effect upon California red scale crawlers causing them to move more rapidly in the majority of his tests, but no directional effect was demonstrated. Tests with purple scale crawlers failed to demonstrate any chemotactic reaction to juices of orange pulp and orange rind (59).

Male and female crawlers of the same species often exhibit different patterns of behavior which result ultimately in separation, or at least nonrandom mixtures, of the sexes on the host. Behavioral differences between sexes have been observed in phototactic and thigmotactic responses, as well as in duration of wandering behavior. Priesner (84) suggested that a differential response to light resulted in the preponderance of male Florida red scales on upper surfaces of leaves where this surface was directed toward light. Geier (48) found that female crawlers of *Epidiaspis leperii* were negatively phototactic and sought sheltered or hidden settling sites in the immediate vicinity of the mother, whereas male crawlers dispersed and eventually settled on young shoots and fruit. In the white peach scale, male crawlers tend to settle quickly after leaving the mother, as soon as a suitable locality is encountered; whereas female crawlers undergo a definite wandering period lasting several hours (11, 12). Brown (21) showed that the average distance traveled before settling was four times as great with female pine needle scales than with males. This explained why nearly all male scales occurred on old growth, while most of the females were on the new growth. Benassy (11) in controlled experiments on white peach scale crawlers showed that females tended to settle in a random manner on the surface of smooth potato tubers, although showing preference for eyes or other surface irregularities. Male crawlers, on the other hand, tended to form definite aggregations. He suggested that after one male crawler settles (usually a thigmotactic response to some surface irregularity), others aggregate with it because of some gregarious instinct. Aggregations of male scales occur in many diaspidids, particularly in genera such as *Phenacaspis, Pinnaspis,* and related forms. The nature of the gregarious instinct of these males is unknown. The possibility that pheromones may be involved has not been investigated. Perhaps aggregation results merely from the thigmotactically induced settling of males as, by chance, they encounter previously settled males.

Hulley (59) divided the usual settling behavior of purple scale crawlers on orange leaves into five stages. During stage one, crawlers walked fast with little turning. At the onset of stage two, there was a sudden decrease in speed often accompanied by increased turning. Usually after a short period of stage two crawlers began to walk in a peculiar manner in which the tarsi were put down at almost the same spot where they were lifted. This behavior was accompanied by much turning and little progress in any direction (stage three). After a short time, crawlers came to a complete halt, making only slight pawing movements (stage four). In stage five, final settling, the antennae were folded back against the body and the crawler flattened itself against the leaf surface. The legs were then withdrawn beneath the body and the crawler remained motionless. Within 50 min to 3 hr characteristic wax threads, which indicated the beginning of scale formation, began to appear. Sometimes this sequence was broken at some point with a reversion to an earlier stage of behavior. Thus, even after flattening (stage five), reversion to stage one occasionally occurred.

Wind Dispersal

Air currents have long been recognized as being of probable importance in the dispersal of armored scale crawlers, at least over short distances (as within orchards)

(16, 59, 63). Crawlers of virtually all Coccoidea possess a broad, flat body form and long fine caudal setae. These features appear to be morphological adaptations for maximum buoyancy in air currents, although armored scale crawlers have been observed to use the caudal setae to right themselves when overturned (59). Even though positive trapping data indicated crawlers of other Coccoidea, such as mealybugs and soft scale, (87, 98) were wind dispersed, relatively little attention had been given to this mode of spread in armored scales until recently. In wind tunnel experiments Brown (21) found that crawlers of the pine needle scale traveled a maximum horizontal distance of 72 inches while falling through an average vertical distance of 7.5 inches in a 4.0 km/hr wind. Using the formula for small bodies: distance traveled = (horizontal wind speed/terminal velocity) X initial height above surface, he calculated the terminal velocity as 0.42 km/hr (0.12 m/sec). A verticle component of wind velocity in excess of this value would keep such a body aloft indefinitely. In field trapping experiments large numbers of crawlers were carried for short distances (up to 30 feet) showing the importance of wind in intrastand dispersal of pine needle scale. Small numbers of crawlers were trapped up to 2.8 km (the maximum distance tested) downwind from the nearest source. Timlin (111) using spaced greased plates to trap windborne crawlers showed that *Parlatoria pittospori* was spread downwind into apple orchards for 250 yards from infested shelter belts of *Pinus radiata.* Data from trapping tests designed to measure aerial density of crawlers of the sugar cane scale, *Aulacaspis tegalensis,* in East Africa (54) showed that crawler distribution conformed to the pattern expected from similar experiments with airborne aphids and the like, i.e. log density was inversely proportional to log height above crop. Below the upper level of the crop the pattern of crawler density was complicated by eddy currents, but tended to approach expected values at higher wind velocities. Crawler density increased with increasing wind speed up to about 2.0 m/sec. Observed departures from expected velocity-density relationships were attributed to reduced crawler activity during periods of very high humidity. Densities of airborne crawlers below the upper surface of the crop ranged between 0.03 and 5.0/m^3 and the author concluded that such densities were more than sufficient to explain the spread of scales across roads between blocks of sugar cane. Densities at the crop surface were in the range of 0.1–2.0/m^3 and it was calculated that the crawler emission rate from the crop surface ranged between 500 and 2.5 X 10^6/m^2/hr. The previously cited formula employed by Brown to estimate the distance a crawler will be carried does not take into account vertical components of wind velocity. Greathead stated that with turbulent air conditions characteristic of East Africa *A. tegalensis* crawlers, which have a low terminal velocity of 0.12 m/sec, could be carried upward and eventually become distributed to the level of inversion close to the cloud base. He postulated that under certain atmospheric conditions, such as in converging winds associated with storm fronts, airborne crawlers might be concentrated and deposited over a relatively small area some distance from their source. He suggested that such long-range dispersal of airborne crawlers may have initiated *A. tegalensis* infestations at localities 150 and 260 km inland from known infestations on the Kenya Coast. Crawlers of this species appear to possess a distinctive behavior pattern which favors wind dispersal. The crawlers, most of which emerge during the morning, move upward from infested stalks to cane leaves where part of the population is detached by wind. The remaining crawlers move downward at night and settle on the stalks beneath leaf sheaths. Greathead suggested that behavior patterns favoring wind dispersal may occur in other diaspidid crawlers, particularly those which infest relatively short-lived hosts such as grasses and herbs.

THE SCALE COVERING

A major advance in the evolution of the armored scales was the development of the pygidium, the organ responsible for the formation of the external scale covering (26). The pygidium is composed of the more or less fused and sclerotized posterior abdominal segments. This structure bears the openings of secretory glands of various sizes and shapes, the anus, and usually, variously shaped marginal lobes and special marginal gland spines or flattened fringe plates. All of these structures function in the construction of the external scale. The number, type, and arrangement of these pygidial structures are of paramount importance in the taxonomy of the diaspidids (110). A well-developed pygidium is present in nearly all armored scale insects, although occasionally it is reduced or largely unsclerotized, and sometimes lacks marginal lobes, gland spines, or fringe plates, as in *Protodiaspis agrifoliae* which produces an external scale of unconsolidated filaments. Normally the pygidium includes the fifth abdominal segment (sometimes also the fourth) and those posterior to it, but occasionally it contains no segments anterior to the sixth (e.g. *Ancepaspis* spp.) (45).

The scale coverings of female diaspidids are formed from three basic materials: 1. loose fibers secreted mainly by the pygideal glands, 2. a fluid material discharged from the anus which is thought to bind the fibers together, and 3. the two larval exuviae which are incorporated into the scale at each molt (39). Sometimes fragments of the host plant epidermis also may become incorporated into the scale around its margins (38). The so-called mining scales, such as *Howardia biclavus* and *Psuedaonidia claviger,* bury themselves almost completely beneath the epidermal layer of host twigs and the host epidermis becomes intimately associated with the dorsal scale covering (37). In most species the mature female scale is composed of three distinct segments corresponding to the three instars of the female developmental cycle. The first and second segments consist of secretory products plus the dorsal exuviae of the first and second instars, respectively. The third or outer segment, laid down by the adult female, is composed entirely of secretory materials.

In addition to the dorsal scale most diaspidids also produce a membranous ventral scale which separates the insect from direct contact with the plant surface. The ventral scale is formed of secretions from ventral pygideal glands, plus incorporated ventral exuvial residues. In some Lepidosaphini, the ventral scale forms an internal fold which serves as an incubation chamber (86). In some other diaspidids (e.g. Odonaspidini), it is nearly as thick as the dorsal portion, but in many scales it is extremely delicate or not detectable.

Soon after a diaspidid crawler has settled and inserted its stylets, secretory filaments begin to issue from gland orifices on the pygidium and elsewhere on the body. In the red scale, the material from the pygideal glands is forced out by periodic contractions of the abdomen (9). In the oystershell scale the newly settled crawler produces a mass of thread-like fluff which exudes from abdominal gland spines and pores, and from wax pores in the region of the head. When this secretion is complete only the head region of the insect remains uncovered (92). Subsequently, the body of the insect becomes flattened and the dorsal skin hardens and thickens to form the first scale. Observations on newly settled purple scales indicated that this species first produces two long tangled threads extending from the anterior part of the body, over and around the insect. These threads (presumably produced from a pair of large cephalic ducts which are present in the first instar in the Lepidosaphini and some other tribes) serve to support the more compact covering which is produced later (86).

In describing the formation of the dorsal shield of the red scale, Dickson (38) stated "The insect uses its pygidium much as a plasterer uses a trowel, in this case a self-feeding trowel that continually produces its own plaster." The scale covering is enlarged by the insect "pressing the tip of the pygidium against an area at the edge and slowly shoving outward, building a flatly triangular extension the size and shape of the pygidium. The pygidium is then drawn back, the body rotates, the pygidium is reapplied, and another portion of the edge of the covering extended."

Except for minor modifications due to crowding, irregularities in the substrate, and the like, the form of the external covering is quite uniform within a given species. In general, the shape of the female body is reflected in the form of the external scale, i.e. species with elongate bodies tend to produce elongate scales, whereas short-bodied species tend to produce circular or oval scales. There are exceptions, however (e.g. the circular scale produced by *Aulacaspis greeni,* an elongate-bodied species) (101).

Several workers have discussed the movements executed by the pygidium during the formation of the secretory portions of the scale covering (9, 38, 39, 72, 101). In general these movements appear to be oscillatory in nature, i.e. the pygidium is moved through an arc as the insect pivots around the inserted stylets and the scale is built up of concentric arcuate strips added to the margin. The length of the arcs described by the pygidium and their alignment determines the shape of the scale. In species with long, narrow scales the pygidium describes narrow arcs which are generally aligned in one direction. In the case of circular scales with central or subcentral exuviae, the process of formation has been called rotation by some authors (72), although the insects do not rotate continuously in one direction. Such scales are formed of a series of arcuate segments. Disselkamp (39), describing the formation of the dorsal shield of the female San Jose scale, stated that the pygidium is first rotated through an arc of approximately 320° during which a strip of loose fibers from the pygideal glands is laid down. The insect then agglutinates these fibers with the anal discharge while rotating the pygidium in the opposite direction. During this process one droplet of anal secretion is deposited every one or two minutes. The sequence is then repeated to lay down additional arcs of scale material. By shifting the starting point of successive arcs, a circular or oval scale eventually is formed. This method of scale formation is somewhat different than the deposition of discontinuous segments described for red scale (38). Takagi & Tippins (101) postulated an evolutionary trend toward forms which produce circular scales by rotational types of pygideal movements. Scales of this type are characteristic of the advanced tribe Aspidiotini but have evolved independently in a few species in other tribes.

The scales of male diaspidids are formed during the feeding stages, which are the first and second instars in that sex. Therefore, only the first stage exuviae are incorporated into the scales. Male scales are smaller and usually distinctly different in shape, and often in color and texture, from those of females of the same species. For example, in many Diaspidini, Aspidiotini, and Parlatorini in which the female scales are circular or oval, the male scales are more or less elongate. In some genera (e.g. *Chionaspis, Diaspis*) the surface of the male scales are frequently decorated with one or more longitudinal raised ridges. Disselkamp (39) described differences in the pygideal movements in second instar San Jose scales which resulted in sexual differences in scale shape. In male scales the first bands of secretory material laid down are broad arcs forming a complete ring, as in the female, but later bands cover an arc of only about 180° so that the final resulting scale is oblong.

Information on the chemical composition of diaspidid scale covering is still meager. Many early workers referred to the waxy nature of these coverings. Green (55) for example described the scales as formed of "waxy, fibrous or resinous matter." Manlik (68), while attempting to use wax solvents to improve insecticides for scale control, tested various solvents and concluded the scale covering of *Lepidosaphes ulmi* probably was not a wax. Disselkamp (39) found that the secretory fibers of the shield of San Jose scale were composed of extremely resistant chitin-like secretion products which were insoluble in organic solvents such as xylene, methanol, ether, acetone, and chloroform. These fibers did dissolve almost completely in hot concentrated sulfuric and nitric acids. By paper chromatography, the presence of serine and tyrosine in the fibers was established.

It appears that the scales of most diaspidids do contain at least some waxy materials. Chloroform-soluble substance in the covering of the white peach scale amounted to about 35% (104), and approximately 50% in the Florida red scale (41). Metcalf & Hockenyos (77) extracted the scale coverings of a number of species with hot carbon tetrachloride and found that waxes accounted from 31 to 58% of the weight of the scales tested. Dickson (38) reported that the covering of California red scale was about 45% waxes, 47% proteinaceous material, and 8% exuviae.

The nonwaxy fraction has been found to be proteinaceous in nature and does not contain chitin, except for the exuviae incorporated into it. Spectrophotometric studies in Florida red scale showed that the scale material contained 3.7% nitrogen, and amino acids accounted for only 15% of the total weight. The nonwaxy component did not contain DNA, RNA, or carbohydrate. The bulk of this fraction was composed of material with a molecular weight of at least 200,000, and its chemical composition was similar to that of melanin and polyphenol polymers. The possible identification of a polyphenol polymer suggests that hardening of the shield may result from enzymatic polymerization of tyrosine (41).

The amount of polyphenol polymers must undoubtedly vary among the various species, as reflected by the different quality and texture of the shield. Very little information exists on the density and permeability of the scale covering. The production of waxy filaments coming through the shield in young stages is considered to be a result of the scale permeability. The diffusion of plant volatile materials through the shield of the fir scale, *Dynaspidiotus abietis,* has been reported by Goidanich (52).

PUPILLARIAL FORMS

In several specialized genera (e.g. *Ancepaspis*), females do not shed the exuviae during the second molt, but remain enclosed within the hardened second stage skin. Such species are referred to as pupillarial or exuvial. Glandular structures associated with scale formation are often much reduced or absent in pupillarial species, except in the second stage males which produce a scale of normal form (26). It appears that in most pupillarial species the second stage exuvium is ruptured along its posterior margin by the mature female within to accommodate mating and the egress of the crawlers. In several species of *Ancepaspis,* the pygidia of mature females are exerted from the second stage exuviae. In such forms the posterior parts of the pygidia are strongly sclerotized with pronounced, tooth-like, apical lobes (26). In species of *Fiorinia* the pupillarial females generally retain a few glandular pygideal structures and may make some secretory additions to the second stage exuviae (44). Species of *Protodiaspis* are semiexuvial, as the adult female is partly, but never completely, enclosed within the second stage molt which fragments or ruptures in the normal fashion (26).

Brown & McKenzie (26) presented evidence that pupillarial or partly pupillarial forms such as *Ancepaspis* and *Protodiaspis* represent a specialized, derived condition rather than a primitive ancestral one.

Literature Cited

1. Agarwal, R. A., Sharma, D. P. 1961. Studies of some epidermal characteristics and hardness of sugar cane stem in relation to the incidence of scale insects *Melanaspis glomerata* (Green). *Indian J. Entomol.* 22:197–203
2. 1964. Distribution maps of pests. Series A (Agricultural), Map No. 187. *Commonw. Inst. Entomol. London*
3. 1967. Distribution maps of pests. Series A (Agricultural), Map No. 7 (revised). *Commonw. Inst. Entomol. London*
4. 1968. Distribution maps of pests. Series A (Agricultural), Map No. 2 (revised). *Commonw. Inst. Entomol. London*
5. Bacetti, B. 1970. I tubi Malpighiani nella femmina adulta dei Diaspidini. *Proc. Int. Congr. Entomol., 10th, Wein* 1:752–58
6. Balachowsky, A. 1932. Etude biologique des Coccides du bassin occidental de la Méditerranée. *Encycl. Entomol. Ser. A,* Vol. 15
7. Balachowsky, A. 1937. Les cochenilles de France, d'Europe, du Nord de l'Afrique et du bassin Méditerranéen. II: Caractères generaux des cochenilles. Morphologie interne. *Actual. Sci. Ind. Entomol. Appl.* 564:73–129
8. Balachowsky, A. 1948. Les cochenilles de France, d'Europe, du Nord de l'Afrique, et du bassin Méditerranéen. IV. Monographie des Coccoidea; classification—Diaspidinae (Première Par-

tie). *Actual. Sci. Ind. Entomol. Appl.* 1054:243–394
9. Baranyovits, F. 1953. Some aspects of the biology of armoured scale insects. *Endeavour* 12(48):202–9
10. Beardsley, J. W. 1970. *Aspidiotus destructor* Signoret, an armored scale pest new to the Hawaiian Islands. *Proc. Hawaii. Entomol. Soc.* 20:505–8
11. Bénassy, C. 1961. Contribution a l'étude de l'influence de quelques facteurs écologiques sur la limitation des pullulations de cochenilles-diaspines. *Inst. Nat. Rech. Agr. Ser. A,* No. 3747. 165 pp.
12. Bennett, F. D., Brown, S. W. 1958. Life history and sex determination in the diaspine scale *Pseudaulacaspis pentagona* (Tar.). *Can. Entomol.* 90: 317–25
13. Berlese, A. 1896. Cocciniglie italiane viventi sugli agrumi. Part. III: I Diaspiti, anno IV et V. *Riv. Patol. Veg.* 4:195–292; 5:3–73
14. Bielenin, I., Weglarska, B. 1967. Anatomy and histology of female and male alimentary canal of *Quadraspidiotus ostraeiformis* (Curt.) including cytology of mid-gut. *Acta Biol. Cracov. Ser. Zool.* 10:105–21
15. Bliss, C. I., Cressman, A. W., Broadbent, B. M. 1935. Productivity of the camphor scale and the biology of its eggs and crawler stages. *J. Agr. Res.* 50:243–66
16. Bodenheimer, F. S. 1951. *Citrus Entomology in the Middle East with Special References to Egypt, Iran, Irak, Palestine, Syria, Turkey.* The Hague: Junk. 633 pp.
17. Bodenheimer, F. S., Harpez, A. 1951. Holometabolic development in the males of Coccoidea. *Isr. Res. Counc. Bull.* 1:133–35
18. Boratynski, K. L. 1953. Sexual dimorphism in the second instar of some Diaspididae. *Trans. Roy. Entomol. Soc. London* 104(Part 12):451–79
19. Borchsenius, N. S. 1966. *A Catalogue of the Armoured Scale Insects (Diaspidoidea) of the World.* Moscow and Leningrad: Academy of Sciences of the USSR, Zoological Institute. 449 pp.
20. Boyden, B. L. 1941. Eradication of the parlatoria date scale in the United States. *US Dep. Agr. Misc. Publ.* 443. 62 pp.
21. Brown, C. E. 1958. Dispersal of the pine needle scale, *Phenacaspis pinifoliae* (Fitch). *Can. Entomol.* 90:685–90
22. Brown, S. W. 1963. The Comstockiella system of chromosome behavior in the armored scale insects. *Chromosoma* 14:360–406
23. Brown, S. W. 1965. Chromosomal survey of the armored and palm scale insects. *Hilgardia* 36:189–294
24. Brown, S. W., Bennett, F. D. 1957. On sex determination in the diaspine scale *Pseudaulacaspis pentagona* (Targ.). *Genetics* 42:510–23
25. Brown, S. W., DeLotto, G. 1959. Cytology and sex ratio in an African species of armored scale insect. *Am. Natur.* 93:369–79
26. Brown, S. W., McKenzie, H. L. 1962. Evolutionary patterns in the armored scale insects and their allies. *Hilgardia* 33:133–70A
27. Busgen, M. 1891. Der Honigtau. Biologische Studien an Pflanzen und Pflanzenlausen. *Jena. Z. Naturwiss.* 25, Band. 18
28. Carnegie, A. M. J. 1957. Observations on the behavior of the crawlers of *Lepidosaphes beckii* (Newm.). *J. Entomol. Soc. S. Afr.* 20:164–69
29. Carter, W. 1962. *Insects in Relation to Plant Disease.* New York: Interscience. 705 pp.
30. Childs, L. 1914. The anatomy of the Diaspinine scale insect *Epidiaspis pyricola* (Del Guerc.). *Ann. Entomol. Soc. Am.* 7:47–57
31. Cressman, A. W. et al 1935. Biology of the camphor scale and a method for predicting the time of appearance of stages in the field. *J. Agr. Res.* 50: 267–83
32. Dale, J. L., McCoy, C. E. 1964. The relationship of scale insects to death of Bermuda grass in Arkansas. *Plant Dis. Rep.* 48:228
33. Danzig, E. M. 1959. Concerning the biological forms of the apple comma scale, *Lepidosaphes ulmi* (L). *Zool. Zh.* 38:879–86. (In Russian, English summary)
34. Danzig, E. M. 1970. Synonymy of some polymorphous species of coccids. *Zool. Zh.* 49:1015–24. (In Russian, English summary)
35. DeBach, P. 1951. The necessity for an ecological approach to pest control on citrus in California. *J. Econ. Entomol.* 44:443–47
36. DeBach, P., Fisher, T. W. 1956. Experimental evidence for sibling species in the oleander scale, *Aspidiotus hederae* (Vallot). *Ann. Entomol. Soc. Am.* 49: 235–39
37. Deckle, G. W. 1962. Camelia mining scale, *Pseudoaonidia clavigera* (Ckl.) *Fla. Dep. Agr. Entomol. Circ.* No. 1

38. Dickson, R. C. 1951. Construction of the scale covering of *Aonidiella aurantii* (Mask.). *Ann. Entomol. Soc. Am.* 44:596–602
39. Disselkamp, C. 1954. The scale formation of the San Jose scale (*Quadraspidiotus perniciosus* Comst.). *Hofchen-Briefe* 7:105–51
40. Ebling, W. 1959. *Subtropical Entomology.* San Francisco: Lithotype Process Co. 747 pp.
41. Ebstein, R. P., Gerson, U. 1970. The non-waxy component of the armored scale insect shield. *Biochem. Biophys. Acta* 237:550–55
42. Enser, K. 1941. Histologische untersuchungen der Saugstiche von *Aspidiotus perniciosus* (Comst.). *Mitt. Biol. Reichsanst.* 65:96–98
43. Ezzat, Y. M. 1957. Biological studies on the olive scale (*Parlatoria oleae* Colvee). *Bull. Soc. Entomol. Egypt* 41:351–63
44. Ferris, G. F. 1937. *Atlas of Scale Insects of North America,* Series I. Stanford, Calif.: Stanford Univ. Press. 275 pp.
45. Ibid 1942. Ser. IV
46. Flanders, S. E. 1956. Struggle for existence between red and yellow scale. *Calif. Citrog.* 41:396–403
47. Froggatt, W. W. 1915. A descriptive catalogue of the scale insects ("Coccidae") of Australia. Part I. *Dep. Agr. N. S. W. Sci. Bull.* No. 14. 64 pp.
48. Geier, P. W. 1949. Contribution a l'etude de la cochenille rouge du poirier (*Epidiaspis leperii* Sign.) en Suisse. *Rev. Pathol. Veg. Entomol. Agr. Fr.* 28:177–261
49. Gentile, A. G., Summers, F. M. 1958. The biology of the San Jose scale on peaches with special reference to the behavior of males and juveniles. *Hilgardia* 27:269–85
50. Gerson, U. 1967. Interrelationships of two scale insects on citrus. *Ecology* 48:872–73
51. Ghauri, M. S. K. 1962. *The Morphology and Taxonomy of Male Scale Insects.* London: British Mus. (Natur. Hist.). 221 pp.
52. Goidanich, A. 1956. Sulla permeabilita del sollicolo sericco in alcum Diaspididi. *Mem. Soc. Entomol. Ital.* 25: 207–24
53. Goidanich, A. 1960. Specializzaciones ecologica e nomenclatura delle *Carulaspis* del viscum e delle Cuppressaceae. *Boll. Ist. Entomol. Univ. Bologna* 24: 1–38
54. Greathead, D. J. 1972. Dispersal of the sugar-cane scale *Aulacaspis tegalensis* (Zhnt.) by air currents. *Bull. Entomol. Res.* 61:547–58
55. Green, E. E. 1896. *The Coccidae of Ceylon,* Part I. London: Dulau & Co. 103 pp.
56. Heriot, A. D. 1934. The renewal and replacement of the stylets of sucking insects during each stadium, and the method of penetration. *Can. J. Res.* 11:602–12
57. Huffaker, C. B., Kennett, C. E., Finney, G. L. 1962. Biological control of olive scale, *Parlatoria oleae* (Colvee), in California by imported *Aphytis maculicornis* (Masi). *Hilgardia* 32:541–636
58. Hughes-Schraeder, S. 1948. Cytology of coccids. *Advan. Genet.* 2:127–203
59. Hulley, P. 1962. On the behavior of the crawlers of the citrus mussel scale, *Lepidosaphes beckii* (Newm.). *J. Entomol. Soc. S. Afr.* 25:56–72
60. Ishaaya, I., Swirski, E. 1970. Invertase and amylase activity in the armoured scales *Chrysomphalus aonidum* and *Aonidiella aurantii. J. Insect Physiol.* 16:1599–1606
61. James, H. C. 1938. The effect of the humidity of the environment on sex ratios from over-aged ova of *Pseudococcus citri* Risso. *Proc. Roy. Entomol. Soc. Ser. A.* 13:73–79
62. Jenser, G., Sheta, I. B. 1969. Investigation of the resistance of a few Hungarian sour-cherry hybrids against the San Jose scale. *Acta Phytopathol. Acad. Sci. Hung.* 4:313–15
63. Jones, E. P. 1936. The bionomics and ecology of the red scale—*Aonidiella aurantii* Mask.—in Southern Rhodesia. *Brit. S. Afr. Co. Mazoe Citrus Exp. Sta. Publ. 5*
64. Klemm, M. 1944. San Jose Schildlaus in Nordkaukasus. *Arb. Physiol. Angew. Entomol. Berlin Dahlem* 11:1–24
65. Ludicke, M. 1950. Uber biologische Besonderheiten der San Jose Schildläus im Zusammenhang mit der Wirking von Phosphorsaureestern. *Hofchen-Briefe* 3:17–32
66. Lupo, V. 1943. II *Mytilococcus ficifoliae* (Berlese) e una forma estiva del *M. conchiformis* (Gmelin). *Boll. Lab. Entomol. Agr. Portici* 5:197–205
67. Lupo, V. 1957. Precisazione sulla posizione e rotura delle esuvie larval delle Diaspine. *Boll. Acad. Gioenia Sci. Natur. Catania Ser. 4* 3:419–29
68. Manlik, S. 1917. Solubility of the scale, *Lepidosaphes ulmi* (L). *Bull. Entomol. Res.* 7:267–69
69. Marek, J. 1952. Über passive Verbreitung der San Jose Schildlaus druch

einheimesche Ameisen. *Z. Pflanzenbau Pflanzenschutz* 3:254–63
70. Marlatt, C. L. 1906. The San Jose or Chinese scale. *US Dep. Agr. Bur. Entomol. Bull.* 62. 89 pp.
71. Mathis, J. 1947. Biology of the Florida red scale in Florida. *Fl. Entomol.* 29: 13–35
72. Matsuda, M. 1927. Studies on the rotatory movements necessary for the formation of the scale in *Chrysomphalus aonidum. Trans. Natur. Hist. Soc. Formosa* 17:391–417. (In Japanese, English summary)
73. McKenzie, H. L. 1952. Distribution and biological notes on the olive parlatoria scale, *Parlatoria oleae* (Colvee), in California. Scale studies—Part X. *Bull. Calif. Dep. Agr.* 41:127–38
74. McLaren, I. W. 1971. A comparison of the population growth potential in California red scale, *Aonidiella aurantii* (Maskell), and yellow scale, *A. citrina* (Coquillet) on citrus. *Aust. J. Zool.* 19:189–204
75. Melis, A. 1943. Contributo alla conocenza dell' *Aspidiotus perniciosus* Comst. *Redia* 29:1–170
76. Melis, A. 1951. Precisazioni Morphobiologiche sull' *Aspidiotus perniciosus* Comst. *Redia* 36:1–91
77. Metcalf, C. L., Hockenyos, G. L. 1930. The nature and formation of scale insect shells. *Trans. Ill. State Acad. Sci.* 22:166–84
78. Moreno, D. S. 1972. Location of the site of production of the sex pheromone in the yellow scale and the California red scale. *Ann. Entomol. Soc. Am.* 65: 1283–86
79. Moreno, D. S., Carman, G. E., Rice, R. E., Shaw, J. G., Bain, N. S. 1972. Demonstration of a sex pheromone of the yellow scale, *Aonidiella citrina. Ann. Entomol. Soc. Am.* 65:443–46
80. Nel, R. G. 1933. A comparison of *Aonidiella aurantii* and *Aonidiella citrina,* including a study of the internal anatomy of the latter. *Hilgardia* 7: 417–66
81. Oda, T. 1963. Studies on the dispersion of the mulberry scale *Pseudaulacaspis pentagona. Jap. J. Ecol.* 13:41–46. (In Japanese, English summary)
82. Parr, T. 1940. *Asterolecanium variolosum* Ratzeburg, a gall forming coccid, and its effect upon the host tree. *Yale Univ. Sch. Forest. Bull. 46* 49 pp.
83. Pesson, P. 1944. *Contribution a l'etude morphologique des femelles de Coccides.* Thesis Fac. Cienc. Univ. Paris. No. 2899, Ser. A., 266 pp.
84. Priesner, H. 1931. On the biology of *Chrysomphalus ficus* Ril. with suggestions on the control of this species in Egypt. *Egypt Min. Agr. Tech. Sci. Serv. Bull.* 117. 19 pp.
85. Quayle, H. J. 1911. Locomotion of certain young scale insects. *J. Econ. Entomol.* 4:301–6
86. Quayle, H. J. 1912. The purple scale. *Univ. Calif. Bull.* 226:319–40
87. Quayle, H. J. 1916. Dispersion of scale insects by the wind. *J. Econ. Entomol.* 9:486–92
88. Reh, L. 1901. Über die postembryonale Entwickelung der Schildläuse und Insekten Metamorphose. *Allg. Z. Entomol.* 6:51–89
89. Rice, R. E., Moreno, D. S. 1969. Marking and recapture of California red scale for field conditions studies. *Ann. Entomol. Soc. Am.* 62:558–60
90. Rice, R. E., Moreno, D. S. 1970. Flight of male California red scale. *Ann. Entomol. Soc. Am.* 63:91–96
91. Richards, A. M. 1962. Damage to apple crops infested with San Jose scale. *N. Z. J. Agr. Res.* 5:479–84
92. Samarasinghe, S., Leroux, E. J. 1966. The biology and dynamics of the oystershell scale, *Lepidosaphes ulmi* (L) on apples in Quebec. *Ann. Entomol. Soc. Quebec* 11:206–59
93. Schmutterer, H., Kloft, W., Lüdicke, M. 1957. Coccoidea, Schildäuse, scale insects, cochenilles. In *Handbuch der Pflanzenkrankh,* ed. P. Sorauer, 5:403–520. Berlin: Paul Parey
94. Schweig, C., Grunberg, A. 1936. The problem of black scale, *Chrysomphalus ficus* (Ashm.) in Palestine. *Bull. Entomol. Res.* 27:677–714
95. Snodgrass, R. E. 1935. *Principles of Insect Morphology.* New York: McGraw-Hill. 667 pp.
96. Stafford, E., Barnes, D. F. 1948. The biology of the fig scale. *Hilgardia* 18:567–98
97. Stanaard, L. J. 1965. Polymorphism in the Putnam's scale, *Aspidiotus ancylus. Ann. Entomol. Soc. Am.* 58:573–76
98. Strickland, A. H. 1950. The dispersal of Pseudococcidae by air currents in the Gold Coast. *Proc. Roy. Entomol. Soc. London, Ser. A* 25:1–9
99. Takagi, S. 1969. Diaspididae of Taiwan based on material collected in connection with the Japan-U.S. Cooperative Science Programme, 1965. *Insecta Matsumurana* 32:1–110
100. Takagi, S., Kawai, S. 1967. The genera *Chionaspis* and *Pseudaulacaspis* with a

criticism on *Phenacaspis. Insecta Matsumurana* 30(1):29–43

101. Takagi, S., Tippins, H. H. 1972. Two new species of the Diaspididae occurring on Spanish moss in North America. *Kontyu* 40(3):180–86
102. Takahashi, R. 1953. Dimorphism in some species of *Chionaspis* and *Phenacaspis. Boll. Lab. Zool. Gen. Agr. Portici* 33:48–56
103. Takezagua, H., Uchida, M. 1969. Relationship between the ovarial development and the appearance of nymphs in the arrowhead scale, with special reference to its application to forecasting. *Jap. J. Appl. Entomol. Zool.* 13(1):31–39. (In Japanese, English summary)
104. Tamaki, Y., Kawai, S. 1969. X-ray diffraction studies on waxy covering of scale insects. *Appl. Entomol. Zool.* 4(2): 79–86
105. Tashiro, H., Beavers, J. B. 1968. Growth and development of the California red scale, *Aonidiella aurantii. Ann. Entomol. Soc. Am* 61:1009–14
106. Tashiro, H., Chambers, D. L. 1967. Reproduction in the California red scale, *Aonidiella aurantii.* I. Discovery and extraction of a female sex pheromone. *Ann. Entomol. Soc. Am.* 60:1166–70
107. Tashiro, H., Moffitt, C. 1968. Reproduction in the California red scale, *Aonidiella aurantii.* II. Mating behavior and postinsemination female changes. *Ann. Entomol. Soc. Am.* 61:1014–20
108. Teodoro, G. 1915. Osservazioni sulla ecologia dell Cocciniglie con speciale riguardo alla morfologia e alla fisiologia di questi insetti. *Redia* 11:129–209
109. Thiem, H. 1933. Uber Ein-und Zweigeschlechtliche Kommaschildläuse (*Lepidosaphes ulmi* unisexualis und bisexualis, *L. rubri* und *L. newsteadi*) der deutchen Coccidenfauna. *Z. Pflanzenkr.* 43:638–57
110. Thiem, H., Gerneck, R. 1934. Untersuchungen an deutschen Austernschildlaüsen (Aspidiotini) im Vergleich mit der San Jose-Schildlaus. *Arb. Physiol. Angew. Entomol. Berlin Dahlem* 1:130–58, 208–38
111. Timlin, J. S. 1964. The biology bionomics, and control of *Parlatoria pittospori* Mask.: a pest on apples in New Zealand. *N. Z. J. Agr. Res.* 7:536–50
112. Tremblay, E. 1958. Ovoviparita, comportamento della femmine vergini, sesso delle larve e ghiandole cefaliche larvali della *Diaspis pentagona* Targ. *Boll. Lab. Entomol. Agr. Portici* 16:215–46
113. Van Dinther, J. B. M. 1950. Morphologie en Biologie van de Schildluis *Chionaspis salicis* L. *Tijdschr. Plantenziekten* 56:173–252
114. Vasseur, R., Schvester, D. 1957. Biologie et ecologie du pou de San Jose (*Quadraspidiotus perniciosus* Comst.) en France. *Ann. Epiphyt. Pathol. Veg. Zool. Agr. Phytopharm.* 8:5–66
115. Vayssiere, P. 1926. Contribution a l'étude biologique et systematique des coccidae. *Ann. des Epiphyties* 12:187–382
116. Waterson, J. M. 1947. *Report of the Plant Pathologist* (*Bermuda*) *for the year 1946.* 18 pp. Hamilton Dep. Agr.
117. Weber, H. 1930. *Biologie der Hemiptera.* Berlin: Springer. 543 pp.
118. Williams, J. R. 1970. Studies on the biology, ecology and economic importance of the sugar-cane scale insect, *Aulacaspis tegalensis* (Zhnt.) in Mauritius. *Bull. Entomol. Res.* 60:61–95

PHYSIOLOGY OF TREE RESISTANCE TO INSECTS[1]

❖6081

James W. Hanover[2]
Department of Forestry, Michigan State University, East Lansing, Michigan 48823

The economic impact of insect damage to trees is so great that research efforts aimed at controlling insect pests are easily justified and research accomplishments are readily measured. As a result of past research on many insect-tree combinations, progress has been made towards the discovery and development of control methods including insecticides, silvicultural techniques, attractants, and the use of parasites, predators, and diseases. Remarkably little use has been made of tree resistance for control of pest populations despite the fact that this is generally regarded as the ideal method (10, 66).

Although tree losses are being reduced by nongenetic means and will continue to be reduced as methods are refined, insect damage to our forest and ornamental trees remains enormous. Genetic methods have proved successful in combating many crop insects and offer as much or more potential for alleviating damage to trees (77).

Development of host resistance as an effective tool for long-term control of insect pests requires a multidisciplinary approach by geneticists, physiologists, and entomologists. Unfortunately, one or more of the disciplines is usually missing and this has partially contributed to the general lack of progress in this area. This has recently been discussed by Philogene (79) with regard to the need for plant physiologists to work on host-insect relationships, but it is equally true with respect to forest geneticists.

Stark (105) covered several facets of tree resistance to insects in his 1965 review of forest entomology. The book edited by van Emden (25) also contains much information pertinent to the topic. The objective of the following discussion is to critically examine the concept of tree resistance to insects and to review our knowledge of the physiological mechanisms involved in resistance. Hopefully, this may at least stimulate more discussion of the problem and perhaps productive research among the three disciplines.

[1]Michigan Agricultural Experiment Station Article No. 6142.
[2]The survey of the literature pertinent to this review was concluded in February 1974.

CONCEPT OF TREE RESISTANCE

Definition of Resistance

Attempts have been made to devise a framework of terminology within which the developmental interrelationships between host and insect may be described (10, 76, 102). At one of the two extremes in this relationship lies complete success of an insect infestation leading to death of a host species, population, or individual. At the other extreme lies immunity of the host species, population, or individual to damage by the insect.

Host resistance lies in the broad area between these extremes and has been defined by Beck (10) as "the collective heritable characteristics by which a plant species, race, clone, or individual may reduce the probability of successful utilization of that plant as a host by an insect species, race, biotype, or individual." Obviously, this concept of resistance is broad and reflects the complexity of the phenomenon itself.

A further insight into the complexity of the host-insect relationship is given in Figure 1 in which both insect and host behavior related to tree resistance are summarized. Implicit in the concept of resistance is that the factors conferring resistance upon a host are genetically codetermined in the insect and host; therefore these factors are able to be perpetuated by the geneticist. Omitted from this concept is pseudoresistance—escape from insect attack or lack of insect success on a host due to temporarily induced resistance. Environmental conditions may lead to pseudoresistance. However, if the host and insect genotypes provide the basic potential for a host-resistance response, some environmental variables may modify the expression of genetic resistance and occasionally increase host susceptibility to a degree unsuitable for management objectives. Unstable or transient host responses such as pseudo-resistance and increased susceptibility of a resistant host are often encountered and are no doubt important in integrated control programs. However, my concern here is with the more stable types of resistance that will be useful in a genetic improvement program. This includes resistance mechanisms that are relatively stable to fluctuations in insect behavior resulting from genetic and environmental changes.

It is interesting to note that in their basic text on forest entomology, Graham & Knight (34) define "environmental resistance" as including genetic resistance mechanisms expressed by the host tree (evidently because these are part of the insect's environment). Such a definition is confusing and unacceptable because it combines genetic and environmentally conditioned resistance responses when they must be separated to study the physiology of resistance.

Demonstration of Resistance

Before attempting an investigation into the physiological basis for insect resistance, it is first necessary to find a resistant tree, as Painter (77) succinctly points out. In other words, there must be demonstrable genetic variation in physiological traits so that resistant and susceptible individuals can be handled experimentally.

The search for resistant trees is a most crucial phase of resistance physiology studies and requires the joint efforts of entomologist and physiologist. The en-

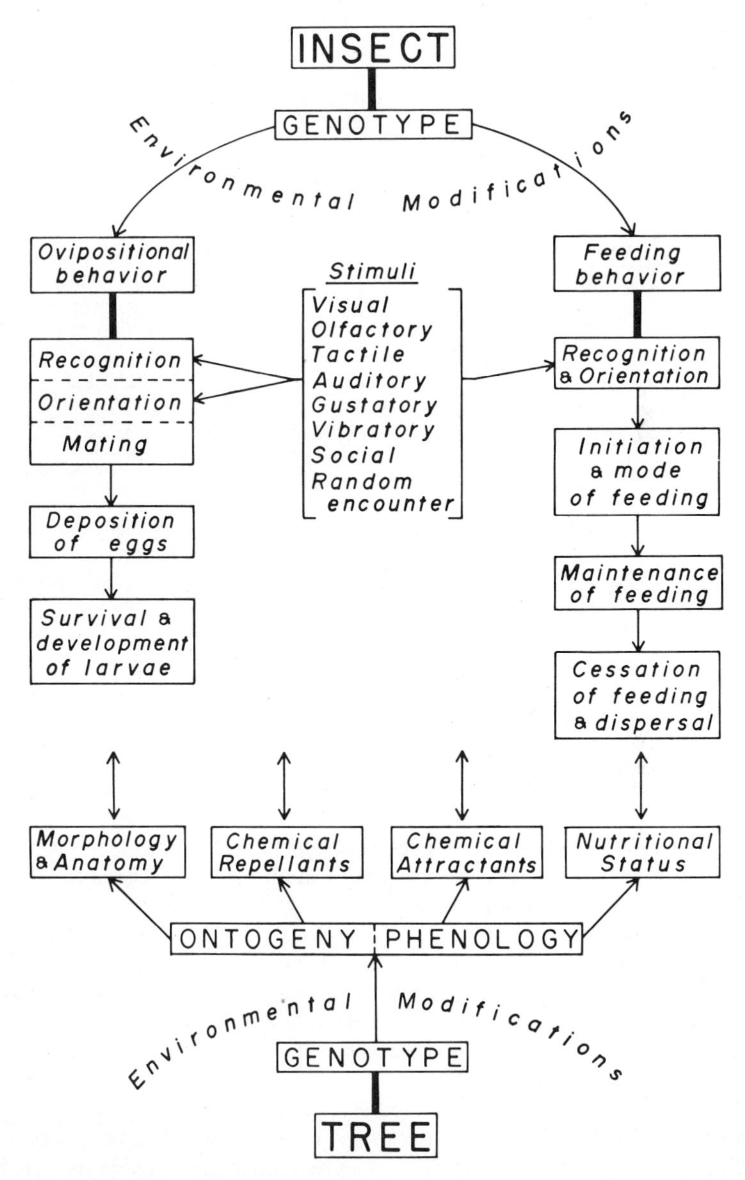

Figure 1 Diagrammatic summary of the host-insect complex emphasizing the potential mechanisms for tree resistance to an insect.

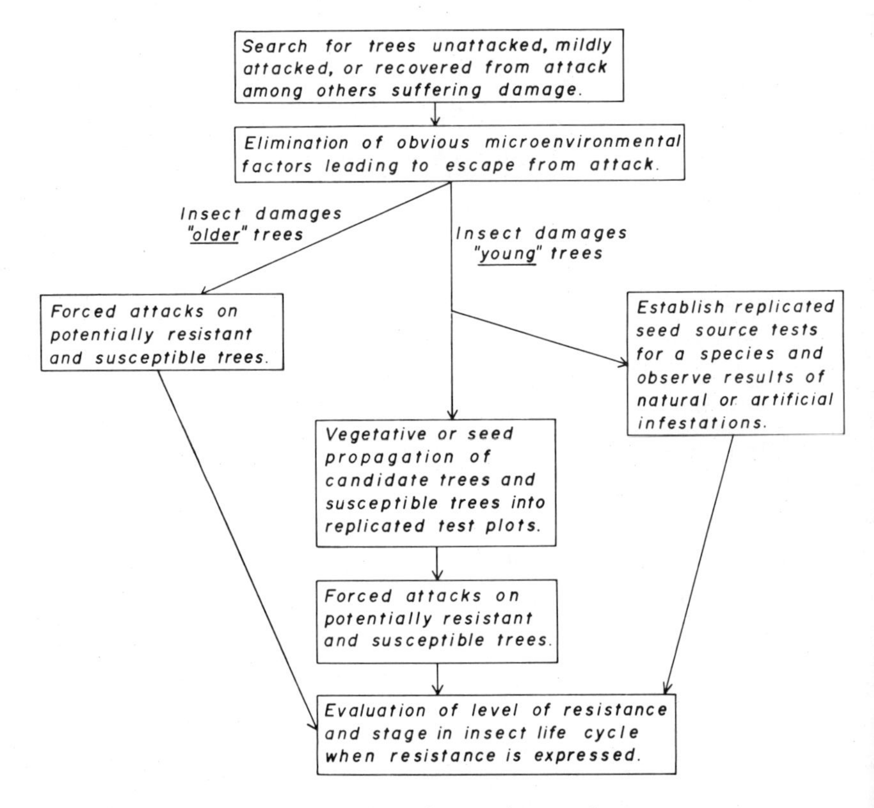

Figure 2 Procedures for demonstrating resistance of trees to insects.

tomologist must know the bionomics and population dynamics of the insect, whereas the physiologist must be familiar with all aspects of the ontogeny and phenology of the tree species.

General procedures for detecting, testing, and propagation of insect resistant trees are discussed in a symposium publication edited by Gerhold et al (29). The main steps involved in demonstrating heritable resistance to an insect are outlined in Figure 2. Relatively few cases of such a rigorous demonstration of resistance to insects exist within tree species, but they are the prerequisite to physiological investigations into resistance mechanisms.

Probably the most productive method used for revealing intraspecific resistance to an insect, which is our main concern here, has been from observations made in replicated seed source tests established for a given species. In recent years there has been a great proliferation of seed source tests and some results of differential insect attack have been reported. For example Bouvarel & Lemoine (17) observed differences in resistance of Norway spruce (*Picea exelsa*) provenances to *Lygaeonematus abietinus* in France. Teucher (107) demonstrated that *virdis* and *caesia* types of

Douglas fir were more susceptible to the Sitka spruce gall aphid than were *glauca* types. Rudolf & Patton (91) summarized the results of observations of seed source-related differences in injury to insects by many species in genetic plantations. More recent reports of apparent resistance by certain seed sources include Scotch pine to European pine sawfly (*Neodiprion sertifer*) (125) and the pine root collar weevil (*Hylobrius radicis*) (126), and white and Norway spruces to the eastern spruce gall aphid (*Adelges abietis*) (108).

MECHANISMS OF TREE RESISTANCE

Referring again to Figure 1, it is readily apparent that there are innumerable possibilities for an unfavorable condition to exist in a tree that would reduce the possibility of successful utilization of the tree by an insect. This reduction in utilization of the tree, or host resistance, may be manifested in single, scattered trees or in groups or populations of trees. The unfavorable or less favorable condition in the tree adversely affects the insect's feeding or reproductive behavior to the benefit of the tree.

Imbalances in the insect-tree relationship leading to tree resistance may be caused by variation in any of the four basic host characteristics shown in Figure 1: 1. morphology and anatomy of the host; 2. chemical repellents produced by the host; 3. chemical attractants produced by the host; and 4. nutritional status of the host.

These four broad categories of host behavior are also dependent upon the phenological and ontogenetic patterns of variation, e.g. variation associated with season and age of the tree. Painter (76) divided plant resistance mechanisms into three categories: (*a*) preference or nonpreference, (*b*) antibiosis, and (*c*) tolerance. This is a widely quoted but somewhat vague statement of mechanisms. The four categories I am using here are perhaps more specific and more host-oriented for the purposes of this discussion.

As yet there are remarkably few cases in which specific resistance mechanisms in trees have been defined. However, there are a number of instances producing reasonably good evidence to illustrate the four general ways host resistance may be expressed. It should be emphasized that the actual host resistance mechanism is most likely a combination or interaction of conditions either within the host itself or between host and insect. Some examples of each of the resistance categories follow.

Morphology and Anatomy

All developmental changes in the physical characteristics of the tree through its ontogeny or phenology are included in the morphological and anatomical category. Important biochemical changes can also be included with physical characteristics, but they are of secondary importance in this category. For example, Bennett (11) correlated the level of attack by the pine needle miner, *Exoteleia pinifoliella,* with number and size of the resin ducts in several hard pine species. Similarly, Stroh & Gerhold (106) demonstrated that white pine weevils (*Pissodes strobi*), feeding in the cortex of the leaders, consistently avoid resin ducts and cease feeding if they cannot circumvent the ducts. Plank & Gerhold (80) report that western white pine which is resistant to the weevil has a greater number of outside cortical resin ducts than eastern white pine which is susceptible. Thus, anatomical differences among trees may lead to differences in resistance which are ultimately chemical in nature. In these cases the characteristics for which a geneticist would breed to develop resistant varieties are anatomical or morphological and this determines into which category the resistance mechanism is classified.

One of the most obvious cases of tree resistance due to phenological differences between hosts is that of the spruce budworm, *Choristoneura fumiferana,* which feeds on foliage of conifers of several genera. In eastern Canada and the United States black spruce is relatively immune to budworm damage compared with the preferred hosts white spruce and balsam fir. A partial explanation for the resistance of black spruce is that it opens its buds much later than white spruce or balsam fir (16). There may also be a chemical basis for budworm aversion to black spruce as I will point out in a moment. Budworm larvae development and survival are dependent on the availability of food in the form of staminate flowers or newly emerging vegetative shoots at the time of larval emergence in the spring. Asynchrony between food availability and larval emergence as an effective resistance mechanism has been well demonstrated by Eidt & Cameron (23) within balsam fir trees and by Eidt & MacGillivray (24) between seven species of fir.

Other examples of anatomical or phenological bases for insect resistance are cited by von Schönborn (97).

Chemical Repellents

The presence of certain chemicals in tree tissues can make them such undesirable hosts that the insect is prevented from completing a phase of its development on that host. Chemical repellents may exert their effect in the volatile state causing an insect to avoid the tree completely or they may repel the insect after it contacts the tree or ingests tissue. According to Wright et al (127) substances in the host may also act as antiattractants. Such substances have no intrinsic repellency themselves but can diminish the attractiveness of other compounds.

Referring again to the spruce budworm, Manley & Fowler (64) have shown that in New Brunswick resistant black spruce and susceptible red spruce do not differ sufficiently in their phenology to explain resistance, as was the case with black and white spruces. Heron (47) found that the glycoside pungenin has a deterrent effect on budworm larvae feeding in laboratory bioassays. Pungenin also occurs naturally in very low levels in the preferred tissues, staminate flowers, and new vegetative shoots, and in high levels in mature needles which the budworm avoids. Wilkinson (119) found that a compound almost identical to pungenin constitutes a major chemical difference between the foliage of red spruce which lacks it and black spruce which contains large amounts. Species hybrids have varying levels of both pungenin and resistance to budworm attack, although statistical correlations have not been done for these two traits. Other chemicals undoubtedly affect budworm larvae feeding behavior but pungenin has significant repellent characteristics in spruces which could be useful in genetic improvement for insect resistance.

From the work of Norris (31, 73) and his associates, we know that the elm bark beetle, *Scolytus multistriatus,* is deterred from feeding on nonhost species such as hickory (*Carya* spp.) by the compound juglone and other 1,4-naphthoquinones. Without a doubt the most commonly reported example of chemical repellency involves the components and characteristics of conifer oleoresin. Though the role of resin in resistance to bark beetle attack has been hypothesized for at least 70 years (49), confirmatory experimental evidence has come only recently for bark beetles and other insects as well. Yet how it acts (other than to say it suffocates) remains unresolved.

Smith (98, 99, 101) has associated toxicity of resin vapors of nonhost species with the host specificity of bark beetles attacking hard pines. Unfortunately, the results did not hold true for soft pines, although Anderson & Fisher (4) earlier had demonstrated repellency of white pine weevil by volatile oils extracted from white pine and spruce. Rudinsky (88) suggests that oleoresin vapors, specifically monoterpenes, act as repellents to bark beetles after initially attracting them to host stands.

From results of olfactometer studies, Heikkenen & Hrutfiord (45) proposed that since β-pinene is a repellent to the Douglas fir bark beetle (*Dendroctonus pseudotsugae*), and α-pinene is an attractant, tissue specificity shown by this insect may be related to the ratio of α- to β-pinene in a given tissue.

As we shall see later, most of the research on chemicals in a vapor state has focused on their attracting rather than repelling properties. It is quite probable that the individual components of oleoresin may have different repellent properties for different insects. This should be rigorously established to open the way to breeding work for chemical composition.

Any insect that has selected a conifer as a host must be able to tolerate or avoid the detrimental effects of the host oleoresin system. Since trees vary greatly in all aspects of these systems, it is not surprising that a most effective resistance mechanism involves the repellent or toxic properties upon contact. Individual tree or species resistance resulting from the insect's inability to avoid direct contact with oleoresin has been strongly implicated in many studies, some of which are listed in Table 1.

The complexity and diversity of oleoresin systems, especially their close relationship with the water status of trees, provide a logical basis for the observed selectivity of hosts by insects. Much of this selectivity is stimulated by environmental conditions, e.g. the widely observed susceptibility of trees or portions thereof in a stressed or weakened condition that reduces the resin pressure system to a level that makes the resin innocuous to invading insects (113). However, genetic variation in resin properties, although far less studied, is also substantial and should be exploited to develop resistant strains of trees.

Chemical Attractants

The subject of chemical insect attraction is a most vigorous area of research activity with trees and other plants. This situation will no doubt continue as techniques for biological control of insects are developed on the basis of attractant principles developed (13, 53–55, 114). Insects possess extremely sensitive mechanisms for

Table 1 Selected cases in which direct contact with oleoresin has been associated with host resistance to insects

Principle host	Insect	Authors	
Pinus ponderosa	*Ips confusus*	Vitè	(113)
	Dendroctonus brevicomis	Vitè	(113)
Pinus radiata	*Sirex noctilio*	Coutts & Dolezal	(21)
Pinus radiata	*Sirex noctilio*	Titze & Mucha	(110)
Pseudotsuga menziesii	*Dendroctonus pseudotsugae*	Rudinsky	(87)
Pinus contorta	*Dendroctonus ponderosae*	Reid, Whitney & Watson	(82)
Abies grandis	*Scolytus ventralis*	Berryman & Ashraf	(15)
Pinus sylvestris	*Rhyacionia buoliana*	Harris	(43, 44)
Pinus taeda	*Ips avulsus*	Mason	(65)
	Ips grandicollis	Mason	(65)

perceiving their hosts. Olfactory organs detect chemical messages emitted by both the host and other insects (75, 95, 119, 120).

Much of the past work on chemical attraction has focused on secondary attraction, i.e. attractants produced following initial attack of a new host by "pioneer" insects. Primary attraction of an insect to its host may be by any of the mechanisms shown in Figure 1, but olfactory responses are probably most common.

Host resistance mechanisms based on the principle of chemical attraction must obviously result from inability of the host to produce the attractant and thereby avoid both initial and/or mass attacks. There are only a few examples of this mechanism in the literature. Austin, Yuill & Brecheen (5) found that ponderosa pines lacking viscid or resinous new shoots were relatively undamaged by the resin midge (*Retinodiplosis* sp.), compared with trees having glabrous or glaucous shoots. However, it is not certain that the resinous shoots were actually attractive to the midge.

It has often been observed that vigorous, standing trees of many species are much more resistant to attack by bark beetles than freshly cut or lightning struck trees (15, 48, 56, 57, 87, 94). Furniss (28) has reported that standing western larch are virtually immune to beetle attack, but are attacked immediately after being cut down. Anderson & Anderson (3) studied *Ips* bark beetle attack in relation to the physiological decline of a loblolly pine (*Pinus taeda* L.) that had been struck by lightning. They concluded that release of host volatiles soon after injury provides the primary attraction for the beetles. Ferrell (27) reached the same conclusion with regard to the greater attraction of girdled versus nongirdled white fir (*Abies concolor*) to the fir engraver bark beetle, *Scolytus ventralis.* Similar examples of primary attraction after girdling occur for *Scolytus* spp. attacking hardwoods (32, 70). Most investigators in this area of research agree that the lack of sufficient host attractants in standing trees is probably responsible for their resistance.

Analogous situations also exist in hardwood trees because Riddiford & Williams (84) have clearly demonstrated that a chemical released by oak leaves is necessary for the mating of the polyphemus moth (*Antheraea polyphemus*). Chemicals found to be effective in attracting insects include monoterpenes (83, 88), fatty acids (2), benzoic acid (128), and ethanol (19).

Thus, there seems to be ample evidence that host chemicals act as primary attractants to their associated insect pests. It logically follows that a modification in or suppression of these chemicals through breeding could provide useful resistance.

Knerer & Atwood (59) have described some interesting relationships among strains of diprionid sawflies and their preferred conifer hosts that appear to involve both chemical attractiveness to females and host toxicity to larvae. The *Neodiprion abietes* complex contains strains that feed on species of fir (*Abies*), spruce (*Picea*), Douglas fir (*Pseudotsuga*), and hemlock (*Tsuga*). Within this complex there are three distinct subdivisions based on seasonal appearance: those preferring balsam fir, those preferring white fir, and a large conglomerate of all the spruce strains plus a larger balsam form. Females of three *N. abietis* strains accept only their respective hosts, balsam fir, black spruce, and white spruce, for oviposition and die after several days if offered a different host. Hybrids between the balsam and spruce strains of the sawflies chose balsam fir for oviposition which led Knerer and Atwood to suggest a simple genetic control system over host specificity among the sawflies. These authors also point out that host toxicity to larval development is also a factor in the occurrence of sawfly damage. For example, the spruce strain avoids balsam fir because of its high toxicity to the larvae whereas larvae of the balsam strain actually prefer spruce over balsam fir if given a choice. However, under natural conditions the larvae have little choice and must usually feed where they hatch. Ghent (30) has also demonstrated that the larvae of *Neodiprion pratti banksiannae* which feed on jack pine (*Pinus banksiana*) are highly responsive to the odor of pine foliage in their feeding habits.

Host resistance may also reside in some facet of secondary attraction because, escaping the mass attack, trees may suffer little damage and recover. A deficiency in the production of a secondary attractant could adversely affect the insect's activity, although to my knowledge there is no evidence for this occurring. Attraction models have been proposed by Renwick & Vité (83) for *Dendroctonus ponderosae, D. frontalis,* and *D. brevicomis* which postulate regulatory roles in primary and secondary attraction for individual terpenes produced by the hosts. If these models prove valid we will then also have a firm basis for a host role in secondary attraction and therefore a potential resistance mechanism. Of course, before either primary or secondary attraction can be exploited to develop genetic resistance in a tree species, more definitive physiological studies are needed to provide the basis for selection and breeding.

Nutritional Status

The fourth category of resistance mechanisms is the nutritional status of the host. Although any green leaf probably provides the food materials that foliage-infesting insects require, there are needs for nutrient balance (52) and different insects do have different nutritional requirements (50, 51), judging from work with nontree insect pests. Moreover, wood itself is a poor nutritional substrate for insects, being deficient in vitamins, sterols, and other growth factors. Painter (77) has enumerated the varied effects of an insect feeding on a resistant plant including death of larvae, lowered fecundity of female, small size, low weight, abnormal length of life, low food reserves affecting hibernation, death before adult stage, and physiological aberrations. The actual causes of these deleterious effects include the presence in the host of toxins; growth or reproduction inhibitors; lack of a nutritional element, e.g. a vitamin; deficiency of a nutritional element; and imbalance in available nutrients, especially carbohydrate-protein ratios. Also included here would be the presence of certain feeding deterrents in host tissue.

Several studies have shown that the nutritional status of trees may be a source of genetic resistance to insects. The extensive work of Norris and Baker in Wisconsin with bark beetles feeding on elm and other hardwood species reveals that phenols and other chemicals in the bark stimulate feeding by the beetles (6, 7, 73).

But host specificity is apparently governed more by a lack of chemicals which deter feeding and thus a balance between stimulating and deterring substances (74).

The feeding behavior of spruce budworm larvae varies according to host nutritional balance in addition to the concentration of pungenin mentioned earlier (47).

The feeding habits of aphids make them especially prone to both anatomical or nutritional variation in the host, thereby preventing or deterring feeding. Thielges & Campbell (108) have recently given evidence for variation in resistance to the eastern spruce gall aphid (*Adelges abietis*) in Ohio plantations of both white and Norway spruces. They suggest that resistance is due either to physical barriers or to an adverse chemical environment in host foliage, because there are no phenological differences between resistant and susceptible trees.

Although resistance mechanisms involving host nutritional status may actually be quite prevalent, they are most difficult to prove because the net effect is likely to be more quantitative or subtle than that of the other resistance types. In addition, nutritional status is closely tied to and perhaps inseparable from developmental changes in host anatomy or the presence of chemical repellents and attractants.

As stated earlier the actual relationship between host and insect that leads to expression of a level of resistance may often be a combination or interaction of conditions associated with either the host or insect. Examples of this are found with the sawflies (59), the apparent need for fungal associations with the bark beetles (14, 124), and probably many other situations.

PHYSIOLOGY OF SECONDARY CHEMICAL CONSTITUENTS IN RELATION TO HOST RESISTANCE

From the preceding it is evident that the so-called secondary chemical constituents in plants play a predominant role in tree resistance to insects. In fact, these chemicals along with proteins form the basis for individuality and the chemical diversity among all plants. Despite this fact, there is a dearth of information about the physiology of these substances.

Beck (10) has stated that the rate of progress in development of resistant plant varieties will be closely correlated with the rate of accumulation of fundamental biological and biochemical knowledge concerning the complex interactions between insects and their host plants.

In my judgment, forest entomologists have led the way by accumulation of information on the biology of important insects while tree physiologists have ignored the physiology of secondary chemical constituents. It behooves us to learn more about the complex chemical environment that an insect encounters when it feeds or reproduces on a tree or when it is naturally prevented from doing so.

It may be helpful at this point to briefly review the biochemical environment of a tree and point out some of the important characteristics of one important group, the terpenes.

Major Chemicals and Pathways

The major chemical constituents in trees and their overall metabolic relationships are shown in Figure 3. The biosynthesis of secondary compounds represents a series of offshoots from primary metabolic pathways. In many instances combinations of primary and secondary compounds contribute to even greater chemical diversity in the tree, e.g. phenolic glycosides, in which a sugar is joined to a phenolic compound to form important pigments and other substances.

The large number of secondary compounds (probably close to 1000) in an infinite number of combinations in a given tree, plus their stability to environmental modification compared with primary metabolic products, make them the major source of

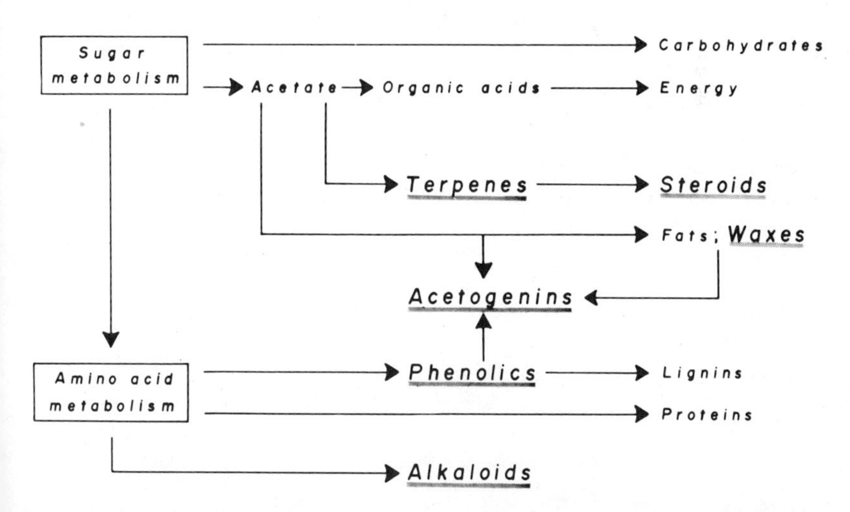

Figure 3 Diagrammatic summary of the major groups of chemicals in trees and their metabolic relationships. Secondary compounds are in large type and underlined [modified from Whittaker & Feeny (118)].

chemical diversity to which the majority of forest insect pests have adapted in selecting their preferred hosts. Wilkinson (119) detected more than 100 compounds in the phenolic group alone by chromatographing extracts from the foliage of a single conifer (black spruce). Phenolic compounds are proving very useful in chemical systematic work (42, 118), especially in relation to insect resistance, as the work of Norris and others is revealing.

Acetogenins are also phenolic derivatives and include such compounds as stilbenes which are toxic to insects (121), quinones (such as juglone from walnut), and anthocyanidin pigments. The report of a highly toxic compound, nagilactone, isolated from leaves of *Podocarpus nivalis,* is an example of a rather unique secondary constituent having inhibitory effects on insect development (93). Tannins are complex polymers of phenol compounds which occur in relatively high concentrations in woody plant tissues. The work of Feeny (26) has demonstrated how these substances in oak leaves can regulate growth and development of lepidopterous larvae such as the winter moth (*Operophtera brumata*). The presence of hydroxylactones and aldehydes in leaves of the ginkgo tree (*Ginkgo biloba*) is believed to be responsible for the apparent immunity of this ancient species to insect attack (63).

The waxes present in or on trees are an example of not only chemical diversity, but also physical diversity, another factor with which an insect must contend. Eglinton & Hamilton (22) have reviewed the chemistry and distribution of waxes in plants, but there are only a few biochemical studies involving tree waxes (12, 46, 89). The scanning electron photomicrographs published by Hanover & Reicosky (41) reveal the physical nature of waxes on foliage of various tree species.

Alkaloids are a very large and structurally diverse group of secondary substances characterized by having nitrogen. To my knowledge there have been no reported studies of alkaloids in trees, although it would be surprising if these substances did not occur.

Terpenes have received by far the most attention by those working on insect-tree relationships. As suggested earlier, our understanding of the physiology of terpenoid compounds still leaves much to be desired, particularly in relation to insect problems. To conclude this review, I will briefly summarize some pertinent facts about the genetics and physiology of terpenes as they may relate to tree resistance to insects.

Physiology of Terpenes in Trees

Terpenes are also a diverse and ubiquitous group of chemicals that includes rubber, the highly aromatic monoterpenes, resin acids, carotenoid pigments, sterols, and the growth regulatory gibberellins and abscisic acid (Figure 4). However, all terpenes are basically increasingly complex polymers of the 5-carbon isoprene unit:

$$CH_3 \cdot C(:CH_2) \cdot CH(:CH_2)$$

From the standpoint of insect resistance, most of the interest has centered around the volatile fraction of conifer oleoresin, mainly monoterpenes (Figure 5). Monoterpenes occur not only in conifer oleoresin, but are found in just about every type of

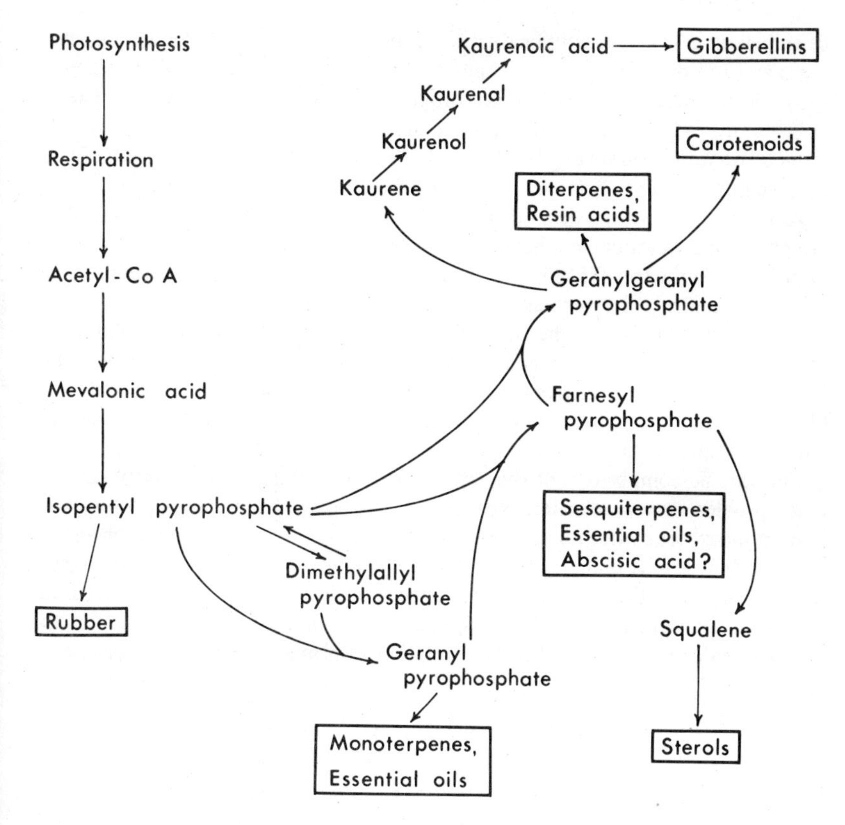

Figure 4 Biogenetic pathways of terpenes in plants.

plant analyzed (71). In addition to monoterpenes, the nonvolatile and often predominant terpene fraction of oleoresin consists of the diterpenes or resin acids some of which are shown in Figure 6.

Oleoresin may also contain varying amounts of sesquiterpenes, terpene alcohols, and fatty acids. Especially in the conifers, resinous systems have evolved to such a degree that they comprise a remarkable model of genetically coordinated biochemistry, physiology, and anatomy.

There is a large body of knowledge accumulating on the subject of terpenoids in plants and especially the resinous systems in conifers (8, 20, 33, 58, 72, 78, 81, 92, 109, 115, 116). Although space will not allow a detailed examination of the subject here, some facts of particular importance to the physiology of insect resistance deserve mention.

Of fundamental importance to potential tree insect resistance mechanisms is the fact that oleoresin has a very wide range of variation in chemical and physical properties which in combination gives almost every tree a unique individuality (35, 39, 96, 100). Both the monoterpenes and diterpene resin acids are under strong genetic control which greatly facilitates the identification of desired chemical genotypes and breeding for specific compositions (9, 35–38, 69, 86, 103, 104, 112, 122, 123). Thus, if some of the mechanisms proposed by Renwick & Vité (83), Heikkenen & Hrutfiord (45), and Rudinsky (87) for the primary attraction of insects to their hosts prove to be operative, geneticists may be able to modify attractiveness of trees by breeding different chemical varieties of the species.

Along with the strong genetic control of oleoresin characteristics in trees there are predicable variations in chemical composition associated with different tissues or positions in the tree (36, 85, 111, 117). An example of how monoterpene concentrations change according to age of tissue, position within the same aged tissue or internode, and changes in the concentrations of other monoterpenes is shown for *Pinus strobus* in Table 2. Much greater changes in composition may be found between oleoresin sampled from wood, cortex, bark blisters, foliage, buds, and cones (36, 111). Within the same tissue there are also relatively small but predicable changes in monoterpene levels between seasons of the year (1, 40, 61, 90, 111).

The specific composition of the vapors emitted by trees is particularly relevant to the problem of host selection by an insect. Obviously, the terpene vapors released to the atmosphere by tree foliage are qualitatively determined by the internal composition of the various tissue systems. However, Hanover (39) has shown that there can be a wide quantitative discrepancy between the internal and external concentrations of monoterpenes of different species. More information is needed on the chemical halos that surround trees in relation to primary attraction of insects.

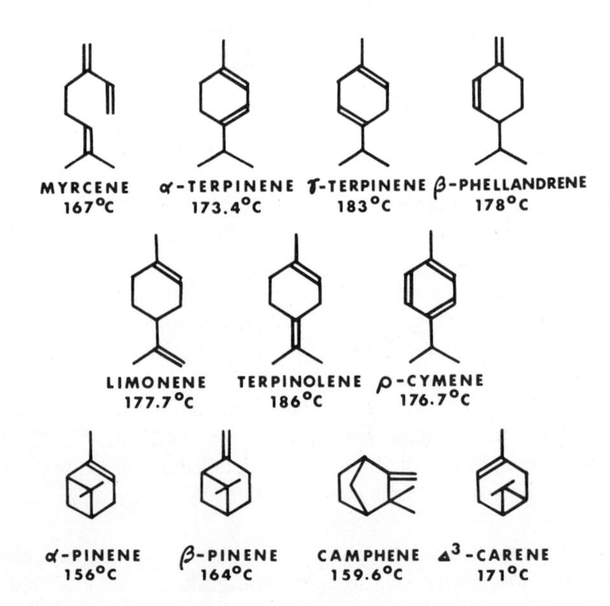

Figure 5 Some monoterpenes commonly found in conifer oleoresin. Boiling points are shown below each compound.

Labdic Type Resin Acids

COOH
Anticopalic acid

COOH
Communic acid

COOH
Strobic acid

Abietic Type Resin Acids

COOH
Levopimaric acid

COOH
Abietic acid

COOH
Palustric acid

COOH
Neoabietic acid

COOH
Dehydroabietic acid

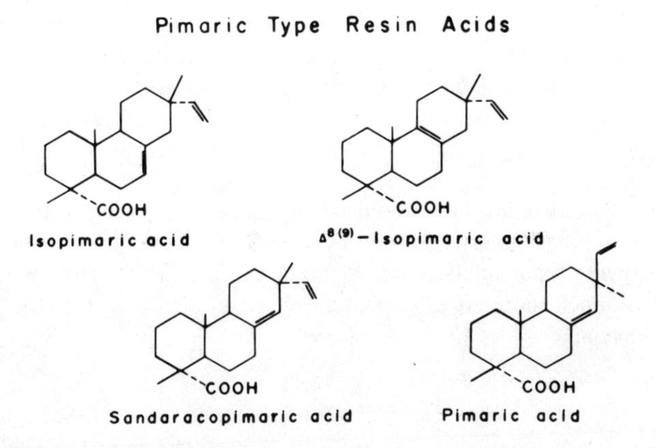

Figure 6 Some resin acids commonly found in conifer oleoresin.

Table 2 Within-tree variation of the principal monoterpenes in oleoresin sampled from young *Pinus strobus*

3–Carene concentration	Tissue age (yrs.) and internodal position[a]	Monoterpene		
		α–pinene	β–pinene	3–Carene
			percent of oleoresin	
High	1 U	3.28	9.51	13.56
	1 L	3.66	9.79	13.34
	2 U	3.90	7.45	12.24
	2 L	4.51	7.56	11.21
	3 U	5.25	7.05	10.87
	3 L	6.60	8.57	11.47
Low	1 U	14.38	15.73	0.05
	1 L	14.92	16.13	0.09
	2 U	15.83	13.23	0.03
	2 L	16.59	13.03	0.04
	3 U	14.64	13.93	0.05
	3 L	15.53	14.06	0.04

[a] U = upper part of internode; L = lower part of internode. Analyses of variance showed highly significant differences between internodes for all three terpenes and significant intranodal differences for α–pinene only (117).

As mentioned earlier in this review, direct contact with oleoresin has been shown to be a primary factor in insect resistance by tree species (Table 1). Presumably, the physical and/or chemical characteristics of the oleoresin are responsible for resistance, although the exact mechanism is apparently unknown. No studies have been done on the total composition of oleoresin for an individual tree. It is known that the relative proportions of monoterpene and diterpene resin acids can vary substantially between trees, seed sources, and species (117, 123). Stark (105) gave a detailed review of the role of oleoresin in pines as a resistance mechanism to attack by insects. He emphasized the physical characters of resin that contribute to resistance, including rate and duration of flow, quantity produced, viscosity, rate of crystallization, exudation pressure, and resin duct anatomy. Many of these physical characteristics appear to be genetically regulated, whereas others are sensitive to environmental changes. For example, total oleoresin yield (68, 87), viscosity (67, 69), color (60), and rate of crystallization (18) would be amenable to breeding in trees, whereas exudation pressure is mainly regulated by the water status of the tree (62, 87, 113) and the degree of tissue lignification. As Stark points out, the oleoresin systems of trees deserve much more attention by tree physiologists and geneticists than they are now receiving.

Literature Cited

1. Adams, R. P. 1970. Seasonal variation of terpenoid constituents in natural populations of *Juniperus pinchotii* Sudw. *Phytochemistry* 9:397–402
2. Adlung, K. G. 1960. Über die Ergebnisse der Schwarzwald 1958 and 1959 durchgeführten Freilundversuche zur Anlockung von Borkenkäfern mit Lockstoffen. *Z. Angew. Entomol.* 45: 430–35
3. Anderson, N. H., Anderson, D. B. 1968. Ips bark beetle attacks and brood development on a lightning-struck pine in relation to its physiological decline. *Fla. Entomol.* 51:23–30
4. Anderson, J. M., Fisher, K. C. 1960. The response of the white pine weevil to naturally occurring repellents. *Can. J. Zool.* 38:547–64
5. Austin, L., Yuill, J. S., Brecheen, K. G. 1945. Use of shoot characters in selecting ponderosa pines resistant to resin midge. *Ecology* 26:288–96
6. Baker, J. E., Norris, D. M. 1968. Behavioral responses of the small European elm bark beetle, *Scolytus multistriatus,* to extracts of non-host tree tissues. *Entomol. Exp. Appl.* 11:464–69
7. Baker, J. E., Rainey, D. P., Norris, D. M., Strong, F. M. 1968. p-Hydroxybenzaldehyde and other phenolics as feeding stimulants for the smaller European bark beetle. *Forest Sci.* 14:91–95
8. Banthorpe, D. V., Charlwood, B. V., Francis, M. J. O. 1972. The biosynthesis of monoterpenes. *Chem. Rev.* 72:115–55
9. Baradat, P., Bernard-Dagan, C., Fillon, C., Marpeau, A., Pauly, G. 1972. Les terpenes du pin maritime: aspects biologiques et genetiques. II. Hérédité de la teneur en monoterpenes. *Ann. Sci. Forest.* 29:307–34
10. Beck, S. D. 1965. Resistance of plants to insects. *Ann. Rev. Entomol.* 10:207–32
11. Bennett, W. H. 1954. The effect of needle structure upon the susceptibility of hosts of the pine needle miner (*Exotleia pinifoliella* Chamb.) *Can. Entomol.* 86:49–54
12. Beri, R. M., Lemon, H. W. 1970. Chemical examination of the wax from needles of black spruce and balsam fir. *Can. J. Chem.* 48:67–69
13. Beroza, M. 1971. Insect sex attractants. *Am. Sci.* 59:320–25
14. Berryman, A. A. 1972. Resistance of conifers to invasion by bark beetle-fungus associations. *Bioscience* 22:598–602
15. Berryman, A. A., Ashraf, M. 1970. Effects of *Abies grandis* resin on the attack behaviour and brood survival of *Scolytus ventralis. Can. Entomol.* 102:1229–36
16. Blais, J. R. 1957. Some relationships of the spruce budworm, *Choristoneura fumiferana* (Clem.) to black spruce, *Picea mariana* (Moench) Voss. *Forest. Chron.* 33:364–72
17. Bouvarel, P., Lemoine, M. 1957. L'experience internationale sur les provenances d'Epicea (*Picea excelsa* Link.) *Silvae Genet.* 6:91–97
18. Buijtenen, J. P. van, Santamour, F. 1972. Resin crystallization related to weevil resistance in white pine (*Pinus strobus*). *Can. Entomol.* 104:215–19
19. Cade, S. C., Hrutfiord, B. F., Gara, R. I. 1970. Identification of a primary attractant for *Gnathotrichus sulcatus* isolated from western hemlock logs. *J. Econ. Entomol.* 63:1014–15
20. Clayton, R. B. 1970. The chemistry of non-hormonal interactions: terpenoid compounds in ecology. In *Chemical Ecology,* ed. E. Sondheimer, J. B. Simeone, 235–80. New York: Academic. 336 pp.
21. Coutts, M. P., Dolezal, J. E. 1966. Polyphenols and resin in the resistance mechanism of *Pinus radiata* attacked by the wood wasp, *Sirex noctilio,* and its associated fungus (*Amylostereum* sp.) *Leafl. Forest. Timber Bur. Aust.* No. 101, 19 pp.
22. Eglinton, G., Hamilton, R. J. 1967. Leaf epicuticular waxes. *Science* 156:1322–34
23. Eidt, D. C., Cameron, M. D. 1971. Delayed budbreak and spruce budworm survival. *Bimon. Res. Notes* 27:28–29
24. Eidt, D. C., MacGillivray, H. G. 1972. Resistance of seven fir species to spruce budworm and other insects. *Bimon. Res. Notes* 28:2–3
25. Emden, H. F. van, Ed. 1973. *Insect/Plant Relationships.* New York: Wiley. 215 pp.
26. Feeny, P. P. 1968. Effect of oak leaf tannins on larval growth of the winter moth *Operophtera brumata. J. Insect. Physiol.* 14:805–17
27. Ferrell, G. T. 1971. Host selection by the fir engraver, *Scolytus ventralis:* preliminary field studies. *Can. Entomol.* 103:1717–25
28. Furniss, M. M. 1972. *Observations on resistance and susceptibility to Douglas-fir beetles. Proc. Forest Biol. Workshop, 2nd, Corvallis, Ore. July 31-August 4, 1972*

29. Gerhold, H. D., Schreiner, E. J., McDermott, R. E., Winieski, J. A., Eds. 1966. *Breeding Pest-Resistant Trees.* London: Pergamon. 505 pp.
30. Ghent, A. W. 1960. A study of the group-feeding behaviour of larvae of the jack pine sawfly, *Neodiprion pratti banksianae* Roh. *Behaviour* 16:110–48
31. Gilbert, B. L., Norris, D. M. 1968. A chemical basis for bark beetle (*Scolytus*) distinction between host and non-host trees. *J. Insect. Physiol.* 14:1063–68
32. Goeden, R. D., Norris, D. M. Jr. 1964. Attraction of *Scolytus quadrispinosus* to *Carya* spp. for oviposition. *Ann. Entomol. Soc. Am.* 57:141–46
33. Goodwin, T. W. 1971. *Aspects of Terpenoid Chemistry and Biochemistry.* New York: Academic. 441 pp.
34. Graham, S. A., Knight, F. B. 1965. *Principles of Forest Entomology.* New York: McGraw-Hill. 4th ed. 417 pp.
35. Hanover, J. W. 1966. Genetics of terpenes. I. Gene control of monoterpene levels in *Pinus monticola* Dougl. *Heredity* 21:73–84
36. Hanover, J. W. 1966. Environmental variation in the monoterpenes of *Pinus monticola* Dougl. *Phytochemistry* 5: 713–17
37. Hanover, J. W. 1966. Inheritance of 3-carene concentration in *Pinus monticola. Forest. Sci.* 12:447–50
38. Hanover, J. W. 1971 Genetics of terpenes. II. Genetic variances and interrelationships of monoterpene concentrations in *Pinus monticola. Heredity* 27:237–45
39. Hanover, J. W. 1972. Factors affecting the release of volatile chemicals by forest trees. *Mitt. Forstl. Bundes-Versuchsanstalt. Wein* No. 97:625–44
40. Hanover, J. W., Furniss, M. M. 1966. Monoterpene concentration in Douglas-fir in relation to geographic location and resistance to attack by the Douglas-fir beetle. *US Forest Serv. Res. Pap.* NC-6, pp. 23–28
41. Hanover, J. W., Reicosky, D. A. 1971. Surface wax deposits on foliage of *Picea pungens* and other conifers. *Am. J. Bot.* 58:681–87
42. Hanover, J. W., Wilkinson, R. C. 1970. Chemical evidence for introgressive hybridization in *Picea. Silvae Genet.* 19:17–22
43. Harris, P. 1960. Natural mortality of the pine shoot moth *Rhyacionia buoliana* (Schiff) in England. *Can. J. Zool.* 38:755–68
44. Harris, P. 1960. Production of pine resin and its effect on survival of *Rhyacionia buoliana* (Schiff). *Can. J. Zool.* 38:121–30
45. Heikkenen, H. J., Hrutfiord, B. F. 1965. *Dendroctonus pseudotsugae:* a hypothesis regarding its primary attractant. *Science* 150:1457–59
46. Herbin, G. A., Sharma, K. 1969. Studies on plant cuticular waxes-V. The wax coatings of pine needles: a taxonomic survey. *Phytochemistry* 8:151–60
47. Heron, R. J. 1965. The role of chemotactic stimuli in the feeding behavior of spruce budworm larvae on white spruce. *Can. J. Zool.* 43:247–69
48. Hodges, J. D., Pickard, L. S. 1971. Lightning in the ecology of the southern pine beetle, *Dendroctonus frontalis. Can. Entomol.* 103:44–51
49. Hopkins, A. D. 1902. Insect enemies in the Black Hills forest reserve. *US Dep. Agr. Div. Entomol. Butl.* 32:24 pp.
50. House, H. L. 1961. Insect nutrition. *Ann. Rev. Entomol.* 6:13–26
51. House, H. L. 1962. Insect nutrition. *Ann. Rev. Biochem.* 31:653–72
52. House, H. L. 1966. Effects of varying the ratio between the amino acids and the other nutrients in conjunction with a salt mixture on the fly *Agria affinis* (Fall). *J. Insect Physiol.* 12:299–310
53. Jacobson, M. 1965. *Insect Sex Attractants.* New York: Wiley-Interscience. 154 pp.
54. Jacobson, M. 1966. Chemical insect attractants and repellents. *Ann. Rev. Entomol.* 11:403–22
55. Jacobson, M., Beroza, M. 1963. Chemical insect attractants. *Science* 140: 1367–72
56. Jantz, O. K. 1965. *Studies on the olfactory behavior of the Douglas-fir-beetle, Dendroctonus pseudotsugae Hopkins.* Ph.D. Thesis, Oregon State University, Corvallis, Ore.
57. Johnson, P. C. 1966. Attractiveness of lightning-struck ponderosa pine trees to *Dendroctonus brevicomis. Ann. Entomol. Soc. Am.* 59:615
58. Keng, H., Little, E. L. Jr. 1961. Needle characteristics of hybrid pines. *Silvae Genet.* 10:131–46
59. Knerer, G., Atwood, C. E. 1973. Diprionid sawflies: polymorphism and speciation. *Science* 179:1090–99
60. Kraus, J. F., Squillace, A. E. 1964. Inheritance of yellow oleoresin and virescent foliage in slash pine. *Silvae Genet.* 13:114–16
61. Levinson, A. S., Lemoine, G., Smart, E. C. 1971. Volatile oil from foliage of *Sequoiadendron giganteum.* Change in

composition during growth. *Phytochemistry* 10:1087–94
62. Lorio, P. L. Jr., Hodges, J. D. 1968. Microsite effects on oleoresin exudation pressure of large loblolly pines. *Ecology* 49:1207–10
63. Major, R. T. 1967. The ginkgo, the most ancient living tree. *Science* 157:1270–73
64. Manley, S. A. M., Fowler, D. P. 1969. Spruce budworm defoliation in relation to introgression in red and black spruce. *Forest. Sci.* 15:365–66
65. Mason, R. R. 1967. *Dynamics of Ips populations after summer thinning in a loblolly pine (Pinus taeda) plantation: with special reference to host tree resistance.* Abstr. of thesis, in Dissert. Univ. of Michigan, Ann Arbor, Mich. Abstr. 27B, 7, 2215
66. Maxwell, F. G., Jenkins, J. N., Parrott, W. L. 1972. Resistance of plants to insects. *Advan. Agron.* 24:187–265
67. McReynolds, R. D., Lane, J. M. 1971. Adapting the bubble-time method for measuring viscosity of slash pine oleoresin. *US Dep. Agr. Forest Serv. Res. Note* No. 147, 6 pp.
68. Mergen, F. 1953. Gum yields in longleaf pine are inherited. *Southeast. Forest. Exp. Sta. Res. Notes* No. 29, 2 pp.
69. Mergen, F., Hoekstra, P. E., Echols, R. M. 1955. Genetic control of oleoresin yield and viscosity in slash pine. *Forest. Sci.* 1:19–30
70. Meyer, H. J., Norris, D. M. 1967. Behavioral responses by *Scolytus multistriatus* to host (*Ulmus*) and beetle-associated chemotactic stimuli. *Ann. Entomol. Soc. Am.* 60:642–47
71. Nicholas, H. J. 1963. The biogenesis of terpenes in plants. In *Biogenesis of Natural Compounds,* ed. P. Bernfield, 641–91. New York: Pergamon
72. Nicholas, H. J. 1973. *Terpenes.* In *Phytochemistry,* ed. L. P. Miller, 2:254–309. New York: Van Nostrand. 445 pp.
73. Norris, D. M. 1970. Quinol stimulation and quinone deterrency of gustation by *Scolytus multistriatus. Ann. Entomol. Soc. Am.* 63:476–78
74. Norris, D. M. 1972. Quinones in hardwoods: resistance factors to scolytids. *Proc. Forest Biol. Workshop, 2nd, Corvallis, Ore., 1972*
75. Norris, D. M., Baker, J. E., Borg, T. K., Ferkovich, S. M., Rozental, J. 1970. An energy-transduction mechanism in chemoreception by the bark beetle, *Scolytus multistriatus. Contrib. Boyce Thompson Inst.* 24:263–74
76. Painter, R. H. 1958. Resistance of plants to insects. *Ann. Rev. Entomol.* 3:267–90
77. Painter, R. H. 1966. Lessons to be learned from past experience in breeding plants for insect resistance. See Ref. 29, pp. 349–55
78. Pauly, G. 1963. The resin ducts of *Pinus pinaster. Trav. Lab. Forest. Toulouse* 1:Art. 18, 32 pp.
79. Philogène, B. J. R. 1972. Physiological studies and pest control. *Bioscience* 22:715–18
80. Plank, G. H., Gerhold, H. D. 1965. Evaluating host resistance to the white pine weevil, *Pissodes strobi,* using feeding preference tests. *Ann. Entomol. Soc. Am.* 58:527–32
81. Pridham, J. B., Ed. 1967. *Terpenoids in Plants.* New York: Academic. 257 pp.
82. Reid, R. W., Whitney, H. S., Watson, J. A. 1967. Reactions of lodgepole pine to attack by *Dendroctonus ponderosae* Hopkins and blue stain fungi. *Can. J. Bot.* 45:1115–26
83. Renwick, J. A. A., Vité, J. P. 1970. Systems of chemical communication in *Dendroctonus. Contrib. Boyce Thompson Inst.* 24:283–92
84. Riddiford, L. M., Williams, C. M. 1967. Volatile principle from oak leaves: role in sex life of Polyphemus moth. *Science* 155:589–90
85. Roberts, D. R. 1970. Within tree variation of monoterpene hydrocarbon composition of slash pine oleoresin. *Phytochemistry* 9:809–15
86. Rottink, B. A., Hanover, J. W. 1972. Identification of blue spruce cultivars by analysis of cortical oleoresin monoterpenes. *Phytochemistry* 11:3255–57
87. Rudinsky, J. A. 1966. Host selection and invasion by the Douglas-fir beetle, *Dendroctonus pseudotsugae* Hopkins, in coastal Douglas-fir forests. *Can. Entomol.* 98:98–111
88. Rudinsky, J. A. 1966. Scolytid beetles associated with Douglas fir: response to terpenes. *Science* 152:218–19
89. Rudloff, E. von 1959. The wax of the leaves of *Picea pungens* (Colorado spruce). *Can. J. Chem.* 37:1038–42
90. Rudloff, E. von 1972. Seasonal variation in the composition of the volatile oil of the leaves, buds, and twigs of white spruce (*Picea glauca*). *Can. J. Bot.* 50:1595–1603
91. Rudolf, P. O., Patton, R. F. 1966. Genetic improvement of forest trees for disease and insect resistance in the lake states. See Ref. 29, pp. 63–68

92. Runeckles, V. C., Mabry, T. J., Eds. 1973. *Terpenoids: Structure, Biogenesis, and Distribution.* New York: Academic. 241 pp.
93. Russell, G. B., Fenemore, P. G., Singh, P. 1972. Insect-control chemicals from plants. Nagilactone C, a toxic substance from the leaves of *Podocarpus nivalis* and *P. hallii. Aust. J. Biol. Sci.* 25:1025–29
94. St. George, R. A. 1930. Drought-affected and injured trees attractive to bark beetles. *J. Econ. Entomol.* 23:825–28
95. Schantz, M. von, Juvonen, S. 1966. Chemotaxonomische Untersuchungen in der Gattung *Picea. Acta Bot. Fenn.* 73:1–51
96. Schneider, D. 1969. Insect olfaction: deciphering system for chemical messages. *Science* 163:1031–37
97. Schönborn, A. von 1966. The breeding of insect resistant forest trees in central and northwestern Europe. See Ref. 29, pp. 25–27
98. Smith, R. H. 1961. The fumigant toxicity of three pine resins to *Dendroctonus brevicomis* and *D. jeffreyi. J. Econ. Entomol.* 54:365–69
99. Smith, R. H. 1963. Toxicity of pine resin vapors to three species of *Dendroctonus* bark beetles. *J. Econ. Entomol.* 56:827–31
100. Smith, R. H. 1964. Variation in the monoterpenes of *Pinus ponderosa* Laws. *Science* 143:1337–38
101. Smith, R. H. 1966. Resin quality as a factor in the resistance of pines to bark beetles. See Ref. 29, pp. 189–96
102. Snelling, R. O. 1941. Resistance of plants to insect attack. *Biol. Rev.* 7:543–86
103. Squillace, A. E. 1971. Inheritance of monoterpene composition in cortical oleoresin of slash pine. *Forest. Sci.* 17:381–87
104. Squillace, A. E., Hedrick, G. W., Green, A. J. 1971. Variation and inheritance of levopimaric acid content and its relationship to oleoresin yield in slash pine. *Silvae Genet.* 20:90–91
105. Stark, R. W. 1965. Recent trends in forest entomology. *Ann. Rev. Entomol.* 10:303–24
106. Stroh, R. C., Gerhold, H. D. 1965. Eastern white pine characteristics related to weevil feeding. *Silvae Genet.* 14:160–69
107. Teucher, G. 1955. Die Anfalligkeit der Douglasienrassen gegenüber der Douglasien Wollaus (*Gilletteella cooleyi*) (Gill) C. B. *Forst Jagd* 5:297–99, 337–42
108. Thielges, B. A., Campbell R. L. 1972. Selection and breeding to avoid the eastern spruce gall aphid. *Am. Christmas Tree J.* May 1972:3–6
109. Thomas, B. R. 1970. Modern and fossil plant resins. In *Phytochemical Phylogeny,* ed. J. B. Harborne, 59–79. New York: Academic. 335 pp.
110. Titze, J. F., Mucha, S. 1965. Testing of vigorous regrowth trees for resistance to *Sirex* by infestation with caged insects. *Aust. Forest. Res.* 1:14–22
111. Tobolski, J. J. 1968. *Variations in monoterpenes in Scotch pine (Pinus sylvestris* L.). PhD thesis, Michigan State University, East Lansing, 125 pp.
112. Tobolski, J. J., Hanover, J. W. 1971. Genetic variation in the monoterpenes of Scotch pine. *Forest. Sci.* 17:293–99
113. Vité, J. P. 1961. The influence of water supply on oleoresin exudation pressure and resistance to bark beetle attack in *Pinus ponderosa. Contrib. Boyce Thompson Inst. Plant Res. Prof. Pap.* 21:37–66
114. Vité, J. P., Pitman, G. B. 1970. Management of western pine beetle populations: use of chemical messengers. *J. Econ. Entomol.* 63:1132–35
115. Weissmann, G. 1966. *The distribution of terpenoids. Comparative Phytochemistry,* ed. T. Swain, New York: 97–120. Academic. 360 pp.
116. Werker. E., Fahn, A. 1969. Resin ducts of *Pinus halepensis* Mill.—Their structure, development and pattern of arrangement. *Bot. J. Linn. Soc.* 62:379–411
117. Westfall, R. D. 1972. *Developmental and genetic variation in the cortical terpenes of species of Pinus and Picea.* PhD thesis, Michigan State University, East Lansing, 102 pp.
118. Whittaker, R. H., Feeny, P. P. 1971. Allelochemics: chemical interactions between species. *Science* 171:757–69
119. Wilkinson, R. C. 1970. *Chemical analysis of species relationships in northeast American spruces.* PhD Thesis, Michigan State University, East Lansing, Mich. 111 pp.
120. Wilkinson, R. C., Hanover, J. W. 1972. Geographical variation in the monoterpene composition of red spruce. *Phytochemistry* 11:2007–10
121. Wilkinson, R. C., Hanover, J. W., Wright, J. W., Flake, R. H. 1971. Genetic variation in the monoterpene composition of white spruce. *Forest. Sci.* 17:83–90

122. Wilson, D. M. 1972. Genetic and sensory mechanisms for locomotion and orientation in animals. *Am. Sci.* 60:357–65
123. Wilson, E. O. 1970. Chemical communication within animal species. See Ref. 20, pp. 133–55
124. Wood, D. L. 1973. Selection and colonization of ponderosa pine by bark beetles. See Ref. 25, pp. 101–17
125. Wright, J. W., Wilson, L. F., Randall, W. K. 1967. Differences among Scotch pine varieties in susceptibility to European pine sawfly. *Forest. Sci.* 13:175–81
126. Wright, J. W., Wilson, L. F. 1972. Genetic differences in Scotch pine resistance to pine root collar weevil. *Mich. State Univ. Agr. Exp. Sta. Res. Rep.* 159, 6 pp.
127. Wright, R. H., Chambers, D. L., Keiser, I. 1971. Insect attractants, anti-attractants, and repellents. *Can. Entomol.* 103:627–30
128. Yasunaga, K., Oshima, Y., Kuwatsuka, S. 1962. Attractants of the pine bark beetles. I. Isolation of an attractant, benzoic acid, from the red pine bark. *J. Agr. Chem. Soc. Jap.* 36:802–4

THE STATUS OF VIRUSES PATHOGENIC FOR INSECTS AND MITES

❖6082

W. A. L. David
Glasshouse Crops Research Institute, Worthington Road, Littlehampton, England BN16 3PU

Several very different reviews could be presented under this title, but here *status* is taken to mean condition of things, and the emphasis is on recent investigations—chiefly on those which have some relevance to the exploitation of viruses for pest control. As Tarasevich (190) has emphasized this aim involves the solution of many fundamental problems, such as those relating to the characterization, entry, multiplication, and epizootiology of viruses. It is a testimony to the present interest in insect viruses that for the purpose defined, it was necessary to consider over 400 papers published since January 1970. Many of these have had to be omitted from the list of references, and the results, as for earlier publications, are referred to only through later papers, reviews, and textbooks (31, 36, 57, 89, 93, 115, 131, 132, 170, 171, 194, 200). Aspects of the ecology and epizootiology of viruses have been reviewed or discussed (169, 183, 186, 187, 228), and also insect tissue cultures and the part they can play in the study and production of insect viruses have been reviewed (26, 80, 199, 210).

The characterization, nomenclature, and classification of insect viruses are all matters for active research and debate (128). At the present time, a system being worked out by the International Committee for the Nomenclature of Viruses seems to be favored and has been adopted in this review. An important feature of this system, based mainly on the properties of the virus particle, is the cryptogram which embodies in four pairs of symbols certain biochemical, biophysical, and morphological features of viruses or groups of viruses (221). As knowledge advances it may appear desirable to give less emphasis in the classification to the attributes of viruses in their dormant form (the virions) and to include certain dynamic criteria (179, 222). So far, insect viruses are represented in six genera and one group ("group" is a provisional term), but there are also many unclassified viruses.

Up-to-date information about the distribution of the viruses in the different host orders has been kindly provided by M. E. Martignoni (personal communication) from a computer-based catalog (135) (see Table 1). Some relatively familiar, but

Table 1 Number of species of insects and mites reported to have viral diseases[a]

Orders of insects and mites	DNA viruses					RNA viruses				Unclassified viruses
	Baculovirus		*Iridovirus*	*Poxvirus*	*Parvovirus*	Cytoplasmic polyhedrosis viruses	*Rhabdo-virus*	*Entero-virus*	Other RNA viruses	
	Nuclear poly-hedrosis viruses	Granulosis viruses		Entomopox-viruses	Densonu-cleosis viruses					
Acarina	–	–	–	–	–	–	–	–	–	5
Orthoptera	4	–	–	1	–	–	–	–	9	4
Isoptera	–	–	–	–	–	–	–	–	–	3
Hemiptera	–	–	3	–	–	–	–	–	–	1
Neuroptera	2	–	–	–	–	–	–	–	–	1
Trichoptera	1	–	–	–	–	–	–	–	–	–
Lepidoptera	243	65	12	7	6	141	–	–	1	69
Coleoptera	4	–	4	9	1	–	–	–	5	3
Hymenoptera	21	–	2	–	1	1	–	6	2	12
Diptera	9	–	28	2	–	5	8	–	–	20

[a]This table was compiled from a February 1974 scan of MART-FAM, a computer-based insect virus catalog (M. E. Martignoni, personal communication). Classification according to Wildy (221). Generic names on line one, group or subgroup names on line 2.

as yet unclassified, viruses have been omitted from this review, e.g. chronic bee paralysis and infectious flacherie of the silkworm.

BACULOVIRUSES

This genus comprises the best known insect viruses—the nuclear polyhedroses (NPVs) and the granuloses (GVs). The virions are rod-shaped and contain DNA. Recently, viruses believed to belong to this genus, but which do not fit into either of the two subgroups above, have been found in the Coleoptera, *Gyrinus* (59), and in the Diptera, *Aedes* (50). The virus known as *Rhabdionvirus orcytes,* from *Oryctes rhinoceros,* the coconut beetle, is also now considered to be a member of this genus (136). Altogether NPVs have been found in > 200 Lepidoptera, ca 20 Hymenoptera, and ca 9 Diptera, and more rarely in Orthoptera, Coleoptera, and Trichoptera (*Neophylax* sp.) (61, 91, 221; Table 1). GVs are known from ca 65 Lepidoptera but from no other orders.

The general features of the NPVs and GVs have been reviewed (171, 174) and the reasons for grouping them together have been discussed (15). There are simple methods for recognizing polyhedra and for distinguishing between the nuclear and cytoplasmic types (171, 134). The shapes and surface structures of polyhedra have been studied by scanning electron microscopy (201).

Baculoviruses (BVs) can be obtained in quantity and purified by methods suitable for large-scale production (25). Studies of the polyhedra and capsules show that, by treatment with alkali, they can be separated into three components—the inclusion body protein, the enveloped virus particles, and the inclusion body envelope. The inclusion bodies contain two types of protein which differ in amino acid composition and are not identical serologically. Only a small quantity of one is present. This is relatively insoluble in alkali and, in the GVs of *Pieris brassicae,* it occurs only on the surface of the inclusion bodies and the virions (40, 66, 127).

By treating the polyhedra of *Bombyx mori* NPV with Na_2CO_3 under various conditions, it has been shown that the released virions consist of an outer membrane (the envelope which contracts into a sphere when the contents are released), an inner membrane (which has cross striations on the surface), and the core (which breaks up when exposed to the alkali to give subunits of 20–25 nm diameter) (112). In *Lymantria (Porthetria) dispar,* the outer membrane envelops a virus bundle and appears to consist of three layers, one structureless and the other two composed of hexagonally packed subunits. Inside are the virus capsids (inner membranes) each enclosing a hollow core (67).

From investigations of the morphogenesis of NPVs in Lepidoptera it seems that the chromatin strands within the hypertrophied nucleus (virogenic stroma) produce all the components of the nucleocapsids including the DNA. The nucleocapsids develop at the surface of the stromal strands, an envelope forms around each, and subsequently polyhedron protein with lattice structure begins to appear between them (68, 96, 182, 214). A similar sequence of events has been described for the GV of *Hyphantria cunea* (216). The source of the envelope is uncertain, but there is evidence that it may arise de novo within the nucleus and independent of the inner nuclear membrane (177).

Although the inclusion bodies of BVs are proteinaceous and proteolytic enzymes occur in the mid-gut of insects, the dissolution of ingested inclusions is believed to be primarily attributable to pH, though various alcohols and hydrogen bond cleaving agents also promote solubilization (40, 47, 104, 140).

The envelope, which surrounds the nucleocapsid as released from the occlusion body, is probably involved in the fusion between the virion and the microvilli of the columnar cells of the mid-gut epithelium (65, 105, 180). After the nucleocapsid has entered the columnar cells, virus is assembled within the nucleus, but usually no occlusion bodies are seen (88, 188). The newly formed nucleocapsids apparently become enclosed in vesicular transport membranes as they traverse the plasma membranes, after which they infect the other tissues and undergo further cycles of multiplication, during which occlusion bodies are formed (65, 177, 180) until protein synthesis becomes defective (215). Once in the hemocoel, virus particles with the envelope intact seem to be no more infective than virus particles in which only the inner membrane is present (111).

Aizawa (1) has discussed the virulence of insect viruses and has summarized results which show that isolates differing in virulence occur, and that changes in level may take place following serial passage through the normal hosts, or another susceptible species. Recently, the virulence of a NPV of *Spodoptera frugiperda* has been increased by chemical mutagens (153).

The resistance of insects to virus diseases has been reviewed (35, 170), and only a few recent papers are referred to here. *Malacosoma disstria* larvae, which have received a sublethal dose of NPV, are not protected against a further dose given 5 days later (149). Strains of some insects, which show a maintained difference in the level of resistance under the same conditions, are known to occur naturally. For example, two strains of *Plodia interpunctella* (Indian meal moth) showed a sevenfold difference in resistance to a granulosis virus (87). How common it is for insect stocks to become resistant to viruses as a result of exposure in successive generations is not clear. When the possibility was critically tested, using *Heliothis zea* and its NPV, no resistance developed in 25 laboratory generations (93).

Various factors are known to affect the susceptibility of insects to viruses, e.g. *Pseudaletia unipuncta* (Noctuidae) larvae held at a high temperature (37°C) after exposure to NPV given orally or by injection did not succumb, though some cells became infected (217). Also, food factors may influence the effect of a given dose of virus (44, 219, 225).

The specificity of an insect virus is of interest in relation to the range of pests it might control, the effect on beneficial insects, and, more broadly, in relation to hazards for man and higher animals. This subject has been extensively reviewed (92) and the conclusions for the BVs can be summarized as follows: the GVs are the most specific of all insect viruses; they are followed by the NPVs. So far as is known, vertebrates and other invertebrates are not susceptible to the BVs used to control insects, when administered by all the usual routes. However, a BV type virus has recently been found in a shrimp (39). In long-term feeding tests, the NPV of *H. zea* was not carcinogenetic in rats (92), and when fed during the fifth to fourteenth day of gestation it produced no detectable effect on the number or weight of fetuses nor any teratologies (94). Certain NPVs (but not GVs) may multiply in vertebrate cell lines, but *H. zea* NPV failed to develop or to produce any other detectable effects after three serial passages through four primate cell types (92).

From many reports, Ignoffo (92) concluded inter alia that: "commercial microbial insecticides tested to date are virtually harmless to vertebrates and forms of life other than target pests." Each pathogen which is proposed for use must be individually tested. Absolute "safety cannot be guaranteed in all living systems for all time" and a decision to use a pathogen "must be based on a prudent consideration of benefits to be obtained, balanced against potential risks." There seems to be no reason for disputing these conclusions, though attention should be drawn to the more cautious attitude of Kurstak (117). Very germain to discussions about safety are the observations that natural epizootics of virus disease often release much more of the pathogen than would be applied in a biological control program, and that much of the virus is present when the crop is marketed, e.g. the NPV of *Trichoplusia ni* on cabbage (76).

Too high a specificity can be a disadvantage in a candidate biological control agent. A virus which will control several pests is a more attractive proposition for producer and user. Such is the NPV of *Autographa californica,* which produced typical symptoms in five lepidopterous pests and is not altered in infectivity by passage through two of these alternative hosts, and so, could be produced in any of the three (205, 207).

There is no conclusive evidence that NPVs or GVs are transmitted in the eggs, though in some cases the circumstantial evidence is strong (45, 166, 170, 185), while in others, transmission within the eggs is said not to occur (138, 203). What is needed are detailed studies similar to those carried out with other systems (4, 164), using several strains of hosts and viruses. If transmitting strains could be established they might be valuable agents in biological control programs.

BVs, especially the NPVs, have been more tested and used for biological control than viruses in other orders, and numerous observations have been made on their persistence in the ecosystem (186). In soil they are resistant to leaching and sufficiently stable to persist from one season to the next. The NPV of *T. ni* is resistant to the action of many soil organisms (186) but its persistence is affected by pH (193). By means of a scanning electron microscope, the inclusion bodies can be seen adhering to the soil particles and certain types of inclusion bodies can be distinguished (84). When exposed on foliage NPVs and GVs are rapidly inactivated by the UV and near-UV radiation present in sunlight (228). For field use, the viruses have been formulated with various additives which screen out UV (98) or increase their effectiveness in other ways (219). The most successful additive to date seems to be activated carbon: it extended the half-life of NPV from < 1 day to ca 3 days on corn silks (95) and to ca 5 days on cotton (93).

Until the present time all the viruses used in field trials or sold commercially have been produced in living insects (93, 226). Frequently, such procedures are not difficult to manage, though industrialization presents problems (162).

A system for producing the viruses in tissue culture cell lines would lend itself more readily to industrialization. The problems are to find an inexpensive medium in which to grow the cells, to maintain the virulence of the virus to the host, to obtain a high yield of occluded virus, and to make the process continuous. Several authors report recent progress in this field, including the production of good yields of polyhedra in established cell lines (77, 148, 172, 206).

Attempts to replicate viruses of higher animals in bacteria and yeasts have met with some success (121, 220) but with insect viruses only a few infective particles resembling virions and no polyhedra have been obtained (220).

Undoubtedly, the most important and advanced study on the use of viruses for biological control has been carried out on the NPV of *Heliothis.* This work, which has been summarized by Ignoffo (93), makes history in the development of viral insecticides, since it was the first such product to be granted a temporary exemption from the requirement of a tolerance level by the U.S. Food and Drug Administration (197). At present the registration applies to use on cotton, but products containing the virus (Virion/H and Biotrol VHZ) have been extensively tested on *Heliothis* species attacking other crops. The results have not always been entirely satisfactory, and it is probably too early to decide to what extent they can be improved by right formulation and application (7, 46, 93).

Various BVs are being developed for use against pests of other field crops. Notable among these are the NPVs of *T. ni* (99, 173, 208) and of *A. californica* (208) and the GVs of *Pieris* spp. (3, 99) and of the potato moth, *Phthorimaea operculella,* in Australia (152). The multiple embedded virus of *A. californica* seems to be a likely candidate for commercialization since it infects several other lepidopterous hosts (*T. ni, Spodoptera exigua, Estigmene acrea, H. zea, Bucculatrix thurberiella,* and *Plutella xylostella*) (205).

The NPVs have been extensively used in controlling pests on forest trees (175, 218) and investigations are continuing (227). GVs have had some success in forests, e.g. against the fir budworm, *Choristoneura murinana* (163), and on fruit trees for controlling the codling moth (49, 107) and the tortricid, *Adoxophyes orana,* in Japan (165).

The virus which occurs in *Oryctes rhinoceros* (formerly considered to be a rhabdovirus) has been studied, especially in relation to the control of this serious coconut pest (13, 133, 230, 231). In adults it causes the epithelium of the mid-gut to proliferate, so that the lumen is filled with a mass of cells in which virus is multiplying actively. The adults thus become flying reservoirs of virus which they excrete into the host habitat (82). For field application, a strained suspension of ground up, infected larvae mixed with sawdust has been used. A fresh preparation causes a lethal infection in larvae of *O. rhinoceros* and the related *Scapanes australis grossepunctatus.* The early instars succumb more rapidly than late instars. After one week, 99% of the viral activity was lost under normal, damp conditions and drying or raising the temperature increased the rate of inactivation (13, 230). No broad study on the specificity or safety of this virus seems to have been made.

Much further work is required on the ecology and epizootiology of BVs, especially in relation to particular crops, so that epizootics can be understood and, if possible, fostered (48). Only the most obvious consequence of infection, death of the larvae, is usually studied, but in terms of crop protection and population dynamics other effects need to be considered. Among these are reduction in food consumption by the larvae (64), death of pupae, and debilitation of adults—though effects on the adult are sometimes minimal (203). Regarding the distribution of virus from one area to another, the fate of the viruses in the alimentary canal of small mammals, birds, and predatory insects is important (58, 81).

POXVIRUSES

The virions of entomopoxviruses (EPVs) resemble those of vertebrate poxviruses (VPVs) in general structure and consist of a beaded lipoprotein envelope, a plate-like core, and either one or two lateral bodies. The core contains the double-stranded DNA viral genome enclosed in a proteinaceous coat (16, 18, 60a, 160). However, EPVs differ from VPVs in being occluded in large (up to 20 μ maximum diameter) proteinaceous bodies. The virions are either oval or brick shaped and vary in size from 150 to 400 nm in length and from 165 to 250 nm in width, being largest in the Coleoptera and smallest in the Diptera (17). Sometimes, there are antigenically distinct spindle-shaped bodies associated with them in the cells (41, 42). Since the reviews quoted, other EPVs have been found in Coleoptera (18, 124), Lepidoptera (184), and Diptera (83, 178).

Purification of EPVs presents no special problems, except that, for dissolving the inclusion bodies, it is essential to use a mixture of alkali and a reducing compound, e.g. sodium carbonate and sodium thioglycollate at high pH (10.7–11.0) (200), probably because of the high cystine content (19). The action is stopped by adding potassium dihydrogen phosphate to reduce the pH to 8.5–9.0 (161). The viruses can then be banded by rate zonal sucrose density gradient centrifugation.

When examined by sodium dodecyl sulphate polyacrylamide gel, electrophoresis solubilized inclusion body protein from *Amsacta moorei* virus, produced in *E. acrea,* gave three major protein bands and the purified virions four major bands (161). The virions are reported to contain 1.4% DNA, which is low compared with poxviruses in general (161, 221). The molecular weight of the DNA has been put at around 160×10^6 daltons (161). Like a minority of VPVs, the EPVs contain enzymes in the core and four have now been found in the *Amsacta* virus (146).

Entry of an EPV into a host insect involves fusion of the viral envelope with the plasma membrane of the microvilli of the gut epithelium at their tips, followed by release of the core which then migrates into the body of the cell (60).

During the course of multiplication the virus is only seen in the cell cytoplasm but the nuclei of infected cells undergo certain changes (83, 178). Unlike the situation with VPVs in vertebrate cell cultures, the replicative cycle of EPVs inoculated into living hosts (the only possible technique to date) is not synchronous. It has therefore to be deduced from a vertebrate model and a summary of the events for the poxvirus of *Melolontha melolontha* has been given (17). More recently the morphogenesis and structure of the poxvirus of the midge, *Chironomus attenatus,* has been described (178).

IPVs develop in the cytoplasm of the fat body cells and/or hemocytes. Diseased *M. melonontha* larvae grow slowly and may pupate, but seldom, if ever, survive to become adults (23, 159). The diseased larvae assume a chalky white appearance due to virus in the fat body, which is somewhat distended, although the fat content is reduced. In the midge, *C. attenatus*, the virus multiplies only in the hemocytes, but nevertheless gives rise to irregularly distributed white masses just below the integument. The size and activity of the larvae appear to be normal but they die in the late fourth instar (178). Also, in the moth *Choristoneura conflicata* development of diseased larvae is much prolonged (42).

Preliminary studies show no common antigens between the EPVs and VPVs (17). The EPVs examined tend to infect related hosts. As is usually the case, they are more specific when fed as intact inclusions than when injected as free virions, e.g. the virus originally from *A. moorei* infected *Estigmene acrea* (both Arctiidae), *L. dispar* (Lymantriidae), and *Galleria mellonella* (Pyralidae) when injected, but only the first two by feeding. EPVs of *Choristoneura* spp. are not species specific (42, 161). The *Amsacta* virus did not infect certain *Orthoptera, Coleoptera,* or *Diptera,* and neither it nor the *Choristoneura virus* showed any effect on small mammals or birds (30, 90, 161).

Since EPVs are occluded, they may be expected to be relatively stable under field conditions and, therefore, to be useful for biological control. Evidence so far available confirms this expectation. The virus of *M. melolontha* is stable in wet and dry preparations and at normal temperatures, and would persist from one generation to the next in cadavers in the soil. It is inactivated by short exposures to UV (253 nm) (158). Reasonable success has been reported with *M. melolontha* poxvirus in limited field tests and the opinion was expressed that EPVs should have the same potential for use in biological control as the baculoviruses (BVs) (90). However, further work on the field efficacy and safety of EPVs is required to substantiate these opinions.

IRIDOVIRUSES

Early work on iridescent insect viruses (IVs) has been reviewed (14, 170), and Tinsley & Kelly (195) have proposed an interim nomenclature system (essentially numbering them in order of discovery). Recently Kelly & Robertson (108) discussed IVs in association with other viruses and questioned whether this genus, as defined (221), should have been established. Readers should refer to this review for an evaluation of recent fundamental studies and for comparisons of the IVs with icosahedral cytoplasmic deoxyriboviruses from other phyla.

IVs have been found in Diptera, Coleoptera, and Lepidoptera and are widely distributed. The particles are isometric (diameter 130 to 240 nm), are hexagonal in section, and have icosahedral symmetry. It has been confirmed that the DNA in *Tipula* iridescent virus (TIV) is double stranded (119).

There is some uncertainty as to whether IV virions are normally enveloped. Mosquito iridescent virus (MIV) virions were thought to be naked (144), but recent studies suggest that TIV virions may easily become enveloped in extra membranes from the host cell (229). In iridescent viruses types 2 and 6 (*Sericesthis* and *Chilo*) an internal membrane has been observed and sufficient lipids occur to form a continuous bilayer at this site (110). A recent study of the capsid structure of MIV suggests that the shell consists of a single structural unit membrane, formed of triangular arrays of capsomeres, with a fringe of knobbed fibrils attached (176). Within this is the core membrane which possibly corresponds to the internal membrane observed by Kelly & Vance (110).

Large yields of IVs can often be obtained and purified (43, 109), but this is not so for *Aedes taeniorhynchus* MIV (144). Serological tests show that viruses from the same host genus have antigens in common; those from different genera of the same host family and also those from different host orders may or may not have common antigens (108). It is interesting that two strains of *Aedes* MIV which have different physical characteristics are indistinguishable by immunodiffusion (62). When the profiles of the viral structural proteins of various IVs were examined, using acrylamide gel electrophoresis, it was found that they differed for serologically related viruses, but not so much for unrelated viruses (108). Various enzyme activities have been found to exist in the virions, as in EPVs (108, 146).

The details concerning the uptake of MIV and TIV from the gut have not been elucidated (181). The primary tissues infected are the fat body, epidermis, and imaginal disks, but iridescence may be observed later in many other tissues (63). The virus particles are assembled in the cytoplasm in the presence of virogenic stroma. A study of TIV infected *G. mellonella* larval hemocytes showed that viral structural protein fluorescence first appeared in the cytoplasm, outside the virus-induced viroplasm, one day after infection. Also, viral DNA was located entirely in the viroplasm. Doubt has been expressed concerning the virus assembly process, i.e. whether the capsid or the nucleoid forms first (108, 229), but with TIV it seems that the shell forms first and that the core material is introduced subsequently (170, 229).

When the IVs are fed to their natural hosts (37, 38, 144, 170, 178) or to other species (53, 63, 129, 224) the resulting level of infection is often low. The same viruses are much more infective when injected. In *Tipula* uptake from the gut seems to be the only mode of infection (38), but in *Aedes* infection may occur orally or by transovarian transmission (224). In an attempt to explain the low infectivity of MIV given orally, Stoltz & Summers (178) showed that the particles were destroyed rapidly (process unknown) in the anterior portion of the mid-gut. They also showed that the peritrophic membrane acts as a very effective barrier, since it seems to lack the gaps present in lepidopterous larvae. Presumably, infection takes place when particles pass through rare ruptures in the peritrophic membrane, or it may depend on some infective subviral unit.

This low oral infectivity of the IVs contrasts with their high infectivity and low level of specificity when injected into the hemocoel (170). Although apparently related viruses occur in vertebrates, other invertebrates, and plants, the only tests carried out suggest that insect IVs will not multiply in vertebrate cells (92). Undoubtedly, the low level of infectivity observed in feeding tests has discouraged serious attempts to use these viruses to control their hosts, e.g. *Tipula* and *Aedes*.

PARVOVIRUSES

Parvoviruses (PVs) are isometric, naked viruses, 18–22 nm in diameter, with icosahedral symmetry which contain single-stranded DNA. They are known with certainty in mammals and the lepidopterous insects *G. mellonella* and *Junonia coenia* (118, 196), but recently a possible member has been reported from a laboratory stock of *Aedes aegypti* (122).

Relatively large quantities of *Galleria* densonucleosis virus (DNV) can be produced in the larvae and extracted in a high level of purity. The methods recommended involve two cycles of differential centrifugation with DNase and RNase treatment in between (to eliminate cellular nucleic acids), followed by zonal centrifugation on a sucrose gradient: dialysis of the virus-containing sample against buffer and filtration through millipore filters (118, 211). The virions are resistant to ether and chloroform and retain their infective properties for 10 min at 80°C or for several years at –70°C. In dry deposits they are less stable (118, 157, 221).

Because of their small size, the fine structure of the virions is difficult to determine. It has been concluded that the particles of certain mammalian PVs have 32 capsomeres and that the DNV virion probably has 42 (118), which, if confirmed, would indicate a significant difference.

Studies of the structural proteins using 10% polyacrylamide gel electrophoresis showed that both DNV and *Junonia* virus contained four protein bands, and the profiles for the two were identical. Using a full complement of serological tests a close relationship has been found between these viruses, but no relationship seems to exist between DNV and the mammalian parvoviruses. DNV does not agglutinate the erythrocytes from the animal species tested (118, 196).

The reasons for concluding that the DNA of various parvoviruses is linear and single-stranded have been summarized (118, 196). In certain of these viruses, including DNV, complimentary (+) and (–) forms of DNA, encapsulated separately, are present. Depending on the conditions of extraction, these may persist as such, or they may associate to give double-stranded DNA (118, 120).

The uptake of DNV from the gut has not been studied. Once in the host it infects most tissues and gives rise to strong Feulgen-positive lesions in the nuclei of infected cells. By means of immunofluorescence and immunoperoxidase techniques, it has been shown that early antigens can be detected in the cytoplasm 4–7 hr after infection. Later (8–13 hr at 32°C) structural antigens appear in the nucleus and, at about this time, electron microscope observations reveal the first virions emerging from dense chromatin areas in the nucleus. They continue to multiply, the nuclear membrane is broken, and finally virions may entirely replace the cellular material (118).

DNV is very contagious and kills larvae of *Galleria* in about 4–6 days at 28°C. In the infected larvae, glycogen reserves are depleted (143) and there is a marked increase in uric acid in the fat body, especially when infection occurs during the early instars (168). When fully grown larvae are inoculated with a high concentration of virus, development is stopped. In contrast, if a lower concentration is administered, pupation proceeds and infection spreads from the larval tissues, which are still present, to the developing, imaginal organs, including the gonads of both sexes. The gametes themselves do not show any signs of infection but, in any case, visibly infected larvae and pupae do not become adults, so that there is no possibility of obviously diseased insects transmitting the virus transovarially. However, individuals which show no histopathological changes may possibly carry infection and transmit it (21, 22).

DNV is very host specific and it has not been possible to transmit it even to related Lepidoptera (22). When injected into *Melolontha melolontha,* it persisted for 30 days, but a progressive loss of virus was shown to occur from the eighth day following inoculation (142). On the other hand, the *J. coenia* (Nymphalidae) virus is far less host specific. In feeding tests it readily infected *Aglais urticae* and virus propagated in this host was in turn infective to *Mamestra brassicae, B. mori,* and *L. dispar.* It was not infective to *P. brassicae, Philosamia* spp., or, interestingly, *G. mellonella* (157).

DNV has been tested for the control of *G. mellonella* in bee hives and it has been claimed that the virus is not injurious to mammalian cells. However, although DNV does not replicate or cause disease in mice, in vitro tests have now shown that the virus can infect and destroy up to 30% of mouse L-cells and that others of these cells are transformed. Virions infective to *Galleria* could be extracted from L-cells infected six days earlier but no DNV virions could be seen in the transformed cells and extracts of these cells did not cause typical infection in *Galleria,* though they continued to produce DNV antigens. Rat embryo fibroblasts exposed to DNV transformed in vitro, and the cells contained enhanced numbers of RNA type-C virions, which signals caution, since it has been suggested that genomes of these viruses may act as determinants of DNA virus oncogenicity (117, 118).

CYTOPLASMIC POLYHEDROSES

The classification of the cytoplasmic polyhedroses viruses (CPVs) is unresolved and, provisionally, they are designated a group and placed near the genus *Reovirus.* The naked virions have icosahedral symmetry, are approximately 60 nm in diameter, and contain double-stranded RNA (dsRNA). Many are contained in large crystalline inclusions (5, 170, 171, 221, 223). Simple methods for detecting CPVs in tissue (167) and for distinguishing CPVs from NPVs (134) have been described.

Aruga & Tanada (5) list 168 species of Lepidoptera known to have a CPV infection (but it is doubtful whether all these are different viruses), as well as 2 species of Neuroptera, 3 species of Diptera, and 1 species of Hymenoptera. More recently a CPV has been found in an adult mosquito, *Anopheles stephensi,* where it was also present in the sporogonic stages of a malarial parasite (20). Two have been found in chironomid larvae collected in Florida (51).

CPVs chiefly infect the epithelial cells of the midgut, and free virions, as well as polyhedra, can be isolated by standard techniques (2, 69, 70). In *Bombyx mori* just over half the virions exist free, and those which are occluded can be released, undamaged, by treating the polyhedra with 0.05 M sodium carbonate (73).

The shape of the polyhedra is variable and in *B. mori* CPV four different forms have been observed by scanning electron microscope: regular cube, rhombic dodecahedron, tetragonal tristetrahedron, and globular. The value of shape for the identification of strains has yet to be determined (150). What little is known about the chemical composition of the polyhedra has been summarized by Kawase (106). Ignoring the virions, the polyhedra consist of protein and probably traces of metal ions.

Studies of the virions suggest that the structure proposed for *B. mori* CPV applies also to other viruses (106, 223). This shows two concentric icosahedral shells: the outer consists of 12 subunits (capsomeres) and is connected to the inner shell by 12 tubular structures. A segmented hollow tube terminating in a spherical particle projects from each capsomere (6).

The virions are reported to be resistant to ether, deoxycholate, and alkali (pH 11) at room temperature, but sensitive to chloroform and alcohol (70, 106, 155). However, it is now recognized that free virions may be more sensitive than the possibly more mature virions obtained by dissolving polyhedra (156). Acetone releases the viral genome RNA from virions of *Malacosoma disstria* in high yield and could probably be used for extracting other dsRNAs (71).

The protein and nucleic acid components of the virions have been studied separately (106). Virions of CPV of *B. mori* are serologically distinct from those of *Hemerocampa (Orgyia) leucostigma,* tussock moth, and *M. disstria,* forest tent caterpillar, which are closely related but differ quantitatively in their antigenic composition (116). These conclusions are supported by observations which show that the ratios of basic to acidic amino acids in the proteins extracted from the free virions of *M. disstria* and *H. leucostigma* were similar, and yet different from that of the serologically very distinct *B. mori* (72).

Kawase (106) has summarized the evidence for concluding that the genome of a CPV is a dsRNA which fragments in a nonrandom fashion, producing several pieces. Whether or how these segments are physically linked inside the virion has not been determined (54, 123). It appears that the resulting RNAs fall into long and short groups. For *B. mori* CPV all ten genome segments have the identical 3' terminal structure, carrying cytosine in one RNA chain and uracil in another (55). The electrophoretic patterns of the CPV dsRNAs from *M. disstria* and *B. mori* have been compared. In both, the genome segments separated into two groups composed of either slow- or fast-migrating fragments. In the *M. disstria* DNA six components were found, as against nine [or ten according to Furuichi & Mirau (55)] in *B. mori* DNA (74), which clearly shows that the two viruses are different and supports the results of the serological tests and amino acid analysis just mentioned (72, 116).

Owing to their small size and other difficulties the exact process of CPV formation has not been fully determined. Iwashita and also Kobayashi (97, 114) have summarized what is known, though it is not always easy to fully reconcile these two accounts. It appears that the viral RNA is synthesized in the nucleus and then migrates into the cytoplasm [though some may be synthesized there (75)]. According to Kobayashi, the origin of the viral cores which then form seems to be correlated with either or both of the two abnormal structures which can be seen in the cytoplasm of infected cells. They give rise to small dense granules which diffuse towards the virogenic stroma where they form core-like particles. These are then enclosed in capsidal protein from the virogenic stroma and subsequently the projections of the capsids are believed to be attached. The polyhedra form on the surface of the virogenic stroma as a result of threads of polyhedral protein diffusing among the virus particles and later crystallizing.

Infection of the host takes place from the gut, but little is known of the processes involved in the living insect. It seems probable that in some way intact virions pass through the peritrophic membrane. What happens then has been deduced from observations on midgut epithelial cell cultures (114). It appears that a spherical particle at the end of one of the projections from the virion attaches to the host cell (6). The attachment becomes closer and the virus then appears to pass through the projection into the cell cytoplasm, where the core substance is released, leaving only the empty outer shell on the cell surface.

Strains of CPVs which have polyhedra of more than one shape are well known, and strains which differ in virulence have been reported. After several passages, the virulence of a strain, transferred to a different host, has been observed to increase for the new host, but to decrease for the original host (189).

Commercial strains of silkworms which are genetically resistant to CPV have been selected. For optimum results when breeding, care should be taken to ensure that factors which influence susceptibility, such as quality of the larval food and environmental conditions, are favorable (4, 213). However, exposure to abnormal conditions may have a therapeutic effect. Thus, larvae exposed to a high temperature may resist the disease, probably because the multiplication of the virus is suppressed (32, 213). More use could probably be made of this technique when rearing virus-free insects.

It can now be stated with confidence that CPVs have a wider host range than NPVs. This conclusion holds true whether the virus is administered orally or injected (91, 170, 189). There appears to be little information about the effect of CPVs on other invertebrates or vertebrates. In a limited number of in vivo and in vitro tests with the CPVs of *B. mori* and *Thaumetopoea pityocampa,* pine processionary caterpillar, the results were negative (92). However, as they resemble members of the genus *Reovirus* in several respects, they should be treated with caution.

Summarizing the evidence available in 1971 regarding the CPV of *Bombyx mori,* Aruga (4) concluded that generation-to-generation transmission involved not merely a mechanical contamination of the egg, but also a more complicated genetic mechanism. This conclusion does not necessarily imply the presence of virions of CPV, though they have apparently been observed in the eggs of *H. zea* and *Pectinophora gossypiella* just before hatch (66). In investigations with other bollworms and with cabbage loopers, it appeared that most (33, 202), or all (166), of the virus was on the egg surface and could be removed by suitable washing. It seems likely that the virus frequently occurs on the shell but, also, that it may occur in some state within the egg.

Although CPVs are at least as numerous as NPVs, only that of *T. pityocampa* has been used on any scale for control purposes. In field trials a CPV of *Trichoplusia ni* gave reasonably good control of its host (209).

Further work on the persistence of CPVs in the environment seems desirable. General evidence suggests that CPV virions occluded in polyhedra retain their activity in the natural environment if protected from sunlight UV, but few studies have been made (130). Recently, too, it has been shown that particles of the cubic strain of *B. mori* CPV are adsorbed to soil particles mainly by Coulomb force and that, as with GVs, they are not readily leached out (85).

CPVs tend to act slowly on their hosts; the infected larvae are noticeably backward and may survive for a long time (130, 170, 204). However, since these larvae excrete considerable quantities of virus (which can infect other larvae) (130, 212) and since some species do not produce normal adults with full reproductive potential (34, 130, 202), the effect of the virus on an insect population may be much greater than would appear from a study of the rate and level of larval mortality alone.

ENTEROVIRUSES

Enteroviruses (EVs) are isometric naked viruses, 20–30 nm in diameter, with icosahedral symmetry which contain single-stranded RNA (ssRNA) (27, 221). Many have been found in vertebrates (mammals and birds) and several in invertebrates (insects and mites) (27, 170, 198, 200). Since these reviews were published, other members or probable members of the genus, in addition to those discussed subsequently, have been described from *Antherea* sp. (29), termites (56), crickets (154), and grasshoppers (78). An isometric virus found in citrus red mites, which has probably incorrectly been considered to be the cause of paralysis in these mites, may also be an EV (151).

Methods have been described for obtaining and purifying two bee viruses, acute paralysis virus (ABPV) and sacbrood virus (SBV) (139) and the virus from *Gonometa podocarpi* (GPV) (126). Like other members of the genus these enteroviruses are acid stable and also ether resistant (126, 137, 139). They can be stored at –20°C, but their stability at room temperature and above seems not to have been investigated in detail, though the infectivity of SBV is soon lost in dead larvae (8). Other EVs are inactivated by exposure to 50–60°C for 30 min (170, 221).

Analyses show that GPV has five polypeptides and Nodamura virus (NV) one major polypeptide and possibly small amounts of two others (27, 126, 137). The RNA has been shown to be single-stranded and in ABPV, SBV, and GPV only one species is present, though two occur in NV (both of which appear to be necessary for infection). The base composition of the two bee viruses is similar, but it differs from that of NV and most mammalian viruses (27, 126, 139).

Serological tests with ABPV and SBV failed to show any relationship between them, and GPV virus did not react in immunodiffusion tests with certain other small RNA viruses of insects (102, 200). NV also appears to be unrelated to any known virus (137).

Infection with enteroviruses takes place through the gut, but the process has not been studied in insects and little is known about the morphogenesis of the virus particles (78).

Bailey (8) has reviewed the bee viruses and since then the relationship of ABPV and SBV with adult and larval bees has been further analyzed. ABPV accumulates in the heads, especially in the hypopharyngeal glands and to a lesser extent in the brains, of acutely paralyzed bees. The symptoms are more pronounced in bees kept at 30°C than at 35°C (11). SBV primarily affects larvae. But a high dose will infect young adults, primarily in the heads, especially in the hypopharyngeal glands and brains, without causing symptoms. Such infected bees eat little pollen and live only about 3 weeks, as do bees receiving no pollen; whereas normal bees consume pollen and live 9 weeks. Infected individuals also fly earlier in life but most collect no pollen. Those that do, contaminate it with SBV (9, 10).

Three viruses from *Drosophila* spp. P, iota, and C have been found and shown to have several of the characteristics of enteroviruses. P and iota infections often pass unnoticed, but, in fact, P can reduce fecundity and iota increases sensitivity to CO_2. They infect the cells of the intestine and malpighian tubules. Virus C is much more pathogenic; tracheal cells are most infected but the gut epithelium and the malpighian tubules appear to escape infection. All three viruses multiply in the brain. Serologically, P and iota appear to be closely related, if not identical, and quite distinct from virus C (100, 101, 145, 192).

Nodamura virus is of special interest because it multiplies in some insects without causing symptoms, kills others, and is transmissable to suckling mice in which it causes a fatal infection (12, 137).

Preliminary in vivo tests with vertebrates suggest that ABPV and SBV are not harmful (92). However, bearing in mind the behavior of NV, viruses of this genus would need careful testing.

RHABDOVIRUSES

Rhabdoviruses (RVs) are bullet-shaped or bacilliform (see later) in outline, measuring about 175 X 70 nm if bullet-shaped, but longer if bacilliform. They contain about 2% ssRNA, are sensitive to lipid solvents, and are acid-labile (86, 113, 221).

RVs are known from vertebrate, invertebrate, and plant tissues. Some multiply in the invertebrate vector by which they are transmitted to either a susceptible vertebrate or plant hosts. Such viruses are not considered here and the sigma virus of *Drosophila* is the only accepted member of this genus known to infect insects or mites. It seems possible, however, that the rod-shaped viruses found in *Panonychus ulmi,* European red mite (147), and in *P. citri,* citrus red mite, (151) should be included, though morphogenesis of these viruses occurs exclusively in the nucleus, unlike that of vertebrate RVs (113). Two viruses, one from *Oryctes rhinoceros,* rhinoceros beetle (82) (136), and the other from *Gyrinus natator* (59) have been transferred recently to the *Baculovirus* genus.

The fine structure of the insect RVs has not been studied. In other RVs the RNA is intimately associated with the proteins to form the so-called ribonucleoprotein, and this is surrounded by another layer (the matrix protein). Together they form the nucleocapsid or naked virion, which often appears to be bullet-shaped. There are, however, reasons for thinking that the intact nucleocapsid is bacilliform and that it only appears to be bullet-shaped because one end is labile. In the final stage of maturation the nucleocapsid acquires an envelope by budding through a membrane—thus becoming the virion. The envelope takes the form of a host-derived lipid layer from which protein spikes protrude (113). The insect RVs may be expected to conform to this general pattern; there is recent evidence that sigma virus contains RNA (79).

Mites acquire RVs particularly when they feed on leaves contaminated with recently deposited material; there may be a low level of transovarial transmission (147). Uptake of the virus from the gut has not been studied, but in *P. citri,* virions have been observed in the nuclei of midgut epithelial cells (151). In contrast, fascinating studies of sigma virus show that under natural conditions it is acquired only by transmission through the gametes, though the virus is also infectious when artificially injected. Two types of sigma-infected flies are recognized: the so-called nonstabilized, in which only somatic cells are infected, and the stabilized, in which the gonads are infected. The latter condition is regularly transmitted through maternal inheritance (125, 164).

Sigma virus, which has been known for 30 years or more, is widely distributed in *Drosophila* but sometimes at low frequency (52, 164). Its most notable effect on the host is to cause the adults to become sensitive to CO_2. However, one American strain, which reduced the percentage hatch of eggs, has been described. A strain taken at random was found to reduce the number of eggs laid by both stabilized and nonstabilized infected females (103). Recently, a comparison has been made between the maturation processes in the testes of strains of sigma which are, and which are not, transmitted. In neither case are virions seen in the somatic cells but virions of strains which are transmitted appear in the germ cells. However, the infectivity of testes extracts is independent of these virions and appears to be related, instead, to the content of viral genomes in the cytoplasm (191). Sigma virus can be eliminated from the germ cells of *Drosophila* by suitable heat treatment or by feeding with ethyl-methane-sulphonate (24, 28).

It now seems fairly certain that paralysis and allied symptoms in *P. citri* are caused by the elongated virus and not, as was previously thought, by the isometric virus which is also present in most tissue (151). Most of the field work on the control of citrus mites with the virus disease seems to have been carried out under this misapprehension and is, therefore, difficult to evaluate. However, since control was obtained, it must be supposed that the rod-shaped virus, possibly an RV, was present and that many of the conclusions are valid.

CONCLUDING REMARKS

Viruses pathogenic to insects command attention because of their intrinsic interest, because of the contribution their study can make to virology, but also, undoubtedly, because of their unknown potentialities for biological pest control. It is true that they may involve subtle hazards, but so do chemical insecticides. And, for many generations, man and other animals have been exposed to their influence, since natural epizootics of some of the virus diseases lead to levels of contamination, even of food crops, far greater than would result from an application for biological control.

Some insect viruses are already being successfully employed to control pests and it may be possible to exploit others without danger. Whether this is so, or not, can only be determined by means of detailed, fundamental investigations, such as those reviewed here. Further information is required about the properties of viruses (so that they can be characterized), the complex relationships between the viruses and their hosts, their ecology and epizootiology, the means of producing them in tissue cultures, and the hazards which may be associated with their use.

Acknowledgments

I thank, sincerely, Miss Gwen Marsh for her patient help in preparing and checking the draft typescript and Miss M. E. M. Price (Librarian) for her very willing assistance in obtaining numerous publications.

Literature Cited

1. Aizawa, K. See Ref. 36, 655–72
2. Aizawa, K. See Ref. 5, 23–36
3. Akutsu, K. 1971. *Jap. J. Appl. Entomol. Zool.* 15:56–62
4. Aruga, H. See Ref. 5, 3–21
5. Aruga, H., Tanada, Y., Eds. 1971. *The Cytoplasmic Polyhedrosis Virus of the Silkworm.* Tokyo: Tokyo Univ. Press. 234 pp.
6. Asai, J., Kawamoto, F., Kawase, S. 1972. *J. Invertebr. Pathol.* 19:279–80
7. Atger, P. 1972. *C. R. Acad. Agr. Fr.* 57:1544–48
8. Bailey, L. 1968. *Ann. Rev. Entomol.* 13:191–212
9. Bailey, L. 1969. *Ann. Appl. Biol.* 63:483–91
10. Bailey, L., Fernando, E. F. W. 1972. *Ann. Appl. Biol.* 72:27–35
11. Bailey, L., Milne, R. G. 1969. *J. Gen. Virol.* 4:9–14
12. Bailey, L., Scott, H. A. 1973. *Nature* 241:545
13. Bedford, G. O. 1973. *J. Invertebr. Pathol.* 22:70–74
14. Bellett, A. J. D. 1968. *Advan. Virus Res.* 13:225–46
15. Bellett, A. J. D. 1969. *Virology* 37:117–23
16. Bergoin, M., Dales, S. See Ref. 132, 169–205
17. Bergoin, M., Devauchelle, G. See Ref. 194, 3–7
18. Bergoin, M., Devauchelle, G., Vago, C. 1971. *Virology* 43:453–67
19. Bergoin, M., Veyrunes, J. C., Scalla, R. 1970. *Virology* 40:760–63
20. Bird, R. G., Draper, C. C., Ellis, D. S. 1972. *Bull. WHO* 46:337–43
21. Boemare, N. 1973. *Ann. Zool. Ecol. Anim.* 5:81–97
22. Boemare, N., Brès, N. 1969. *Ann. Zool. Ecol. Anim.* 1:309–20
23. Bonami, J. R. 1970. *Ann. Soc. Entomol. Fr.* 6:451–65
24. Bregliano, J. C. 1973. *Ann. Microbiol.* 124A:393–408
25. Breillatt, J. P. et al 1972. *Appl. Microbiol.* 23:923–30
26. Brooks, M. A., Kurtti, T. J. 1971. *Ann. Rev. Entomol.* 16:27–52
27. Brown, F., Hull, R. 1973. *J. Gen. Virol.* 20:43–60
28. Brun, G., Diatta, F. 1973. *Ann. Microbiol.* 124A:421–38
29. Brzostowski, H. W., Grace, T. D. C. 1970. *J. Invertebr. Pathol.* 16:277–79
30. Buckner, C. H., Cunningham, J. C. 1972. *Can. Entomol.* 104:1333–42
31. Bulla, L. A. 1973. *Ann. NY Acad. Sci.* 217:1–243
32. Bullock, H. R. 1972. *J. Invertebr. Pathol.* 19:148–53
33. Bullock, H. R., Mangum, C. L., Guerra, A. A. 1969. *J. Invertebr. Pathol.* 14:271–73
34. Bullock, H. R., Martinez, E., Stuermer, C. W. 1970. *J. Invertebr. Pathol.* 15:109–12
35. Burges, H. D. See Ref. 36, 445–57

36. Burges, H. D., Hussey, N. W. 1971. *Microbial Control of Insects and Mites.* London: Academic. 861 pp.
37. Carter, J. B. 1973. *J. Invertebr. Pathol.* 21:123–30
38. Ibid 136–43
39. Couch, J. A. 1974. *Nature* 247:229–31
40. Croizier, G., Meynadier, G. 1972. *Entomophaga* 17:231–39
41. Croizier, G., Veyrunes, J. C. 1971. *Ann. Inst. Pasteur* 120:709–15
42. Cunningham, J. C., Burke, J. M., Arif, B. M. 1973. *Can. Entomol.* 105:767–73
43. Cunningham, J. C., Hayashi, Y. 1970. *J. Invertebr. Pathol.* 16:427–35
44. David, W. A. L., Ellaby, S., Taylor, G. 1972. *J. Invertebr. Pathol.* 20:332–40
45. David, W. A. L., Gardiner, B. O. C., Clothier, S. E. 1968. *J. Invertebr. Pathol.* 12:238–44
46. Davidson, A., Pinnock, D. E. 1973. *J. Econ. Entomol.* 66:586–87
47. Egawa, K., Summers, M. D. 1972. *J. Invertebr. Pathol.* 19:395–404
48. Falcon, L. A. 1973. *Ann. NY Acad. Sci.* 217:173–86
49. Falcon, L. A., Kane, W. R., Bethell, R. S. 1968. *J. Econ. Entomol.* 61:1208–13
50. Federici, B. A., Anthony, D. W. 1972. *J. Invertebr. Pathol.* 20:129–38
51. Federici, B. A., Hazard, E. I., Anthony, D. W. 1973. *J. Invertebr. Pathol.* 22:136–38
52. Fleuriet, A. 1972. *C. R. Soc. Biol.* 166:598–601
53. Fukuda, T. 1971. *J. Invertebr. Pathol.* 18:152–54
54. Furuichi, Y., Mirau, K. 1972. *J. Mol. Biol.* 64:619–32
55. Furuichi, Y., Mirau, K. 1973. *Virology* 55:418–25
56. Gay, F. J., Wetherly, A. H. 1970. *Virology* 40:1063–65
57. Gibbs, A. J. 1973. *Viruses and Invertebrates.* Amsterdam: North-Holland. 673 pp.
58. Gitay, H., Polson, A. 1971. *J. Invertebr. Pathol.* 17:288–90
59. Gouranton, J. 1972. *J. Ultrastruct. Res.* 39:281–94
60. Granados, R. R. 1973. *Virology* 52:305–9
60a. Granados, R. R. 1973. *Misc. Publ. Entomol. Soc. Am.* 9(2):73–94
61. Hall, D. W., Hazard, E. I. 1973. *J. Invertebr. Pathol.* 21:323–24
62. Hall, D. W., Lowe, R. E. 1972. *J. Invertebr. Pathol.* 19:317–24
63. Hama, H. 1968. *Jap. J. Appl. Entomol. Zool.* 12:34–39
64. Harper, J. D. 1973. *J. Invertebr. Pathol.* 21:191–97
65. Harrap, K. A. 1970. *Virology* 42: 311–18
66. Ibid 1972. 50:114–23
67. Ibid 124–32
68. Ibid 133–39
69. Hayashi, Y. 1970. *J. Invertebr. Pathol.* 16:442–50
70. Hayashi, Y., Bird, F. T. 1970. *Can. J. Microbiol.* 16:695–701
71. Hayashi, Y., Cunningham, J. C. 1971. *J. Invertebr. Pathol.* 17:433–39
72. Hayashi, Y., Durzan, D. J. 1971. *J. Invertebr. Pathol.* 18:121–26
73. Hayashi, Y., Kawarabata, T., Bird, F. T. 1970. *J. Invertebr. Pathol.* 16:378–84
74. Hayashi, Y., Krywienczyk, J. 1972. *J. Invertebr. Pathol.* 19:160–62
75. Hayashi, Y., Retnakaran, A. 1970. *J. Invertebr. Pathol.* 16:150–51
76. Heimpel, A. M., Thomas, E. D., Adams, J. R., Smith, L. J. 1973. *Environ. Entomol.* 2:72–75
77. Henderson, J. F., Faulkner, P., Mackinnon, E. A. 1974. *J. Gen. Virol.* 22:143–46
78. Henry, J. E., Oma, E. A. 1973. *J. Invertebr. Pathol.* 21:273–81
79. Herforth, R. S. 1973. *Virology* 51:47–55
80. Hink, W. F. 1972. *Advan. Appl. Microbiol.* 15:157–214
81. Hostetter, D. L. 1971. *J. Invertebr. Pathol.* 17:130–31
82. Huger, A. M. 1973. *Z. Angew. Entomol.* 72:309–19
83. Huger, A. M., Krieg, A., Emschermann, P., Gotz, P. 1970. *J. Invertebr. Pathol.* 15:253–61
84. Hukuhara, T. 1972. *J. Invertebr. Pathol.* 20:375–76
85. Hukuhara, T., Wada, H. 1972. *J. Invertebr. Pathol.* 20:309–16
86. Hummeler, K. See Ref. 132, 361–86
87. Hunter, D. K., Hoffmann, D. F. 1973. *J. Invertebr. Pathol.* 21:114–15
88. Hunter, D. K., Hoffmann, D. E., Collier, S. J. 1973. *J. Invertebr. Pathol.* 21:91–100
89. Hurpin, B. 1971. *Ann. Parasitol. Hum. Comp.* 46:243–76
90. Hurpin, B., Robert, P. H. 1969. *Entomophaga* 14:349–57
91. Ignoffo, C. M. See Ref. 131, 129–67
92. Ignoffo, C. M. 1973. *Ann. NY Acad. Sci.* 217:141–64
93. Ignoffo, C. M. 1973. *Exp. Parasitol.* 33:380–406
94. Ignoffo, C. M., Anderson, R. F., Woodard, G. 1973. *Environ. Entomol.* 2:337–38

95. Ignoffo, C. M., Parker, F. D., Boening, O. P., Pinnell, R. E., Hostetter, D. L. 1973. *Environ. Entomol.* 2:302–3
96. Injac, M., Duthoit, J. L., Amargier, A. 1973. *Ann. Zool. Écol. Anim.* 5:99–110
97. Iwashita, Y. See Ref. 5, 79–101
98. Jacques, R. P. 1972. *Can. Entomol.* 104:1985–94
99. Ibid 1973. 105:21–27
100. Jousset, F. X. 1972. *Ann. Inst. Pasteur* 123:275–88
101. Jousset, F. X., Plus, N., Croizier, G., Thomas, M. 1972. *C. R. Acad. Sci. Paris Ser. D* 275:3043–46
102. Juckes, I. R. M., Longworth, J. F., Reinganum, C. 1973. *J. Invertebr. Pathol.* 21:119–20
103. Jupin, N., Plus, N., Fleuriet, A. 1968. *Ann. Inst. Pasteur* 114:577–94
104. Kawanishi, C. Y., Egawa, K., Summers, M. D. 1972. *J. Invertebr. Pathol.* 20:95–100
105. Kawanishi, C. Y., Summers, M. D., Stoltz, D. B., Arnott, H. J. 1972. *J. Invertebr. Pathol.* 20:104–8
106. Kawase, S. See Ref. 5, 37–59
107. Keller, S. 1973. *Z. Angew. Entomol.* 73:137–81
108. Kelly, D. C., Robertson, J. S. 1973. *J. Gen. Virol.* 20:17–41
109. Kelly, D. C., Tinsley, T. W. 1972. *J. Invertebr. Pathol.* 19:273–75
110. Kelly, D. C., Vance, D. E. 1973. *J. Gen. Virol.* 21:417–23
111. Khosaka, T., Himeno, M. 1972. *J. Invertebr. Pathol.* 19:62–65
112. Khosaka, T., Himeno, M., Onodera, K. 1971. *J. Virol.* 7:267–73
113. Knudson, D. L. 1973. *J. Gen. Virol.* 20:105–30
114. Kobayashi, M. See Ref. 5, 103–28
115. Krieg, A. 1973. *Arthropodenviren.* Darmstadt, German Federal Republic: Biol. Bundesdanst. Land-und Forstwirt. 328 pp.
116. Krywienczyk, J., Hayashi, Y., Bird, F. T. 1969. *J. Invertebr. Pathol.* 13:114–19
117. Kurstak, E. 1971. *Ann. Parasitol. Hum. Comp.* 46:277–78
118. Kurstak, E. 1972. *Advan. Virus Res.* 17:207–41
119. Kurstak, E., Garzon, S., Onji, P. A. 1972. *Arch. Ges. Virusforsch.* 36:324–34
120. Kurstak, E., Vernoux, J. P., Brakier-Gingras, L. 1973. *Arch. Gesamte Virusforsch.* 40:274–84
121. Lavroushin, A., Treagan, L. 1972. *Wasmann J. Biol.* 30:87–92
122. Lebedeva, O. P., Kuznetsova, M. A., Zelenko, A. P., Gudz-Gorban, A. P. 1973. *Acta Virol.* 17:253–56
123. Lewandowski, L. J., Millward, S. 1971. *J. Virol.* 7:434–37
124. Lipa, J. J., Bartkowski, J. 1972. *J. Invertebr. Pathol.* 20:218–19
125. L'Héritier, P. 1971. *Ann. Parasitol. Hum. Comp.* 46:173–78
126. Longworth, J. F., Payne, C. C., Macleod, R. 1973. *J. Gen. Virol.* 18:119–25
127. Longworth, J. F., Robertson, J. S., Payne, C. C. 1972. *J. Invertebr. Pathol.* 19:42–50
128. Lwoff, A., Tournier, P. See Ref. 132, 1–42
129. McLaughlin, R. E., Scott, H. A., Bell, M. R. 1972. *J. Invertebr. Pathol.* 19:285–90
130. Maleki-Milani, H. 1970. *Entomophaga* 15:315–25
131. Maramorosch, K. 1968. Insect viruses. *Curr. Top. Microbiol. Immunol.* 42: 192 pp.
132. Maramorosch, K., Kurstak, E. 1971. *Comparative Virology.* New York: Academic. 584 pp.
133. Marschall, K. J. 1970. *Nature* 225:288–89
134. Martignoni, M. E. 1972. *J. Invertebr. Pathol.* 19:281–83
135. Martignoni, M. E., Williams, P., Reineke, D. E. 1973. *J. Invertebr. Pathol.* 22:100–7
136. Monsarrat, P., Meynadier, G., Croizier, G., Vago, C. 1973. *C. R. Acad. Sci. Ser. D* 276:2077–80
137. Murphy, F. A., Scherer, W. F., Harrison, A. K., Dunne, H. W., Gary, G. W. 1970. *Virology* 40:1008–21
138. Neilson, M. M., Elgee, D. E. 1968. *J. Invertebr. Pathol.* 12:132–9
139. Newman, J. F. E., Brown, F., Bailey, L., Gibbs, A. J. 1973. *J. Gen. Virol.* 19:405–9
140. Nordin, G. L. 1972. *Diss. Abstr. B* 32:4648B
141. Oatman, E. R. et al 1970. *J. Econ. Entomol.* 63:415–21
142. Odier, F. 1971. *Entomophaga* 16: 353–58
143. Pascal, A., Baud, L. 1972. *Ann. Soc. Entomol. Fr.* 8:481–84
144. Paschke, J. D., Campbell, W. R., Webb, S. R. See Ref. 194, 20–24
145. Plus, N. 1971. *Ann. Zool. Écol. Anim.* 3:449–53
146. Pogo, B. G. T., Dales, S., Bergoin, M., Roberts, D. W. 1971. *Virology* 43:306–9
147. Putman, W. L. 1970. *Can. Entomol.* 102:305–21
148. Quiot, J. M., Vago, C., Paradis, S. 1970. *Entomophaga* 15:437–43
149. Raheja, A. K., Brooks, M. A. 1971. *J. Invertebr. Pathol.* 17:136–37

150. Rao, C. B. J. 1973. *J. Ultrastruct. Res.* 42:582–93
151. Reed, D. K., Hall, I. M. 1972. *J. Invertebr. Pathol.* 20:272–78
152. Reed, E. M., Springett, B. P. 1971. *Bull. Entomol. Res.* 61:223–33
153. Reichelderfer, C. F., Benton, C. V. 1973. *J. Invertebr. Pathol.* 22:38–41
154. Reinganum, C., O'Loughlin, G. T., Hogan, T. W. 1970. *J. Invertebr. Pathol.* 16:214–20
155. Richards, W. C. 1970. *J. Invertebr. Pathol.* 15:457–58
156. Richards, W. C., Hayashi, Y. 1971. *J. Invertebr. Pathol.* 17:42–47
157. Rivers, C. F., Longworth, J. F. 1972. *J. Invertebr. Pathol.* 20:369–70
158. Robert, P. H. 1968. *Entomophaga* 14:159–75
159. Robert, P. H. 1969. *Ann. Zool. Écol. Anim.* 1:289–307
160. Roberts, D. W., Bergoin, M. 1970. *Proc. Int. Colloq. Insect Pathol., Maryland, USA,* 380–85
161. Roberts, D. W., McCarthy, W. J. See Ref. 194, 7–9
162. Rogoff, M. H. 1973. *Ann. NY Acad. Sci.* 217:200–10
163. Schönherr, J. 1969. *Entomophaga* 14:251–60
164. Seecof, R. See Ref. 131, 59–93
165. Shiga, M., Yamada, H., Oho, N., Nakazawa, H., Ito, Y. 1973. *J. Invertebr. Pathol.* 21:149–57
166. Sikorowski, P. P., Andrews, G. L., Broome, J. R. 1973. *J. Invertebr. Pathol.* 21:41–45
167. Sikorowski, P. P., Broome, J. R., Andrews, G. L. 1971. *J. Invertebr. Pathol.* 17:451–52
168. Smirnoff, W. A., Loiselle, J. M. 1969. *J. Invertebr. Pathol.* 14:421–22
169. Smith, C. E. 1971. *J. Invertebr. Pathol.* 18:i–xi
170. Smith, K. M. 1967. *Insect Virology.* London:Academic. 250 pp.
171. Smith, K. M. See Ref. 132, 479–507
172. Sohi, S. S., Cunningham, J. C. 1972. *J. Invertebr. Pathol.* 19:51–61
173. Splittstoesser, C. M., McEwen, F. L. 1971. *J. Invertebr. Pathol.* 17:194–98
174. Stairs, G. R. See Ref. 131, 1–23
175. Stairs, G. R. 1972. *Ann. Rev. Entomol.* 17:355–72
176. Stoltz, D. B. 1973. *J. Ultrastruct. Res.* 43:58–74
177. Stoltz, D. B., Pavan, C., Cunha da, A. B. 1973. *J. Gen. Virol.* 19:145–50
178. Stoltz, D. B., Summers, M. D. 1972. *J. Ultrastruct. Res.* 40:581–98
179. Subak-Sharpe, J. H. 1971. *Strategy of the Viral Genome,* ed. G. E. W. Wolstenholme, M. O'Connor, 389–94. Edinburgh: Churchill Livingstone
180. Summers, M. D. 1971. *J. Ultrastruct. Res.* 35:606–25
181. Summers, M. D. See Ref. 194, 16–20
182. Summers, M. D., Arnott, H. J. 1969. *J. Ultrastruct. Res.* 28:462–80
183. Surtees, G. 1971. *Int. J. Environ. Stud.* 2:195–201
184. Sutter, G. R. 1972. *J. Invertebr. Pathol.* 19:375–82
185. Swaine, G. 1966. *Nature* 210:1053–54
186. Tanada, Y. 1971. *Entomological Essays to Commemorate the Retirement of Professor K. Yasumatsu,* ed. S. Asahina, J. L. Gressitt, Z. Hidaka, T. Nishida, K. Nomura, 367–79. Tokyo: Hokuryukan
187. Tanada, Y. 1973. *Ann. NY Acad. Sci.* 217:120–30
188. Tanada, Y., Leuthenegger, R. 1970. *J. Ultrastruct. Res.* 30:589–600
189. Tanaka, S. See Ref. 5, 210–7
190. Tarasevich, L. M. 1968. *Proc. Int. Congr. Entomol., Moscow* 2:98–99
191. Teninges, D. 1972. *Ann. Inst. Pasteur* 122:1183–92
192. Teninges, D., Plus, N. 1972. *J. Gen. Virol.* 16:103–9
193. Thomas, E. D., Reichelderfer, C. F., Heimpel, A. M. 1973. *J. Invertebr. Pathol.* 21:21–25
194. Tinsley, T. W., Harrap, K. A. 1972. *Moving Frontiers in Virology.* Monogr. Virol., Vol. 6. Basel: Krager. 66 pp.
195. Tinsley, T. W., Kelly, D. C. 1970. *J. Invertebr. Pathol.* 16:470–72
196. Tinsley, T. W., Longworth, J. F. 1973. *J. Gen. Virol.* 20:7–15
197. Upholt, W. M., Engler, R. A., Terbush, L. E. 1973. *Ann. NY Acad. Sci.* 217:234–37
198. Vago, C. See Ref. 131, 24–37
199. Vago, C. 1972. *Methods in Invertebrate Tissue Culture,* 2 vols. New York: Academic. 410 + 426 pp.
200. Vago, C., Bergoin, M. 1968. *Advan. Virus Res.* 13:247–303
201. Vago, C. et al 1970. *C. R. Acad. Sci.* 271:1053–55
202. Vail, P. V., Gough, D. 1970. *J. Invertebr. Pathol.* 15:397–400
203. Vail, P. V., Hall, I. M. 1969. *J. Invertebr. Pathol.* 13:358–70
204. Vail, P. V., Hall, I. M., Gough, D. 1969. *J. Invertebr. Pathol.* 14:237–44
205. Vail, P. V., Jay, D. L. 1973. *J. Invertebr. Pathol.* 21:198–204
206. Vail, P. V., Jay, D. L., Hink, W. F. 1973. *J. Invertebr. Pathol.* 22:231–37
207. Vail, P. V., Jay, D. L., Hunter, D. K. 1973. *J. Invertebr. Pathol.* 21:16–20

208. Vail, P. V., Soo Hoo, C. F., Seay, R. S., Killinen, R. G., Wolf, W. W. 1972. *Environ. Entomol.* 1:780–85
209. Vail, P. V., Whitaker, T., Toba, H., Kishaba, A. N. 1971. *J. Econ. Entomol.* 64:1132–36
210. Vaughn, J. L. See Ref. 131, 108–28
211. Vernoux, J. P., Kurstak, E. 1972. *Arch. Gesampte Virusforsch.* 39:190–95
212. Watanabe, H. 1968. *J. Sericult. Sci. Jap.* 37:385–89
213. Watanabe, H. See Ref. 5, 169–84
214. Watanabe, H. 1972. *J. Invertebr. Pathol.* 20:223–25
215. Watanabe, H., Imanishi, K. 1972. *Jap. J. Appl. Entomol. Zool.* 16:193–201
216. Watanabe, H., Kobayashi, M. 1970. *J. Invertebr. Pathol.* 16:71–79
217. Watanabe, H., Tanada, Y. 1972. *Appl. Entomol. Zool.* 7:43–51
218. Wellenstein, G. 1973. *OEPP/EPPO Bull.* No. 9:43–52
219. Wellenstein, G., Lühl, R. 1972. *Naturwissenschaften* 59:517
220. Wells, F. E., Heimpel, A. M. 1970. *J. Invertebr. Pathol.* 16:301–4
221. Wildy, P. 1971. *Classification and Nomenclature of Viruses.* Monogr. Virol., Vol. 5. Basel: Karger. 81 pp.
222. Wildy, P. 1973. *J. Gen. Virol.* 20:1–5
223. Wood, H. A. 1973. *J. Gen. Virol.* 20:61–85
224. Woodard, D. B., Chapman, H. C. 1968. *J. Invertebr. Pathol.* 11:296–301
225. Yadava, R. L. 1971. *Z. Angew. Entomol.* 69:303–11
226. Yamada, H., Oho, N. 1973. *J. Invertebr. Pathol.* 21:144–48
227. Yearian, W. C., Young, S. Y., Livingston, J. M. 1973. *J. Invertebr. Pathol.* 22:34–37
228. Yendol, W. G., Hamlen, R. A. 1973. *Ann. NY Acad. Sci.* 217:18–30
229. Yule, B. G., Lee, P. E. 1973. *Virology* 51:409–32
230. Zelazny, B. 1972. *J. Invertebr. Pathol.* 20:235–41
231. Ibid 1973. 22:122–26

FEDERAL AND STATE PESTICIDE REGULATIONS AND LEGISLATION

❖6083

Errett Deck
Washington State Department of Agriculture, Olympia, Washington 98504

Responsibility for regulating the distribution and use of pesticides in the United States has been shared by state and federal government agencies. Municipal and local governments have played a minor role. Federal authority has emphasized control over labeling, setting of safe residue tolerances, and evaluation of toxicity and safety. State authority has emphasized control over local distribution, storage, use and application, and disposal. Both authorities have cooperated in surveillance, inspection, and analyses of pesticide products.

Since the Federal Environmental Pesticide Control Act of 1972 (PL 92-516) (2) was signed on October 21, 1972, pesticide regulatory programs in the United States have been and are being significantly changed. This Act was adopted as a major amendment to the Federal Insecticide, Fungicide, and Rodenticide Act of 1947. As provided in the bill, Congress intends that the new federal responsibilities delegated in the Act be phased in over a 4-yr period, with final implementation of all provisions by October 21, 1976.

Although there will be major changes in pesticide regulatory programs during this implementation period, the new legislation provides for continuing and expanding cooperative responsibility between federal and state agencies. All states will be required to meet certain minimum standards. However, states will be allowed to have more restrictive standards, and a number of states have implemented legislation which contains additional statutory authority not provided for in the federal statute. Past and future phases of these coordinated programs will be discussed in this article.

FEDERAL PESTICIDE LEGISLATION

Federal Insecticide Act

The first Federal Insecticide Act (4) was passed in 1910 and was intended primarily to protect the buyer (the farmer) against adulterated or misbranded products. Prior to 1910, the purchase of agricultural chemicals was on a buyer beware basis. Dramatic changes have taken place since that time in the field of chemical pesticides (previously termed economic poisons). Synthetic organic compounds, such as the chlorinated hydrocarbons, organophosphates, carbamates, and phenoxy herbicides, have been added to the early arsenal of pesticides such as arsenic, fluorine, copper, and sulfur compounds. Both private and publicly financed research have greatly increased the number of pesticides available and widened the scope of their usefulness. During this 64-yr period, federal laws have been broadened to regulate the labeling and interstate distribution of insecticides, rodenticides, nematocides, herbicides, fungicides, plant growth regulators, defoliants, and desiccants for use by farmers, homeowners, and industry. All of these materials which are intended for preventing, destroying, repelling, or mitigating pests or for the purpose of use as plant regulators, defoliants, or desiccants are defined as pesticides.

Federal Insecticide, Fungicide, and Rodenticide Act of 1947 (5)

This Act added a new concept by placing the burden of proof of acceptability of a product on the manufacturer prior to its being marketed. This Act was oriented to protect the user (the consumer) and the general public from pesticides—some of which were highly toxic and all of which were subject to use limitations as required by labeling. The registrant was required to submit scientific proof of the efficacy and safety of the chemical to accomplish the purpose for which it was to be used when the instructions on the label were followed. The U.S. Department of Agriculture was responsible for the enforcement of this Act until December 2, 1970. The U.S. Environmental Protection Agency was created on that date and the responsibility for enforcement of FIFRA was transferred by Executive Order to the new agency.

In 1948, the Food and Drug Administration began to establish safe tolerance levels of pesticide residues in foods. In no case were these tolerances established at a level higher than that necessary to accommodate the registered pesticide use as approved by the U.S. Department of Agriculture.

Miller Amendment

In 1954, the "Miller Amendment" to the Federal Food, Drug, and Cosmetic Act (3) formalized what had been an informal cooperative arrangement whereby the USDA registered only pesticide uses which resulted in no residues on raw agricultural commodities or residues declared safe by the FDA. This amendment required that the pesticide registrant submit data not only to the Department of Agriculture, proving the efficacy and safety of the chemical, but also to the FDA, proving the safety of measurable residues found in the raw agricultural commodity produced following labeled use of the pesticide. In establishing this level of safety, the FDA required a safety factor of over 100 to 1 of any measurable effect of the pesticide in controlled warm-blooded animal tests. Residue tolerance-setting has been complicated for certain pesticides because of interpretation of the "Delaney Clause" (3) [Section 409 (c) (3) (A) Federal Food, Drug, and Cosmetic Act as Amended] to prohibit food uses of pesticides shown capable of inducing cancer in experimental animals. Although the Delaney Clause does not technically apply to pesticide residues, this interpretation leaves no room for judgment in determining a no effect level for a weak carcinogen.

FDA inspectors regularly examined samples of food products shipped in interstate commerce in order to detect if there were violations of the residue tolerance amendment. This enforcement program proved a significant deterrent to a grower violating the pesticide label directions because if residues were found above the allowable tolerance levels his crop was subject to seizure and he was subject to prosecution.

Beginning in 1961, increased emphasis was placed on longer term public health and ecological effects such as the protection of fish, wildlife, and the environment. The President's Science Advisory Committee, the National Academy of Sciences-National Research Council, and various subcommittees were instrumental in reshaping the organization and emphasis of pesticide regulation procedures. A federal interdepartmental agreement was developed in 1964 which gave the Department of Health, Education, and Welfare and the Department of the Interior increased pesticide responsibilities in the areas of health and protection of fish and wildlife. Both the Jensen Report of 1969 (7) (by the National Research Council's Committee on Persistent Pesticides) and the Mrak Report of 1969 (9) (by HEW's Commission on Pesticides) expressed concern about the continued use of persistent pesticides and recommended additional restrictions of pesticide uses deemed to be hazardous by any of the three federal agencies.

Environmental Protection Agency

This Agency was created December 2, 1970, by a reorganization plan issued by the President. This action brought together the major environmental control programs of the federal government into a single agency. The pesticide regulatory functions of the Departments of Agriculture, Interior, and Health, Education, and Welfare were transferred to this new agency. These regulatory functions included registration of pesticides as required under FIFRA; setting of tolerances as required by the Miller Amendment to the Federal Food, Drug, and Cosmetic Act; and many pesticide research and monitoring programs previously conducted by the three departments. In addition, the EPA assumed the functions of the Federal Water Quality Administration (formerly under the Department of Interior); the National Air Pollution Control Administration (formerly under the Department of Health, Education, and Welfare); and the nuclear radiation controls (formerly under the Atomic Energy Commission).

Following the creation of this new agency and the required reorganization and consolidation of functions, there was a need to amend prior legislation covering the transferred responsibilities. Since 1970, major amendments have been passed affecting water quality, air pollution control, and pesticide regulation responsibilities. Details of the new federal pesticide legislation will be covered in this article. Several specialized areas of pesticide responsibility are retained by other federal agencies.

Food and Drug Administration

The Food and Drug Administration still retains the responsibility of monitoring food for humans and feed for animals. In addition, any products violating pesticide residue tolerances now established by the EPA are subject to seizure by the FDA under the authority of the Federal Food, Drug, and Cosmetic Act.

Department of Transportation

The Department of Transportation has rules and regulations governing the transportation of pesticides under the Code of Federal Regulations (CFR) Title 49-Transportation, Parts 100–199, revised as of January, 1974. Certain pesticides are classified as Class B poisons and are the pesticides of primary concern to the Department of Transportation. Pesticide products designated by the EPA as highly toxic formulations are generally designated by the DOT as Class B poisons for transportation purposes. All labeling requirements and package specifications, including maximum quantities permitted to be shipped in a single outside container, are listed in the DOT regulations for those commodities classified as Class B poisons. Those pesticides not classified as Class B poisons, and Class B poisons moved intrastate by an intrastate carrier not possessing ICC interstate authority, do not fall under DOT authority.

Federal Aviation Administration

The Federal Aviation Administration, under the Federal Aviation Regulation, Part 137-Agricultural Aircraft Operations, of January 1, 1966, regulates the dispensing of pesticides by aircraft. Under these regulations, as amended, it is a violation for an aerial applicator to apply any pesticide in a manner inconsistent with the directions for use on the federally registered label.

Department of Agriculture

The Department of Agriculture, through its Consumer and Marketing Services, Meat and Poultry Inspection Program, monitors meat and poultry products including analyses for possible pesticide residues. In addition to the authority to seize contaminated products, the Department identifies, where possible, the source or origin of the contaminated meat with follow-up inspections in order to prevent further marketing of such products.

STATE PESTICIDE LEGISLATION

By 1901, five states had enacted insecticide laws to protect the farmer against adulteration of arsenical and sulfur spray materials. In the 70-yr period since that time, all states have adopted pesticide legislation which requires the registration of pesticide products distributed in their respective states. State pesticide registration laws closely coordinate with federal legislation and include authority for inspection and analysis of pesticide products.

Many states have adopted very comprehensive pesticide legislation which covers all areas including the labeling, registration, distribution, storage, transportation, use and application, and disposal of pesticides. Over two thirds of the states have found it necessary to adopt legislation which specifically regulates the use and application of pesticides. This includes the regulation of commercial applicators and broad authority to control certain uses of pesticides which have proven hazardous and are classified for restricted use in their states. In addition, some states have found it advisable to implement legislation which requires the licensing of pesticide dealers; the licensing and examination of pesticide dealer managers and pesticide consultants; the restriction of the sale and use of restricted-use pesticides by a permit system or other means; and the regulation and inspection of application equipment and pesticide storage and disposal.

Legislation in the various states differs greatly because of a difference in need for regulation, due to differences in agricultural and environmental situations, and because of the independent nature of state government. There is a significant difference between the necessary level of pesticide control between the open spaces of Wyoming and Montana and the congested intensive agricultural and urban mix of valleys in California or between certain areas of New York and Alaska. Historically and logically, laws to control the local distribution and the application and use of pesticides were within the realm of inherent state police powers. These responsibilities are essentially intrastate in nature and can best be handled under state legislative authority. Unfortunately, a number of states—some of them important agricultural states—did not choose to accept responsibility for controlling the use of pesticides within their borders. Congress reacted to this lack of control of pesticide use by some states and in 1972 enacted pesticide legislation which gave the federal government the additional authority to regulate the use and registration of pesticides produced and used solely in intrastate commerce. However, Congress recognized the advantage of utilizing state expertise and the state's understanding of local problems and made provisions for cooperation between the EPA and state government agencies. While the Federal act preempts municipal and local governments from regulating pesticides, it provides that states may retain important functions in pesticide regulatory programs, such as the certification of applicators, the registration of pesticides for special local needs, the issuance of experimental use permits, and the enforcement of the act.

The need for uniform national standards for pesticide labeling, residue tolerances, and the scientific evaluation of toxicity hazards was obvious. For this reason, states have adopted the residue tolerances as established by federal agencies and almost uniformly have accepted federal labels, at least as far as safety precautions, general format, and standard directions for use. The states provided for special local needs, either by means of state supplemental labeling to accompany federally registered products or by issuing state registrations for labels of products formulated for use within the state.

Prior to the implementation of the 1972 federal pesticide amendments, state regulations were the only authority filling an important gap between the federal authority to control the interstate shipment of a federally registered pesticide and the interstate shipment of a food or feed grown from a crop treated with that pesticide. Even with the full enactment of the Federal Environmental Pesticide Control Act of 1972, there are several almost mandatory responsibilities and a number of permissive responsibilities which states should accept to assure an efficient and effective pesticide control program throughout the United States.

State government was first to adopt and implement pesticide legislation in this country. There was an early attempt to guide states in adopting, as appropriate, uniform legislation. The Uniform State Insecticide, Fungicide, and Rodenticide Act (13) was published by the Council of State Governments in 1946 as a guide to state governments for the registration of pesticides. This proposed state legislation was designed to work in harmony with the Federal Insecticide, Fungicide, and Rodenticide Act, passed by Congress the following year. The 1946 state act (designated as the "Suggested State Pesticide Control Act") was amended in 1971 and again in 1973 to include provisions for: licensing pesticide dealers; licensing pest management consultants; establishing a pesticide advisory board; adopting regulations to control storage, distribution, and disposal of pesticides; and providing for cooperative agreements with the EPA or other agencies. The 1973–1974 Official Publication of the Association of American Pesticide Control Officials includes the latest draft of the "Suggested State Pesticide Control Act" (11).

During the 1940s, a number of states adopted legislation to control the use and application of pesticides. The Association of American Pesticide Control Officials developed the Model Custom Application of Pesticides Act, which was published as suggested state legislation by the Council of State Governments in 1951. This suggested legislation to regulate use was updated again in 1970, and this Model Pesticide Use and Application Act (8), as adopted by AAPCO and the National Association of State Departments of Agriculture (NASDA), was published in the Council of State Governments' publication entitled, "1971 Suggested State Legislation." In order that this suggested state legislation might closely coordinate with the new Federal Act, which now relates to use, it was again updated in 1973 and the "Suggested State Pesticide Use and Application Act" (12) is published in the Council of State Governments' "1974 Suggested State Legislation" (Volume XXXIII, p. 80). These model pesticide registration and application bills, developed by AAPCO, have helped many states to adopt relatively uniform and effective legislation to cover their responsibilities.

The Association of American Pesticide Control Officials has been an active organization in coordinating state and federal legislation. Members include federal and state officials regulating pesticides, individuals inspecting and analyzing pesticides, and research workers investigating pesticides. Members supported a number of proposals to improve the preliminary drafts of the Federal Act and congressional committees were responsive. The Association's purpose, as filed with the Recorder of Deeds, Washington, D.C., emphasizes the promotion of uniform and effective legislation.

Pesticides applied according to label directions have not caused significant health hazards or illnesses. The major problems have resulted from misuse—either from ignorance or by accident. Most pesticide incidents have resulted from misuse in the distribution, storage, transportation, or disposal, or by the misapplication by the ultimate user, whether he be a small private user, a large commercial grower, or a commercial applicator. For this reason, the role of states in education and regulation is important. The full implementation of the new federal regulations will encourage states that have not implemented comprehensive control programs to do so, otherwise their agricultural producers may not be able to continue to use certain restricted-use pesticides necessary for successful crop production.

During the past two years, there has been a flurry of activity in initiating or amending state pesticide legislation. Until the final regulations are adopted by the EPA, it is difficult to determine whether a particular state's authority adequately allows for full participation in all the federal-state programs provided for in the federal act. A number of states have either an effective program which should be acceptable, or broad enough legislative authority that their state-developed regulations should satisfy the EPA regulatory requirements.

While approximately 35 states (1) have some type of regulatory control over commercial pesticide applicators, very few have broad enough statutory authority to certify private applicators (farmers) and no state has fully implemented such a certification program. Even though the EPA has not completed specific guidelines for state plans and other state responsibilities covered by the federal act, those states with no experience in licensing or certifying applicators should be developing and introducing basic enabling legislation. New programs cannot be effective on a crash basis and the action dates of October 21, 1974, 1975, and 1976 may pass without successful implementation of all requirements.

FEDERAL ENVIRONMENTAL PESTICIDE CONTROL ACT OF 1972

After two years of intensive study by the House Agriculture Committee, the Senate Agriculture and Forestry Committee, and the Senate Commerce Committee, Congress enacted PL 92-516, which was a major amendment to the Federal Insecticide, Fungicide, and Rodenticide Act of 1947. This bill was signed by President Nixon on October 21, 1972. Committee members, through input at public hearings and volumes of written testimony, considered all affected segments of our society in developing this legislation. There was extensive testimony from environmental groups, chemical industry associations, consumer interest groups, agricultural producer associations, and many governmental agencies, both federal and state. In total, over 15 drafts of this legislation were proposed and evaluated before the final draft was submitted by the Joint House/Senate Committee of Conference (Report No. 92–1540) (6). This process required thousands of hours of input and resulted in the elimination of hundreds of unworkable proposals. Many were submitted by individuals with no experience in enforcing pesticide legislation and with little understanding of the problems their proposals would create for the federal enforcing agency (EPA), cooperating state control officials, agriculture, other users of pesticides, the chemical industry, and eventually the public welfare. As is true with all compromise legislation involving a controversial issue, no one is completely satisfied with the act. There were many vitally interested groups, some constructive and some polarized on opposite extremes. Some may question the need for certain provisions in the act or the lack of certain authorities, however, at this point such deliberation is academic. The legislation does grant the EPA broad and flexible authority to regulate pesticides in order to provide for the protection of man and his environment. The successful implementation of the many provisions in the act will depend upon the responsible development of regulations and enforcement policies by the EPA. The EPA could accomplish the purposes of the act and carry out the intent of Congress without undue interference in the country's ability to control pests or without creating hardship on the producers or users of pesticides.

Major New Responsibilities of the EPA

The Federal Environmental Pesticide Control Act of 1972 increases the responsibility and authority of the EPA substantially. Some of the major new provisions are: (*a*) Regulate the use of all pesticides (some previously covered only by state authority). (*b*) Extend Federal pesticide regulations to actions entirely within a single state (intrastate registration, distribution, and use). (*c*) Prohibit the use of any pesticide in a manner inconsistent with its labeling. In Senate Report No. 92-838 (10), the committee indicated that administrative interpretation of this provision should not prohibit a use "which has only beneficial effects on man and his environment. . . . Further, it is the belief of the Committee that the use of the word 'inconsistent' should be read and administered in a way so as to visit penalties only upon those individuals who have disregarded instructions on a label that would indicate to a man of ordinary intelligence that use not in accordance with such instructions might endanger the safety of others or the environment." (*d*) Require pesticides and their uses to be classified for general or restricted use (explained below under heading: Classifications of Pesticides and Pesticide Uses). (*e*) Require the registration of all pesticide-producing establishments (explained below under heading: Registration of Establishments). (*f*) Authorize stop sale, use, or removal orders and provide for civil penalties. While criminal provisions may be used where circumstances warrant, the flexibility of having civil remedies available provides an appropriate means of enforcement without subjecting a person to criminal sanctions. (*g*) Authorize the establishment of packaging standards and the regulation of pesticide and container disposal. (*h*) Give applicants for registration proprietary rights in their test data. (*i*) Authorize cooperation with the states in enforcement, registration of pesticides for special local needs, training and certification of applicators (explained below under heading: Certification of Applicators), research and monitoring, and issuing experimental use permits.

Section 2 of Public Law 92-516 contains all the amendments to the Federal Insecticide, Fungicide, and Rodenticide Act. It consists of 27 sections which may be cited as the "Federal Environmental Pesticide Control Act of 1972." The act is now commonly being referred to as "Amended FIFRA." Section 3 of PL 92-516 briefly amends the Federal Hazardous Substances Act, the Poison Prevention Packaging Act, and the Federal Food, Drug, and Cosmetic Act. Section 4 of PL 92-516 provides the effective dates for implementation of the various provisions of the act. While it was provided that the amendments to FIFRA would take effect at the close of the date of enactment of the act, the exemptions and timing for their implementation are very important. All procedures and established regulations which are consistent with the amendments are to remain in force until amended or superseded. New regulations were necessary for implementing some provisions of the act that were to become effective on date of enactment. Those regulations were to be promulgated and become effective within 90 days of enactment. This was a difficult requirement and in a number of instances the EPA was not able to comply with the time frame.

In order to allow everyone an opportunity to comply with the provisions of the new regulations, no criminal or civil penalties are to be imposed for any act or failure to act until 60 days after the effective regulations have been published in the Federal Register. For the purposes of this article, it is not appropriate that details of implementation of each section of the 27 sections of the amendments be listed, but it is important to outline the annual bench-marks in implementation that are to be met.

Certification of Applicators

By October 21, 1973, the Administrator of the EPA was to have prescribed standards for the certification of applicators. The law states that certified applicators are to be individuals who are certified as competent and authorized to use or supervise the use and handling of any pesticide classified for restricted use. This certification is to be accomplished by states who are approved by the Administrator to implement this function under an approved state plan. The complexity of establishing these standards and the considerable controversy over how specific, comprehensive, or restrictive the standards should be resulted in delay. The proposed regulation was not published in the Federal Register until February 22, 1974. There was tremendous input from all interested and affected persons in the development of these proposed regulations and the final regulations should be adopted by June 1974. These regulations can serve as a guide for states in developing their state plan and in developing and testing programs for certifying applicators to cover the requirements of the federal act. By October 1, 1975, each state that desires to certify applicators must submit a state plan indicating how it proposes to accomplish this responsibility. Regulations or guidelines for these state plans (developed at regional meetings between state control officials and the EPA staff) are now being prepared for publication in the Federal Register.

By October 21, 1976, states must have certification plans in operation in order that commercial or private applicators may use pesticides which are restricted for use only by certified applicators. At this time, the EPA has no plans to set up a federal applicator certification program and indeed the federal act does not prescribe a mechanism for setting up such a program. According to the EPA the availability of restricted use pesticides in a state for its legitimate users could depend upon the success of a state in submitting, getting approved, and implementing a state plan for certifying applicators.

Registration of Establishments

The act required that, by October 21, 1973, the EPA promulgate effective regulations relating to the registration of establishments. On November 6, 1973, the EPA published adopted regulations regarding the registration of pesticide-producing establishments and the submission of pesticide reports and labeling by the pesticide producer. Under this legislation, an establishment is a location in which any pesticide subject to the act is produced. Under this establishment registration requirement, the producer who operates the establishment must apply for registration and must inform the EPA of the types and amounts of pesticides he is currently producing, the amount he produced during the past year, and the type and amount he has sold or distributed during the past year. After the initial registration application, the producer is required to keep a current record and submit this record annually to the Administrator of the EPA. The timetable as developed in the regulations requires registration by December 24, 1973, of all establishments producing pesticides distributed in interstate commerce and by October 21, 1974, for all establishments producing pesticides solely for intrastate commerce. Each location will be assigned an establishment registration number and that number must appear on the container or label of each pesticide product produced at that location.

Classification of Pesticides and Pesticide Uses

Two years after enactment of the act, the EPA must have final regulations adopted which provide for the classification of pesticide uses. After that time all new applications for registration must be classified either restricted use or general use. Between October 21, 1974 and October 21, 1976 all prior registered pesticide uses will have to be reclassified and designated as either restricted or general use pesticides. The federal act provides that:

> (B) If the Administrator determines that the pesticide, when applied in accordance with its directions for use, warnings and cautions and for the uses for which it is registered, or for one or more of such uses, or in accordance with a widespread and commonly recognized practice, will not generally cause unreasonable adverse effects on the environment, he will classify the pesticide or the particular use or uses of the pesticide to which the determination applies, for general use.

In the same manner, the federal act provides that:

> (C) If the Administrator determines that the pesticide, when applied in accordance with its directions for use, warnings and cautions and for the uses for which it is registered, or for one or more of such uses, or in accordance with a widespread and commonly recognized practice, may generally cause, without additional regulatory restrictions, unreasonable adverse effects on the environment, including injury to the applicator, he shall classify the pesticide, or the particular use or uses to which the determination applies, for restricted use.

These two very general definitions leave considerable flexibility for judgment. The standards developed to implement this requirement are of critical importance to the federal scheme. Many of the regulatory provisions of the act hinge on the classification scheme. For many years, the basic consideration for safety in pesticide use was "Before Using Any Pesticide—STOP—Read the Label—Follow Label Directions." Because of the comprehensive requirements that must be met before a pesticide use and its labeling can be registered, there have been few incidents of illegal residues on food crops and few illnesses of users and the public when labeling directions and precautions have been followed. Those uses which present a special hazard to the applicator or the environment, for which labeling restrictions are not sufficient protection, can be further regulated by the restricted use concept. An example would be sodium fluoroacetate (compound 1080), an effective but hazardous rodenticide which has been strictly regulated because of its high toxicity and potential for secondary poisoning.

Within the restricted use classification there will be those pesticide uses which, because of acute dermal or inhalation toxicity, present a hazard to the applicator and for that reason may be applied only by or under the direct supervision of a certified applicator. Other pesticides or pesticide uses may cause unreasonable adverse effects on the environment. These pesticides or pesticide uses may be restricted to use by, or under the direct supervision of, a certified applicator, or may be subject to other restrictions as provided by the EPA regulations. The final date for requiring the classification of all pesticides and their uses as general use or restricted use and for the certification of those persons who need to apply restricted use pesticides is coordinated in the one date for full implementation of the act—October 21, 1976.

Registrations for Special Local Needs

After October 21, 1974, not only must all new pesticide registrations be classified as general or restricted use, but all intrastate pesticides must be registered under provisions of the act. The Conference Report No. 92-1540, page 33, point (47), clarifies the issue of the validity of state registrations of pesticides which are distributed only in intrastate commerce: "Section 4(c)(1) of this bill gives the Administrator up to two years to promulgate regulations providing for registration of pesticides under provisions of H.R. 10729. This provision of section 4(d) makes it clear that state registered pesticides moving only in intrastate commerce would be provided an opportunity to register under the federal law before their distribution would be prohibited."

Section 24(c) of the act provides for states to continue to register pesticides for special local needs. A copy of the subsection follows:

> Sec. 24.(c) A State may provide registration for pesticides formulated for distribution and use within that State to meet special local needs if that State is certified by the Administrator as capable of exercising adequate controls to assure that such registration will be in accord with the purposes of this Act and if registration for such use has not previously been denied, disapproved, or canceled by the Administrator. Such registration shall be deemed registration under Section 3 for all purposes of this Act, but shall authorize distribution and use only within such State and shall not be effective for more than 90 days if disapproved by the Administrator within that period.

The EPA regulations setting standards for these registrations must be published with adequate lead time so that states can be certified well in advance of October 21, 1974. States will be able to assist the EPA in the registration of these locally formulated products for local or minor uses, pesticide-fertilizer mixes, or special package mixes. Other special local needs include supplemental labeling to accompany a product. This labeling may contain minor changes or additions to directions for use which will accommodate special situations but which are within the scope of federally approved uses.

The Congressional intent is expressed in Senate Report No. 92-838: "The purpose of this subsection is to give a State the opportunity to meet expeditiously and with less cost and administrative burden on the registrant the problem of registering for local use a pesticide needed to treat a pest infestation which is a problem in such State but is not sufficiently widespread to warrant the expense and difficulties of Federal registration."

SUMMARY

The combined research, education, and regulatory efforts of federal and state governments have made possible an impressive record of abundant production while imposing minimal harm to human health and the environment. Considering that approximately one billion pounds of pesticides are being applied in the United States annually to control about 2000 pest species, the safety record is remarkable. The increased public concern about environmental values and long-term health effects is constructive and is resulting in some of the regulatory changes discussed in this article. The Federal Environmental Pesticide Control Act can result in an improved and expanded joint effort by federal and state governments to limit the misuse of pesticides while permitting their use to facilitate increased production of food, feed, and fiber; provide protection from spoilage; improve health; and control nuisance insects and unwanted plants.

Implementation of current regulations, plus those in the process of being enacted under authority of the new federal act, will be complex and costly. Regulations will control the registration, labeling, distribution, storage, use and application, and disposal of pesticides.

A small segment of our society would like to have the use of all chemicals banned. A few individuals oppose all government controls. However, the vast majority of agricultural producer groups, representatives of industry, and the public support the need for pesticide regulations. Only through continued legal controls over pesticides may we hope to retain their use. Our society has gained tremendous benefits from the use of pesticides to prevent disease and to increase the production of food and fiber. Our need to use pesticides will continue to increase for the foreseeable future. Government agencies responsible for protecting human health and the environment must make sound judgments on an individual basis in evaluating the evidence concerning both the benefits and risks of using pesticides.

Literature Cited

1. Baker, E. R. 1973. *Digest of State Pesticide Use and Application Laws.* EPA Operations Division. 1a-55j
2. Federal Environmental Pesticide Control Act of 1972. Public Law 92-516. Amendment to Federal Insecticide, Fungicide, and Rodenticide Act of 1947
3. Federal Food, Drug, and Cosmetic Act, 52 Stat. 1040, 21 U.S.C. 301 et/seq. 1938. Miller Amendment, 68 Stat. 511; 21 U.S.C. 346a. 1954. As amended 1958. Delaney Clause, 21 U.S.C. 348(c)(3)(A)
4. Federal Insecticide Act, 36 Stat. 335; 7 U.S.C. 121. 1910
5. Federal Insecticide, Fungicide, and Rodenticide Act, 7 U.S.C. 135–135K. 1947
6. House of Representatives Conference Report No. 92-1540. October 5, 1972. Federal Environmental Pesticide Control Act
7. Jensen, J. H. 1969. *Report of Committee on Persistent Pesticides of the National Research Council's Division of Biology and Agriculture,* 1–34
8. Model Pesticide Use and Application Act. 1971. Council of State Government's Suggested State Legislation. 1971. 30:185–200
9. Mrak, E. M. 1969. *Report of the Secretary's Commission on Pesticides and Their Relationship to Environmental Health* U.S. Dep. Health, Education and Welfare
10. Senate Report No. 92-838. June 7, 1972. Report to Accompany H.R. 10729. 16
11. Suggested State Pesticide Control Act. 1973. The 1973–74 Official Publication of the Association of American Pesticide Control Officials, 20–39
12. Suggested State Pesticide Use and Application Act. 1973. Council of State Government's Suggested State Legislation. 1974. 33:80–100
13. Uniform State Insecticide, Fungicide, and Rodenticide Act. 1946. Council of State Government's Suggested State Legislation, Program for 1947. 1946. A89–109

NEUROSECRETION AND THE CONTROL OF VISCERAL ORGANS IN INSECTS [1,2]

❖6084

Thomas A. Miller
Division of Toxicology and Physiology, University of California,
Riverside, California 92502

Since Van der Kloot's review on neurosecretion in insects in 1960 (116), the structures of juvenile hormone and ecdysone, the molting hormone, have been published. As predicted by Van der Kloot these events have vastly increased the number of experimental approaches at hand and have expanded knowledge accordingly. In the interim, the subject of neurosecretion in insects and related animals has been reviewed on numerous occasions (5, 7, 26, 31, 38, 41, 61, 66, 67, 95, 105, 107, 119, 120). As of this writing there are no less than four reviews on various aspects of neurosecretion in insects which are completed or in preparation (34, 44, 68, 90).

Much of the work on neurosecretion has centered upon the release of neurohormones from neurohemal organs and their effect upon bodily processes. Hormonal control of visceral muscles and scattered reports of innervation of glands and organs by neurosecretory axons have not received as much research attention as other areas of neuroendocrinology, and only very recently have electrical properties of insect neurosecretory neurons been investigated in detail. These latter subjects will be reviewed below along with a few other selected topics in neurosecretion.

INTRODUCTION

Many nerve structures have been called neurosecretory because the concept of neurosecretion has been defined according to different criteria. Some prefer the more vague definition of neurosecretory cells as those showing cytological evidence of secretion and bearing resemblance to endocrine cells (66, 67, 116). Others are more restrictive by specifying that the neurosecretory axon must release material at a blood sinus structure and not be impinging directly on an effector cell (62). Still

[1]Survey of literature was completed in January 1974.

[2]Supported in part by grants from National Institute of Environmental Health Sciences.

others are guided largely by the presence of neurosecretory granules and define neurosecretory nerves as those containing membrane-bound electron-dense granules of a size generally between 1000 and 3000 Å in diameter (32). Motor axons in insects release neurohumors (also called neurotransmitters) at neuromuscular junctions. Where the distinction between an ordinary motor neuron and a neurosecretory neuron becomes unclear, the term neurosecretomotor has been applied (7, 96).

The difficulty of defining terminology was very adequately aired by Bern (7). Knowles (61) generalized the categorization of neurosecretory fibers into Type A fibers which are thought to secrete peptides and are characterized by electron-dense granules of diameter greater than 1000 Å. Type B fibers contain monoamines in abundance with electron-dense granules usually smaller than 1000 Å diameter. While both types of nerve fiber have been termed neurosecretory, evidence is accumulating that the B-type fiber is involved in neurosecretomotor structures in insects.

To date there is a vast profusion of types of granules reported in neurosecretory axons. These are summarized for corpora cardiaca by Goldsworthy & Mordue (44). Authors have reported some difficulty in classing neurosecretory materials according to a central scheme. As a result many species contain unique labels for neurosecretory materials (43, 109, 115).

Histological staining of neurosecretory neurons or their ultrastructural appearance must now be considered with due caution. Gabe (39) found that materials in the same axon stained differently in different regions. Normann (86) considered different ultrastructural appearance of granules to be partly due to osmotic stress. And very recently, the appearance of granules in *Carausius morosus* was found to change drastically on starvation (34).

It appears, therefore, as though slightly different fixation or handling procedures might produce granules of differing electron density in one species. In fact, different research groups have described different granules for structures in the same animal (14, 16). Given these problems of inconsistency, perhaps the final appearance of the granules is less important when compared between species, as it is in comparing the granules in axons from the same section of tissue.

GENERAL

The best understood neurosecretory system in insects is the brain and corpora cardiaca complex, probably because it is amenable to surgical manipulation. Since this complex plays a vital role in growth and metamorphosis, cauterization or implantation can produce dramatic effects (120). Except for primitive orders (38), the brain of insects always contains medial neurosecretory neurons associated with the pars intercerebralis. The medial cells give axons off which track via nervi corpori cardiaci (NCC I) to the corpora cardiaca, a neuroendocrine ganglion. Medial cell axons are also believed to traverse the cardiaca and enter the corpora allata, organs of non-nervous origin which are responsible for the manufacture of juvenile hormone (120). The medial cells may also give branches to the aorta where in Hemiptera, an extensive neurohemal organ is formed which is thought to represent that neurohemal structure ordinarily found in corpora cardiaca of other insects (115).

A second group of cells, the lateral neurosecretory cells, are also located in the protocerebrum. The lateral cells may be interconnected with the medial cells, but they do send branches to the corpora cardiaca along identified tracts (NCC II). While the lateral neurosecretory cells have long been associated with the protocerebral system involved in molting, their exact role has largely remained unresolved (cf. 44, 106, 109, 116).

Besides units in the brain and corpora cardiaca, neurosecretory cells are thought to occur in the ventral ganglia (27, 116, 120), subesophageal ganglia (93), frontal ganglia (10), ingluvial ganglia (3, 17), and the median (unpaired) nervous system which has been classified as a special neurohemal structure, termed by Raabe the perisympathetic organs (11, 32, 98, 117). (Vincent uses the spelling parasympathetic organ.)

The segmental nerves of some Orthoptera carry granule-filled axons to the lateral cardiac nerve cords where extensive neurohemal organs are situated (57, 77, 114). In addition the segmental vessels of *Periplaneta americana* appear to be a connective tissue matrix impregnated with neurosecretory axon endings forming release sites (T. A. Miller, unpublished observations; T. M. Beattie, personal communication; 76).

Not all insects have segmental vessels (91) nor lateral cardiac nerve cords, and it is unlikely that the cardiac neurohemal organs are universal among insects. However, these extensive structures have been overlooked in the past and they undoubtedly contribute to the overall neurohormonal milieu in the hemolymph. In particular, estimates of neurohormonal flux from corpora cardiaca alone risk ignoring a substantial contribution from peripheral neurosecretory structures, if the same material is elaborated from both. This is especially true of *P. americana, Locusta migratoria,* and *Schistocerca gregaria.*

There will undoubtedly be more neurosecretory structures described in the future which compliment those already described. Of special significance to be mentioned here are the neurosecretory neurons of the optic lobes of *P. americana* (6, 30), the cells described in the ocellar nerve of *Sarcophaga falculata* (108) which are thought to control postemergence growth, and the peripheral neurons reported by Finlayson & Osborne (32). The peripheral region of the caudal nervous system (120), including the posterior alimentary canal, has been relatively untouched in this regard.

Assigning or demonstrating a physiological role for the release of neurohormones from many of the neurohemal organs now known is difficult. Surgical procedures on the perisympathetic organs especially would be painstaking in many cases. Extracts of these organs show equivocal results for bursicon activity in orthopteroid median nervous system (117). Raabe et al (97) have reported diuretic hormone and cardioaccelerating properties for perisympathetic organs of insects.

While fairly amenable to surgical operation, a role for neurosecretory cells in the frontal ganglion is also difficult to establish if indeed one exists. Two of the largest neuron cell bodies (40 to 45 μ diameter) were considered to be neurosecretory of the A type, synthesizing protein-filled granules in the frontal ganglion of *Manduca sexta* (10). A cycle of synthesis was described; and immature granules accumulated in these cells in diapause pupae, as might be expected.

Reports of the lack of neurosecretory cells in the frontal ganglion of other insects, which are based upon histochemical evidence, must be carefully considered in light of the number of neurosecretory structures which do not stain with many of the common staining procedures (44). However, it is premature to assume that neurosecretion occurs in the frontal ganglion of all insects. While a possible role of neurosecretory units in the frontal ganglion is not obvious, a link between the frontal ganglion and neurohormonal activity in the protocerebrum has been inferred (94). This may be largely due to the frontal ganglion simply lying astride sensory afferents from the foregut sensory receptors which supply higher centers with osmotic information, for example.

Frontal ganglionectomy in insects normally causes a restriction in food passage along the gut (20, 29). This can be accompanied by reduction or inhibition in egg development in adult females. In *Locusta* which are frontal ganglionectomized at the fourth or fifth instar, the protocerebral neurosecretory cells of young adults appear normal. Neurosecretory material piles up in the NCC I nerve pathway from the medial neurosecretory cells to the corpora cardiaca from which release is reduced, and the corpora allata degenerate (19). There is a reduced protein synthesis and growth ceases.

In general, some of the conditions in frontal ganglionectomized animals resemble those produced by starvation, and in fact many of the effects could be attributed to a reduction in the supply of food and nutrients which would be expected if the frontal ganglion were a major motor and relay center for the gut. Injection of corpora cardiaca extracts into frontal ganglionectomized locusts reverses many of the changes induced by the surgery (19). Evidently the frontal ganglion enters the neuroendocrinological interplay of bodily processes in some way.

A situation similar to the frontal ganglion also holds for the ingluvial ganglion. The ingluvial ganglion in *Blabera craniifer* (17) and *Periplaneta americana* (3) contains ordinary neurons and neurons with electron-dense granules suspected of being neurosecretory. The stomatogastric system has been shown to stain positive for aminergic neurons (18, 60). Therefore it is reasonable to assume that some of the neurosecretory neurons in the ingluvial ganglion may innervate visceral muscles in the foregut in analogy with several other cases (see section on neurosecretomotor innervation). Chanussot concluded that the ingluvial ganglion possessed neurohemal and relay functions (17).

Where neurosecretion and neurohemal organs were once considered to be mostly anterior, concentrated in the hypocerebral complex, any new theories of the interplay of neurohormones and their receptors must take into account the possibility of many release sites distributed throughout the body. Furthermore, neurosecretory materials may be transported considerable distances in their axons and released locally near target tissues (58, 110).

NEUROSECRETOMOTOR INNERVATION

The hearts of *Periplaneta americana* (57) and *Calliphora erythrocephala* (87) are innervated by granule-filled axons of at least two types. Normann (87) suggested that one of these may be inhibitory innervation, but the ultrastructural appearance of the other suggests aminergic innervation of the B type. The alary muscles of *P. americana* (1) and *Cyclocephala pasadenae* (74, 76), the segmental vessel valve muscle (76), and the hyperneural muscle (75) of *P. americana* are also innervated by granule-filled axons whose electron-dense granules of less than 1500 Å diameter again suggest B-type fibers. Chemical neuromuscular transmission appears unequivocally demonstrated for the fibers innervating heart and hyperneural muscle of *P. americana* (75, 76). Spontaneous miniature postsynaptic potentials, neurally evoked excitatory postsynaptic potentials, and synaptic delays were recorded. Similar neurally evoked junctional potentials were reported in rectal longitudinal muscles of *P. americana* (80, 81).

Axons containing electron-dense granules typical of B fibers were found innervating the muscles investing the spermatheca of *P. americana* (48) and *Sitophilus granarius* (112). In addition, granule-filled axons were found closely apposed to the gland surface of *P. americana,* but not in *Sitophilus.*

The rectum of the blowfly, *C. erythrocephala,* is innervated from the thoracic ganglion mass. Some of the axons contain electron-dense granules of a diameter between 1000 and 3000 Å, while others were 1000 to 1500 Å in diameter (47). The neurosecretory axons branched and terminated near the medulla of the rectum, but no synaptoid structures or synaptic vesicles were reported. The axons were thought to control rectal function by release of antidiuretic hormones, presumably consistent with A-type fibers and following the pattern suggested of transport of materials to target organs (58).

Similar axons were reported to innervate the epidermis of *Rhodnius prolixus* in a solitary report (65). The epidermal glands of some insects were shown to receive neurosecretory axons (84a, 96). Granule-filled axons were found to form release sites near the skeletal muscles of *Schistocerca gregaria, Carausius morosus,* and *Phormia terrae-novae* (92). No synaptoid structures were observed, but apparent synaptic vesicles were present, as well as electron-dense granules generally less than 1500 or 1600 Å in diameter. This would qualify as a B-type fiber according to Knowles (61). These axons were suggested as contributing a trophic effect to the muscles or tissues near the release sites (92). Axons containing electron-dense granules of diameter less than 2500 Å were found to innervate the ventral intersegmental muscles of *R. prolixus* at ordinary synaptic junctions. The same axons also formed release sites with apparent exocytosis of the granules adjacent to the hemolymph (3a).

Hoyle et al (55a) have recently reported a dorsal unpaired median neurosecretory neuron in the metathoracic ganglion of *Schistocerca gregaria.* Branches descend to those muscle fibers in the left and right extensor tibiae which receive only fast innervation. It was concluded that this unit has a trophic function since stimulation caused no obvious change in the ordinary stimulus response behavior of the fast motor units. Again no actual synaptoid connections were found between the neurosecretory unit and the muscle fibers.

The midgut and hindgut of insects have variously been reported to receive granule-filled axons from the last abdominal ganglion (12, 84) and from the corpora cardiaca directly (56). In the cockroach specifically, axons of the B type innervate the rectal longitudinal muscles, where characteristics of chemically transmitting neurosecretomotor junctions seem strongly indicated (80).

The rectal longitudinal muscles of *Periplaneta americana* are said to be innervated by axons containing electron-dense granules on the order of 2000 Å in diameter (Brown and Graham, cited in 12). Extracts of the foregut or hindgut of *P. americana* or of nerves supplying these areas of the gut are said to duplicate neurally evoked contractions when assayed onto the whole hindgut or the rectum of *P. americana* (12). Two active fractions sedimented by centrifugation at 1000 G for 10 min and 127,000 G for 15 min were thought to contain a visceral neurotransmitter, but neither 5-HT, acetylcholine, adrenaline, norodrenaline, γ-aminobutyric acid, nor glutamate behaved in assay in a manner consistent with their participation (12).

More recently Holman & Cook (53) isolated four active substances from the midgut of *Leucophaea maderae* and *P. americana,* including glutamate and aspartate. They contended that Brown's fraction contained a combination of factors which complicated his bioassay responses. However, Brown (12) reported that glutamate did not duplicate neurally evoked contraction of the hindgut preparation of *P. americana,* while Holman & Cook (53) reported that glutamate or aspartate did produce contractions similar to neurally evoked responses in hindgut of *L. maderae.* Obviously more detailed work is called for on neural control of insect gut movement.

Axons with electron-dense granules have been reported innervating the salivary glands of *P. americana* (118), *Nauphoeta cinerea* (54), and *Manduca sexta* (64). In these cases the axons resemble B-type fibers. House has reported a slow, large hyperpolarizing potential in the gland cell of *N. cinerea* salivary gland in response to single shocks of the axons running to the gland along the salivary duct (54). A rather long delay of 1 sec followed the shock before responses were recorded which is consistent with similar measurements from salivary glands of other animals.

Spontaneous hyperpolarizations of the gland cells whose time course duplicated neurally evoked responses were also recorded intracellularly (54). While the spontaneous potentials resemble the normal release of synaptic neurotransmitter qualitatively, the time intervals between events were not entirely random. However, perfect randomness may not be a realistic expectation for spontaneous neurotransmitter release (99), and the analogy to ordinary synaptic transmission is compelling.

Corpora allata in *Oncopeltus fasciatus* are innervated by neurosecretory fibers of unknown origin (115). The allata also receive ordinary fibers which do not contain granules, but no synaptoid structures were observed. The corpora allata of *Calliphora erythrocephala* were found to be innervated by axons containing electron-dense granules of diameters less than 2000 Å and synaptic vesicles (85, 111). Synaptic structures were observed with some evidence for exocytosis.

Neurosecretory neurons of the B type are described in the hypocerebral ganglion of *Schizodactylus monstrosus* (59). It is not clear how the axons of these cells are disposed. The intrinsic cells of the corpora cardiaca of *C. erythrocephala* are innervated by axons containing electron-dense granules of a diameter between 800 and 1400 Å (86). The intrinsic cells contain granules up to 3000 Å diameter so the innervating cells are distinctive. The presynaptic axons were thought to be B fibers whose cell bodies reside in the brain (86).

The type PIC IV cells (pars intercerebralis IV) in the brain of *O. fasciatus,* which also apparently occur in the metathoracic ganglion mass of *Rhodnius* (115), resemble B-type granules of vertebrates (61). These axons in *O. fasciatus* enter the neuropile of the brain where they make synaptic contacts.

Neurons staining positive with the fluorescence technique occur in the brain and subesophageal ganglion of *P. americana* (18, 36, 37, 71). These are thought to contain catecholamines. Chanussot (17) found neurosecretory granules in neurons of the ingluvial ganglion of *Blabera craniifer.* Certain of these cells give histochemical evidence of monoamines (18). Histochemical evidence for monoaminergic innervation of foregut in *Schistocerca gregaria* was also reported (60).

5-Hydroxytryptamine (5-HT) is known to increase the rate of rhythmic contraction of cockroach and locust hearts (74) and cockroach foregut (21); to increase the rate of coiling and uncoiling of Malpighian tubules of cockroach and *Carausius morosus,* but not locust (35); to increase the flow of Malpighian tubule secretion in *Rhodnius* and *C. morosus* (70); to duplicate the neurally evoked responses of *Nauphoeta cinerea* salivary glands (55); to induce secretion in the salivary glands (7, 118); and to inhibit the activation of phosphorylase in the ventral nerve cord of *Periplaneta americana* (49).

5-HT is active in pharmacological concentrations, in most cases 10^{-7} *M* being needed to induce the stated effect. A role for 5-HT as a neurohormone or neurotransmitter is suggested by these responses, and by the fact that 5-HT has been found in many insect nervous tissues (52). However, 5-HT has not been found to play a role in physiological processes as yet. 5-HT is not present in detectable amounts in extracts of corpora cardiaca of cockroach (82) and the cardiaca do not fluoresce when treated for demonstration of biogenic amines (82). 5-HT is not present among extracts showing diuretic hormone activity in *Rhodnius* (4). Pharmacological responses of salivary glands in *Nauphoeta* are more consistent with dopamine receptors than 5-HT receptors (9, 55). 5-HT acts on the Malpighian tubule muscle of *P. americana* to increase the rate of writhing, although the area of the tubule affected is known not to be innervated (25). These examples rather suggest nonspecific effects by 5-HT upon visceral organs and a neuromuscular transmitter role for 5-HT will be difficult to clarify in view of the actions on Malpighian tubule.

The uptake of tritiated noradrenaline and 5-HT into neurons of the corpora cardiaca of *Locusta migratoria* from saline was autoradiographically found to be associated with three types of axon based on the appearance of granules (15). Tritiated amines specifically localized in neurosecretory axons may reflect a nonspecific sequestration since certain nervous structures are known to accumulate supra-vital stains (Miller, unpublished observations).

Little can be concluded from these uptake experiments alone. The Victoria blue preferential staining of cystine–cysteine could mean these sulfhydryl moieties are part of the carrier protein material in A fibers (72, 79). Similarly, tritiated amines may represent an interaction with carrier structures. The inference that neurosecretory granules constantly sequester amines during axoplasmic transport implies that the granules are incomplete upon manufacture at the Golgi complex in the perikarya or that they are in equilibrium with the constituents of the hemolymph. One wonders how many different types of amines can be sequestered by neurosecretory neurons and whether these amines once sequestered can be released by potassium-induced depolarization of neurohemal terminals (4, 45, 68, 69, 89).

In addition to the cases of innervation described above, axons leave the corpora cardiaca of *Rhodnius* and terminate on pericardial cells (C. G. H. Steel, unpublished observations). Other than noticing that the pericardial cells show cytological changes which are correlated with cycles in the neurosecretory cells of the brain, a role for this innervation is unknown.

A neuron containing electron-dense granules near Golgi apparati and in its axons was described as attached to the abdominal fat body of *C. morosus* (32). Axon terminations in the fat body were described, but no single feature provided compelling evidence for categorizing this as a neurosecretory neuron (32).

In *Leucophaea madera* the proctodeal nerve contains large neurosecretory cells located just at the bifurcation of the nerve near the rectal sphincter (B. J. Cook, personal communication).

In *Calliphora erythrocephala* the thoracic glands were found to be innervated by at least two types of axon distinguishable by their axoplasmic granules (85). One of the axon profiles was chacteristic of those A cells described for medial neurosecretory cells or intrinsic neurons of the corpora cardiaca. The other axon profiles, containing electron-dense granules of 1000 Å or less diameter (or more similar to B fibers), were not characteristic of cells in the corpora cardiaca nor medial brain group, and the origin of these fibers is unknown (85). Like many of the reports of innervation of the visceral organs of insects the description reviewed above is from a study not directly concerned with the thoracic glands. This has occurred with other viscera (24, 84a, 96, 122).

The identity of neurosecretory materials most naturally falls into two categories: those materials released to the hemolymph from neurohemal organs and other chemicals released near organs or at neurosecretomotor junctions. The former materials, normally considered to be peptides (44), are typified by recent work on the diuretic hormone of *Rhodnius.* Using improved methods of collecting neurohemal products with high potassium treatment, Aston & White (4) have found an extremely labile component of less than 2000 mol wt. Large molecular weight fragments with diuretic activity which are extracted from the metathoracic ganglion mass are not found in the blood of fed *Rhodnius* when diuretic hormone is normally present, nor in high potassium perfusates of the ganglion mass which is known to release diuretic hormone (4). It is possible that the large molecular weight fractions are particulate or bound neurohormones.

The second category of materials—those involved in neurally evoked stimulation of organs or visceral muscles—is quite distinct from the former peptide chemicals, and indeed this distinction may eventually serve to unravel the meaning of the term neurosecretion. Candidate neurotransmitters in insects include 5-HT, noradrenaline, dopamine, synephrine, and octopamine, primarily, because these molecules are either known to be present or insect tissues have the ability to synthesize the compounds from their precursors (28, 52, 63, 83, 95, 101, 113).

Since it is important in cuticular hardening and darkening, dopamine is thought to be present in high concentration at various times during growth. However, since it is not entirely clear yet where the tanning precursors are synthesized, the actual concentration of dopamine in the blood may be minimal, with only tyrosine and N-acetyl dopamine being exposed to the hemolymph (44).

Adrenergic drugs induced luminescence in the extirpated lantern organ of firefly larvae of the genus *Photuris.* The order of potency was synephrine > octopamine > adrenaline > noradrenaline > dopamine and tyramine (13). Amphetamine and high potassium saline induced intense luminescence. Prior reserpine injection reduced these responses greatly. By virtue of similarities between the action of synephrine in production of a glow in the light organ and neurally evoked glowing, synephrine was implied as a possible neurotransmitter.

Several recent reports indicated that cyclic-AMP (adenosine 3':5'-cyclic monophosphate) may play a role in the action of neurohormones in insects (8, 100). This is based on the wider concept of the emerging role of cyclic-AMP in directing cellular activities in animals (46). Components of the cyclic-AMP system are present in the nervous systems of insects in the few reports which have appeared (102, 103).

Octopamine, dopamine, and 5-HT each activate specific and different adenylate cyclases in breis of the thoracic ganglion of *Periplaneta americana* (83). This is assumed to be indirect evidence that the three amines are neurotransmitters in the cockroach central nervous system (83). If biogenic amines are serving as central neurotransmitters, their role at the visceral neurosecretomotor junctions takes on added importance.

CARDIOACCELERATORS

There has been an assumption in certain work over the past 20 years that neurohormones are released from neurohemal organs (especially the corpora cardiaca), are circulated, and ultimately cause an overall increase in heartbeat rate (26). The hypothetical phenomenon of an increase in heartbeat brought about by neurohormones released from one central location has been termed cardioacceleration and the materials involved are cardioaccelerators. In use these terms have implied a real physiological function.

The strongest evidence for their role in insects comes from the semi-isolated heart preparation, and *P. americana* has been used extensively in the bioassay of cardioaccelerators. The difficulties of interpreting the mode of action of agents on the semi-isolated heart of *P. americana* have been reviewed in depth elsewhere (74).

The key problem appears to be in exactly how much inference may be drawn from in vitro responses of the semi-isolated heart preparation. Mordue & Goldsworthy (79) reported that the glandular and storage lobes of the corpora cardiaca of *Locusta* produced 53 and 37% increase in heartbeat rate, respectively, in locust semi-isolated heart preparations. The amount applied was equivalent to 2 pairs of corpora cardiaca. Extracts injected into the whole animal did not cause equivalent effects even when correcting for dilution. The responses of the semi-isolated heart were considered to be artifactual, since the amount of in vitro heartbeat increase brought about by a given amount of corpora cardiaca extract was disproportionately greater when compared to the effects of the same amount of extract on amaranth excretion through Malpighian tubules or activation of phosphorylase in the fat body (79).

Hertel (51) reported that 0.1 to 0.2 pairs of corpora cardiaca, when extracted and injected into the abdomen of whole *P. americana,* caused an increase in heartbeat rate. He found that the effect was not consistent from one animal to another, but always produced stimulation or heart block (Herzblock). Regardless of the response, the effect lasted only 10 min after which the heartbeat rate returned to near the original value. It must be assumed that saline injections caused no such effect; however, *P. americana* is known to respond to disturbances by a brief increase in heartbeat rate under ordinary circumstances.

These responses to injected extracts of corpora cardiaca were not found by W. Mordue (personal communication) who obtained no response of heartbeat upon injection of up to the equivalent of 1 pair of the glands into *P. americana, Schistocerca gregaria,* or *Locusta migratoria.* Roussel & Cazal (104) reported a 14% increase in heartbeat of *L. migratoria* which persisted up to 2 hr following injection of cardiaca extract. Significantly, however, both cardiaca extracts and control injections of muscle extract caused an immediate peak of heartbeat rate increase over the first 15 min with return of the heartbeat rate to original value in an hour following injection; 25 μ were injected into a reported blood volume of 125 μ.

At present, the case for neurohormonal control of heartbeat seems to fall back on equivocal evidence in *P. americana,* and the early work on *Chaoborus* larvae (40). *Calliphora erythrocephala* does not appear to possess a neurohormone acting as a cardioaccelerator (87). The intimate relationship between heartbeat and temperature control in *Manduca sexta* shown by Heinrich (50) seems to preclude any hormonal control with its obvious time lags. If other insects are found to have an intimate relationship between circulation and temperature control, then perhaps they too must rely exclusively on the more rapid response provided by nervous control alone.

While there is no substantial measurement in whole insects which demonstrates cardioacceleration, this is not to imply that some form of hormonal control of heartbeat will not be found in insects. The effects of extracts on the activity of semi-isolated insect hearts have so far been valuable in distinguishing between various fractions as pointed out by Goldsworthy & Mordue (44). However, the response of the semi-isolated heart cannot be used by itself to assume a role in the whole animal, and in fact, there is more evidence against control of heartbeat by circulating neurohormones than there is in favor of such a scheme.

ELECTRICAL PROPERTIES OF NEUROSECRETORY CELLS

For some time now, neurosecretory cells have been thought to have the properties of ordinary neurons, that is, to conduct impulses, respond to synaptic input, and perhaps synapse onto other cells. In particular, release of neurohormones has been thought to be correlated with electrical activity in the neurosecretory axon (85, 116).

The electrical properties of medial neurosecretory cells (22), of intrinsic neurons of the corpora cardiaca (88), of perisympathetic organs (33), and of the lateral cardiac neurohemal organs (78) have been investigated comparatively recently. These studies and others (41, 42, 45) have confirmed the previous contentions that release of neurohormones can be induced by neurally evoked stimulation. Indeed high potassium, which depolarizes ordinary neurons, also causes depletion of neurosecretory material (4, 45, 69, 89), with some dependence on calcium.

Attempts to record intracellular potentials from medial neurosecretory cells of *Periplaneta americana* met with success in 4% of the penetrations attempted. Of the successful recording, a potential of about 20 mV was found across the sheath layer and cellular membrane potentials of up to 50 mV were recorded (22). The largest cellular potentials did not coincide with spontaneous activity, whereas transcellular potentials of around 30 mV sometimes exhibited spontaneous overshooting potentials of 50 mV absolute amplitude. There was some uncertainty about whether recordings were taken from cell bodies or axons (22), and changes in ambient light showed some effect on the activity recorded from neurosecretory cells. However, the amount and type of light was not described.

Depolarizing pulses caused an increase and hyperpolarizing pulses a decrease in spontaneous firing rate of some neurosecretory cells, hinting at the presence of electrically excitable pacemaker mechanisms (22). Records suggest generator potentials are present, also indicating a pacemaker (121).

The excitable membrane in these cases is not necessarily the cell body. In fact the somata of neurons in the corpora cardiaca of *C. erythrocephala* appear to be electrically inexcitable (88). The shape of recorded potentials gave an indication of the position of recording microelectrodes. Biphasic potentials suggested that action potentials partially invaded the cell somata, but other monophasic records suggested electronic spread of residual action potential activity across the somata (121).

Inhibitory postsynaptic potentials were recorded or implied by the behavior of endogenous spike trains. By arguments of the geometry of the neuron model from Adiyodi & Bern (2), it was concluded that inhibitory (or excitatory) synaptic inputs to the neurosecretory cell occur near the perikarya (121).

When recorded intracellularly (22, 88) or extracellularly (33, 78) impulses in neurosecretory axons appear to be of slightly longer duration, smaller amplitude, and lower propagation velocity when compared to ordinary nervous impulses. Action potentials recorded from the neurosecretory cells of the medial group from the brain of *Sarcophaga bullata* were 2–20 msec in duration compared with durations of 1 msec recorded from ordinary brain neurons (121). Propagation velocities of 0.5–0.25 m/sec have been recorded for crab neurosecretory axons (23) and similar values were reported for *P. americana* segmental neurosecretory axons and lateral cardiac nerve cords (78).

These properties may be partly due to the extremes in diameter which are characteristic of neurosecretory axons. Normann (88) measured a mean diameter of 0.81 μ from 564 measurements of axons in the corpora cardiaca of *Calliphora erythrocephala.* Smaller diameters of the varicosities in neurosecretory axons could collectively contribute to a high internal resistance and thus a low propagation velocity.

Neurosecretory neurons may be filled with cobalt for neuroanatomical mapping in analogy to ordinary neurons (73). Thus the similarity between ordinary neurons and neurosecretory neurons is further strengthened.

CONCLUSIONS

Neurosecretion means different things to different people. As a broad generality based on the ideas of Knowles (61), those cells forming neurohemal organs in insects are most often associated with A fibers which are thought to release peptides. Many of the axons innervating visceral muscles and organs belong to the B fiber category which often stain for the presence of aromatic amines.

The latter class of neurosecretory neurons deserves the imperfect name neurosecretomotor, and probably does not belong to the classical concept of neurosecretory neurons. Rather, it may perform a role intermediate between that of an ordinary motor neuron and a neuron performing a neurohemal role. Undoubtedly other categories of neurosecretory neurons will emerge as more is learned of the insect vegetative nervous system.

Very little is known about the possible hormonal control of the activity of visceral muscles including the heartbeat. Much evidence for cardioaccelerators is indirect, based on the semi-isolated heart preparation.

Acknowledgments

The preparation of this review was aided immeasurably by W. Mordue and G. J. Goldsworthy, T. C. Normann, R. J. Aston and A. F. White, and E. P. Marks who provided copies of papers or reviews in press. The author is grateful to I. Graham-Bryce of Rothamsted Experimental Station and Leo Stones and Paul Capstick of the Wellcome Research Laboratories, Berkhamsted for providing facilities.

Literature Cited

1. Adams, M. E., Miller, T., Thomson, W. W. 1973. Fine structure of the alary muscles of the American cockroach. *J. Insect Physiol.* 19:2199–2207
2. Adiyodi, K. G., Bern, H. A. 1968. Neuronal appearance of neurosecretory cells in the pars intercerebralis of *Periplaneta americana* (L.). *Gen. Comp. Endocrinol.* 11:88–91
3. Aloe, L., Levi-Montalcini, R. 1972. Interrelation and dynamic activity of visceral muscle and nerve cells from insect embryos in long-term cultures. *J. Neurobiol.* 3:3–23

3a. Anwyl, R., Finlayson, L. H. 1973. The ultrastructure of neurons with both a motor and a neurosecretory function in the insect, *Rhodnius prolixus. Z. Zellforsch. Mikrosk. Anat.* 146:367–74

4. Aston, R. J., White, A. F. 1974. Isolation and purification of the duiretic hormone from *Rhodnius prolixus. J. Insect Physiol.* In press
5. Barth, R. H., Lester, J. J. 1973. Neurohormonal control of sexual behavior in insects. *Ann. Rev. Entomol.* 18:445–72
6. Beattie, T. M. 1971. Histology, histochemistry, and ultrastructure of neurosecretory cells in the optic lobe of the cockroach, *Periplaneta americana. J. Insect Physiol.* 17:1843–55
7. Bern, H. A. 1966. On the production of hormones by neurons and the role of neurosecretion in neuroendocrine mechanisms. *Soc. Exp. Biol. Symp.* XX:325–44
8. Berridge, M. J., Prince, W. T. 1972. The role of cyclic AMP and calcium in hormone action. *Advan. Insect Physiol.* 9:1–49
9. Bland, K. P., House, C. R., Ginsborg, B. L., Laszlo, I. 1973. Catecholamine

transmitter for salivary secretion in the cockroach. *Nature New Biol.* 344: 26-27
10. Borg, T. K., Bell, R. A., Picard, D. J. 1973. Ultrastructure of neurosecretory cells in the frontal ganglion of the tobacco hornworm, *Manduca sexta* (L.) *Tissue Cell* 5:259–67
11. Brady, J., Maddrell, S. H. P. 1967. Neurohaemal organs in the medial nervous system of insects. *Z. Zellforsch. Mikrosk. Anat.* 76:389–404
12. Brown, B. E. 1967. Neuromuscular transmitter substance in insect visceral muscle. *Science N.Y.* 155:595–97
13. Carlson, Albert D. 1972. A comparison of transmitter and synephrine of luminescence induction in the firefly larva. *J. Exp. Biol.* 57:737–43
14. Cassier, P., Fain-Maurel, M. A. 1970. Contribution a l'étude infrastructurale due systeme neurosecreteur retrocerebral chez, *Locusta migratoria migratorioides* (R. et F.). I. Les corpora cardiaca. *Z. Zellforsch. Mikrosk. Anat.* 111: 471–82
15. Cazal, M. L., Calos, A., Bosc., S. 1973. Capture et retention de monoamines tritiees dans les corpora cardiaca de *Locusta migratoria* L. étude in vitro par radioautographie à haute résolution. *J. Microsc.* 17:223–26
16. Çazal, M., Joly, L., Porte, A. 1971. Étude ultrastructurale des corpora cardiaca et de quelques formations annexes chez *Locusta migratoria* L. *Z. Zellforsch. Mikrosk. Anat.* 114:61–72
17. Chanussot, B. 1972. Étude histologique et ultrastructurale du ganglion ingluvial de *Blabera craniifer* Burm. *Tissue Cell* 4:85–97
18 Chanussot, B., Dando, J., Moulins, M., Laverack, M. S. 1969. Mise en évidence d'une amine biogène dans le système nerveux stomatogastrique des insectes. Étude histochimique et ultrastructurale. *C. R. Acad. Sci. Ser. D* 268: 2101–4
19. Clarke, K. U., Langley, P. A. 1963. Studies on the initiation of growth and moulting in *Locusta migratoria migratorioides,* R. & F. IV. The relationship between the stomatogastric nervous system and neurosecretion. *J. Insect Physiol.* 9:423–30
20. Clarke, K. U., Anstee, J. H. 1971. Effect of the removal of the frontal ganglion on cellular structure in *Locusta. J. Insect Physiol.* 17:929–43
21. Cook, B. J., Eraker, J., Anderson, G. R. 1969. The effect of various biogenic amines on the activity of the foregut of the cockroach, *Blaberus giganteus. J. Insect Physiol.* 15:445–55
22. Cook, D. J., Mulligan, J. V. 1972. Electrophysiology and histology of the medial neurosecretory cells in adult male cockroaches, *Periplaneta americana. J. Insect Physiol.* 18:1197–1214
23. Cooke, I. M. 1964. Electrical activity and release of neurosecretory material in crab pericardial organs. *Comp. Biochem. Physiol.* 13:353–66
24. Crossley, A. C., Waterhouse, D. F. 1969. The ultrastructure of a pheromone-secreting gland in the male scorpion-fly *Harpobittacus australis. Tissue Cell* I:273–94
25. Crowder, L. A., Shankland, D. L. 1972. Pharmacology of the Malpighian tubule muscle of the American cockroach, *Periplaneta americana. J. Insect Physiol.* 18:929–36
26. Davey, K. G. 1964. The control of visceral muscles in insects. *Advan. Insect Physiol.* 2:219–45
27. Delphin, F. 1965. The histology and possible functions of neurosecretory cells in the ventral ganglia of *Schistocerca gregaria* Forskal. *Trans Roy. Entomol. Soc. London* 117:167–214
28. Dewhurst, S. A., Croker, S. G., Ikeda, K., McCaman, R. E. 1972. Metabolism of biogenic amines in *Drosophila* nervous tissue. *Comp. Biochem. Physiol.* 43B:975–81
29. Dogra, G. S. 1967. Studies on the neurosecretory system and the functional significance of the NSM in the aorta wall of the bug, *Dysdercus koenigii. J. Insect Physiol.* 13:1895–1906
30. Elofsson, R., Klemm, N. 1972. Monoamine-containing neurons in the optic ganglia of crustaceans and insects. *Z. Zellforsch. Mikrosk. Anat.* 133:475–99
31. Engelmann, F. 1970. *The Physiology of Insect Reproduction.* New York: Academic
32. Finlayson, L. H., Osborne, M. P. 1968. Peripheral neurosecretory cells in the stick insect (*Carausius morosus*) and the blowfly larva (*Phormia terrae-nova*). *J. Insect Physiol.* 14:1793–1801.
33. Finlayson, L. H., Osborne, M. P. 1970. Electrical activity of neurohaemal tissue in the stick insect, *Carausius morosus. J. Insect Physiol.* 16:791–800
34. Finlayson, L. H., Osborne, M. P. 1974. Secretory activity of neurons and related electrical activity. *Advan. Comp. Physiol. Biochem.* In press
35. Flattum, R. F., Watkinson, I. A., Crowder, L. A. 1973. The effect of insect "Autoneurotoxin" on *Periplaneta*

americana (L) and *Schistocerca gregaria* (Forskal) malpighian tubules. *Pestic. Biochem. Physiol.* 3:237–42
36. Frontali, N. 1968. Histochemical localisation of catecholamines in the brain of normal and drug treated cockroaches. *J. Insect Physiol.* 14:881–86
37. Frontali, N., Häggendal, J. 1969. Noradrenaline and dopamine content in the brain of the cockroach *Periplaneta americana. Brain Res.* 14:540–42
38. Gabe, M. 1966. *Neurosecretion.* London: Pergamon
39. Gabe, M. 1972. Données histochimiques sur l'évolution du produit de neurosécrétion protocéphaligne des insectes ptérygotes au cours de son cheminement axonal. *Acta Histochem.* 43(1):168–83
40. Gersch, M. 1958. Neurohormonale Beeinflüssung der Herztätigkeit bei der Larve von *Corethra. J. Insect Physiol.* 2:281–97
41. Gersch, M. 1969. Neurosecretory phenomena in invertebrates. *Gen. Comp. Endocrinol. Suppl.* 2:553–64
42. Gersch, M. 1972. Experimentelle Untersuchungen zum Freisetzungsmechanismus von Neurohormonen nach elektrischer Reizung der Corpora Cardiaca von *Periplaneta americana in vitro. J. Insect Physiol.* 18:2425–39
43. Gillott, C., Yin, G. M. 1972. Morphology and histology of the endocrine glands of *Zootermopsis angusticollis* Hagen. *Can. J. Zool.* 50:1537–45
44. Goldsworthy, G. J., Mordue, W. 1974. Neurosecretory hormones in insects. *J. Endocrinol.* In press
45. Gosbee, J. L., Milligan, J. V., Smallman, B. N. 1968. Neural properties of the protocerebral neurosecretory cells of the adult cockroach *Periplaneta americana. J. Insect Physiol.* 14: 1785–92
46. Greengard, P., McAfee, D. A. 1972. Adenosine 3',5'-cyclic monophosphate as a mediator in the action of neurohumoral agents. *Biochem. Soc. Symp.* 36:87–102
47. Gupta, B. L., Berridge, M. J. 1966. Fine structural organisation of the rectum in the blowfly, *Calliphora erythrocephala* (Meig.) with special reference to connective tissue, tracheae and neurosecretory innervation in the rectal papillae. *J. Morphol.* 120:23–82
48. Gupta, B. L., Smith, D. S. 1969. Fine structural organisation of the spermatheca in the cockroach, *Periplaneta americana. Tissue Cell* 1(2):295–324
49. Hart, D. E., Steele, J. E. 1969. Inhibition of insect nerve cord phosphorylase action by 5-hydroxytryptamine. *Experientia* 25:243
50. Heinrich, B. 1971. Temperature regulation of the sphinx moth, *Manduca sexta.* II. Regulation of heat loss by control of blood circulation. *J. Exp. Biol.* 54:153–66
51. Hertel, W. 1971. Untersuchungen zur neurohormonalen Steuerung des Herzens der Amerikanischen Schabe *Periplaneta americana* (L.). *Zool. Jahrb. Physiol. Bd.* 56:152–84
52. Hiripi, L., Salanki-Rozsa, K. 1974. Fluorimetric determination of 5-hydroxytryptamine and catecholamines in the CNS and heart of *Locusta migratoria migratorioides. J. Insect Physiol.* 19:1481–85
53. Holman, G. M., Cook, B. J. 1970. Pharmacological properties of excitatory neuromuscular transmission in the hindgut of the cockroach, *Leucophaea maderae. J. Insect Physiol.* 16:1891–1907
54. House, C. R. 1973. An Electrophysiological study of neuroglandular transmission in the isolated salivary glands of the cockroach. *J. Exp. Biol.* 58:29-43
55. House, C. R., Ginsborg, B. L., Silinsky, E. M. 1973. Dopamine receptors in cockroach salivary gland cells. *Nature New Biol.* 245:63
55a. Hoyle, G., Dagan, D., Moberly, B., Colquhoun, W. 1974. Dorsal unpaired median insect neurons make neurosecretory endings on skeletal muscles. *J. Exp. Zool.* 187:159–65
56. Johnson, B. 1963. A histological study of neurosecretion in aphids. *J. Insect Physiol.* 9:727–79
57. Johnson, B. 1966. Fine structure of the lateral cardiac nerves of the cockroach, *Periplaneta americana* (L.). *J. Insect Physiol.* 12:645–53
58. Johnson, B., Bowers, B. 1963. Transport of neurohormones from the corpora cardiaca in insects. *Science* 141:264–66
59. Khattar, N. 1968. The neurosecretory system of *Schizodactylus monstrosus* (Dury). *Bull. Soc. Zool. Fr.* 93:225–32
60. Klemm, N. 1972. Monoamine-containing nervous fibers in foregut and salivary gland of the desert locust, *Schistocerca gregaria* Forskal. *Comp. Biochem. Physiol.* 43A:207–11
61. Knowles, F. G. W., 1967. Neuronal properties of neurosecretory cells. In *"Neurosecretion" IV International Sym-*

posium on Neurosecretion, ed. F. Stutinsky, 8–19. Berlin: Springer
62. Knowles, F. G. W., Carlisle, D. B. 1956. Endocrine control in the crustacea. *Biol. Rev.* 31:396–473
63. Lake, C. R., Mills, R. R., Brunet, P. C. J. 1970. Hydroxylation of tyramine by cockroach hemolymph. *Biochim. Biophys. Acta* 215:226–28
64. Leslie, R. A., Robertson, H. A. 1973. The structure of the salivary gland of the moth. *Z. Zellforsch. Mikrosk. Anat.* 146:553–64
65. Maddrell, S. H. P. 1965. Neurosecretory supply to the epidermis of an insect. *Science* 150:1033–34
66. Maddrell, S. H. P. 1967. Neurosecretion in insects. In *Insects and Physiology,* ed. J. E. Treherne, 103–18. Edinburgh: Oliver & Boyd
67. Maddrell, S. H. P. 1970. Neurosecretory control systems in insects. In *Insect Ultrastructure,* ed. A. C. Neville, 101–16. Oxford: Blackwell Sci. Publ.
68. Maddrell, S. H. P. 1974. Neurosecretion. In *Insect Neurophysiology,* ed. J. E. Treherne. Amsterdam: North-Holland. In press
69. Maddrell, S. H. P., Gee, J. D. 1974. Release of the diuretic hormones of *Rhodnius prolixus* and *Glossina austeri* induced by K-rich solution: Ca dependence, time course of release and localization of neurohaemal areas. *J. Exp. Biol.* In press
70. Maddrell, S. H. P., Klunsuwan, S. 1973. Fluid secretion by *in vitro* preparations of the Malpighian tubules of the desert locust *Schistocerca gregaria. J. Insect Physiol.* 19:1369–76
71. Mancini, G., Frontali, N. 1970. On the ultrastructural localization of catecholamines in the beta lobes (corpora pedunculata) of *Periplaneta americana. Z. Zellforsch. Mikrosk. Anat.* 103: 341–50
72. Marks, E. P., Holman, G. M., Borg, T. K. 1973. Synthesis and storage of a neurohormone in insect brains *in vitro. J. Insect Physiol.* 19:471–77
73. Mason, C. A. 1973. New features of the brain-retrocerebral neuroendocrine complex of the locust *Schistocerca vaga* (Scudder). *Z. Zellforsch. Mikrosk. Anat.* 141:19–32
74. Miller, T. A. 1974. Visceral muscle. In *Insect Muscle,* ed. P. N. R. Usherwood. London: Academic. In press
75. Miller, T., Adams, M. E. 1974. Ultrastructure and electrical properties of the hyperneural muscle of *Periplaneta americana. J. Insect Physiol.* In press
76. Miller, T., Rees, D. 1973. Excitatory transmission in insect neuromuscular systems. *Am. Zool.* 13:299–313
77. Miller, T. A., Thomson, W. W. 1968. Ultrastructure of cockroach cardiac innervation. *J. Insect Physiol.* 14:1099–1104
78. Miller, T., Usherwood, P. N. R. 1971. Studies of cardio-regulation in the cockroach *Periplaneta americana. J. Exp. Biol.* 54:329–48
79. Mordue, W., Goldsworthy, G. J. 1969. The physiological effects of corpus cardiacum extracts in locusts. *Gen. Comp. Endocrinol.* 12:360–69
80. Nagai, T. 1973. Insect visceral muscle. Excitation and conduction in the proctodeal muscles. *J. Insect Physiol.* 19: 1753–64
81. Nagai, T., Brown, B. E. 1969. Insect visceral muscle. Electrical potentials and contraction in fibres of the cockroach proctodeum. *J. Insect Physiol.* 15:2151–67
82. Natalizi, G. M., Frontali, N. 1966. Purification of insect hyperglycaemic and heart accelerating hormones. *J. Insect Physiol.* 12:1279–87
83. Nathanson, J. A., Greengard, P. 1973. Octopamine-sensitive adenylate cyclase: evidence for a biological role of octopamine in nervous tissue. *Science* 180:308–10
84. Nayar, K. K. 1954. The neurosecretory systems of the fruitfly, *Chaetodocus cucurbitae.* Cof. I. Distribution and description of the neurosecretory cells in the adult fly. *Proc. Indian Acad. Sci. Sect. B.* 40:138–44
84a. Noirot, C., Quennedey, A. 1974. Fine structure of insect epidermal glands. *Ann. Rev. Entomol.* 19:61–80
85. Normann, T. C. 1965. The neurosecretory system of the adult *Calliphora erythrocephala.* I. The fine structure of the corpus cardiacum with some observations on adjacent organs. *Z. Zellforsch. Mikrosk. Anat.* 67:461–501
86. Normann, T. C. 1970. The mechanism of hormone release from neurosecretory axon endings in the insect *Calliphora erythrocephala.* In *Aspects of Endocrinology,* ed. W. Bargmann, B. Scharrer, 30–42. Berlin: Springer
87. Normann, T. C. 1972. Heart activity and its control in the adult blowfly, *Calliphora erythrocephala. J. Insect Physiol.* 18:1793–1810
88. Normann, T. C. 1973. Membrane potential of the corpus cardiacum neurosecretory cells of the blowfly, *Cal-*

liphora erythrocephala. J. Insect Physiol. 19:303–18

89. Normann, T. C. 1974. Calcium-dependence of neurosecretion by exocytosis. *J. Exp. Biol.* In press
90. Normann, T. C. Neurosecretion by Exocytosis. Manuscript in preparation
91. Nutting, W. L. 1951. A comparative anatomical study of the heart and accessory structures of the orthopteroid insects. *J. Morphol.* 89:501–97
92. Osborne, M. P., Finlayson, L. H., Rice, M. J. 1971. Neurosecretory endings associated with striated muscles in three Insects (*Schistocerca, Carausius,* and *Phormia*) and a frog (*Rana*). *Z. Zellforsch. Mikrosk. Anat.* 166:391–404
93. Park, K. E. 1973. Fine structure of the diapause-regulation cell in the suboesophageal ganglion in the silkworm, *Bombyx mori. J. Insect Physiol.* 19:293–302
94. Penzlin, H. 1971. Zur Rolle des Frontalganglions bei Larven der Schabe, *Periplaneta americana. J. Insect Physiol.* 17:559–73
95. Pitman, R. M. 1971. Transmitter substances in insects: a review. *Comp. Gen Pharmacol.* 2:347–71
96. Quennedey, A. 1969. Innervation de type neurosécréteur dans la glande sternale de *Kalotermes flavicollis.* Étude ultrastructurale. *J. Insect Physiol.* 15: 1807–14
97. Raabe, M., Baudry, N. Grillot, J. P., Provansol, A. 1972. The perisympathetic organs of insects. *Int. Congr. Entomol. Canberra, Australia,* 14th abstracts, pp. 131–32
98. Raabe, M., Cazal, M., Chalaye, D., Besse, N. 1966. Action cardioaccélératrice des organes périsympathiques ventraux de quelques insectes. *C. R. Acad. Sci.* 263:2002–5
99. Rees, D. 1974. The spontaneous release of transmitter from insect nerve terminals as predicted by the negative binomial theorem. *J. Physiol.* 236: 129–42
100. Robertson, H. A., Steele, J. E. 1972. Activation of insect nerve cord phosphorylase by octopamine and adenosine 3',5'-monophosphate. *J. Neurochem.* 19:1603–6
101. Robertson, H. A., Steele, J. E. 1973. Octopamine in the insect central nervous system: distribution, biosynthesis and possible physiological role. *J. Physiol.* 237:34–35
102. Rojakovick, A. S., March, R. B. 1972. The activation and inhibition of adenylcyclase from the brain of the Madagascar cockroach (*Gromphadorhina portentosa*). *Comp. Biochem. Physiol.* 43B: 209–15
103. Rojakovick, A. S., March, R. B. 1973. Characteristics of cyclic 3',5'-nucleotide phosphodiesterase from the brain of the madagascar cockroach (*Gromphadorhina portentosa*). *Comp. Biochem. Physiol.* 47B:189–99
104. Roussel, J. P., Cazal, M. 1969. Action, *in vivo,* d'extraits de corpora cardiaca sur le rythme cardiaque de *Locusta migratoria* L. *C. R. Acad. Sci.* 268:581–83
105. Sakharov, D. A. 1970. Cellular aspects of invertebrate neuropharmacology. *Ann. Rev. Pharmacol.* 15:335–52
106. Sandifer, J. B., Tombes, A. S. 1972. Ultrastructure of the lateral neurosecretory cells during reproductive development of *Sitophilus granarius* (L.). *Tissue Cell* 4:437–46
107. Scharrer, B., Weitzman, M. 1970. Current problems in invertebrate neurosecretion. In *Aspects of Neuroendocrinology,* ed. W. Bergmann, B. Scharrer, 1–23. Berlin: Springer
108. Schein, Y. 1972. Postemergence growth in the fly, *Sarcophaga falculata,* initiated by neurosecretion from the ocellar nerve. *Nature New Biol.* 236:217–19
109. Steel, C. G. H., Harmsen, R. 1971. Dynamics of the neurosecretory system in the brain of an insect, *Rhodnius prolixus,* during growth and molting. *Gen. Comp. Endocrinol.* 17:125–41
110. Thomsen, E. 1954. Studies on the transport of neurosecretory materials in *Calliphora erythrocephala* by means of ligaturing experiments. *J. Exp. Biol.* 31: 322–30
111. Thomsen, E., Thomsen, M. 1970. Fine structure of the corpus allatum of the female blow-fly *Calliphora erythrocephala. Z. Zellforsch. Mikrosk. Anat.* 110:40–60
112. Tombes, A. S., Roppel, R. M. 1972. Ultrastructure of the spermatheca of the granary weevil, *Sitopholus granarius* (L.) *Int. J. Insect Morphol. Embryol.* 1:141–52
113. Tunnicliff, G., Rick, J. T., Connolly, K. 1969. Locomotor activity in *Drosophila*-V. A comparative biochemical study of selectively bred populations. *Comp. Biochem. Physiol.* 29:1239–45
114. Tyrer, N. M. 1971. Innervation of the abdominal intersegmental muscles in the grasshopper. *J. Exp. Biol.* 55: 305–14
115. Unnithan, G. C., Bern, H. A., Nayar, K. K. 1971. Ultrastructural analysis of the neuroendocrine apparatus of *Oncopeltus fasciatus. Acta Zool.* 52:117–43

116. Van der Kloot, W. G. 1960. Neurosecretion in insects. *Ann. Rev. Entomol.* 5:35–52
117. Vincent, J. F. V. 1972. The dynamics of release and the possible identity of bursicon in *Locusta migratoria migratorioides. J. Insect Physiol.* 18:757–80
118. Whitehead, A. T. 1971. The innervation of the salivary gland in the American cockroach: Light and electron microscopic observations. *J. Morphol.* 135: 483–506
119. Wigglesworth, V. B. 1964. The hormonal regulation of growth and reproduction in insects. *Advan. Insect Physiol.* 2:247–336
120. Wigglesworth, V. B. 1970. *Insect Hormones.* London: Methuen
121. Wilkens, J. L., Mote, M. I. 1970. Neuronal properties of the neurosecretory cells in the fly *Sarcophaga bullata. Experientia* 26:275–76
122. Wright, R. D., Sauer, J. R., Mills, R. R. 1970. Midgut epithelium of the American cockroach: Possible sites of neurosecretory release. *J. Insect Physiol.* 16:1485–92

NEUROMUSCULAR PHARMACOLOGY OF INSECTS

❖6085

T. J. McDonald
Transidyne General Corporation, Ann Arbor, Michigan 48106

The pharmacology of the insect neuromuscular mechanism has been the subject of previous reviews. In 1965, Hoyle (35) presented a comprehensive compendia on the neural control of insect skeletal muscle. Although little had been clearly established on insect neuromuscular pharmacology until that time, it is a prime source for references to previous works. More recently, Aidley (1), Boistel (8), and Usherwood (72, 76) reviewed the general features of insect neuromuscular transmission. Usherwood (74) prepared an excellent monograph on the electrochemistry of insect muscle. Phillis (62) authored a book entitled *The Pharmacology of Synapses* which is a general overview of synaptic pharmacology for both vertebrate and invertebrate tissues. Huddart (36) not only reviews the features of insect muscle contraction, but presents detailed experimental procedures for investigation of insect nerve-muscle preparations. The transmitter substances of insects have been recently reviewed (64).

In the light of previous reviews, the present one is limited to the pharmacology of excitatory neuromuscular synapses of insect skeletal muscle. This focus permits a more detailed discussion of the considerable progress in insect excitatory neuromuscular pharmacology and draws attention to problems yet to be solved. In spite of their importance and interest, other aspects of insect neuromuscular pharmacology are not the subject of this review. For example, the pharmacology of skeletal inhibitory neuromuscular synapses, innervated heart, gut, light organs of fireflies, and neurosecretory structures of insects are not dealt with in this review.

NEUROMUSCULAR MORPHOLOGY

Since the excitatory neuromuscular synaptic morphology is important to the discussion of its pharmacology, a brief presentation of it is given. Figure 1 shows the three major cell types associated with the synapse: the axon, glial, and muscle fiber. Frequently, tracheole cells are associated with the axon and have similar structure to unmyelinated sheath or glial cells (24). The nerve terminal is identified by the aggregation of synaptic vesicles believed to contain the transmitter substance. The most widely held hypothesis for transmitter release from the presynaptic axon is that

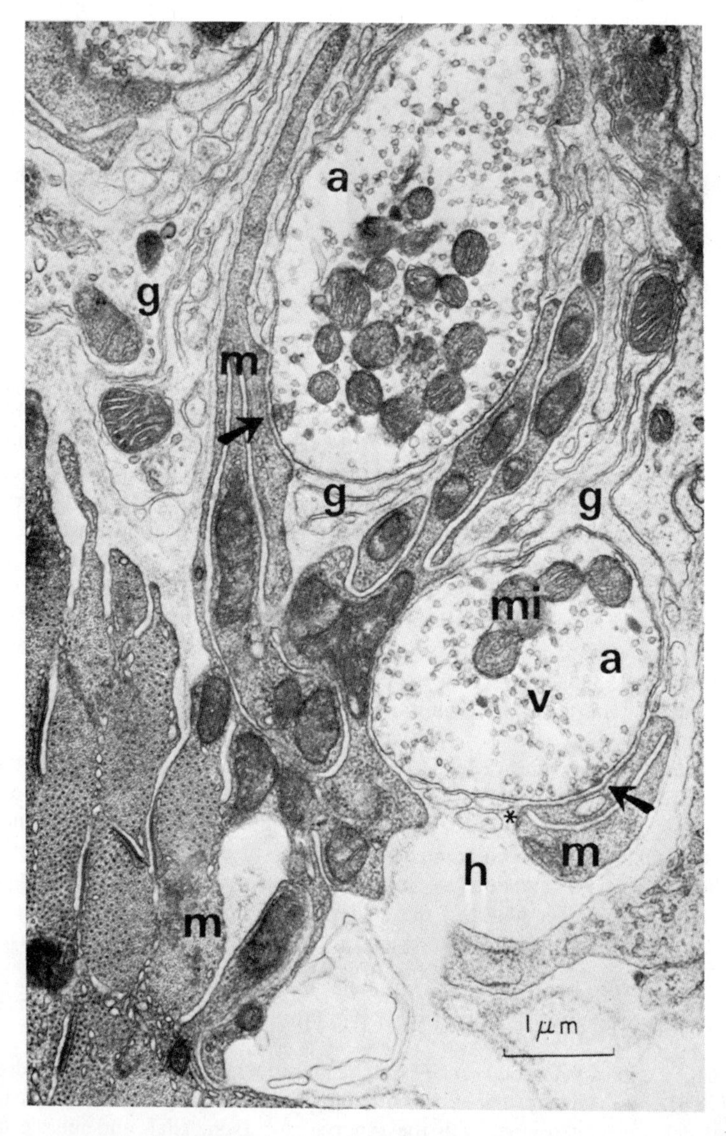

Figure 1 Neuromuscular junctions of body wall muscle of *Carausius morosus.* Two axons (a) make synaptic contact with projections from the surface of the muscle cell (m) at points marked by arrows. Unapposed areas of the axons are capped by glial cell (g). Inclusions in the axoplasm are synaptic vesicles (v) and mitochondria (mi). In some places (*) the axons are poorly isolated from the hemolymph (h) [by permission (61)].

the synaptic vesicles fuse with the presynaptic membrane and liberate the transmitter substance through a rupture at the point of fusion of the presynaptic and vesicular membranes. This packet or quantum of transmitter diffuses across the synaptic cleft, and the transmitter molecule interacts with the postsynaptic receptors. This interaction results in a postsynaptic muscle membrane depolarization (61). The quantal nature of transmitter release is considered to give rise to the miniature endplate potentials detected during microelectrode recordings (66, 75).

The insect muscle ultrastructure is similar to vertebrate muscle (72). Like vertebrate striated muscle, insect muscle has interdigitating thick and thin myofilaments (Figure 2). The sarcolemma invaginates to form the transverse tubular system, which is believed to conduct the ionic current from the cell surface into the interior of the muscle fiber (68). The transverse tubules form dyadic junctions with the sarcoplasmic reticulum—the probable storage site for intracellular calcium (36, 69). The surface membrane action currents conducted into the muscle cell through the transverse tubular system are considered to cause the sarcoplasmic reticulum to release calcium ions which trigger myofilament contractile movement (68).

Since an isolated nerve-muscle preparation consists of axon, glial, tracheole, and muscle cells, a drug perfused over the tissue could act on any or all of these cells. Therefore, care must be exercised when interpreting the tissue response to a particular drug action. For example, if the neurally evoked muscle contractions were diminished in response to drug perfusion, the chemical agent might have blocked axonal conduction, transmitter secretion, receptor activation, muscle membrane depolarization, or the contractile mechanism. However, the use of an isolated nerve-muscle preparation eliminates a number of possible drug action sites compared to the number of possible sites of action in whole insect preparations. An illustration in point is a series of reports dealing with the d-tubocurarine chloride effect on whole insects. Curare injected into *Calliphora erythrocephala* caused flaccid paralysis (43). Based on data from a whole fly preparation of *Sarcophaga bullata,* McCann (48, 49) indicated that curare was a neuromuscular blocking agent similar to that described for vertebrates. Flattum (25) investigated the action of injected curare on the cricket, *Acheta domesticus,* and concluded that the nonvertebrate behavioral response to the injected drug did not indicate a neuromuscular mode of action. It was considered that the coordinating centers of the cricket central nervous system were influenced by the curare (25). Friedman & Carlson (26, 27), on the other hand, presented evidence that curare blocked axonal conduction. The above series of investigations employed either whole insects or large complex preparations; the drug had the possibility of interacting with a number of potential target sites, making definitive conclusions regarding the mode of action difficult. As discussed below, it seems unlikely that curare is an insect excitatory neuromuscular blocking agent.

NEUROMUSCULAR PHARMACOLOGY

Acetylcholine

It has been known for a considerable time that acetylcholine is the chemical transmitter substance at the vertebrate skeletal neuromuscular junction (11). As a result,

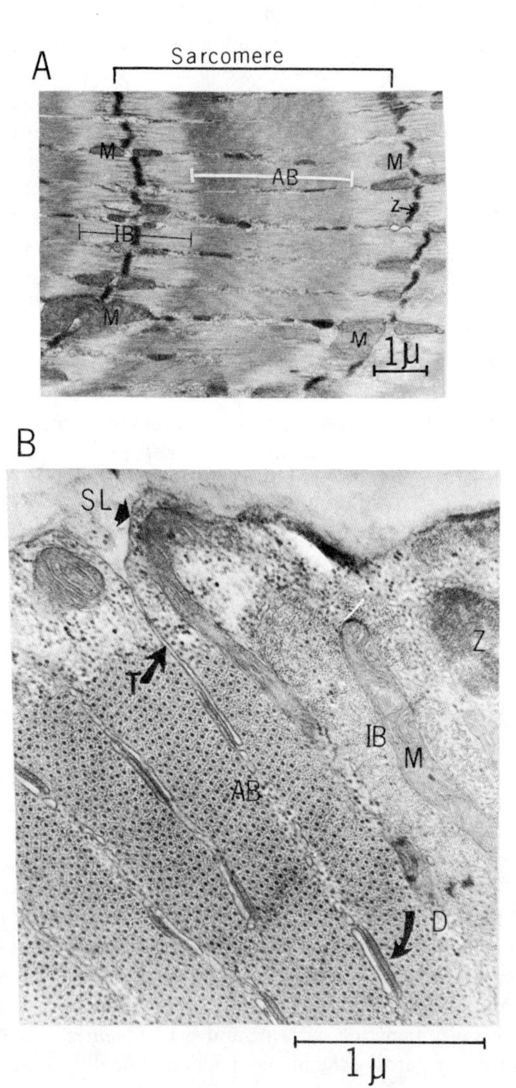

Figure 2 A: Longitudinal section of the retractor unguis muscle of *Romalea microptera* illustrating the A-band (AB) and I-band (IB). B: Cross section of the muscle showing an invagination of the sarcolemma (SL) forming a transverse tubule (T). At the dyad (D), the tubule is electron transparent and the sarcoplasmic reticulum cisternae is opaque (McDonald, unpublished observation).

attempts have been made to show a similar role for acetylcholine at the insect neuromuscular synapse.

Wigglesworth (85) demonstrated the absence of acetylcholine esterase histochemically for muscle tissue in the assassin bug, *Rhodnius prolixus.* Hamori (28) observed similar results for a number of different species. Mandel'shtam (57) used long incubation times to show the presence of cholinesterase at the leg muscle synapses of *Locusta migratoria.* The long incubation times would indicate either low or nonspecific enzyme activity. More recently Booth & Lee (9) reported a highly active cholinesterase at the motor endplates in the head of the cricket, *A. domestica.* The substrate used in the histological procedure was a specific substrate for cholinesterase, viz, acetylthiocholine. Utilizing identical histological conditions, the head muscle of *Periplaneta americana* did not yield a cholinesterase hydrolysis product (9). When 10^{-6} *M* physostigmine was added to the incubation media, the esterase activity of the cricket muscle was inhibited completely. Further studies will be necessary to determine the role, if any, of the enzyme in this cricket tissue. Of the insects examined histochemically, cholinesterase enzyme activity is generally not noted in insect muscle (9). Biochemical studies demonstrate that insect muscle lacks the cholinesterase, acetylcholine, and choline acetylase (16).

Drug perfusion studies support a noncholinergic mechanism for nerve-muscle transmission in insects. Roeder & Weiant (67) found that 10^{-3} *M* curare applied to a cockroach nerve-muscle preparation had no effect upon neuromuscular transmission. For the locust leg preparation, Harlow (32) also found that curare, decamethonium, and prostigmine had no effect on the neurally elicited contractions. However, 10^{-5} *M* to 10^{-1} *M* acetylcholine sometimes caused muscle contractions followed by brief tetanus (32). More recently, a number of cholinergic drugs were tested on several insect nerve-muscle preparations, and the results indicated that none of the effects observed could be traced to an action on neuromuscular transmission (23, 60). McDonald, Farley & March (54) tested a wide variety of cholinergic and anticholinergic drugs on the retractor unguis muscle of locust and grasshopper species. Both ionized and un-ionized compounds were perfused to determine whether or not ionization influenced the cholinergic drug action. Since none of the compounds reduced the neurally evoked response unless 10^{-3} *M* or higher concentration were perfused, they concluded that the insect neuromuscular junction was noncholinergic.

Biogenic Amines

Those biogenic amines tested so far require high concentrations (10^{-2} *M*) to alter significantly the insect neurally evoked muscle response, indicating that the excitatory neurohumor is not a biogenic amine. In proof, Harlow (32) reported that adrenaline was inactive on the locust leg preparation. Noradrenaline (the adrenergic transmitter) and harmine (a monoamine oxidase inhibitor) had no effect on the cockroach leg muscle (60). Hill & Usherwood (33) reported that tryptamine, 5-hydroxytryptamine, lysergic acid diethylamide, bromolysergic acid diethylamide, 5,6-dimethoxytryptamine, and 3-(pyrrolidinomethyl)-thionaphthene diminished the contractile force of the locust leg muscle only at high concentrations.

The neurally evoked response of the retractor unguis muscle of locust and grasshopper species was reduced by high concentrations of norepinephrine, epinephrine, 3-hydroxytyramine, 5-hydroxytryptamine, and DL-amphetamine (54), supporting previous observations (33). Although high concentrations of biogenic amines appear necessary to be active on insect tissue, their site of action seems to be the neuromuscular junction. Hill & Usherwood (33) found that the nerve electrical activity and muscle fiber electrical properties were unaltered during drug perfusion. Since neural stimulation failed to elicit a muscle response under these conditions, it was logically concluded that synaptic transmission was impeded. L-Glutamate is also not able to evoke a phasic muscle contraction during blockage by either tryptamine or 5-hydroxytryptamine (80). Possibly these drugs interact with the hypothesized glutamate receptor, so that applied glutamate is not able to elicit a muscle response.

Amino Acids

There is a considerable and growing body of evidence which indicates that L-glutamic acid is the excitatory transmitter at insect skeletal neuromuscular junctions. Various aspects of the transmitter role for L-glutamate have been examined. The amino acid is able to mimic the natural transmitter substance in a number of insect species (22, 34, 40, 41, 56, 76). Iontophoretic applications have shown that the amino acid has its site of action at the synaptic cleft (6, 41, 73, 75, 78, 80). The amino acid is released from nerve-muscle tissues during electrical stimulation of the nerve (39, 81), and possibly inactivated by an uptake mechanism (24).

At high concentration, glutamate desensitizes the insect neuromuscular junctions (56, 80), a phenomenon noted for cholinergic preparations (3). A molecular mechanism has been proposed recently for the phenomenon of desensitization by Eldefrawi & O'Brien (20). In brief, they suggest that receptor desensitization occurs when the transmitter at high concentration binds to a proposed regulatory site on the receptor moiety, resulting in a subsequent receptor conformational change which no longer allows further ionic conductance across the membrane. Thus, high concentrations of transmitter autoinhibits the depolarizing action of the usual receptor-transmitter complex. It is useful to note that the desensitization phenomenon may aid in identifying synapses where glutamate is the suspected transmitter. For example, when a drop of 10^{-3} *M* glutamate was added to the saline flowing over a semi-isolated heart preparation of *Schistocerca gregaria,* the contractile response of the alary muscles were selectively abolished (Figure 3). The heart muscle contractions were not influenced by glutamate, but the alary muscle contractions were completely abolished while glutamate was present (T. J. McDonald, unpublished observations).

A major obstacle in accepting the hypothesis that L-glutamate is the excitatory transmitter is that insect hemolymph has an amino acid titer sufficient to desensitize the neuromuscular synapse (24, 76, 80). As pointed out by Usherwood & Machili (80): "It is not possible to invoke either structural or enzymic barriers around the neuromuscular junctions to protect the synapses from the glutamate in the blood, since the postsynaptic receptors are excited by very low concentrations of topically and iontophoretically applied glutamate." In contrast, Faeder & Salpeter (24) argue that the sheath and tracheole cells can provide an effective barrier against hemolymph glutamate. Why these same cells do not protect isolated tissues against the glutamate of perfused solutions has not been clarified. In addition, morphological evidence does not always support synapse protection from the hemolymph by sheath cells. Thus Figure 1 shows an axon and synapse very poorly isolated from the hemolymph.

It has been speculated that there is actually little free glutamate in the insect hemolymph and that the glutamate detected by biochemical means is actually in a bound form (80). In consideration of bound and free glutamate, there is an important thermodynamic consideration which should not be overlooked, viz, glutamate must be bound more tightly to the hemolymph component than to the postsynaptic receptors. In addition, the hemolymph component must not become saturated with glutamate, because then the excess glutamate would be free to interact with the synaptic receptors. Data of Lunt (47) indicates that an insect muscle component which possibly represents a constituent part of the glutamate receptor has an apparent dissociation constant of approximately 5×10^{-7} *M*. If this indeed represents the dissociation constant between glutamate and the receptor, the binding component of the hemolymph must possess even higher affinity for glutamate. Possibly there is a mechanism which limits the rate of diffusion of glutamate to the receptor sites. This would satisfy the thermodynamic considerations. Usherwood (76) suggested that hemocytes possibly sequester glutamate. An amino acid active transport mechanism across the hemocyte cell membrane could satisfy the thermodynamic requirements. Further studies on the problem of hemolymph glutamate are needed to understand the role of glutamate in insect excitatory neuromuscular transmission.

Studies of amino acids have shown that nerve-muscle preparations are most sensitive to L-glutamate (50, 56, 76, 80, 84). The receptor is stereospecific for the L-isomer (44–46, 50, 53, 56, 80, 82). N-Acetyl-L-glutamic acid, N-carbamyl-L-glutamic acid, L-glutamic acid-α-methyl ester, L-glutamic acid-γ-methyl ester, L-glutamic acid dimethyl ester, and L-glutamic acid-γ-hydrazide are substituted glutamate analogs which possessed little activity on the retractor unguis preparation at concentrations as high as 10^{-3} *M* (50). Since substitution eliminates the ionized charge of the functional group, these results indicate that each functional group must be ionized (56).

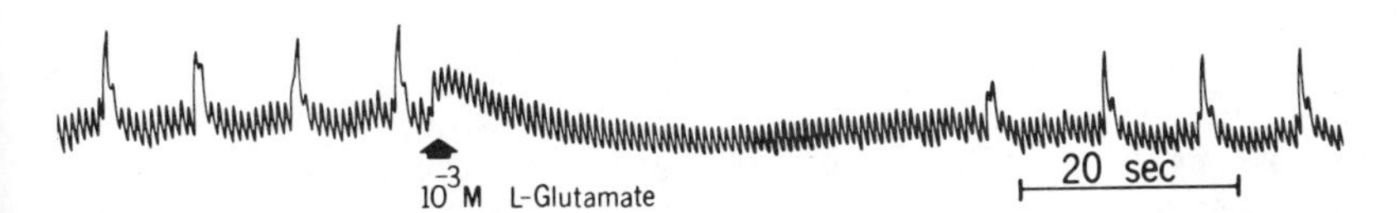

Figure 3 Myogram of spontaneously beating heart of *Schistocerca gregaria*. The large amplitude contractions of the alary muscle were selectively diminished by one drop of 10^{-3} *M* glutamate added to perfusion saline flowing at 2 ml/min. The lower amplitude heart muscle contractions were not reduced (McDonald, unpublished observation).

Since glutamate was suspected to be degraded by L-glutamic acid decarboxylase (GAD) [an enzyme requiring pyridoxal phosphate (52, 79)], studies of enzyme inhibitors have been conducted. Usherwood & Machili (80) showed that hydroxylamine, *p*-phenylenediamine, semicarbazide, and thiosemicarbazide had little or no activity on the retractor unguis preparation. In a further investigation, McDonald (50) found that O-diazoacetyl-L-serine (Azaserine), DL-methionine-DL-sulfoximine (which is a convulsant in vertebrates) (42), DL-α-methyl-glutamic acid, amino-oxyacetic acid (83), cycloserine, and 4-iodoacetamidosalicylic acid were inactive. In contrast to these compounds with low activities, Usherwood & Machili (79) reported high activity for phenylhydrazine and speculated that the potentiation observed during perfusion with the drug may have been caused by inhibition of the proposed degradative enzyme, L-glutamic acid decarboxylase. Hydrazine pharmacology will be discussed in greater detail below. The observation that the above enzyme inhibitors, with the exception of phenylhydrazine, only slightly altered the force of the neurally evoked contractions is disappointing. Particularly, when one considers the usefulness of cholinesterase inhibitors used to investigate cholinergic junctions.

So far, covalently bonding GAD inhibitors have not been found active on insect nerve-muscle preparations. This lack of activity is evidence that the coenzyme, pyridoxal phosphate, is not involved in the immediate events of synaptic transmission. This is indirect support for the hypothesis that glutamate is inactivated by an uptake mechanism rather than enzymatic decarboxylation (24). This is further supported by McDonald & O'Brien (56), as they were not able to confirm the report of Usherwood & Machili (79) that GAD diminished the mechanical response of the retractor unguis muscle to nerve stimulation. Although Usherwood (76) considered that the effect of perfused GAD was the result of altering the metabolism of intraneuronal glutamate, the effect may have been due to reduced calcium ion concentration (56).

Hydrazines

As pointed out above, Usherwood & Machili (79, 80) reported that phenylhydrazine potentiated the neurally evoked contractions of the retractor unguis muscle of the locust, *Schistocerca gregaria.* They considered that the hydrazine inhibited an enzyme involved in the inactivation of the transmitter at the synapse, allowing prolonged transmitter-receptor interaction. Holman & Cook (34) also observed that phenylhydrazine potentiated the contractions of the hindgut of the cockroach, *Leucophaea maderae,* and speculated that its mode of action was similar to that first proposed by Usherwood & Machili (79). More recently, Dowson & Usherwood (18) concluded that phenylhydrazine action was too complex to be solely due to enzyme inhibition. This would support the free radical mode of action for phenylhydrazine proposed by McDonald (51, 52).

Using the retractor unguis preparation of the grasshopper, *Romalea microptera,* it was demonstrated that only after oxidation of phenylhydrazine solutions had occurred was potentiation of the neurally evoked contractions observed (51, 52).

$$C_6H_5{-}NHNH_2 + O_2 \longrightarrow [C_6H_5{-}N{=}\dot{N}H] + H_2O_2$$

Phenylhydrazine → Phenyldiazene radical

$$[C_6H_5\bullet] + N_2 + H_2O_2 \longrightarrow C_6H_6 + N_2 + H_2O_2$$

Phenyl radical → Benzene

Since benzene is the major oxidation product of phenylhydrazine, the potentiation effect was considered due to benzene.

In an aqueous environment, phenylhydrazine auto-oxidizes via phenyldiazine (phenyldiimide) and phenyl free radicals to benzene, nitrogen, and hydrogen peroxide. Minor products are azobenzene, biphenyls, and hydrazobenzene. Cupric ions catalyze the oxidation (4, 19, 31, 70).

It was further observed that cupric ions lessen the time required for phenylhydrazine to diminish the contractile response of the retractor unguis muscle (52). When ascorbic acid and EDTA were added to the phenylhydrazine solution to prevent its oxidation, the muscle response was not abolished (Figure 4). This indicated that the hydrazine itself was inactive. Only during the course of phenylhydrazine oxidation was the contractile response reduced, indicating that the free radicals of oxidation were responsible for the ability of oxidizing solutions to abolish the muscle response. Although the hydrazine pharmacology does not support a glutamate enzymatic degradation hypothesis, it is an interesting example of a free radical mode of action for a drug.

Aromatic hydrazines have been observed to abolish the contractile response of retractor unguis muscle of *R. microptera* (McDonald, unpublished observations). For this study, the activities of *para*- and *meta*-substituted phenylhydrazines were correlated with their Hammett sigma values. Figure 5 is a graph showing the activities of a number of *p*- and *m*-substituted hydrazines plotted vs their Hammett sigma values (σ). The equipotent concentrations plotted were the concentrations required to diminish the contractile response by 50% in 30 min. As indicated on the graph of Figure 5, the most active hydrazine was *p*-methoxyphenylhydrazine; and the least active hydrazine was *m*-nitrophenylhydrazine. As calculated according to the method of least squares, the best equation is,

$$\text{Log } 1/C = 1.917 \text{ Log } \sigma - 4.555$$

with a correlation coefficient of 0.9177. This physicochemical analysis of the drug activities is evidence that the activity of the hydrazine is proportional to the electronic contribution of the ring substituent, with the electron donating substituents favoring high activity (14).

Alcohols

The action of alcohols on the neurally evoked response of the retractor unguis muscle of *Schistocerca gregaria* was recently investigated by McDonald, Farley &

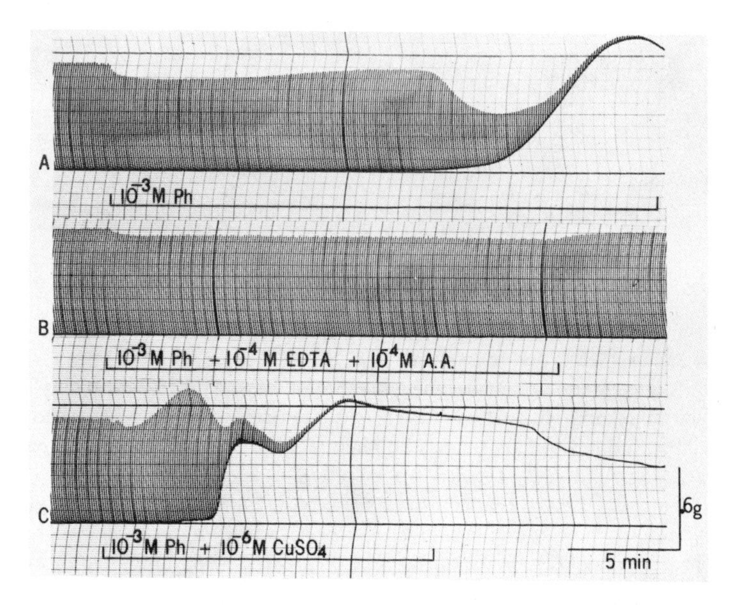

Figure 4 Effect of various 10^{-3} *M* phenylhydrazine solutions (Ph) on the neurally evoked contractions of the retractor unguis muscle of *Romalea microptera.* A: 10^{-3} *M* phenylhydrazine itself, B: hydrazine with 10^{-4} *M* EDTA and ascorbic acid (AA) which prevent phenylhydrazine oxidation. C: the hydrazine with 10^{-6} *M* $CuSO_4$ which catalyzed phenylhydrazine oxidation (52).

March (55). They observed that a series of alcohols potentiated the contractile force of the muscle. Figure 6 is a plot of equipotent concentrations with respect to their 1-octanol/water partition coefficient (29). As calculated according to the method of least squares, the best equation is,

$$\text{Log } 1/C = 1.282 \text{ Log P} + 1.684$$

with a correlation coefficient of 0.963.

Since the 1-octanol/water partition coefficient (P) is a measure of the ratio of lipophilicity/hydrophilicity for a chemical substance, Figure 6 would indicate that the more lipophilic the alcohol, the more active it is. The adequate description by a single parameter is evidence that there is a common site of action for the alcohols and that the only limiting factor for action is the free energy of transfer from saline to a hydrophobic site of action (30). Ethanol was reported to potentiate the contractions of the rat phrenic nerve-diaphragm muscle preparation (21), which indicates that the potentiation effect is not specific for insect muscle tissues. Additional research is needed to indicate mode of action of ethanol on excitable tissues (7, 65).

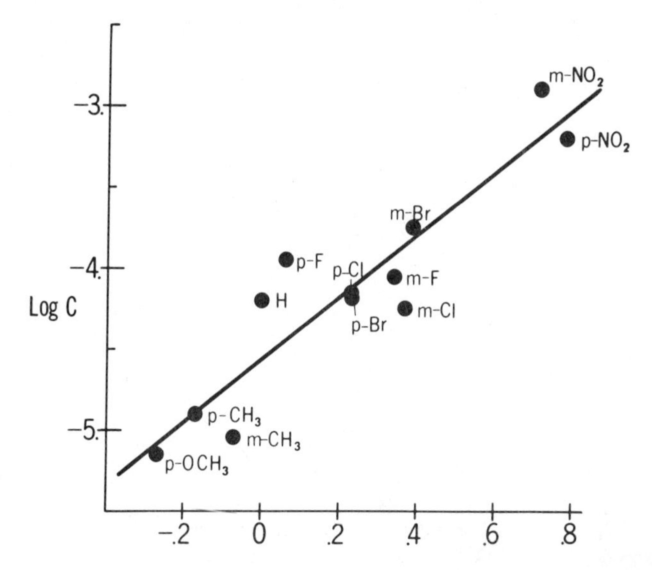

Figure 5 Structure-activity relationship of *para* (*p*) and *meta* (*m*) substituted phenylhydrazines on the retractor unguis muscle of *Romalea microptera.* Equipotent concentrations which reduced the contractile response to 50% in 30 min were plotted versus Hammett sigma values (14), a measure of the electron withdrawing contribution of the substituent (McDonald, unpublished observation).

Miscellaneous Drugs

N-Ethyl maleimide, iodoacetic acid, *p*-chloromercuribenzoic acid, and mersalyl (salyrganic acid) are sulphydryl reagents which have been shown to abolish muscle contractures elicited by L-glutamate perfusion as well as the neurally evoked contractile response of the retractor unguis muscle of *R. microptera* (10). A particularly interesting feature of this study was that blockage caused by *p*-chloromercuribenzoic acid could be reversed completely by postperfusion of dithiothreitol solutions. This was considered evidence for an active free sulphydryl group necessary for muscle contraction, since dithiothreitol causes complete reduction of disulphide linkages to free sulphydryl groups (15). It would be of interest to determine the site of action for this compound, since vertebrate studies have shown that this class of drugs is active at cholinergic junctions (5, 37, 38, 59).

Black widow spider venom was recently shown to increase the spontaneous miniature discharges of the locust extensor tibiae muscle fibers (17). Electron micrographs of treated tissues demonstrated that there was a reduction in the synaptic vesicles contained in a number of the nerve terminals and that the venom caused irregularly shaped membrane bound structures to appear. This was considered evidence that the venom effect observed during microelectrode recordings was also

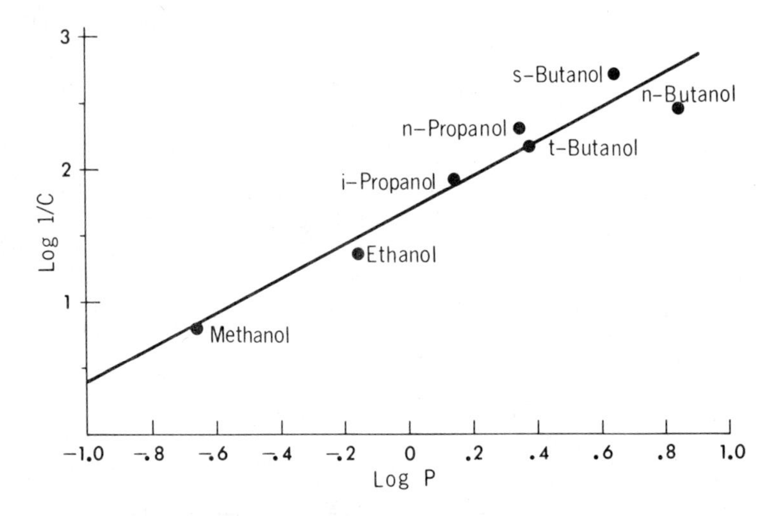

Figure 6 Structure-activity relationship of alcohols on the retractor unguis muscle of *Romalea microptera* (55). The negative logarithm of equipotent concentrations (Log 1/C) are plotted against the logarithm of the 1-octanol/water partition coefficients. The more lipophilic alcohols were more effective in potentiating the contractile response of the muscle.

accompanied with a corresponding change in muscle structure. In another study, it was shown that the venom of the digger wasp, *Philanthus triangulum,* diminished the electrically excitable muscle membrane potential (63).

α-Bungarotoxin, a component of the venom of a Formosan snake, *Bungarus multicinctus,* selectively blocks the depolarizing action of acetylcholine at the vertebrate neuromuscular junction (12, 13, 58). This toxin of the snake venom was perfused over the retractor unguis tissue; it was inactive on the tissue at concentrations as high as 2.5×10^{-5} g/ml (56). This evidence supports the claim that it is specific for cholinergic receptors (13).

CONCLUSION

Considerable progress has been made since Hoyle (35) reviewed the subject of insect neuromuscular pharmacology. Presently, the most likely candidate for the excitatory neuromuscular transmitter substance is L-glutamate. More research is needed on the mode of transmitter inactivation. Pharmacological evidence does not support the hypothesis that L-glutamic acid decarboxylase is involved in the inactivation process. Although there is only one report presenting an uptake mechanism for the insect synapse (24), pharmacological evidence does not challenge this mode of transmitter inactivation. The occurrence of high hemolymph glutamate concentrations remains a problem and requires further investigation. It would be desirable to discover drugs which covalently bond to the receptor and inactivation mechanism. The receptor isolation and characterization will prove a fruitful area of study (47).

Although pharmacological investigations are of considerable importance, they should be done in collaboration with biochemical, physiological, and morphological studies. Only in this manner will insect excitatory neuromuscular transmission be understood.

ACKNOWLEDGMENTS

This review was supported by NIH Training Grant ES 47 and ES 98, Rockefeller Foundation Grant RF 69073, and NIH Grant NB 08110-01.

Literature Cited

1. Aidley, D. J. 1967. The excitation of insect skeletal muscles. *Advan. Insect Physiol.* 4:1–31
2. Atwood, H. L., Jahromi, S. S. 1967. Strychnine and neuromuscular transmission in the cockroach. *J. Insect Physiol.* 13:1065–74
3. Axelsson, J., Thesleff, S. 1958. The "desensitizing" effect of acetylcholine on the mammalian motor end-plate. *Acta Physiol. Scand.* 43:15–26
4. Aylward, J. B. 1969. Oxidation of hydrazine derivatives. Part 1. Oxidation of phenylhydrazine with lead tetra-acetate. *J. Chem. Soc.* (C):1663–65
5. Bartels, E., Rosenberry, T. L. 1971. Snake neurotoxins: Effect of disulfide reduction on interaction with electroplax. *Science* 174:1236–37
6. Beranek, R., Miller, P. L. 1968. The action of iontophoretically applied glutamate on insect muscle fibers. *J. Exp. Biol.* 49:83–93
7. Blum, K., Wallace, J. E., Geller, I. 1972. Synergy of ethanol and putative neurotransmitters: glycine and serine. *Science* 176:292–94
8. Boistel, J. 1968. The synaptic transmission and related phenomena in insects. *Advan. Insect Physiol.* 5:1–64
9. Booth, G. M., Lee, A. H. 1971. Distribution of cholinesterase in insects. *Bull. WHO* 44:91–98
10. Bratkowski, T. A., McDonald, T. J., O'Brien, R. D. 1972. Effects of sulphydryl reagents on an insect nerve-muscle preparation. *J. Insect Physiol.* 18:1949–60
11. Brown, G. L., Dale, H. H., Feldberg, W. 1936. Reactions of the normal mammalian muscle to acetylcholine and eserine. *J. Physiol. London* 87:394–424
12. Chang, C. C., Lee, C. Y. 1963. Isolation of neurotoxins from the venom of *Bungarus multicinctus* and their modes of neuromuscular blocking action. *Arch. Int. Pharmacodyn.* 144:241–57
13. Changeux, J. P., Kasai, M., Lee, C. Y. 1970. Use of a snake venom to characterize the cholinergic receptor protein. *Proc. Nat. Acad. Sci. USA* 67:1241–47
14. Charton, M. 1963. The estimation of Hammett substituent constants. *J. Org. Chem.* 28:3121–24
15. Cleland, W. W. 1964. Dithiothreitol, a new protective reagent for SH groups. *Biochemistry* 3:480–82
16. Colhoun, E. H. 1963. The physiological significance of acetylcholine in insects and observations upon other pharmacologically active substances. *Advan. Insect Physiol.* 1:1–46
17. Cull-Candy, S. G., Neal, H., Usherwood, P. N. R. 1973. Action of black widow spider venom on an aminergic synapse. *Nature* 241:353–54
18. Dowson, R. J., Usherwood, P.N.R. 1973. The mode of action of phenylhydrazine hydrochloride on the locust neuromuscular system. *J. Insect Physiol.* 19:355–68
19. Eberson, L. E., Persson, K. 1962. Studies on monoamine oxidase inhibitors. I. The autoxidation of β-phenylisopropylhydrazine as a model reaction for irreversible monoamine oxidase inhibition. *J. Med. Pharm. Chem.* 5:738–52
20. Eldefrawi, M. E., O'Brien, R. D. 1971. Autoinhibition of acetylcholine binding to *Torpedo* electroplax; a possible molecular mechanism for desensitization. *Proc. Nat. Acad. Sci. USA* 68: 2006–7
21. Etessami, S. 1972. Effect of ethanol on neuromuscular contractions. *Comp. Gen. Pharmacol.* 3:200–4
22. Faeder, I. R., O'Brien, R. D. 1970. Responses of perfused isolated leg preparations of the cockroach, *Gromphadorhina portentosa,* to L-glutamate, GABA, picrotoxin, strychnine and chlorpromazine. *J. Exp. Zool.* 173: 203–14

23. Faeder, I. R., O'Brien, R. D., Salpeter, M. M. 1970. A re-investigation of evidence for cholinergic neuromuscular transmission in insects. *J. Exp. Zool.* 173:187–202
24. Faeder, I. R., Salpeter, M. M. 1970. Glutamate uptake by stimulated insect nerve muscle preparation. *J. Cell Biol.* 46:300–7
25. Flattum, R. F., Friedman, S., Larsen, J. R. 1967. The effects of d-tubocurare chloride on nervous activity and muscular contraction in the house cricket, *Acheta domesticus* (L.). *Life Sci.* 6:1–9
26. Friedman, K. J., Carlson, A. D. 1970. The effects of curare in the cockroach. I. dTC-induced failure of leg contraction. *J. Exp. Biol.* 52:583–92
27. Friedman, K. J., Carlson, A. D. 1970. The effects of curare in the cockroach. II. blockage of nerve impulses by dTC. *J. Exp. Biol.* 52:593–601
28. Hamori, J. 1961. Cholinesterase in insect muscle innervation with special reference to insecticide effects of DDT and DFP. *Bibl. Anat.* 2:194–206
29. Hansch, C. 1968. The use of substituent constants in structure-activity studies. In *Physico-Chemical Aspects of Drug Action,* ed. E. J. Ariens, 141–67. New York: Pergamon
30. Hansch, C., Anderson, S. M. 1967. The structure-activity relationship in barbiturates and its similarity to that in other narcotics. *J. Med. Chem.* 10:745–53
31. Hardie, R. L., Thomson, R. H. 1957. The oxidation of phenylhydrazine. *J. Chem. Soc.* 2512–18
32. Harlow, P. A. 1958. The action of drugs on the nervous system of the locust (*Locusta migratoria*). *Ann. Appl. Biol.* 46:55–73
33. Hill, R. B., Usherwood, P.N.R. 1961. The action of 5-hydroxytryptamine and related compounds on neuromuscular transmission in the locust, *Schistocerca gregaria. J. Physiol. London* 157:393–401
34. Holman, G. M., Cook, B. J. 1970. Pharmacological properties of excitatory neuromuscular transmission in the hindgut of the cockroach, *Leucophaea maderae. J. Insect. Physiol.* 16:1891–1907
35. Hoyle, G. 1965. Neural control of skeletal muscle. In *Physiology of Insecta,* ed. M. Rockstein, 2:407–49. New York: Academic
36. Huddart, H. 1971. Contraction of insect muscle. *Exp. Physiol. Biochem.* 4: 219–88
37. Karlin, A. 1969. Chemical modification of the active site of the acetylcholine receptor. In *Membrane Proteins, Proceedings from a Symposium Sponsored by the New York Heart Association,* D. Nachmansohn (chairman), 245–64. Boston: Little, Brown
38. Karlin, A., Bartels, E. 1966. Effects of blocking sulfhydryl groups and of reducing disulfide bonds on the acetylcholine-activated permeability system of the electroplax. *Biochim. Biophys. Acta* 126:525–35
39. Kerkut, G. A., Leake, L. D., Shapira, A., Cowan, S., Walker, R. J. 1965. The presence of glutamate in nerve-muscle perfusates of *Helix, Carcinus,* and *Periplaneta. Comp. Biochem. Physiol.* 15: 485–502
40. Kerkut, G. A., Shapira, A., Walker, R. J. 1965. The effect of acetylcholine, glutamic acid and GABA on contractions of the perfused cockroach leg. *Comp. Biochem. Physiol.* 16:37–48
41. Kerkut, G. A., Walker, R. J. 1966. The effect of L-glutamate, acetylcholine and gamma-aminobutyric acid on the miniature endplate potentials and contractures of the coxal muscles of the cockroach, *Periplaneta americana* L. *Comp. Biochem. Physiol.* 17:435–54
42. Lamar, C. 1968. The duration of the inhibition of glutamate synthetase by methionine sulfoximine. *Biochem. Pharmacol.* 17:636–40
43. Larsen, J. R., Miller, D. M., Yamamoto, T. 1966. d-Tubocurarine chloride: effect on insects. *Science* 152:225–26
44. Lea, T. J., Usherwood, P. N. R. 1970. Increased chloride permeability of insect muscle fibres on exposure to ibotenic acid. *J. Physiol. London* 211:32–33P
45. Lea, T. J., Usherwood, P. N. R. 1973. The site of action of ibotenic acid and the identification of two populations of glutamate receptors on insect muscle-fibres. *Comp. Gen. Pharmacol.* 4: 333–50
46. Lea, T. J., Usherwood, P. N. R. 1973. Effect of ibotenic acid on chloride permeability of insect muscle-fibres. *Comp. Gen. Pharmacol.* 4:351–63
47. Lunt, G. G. 1973. Hydrophobic proteins from locust (*Schistocerca gregaria*) muscle with glutamate receptor properties. *Comp. Gen. Pharmacol.* 4:75–79
48. McCann, F. V. 1966. Curare as a neuromuscular blocking agent in insects. *Science* 154:1023–24

49. McCann, F. V., Reece, R. W. 1967. Neuromuscular transmission in insects: effect of injected chemical agents. *Comp. Biochem. Physiol.* 21:115–24
50. McDonald, T. J. 1971. *Studies on the physiology and pharmacology of insect neuromuscular transmission.* PhD thesis. Univ California, Riverside, Calif. 236 pp.
51. McDonald, T. J. 1971. Phenylhydrazine action on the insect retractor unguis muscle: a free radical mechanism. *Am. Zool.* 11:660
52. McDonald, T. J. 1972. Free radical mechanism for phenylhydrazine action on the retractor unguis muscle of the grasshopper, *Romalea microptera. Comp. Gen. Pharmacol.* 3:319–26
53. McDonald, T. J., Farley, R. D., March, R. B. 1970. The chemical and physical properties of the retractor unguis muscle from the grasshopper, *Romalea microptera. Am. Zool.* 10:304
54. McDonald, T. J., Farley, R. D., March, R. B. 1972. Pharmacological profile of the excitatory neuromuscular synapses of the insect retractor unguis muscle. *Comp. Gen. Pharmacol.* 3:327–38
55. McDonald, T. J., Farley, R. D., March, R. B. 1974. Alcohol induced potentiation of the neurally-evoked contractions of the retractor unguis muscle of the locust, *Schistocerca gregaria. J. Insect. Physiol.* In press
56. McDonald, T. J., O'Brien, R. D. 1972. Relative potencies of L-glutamate analogs on excitatory neuromuscular synapses of the grasshopper, *Romalea microptera. J. Neurobiol.* 3:277–90
57. Mandel'shtam, Y. E. 1961. On the nature of neuro-muscular transmission in the Asiatic locust. *Dokl. Akad. Nauk. SSSR (Bio. Eng.)* 140:1010–12
58. Miledi, R., Molinoff, P., Potter, L. T. 1971. Isolation of the cholinergic receptor protein of *Torpedo* electric tissue. *Nature* 229:554–57
59. O'Brien, R. D., Eldefrawi, M. E., Eldefrawi, A. T. 1972. Isolation of acetylcholine receptors. *Ann. Rev. Pharmacol.* 12:19–34
60. O'Conner, A., O'Brien, R. D., Salpeter, M. M. 1965. Pharmacology and fine structure of peripheral muscle innervation in the cockroach, *Periplaneta americana. J. Insect. Physiol.* 11:1351–58
61. Osborne, M. P. 1970. Structure and function of neuromuscular junctions and stretch receptors. In *Insect Ultrastructure,* ed. A. C. Neville. *Symp. Roy. Entomol. Soc. London* 5:77–100
62. Phillis, J. W. 1970. *The Pharmacology of Synapses.* New York: Pergamon. 358 pp.
63. Piek, T. 1966. Site of action of the venom of the digger wasp *Philanthus triangulum* F. on the fast neuromuscular system of the locust. *Toxicon* 4:191–98
64. Pitman, R. M. 1971. Transmitter substances in insects: a review. *Comp. Gen. Pharmacol.* 2:347–71
65. Quastel, D. M. J., Hackett, J. T., Cooke, J. D. 1971. Calcium: is it required for transmitter secretion? *Science* 172:1034–36
66. Rees, D. 1974. The spontaneous release of transmitter from insect nerve terminals as predicted by the negative binomial theorem. *J. Physiol.* London 236:129–42
67. Roeder, K. D., Weiant, E. A. 1950. The electrical and mechanical events of neuromuscular transmission in the cockroach, *Periplaneta americana. J. Exp. Biol.* 27:1–13
68. Sandow, A. 1965. Excitation-contraction coupling in skeletal muscle. *Pharmacol. Rev.* 17:265–320
69. Stössel, W., Zebe, E. 1968. Zur intracellulären Regulation der Kontractionsaktivität. *Pfluegers Arch. Gesamte Physiol. Menschen Tiere* 302:38–56
70. Stroh, H. H., Ebert, B. 1964. Über die Autoxydation kernsubstituierter Phenylhydrazine. *Chem. Ber.* 97:2335–41
71. Usherwood, P. N. R. 1962. The action of ryanodine on insect skeletal muscle. *Comp. Biochem. Physiol.* 6:181–99
72. Usherwood, P. N. R. 1967. Insect neuromuscular mechanism. *Am. Zool.* 7:553–82
73. Usherwood, P. N. R. 1969. Glutamate sensitivity of denervated insect muscle fibres. *Nature,* 223:411–13
74. Usherwood, P. N. R. 1969. Electrochemistry of insect muscle. *Advan. Insect Physiol.* 6:205–78
75. Usherwood, P. N. R. 1972. Transmitter release from insect excitatory motor nerve terminals. *J. Physiol.* 227:527–51
76. Usherwood, P. N. R. 1972. Glutamate as an excitatory neuromuscular transmitter in insects. *Neurosci. Res. Program. Bull.* 10:136–43
77. Usherwood, P. N. R. 1973. Action of iontophoretically applied gamma-aminobutyric acid on locust muscle fibres. *Comp. Biochem. Physiol.* 44A:663–64
78. Usherwood, P. N. R., Cochrane, D. G., Rees, D. 1968. Changes in the structural, physiological and pharmacologi-

cal properties of insect excitatory nerve-muscle synapses after motor nerve section. *Nature* 218:589–91

79. Usherwood, P. N. R., Machili, P. 1966. Chemical transmission at the insect excitatory neuromuscular synapse. *Nature* 210:634–36
80. Usherwood, P. N. R., Machili, P. 1968. Pharmacological properties of excitatory neuromuscular transmission in the locust. *J. Exp. Biol.* 49:341–61
81. Usherwood, P. N. R., Machili, P., Leaf, G. 1968. L-Glutamate at insect excitatory nerve-muscle synapses. *Nature* 219:1169–72
82. van Gelder, N. M. 1971. Molecular arrangement for physiological action of glutamic acid and γ-aminobutyric acid. *Can. J. Physiol.* 49:513–19
83. Wallach, D. P. 1961. Studies on the GABA pathway—I. The inhibition of γ-aminobutyric acid-α-ketoglutaric acid transaminase *in vitro* and *in vivo* by U-7524 (amino-oxyacetic acid). *Biochem. Pharmacol.* 5:323–31
84. Walther, C., Usherwood, P. N. R. 1972. Characterisation of the glutamate receptor at the locust excitatory neuromuscular junction. *Verh. Deut. Zool. Ges. Helgoland* 65:309–12
85. Wigglesworth, V. B. 1958. The distribution of esterase in the nervous system and other tissues of the insect, *Rhodnius prolixus*. *Quart. J. Microsc. Sci.* 99:441–50

INHERITED STERILITY IN LEPIDOPTERA

◆6086

David T. North
Metabolism and Radiation Research Laboratory, Agricultural Research Service,
U.S. Department of Agriculture, Fargo, North Dakota 58102

Lepidoptera include some of the most destructive insect pests of agriculture. In the past decade, the increasing resistance to insecticides, along with the more restrictive use of many chemical pesticides, has prompted research for new avenues of insect control. Many researchers have studied the possibilities of population suppression through the release of radiation-sterilized moths, trying to emulate the success with the screwworm fly (*Cochliomyia hominivorax*) (9). These attempts were not resounding successes, though there have been several exceptions. Through the release of sterile moths, control of the codling moth (*Laspeyresia pomonella*) (72, 73) was obtained in isolated orchard tests, and the tobacco hornworm (*Manduca sexta*) (85) was controlled on the island of St. Croix, U.S. Virgin Islands. Both these successes were under somewhat isolated conditions and on a relatively small scale.

Lepidoptera require large doses of radiation to effect sterility compared with most other insects and have often been referred to as being radioresistant (43, 58). Moths are radioresistant when the criterion is induced dominant lethality or sterility. However, they are equally as radiosensitive as the house fly (*Musca domestica*) (60) when lifespan is the criterion.

When Lepidoptera are given sterilizing doses of radiation, induced physiological disturbances (36, 60, 63, 64) are manifested, such as lack of or insufficient sperm transfer, lack of mating, etc. In fact, radiation-induced sterility in a majority of the lepidopteran species investigated is not due primarily to dominant lethality but is more directly related to unfertilized eggs. It was the realization that these debilitating effects would limit the ability of the sterile males to compete with natural males in suppressing a population that prompted the reinvestigation of inherited sterility (58, 60, 101). It was believed that more competitive insects would result from using partially sterilizing doses of radiation.

Proverbs (71), using the codling moth, was the first to observe that the progeny of irradiated males were semi- to completely sterile. The surviving progeny of irradiated parents inherit sufficient genetic material to make them partially or completely sterile; this is not in itself unique (84). It is important, however, that in contrast with other orders, the progeny of irradiated parents are more sterile than their parents. Chromosome translocations were attributed to be the basis for the inherited sterility in Lepidoptera (61). Although the Japanese had used genetic markers to identify translocations (95) in genetic studies with the silkworm (*Bombyx mori*), Bauer (7) was the first to describe the cytogenetic behavior of translocations in Lepidoptera.

There are some unique advantages of using such a phenomenon in controlling or suppressing native insect populations (42). This chapter will deal with what is generally known regarding the efficiency of induction of inherited sterility and the behavior of the F_1 individuals and will then discuss the possible genetic basis of this phenomenon.

INDUCED INHERITED STERILITY

The first report of induced inherited sterility in Lepidoptera showed that the progeny from codling moth males given 30 krad as pupae were highly sterile, and that the majority of the progeny that became adults were male (71, 74). A summarization of research on the effects of radiation on the gypsy moth (*Porthetria dispar*) from 1957 to 1962 (25) showed that: (*a*) mortality among F_1 individuals from irradiated parents (irradiated as pupae) was highest in the larval stage; and (*b*) moths that had been irradiated as fourth- or fifth-instar larvae were partially sterile. Results with the Indian meal moth (*Plodia interpunctella*) and the Angoumois grain moth (*Sitotroga cerealella*) (13) suggested it was plausible to capitalize on the radiation-induced inherited sterility in these species to suppress later generations. Inherited sterility was then induced in the cabbage looper (*Trichoplusia ni*) (58) and the sugarcane borer (*Diatraea saccharalis*) (101), and it has since been induced in over a dozen species (Table 1). Not until 1969 (62) was suppression of a laboratory population tried by a single release of partially sterile males. Hence, the salient factors of induced inherited sterility were presented in the years from 1962 to 1968. Increased efforts have been made in the past six years to explore this possibility as a means of insect control.

Methods of Induction

Radiation was used in the majority of the studies to induce inherited sterility. There is nothing magical about ionizing radiation; probably any mutagen that is an effective chromosome breaker will induce inherited sterility. The general availability of Cobalt-60 and Cesium-137 is responsible for most of the research being done with gamma radiation. Neutrons are more efficient than X or γ rays in causing chromosome breaks (47), and therefore, this would hold true for inducing inherited sterility. Neutron sources, not being as available, have not been used extensively but were shown to be more efficient than ionizing radiation in inducing recessive lethals in the silkworm (55) and other insects (49). Inherited sterility is induced more efficiently by neutrons in cabbage loopers (63, 64) and neutrons enjoy several advantages over gamma irradiation. First, the percentage hatch of the F_1 generation was higher, probably due to a higher degree of fertilized eggs; this means that more

Table 1 Species of Lepidoptera in which inherited sterility has been induced

Family and species	Reference
Crambidae	
Diatraea saccharalis	101
Galleriidae	
Galleria mellonella	57
Gelechiidae	
Sitotroga cerealella	13
Pectinophora gossypiella	31
Noctuidae	
Spodoptera exigua	15
Trichoplusia ni	58
Heliothis zea	63
Heliothis virescens	66
Olethreutidae	
Laspeyresia pomonella	71
Phycitidae	
Ephestia cautella	1, 30
Plodia interpunctella	13
Anagasta kuhniella	H. C. Hofmann and J. G. Riemann, unpublished observations
Paramyelois transitella	39
Pieridae	
Pieris brassicae	7

F_1 individuals could be placed into a native population in a control program. Second, the required dose to the P_1 generation to induce completely sterile F_1 was about one-third less for neutrons than for gamma radiation. More competitive moths would be expected by subjecting later-stage pupae or adults to neutron irradiation. Radiations with a high linear energy transfer (LET) are more effective per unit of absorbed energy in inducing inherited sterility and will most likely give a more competitive P_1 moth than will gamma irradiation. The immobility of nuclear reactors (the usual source of neutrons) makes this impractical for most programs.

There has been very little use of chemosterilants in inducing inherited sterility. Stimmann (86) found that a topical application of tepa in acetone induced both P_1 and F_1 sterility in cabbage loopers. The amount of inherited sterility induced was dose-dependent as it was in irradiation experiments.

Life Stage Irradiated

Historically, it has been a common practice of many workers studying radiation-induced sterility in insects to irradiate all life stages. The irradiation of early stages (prior to pupa) to induce sterility in species that have monokinetic chromosomes has been inefficient because the induced damage is selected against and not included in the newly formed gametes. Species with holokinetic chromosomes (e.g. Lepidoptera) present a different situation since the incidence of cell death would not be expected to be as great.

The adult stage, since it is not undergoing metamorphosis, is considered a better radiation target and does not suffer as many of the debilitating effects of radiation (such as malformed wing, improper eclosion, inability to mate, etc) that plague individuals irradiated earlier (13, 25, 71, 75). Adult irradiation, however, does not totally avoid physiological damage (35, 63, 65). The progeny of moths irradiated as either adults or late pupae exhibit the same amount of inherited sterility (13, 15, 57, 102).

The dose of radiation required to induce inherited sterility is between 12.5 and 22.5 krad as borne out by studies of the 15 species listed in Table 1. The F_1 male progeny from irradiated parents in all these species are more sterile than the F_1 female progeny, regardless of the stage irradiated.

The reproductive capability of cabbage loopers irradiated as pupae was equal to that when adults were irradiated only when the dose was fractionated over several days (96). Female sterility was also higher when fractionated doses were given to pupae. Fractionating the dose also is more efficient in *Galleria mellonella* (100). Pupal irradiation is preferred because the pupal stage is the easiest to handle. Further investigation of dose fractionation of pupal stages would be warranted. Adult codling moths can be chilled before and during irradiation without causing any differences in the amount of F_1 sterility exhibited by the F_1 irradiated moths (106). Chilling does not affect the radiosensitivity of the P_1 generation, and these treatments reduce the injury incurred by restricting the moths' activity.

Larval irradiation proved unsatisfactory because the adults resulting from irradiated larvae usually are not capable of reproduction (2, 15, 19, 20, 25, 76, 102). It was concluded that relatively low doses (3.5 krad) given to fifth-instar larvae of the Indian meal moth induced sufficient genetic damage to spermatogonia to induce sterility (2). The aberrations induced by the low doses are evidently not complex and thus are incorporated into the mature spermatozoa; hence, semisterility is transmitted by the F_1 male. Also, the progeny for several generations following irradiation have been reported to be semisterile (37, 57). First-instar *G. mellonella* larvae given 4 krad of gamma radiation developed into partially sterile adult males. This sterility persisted through at least two generations of the male line, and nearly all of the later populations were male (57). Oogenesis proceeds later than spermatogenesis in Lepidoptera. It is, therefore, understandable that irradiation of early stages (before fifth instar) would completely destroy the developing germ cells of the ovaries (52) and thus produce infecund females. In *Bombyx* (95), the percentage of unfertilized eggs decreases with the age of female pupae irradiated. Irradiation of early cell stages of spermatogenesis in Lepidoptera (2, 81, 82, 88, 98, 99) would not be expected to yield genetic damage that would be inherited in any sufficient amount required to be useful in control. Testes containing only gonial cells (35, 83, 89) irradiated with 5 krad had the definitive gonia killed but the predefinitive gonia were more resistant. Consequently, the testes were repopulated with slightly damaged or undamaged cells. Researchers exploring the possibility of inducing inherited sterility by irradiation of larval and embryonic stages must be able to recognize the target germ cells damaged by the radiation and determine their ultimate fate. Induced genetic damage not incorporated into the sperms or ova because of cell death does not lead to inherited sterility.

Attempts to induce sufficient genetic damage to yield inherited sterility by irradiating embryos have generally been futile (6, 13, 15, 19, 20, 34, 39, 45, 54, 71, 76, 96, 102). The results were varied because heterogeneous embryonic stages were irradiated. In most cases there was no control over such developmental influences as temperature and photoperiod. As the dose of radiation increases, so does embryonic mortality. The resulting adults, however, are often malformed and fail to mate, but some of the moths that do mate are partially sterile. The female appears to be more radiosensitive when irradiated as an embryo. F_1 progeny from moths that were irradiated as eggs are usually fertile. The persistence of sterility for several generations beyond the P_1 of irradiated eggs has been reported (38, 57). These exceptions warrant further exploration of embryonic irradiation as a means of inducing inherited sterility. However, future studies should take into consideration irradiation of precise stages of embryonic development, and control should be exerted over the environment. Temperature and photoperiod during embryonic development play an important role in the subsequent fecundity of codling moths (16).

It should be remembered that an embryo, diploid in chromosome complement, is unlike the haploid gamete, and thus the resulting adult would be heterozygous for chromosomal aberrations. This would make the P_1 of an irradiated embryo the equivalent, genetically, to the F_1 progeny from an irradiated sperm or ovum. Irradiation of embryos, if successful, would allow placing irradiated eggs on noneconomic hosts; this would allow rearing sterile moths in nature. This technique could easily be used for the protection of some field crops.

Male-Female Responses

As stated earlier, the progeny from irradiated male moths are more sterile than their parents. The progeny of irradiated females exhibit some sterility (62, 63, 68) but are more fertile than their irradiated mothers. Irradiation of female sugarcane borers with as little as 2 krad resulted in the subsequent collapse of the laboratory population (103). The female progeny of these irradiated females were reported sterile, and the male progeny were semisterile. F_1 males from irradiated female cabbage loopers (15 krad) were more sterile than the F_1 females (63). However, no difference was found in the sterility between the F_1 males and females of irradiated tobacco budworm females at any dose up to 22.5 krad (68). This is inconsistent with the data reported for the sugarcane borer, though it is difficult to visualize what difference in the species could be responsible for giving such a large variance in the dose required and why the F_1 females are more sterile than F_1 males. F_1 tobacco budworm males from irradiated females did not transfer sperm as well as the untreated males (68). This trait is characteristic of F_1 males from irradiated male parents (4, 63, 66, 78).

The fertility of offspring from irradiated females in most lepidopteran species has an important bearing on the effectiveness of releasing partially sterile females for suppression. This makes it mandatory to release only males if inherited sterility is to be used as a method of suppressing lepidopterous pests. One way to avoid this would be to release both sterile females and partially sterile males. This could be achieved by giving a single dose of radiation to both sexes because females are more radiosensitive than males.

Sex Distortion

It has become generally accepted that irradiated male produce more male than female progeny. This was first shown in the codling moth (71) and later in all species examined (15, 30, 39, 57, 62, 66, 100). Contrary to this, irradiated females produce progeny in a normal sex ratio (68).

In Lepidoptera, the male is homogametic and the female is heterogametic (29, 51, 79, 95, 105). The male is XX (29) or ZZ (79, 95), and the female is XY or ZW (depending on the nomenclature used for the sex chromosomes). It is also possible that in some species the female would be XO or ZO (87). This is neither the time nor the place to get into a lengthy discussion of the symbolism of sex chromosomes. The reason for using the ZZ:ZW scheme is to point out that the female is heterogametic (79). Goldschmidt (26–28) beautifully analyzed the sex determination of *Lymantria dispar,* and I prefer to follow him in retaining the classical genetic sex symbolism of using XX and XY.

The portrayal of sex determination as nothing more than the orderly disjunction of a single chromosome is a gross simplification in Lepidoptera. There are many autosomal factors involved (12, 29, 53, 79, 95, 105) in sex determination in Lepidoptera. The reason that irradiated males tend to produce more male progeny than female was suggested as being due to the induction of recessive lethals on the X chromosome of the male (63, 66). Male cabbage loopers fed ethyl methanesulfonate (EMS), an active mutagen but not known as an effective chromosome breaker, produced more male progeny than female (D. T. North, unpublished observations) without sterility. Individual fertile lines were isolated in the F_2 generation that produced two males to one female. This, along with the fact that the sex ratio of the progeny of the irradiated females (66) is 1:1 is considered evidence that sex distortion is caused by lethals on the X chromosome. When all the progeny from irradiated females were grouped together, the sex distortion was in favor of males, whereas if individual pairs were analyzed, some lines showed distortion toward an excess of females (66). Catcheside & Lea (10) found sex-ratio distortion in irradiated *Drosophila melanogaster* and concluded it was caused by damage to sperms bearing the X chromosome. This gave an excess of males in *Drosophila,* but in Lepidoptera it would reduce the number of XX individuals (males), giving an excess of females. Therefore, it is possible that distortion of both sexes exists in Lepidoptera; however, the majority of the distortion creates an abundance of males. The determination of the beneficial effect of sex distortion on population suppression and manipulation has not been examined except for a unisexual (18) strain of *Estigmene acrea.*

Development of F_1 Progeny

F_1 progeny of irradiated parents develop more slowly than progeny resulting from a mating of untreated individuals (11, 63, 66), and this phenomenon is dose-dependent. The lag in development is also temperature-dependent (67). When progeny were reared at high temperatures (27°C), there was more radiation damage expressed and some progeny appeared to enter diapause. This points out the severe physiological debilitation suffered by these F_1 individuals. The possible hormone imbalances existing in the progeny of irradiated parents have not been studied, but this incident in the budworm (67) shows that the situation exists. This could be a limiting factor under field conditions for achieving population suppression using inherited sterility. The fact that these individuals would be reared in nature (61) has been presented as an argument for their being hardier than the P_1-released, laboratory-reared moths. Certainly the limited data available on development under known conditions would make such a conclusion suspect. Before using inherited sterility as a possible method of suppression, more data are needed on behavior of the F_1 moths in a natural environment.

Sperm Production and Transfer

The male progeny from males given a partially sterilizing dose of radiation often fail to transfer sperm to the spermathecae of the female successfully (11, 30, 44, 64, 66, 102), but no quantitative data on sperm production and transfer are available. The relative amount and type of sperm transferred to the spermatheca by the F_1 male was approximated (11, 44) in the pink bollworm (*Pectinophora gossypiella*). These approximations are crude at best but do serve to illustrate that F_1 males have a problem in sperm production and transfer. It was observed that F_1 progeny of males treated with apholate had few eupyrene sperms (89), although they produced a normal number of apyrene sperms. It has not been determined whether fewer eupyrene sperms are produced or whether eupyrene sperms are unable to reach the spermathecae. It is impossible to differentiate between the two from present data. Now that there are proven methods to determine the number of sperms in the spermathecae (77), quantitative evaluation of sperm transfer by F_1 males is possible and should be studied.

Riemann (78) examined the sperm of male F_1 Mediterranean flour moths, the male parents of which had received 15 krad. Sperms were observed from the spermatheca of an unirradiated female to which the F_1 male had mated or from sperm bundles in the seminal vesicle. The bundles in the seminal vesicles had reduced numbers of cells, and these cells contained a variety of abnormalities. All of the intact sperms in the spermathecae appeared morphologically normal. Apparently only morphologically normal cells are maintained in the spermathecae. The F_1 progeny of Indian meal moths, the male parents of which were irradiated as adults, inherited more ultrastructural abnormalities than those from P_1 males irradiated as first-instar larvae (3). It is interesting to note that the P_1 males irradiated as fifth-instar larvae exhibited a high frequency of flagellar and apical abnormalities of eupyrene sperms. The damage inflicted on the germ cells of the fifth-instar larvae is eliminated when the abnormal sperms are lost. It is questionable, however, whether the abnormalities found in the F_1s from irradiated adults are due to inheritance of the genetic damage or are simply an imbalance of the system, manifesting itself as abnormal sperm. This point definitely needs clarification.

INHERITED STERILITY BY GENETIC MANIPULATION

Hybridization

There are numerous reports of hybridization of species and subspecies in Lepidoptera (79), but their value to genetic understanding has been very limited. Since many hybrids are sterile in the first generation, little genetic information can be derived from hybridization. Although hybridization of lepidopterous species has been studied for over three quarters of a century, this area was not explored as a source of sterility for population suppression until 1959 (17). Hybridization offers a method of inducing sterility without the debilitating effects of radiation, though hormone imbalance, etc is found in hybrids. A hybrid that has limited fertility affords an opportunity to manipulate the genetic material so a vigorous, but sterile, insect would result.

Laster (46) found that the *Heliothis subflexa* X *Heliothis virescens* hybrid could be made in the laboratory. The progeny from such a cross are of interest since the F_1 males are sterile and the F_1 females are fertile when crossed to *H. virescens.* Although the F_1 female is fertile, her male progeny are sterile. This would appear to be an exception to the rule expressed by Haldane (33) that in hybrids of animals, the rare or sterile sex is the heterogametic sex. There is no reason to suspect, however, that the males in *Heliothis* species are not homogametic.

Proshold & LaChance (69) have expanded upon the original work of Laster and were able to make the reciprocal hybrid crosses between *H. subflexa* and *H. virescens.* The F_1 males from both reciprocal crosses were nearly sterile when backcrossed to either parent species. The female F_1s were semisterile, depending upon the interspecific cross. Over 40% of these F_1 females resulting from *H. subflexa* X *H. virescens* entered diapause under conditions which normally do not induce diapause in either species. The nondiapausing females were reluctant to mate with males of either parent species. The male progeny were sterile through three backcross generations of these females X *H. virescens* males. Females from a reciprocal cross, however, produced completely fertile male progeny following three backcrosses to male *H. virescens.* Not all hybrids give sterility (56), but their progeny can inherit chromosomal anomalies.

The male sterility reported in *Heliothis* hybrids (46, 69) was apparently from the lack of fertilization. This conclusion was reached because of the absence of eupyrene sperm in the spermathecae of females mated to hybrid males. Sperm bundles were observed in spermatophores transferred by hybrid males and the eupyrene bundles were described as inactive, usually with scattered sperms (46). This is not necessarily an abnormal situation (36), but failure to recover many eupyrene sperms in a spermatheca would be indicative of problems related to the transport of the eupyrene sperms to the spermathecae. A majority of the eupyrene sperms are morphologically abnormal (R. Richard, personal communication) and, therefore, it is possible that females eliminate these abnormal sperms rather readily from the spermathecae (78). If these observations of the spermathecae are not made soon after the arrival of the sperm, they would be missed in the analysis.

Hybrids afford many types of cytogenetic manipulations to give sterility that can possibly be used for population suppression. One outstanding limitation of using the described *Heliothis* hybrids would be the possible influx of *H. subflexa* genes into *H. virescens* through the fertile hybrid females, assuming the hybrid females mate in the field with *H. virescens* males. The adding of genes that could possibly widen host-plant adaptation and possible insecticide resistance to *H. virescens* could override any possible advantage obtained with the sterile males. It would appear that detailed studies are required in both the areas of quantitative genetics and cytogenetics of these hybrids before considering releasing test insects in the field. With further research, this could prove to be one of the most promising areas where genetic control could be applied to suppression of natural populations.

Translocations

Theoretically, population suppression of insects can be achieved by releasing homozygous translocations (84, 104, 107), but its value in controlling Lepidoptera is doubtful. First, lepidopteran species possess large numbers of chromosomes (50) and they lack size and varied morphology that could be used for cytological identification (7, 61, 95). Second, homozygous translocations, in *Drosophila* at least (8), have proved to be very nonviable. Also, there are no lepidopteran species of economic importance with a sufficient number of mutants to label the chromosomes as has been done in *Bombyx* (95), where over 250 mutants are known. With this extensive knowledge of linkage groups, translocations can be recognized through segregation of these loci in various crosses, and the stock can be easily maintained. It would take a long time to develop such a system for any one economically important species. Bauer (7) conducted a thorough study of translocation behavior in *Pieris brassicae.* North & Holt (61) related heterozygous translocations to radiation-induced inherited sterility. Using the technique of partial castration (40), 24 different heterozygous translocations were examined in the cabbage looper (D. T. North, manuscript in preparation), and there was less than 10% fertility in most cases. Interestingly, there was little decrease in the fecundity of the females to which heterozygous translocation males were mated, inferring adequate sperm transfer. Translocations can be effectively used to sex Lepidoptera (95), though this technique has not been explored for economic species. However, there are enough mutants available in economic species to make this technique worthwhile.

Polyploidy

Polyploids have been induced in *Bombyx* by several methods (95) but have proved impossible to maintain. The tetraploids induced have been females and thus have to be mated with diploid males. This results in triploid offspring. Reports have been made (95) of the establishment of a tetraploid strain from the progeny of a 6N X 2N cross. However, this strain could not be maintained for more than two generations.

Polyploids could be used as a source of sterility in several ways. Sterile triploids can be obtained by crossing 4N X 2N individuals. This would prove impractical because the female line would have to be induced each generation. Triploid individuals could possibly be stockpiled in some species and stored in the diapause condition until needed for release. This would be cumbersome, and the performance of triploids would have to be studied first. Also, there is evidence that the sexual reproductive performance (70) of the moths emerging from diapause is poor.

When eggs are centrifugated prior to undergoing first cleavage division, many triploids and aneuploids result (95). There is the possibility that stable aneuploid lines could be established from these individuals which would be intrafertile and produce no progeny when mated to normal individuals. The ability to remove one testis (40) for cytological examination and then test breed the resulting moth can be used in establishing such strains.

THE CYTOGENETIC BASIS OF INHERITED STERILITY

The lepidopteran karyotype is characteristically small and has symmetrical chromosomes without much morphological variation. There have been many reports on the holokinetic nature of the lepidopteran chromosome (5, 7, 23, 61, 80, 90–94, 99) and the question is still unsettled (105), but these reports serve as the basis to explain radioresistance and inherited sterility in Lepidoptera (43, 58). The small size and lack of distinct shape has not allowed determination of the type of the kinetochore through chromosomal behavioral studies. There are several methods for preparing lepidopteran chromosomes for cytogenetic examination (21, 32, 56, 61).

Transmission electron micrograph studies indicated that lepidopteran chromosomes in gonia and meiotic cells have a localized kinetochore approximately the same size as in monokinetic species (14, 24). During metaphase I they do not appear to have any kinetochore activity (23).

A greater number of translocations (61) could be recovered from species with holokinetic chromosomes than in a monokinetic species since all translocations induced would behave as symmetrical exchanges. Thus, the number of recovered translocations from any one dose of radiation would be double that for a monokinetic species.

Insufficient data are available to make conclusions regarding the role chromosome translocations play in induced inherited sterility; 100% of the progeny of male corn earworms given 20 krad possessed at least one translocation (59), and when their fathers received only 10 krad, 80% possessed at least one translocation. The biggest problem in assessing the amount of sterility contributed per heterozygous translocation is the difficulty of separating fertilized from unfertilized eggs. Eggs laid by females inseminated by F_1 males are often unfertilized because of the abnormal sperm received from the F_1 males (4, 78). It is of more than academic interest that we determine cytologically the event or series of events that lead to abnormal sperm production in the F_1s. It would appear that the cause may be the same in F_1s from irradiated parents as well as for hybrid crosses.

An exciting aspect of the hybridization work with *Heliothis* is the possibility of substituting chromosomes from one species to another. Although 3–7 chromosome pairs in these hybrids are not found as bivalents at metaphase I (69), this is not to infer lack of homology. First, it is not known whether the lack of pairing is asynaptic or desynaptic in origin. There are many factors that can affect either type of pairing relationship; i.e. translocations, inversions, etc (105). However, substitution of one or more of the *H. subflexa* chromosomes into the karotype of *H. virescens* with proper selection could yield an intrafertile strain that would be sterile when crossed back to normal *H. virescens.* It is imperative, therefore, to determine the reason for the lack of pairing observed and the relationship of this to sterility and abnormal sperm production. Even more important, if chromosome substitution lines are developed this would afford isolation against any intragression of the two species under field conditions.

PRACTICAL APPLICATION OF INHERITED STERILITY FOR POPULATION SUPPRESSION

The most obvious drawback to the use of inherited sterility is that the F_1 progeny need to be reared on a host. (If this host is of economic value, this does not present an appealing method to growers.) There are certain insects that appear more suitable for control by inherited sterility than others: the sugarcane borer, where cultural practices are such that damage to early tillers does not necessarily hurt the crop; the gypsy moth, where though the F_1s would cause some defoliation, there should be sufficient sterility of the P_1 generation to lower the original infestation, and the F_1s could effect suppression during the next generation; the corn earworm, particularly in the southeastern U.S., because infested corn usually contains a single larva that does little damage and thus F_1s would be available to suppress the population that infests cotton; and the Asiatic rice borer, *Chilo suppressalis,* where, again, infestation of the early tillers does not harm the crop value. Protection of the southeastern states from cabbage loopers (42) by releasing partially sterile moths in Florida early in the year, and thus minimizing the populations available for northern migration, has been suggested. The idea of using partially sterile moths that also transmit conditional lethals (41) has been suggested. Careful consideration would have to be given to such a plan or the two objectives could cancel each other. For maximum effect of inherited sterility, the P_1 generation should be fertile (22, 62, 102) and the F_1 completely sterile. In order to transmit recessive conditional lethals into a population, the F_1 generation would need to be only semisterile. However, it is a concept worth consideration.

Inherited sterility was effective in suppressing cabbage loopers (97) and sugarcane borers (103) in field cages, but there has been no major assessment of the technique under field conditions. Partially sterile cabbage loopers were released on St. George Island, Florida (48), but there was no evidence that the suppression obtained for one month resulted directly from inherited sterility. Cytological analysis of corn earworms collected on St. Croix, U.S. Virgin Islands, during and following the release of male and female moths given 25 krad, showed their progeny were able to survive (D. T. North and J. W. Snow, manuscript in preparation). As high as 30% of the larvae sampled at one period had chromosome aberrations, indicating that they were the progeny of an irradiated parent. These results provided no data as to the efficiency of infusing F_1s into a natural population, but clearly indicate that F_1s were capable of establishing themselves in a natural population. Based on laboratory studies (62) at least 60% of a population have to be F_1s from an irradiated parent for the inherited sterility concept to be effective. The use of inherited sterility is not the complete answer to suppression of all Lepidoptera, but the advantages it affords outweigh the disadvantages for some species. Greater evaluation of the method is needed at the field level before its total worth as a possible control measure can be assessed.

AKNOWLEDGMENTS

The author is indebted to Jane Houtkooper and Gail Nelson who kept order in the laboratory while this chapter was in preparation. Special thanks are extended to Diane Kastet for deciphering and typing the manuscript and to Fred Proshold for listening. Heartfelt thanks go to Mrs. Vivian Chaska, who once again survived the incompetence of the author and got the job done.

Literature Cited

1. Ahmed, M. S. H., Al-Hakkak, Z., Al-Saqur, A. 1971. Exploratory studies on the possibility of integrated control of the fig moth, *Ephestia cautella* Walk. In *Applications of Induced Sterility for Control of Lepidopterous Populations,* 1–6. Vienna: Int. At. Energy Ag. 169 pp.
2. Ashrafi, S. H., Brower, J. H., Tilton, E. W. 1972. Gamma radiation effects on testes and on mating success of the Indian meal moth, *Plodia interpunctella. Ann. Entomol. Soc. Am.* 65:1144–49
3. Ashrafi, S. H., Roppel, R. M. 1973. Radiation-induced alteration of testes of larvae of the Indian meal moth, *Plodia interpunctella* (Hübner). *Ann. Entomol. Soc. Am.* 66:1324–28
4. Ashrafi, S. H., Roppel, R. M. 1973. Radiation-induced partial sterility related to structurally abnormal sperms of *Plodia interpunctella. Ann. Entomol. Soc. Am.* 66:1309–14
5. Barry, B. D., Guthrie, W. D., Dollinger, E. J. 1967. Evidence of a diffuse centromere in the European corn borer, *Ostrinia nubilalis. Ann. Entomol. Soc. Am.* 60:487–88
6. Bartlett, A. C., Staten, R. T., Ridgway, W. O. 1973. Gamma radiation treatment of pink bollworm eggs. *J. Econ. Entomol.* 66:475–77
7. Bauer, H. 1967. Die kinetische organisation der lepidopteren-Chromosomen. *Chromosoma* 22:102–25
8. Burnham, C. R. 1962. *Discussion in Cytogenetics.* Minneapolis, Minn.: Burgess. 375 pp.
9. Bushland, R. C. 1971. Sterility principle for insect control: Historical development and recent innovations. In *Sterility Principle for Insect Control or Eradication,* 3–14. Vienna: Int. At. Energy Ag. 542 pp.
10. Catcheside, D. G., Lea, D. E. 1945. Induction of dominant lethals in *Drosophila* sperm by x-rays. *J. Genet.* 47:1–9
11. Cheng, W. Y., North, D. T. 1972. Inherited sterility in the F_1 progeny of irradiated male pink bollworms. *J. Econ. Entomol.* 65:1272–75
12. Cockayne, E. A. 1938. The genetics of sex in Lepidoptera. *Biol. Rev.* 13: 107–32
13. Cogburn, R. R., Tilton, E. W., Burkholder, W. E. 1966. Gross effects of gamma radiation on the Indian-meal moth and the Angoumois grain moth. *J. Econ. Entomol.* 59:682–84
14. Danilova, L. V. 1973. An electron microscope study of meiosis in diploid males of the silkworm. *Ontogenez* 4:40–48
15. Debolt, J. W. 1973. Preliminary investigation of the effects of gamma irradiation on the fertility, survival, and mating competitiveness of the beet armyworm, *Spodoptera exigua. Proc. Nat. Symp. Radioecology, 3rd, 1971* 2:1158–61
16. Deseö, K. V. 1973. Side effect of diapause inducing factors on the reproduc-

tive ability of some lepidopterous species. *Nature New Biol.* 242:126–27

17. Downes, J. A. 1959. The gypsy moth and some possibilities of the control of insect by genetical means. *Can. Entomol.* 91:661–64
18. Earle, N. W., MacFarlane, J. 1968. A unisexual strain of the saltmarsh caterpillar, *Estigmene acrea* (Drury). *Ann. Entomol. Soc. Am.* 61:949–53
19. Ercelik, T. M., Holt, G. G. 1972. Sterility inherited by the progeny of male cabbage loopers irradiated in various stages of development. *Environ. Entomol.* 1:592–96
20. El Sayed, E. I., Graves, J. B. 1969. Effects of gamma radiation on the tobacco budworm. III. Irradiation of eggs and larvae. *J. Econ. Entomol.* 62: 296–98
21. Emmel, T. C. 1968. Methods for studying the chromosomes of Lepidoptera. *J. Res. Lepidoptera* 7:23–28
22. Fossati, A., Stahl, J., Granges, J. 1971. Effect of gamma radiation dose on the reproductive performance of the P and F_1 generations of the codling moth, *Laspeyresia pomonella* L. In *Application of Induced Sterility for Control of Lepidopterous Populations,* 41–47. Vienna: Int. At. Energy Ag. 169 pp.
23. Friedlander, M., Wahrman, J. 1970. The spindle as a basal body distributor. A study in the meiosis of the male silkworm moth, *Bombyx mori. J. Cell Sci.* 7:65–89
24. Gassner, G., Klemetson, D. J. 1974. A transmission electron microscope examination of hemipteran and lepidopteran gonial kinetochores. *Can. J. Genet. Cytol.* In press
25. Godwin, P. A., Rule, H. D., Waters, W. E. 1965. Some effects of gamma irradiation of the gypsy moth, *Porthetria dispar. J. Econ. Entomol.* 57:986–90
26. Goldschmidt, R. 1916. The function of the apyrene spermatozoa. *Science* 44: 544–46
27. Goldschmidt, R. 1934. Untersuchungen über Intersexualität. VI. *Z. Indukt. Abstamm. Vererbungsl.* 67:1–40
28. Goldschmidt, R. 1942. Sex determination in *Melandrium* and *Lymantria. Science* 95:120–21
29. Goldschmidt, R. 1955. *Theoretical Genetics.* Berkeley, Calif.: Univ. California Press. 563 pp.
30. Gonen, M., Calderon, M. 1971. Effects of gamma radiation on *Ephestia cautella* (Wlk.) II. Effects on the progeny of irradiated males. *J. Stored Prod. Res.* 7:91–96
31. Graham, H. M., Ouye, M. T., Garcia, R. D., De La Rose, H. H. 1972. Dosages of gamma irradiation for full and inherited sterility in adult pink bollworms. *J. Econ. Entomol.* 65:645–50
32. Guthrie, W. D., Dollinger, E. J., Stetson, J. F. 1965. Chromosome studies of the European corn borer, smartweed borer, and lotus borer. *Ann. Entomol. Soc. Am.* 58:100–5
33. Haldane, J. B. S. 1922. Sex ratio and unisexual sterility in hybrid animals. *J. Genet.* 12:101–9
34. Hathaway, D. O. 1966. Laboratory and field cage studies of the effects of gamma radiation on codling moths. *J. Econ. Entomol.* 59:35–37
35. Holt, G. G. 1968. *Gamma-radiation induced damage to developing male germ cells in a species having holokinetic chromosomes, Trichoplusia ni.* MS Thesis. North Dakota State Univ., Fargo, N.D. 75 pp.
36. Holt, G. G., North, D. T. 1970. Effects of gamma irradiation on the mechanisms of sperm transfer in *Trichoplusia ni. J. Insect Physiol.* 16:2211–22
37. Hossain, M. M., Brower, J. H., Tilton, E. W. 1972. Radiation sensitivity of successively irradiated generations of the Indian meal moth. *J. Econ. Entomol.* 65:673–76
38. Hough, W. S. 1963. Effects of gamma radiation on codling moth eggs. *J. Econ. Entomol.* 56:660–63
39. Husseiny, M., Madsen, H. F. 1964. Sterilization of the navel orange worm, *Paramyelois transitella* (Walker), by gamma radiation. *Hilgardia.* 36:113–37
40. Karpenko, C. P., North, D. T. 1973. Ovipositional response elicited by normal, irradiated, F_1 sons, or castrated male *Trichoplusia ni. Ann. Entomol. Soc. Am.* 66:1278–80
41. Klassen, W., Knipling, E. F., McGuire, J. U. Jr. 1970. The potential for insect-population suppression by dominant conditional lethal traits. *Ann. Entomol. Soc. Am.* 63:238–55
42. Knipling, E. F. 1970. Suppression of pest Lepidoptera by releasing partially sterile males: A theoretical appraisal. *BioScience* 20:465–70
43. LaChance, L. E., Schmidt, C. H., Bushland, R. C. 1967. Radiation-induced sterilization. In *Pest Control: Biological, Physical, and Selected Chemical Methods,* 147–96. New York: Academic. 477 pp.
44. LaChance, L. E., Bell, R. A., Richard, R. D. 1973. Effect of low doses of gamma irradiation on reproduction of

male pink bollworms and their F_1 progeny. *Environ. Entomol.* 2:653–58
45. Lassota, A. 1963. The action of gamma rays on eggs, larvae and pupae of *Bombyx mori. Acta Biochim. Pol.* 10:379–86
46. Laster, M. L. 1972. Interspecific hybridization of *Heliothis virescens* and *H. subflexa. Environ. Entomol.* 1:682–87
47. Lea, D. E. 1956. *Actions of Radiation on Living Cells.* London: Cambridge Univ. Press. 2nd ed. 416 pp.
48. Lingren, P. D., Williamson, D. L., Henneberry, T. J. 1972. Experimentation on population suppression of wild cabbage looper populations through releases of partially sterile adults. *Proc. Joint Meet. Can. US, ESC, and ESA, 2nd, Montreal* (Abstr.)
49. Lightly, P. M. 1971. Neutron irradiation in *Antheraea eucalypti* Scott. *J. Lepidopterists' Soc.* 25:239–40
50. Makino, S. 1956. *A Review of the Chromosome Numbers in Animals.* Tokyo: Hokuryerkan Publ. Co. Ltd. 300 pp.
51. Mittwoch, U. 1967. *Sex Chromosomes.* New York: Academic. 306 pp.
52. Miya, K., Kurihara, M., Tanimura, I. 1970. Electron microscope studies on the oogenesis of the silkworm, *Bombyx mori.* L. III. *J. Fac. Agr. Iwate Univ.* 10:59–84
53. Mosbacker, G. C., Traut, W. 1968. Sex chromatin of *Lymantria dispar* and *Ephestia kuehniella. Naturwissenschaften* 55:349–50
54. Murakami, A. 1969. Comparison of radiosensitivity among different silkworm strains with respect to the killing effect on the embryo. *Mutat. Res.* 8:343–52
55. Murakami, A., Kondo, S., Tazima, Y. 1965. Enhancing effect of fractionated irradiation with 14 MeV neutrons on the induction of visible recessive mutations in silkworm gonia. *Ann. Rep. Nat. Inst. Genet. Jap.* 16:109–10
56. Murakami, A., Imai, H. T. 1973. Studies on chromosomes of the silkworm, *Bombyx mori. Ann. Rep. Nat. Inst. Genet. Jap.* 23:54–56
57. Nielsen, R. A. 1971. *Radiation biology of the greater wax moth: (Galleria mellonella L.): Effects on developmental biology, bionomics, mating-competitiveness, and F_1 sterility.* PhD thesis. Utah State Univ., Logan, Utah. 154 pp.
58. North, D. T. 1967. The cytogenetic basis of radioresistance in lepidopteran species, *Trichoplusia ni.* (Abstr.) *Radiat. Res.* 31:615
59. North, D. T. 1972. Inherited sterility for control of lepidopterous and hemipterous pests. (Abstr.) *Int. Congr. Entomol, 14th, August 22–30, 1972, Canberra.* 356 pp.
60. North, D. T., Holt, G. G. 1968. Genetic and cytogenetic basis of radiation-induced sterility in the adult male cabbage looper, *Trichoplusia ni.* In *Isotopes and Radiation in Entomology,* 391–403. Vienna: Int. At. Energy Ag.
61. North, D. T., Holt, G. G. 1968. Inherited sterility in progeny of irradiated male cabbage loopers. *J. Econ. Entomol.* 61:928–31
62. North, D. T., Holt, G. G. 1969. Population suppression by transmission of inherited sterility to progeny of irradiated cabbage looper, *Trichoplusia ni. Can. Entomol.* 101:513–20
63. North, D. T., Holt, G. G. 1970. Population control of Lepidoptera: The genetic and physiological basis. *Manitoba Entomol.* 4:53–69
64. North, D. T., Holt, G. G. 1971. Inherited sterility and its use in population suppression of Lepidoptera. In *Application of Induced Sterility for Control of Lepidoptera Populations,* 99–111. Vienna: Int. At. Energy Ag. 169 pp.
65. North, D. T., Holt, G. G. 1971. Radiation studies of sperm transfer in relation to competitiveness and oviposition in the cabbage looper and corn earworm. In *Application of Induced Sterility for Control of Lepidoptera Populations,* 87–97. Vienna: Int. At. Energy Ag. 169 pp.
66. Proshold, F. I., Bartell, J. A. 1970. Inherited sterility in progeny of irradiated male tobacco budworms: Effects on reproduction, developmental time, and sex ratio. *J. Econ. Entomol.* 63:280–85
67. Proshold, F. I., Bartell, J. A. 1972. Postembryonic growth and development of F_1 and F_2 tobacco budworms from partially sterile males. *Can. Entomol.* 104:165–72
68. Proshold, F. I., Bartell, J. A. 1973. Fertility and survival of tobacco budworms, *Heliothis virescens,* from gamma-irradiated females. *Can. Entomol.* 105:377–82
69. Proshold, F. I., LaChance, L. E. 1974. Analysis of sterility in hybrids from interspecific crosses between *Heliothis virescens* and *H. subflexa. Ann. Entomol. Soc. Am.* 67:445–49
70. Proshold, F. I., North, D. T. 1973. Survival and reproductive capacity of tobacco budworms irradiated while in diapause. *Proc. N. C. Branch Entomol. Soc. Am.* 28:167

71. Proverbs, M. D. 1962. Progress on the use of induced sexual sterility for the control of the codling moth, *Carpocapsa pomonella. Proc. Entomol. Soc. Ontario* 92:5–11
72. Proverbs, M. D. 1970. Procedures and experiments in population suppression of the codling moth, *Laspeyresia pomonella* in British Columbia orchards by release of radiation sterilized moths. *Manitoba Entomol.* 4:46–52
73. Proverbs, M. D. 1971. Orchard assessment of radiation-sterilized moths for control of *Laspeyresia pomonella* (L.). In *Application of Induced Sterility for Control of Lepidoptera Populations,* 117–33. Vienna: Int. At. Energy Ag. 169 pp.
74. Proverbs, M. D., Newton, J. R. 1962. Some effects of gamma radiation on the reproductive potential of the codling moth, *Carpocapsa pomonella* (L.). *Can. Entomol.* 94:1162–70
75. Qureshi, Z. A., Wilbur, D. A., Mills, R. B. 1967. Sublethal gamma radiation effects on prepupae, pupae, and adults of Angoumois grain moth. *J. Econ. Entomol.* 61:1699–1705
76. Qureshi, Z. A., Wilbur, D. A., Mills, R. B. 1970. Irradiation of early instars of the Angoumois grain moth. *J. Econ. Entomol.* 63:1241–47
77. Raulston, J. R., Graham, H. M. 1974. A technique for determining quantitative sperm transfer by tobacco budworm males. *J. Econ. Entomol.* 67:463–64
78. Riemann, J. G. 1973. Ultrastructure of sperm of F_1 progeny of irradiated males of the Mediterranean flour moth, *Anagasta kuehniella. Ann. Entomol. Soc. Am.* 66:147–53
79. Robinson, R. 1971. *Lepidoptera Genetics.* New York: Pergamon. 687 pp.
80. Roth, L. E., Wilson, H. J., Chakraborty, J. 1966. Anaphase structure in meiotic cells typified by spindle elongation. *J. Ultrastruct. Res.* 14:460–483
81. Rule, H. D., Godwin, P. A., Waters, W. E. 1965. Irradiation effect on spermatogenesis in the gypsy moth, *Porthetria dispar. J. Insect Physiol.* 11:369–78
82. Sado, T. 1963. Spermatogenesis of the silkworm and its bearing on radiation induced sterility. II. *J. Fac. Agr. Kyushu Univ.* 12:387–404
83. Sado, T. 1968. Spermatogenesis of the silkworm and its bearing on radiation induced sterility II. *J. Fac. Agr. Kyushu Univ.* 12:387–404
84. Serebrovskii, A. S. 1940. On the possibility of a new method of control of insect pests. *Zool. Zh.* 19:618–30. English transl.: *Sterile Male Technique for Eradication or Control of Harmful Insects,* 123. Vienna: Int. At. Energy Ag.
85. Snow, J. W., Haile, D. G., Baumhover, A. H., Cantelo, W. W. 1974. Control of the tobacco hornworm on St. Croix, U.S. Virgin Islands using the sterile male technique. *J. Econ. Entomol.* In press
86. Stimmann, M. W. 1971. Inherited sterility among progeny of male cabbage loopers treated with tepa. *J. Econ. Entomol.* 64:784–87
87. Strickberger, M. W. 1968. *Genetics.* New York: MacMillan. 868 pp.
88. Sugai, E., Iijima, T. 1967. Dose rate effect of radiation spermatogonia of the silkworm. *Nature* 213:943–44
89. Sugai, E., Suzuki, S. 1971. Male sterility in the progeny of the male silkworm, *Bombyx mori,* orally administered apholate. *Appl. Entomol. Zool.* 6:126–30
90. Suomalainen, E. 1953. The kinetochore and the bivalent structure in the Lepidoptera. *Hereditas* 39:88–96
91. Suomalainen, E. 1965. On the chromosomes of the geometrid moth genus *Cidaria. Chromosoma* 16:166–84
92. Suomalainen, E. 1966. Achiasmatische oogenese bei Tricopteren. *Chromosoma* 18:201–7
93. Suomalainen, E. 1969. On the sex chromosome trivalent in some Lepidoptera females. *Chromosoma* 28:298–308
94. Suomalainen, E. 1969. Chromosome evolution in the Lepidoptera. *Chromosomes Today Proc. Oxford Chromosome Conf.* 2:33–43
95. Taxima, Y. 1964. *The Genetics of the Silkworm.* London: Logos. 253 pp.
96. Toba, H. H., Kishaba, A. N. 1973. Cabbage loopers: Pupae sterilized with fractionated doses of gamma irradiation. *Environ. Entomol.* 2:118–24
97. Toba, H. H., Kishaba, A. N., North, D. T. 1972. Reduction of population of caged cabbage loopers by release of irradiated males. *J. Econ. Entomol.* 65:409–11
98. Vinson, S. B., Landano, R. L., Bartlett, A. C. 1969. Effect of gamma radiation on tissues of the tobacco budworm, *Heliothis virescens. Ann. Entomol. Soc Am.* 62:1340–47
99. Virkki, N. 1963. Gametogenesis in the sugarcane borer moth, *Diatraea saccharalis* (F.). *J. Agr. Univ. P. R.* 47:102–37
100. Walker, D. W. 1973. *Insect Sterility Program Techn. Rep. No. 7.* Puerto

Rico Nuclear Center, Mayaguez, P. R. 35 pp.

101. Walker, D. W., Quintana-Munez, V. 1968. Mortality staging of dominant lethals induced in the F_1 generation of the sugarcane borer, *Diatraea saccharalis* (F.). *Radiat. Res.* 26:138–43

102. Walker, D. W., Quintana-Munez, V., Padovani, F. 1971. Effect of gamma irradiation on immature sugarcane borers. In *Sterility Principle for Insect Control or Eradication,* 513–24. Vienna: Int. At. Energy Ag. 542 pp.

103. Walker, D. W., Quintana, V., Torres, J. 1971. Genetic collapse of insect populations. I. Extinction of inbred and outbred lines in laboratory populations of the sugarcane borer. *J. Econ. Entomol.* 64:660–67

104. Waterhouse, D. F., LaChance, L. E., Whitten, M. J. 1974. Use of autocidal methods. In *The Theory and Practice of Biological Control.* New York: Academic P. In press

105. White, M. J. D. 1973. *Animal Cytology and Evolution.* London: Cambridge Univ. Press. 3rd ed. 961 pp.

106. White, L. D., Hutt, R. B. 1972. Effects of treating adult codling moths with sterilizing and substerilizing doses of gamma irradiation in a low-temperature environment. *J. Econ. Entomol.* 65:140–43

107. Whitten, M. J. 1971. Insect control by genetic manipulation of natural populations. *Science* 171:682–84

EVOLUTION AND CLASSIFICATION OF THE LEPIDOPTERA[1]

❖6087

I. F. B. Common
Division of Entomology, CSIRO, Canberra 2601, Australia

Our understanding of the origin and relationships of the Lepidoptera is far from adequate, although it has progressed well beyond the stage when the superficial subdivisions Rhopalocera and Heterocera (30, 65) or Microlepidoptera and Macrolepidoptera were in vogue. Unfortunately, the fossil record has contributed little, and many controversial aspects will persist until we have a much better knowledge of comparative morphology.

Many so-called phylogenetic classifications have been proposed but since 1890 those of Meyrick (51, 53), Comstock (18, 19), Tillyard (68, 70, 73), and Börner (1) have probably had the most influence. Meyrick initially divided the order into nine major groups ranging from the Micropterygina, which included all the homoneurous forms, to the Caradrinina or noctuoid families. Later he increased the number of these groups to twelve. Comstock proposed a division into suborders Jugatae and Frenatae, depending mainly on whether their wing-coupling involves a jugum or frenulum. Tillyard substituted Homoneura and Heteroneura for these primary groups, believing that the venation provides a sounder basis for their higher classification. Börner based his major subdivisions, Monotrysia and Ditrysia, on the structure of the female genitalia.

Essentially the more recent classifications have been variants of one or more of these four systems. Dyar (23) and Fracker (28) based their larval classifications on Comstock's Jugatae and Frenatae. Forbes (26) also accepted Comstock's primary subdivision, but attempted to correlate data from all developmental stages. Imms (39) adopted Tillyard's Homoneura and Heteroneura, but divided the latter into the eight heteroneurous groups of Meyrick (51). Tillyard (73), Bourgogne (3), and Remington (61) also adopted the suborders Homoneura and Heteroneura. Börner's (1) view, shared by Busck (8), that female genital structure provides the best criteria for delineating major categories, has gradually gained acceptance. Hinton (34)

[1]The survey of the literature pertaining to this review was concluded in December 1973.

adopted Börner's Ditrysia, but removed the Micropterigidae from his Monotrysia and established it as a separate order, Zeugloptera. He also placed the Eriocraniidae and Mnesarchaeidae in another suborder, Dacnonypha. Richards & Davies (63) also accepted Börner's Monotrysia and Ditrysia, but distinguished Hinton's Zeugloptera as a third suborder. Common (16) followed Hinton in recognizing four primary groups, but he treated all four as suborders of the Lepidoptera.

Most classifications have attempted to reflect the supposed evolution of the Lepidoptera from primitive homoneurous forms to highly derivative heteroneurous forms. The lack of fossil evidence has necessitated a dependence on the comparative morphology of recent forms, but, in the past, morphological knowledge has been far too superficial or fragmentary for the task, and indeed this is still so of many groups. Some of the more controversial but critical taxa are also known from few specimens, and properly preserved examples of their early stages are seldom available for study. Until these basic problems are overcome, a full understanding of the evolution of the Lepidoptera will not be possible, and there will be no generally accepted classification.

FOSSIL HISTORY

There seems to be no satisfactory explanation for the paucity of fossil Lepidoptera from the more productive insect fossil beds. The wings described by Tindale (74) from the Upper-Triassic of Queensland, and thought to represent an early relative of the Lepidoptera, are now believed to be mecopteroid (64). The earliest indubitably lepidopterous fossil is the head capsule of a ditrysian larva in Cretaceous Canadian amber (50). Representatives of more primitive recent families, including Micropterigidae, Adelidae, Tineidae, Psychidae, Lyonetiidae, Yponomeutidae, Oecophoridae, Gelechiidae, Tortricidae, Cochylidae, and Pyralidae, are known from mid-Tertiary Baltic amber (49). Other Tertiary fossils are known in the families Cossidae, Geometridae, Hesperiidae, Papilionidae, Pieridae, Nymphalidae, and Lycaenidae (43, 49, 78). Most Tertiary fossils differ little from existing genera.

Clearly the haustellate Lepidoptera were well established by the Cretaceous, and many modern families had appeared by the Oligocene. It is likely, therefore, that a simple type of proboscis had already evolved by the time angiosperms appeared, although the major radiation of ditrysian forms may well have paralleled that of the angiosperms. Opler (57) identified the mines of various modern leaf-mining genera in Miocene fossil leaves similar to those of existing North American trees. Thus at least some of the intimate specific associations between existing genera of endophagous Lepidoptera and angiosperms have persisted since the mid-Tertiary.

EVOLUTIONARY TRENDS IN THE MORPHOLOGY OF LEPIDOPTERA

Egg

Chapman (11) discussed the phylogenetic significance of the eggs in Lepidoptera. He recognized two basic types: "flat" eggs, with micropylar axis more or less parallel to the substrate surface, and "upright" eggs, with the micropylar axis vertical to the substrate. He noted that one type of egg occurred throughout any one family, and the flat egg is probably the more primitive. Whereas it was possible for a group that lays upright eggs to have originated in a group laying flat eggs, the reverse was improbable. In advanced Lepidoptera, Chapman recognized that upright eggs are laid by noctuoids and butterflies, whereas flat eggs are characteristic of geometroids and bombycoids.

Larva

HEAD The most useful comparative account of the morphology of the larval head is due to Hinton (36, 38). The head capsule in most lepidopterous larvae is relatively uniform in structure. In the more primitive panorpoid larvae, including Micropterigidae, the epicranial suture is a λ-shaped line of weakness, all or part of which splits at ecdysis. In Eriocraniidae and other Lepidoptera the dorsal cleavage line is Λ-shaped and is situated lateral to a λ-shaped internal thickening of the cranium known as the adfrontal sutures or adfrontal ridge.

The maximum number of ocelli in lepidopterous larvae is six on each side; ocelli are absent in some internal-feeding larvae. The antennae (20) are usually three-segmented and remarkably uniform. They are reduced in certain internal feeders, such as Agathiphagidae, Nepticulidae, Opostegidae, Phyllocnistidae, and Gracillariidae; and they are exceptionally long in Micropterigidae and Limacodidae in which the head is retracted into the prothorax. In aquatic Lepidoptera they are rather longer than normal, in contrast to aquatic Trichoptera in which they are greatly reduced.

The maxilla of Micropterigidae is the least specialized of panorpoid larvae and has separate and distinct lacinia and galea, unlike Eriocraniidae and other Lepidoptera, Trichoptera, and some Mecoptera (38). In all these groups the cardo and stipes are separate and the latter is transversely divided into basistipes and dististipes. In Adelidae and more advanced Lepidoptera the dististipes is divided into two sclerites, whereas it is undivided in Eriocraniidae, Hepialidae, and Tischeriidae.

The postmentum is large in Micropterigidae and other Lepidoptera, as it is in Trichoptera and some Mecoptera. It is clearly divided into mentum and submentum in Trichoptera, less clearly divided in higher Lepidoptera, and not divided in Micropterigidae. In all free-living larvae of higher Lepidoptera a spinneret is developed, but in Micropterigidae, Trichoptera, and some internal-feeding Lepidoptera it is absent.

TRACHEAL SYSTEM The tracheal systems of panorpoid larvae were discussed by Hinton (38). The terrestrial larvae of nearly all Lepidoptera, and the final instar of some aquatic Pyralidae, are peripneustic, with functional spiracles on the prothorax and first eight abdominal segments. In other aquatic Pyralidae the last instar may have functional spiracles only on the first three abdominal segments. All instars of *Acentropus* and the earlier instars of other aquatic pyralids are apneustic. The larva of the micropterigid *Sabatinca* is peripneustic, as Tillyard (71) stated, whereas that of *Micropterix* is holopneustic, with functional spiracles on pro- and metathorax, and on the first seven abdominal segments.

THORACIC LEGS The five-segmented thoracic legs of most lepidopterous larvae are only slightly more specialized than the most primitive panorpoid type found in Trichoptera (38). In Micropterigidae the thoracic legs have been reduced to three segments including the pretarsus. Further reduction or loss of the thoracic legs has occurred independently in many internal-feeding Lepidoptera such as Agathiphagidae, but especially in leaf miners.

ABDOMINAL PROLEGS The ventral and anal prolegs of the abdomen are hollow outgrowths of the body wall, closed apically by the planta, which bears the crochets and in which the retractor muscles are inserted. The conical ventral prolegs of Micropterigidae end in a sclerotized point; they are present on the first eight abdominal segments and are without muscles. Abdominal prolegs occur on segments 3 to 6 and on 10 in most Lepidoptera, but in free-living or leaf-mining larvae may be modified, reduced, or lost. The prolegs are never retained when the thoracic legs are lost. Reduction of their number in free-living larvae proceeds posteriorly from segment 3.

The crochets of the anal prolegs are always in a single band or crescent, never in a complete circle. The most primitive crochet pattern on the ventral prolegs is a single or multiple series of more or less uniform (uniordinal) crochets. This arrangement suggests that crochets originated as modified spinules of the integument. Evolution of the crochet pattern has proceeded in several ways. By the loss of lateral crochets the simple circle may become a pair of curved or straight transverse bands, or with further loss a single transverse band of crochets. Alternatively there may be a progressive increase in the number of crochets and diversification of crochet size, from uniordinal to biordinal or multiordinal. Finally, the inner crochets of each circle may be lost and the outer crochets form a penellipse or a single longitudinal band or mesoseries. In some butterflies the development of the mesoseries from a simple circle of crochets is demonstrated ontogenetically in successive larval instars. The evolution of the clasping type of proleg, with crochets in a mesoseries, was associated with the development of an exophagous arboreal life. Clearly it evolved from the more simple type of proleg several times (37).

CHAETOTAXY Dyar (23) made the first serious attempt to homologize and name the setae of lepidopterous larvae. Fracker (28) made a more detailed study and proposed a new setal nomenclature. Earlier work was reviewed by Hinton (34), who reexamined setal homologies and proposed a system of nomenclature that is now widely accepted.

The body processes of Micropterigidae do not appear to be homologues of the setae in Eriocraniidae and other Lepidoptera. In the latter Hinton (34) distinguished three types of tactile setae: primary, subprimary, and secondary. Primary setae alone are present in the first instar of all but the most specialized larvae; subprimary setae are absent in the first instar but are found in the later instars even of the more primitive families. Secondary setae are additional setae found in the more advanced families and are usually acquired only in the second and later instars. These three types of setae act as receptors for external stimuli. In addition, certain microscopic setae of the thorax and abdomen, and a few of the smallest primary setae of the head, are believed to act as proprioceptors. Hinton (34) considered that primary setae were probably present in the ancestors of the Eriocraniidae and higher Lepidoptera before the group was differentiated into suborders, whereas secondary setae were acquired more recently. Indeed they seldom occur in the more primitive families, but a few occur in certain Gelechioidea, and in Cossidae, Castniidae, and Zygaenidae. In higher groups secondary setae are often numerous and many modifications in their form and function have evolved.

Pupa

With the evolution of a relatively immobile and vulnerable pupal stage in Lepidoptera came the larval habit of preparing a protective cell or cocoon. Evolution in the lepidopterous pupa, therefore, must have been profoundly influenced by the practical problem of how the adult can escape from the cocoon (9, 11, 35).

Since articulated mandibles are found in the pupae of Megaloptera, Neuroptera, Mecoptera, and Trichoptera, their presence in the decticous pupae of Micropterigidae, Eriocraniidae, and Agathiphagidae indicates the primitive nature of these groups. Operated by adult muscles, pupal mandibles enable the pharate adult to rupture the cocoon or cell, and assist its passage through the material in which the cocoon may be embedded (35). The whole escape process in these primitive forms is greatly aided by the thin and flexible cuticle, with separately ensheathed appendages, and the mobile abdomen. In *Agathiphaga,* the pupal tarsi have claws which may aid the pharate adult to escape from the hard pupal cell. Pupal claws are also found in certain Trichoptera and assist the enclosed adult to crawl out of the water (35).

Evolution in Lepidoptera has no doubt proceeded from the decticous to the adecticous type of pupa in which mandibles are absent or reduced to nonfunctional vestiges. Adecticous pupae, found in all but Micropterigidae, Eriocraniidae, and Agathiphagidae, show a clear evolutionary trend towards increased sclerotization and fusion of their various parts (54). Most are protected by a cocoon or cell and, as in decticous pupae, provision for the adult to escape from this shelter has deeply influenced their evolution.

Most primitive pupae have freedom of movement between head and thorax, between the thoracic segments, and between the first seven abdominal segments. The head may have a sharp anterior point or ridge, the cocoon-cutter, which may be the functional equivalent of the mandibles of decticous pupae (35). Posteriorly directed abdominal spines ensure that, just before ecdysis, the pupa is forced forwards and partly out of the cocoon when the abdomen is wriggled. Anteriorly directed hooks or spines, or a silken cable attaching the cremaster to the inside of the cocoon, prevent the pupa from falling out entirely.

In Lepidoptera all adecticous pupae are obtect, with appendages more or less glued down, first, in the course of evolution, to one another and then to the body wall. In more advanced pupae the degree of movement of the abdominal segments decreases, beginning at the base of the abdomen, until, in the most specialized, movement is possible between only two segments or not at all (54). Only in forms having at least three movable segments in the male and two in the female is the pupa protruded from the cocoon before ecdysis. The cocoon-cutter and abdominal spines are almost entirely restricted to these forms. If the adult sheds the pupal cuticle within the cocoon, various adaptations aid its escape from the cocoon (35). A hardened pupal cuticle provides a rigid structure from which the adult can push its way out of the cocoon, sometimes through a lidlike opening or other area of weakness, through a lobster-pot type of exit, or through one end of the cocoon after it has been softened by an oral secretion. Specialized protective adaptations in structure and coloration occur in Papilionoidea, and in a few Lyonetiidae, Cycnodiidae, Oecophoridae, and Pterophoridae, which do not spin cocoons.

Pupal specialization may be signified by gradual loss of cranial sutures, changes in the relative length of the thoracic and abdominal segments, lengthening of the pro- and mesothorax, and shortening of the metathorax and abdominal segments. The labial palpi become concealed by the maxillae and the maxillary palpi decrease in size until they are finally lost. However, maxillary palpi may be indicated by a distinct sclerite even when they are greatly reduced in the adult. Finally, the femur of the prothoracic leg is progressively concealed by the prothoracic tibia and tarsus (54). Differences between primitive and derived pupae are also found in the manner in which the pupal cuticle is shed at ecdysis.

Adult

Most comparative work on Lepidoptera has been done on the adult, and discussions on their evolution and classification usually reflect this imbalance. Nevertheless, the value of adult structures to indicate phylogenetic relationships should not be underrated.

HEAD Derivative trends in the adult head include reduction or loss of cranial sulci; loss of the subgenal process found in Mecoptera, Trichoptera, Micropterigidae, Eriocraniidae, and Neopseustidae (47); enlargement of the eyes; modification of the moniliform or filiform antennae; and reduction and loss of Eltringham's organs and lateral ocelli. The median ocellus is absent. In certain groups chaetosemata (40) have evolved, probably on several occasions. Their function is unknown. The evolution of the mouthparts shows a steady trend from the primitive mandibulate type of Micropterigidae and Agathiphagidae to the haustellate type of Eriocraniidae and higher Lepidoptera. The most primitive lepidopterous labrum, found in Micropterigidae, Agathiphagidae, and Eriocraniidae, is a movable, more or less pentagonal lobe. Reduction of the labrum occurs in other Lepidoptera, but lateral groups of setae (pilifers) are found in Micropterigidae, Agathiphagidae, and most higher Lepidoptera (47).

In Micropterigidae the asymmetrical mandibles, in conjunction with the modified hypopharynx, are adapted to grind vegetable matter such as pollen (72). Agathiphagidae also have asymmetrical, slightly dentate, articulating mandibles, but their habits are unknown. Perhaps they too grind pollen grains, because the concave inner surfaces of the bulbous galeae, figured by Common (16), could be used to manipulate such particles preparatory to grinding. The mandibles of Eriocraniidae are reduced to small lobes furnished with adductor and abductor muscles used by the adult only before ecdysis to operate the hypertrophied pupal mandibles (35, 47). After ecdysis the muscles undergo partial histolysis. The mandibular lobes of Lophocoronidae (17) may have a similar function. Apparently functionless mandibular lobes persist in Mnesarchaeidae, Neopseustidae, and Hepialidae, and in many higher Lepidoptera such as Ethmiidae and Yponomeutidae, all of which have adecticous pupae.

Micropterigidae and Agathiphagidae have the most primitive lepidopterous maxilla, with more or less simple galea and lacinia. In all other Lepidoptera, the laciniae are reduced or lost and the galeae are modified to form the two halves of a tube-like proboscis. Occasionally the proboscis is secondarily reduced or lost. Eriocraniidae retain a vestigial lacinia furnished with a cranial muscle and have a short, functional proboscis, with simple linking mechanism but no internal muscles or septa (45). The much more complex proboscis found in butterflies, for example, has an internal musculature and its two halves have a more intricate linking mechanism, sealed by a glandular secretion. The most primitive maxillary palpi are five-segmented and are found in Micropterigidae, Agathiphagidae, Eriocraniidae, Neopseustidae, Nepticuloidea, Adeloidea, certain Tineoidea, and occasionally elsewhere. Reduction in the size of the palpi and the number of segments is characteristic of most Lepidoptera (60), and in the most advanced forms the palpi may be vestigial or absent.

The labium in Micropterigidae is of the primitive type with well-developed mentum, submentum, glossae, and paraglossae. The postmentum may carry an Eltringham's organ. The labial musculature is said to be the most complete of any endopterygote insect (47). The palpi are three-segmented with a sensory invagination near the apex of the third segment. The labium is more or less reduced in other Lepidoptera, but palpi with apical sensory area are usually retained, although occasionally the palpi may be reduced to two segments and the sensory area lost. Specialization in the Trichoptera has followed an entirely different pattern (47).

THORAX The comparative skeletal anatomy of the thorax in many Lepidoptera is still not well understood. Brock (6) studied many of the more advanced families, selecting features which he thought should indicate phylogenetic trends. These included the sternopleural region of the mesothorax, the metathoracic furca, the metathoracic scutum and scutellum, and the dorsal and ventral articulation of the thorax and abdomen. He traced a progressive modification of these structures through the higher Lepidoptera.

In the sternopleural region of the mesothorax, sclerotization of the parepisternal suture and its gradual movement towards the anapleural cleft, the gradual disappearance of the upper sector of the precoxal suture, and the elongation and strengthening of the marginopleural suture, appear to be derivative conditions. The metathoracic furca in more primitive forms is relatively simple, with an elongate furcal stem and a pair of dorsal laminae extending along the secondary furcal arms. In more advanced forms anterior and posterior ventral laminae first become associated with the apophyses and later the furcal stem tends to shorten. A postfurcal sclerite may appear and a wide lamella may be present along the median invagination to the furcal stem. In primitive Lepidoptera the metascutum is well exposed medially, but in more advanced forms it is progressively overlaid by the scutellum.

Brock stated that the dorsal articulation of the thorax and abdomen is similar in Trichoptera, Micropterigidae, and Eriocraniidae, in which the anterolateral angles of the laterophragmata of the metathorax are fused with the first abdominal tergite. A number of modifications are introduced in the Hepialidae which persist throughout the remaining Lepidoptera. Here the laterophragmata are greatly enlarged and are separated from the epimeron by a strong suture; dorsally they have a narrow articulation with the anterior rim of the first abdominal segment, which is heavily sclerotized and otherwise modified. The Adeloidea show some further modification of the hepialid type with the tergal phragma tending to form a complete sclerite between the laterophragmata. Fusion of the latter to form a complete transphragma is characteristic of all higher Lepidoptera, with a reduction or loss of the wide tergal phragma found in Adeloidea. In advanced families a well-developed phragma is found on the tergal rim, and in the most specialized a supraphragmal sclerite as well.

The ventral articulation of thorax and abdomen also shows features useful in the study of evolution (1). Brock (6) found that it was very similar in Trichoptera, Micropterigidae, Eriocraniidae, and Hepialidae, in which the second abdominal sternum is ventral to the second tergum, and is preceded by sclerotized areas thought to represent the first sternum. In these groups and in the Adeloidea, a tergosternal sclerite extends from near the anterolateral angle of the second sternite to near the apex of the first tergite. In Adeloidea, however, the second sternum extends forward beneath the first tergum and it is doubtful if any remnant of the first sternite remains. In Tineoidea, Yponomeutoidea, and Gelechioidea the second sternite extends even farther forward and the first sternite is lost or possibly fused with the second; there is no tergosternal sclerite. In these groups the second sternite usually has a pair of slender ventrolateral rods which project as apodemes beyond its anterior margin. In Copromorphidae and Tortricoidea the anterior angles of the second sternite are produced. Apodemes persist in both, but the ventrolateral rods are generally lost in Tortricoidea, Cossoidea, Sesioidea, Zygaenoidea, Castnioidea, and Bombycoidea. The second sternite is rather more modified in the remaining advanced superfamilies.

TYMPANAL ORGANS Metathoracic tympanal organs are known only in Notodontoidea and Noctuoidea (6, 42, 62). Those of Notodontidae show the least modification and always have a metascutal bulla. In Noctuoidea they are rather more complex. The metepimeron is least modified in Lymantriidae and, as in Arctiidae, there is a prespiracular counter-tympanal hood at the base of the abdomen. In other Noctuoidea the counter-tympanal hood is postspiracular. Tympanal organs have been secondarily reduced or lost in Amatidae (Ctenuchidae).

Abdominal tympanal organs occur in Pyralidae, Geometroidea, and in the isolated genus *Dudgeonea* (13, 41, 55). Both sexes of Geometridae, and female Uraniidae and Epiplemidae, have tympanal organs associated with the second sternite at the base of the abdomen. In males of the last two families posteriorly directed tympanal organs are invaginated beneath the second tergite. In Thyatiridae and some Drepanidae the tympanal organs are also associated with the second sternite, but unlike Geometridae the opening has a thick sclerotized margin. Clench (13) recorded abdominal tympanal organs in *Dudgeonea* and rudimentary tympanal organs in *Chilecomadia* from Chile and *Pseudocossus* from Madagascar (14). He referred these genera to the Cossidae, but Brock (6) placed *Dudgeonea* provisionally in the Sesioidea.

WINGS Without doubt the wings of Micropterigidae and Eriocraniidae, which are similar to those of certain primitive Trichoptera, represent the primitive lepidopterous form. Their surfaces are densely covered with microtrichia or aculei, the shape is broadly elliptical, and a well-differentiated jugum projects in the forewing. Aculeate wings occur in all homoneurous Lepidoptera and in Adeloidea and Nepticuloidea, and a few aculei are found in some other groups. In many families there is a patch of minute spines or enlarged aculei, thought by Braun (4) to have a wing-coupling function, beneath the forewing near the base. When the wings are folded these spines usually interlock with similar spines on the metascutum and hold the wings in place (15). Both zones are already differentiated in homoneurous Lepidoptera and in some other endopterygotes. They are nearly always present in those forms which fold their wings in the primitive roofwise or stegopterous (70) position. Those with modified resting postures usually lack spined zones, but they are retained throughout the Noctuoidea, even in those that spread the wings out flat in repose.

Sharplin (66) showed that the differentiated jugum of most homoneurous Lepidoptera is present in more restricted form in many primitive heteroneurous groups. It is always folded beneath the forewing when the wings are folded, and lies above the hindwing when the wings are expanded. Sharplin found in all Lepidoptera, including Micropterigidae, a specialized tergopleural apodeme of the mesothorax, not observed in other insects. She also noted that Micropterigidae, Eriocraniidae, and Hepialidae have a lateral extension of the metaprescutum not found elsewhere. In several other characters the most primitive condition found in Trichoptera or other endopterygotes also occurs in Micropterigidae and Eriocraniidae, less frequently in Hepialidae, Adeloidea, and Nepticuloidea, and in a few of the more advanced Lepidoptera.

Tillyard (70) considered that no modern species has retained all of the humoneurous venation of ancestral Lepidoptera, but that of recent homoneurous families can be derived from it by reduction. No venation is known that is transitional between homoneurous and heteroneurous venation, although Busck (7) and Braun (5) thought that the two-branched Rs occasionally found in the hindwing of Gracillariidae and Cycnodiidae may indicate the means by which Rs has been reduced to a single vein. Most authors (19, 70, 77) have maintained that specialization by reduction is the rule in heteroneurous venation, usually by coalescence but sometimes by atrophy and loss. This accounts for the reduction and disappearance of the humeral vein, the stem of M, CuP, and certain anal veins; forking or loss of peripheral veins may also be due to coalescence. Such reduction often accompanies the narrowing of the wings, a process reaching its ultimate effect in the tineoid superfamilies. Headlee (31) claimed that the radial system of the butterfly forewing had departed from the dichotomously branched system of primitive Lepidoptera, in which R_1 and Rs diverged first, followed by the separation of R_{2+3} and R_{4+5}. In *Danaus* and other butterflies he noticed that R_{4+5} arose near the base of R, well before the base of R_1, and he noted this even in the tracheation of pupal (pharate adult) wings. He found nothing similar in other heterocerous wings and, like Tindale (75), thought this a unique feature of butterflies, justifying their basic separation from moths. Tillyard (70) showed that similar radial tracheation in the pupal wings of the cossid *Xyleutes* and the hepialid *Zelotypia* resulted in two distinct radial systems in the adults. In *Zelotypia* the junction of R_{4+5} remained in its primitive position beyond that of R_1, whereas in *Xyleutes* R_{4+5} arose before the base of R_1. He stated that a similar process of "splitting-back" of R_{4+5} could be seen in many families, notably butterflies. Brock (6) claimed that an inward splitting-back of R_5, and often of M_1 and R_4 also, can be recognized in the more specialized members of most of the advanced superfamilies as well as butterflies. However, plausible alternative explanations for the existing venation in some of Brock's examples may cast doubt on the validity of such a general application of the splitting-back hypothesis.

LEGS Except in certain Psychidae and advanced Lepidoptera, the adults have three pairs of functional legs. The epiphysis of the fore tibia is present in Micropterigidae and most other Lepidoptera, but is absent in Trichoptera. Agathiphagidae are exceptional in having one spur on each fore tibia and two pairs on each of the mid and hind tibiae. Other families have a maximum of one pair on the mid tibia and two pairs on the hind tibia. Reduction in this number occurs in many advanced families. Tarsal modification of the forelegs increases progressively within the butterflies.

TRACHEAL SYSTEM Paired spiracles on the prothorax and first seven abdominal segments are usual in both sexes of adult Lepidoptera, including Micropterigidae. Functional or rudimentary spiracles are sometimes present on the eighth abdominal segment in females of Eriocraniidae, Agathiphagidae, Hepialidae, Neopseustidae, Mnesarchaeidae, Nepticuloidea, Adeloidea (21), and some Yponomeutoidea and Gelechioidea (K. Sattler, personal communication).

GENITALIA Homologies of the male genitalia and their bearing on the phylogeny of Lepidoptera have been discussed by Eyer (24, 25), Forbes (27), and Stekol'nikov (67). Stekol'nikov stated that the most characteristic structure is the uncus, thought to be derived from the tenth tergite. In its primitive form it is a paired organ, homologous with the tegminal lobes in *Micropterix,* becoming an articulated process in advanced Lepidoptera; occasionally it is lost. In his view, the function of the uncus and valvae in primitive groups is to hold the female during copulation, whereas in more advanced groups the uncus may have a seizing function. The gnathos is a secondary development and, when the uncus is reduced or lost, may assume its function. Other genital structures may have similar functions. In view of the importance attached to male genitalia in classification, their function deserves thorough examination.

The structure and homologies of the female genitalia have been outlined by Klots (44). The external genitalia of several primitive families were examined by Mutuura (56) and the internal organs by Dugdale (21). The exoporian genitalia of Hepialidae were described by Bourgogne (2). Contrary to the views of Börner (1), Stekol'nikov (67) thought that the single terminal aperture is primitive and the ditrysian type of genitalia with two genital openings was derived from it. The exoporian type he considered to be intermediate, whereas Friese (29) regarded it as the most primitive. Here the oviporus and copulatory aperture are connected by a seminal groove, the closing of which would produce an internal ductus seminalis joining the bursa copulatrix and oviductus communis. Mutuura stressed the phylogenetic importance of modes of oviposition as well as external female genitalia, but Stekol'nikov pointed out that modifications in the external morphology and musculature of female terminalia are generally adaptive. Adaptation for oviposition in crevices or cracks results in the elongation of the muscle-bearing apophyses, the terminal segments and the intersegmental membranes. Similarly the apophyses may be reduced in forms that freely disseminate their eggs.

In Micropterigidae, Eriocraniidae, Agathiphagidae, Nepticuloidea, and Adeloidea, Dugdale (21) found that the oviductus communis enters the bursa copulatrix ventrally, whereas in other Lepidoptera with two genital openings it is dorsal to the bursa. He noted that a cloaca, with an external aperture serving as anus, oviporus, and copulatory opening, occurs in Micropterigidae, Eriocraniidae, Agathiphagidae, and some Nepticuloidea and Adeloidea. An incomplete cloaca, with external aperture serving as anus and oviporus, occurs in certain other primitive Lepidoptera. A complex spermatheca is also characteristic of Lepidoptera, and in some of the more primitive there is a prominent spermathecal papilla on the vagina. Ducted colleterial glands are absent in exoporian genitalia.

ZEUGLOPTERA

The Zeugloptera contain the single family Micropterigidae. Most early lepidopterists recognized their relatively archaic nature, but associated them with the Eriocraniidae. Chapman (10) first distinguished between the two after studying the larva of *Micropterix.* Packard (58) separated the Micropterigidae from all other Lepidoptera as "Lepidoptera laciniata." Sharp (65) also distinguished the Micropterigidae and Eriocraniidae, but Meyrick (52) and many later authors combined the two.

Chapman (12) accepted Packard's isolation of Micropterigidae based on mouthparts, but noted that the female has a single genital opening at the end of the tenth abdominal segment; other Lepidoptera he believed had two genital openings. He therefore proposed a separate order Zeugloptera for the family. Comstock (19) thought that the close similarity of the wings of Micropterigidae and Eriocraniidae to those of Trichoptera necessitated their inclusion in that order. Accordingly, he combined Micropterigidae, Eriocraniidae, and Mnesarchaeidae in one family Micropterigidae, which he called terrestrial Trichoptera.

Braun (4) and Tillyard (69) opposed Chapman's and Comstock's views and accepted Micropterigidae as true Lepidoptera. Tillyard based his conclusion on wing structure which he claimed was in no way inconsistent with the inclusion of the family in the Lepidoptera. In his view Micropterigidae resemble Lepidoptera and differ from Trichoptera by the absence of M_4 as a separate vein in the forewing, the absence of a wing-spot between veins R_4 and R_5 (found in both wings of nearly all Trichoptera), the possession of a complete set of tracheae in the pupal wing (instead of only two tracheae as in Trichoptera), and the presence of broad specialized wing scales with numerous striae (instead of scattered scales with few striae occasionally found in Trichoptera).

Tillyard (71) stated that the larva and pupa of the micropterigid *Sabatinca* exhibit more archaic features than any existing panorpoid insects except the Mecoptera. He recognized the wide divergence of the micropterigid larval type from that of Eriocraniidae and other Lepidoptera, but insisted that Micropterigidae are archaic Lepidoptera which branched off along a separate line from the very base of the order. Nevertheless, Tillyard (73) included Micropterigidae and Eriocraniidae in the one superfamily.

Börner (1) proposed a subordinal division of Lepidoptera on differences in female genitalia, but referred Micropterigidae and Eriocraniidae to the one group Micropterigoidea.

Hinton (34) reestablished Chapman's order Zeugloptera for Micropterigidae, which he regarded as much more archaic than either Trichoptera or Lepidoptera. He listed several primitive attributes of adults and larvae shared by all three orders, and other attributes which he found only in Zeugloptera. Some of the latter were later shown (22) to occur also in *Agathiphaga,* a genus that Hinton (38) accepted as undoubtedly lepidopterous. The ordinal status of Zeugloptera was then sustained by Hinton on the basis of larval characters. He stated that he knew of no reason to separate Zeugloptera from Lepidoptera on adult or pupal morphology. Hinton listed twelve attributes of larval Zeugloptera which distinguish them from Lepidoptera: (*a*) the cranium lacks an adfrontal ridge or adfrontal sutures; (*b*) the tentorial bridge is short and thick; (*c*) the anterior tentorial arms arise mesad of the antennal bases instead of well behind them; (*d*) the maxilla has separate lacinia and galea; (*e*) a cranial flexor of the dististipes is present; (*f*) median and lateral labral retractors are present; (*g*) a pair of submedian retractors are inserted on the labrum (possibly cibarial muscles); (*h*) a distinct spinneret is absent; (*i*) the thoracic legs are reduced by fusion of the coxa, trochanter, and femur; (*j*) the ventral abdominal prolegs lack muscles; (*k*) the metathoracic spiracles are functional; and (*l*) the chaetotaxy of thorax and abdomen is of quite a different type. Two of these differences no longer hold, for the larva of *Agathiphaga* lacks distinct adfrontal sutures, and the tracheal system in the micropterigid *Sabatinca* is peripneustic, as in most Lepidoptera. Of the remaining ten characters, five are expressed in their primitive form in Zeugloptera and in their derivative form in Eriocraniidae and other Lepidoptera. Four of these five also appear in their primitive form in Trichoptera. Hence Zeugloptera and Trichoptera are more archaic than Eriocraniidae and other Lepidoptera, a contention supported by Tillyard (71) and not contested by most authors.

There remain, however, profound differences between the larvae of Zeugloptera and Lepidoptera, as Hinton stated. Some of these may represent specialization in free-living larvae of Zeugloptera after the group diverged from the main lepidopterous stem, but others have not yet been explained adequately.

Although there are, in Hinton's view, no grounds for separating Zeugloptera from Lepidoptera on the basis of adult or pupal morphology, one must not dismiss adult and pupal characters as of no phylogenetic consequence. Indeed, Hinton (38) stated that he had not studied panorpoid larvae because they reveal phylogenetic relationships more clearly than other stages, but because much less was known about larvae. The number of apomorphic attributes of the adult, which suggest a closer relationship of Zeugloptera to Lepidoptera rather than to Trichoptera, is steadily mounting. Kristensen (48) listed 17 advanced attributes, shared by the adults of Zeugloptera and higher Lepidoptera but not Trichoptera, which suggest a common lineage of Micropterigidae and Eriocraniidae after the divergence of Trichoptera. To these should be added the presence in both Zeugloptera and Lepidoptera of a cloaca and a prominent spermathecal papilla on the vagina (21). Mutuura (56) supported an ordinal status for Zeugloptera because the female abdomen lacks apophyses, has separate ninth and tenth segments, and features a simple arrangement of tergite and sternite on the eighth segment. These may well be plesiomorphic attributes, once more indicating that in these respects Zeugloptera may be more primitive than Eriocraniidae and other Lepidoptera.

DACNONYPHA

The suborder was established by Hinton (34) to include Eriocraniidae and Mnesarchaeidae, to which he later added *Agathiphaga* (38). Packard (58) placed Eriocraniidae in his Paleolepidoptera, one of two major groups of his "Lepidoptera haustellata." Comstock (18) included them with Micropterigidae in his Jugatae, but later (19) transferred both families to his suborder Micropterygina of the Trichoptera. Tillyard (70) included Eriocraniidae, with Micropterigidae and Mnesarchaeidae, in his division Jugo-frenata of the Homoneura, for which he later substituted the superfamily Micropterigoidea (73). Börner (1) associated Eriocraniidae, Adelidae, Heliozelidae, and Prodoxidae in a group Incurvariina, largely because of the greatly enlarged apophyses of the female abdomen. He recognized the primitive position of Eriocraniidae in this group, indicated by wing structure and the presence of spiracles on the eighth segment of the female abdomen. Börner omitted Mnesarchaeidae, which were claimed by Busck (8) to have only one genital opening.

Kristensen (45–47) accepted Hinton's suborder Dacnonypha and included in it Agathiphagidae, Eriocraniidae, Neopseustidae, and Mnesarchaeidae. He regarded Agathiphagidae as the most primitive because of its laciniate mouthparts. The Mnesarchaeidae and Neopseustidae he thought might be an advanced monophyletic group. Common (17) accepted Kristensen's concept of the Dacnonypha and added a new family Lophocoronidae. However, Mutuura (56) noticed that *Neopseustis* has exoporian female genitalia and thought, with Philpott (59), that *Mnesarchaea* probably has two genital openings also. Dugdale (21) confirmed this and removed

Neopseustidae and Mnesarchaeidae from Dacnonypha and treated them, with Hepialidae, as primitive Ditrysia. Dacnonypha have thereby been restricted to Agathiphagidae, Eriocraniidae, and Lophocoronidae, although the position of Lophocoronidae must remain doubtful until the female or early stages are discovered.

MONOTRYSIA

Börner (1) placed in his suborder Monotrysia all homoneurous Lepidoptera, including Micropterigidae, and heteroneurous forms with a single genital aperture in the female. Hinton (34) modified Börner's concept by excluding the Zeugloptera and Dacnonypha. The Hepialoidea, he thought, branched off well before the divergence of ancestral Adeloidea and Nepticuloidea.

The apparent weakness in Hinton's concept is that it retains Adeloidea and Nepticuloidea in the one suborder with the very different Hepialoidea. Neither Börner nor Hinton was aware that female Hepialidae have two genital openings and lack colleterial glands. Little evidence now remains to support the close association of the three superfamilies. Certainly the larvae of Hepialidae and Adelidae are exceptional in having two punctures near the fifth ocellus (34), but in many ways larval Hepialidae have developed independently of other Lepidoptera. The frontal puncture of the head is very close to, but laterad of F1, instead of mesad of F1; punctures Oa and Ga are absent; and seta MXD1 of the prothorax has been lost. The disposition of the cranial setae in Hepialidae appears to differ greatly from that in other Lepidoptera, including Adelidae and Nepticulidae (34). Friese (29) listed several plesiomorphic features of adult Hepialidae which he thought indicated that the family was even more primitive than Micropterigidae. However, he did not give due regard to the decticous pupae and many other primitive features of Micropterigidae, or the several apomorphic attributes of hepialid pupae that are comparable with those of many advanced Lepidoptera (54).

Dugdale (21) found that in exoporian genitalia the oviductus communis is dorsal to the bursa copulatrix, as in ditrysian genitalia. In the more variable monotrysian genitalia of Adeloidea and Nepticuloidea, the oviductus communis is ventral to the bursa, as in Trichoptera, Zeugloptera, and Dacnonypha. As exoporian forms share few derived features with Adeloidea and Nepticuloidea, it is unlikely that Hinton's Monotrysia is monophyletic. Dugdale's proposal to restrict the Monotrysia to the Adeloidea and Nepticuloidea, therefore, seems sound, but his decision to treat the Hepialoidea and Mnesarchaeoidea as exoporian Ditrysia may be less firmly based. The dorsal position of the oviductus communis in these groups could have arisen independently in exoporian and ditrysian forms, and exoporian genitalia need not be intermediate between monotrysian and ditrysian as Stekol'nikov (67) supposed. Clearly Hepialoidea developed many apomorphic attributes after they departed from the main stream of development. If they branched off before the Monotrysia, along a separate line, there would be some merit in treating them as a separate suborder Exoporia.

The persistence in the exoporian groups of several plesiomorphic features not found in heteroneurous Lepidoptera may in fact imply their divergence before the Monotrysia. Of note are the homoneurous venation, the projecting jugum, the mandibular lobes, the separate follicles of the male testes and the abdominal ganglia (29), the lateral extension of the metaprescutum, a free tergopleural apodeme in the forewing, and several other basal structures of the wings (66). In addition Adeloidea and Nepticuloidea share several apomorphic features with Ditrysia, including their heteroneurous venation; frenulate wing coupling; reduced jugum; the presence of patagia, parapatagia, and dorsal pronotal plates; and the fusion of the anterolateral angles of the prothoracic sternum to the pleuron (6).

Hinton (34) regarded Prototheoridae and Palaeosetidae as specialized offshoots of a hepialid stock, and Dugdale (21) confirmed their close relationship to the Hepialidae. Mutuura (56) thought Neopseustidae are related to Prototheoridae, and Dugdale included them in Hepialoidea. However, similarities in adult head structure in Neopseustidae and Mnesarchaeidae led Kristensen (46) to suggest that the two may constitute a monophyletic group, but he did not study Hepialidae.

The aculeate heterocerous families Adelidae, Incurvariidae, Nepticulidae, Opostegidae, and Tischeriidae were associated by Busck (7), and Heinrich (32) grouped the last three. Forbes (26) erected the superfamilies Nepticuloidea for Nepticulidae and Incurvarioidea for Incurvariidae. Busck (8) added Opostegidae and Tischeriidae to Nepticuloidea, a group with soft nonpiercing ovipositors, and included Heliozelidae in Incurvarioidea, a group with piercing ovipositors. This division has been accepted by most recent authors, and the female genitalia support it (21). However, Mutuura (56) thought that the otherwise primitive Tischeriidae have two genital openings and suggested a transitional place for them between forms with one and two genital apertures. Dugdale (21), on the contrary, found that *Tischeria complanella* has only a single genital opening and that the internal female genitalia confirm its relationship to the Nepticuloidea.

DITRYSIA

The suborder was proposed by Börner (1) to embrace all heteroneurous Lepidoptera with two genital openings in the female, and this concept was accepted by Hinton (34). With the discovery of exoporian genitalia in Hepialidae (2), Börner's suborder really included those ditrysian Lepidoptera with an internal ductus seminalis. Dugdale (21) broadened the definition to include all forms with two genital openings in the female, both endoporian and exoporian, but this has the disadvantage of associating relatively primitive homoneurous families with the more advanced heteroneurous forms.

Several families of Ditrysia (s.s.) share an array of archaic features which make it difficult to work out their phylogeny. Mainly on the basis of wing venation, Turner (76) suggested that the heteroneurous Lepidoptera had descended from a hypothetical cossid-like ancestor which he called Protocossidae. From this stock he believed the modern Cossidae, Tortricoidea, and tineoid families first arose and, separately, the four groups (*a*) Castniidae; (*b*) Zygaenidae, Limacodidae, Psychidae, and Pyralidae; (*c*) Lasiocampidae; and (*d*) Noctuidae, Arctiidae, Lymantriidae, and Notodontidae, and probably also Geometridae and Thyatiridae. Turner's views on the broad family groupings were largely preserved by Tillyard (73). Nevertheless, there were several major difficulties in Turner's system, notably the inclusion of the three families Psychidae, Zygaenidae, and Limacodidae in the one superfamily Psychoidea, an inheritance from Meyrick (51); the wide separation of Lasiocampidae from Bombycoidea and Uranioidea from Geometroidea; and the association of Notodontidae with Sphingidae and Geometridae. Turner also placed the butterflies in a division Rhopalocera, equal in value to the whole of the remaining heteroneurous Lepidoptera.

Some of these difficulties were also present in Forbes' (26) system. He gave the butterflies subordinal status, separated Sphingoidea widely from Bombycoidea, and associated Uranioidea with Bombycoidea. Richards & Davies (63) overcame most of the earlier difficulties, but the heterogeneous superfamily Psychoidea persisted, and Sphingidae remained widely separated from Bombycoidea.

Common (16) accepted Cossoidea, Tortricoidea, and Tineoidea (including Psychidae) as amongst the most primitive of his 17 superfamilies of Ditrysia and thought that they probably evolved from a common ancestor, the more archaic forms retaining a primitive venation and mouthparts with five-segmented maxillary palpi. In his system Tineoidea gave rise to Yponomeutoidea, Gelechioidea, and Copromorphoidea, with evolutionary trends towards loss of ocelli, reduction of mouthparts and venation, specialization of proboscis, acquisition of a specialized vestiture, loss of subprimary seta L3 of the larval prothorax, and an increase in the number and diversity of size in larval crochets and a progressive specialization of the pupa. He also regarded the Castnioidea and Zygaenoidea as relatively primitive, but outlined only briefly the interrelationships of the higher superfamilies.

Brock's (6) analysis of the more advanced Ditrysia has clarified some of the problems encountered by earlier authors. He accepted 14 superfamilies, each of which he believed was monophyletic. He distinguished Sesioidea, including Sesiidae and Choreutidae, previously placed in Yponomeutoidea, and transferred Limacodidae, Megalopygidae, and Chrysopolomidae from Zygaenoidea to Cossoidea. The Zygaenoidea he restricted to Zygaenidae, removing Epipyropidae and Heterogynidae to Tineoidea. He also included Pterophoridae and Alucitidae in Pyraloidea, Hesperiidae in Papilionoidea, Callidulidae and Pterothysanidae in Geometroidea, and Notodontidae in Noctuoidea, believing that the separation of these respective groups obscured their evolutionary importance.

Brock concluded that Tineoidea are the most primitive Ditrysia, and tineoid stock gave rise to Yponomeutoidea, Gelechioidea, and Copromorphoidea. Tortricoidea, Sesioidea, Cossoidea, and Castnioidea, he believed, have many primitive features in common, and probably also originated in the tineoid complex. The last three and Pyraloidea he derived on a common stem from a tortricoid-like ancestor. Cossoidea, Castnioidea, and Pyraloidea each display strong tendencies towards the attainment of derivative venation and other attributes of the advanced exophagous superfamilies. Accordingly, Brock derived Bombycoidea from cossoid stock, Papilionoidea from castnioid stock, and both Geometroidea and Noctuoidea from pyraloid stock. The change from endophagous to exophagous larval habits has been a prime biological factor in the evolution of advanced Lepidoptera, and the selective pressures imposed by this new way of life have resulted in the independent evolution of similar adaptations in several superfamilies.

Few authors would contest Brock's broad conclusions, but many matters of detail remain to be clarified. For example, the derivation of Papilionoidea from a castnioid-like ancestor is not supported by Hessel's (33) study of the dorsal vessel in Ditrysia. Castniidae have a tineoid-type arrangement of the dorsal vessel, whereas Cossidae alone amongst heterocerous families have a horizontal aortal chamber in the thorax similar to Papilionoidea, an observation suggesting a cossid-like rather than a castniid-like ancestor for butterflies. Further, the structure of the immature stages of Limacodidae and Megalopygidae at once introduces doubts about Brock's treatment of them as advanced Cossoidea. Certainly their exophagous larvae have diverged greatly from any endophagous cossoid ancestor, and Mosher (54) thought that their pupae followed a very different line of development. Unlike Cossidae they have retained several primitive features, such as a thin flexible cuticle, freedom of movement between all abdominal segments in advance of the eighth, and visible spiracles on the first abdominal segment.

Brock's inclusion of Pterophoridae in Pyraloidea, Hesperiidae in Papilionoidea, Callidulidae in Geometroidea, and Notodontidae in Noctuoidea will probably not receive serious opposition, despite the traditional but relatively slight distinction drawn between skippers and butterflies. His placement of a few small families, such as Hyblaeidae, Epipyropidae, and Alucitidae, as well as such genera as *Dudgeonea,* will certainly be questioned. Much further anatomical work is needed to clarify these and similar problems, especially the relationships of the families of Tineoidea, Yponomeutoidea, and Gelechioidea.

CONCLUSIONS

If Zeugloptera are to be accepted as Lepidoptera, and the evidence seems to favor this, they are clearly the most primitive. No phylogeny or classification that fails to recognize their basic position is tenable. The many derivative attributes of the adult and pupa that Zeugloptera share with Eriocraniidae and other Lepidoptera, but not with other panorpoid orders, suggest that they branched off from the lepidopterous lineage after Trichoptera, perhaps as early as the Triassic. Nevertheless, several larval peculiarities in Zeugloptera raise justifiable doubts about this conclusion. Further detailed study of the larvae of primitive families such as Agathiphagidae may help to resolve the question.

Dacnonypha, including Eriocraniidae, Agathiphagidae, and Lophocoronidae in the one superfamily Eriocranioidea, are a primitive group which diverged from the lepidopterous stem probably in the Jurassic or a little earlier. Their pupae are decticous, as in Zeugloptera, and their functional mandibles are hypertrophied. Their endophagous larvae have lost the thoracic legs and abdominal prolegs of the ancestral dacnonyphous larva, but the spinneret and primary setae have persisted.

The adult maxillae have unspecialized galeae and laciniae, as in *Agathiphaga,* or the galeae have been modified to form a primitive but functional proboscis as in Eriocraniidae and Lophocoronidae. Functional mandibles are present in adults of Agathiphagidae and Eriocraniidae, probably used in the former for feeding, but in the latter only to operate the pupal mandibles before ecdysis. The evolution of the proboscis would have permitted these insects, should they live in relatively dry places, to replenish water loss more effectively from moisture droplets or from fluids secreted by plants.

It is not certain whether the exoporian Lepidoptera, including Mnesarchaeoidea and Hepialoidea, evolved from a monotrysian ancestor before or after the common ancestor of Adeloidea and Nepticuloidea. However, few apomorphic attributes are shared by the exoporian families and Adeloidea or Nepticuloidea. As Mnesarchaeoidea and Hepialoidea have several plesiomorphic attributes found also in Dacnonypha and Zeugloptera, it seems probable that they diverged before the heteroneurous Monotrysia. This may justify the separation of the exoporian superfamilies as a suborder Exoporia.

Ditrysia contain the great majority of the Lepidoptera. The suborder was apparently well established by the Cretaceous, and its major radiation probabably paralleled that of the angiosperms. The ancestral form must have combined a heteroneurous venation (similar to the most primitive Tineoidea), five-segmented maxillary palpi, lateral ocelli, a long exposed metathoracic scutum, an endophagous larva with six ocelli, primary setae and crochets of the ventral prolegs arranged in a simple circle, and a mobile pupa with dorsal spines and with the first seven segments in the male abdomen being movable. From this evolved, on the one hand, the Tineoidea, Yponomeutoidea, Gelechioidea, and Copromorphoidea and, on the other, a tortricoid form which in turn gave rise to the Cossoidea, Sesioidea, Castnioidea, Zygaenoidea, and Pyraloidea. If Brock's (6) views are correct, then a cossoid ancester gave rise to the Bombycoidea, a castnioid ancestor to the Papilionoidea, and a pyraloid ancestor to the Geometroidea and Noctuoidea. Alternatively a cossoid ancestor may have provided the origin for both Papilionoidea and Bombycoidea.

Acknowledgments

Thanks are due to Mr. J. S. Dugdale, DSIR, Auckland, New Zealand; Dr. K. Sattler, British Museum (Natural History), London; and to several other colleagues of the British Museum and the Division of Entomology, CSIRO, Canberra, for their stimulating comments on a draft of this review.

Literature Cited

1. Börner, C. 1939. Die Grundlagen meines Lepidopterensystems. *Verh. Int. Kongr. Entomol., 7th, Berlin* 2:1372–1424
2. Bourgogne, J. 1949. Un type nouveau d'appareil génital femelle chez les Lépidoptères. *Ann. Soc. Entomol. Fr.* 115:69–80
3. Bourgogne, J. 1951. Ordre des Lépidoptères. In *Traité de Zoologie,* ed. P. Grassé, 10:174–448. Paris: Masson. 975 pp.
4. Braun, A. F. 1919. Wing structure of Lepidoptera and the phylogenetic and taxonomic value of certain persistent

trichopterous characters. *Ann. Entomol. Soc. Am.* 12:349–66
5. Braun, A. F. 1933. Pupal tracheation and imaginal venation in Microlepidoptera. *Trans. Am. Entomol. Soc.* 59:229–68
6. Brock, J. P. 1971. A contribution towards an understanding of the morphology and phylogeny of the Ditrysian Lepidoptera. *J. Natur. Hist.* 5:29–102
7. Busck, A. 1914. On the classification of the Microlepidoptera. *Proc. Entomol. Soc. Wash.* 16:46–54
8. Busck, A. 1932. On the female genitalia of the Microlepidoptera and their importance in the classification and determination of these moths. *Bull. Brooklyn Entomol. Soc.* 26:199–211
9. Chapman, T. A. 1893. On some neglected points in the structure of the pupae of heterocerous Lepidoptera, and their probable value in classification; with some associated observations on larval prolegs. *Trans. Entomol. Soc. London* 1893:97–119
10. Chapman, T. A. 1894. Some notes on the Micro-Lepidoptera whose larvae are external feeders, and chiefly on the early stages of *Eriocephala calthella. Trans. Entomol. Soc. London* 1894:335–50
11. Chapman, T. A. 1896. On the phylogeny and evolution of the Lepidoptera from a pupal and oval standpoint. *Trans. Entomol. Soc. London* 1896: 567–87
12. Chapman, T. A. 1917. *Micropteryx* entitled to ordinal rank; Order Zeugloptera. *Trans. Entomol. Soc. London* 1916:310–14
13. Clench, H. K. 1957. Cossidae from Chile. *Mitt. Münchner Entomol. Ges.* 47:122–42
14. Clench, H. K. 1959. On the unusual structure and affinities of the Madagascan genus *Pseudocossus. Rev. Fr. Entomol.* 26:44–50
15. Common, I. F. B. 1969. A wing-locking or stridulatory device in Lepidoptera. *J. Aust. Entomol. Soc.* 8:121–25
16. Common, I. F. B. 1970. Lepidoptera (Moths and butterflies). In *The Insects of Australia,* 765–866. Melbourne: Melbourne Univ. Press. 1029 pp.
17. Common, I. F. B. 1973. A new family of Dacnonypha based on three new species from southern Australia, with notes on the Agathiphagidae. *J. Aust. Entomol. Soc.* 12:11–23
18. Comstock, J. H. 1892. The descent of the Lepidoptera. An application of the theory of natural selection to taxonomy. *Proc. Am. Assoc. Advan. Sci.* 41: 199–201
19. Comstock, J. H. 1918. *The Wings of Insects.* Ithaca, NY: Comstock. 430 pp.
20. Dethier, V. G. 1941. The antennae of lepidopterous larvae. *Bull. Mus. Comp. Zool. Harvard Univ.* 87:455–507
21. Dugdale, J. S. 1974. Female genital configuration in the classification of Lepidoptera. *N. Z. J. Zool.* 1:127–46
22. Dumbleton, L. J. 1952. A new genus of seed-infesting micropterygid moths. *Pac. Sci.* 6:17–29
23. Dyar, H. G. 1894. A classification of lepidopterous larvae. *Ann. NY Acad. Sci.* 8:194–232
24. Eyer, J. R. 1924. The comparative morphology of the male genitalia of the primitive Lepidoptera. *Ann. Entomol. Soc. Am.* 17:275–328
25. Eyer, J. R. 1926. Characters of family and superfamily significance in the male genitalia of Microlepidoptera. *Ann. Entomol. Soc. Am.* 19:237–46
26. Forbes, W. T. M. 1923. The Lepidoptera of New York and neighboring States. *Mem. Cornell Univ. Agr. Exp. Sta.* 68:1–729
27. Forbes, W. T. M. 1939. The muscles of the lepidopterous male genitalia. *Ann. Entomol. Soc. Am.* 32:1–10
28. Fracker, S. B. 1915. The classification of lepidopterous larvae. *Ill. Biol. Monogr.* 2:1–169
29. Friese, G. 1970. Zur Phylogenie der älteren Teilgruppen der Lepidopteren. *Tagungsber. Deut. Akad. Landwirtschaftswiss. Berlin* 80:203–22
30. Hampson, G. F. 1892. Moths.—Vol. I. In *The Fauna of British India.* London: Taylor & Francis. 527 pp.
31. Headlee, T. J. 1907. A study in butterfly wing-venation, with special regard to the radial vein of the front wing. *Smithson. Misc. Collect.* 48:284–96
32. Heinrich, C. 1918. On the lepidopterous genus *Opostega* and its larval affinities. *Proc. Entomol. Soc. Wash.* 20:27–38
33. Hessel, J. H. 1969. The comparative morphology of the dorsal vessel and accessory structures of the Lepidoptera and its phylogenetic implications. *Ann. Entomol. Soc. Am.* 62:353–70
34. Hinton, H. E. 1946. On the homology and nomenclature of the setae of lepidopterous larvae, with some notes on the phylogeny of the Lepidoptera. *Trans. Roy. Entomol. Soc. London* 97:1–37
35. Hinton, H. E. 1946. A new classification of insect pupae. *Proc. Zool. Soc. London* 116:282–328

36. Hinton, H. E. 1948. The dorsal cranial areas of caterpillars. *Ann. Mag. Natur. Hist.* (11)14:843–52
37. Hinton, H. E. 1955. On the structure, function, and distribution of the prolegs of the Panorpoidea, with a criticism of the Berlese-Imms Theory. *Trans. Roy. Entomol. Soc. London* 106:455–545
38. Hinton, H. E. 1958. The phylogeny of the panorpoid orders. *Ann. Rev. Entomol.* 3:181–206
39. Imms, A. D. 1925. *A General Textbook of Entomology.* London: Methuen. 698 pp.
40. Jordan, K. 1923. On a sensory organ found on the head of many Lepidoptera. *Nov. Zool.* 30:155–58
41. Kennel, J., Eggers, F. 1933. Die abdominalen Tympanalorgane der Lepidopteren. *Zool. Jahrb. Anat.* 57:1–104
42. Kiriakoff, S. G. 1963. The tympanic structures of the Lepidoptera and the taxonomy of the order. *J. Lepid. Soc.* 17:1–6
43. Kozhanchikov, I. V. 1957. A new representative of the family Cossidae from Miocene deposits of the Caucasus. *Dokl. Akad. Nauk. SSSR* 113:675–77
44. Klots, A. B. 1970. Lepidoptera. In *Taxonomist's Glossary of Genitalia in Insects,* ed. S. L. Tuxen. Copenhagen: Munksgaard. 359 pp.
45. Kristensen, N. P. 1968. The morphological and functional evolution of the mouthparts in adult Lepidoptera. *Opusc. Entomol.* 33:69–72
46. Kristensen, N. P. 1968. The skeletal anatomy of the heads of adult Mnesarchaeidae and Neopseustidae. *Entomol. Meddr.* 36:137–51
47. Kristensen, N. P. 1968. The anatomy of the head and the alimentary canal of adult Eriocraniidae. *Entomol. Meddr.* 36:239–315
48. Kristensen, N. P. 1971. The systematic position of the Zeugloptera in the light of recent anatomical investigations. *Proc. Int. Congr. Entomol., 13th, Moscow.* 1:261
49. Laurentiaux, D. 1953. Classe des Insectes. In *Traité de Paléontologie,* ed. J. Piveteau, 3:397–527. Paris: Masson. 1063 pp.
50. MacKay, M. R. 1970. Lepidoptera in Cretaceous amber. *Science* 167:379–80
51. Meyrick, E. 1895. *A Handbook of British Lepidoptera.* London: Macmillan. 843 pp.
52. Meyrick, E. 1912. Lepidoptera Heterocera Fam. Micropterygidae. *Genera Insect.* 132:1–9
53. Meyrick, E. 1928. *A Revised Handbook of British Lepidoptera.* London: Watkins & Doncaster. 914 pp.
54. Mosher, E. 1916. A classification of the Lepidoptera based on characters of the pupa. *Bull. Ill. State Lab. Natur. Hist.* 12:17–159
55. Mullen, M. A., Tsao, C. H. 1971. Tympanic organ of Indian meal moth, *Plodia interpunctella* (Hübner), almond moth, *Cadra cautella* (Walker) and tobacco moth, *Ephestia elutella* (Hübner). *Int. J. Insect Morphol. Embryol.* 1:3–10
56. Mutuura, A. 1972. Morphology of the female terminalia in Lepidoptera, and its taxonomic significance. *Can Entomol.* 104:1055–71
57. Opler, P. A. 1973. Fossil lepidopterous leaf mines demonstrate the age of some insect-plant relationships. *Science* 179:1321–23
58. Packard, A. S. 1895. Monograph of the bombycine moths of America north of Mexico. Part I. Notodontidae. *Mem. Nat. Acad. Sci. Wash.* 7:1–390
59. Philpott, A. 1927. The genitalia of the Mnesarchaeidae. *Trans. N. Z. Inst.* 57:710–15
60. Philpott, A. 1927. The maxillae in the Lepidoptera. *Trans. N. Z. Inst.* 57: 721–46
61. Remington, C. L. 1954. Order Lepidoptera. In *Classification of Insects,* C. T. Brues, A. L. Melander, F. M. Carpenter, 226–305. *Bull. Mus. Comp. Zool. Harvard Univ.* 108:1–917
62. Richards, A. G. 1933. Comparative skeletal morphology of the noctuid tympanum. *Entomol. Am.* 13:1–43
63. Richards, O. W., Davies, R. G. 1957. In *A General Textbook of Entomology,* A. D. Imms. London: Methuen. 9th ed. 886 pp.
64. Riek, E. F. 1970. Fossil history. In *The Insects of Australia,* 168–86. Melbourne: Melbourne Univ. Press. 1029 pp.
65. Sharp, D. 1909. Insects, Part II. In *The Cambridge Natural History,* ed. S. F. Harmer, A. E. Shipley, 6. London: Macmillan. 626 pp.
66. Sharplin, J. 1963–1964. Wing base structure in Lepidoptera I. Fore wing base; II. Hind wing base; III. Taxonomic characters. *Can. Entomol.* 95:1024–50, 1121–45; 96:943–49
67. Stekol'nikov, A. A. 1967. Functional morphology of the copulatory apparatus in the primitive Lepidoptera and general evolutionary trends in the

genitalia of the Lepidoptera. *Entomol. Rev.* 46:400–9

68. Tillyard, R. J. 1917. The wing venation of Lepidoptera. (Preliminary report). *Proc. Linn. Soc. N. S. W.* 42:167–74
69. Tillyard, R. J. 1919. On the morphology and systematic position of the family Micropterygidae (sens. lat.). *Proc. Linn. Soc. N. S. W.* 44:95–136
70. Tillyard, R. J. 1919. The panorpoid complex. Part 3: The wing-venation. *Proc. Linn. Soc. N. S. W.* 44:533–718
71. Tillyard, R. J. 1923. On the larva and pupa of the genus *Sabatinca. Trans. Entomol. Soc. London* 1922:437–53
72. Tillyard, R. J. 1923. On the mouthparts of the Micropterygoidea. *Trans. Entomol. Soc. London* 1923:181–206
73. Tillyard, R. J. 1926. *The Insects of Australia and New Zealand.* Sydney: Angus & Robertson. 560 pp.
74. Tindale, N. B. 1946. Triassic insects of Queensland. 1. *Eoses,* a probable lepidopterous insect from the Triassic beds of Mt. Crosby, Queensland. *Proc. Roy. Soc. Queensl.* 56:37–46
75. Tindale, N. B. 1963. Origin of the Rhopalocera stem of the Lepidoptera. *Proc. Int. Congr. Zool., 16th, Wash.* 1:304
76. Turner, A. J. 1918. Observations on the lepidopterous family Cossidae and on the classification of the Lepidoptera. *Trans. Entomol. Soc. London* 1918: 155–90
77. Turner, A. J. 1947. A review of the phylogeny and classification of the Lepidoptera. *Proc. Linn. Soc. N. S. W.* 71:303–38
78. Zeuner, F. E. 1961. Notes on the evolution of the Rhopalocera. *Verh. Int. Kongr. Entomol., 11th, Wien* 1:310–13

RECENT DEVELOPMENTS IN INSECT STEROID METABOLISM[1,2]

❖6088

J. A. Svoboda, J. N. Kaplanis, W. E. Robbins, and M. J. Thompson
Insect Physiology Laboratory, Agricultural Research Service, U.S. Department of Agriculture, Beltsville, Maryland 20705

It is generally accepted that insects lack the capacity for de novo biosynthesis of the steroid nucleus, and to date all insects that have been examined critically have been shown to require an exogenous source of sterol to achieve normal growth, development, and reproduction—except in those cases where associated symbiotes may provide a source of sterol. In all but two known exceptions, which will be discussed elsewhere in this chapter, this critical need for sterol can be satisfied by dietary cholesterol or some sterol that can be converted to cholesterol by the insect. Several reviews published since 1970 (17, 39, 42, 54, 56) provide detailed coverage on virtually every facet of the utilization and metabolism of steroids in insects. As the title implies, this chapter will be primarily devoted to a discussion of progress that has been made in this area of insect biochemistry since the last reviews on the subject were written. In the present chapter we will repeat that material covered in previous reviews only when it is required for background information, clarification, discussion, or to emphasize the importance of an aspect that is being considered in detail.

This review will be concerned with three major areas of insect steroid research: the metabolism and utilization of phytosterols and other neutral sterols; the in vivo metabolism of the ecdysones; and the in vitro metabolism of ecdysones. These were selected since most of the recent developments and advances in insect steroid metabolism have been in these three areas.

THE UTILIZATION AND METABOLISM OF PHYTOSTEROLS AND OTHER NEUTRAL STEROLS

During the past seven years detailed metabolic studies utilizing, as the primary experimental animal, the tobacco hornworm, *Manduca sexta,* have resulted in the

[1]The literature survey for this review was completed March, 1974.

[2]The following abbreviations are used in this chapter: NADPH (nicotinamide adenine dinucleotide phosphate, reduced form); ATP (adenosine triphosphate); DEAE (diethyl amino ethyl); GSH (glutathione, reduced form).

accumulation of a considerable body of knowledge (Figure 1) concerning the intermediary metabolism of phytosterols in insects (42, 54, 56). As a result of this and corollary research with other phytophagous species and certain omnivorous insects, we now know a number of intermediate steps involved in the dealkylation and conversion of C_{28} and C_{29} plant sterols to cholesterol in insects. Considerable similarity has been found to exist between insect species, with respect to their ability to utilize and metabolize various plant sterols, but as we expand our knowledge of this area it has become increasingly obvious that generalizations concerning sterol metabolism in insects must be made with care. In this portion of the review we will first outline the various pathways of phytosterol metabolism in insects as we now know them and then discuss some of the interesting differences that have been found to occur between species. As with our previous reviews (42, 54, 56), the experimental data covered will be limited to those studies in which adequate specific biochemical and analytical techniques have been employed.

It has been known for a number of years, from biochemical research utilizing semidefined diets in conjunction with either highly purified or radiotracer-labeled sterols, that sitosterol is readily converted to cholesterol in a number of phytopha-

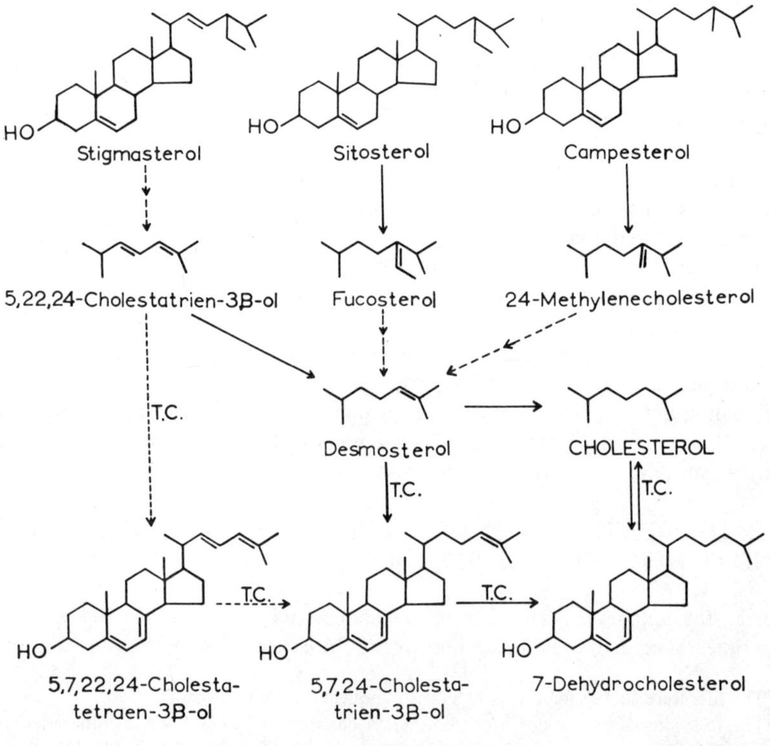

Figure 1 Pathways of phytosterol metabolism in the tobacco hornworm and *Tribolium confusum*. Conversions labeled T. C. have been demonstrated only in *Tribolium*.

gous and omnivorous species. These include the German cockroach (*Blattella germanica*) (41), the pine sawfly (*Neodiprion pratti*) (43), the silkworm (*Bombyx mori*) (20), the boll weevil (*Anthonomus grandis*) (8), the tobacco hornworm (50), and the cockroach (*Eurycotis floridana*) (40). In all of these species, cholesterol was normally the major sterol metabolite present in the insect. In addition to sitosterol, the tobacco hornworm also efficiently converts campesterol, stigmasterol, brassicasterol, 22,23-dihydrobrassicasterol, 24-methylenecholesterol, and fucosterol to cholesterol (Figure 1), with cholesterol being the major insect sterol when each of these sterols is used as the sole added dietary sterol (47).

The first known intermediate involved in the conversion of a plant sterol to cholesterol was isolated and identified as desmosterol (24-dehydrocholesterol) in a study on the metabolism of ^{3}H-sitosterol in the tobacco hornworm (50). Subsequently, this Δ^{24}-sterol was shown to be the normal terminal intermediate in the conversion of campesterol, stigmasterol, brassicasterol, 22,23-dihydrobrassicasterol, 24-methylenecholesterol, and fucosterol to cholesterol in this insect (47). In comparative studies, desmosterol was also identified as an intermediate in the dealkylation and conversion of sitosterol to cholesterol in the firebrat (*Thermobia domestica*), the German cockroach, the American cockroach (*Periplaneta americana*), the corn earworm (*Heliothis zea*), the fall armyworm (*Spodoptera frugiperda*) (48), and the confused flour beetle (*Tribolium confusum*) (49). These findings were facilitated through the use of several inhibitors of sterol metabolism developed in vertebrate biomedical research but which have also been useful in insect sterol metabolism studies (46, 48). These inhibitors, including triparanol (MER-29) and a number of azasteroids, block the Δ^{24}-sterol reductase, an enzyme responsible for converting desmosterol to cholesterol in both vertebrates and insects. These inhibitory compounds have proven to be most useful tools in metabolic studies on plant sterols in insects.

The discovery of desmosterol as the terminal intermediate in the dealkylation of plant sterols to cholesterol led to the development of the first in vitro system available for the study of neutral sterol metabolism in insects. This system, prepared from midgut tissue of the tobacco hornworm (53), efficiently converts ^{14}C-desmosterol to cholesterol, and has been used to study both normal sterol metabolism and the action of the sterol inhibitors. To date, it is still the only in vitro system available for examining a specific enzyme activity involved in neutral sterol metabolism in an insect.

In subsequent studies using radiolabeled sterols, it was determined that fucosterol was a normal intermediate in the conversion of sitosterol to cholesterol in the tobacco hornworm (51) and in *Tribolium* (49). Fucosterol has also been implicated as an intermediate in this pathway through its tentative identification from the German cockroach and *E. floridana* (40). An analogous pathway exists in the case of the metabolism of ^{3}H-campesterol to ^{3}H-cholesterol, in which case ^{3}H-24-methylenecholesterol was shown to be an intermediate in the tobacco hornworm (52, 56). Similarly, 24-methylenecholesterol is also involved as an intermediate in the conversion of dihydrobrassicasterol (the 24S-methyl isomer of campesterol) to cholesterol in the hornworm (52).

Recent related studies (37) have provided some extremely interesting information on the probable mechanism of dealkylation of phytosterols to cholesterol in insects. When radiolabeled 28-isofucosterol (prepared biosynthetically by incorporation of [2-^{14}C-(4*R*)-4-^{3}H]-mevalonate by larch leaves) was fed to young larvae of *Tenebrio molitor,* it was largely converted to cholesterol. By examining the ^{3}H:^{14}C atomic ratio of the cholesterol isolated from the insect, it was found that all tritium administered in the isofucosterol was retained after dealkylation, indicating that the C-25 hydrogen atom must migrate to C-24 during dealkylation. The same research group (38) had previously shown that the transformation in plants of Δ^{24}-sterol into a 24-ethylidene compound involves hydrogen migration from C-24 to C-25, thus supporting the previous suggestion (47) that the process of C-24 dealkylation of sterols in insects may be, at least in part, the reverse of sterol C-24 alkylation in plants. The chemical conversion of fucosterol-24,28-epoxide into desmosterol by BF_3-etherate (19) has been interpreted in terms of a similar mechanism.

Fucosterol-24,28-epoxide has been implicated as a probable intermediate in the conversion of sitosterol to cholesterol in the silkworm (Figure 2) (34). When 2,4-^{3}H-fucosterol-24,28-epoxide was administered orally to *Bombyx* larvae, labeled cholesterol was isolated and identified from the insect. In addition some labeled epoxide, as well as cholesterol, was recovered after ^{3}H-fucosterol was administered in a similar manner. Larvae fed 28-oxo-, 24,28-dihydroxy- or 24-hydroxy-28-oxo-sitosterol, and 24-oxo- or 24-hydroxy-cholesterol as dietary sterol died in the first instar. Fucosterol-24,28-epoxide also appears to be an intermediate in the same metabolic pathway in the locust, *Locusta migratoria* (1). 3-^{3}H-Fucosterol-24,28-epoxide propionate was converted to cholesterol when injected into the abdomen of this insect, and, as in the silkworm, 28-oxo-sitosterol did not serve as an intermediate for cholesterol production. These two studies are of considerable interest, but positive identification of the very unstable fucosterol-24,28-epoxide as a normal intermediate in the metabolic conversion of sitosterol to cholesterol must await its isolation and characterization from insects that have been administered radiolabeled sitosterol.

A new sterol, 22-*trans*-5,22,24-cholestatrien-3β-ol was identified as an intermediate in the conversion of stigmasterol to cholesterol in the tobacco hornworm, German cockroach, and corn earworm (45) and in *Tribolium* (49). It was established that the Δ^{22}-bond is reduced before the Δ^{24}-bond in both the hornworm and *Tribolium.* The conversion of this trienol to desmosterol and cholesterol can also be effected in the same in vitro system from hornworm gut tissue that was used for studying the reduction of desmosterol to cholesterol (45).

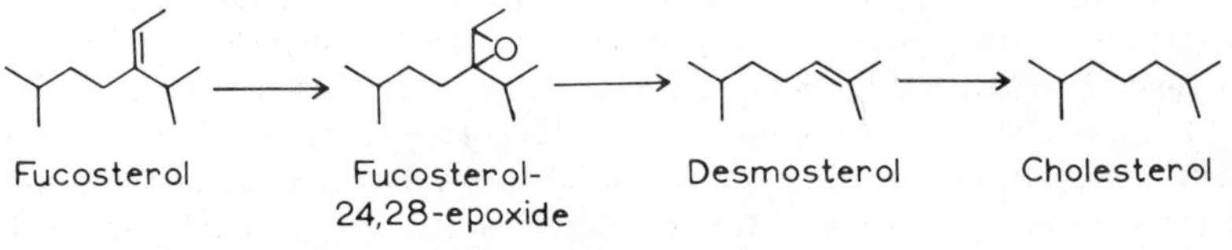

Fucosterol → Fucosterol-24,28-epoxide → Desmosterol → Cholesterol

Figure 2 Probable position of fucosterol-24,28-epoxide as an intermediate in the conversion of sitosterol to cholesterol.

Metabolic studies with *Tribolium* (Figure 1) yielded yet another new sterol, 5,7,24-cholestatrien-3β-ol, which was shown to be a normal intermediate in the conversion of radio-labeled sitosterol, campesterol, stigmasterol, and desmosterol to cholesterol (49). Involvement of this compound as an intermediate in sterol metabolism represents a major variance from sterol metabolism in the hornworm and a number of other species. In addition, a related compound, 5,7,22,24-cholestatetraen-3β-ol, was isolated from *Tribolium* fed ^{3}H-stigmasterol plus an azasteroid inhibitor (49), but it was concluded that this was not a normal intermediate since only trace amounts were found in sterols of normal insects.

Sterols containing a Δ^7-bond have received increased attention in insect biochemistry in the past few years. This emphasis on their importance arose largely because all the insect ecdysones have a Δ^7-bond (18, 26) and Δ^7-sterols have been suggested as precursors of ecdysones (42, 54, 56). Certainly, 7-dehydrocholesterol would appear to be a likely ecdysone precursor, from the standpoint of its relatively high titer in the ecdysial glands (54), as well as the demonstration of its conversion to 20-hydroxyecdysone in *Calliphora stygia* (10).

Although 7-dehydrocholesterol is usually a minor component of the neutral sterols in many species, it has been found to represent as much as 70% of the total sterol in *Tribolium* (2, 49). In this insect there is an equilibrium between cholesterol and 7-dehydrocholesterol, but the significance of this extremely high level of 7-dehydrocholesterol still remains unknown. Studies with labeled sterols have shown that a number of insects readily convert cholesterol to 7-dehydrocholesterol (42, 54, 56), but two species are apparently unable to introduce the Δ^7-bond into the sterol nucleus (5, 14). These two species, *Drosophila pachea* and *Xyleborus ferrugineus,* require a dietary sterol containing either a Δ^7-bond or a $\Delta^{5,7}$-diene system for normal development and reproduction.

Recent studies with the Mexican bean beetle, *Epilachna varivestis,* have shown that Δ^7-sterols also occupy a significant role in sterol metabolism in this species.[3] In bean beetles reared on soybean leaves, lathosterol (Δ^7-cholesten-3β-ol) comprised from 12–16% of the sterols from eggs, prepupae, and adults, and Δ^7-campestenol and Δ^7-stigmastenol (Figure 3) were also identified in all stages, with the highest percentage of these latter two sterols occurring in the egg. To our knowledge, none of these Δ^7-sterols has previously been reported as a normally occurring sterol in insects reared on a natural host plant. However, *D. pachea* readily converts lathosterol to 7-dehydrocholesterol (12) and it may be a normal sterol intermediate in this insect. Also lathosterol is the principal metabolite of cholestanol in two species of cockroaches (6, 32). In addition, the saturated sterols (stanols), cholestanol, campestanol, and stigmastanol (Figure 3) taken together were the major components comprising 54–77% of the total sterols in the egg, prepupa, and adult stages. Perhaps most surprising was the relatively low percentage of cholesterol present in each stage of the bean beetle, which ranged from 2.9–4.5% of the total sterols in

[3]Svoboda, J. A., Thompson, M. J., Elden, T. C., Robbins W. E. Unusual sterol composition of a phytophagous insect, the Mexican bean beetle, reared on soybean plants. Submitted for publication.

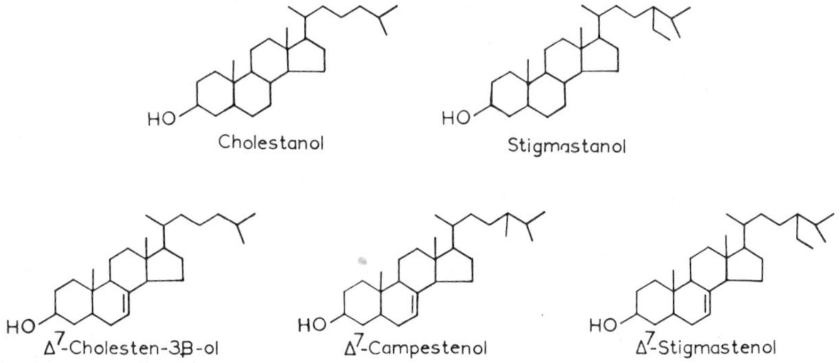

Cholestanol

Stigmastanol

Δ^7-Cholesten-3β-ol

Δ^7-Campestenol

Δ^7-Stigmastenol

Figure 3 Sterols isolated from the Mexican bean beetle.

the stages examined. Analysis of the soybean leaves on which the Mexican bean beetles were reared indicated that over 98% of the soybean leaf sterols were comprised of the normal phytosterols campesterol, stigmasterol, and sitosterol, along with trace amounts of cholesterol. Based on the composition of the sterols of the bean beetle and the soybean leaves, then, the following tentative scheme is proposed for the utilization and metabolism of plant sterols in this insect:

$$\Delta^5\text{-PHYTOSTEROLS} \longrightarrow \Delta^0\text{-PHYTOSTEROLS} \longrightarrow \text{CHOLESTANOL}$$

$$\Delta^0\text{-PHYTOSTEROLS} \downarrow \Delta^7\text{-PHYTOSTEROLS} \qquad \text{CHOLESTANOL} \downarrow \text{LATHOSTEROL}$$

The Mexican bean beetle is, interestingly, a member of the subfamily Epilachninae which contains nearly all of the phytophagous species of the family Coccinellidae, a predominantly predacious family of insects. The unusual spectrum of sterols found in this insect then may result from mechanisms and pathways of phytosterol utilization and metabolism that have secondarily evolved in a phytophagous insect that was originally zoophagous. This then, may account for the differences found in this species as compared to other plant-feeding insects.

Findings such as these should prompt researchers to characterize more thoroughly the sterols isolated from insects, since a number of saturated and unsaturated sterols are not readily resolved by many of the commonly used gas-liquid and thin-layer chromatographic systems. Clearly, more insects from diverse or specialized ecological niches, such as *D. pachea, Tribolium, X. ferrugineus,* and the Mexican bean beetle, should be critically examined with respect to sterol utilization, metabolism, and content. We may well expect to find additional interesting variations which reflect the adaptations of these insects with respect to their utilization and metabolism of dietary sterols to fulfill their requirements for these essential nutrients.

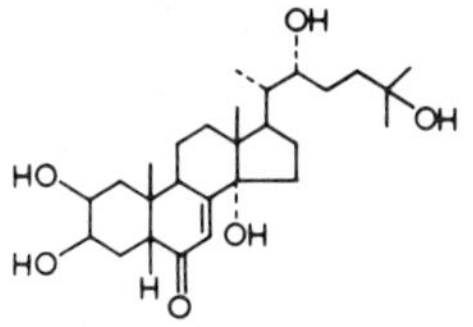

I
α-Ecdysone

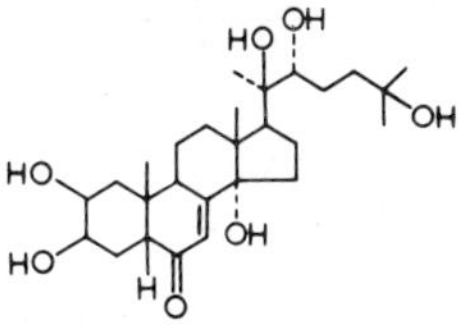

II
20-Hydroxyecdysone

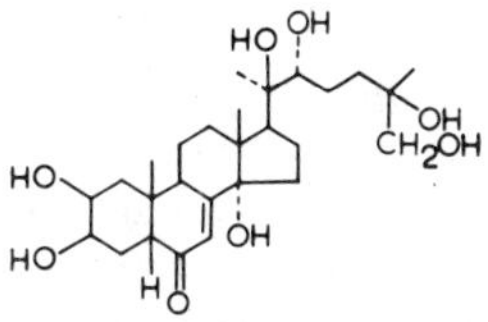

III
20,26-Dihydroxyecdysone

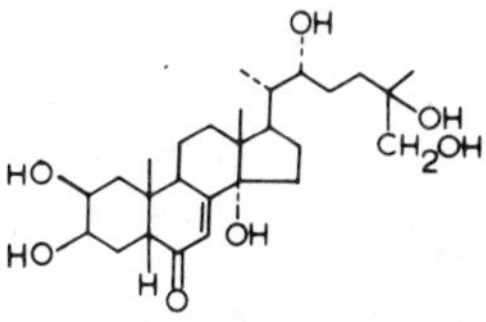

IV
26-Hydroxyecdysone

Figure 4 Insect ecdysones.

IN VIVO METABOLISM OF THE ECDYSONES

The chemistry, biosynthesis, and metabolism, including inactivation, of the insect molting hormones have been previously discussed in detail in a number of comprehensive reviews (11, 17, 39, 42, 54, 56). Salient examples of progress since the latest reviews include the isolation and identification of a new insect molting hormone; a new synthetic steroid that may serve as a precursor for the molting hormones in an insect; the detection, isolation, and identification of apolar metabolites of α-ecdysone or 20-hydroxyecdysone; and further information on the inactivation of the ecdysones and metabolites in insects in vivo. This section will deal briefly with each of these aspects and the reader is referred to the previously cited reviews for background information.

A recent development in molting hormone research is the isolation and identification of the new insect molting hormone, 26-hydroxyecdysone (IV) (22) which brings the total of naturally occurring molting hormones thus far isolated and identified from insects to four (11, 17, 39, 42, 54, 56). This steroid (Figure 4) was found to be the predominant molting hormone in the developing embryo of the tobacco hornworm. In addition, the three previously known molting hormones, α-ecdysone (I), 20-hydroxyecdysone (II), and 20,26-dihydroxyecdysone (III), were also isolated from the tobacco hornworm egg, but these were present in far smaller quantities (22). Based on the qualitative and quantitative nature of the molting hormones isolated from the egg, the following tentative scheme is proposed for the biosynthesis and metabolism of the ecdysones in the developing embryo of the tobacco hornworm:

α-ECDYSONE ⟶ 26-HYDROXYECDYSONE
↓
20-HYDROXYECDYSONE ⟶ 20,26-DIHYDROXYECDYSONE

α-Ecdysone, then, serves as a precursor for both 20-hydroxyecdysone and 26-hydroxyecdysone, with 26-hydroxyecdysone being the major metabolite of α-ecdysone in the developing embryo (48–64 hr old). Although 20-hydroxyecdysone is a minor metabolite of α-ecdysone at this stage of development, it is the major metabolite during pupal-adult development (23). The currently available chemical and biochemical information on the molting hormones, both during embryonic and postembryonic development, indicates the presence of different biosynthetic-metabolic pathways, as well as quantitative and qualitative differences in the ecdysones in the different developmental stages of the tobacco hornworm (22). Different ecdysones then may function at various stages of insect development and the qualitative nature of the molting hormones might well determine the type of molt. In the light of this possibility, certain of the current concepts concerning the role of the ecdysones in the hormonal control of molting and metamorphosis in insects may need to be reassessed.

There is evidence that the biosynthetic-metabolic pathways of the ecdysones may differ in different species of insects as well. Both labeled α-ecdysone and 20-hydroxyecdysone are converted to an as yet unidentified metabolite when these molting hormones are injected into larvae of *Chironomus tentans* (7, 59). However, this metabolite was not produced from either α-ecdysone or 20-hydroxyecdysone in different developmental stages of *Calliphora vicina* (59).

When radiolabeled 2,22,25-trideoxyecdysone was injected into third-instar larvae of *Calliphora stygia* at the time of puparium formation and the insects analyzed 12 hr later, the steroid was converted to 22-deoxyecdysone, α-ecdysone, 20-hydroxyecdysone, and to another compound, present in relatively small quantities and tentatively identified as 2,22-dideoxyecdysone (9). On the basis of these results the sequence of hydroxylation during ecdysone biosynthesis in this insect is suggested to be C-25, C-2, C-22, and then C-20. In the tobacco hornworm, however, the efficient incorporation and conversion of the radiolabeled 22,25-dideoxyecdysone to the insect ecdysones during pupal-adult development (21) strongly suggests that in this insect hydroxylation of the steroid nucleus is completed first prior to any hydroxylation on the side chain. Although candidate precursors such as 22,25-dideoxyecdysone and 2,22,25-trideoxyecdysone are converted to the insect ecdysones, the unequivocal proof as to whether or not one or the other or both are indeed normal intermediates in the biosynthetic-metabolic pathway, must await their direct isolation and identification from insects. Since 22-deoxyecdysone is a metabolite shared by both precursors, it may be the first intermediate to be searched out, isolated, and identified from an insect source. An interesting observation in the above cited study (9) was the conversion of the labeled 2,22,25-trideoxyecdysone analog to both α-ecdysone and 20-hydroxyecdysone in the isolated abdomens of *C. stygia,* indicating that these conversions may occur in the insect independent of the brain-ring gland complex.

An area of increasing interest is the unidentified apolar steroids or metabolites of the molting hormones encountered in vivo from insects. A less polar isomer of 20,26-dihydroxyecdysone was isolated along with this ecdysone from the tobacco hornworm during pupal-adult development. Its structure was speculated to be the 5α-H isomer of 20,26-dihydroxyecdysone (55). Recently, a less polar metabolite of 20-hydroxyecdysone was isolated and identified from the meconium of this insect as the corresponding 3α-hydroxy epimer of this ecdysone.[4] On the basis of this information from an in vivo system and the conversion of both α-ecdysone and 20-hydroxyecdysone to their corresponding 3α-hydroxy epimers in in vitro systems from the hornworm (see in vitro section), it is probable that the compound associated with 20,26-dihydroxyecdysone is the 3α-hydroxy epimer of this ecdysone rather than the 5α-H isomer, as speculated earlier (55).

When radiolabeled α-ecdysone was injected into white prepupae of *C. vicina,* it was oxidized to 3-dehydroecdysone as well as to 3-dehydro-20-hydroxyecdysone (27). Interestingly, the 3-dehydroecdysone was found to be rapidly reduced to α-ecdysone and is thus interconvertible in this insect. The locust, *Locusta migratoria,* (16, 31) also converted radiolabeled α-ecdysone to 20-hydroxyecdysone and to their corresponding 3-dehydro-products (16, 31), indicating similar routes of metabolism for α-ecdysone in these two representatives from two very different orders. In an earlier study (25) it had been suggested that the unidentified radiolabeled apolar metabolite IV (3) and product D (35) detected in the hemolymph of *Antheraea polyphemus* and *Bombyx mori,* respectively, may be identical to 3-dehydroecdysone. In addition to the hemolymph, a number of other tissues including fat body, carcass, and gut were found to possess significant quantities of a less polar metabolite of α-ecdysone in *Hyalophora cecropia* when the pupa of this insect was injected with radiolabeled α-ecdysone (13).

The larva of *Chironomus tentans* converted both radiolabeled α-ecdysone and 20-hydroxyecdysone to an unknown metabolite which, based on its chromatographic behavior, differed from 3-dehydro-20-hydroxyecdysone and ruled out this steroid as the metabolite (59). Based on its chromatographic properties, this metabolite is postulated to be the 5α-H isomer of 20-hydroxyecdysone. When high concentrations of α-ecdysone were injected into *C. tentans,* a different metabolite was produced that migrated ahead of α-ecdysone (59). However, this unidentified compound was not formed from radiolabeled 20-hydroxyecdysone by *Chironomus,* indicating it to be a metabolic product of α-ecdysone only. Neither of these metabolites detected in *C. tentans* were found to occur in the various developmental stages of *C. vicina* (59).

The pupa of *B. mori* has also been reported to produce an unknown metabolite which, when analyzed as a derivative by gas chromatography, has a shorter retention time than the corresponding derivative of 20-hydroxyecdysone, suggesting this metabolite to be less polar (33). Whether or not the unknown metabolites that have

[4]Thompson, M. J., Kaplanis, J. N., Robbins, W. E., Dutky, S. R., Nigg, H. N. 3-Epi-20-hydroxyecdysone from meconium of the tobacco hornworm. Submitted for publication.

been detected are similar or differ in the various species of insects in which they occur, must await their isolation and characterization.

Insects are known to form highly polar metabolites of the molting hormones, such as their glucoside and sulfate conjugates, and these are generally considered to be the end products of ecdysone metabolism (54). Conjugation does not occur to an appreciable extent during embryonic development in the tobacco hornworm since the highly polar fractions, when subjected to enzymic hydrolysis with sulfatase, yielded only small amounts of the 26-hydroxyecdysone (22). No ecdysones were detected in this fraction following hydrolysis with glucosidase. On the other hand, during pupal-adult development sulfates are the predominant conjugates in this insect (54, 56). Recently sulfates and glucuronides have been reported to be the major conjugates of the molting hormones and their metabolites in *Calliphora vicina* (27) and *L. migratoria* (16, 31). In addition to sulfate conjugates, the polar radiolabeled fractions from *Chironomus tentans* and *C. vicina* also contained glucosides as determined by enzymic hydrolysis with glucosidase (59). Thus far, the intact conjugates of the molting hormones from insects have not been separated, purified, isolated, and identified, nor have their roles been clearly defined in insects. Whether or not conjugates have a function in the biosynthesis, metabolism, and regulation of titer of the ecdysones as previously discussed or are simply inactivation products, remains to be determined.

IN VITRO METABOLISM OF THE ECDYSONES

A diversity of in vitro systems including cell-free preparations, intact organs, tissues, and intact isolated abdomens have been used to study the biosynthesis, metabolism, and inactivation of the ecdysones in insects. The following discussion is a compendium of the in vitro studies of the ecdysones in which emphasis will be placed on briefly describing the experimental conditions employed and the metabolic conversions effected. This discussion will also serve to emphasize the paucity of information available on the nature of the enzymes involved in molting hormone biosynthesis and metabolism.

The first in vitro study involving a molting hormone described the presence of an enzyme system in the cytosol from the fat body of *C. vicina* that was capable of inactivating α-ecdysone (24). The enzymic activity was retained after a sixfold purification using centrifugation (100,000 X g) and ammonium sulfate precipitation. The enzyme activity was measured, after incubation with α-ecdysone in phosphate buffer (pH 7.0) for 2 hr at 37°C, by determining the residual molting hormone activity in the *Calliphora* bioassay. With this system, it was demonstrated that the fluctuation in the activity of this inactivating enzyme system was a mirror image of the ecdysone titer curve previously charted for this insect (44).

In studies with the same insect, intact larval fat body tissue incubated for up to 60 min at 25°C in a sucrose solution (containing KCl, $MgCl_2$, $KHCO_3$, at pH 7.6) with ^{3}H-25-deoxy-20-hydroxyecdysone converted this compound to its glycoside (15). 20-Hydroxyecdysone was incubated at pH 7.2 for 6 hr at 37°C in combination with uridine diphosphate-glucose-^{14}C in a soluble, dialyzed enzyme preparation obtained by homogenizing larval fat body of *C. vicina* in Tris/HCl buffer [with NH_4Cl and $(CH_3COO)_2Mg$]. Labeled 3β-(α-glucopyranosido-)-20-hydroxyecdysone was produced in this cell-free system. These results indicated that the inactivation of this molting hormone was catalyzed by the enzyme UDP-Glucose:20-Hydroxyecdysone-Glucosyl-transferase.

An in vitro system from early sixth-instar larval gut tissues of the southern armyworm (*Spodoptera eridania*) was found to have sulfotransferase enzyme activity capable of sulfating α-ecdysone and a number of other steroids (58). The guts were homogenized in 0.05 M sodium phosphate and the 100,000 X g supernatant was used as the enzyme source. The incubations were carried out in sodium phosphate buffer, pH 7.5, containing ATP, K_2SO_4, $MgCl_2$, and GSH for 1 hr at 38°C. This was the first report on sulfoconjugation of steroids in insects by an in vitro enzyme system and the authors propose that this system (conjugation) may be involved in the regulation of molting hormone titer.

In experiments with last-instar larvae of *C. tentans,* labeled 20-hydroxyecdysone was incubated with gut and body wall in modified Cannon's medium, a complex mixture of inorganic salts, sugars, organic acids, amino acids, vitamins, cholesterol, and antibiotics, for 4 hr at pH 7.3 (59). Very polar metabolites, including glucose conjugates plus a less polar compound, were formed by both tissues, with the gut having greater metabolic activity than the body wall.

The first studies involving the metabolic conversion of α-ecdysone to 20-hydroxyecdysone in other than an intact insect were done with isolated abdomens of *C. vicina* and *Bombyx* larvae (35). Following injection of ^3H-α-ecdysone into the isolated abdomens of these insects, ^3H-20-hydroxyecdysone was isolated and identified. This work presented the first evidence that the conversion of α-ecdysone to 20-hydroxyecdysone could occur peripheral to the ecdysial glands and posed the question as to whether or not this conversion occurs in these glands.

Isolated abdomens from 6-day-old fifth-instar silkworm larvae have also been shown to be capable of converting injected ^{14}C-cholesterol to α-ecdysone and 20-hydroxyecdysone (36). These studies provided further proof that molting hormones may be produced in tissues other than the prothoracic glands. However, isolated abdomens from 2-day-old fifth-instar larvae did not convert cholesterol to the molting hormones, thus indicating that active prothoracic glands are somehow related to ecdysone biosynthesis. In more recent studies, the isolated abdomens of *Calliphora stygia* (9) were shown to convert injected ^3H-2,22,25-trideoxyecdysone to α-ecdysone and 20-hydroxyecdysone, indicating the ability to produce ecdysones in tissues other than prothoracic or ring glands in yet another species.

In another study, using intact brain-ring gland complexes from 7-day-old *Calliphora vicina* larvae, some interesting results concerning ecdysone metabolism were obtained from incubation of these complexes with ^{14}C-cholesterol (57). The incubation medium consisted of a phosphate buffer system (pH 7.4) containing KCl, $MgCl_2$, sodium citrate, nicotinamide, streptomycin sulphate, benzylpenicillin-Na, and an NADPH-generating system plus ATP. The incubations were carried out aerobically at 25°C for 12 hr. Neither α-ecdysone nor 20-hydroxyecdysone, either in the free form or as esters or glycosides, was produced in this system, although labeled substances were produced which co-chromatographed with α-ecdysone and 20-hydroxyecdysone, but were shown to be different by co-crystallization. In further in vitro experiments with unlabeled cholesterol, hormonal activity, as indicated by the *Calliphora* test, was found in a fraction less polar than α-ecdysone after treating this fraction with an esterase. Similarly, considerable hormonal activity was found after treatment of a fraction more polar than 20-hydroxyecdysone and also the water soluble fraction with α-glucosidase, but no known insect ecdysones were detected here either.

22,25-Dideoxyecdysone was readily metabolized to 22-deoxyecdysone, α-ecdysone, 20-hydroxyecdysone, and 20,26-dihydroxyecdysone in isolated fat body and Malpighian tubule tissue of *Manduca* prepupae in vitro (28, 29). Ecdysone glucosides also were formed by *Manduca* fat body and Malpighian tubule tissue incubated in vitro with ^{3}H-α-ecdysone and uridine diphosphate- ^{14}C-glucose. These glucosides yielded ^{3}H-α-ecdysone, 20-hydroxyecdysone, and 20,26-dihydroxyecdysone when treated with α-glucosidase. Fat body, Malpighian tubule, gut, and body wall tissues of this species are all able to convert α-ecdysone to 20-hydroxyecdysone in vitro; however, other tissues including the prothoracic gland are unable to effect this conversion. The prothoracic gland was found to metabolize 22,25-dideoxyecdysone to α-ecdysone, but no farther. The incubation system for these studies involved the gentle agitation of intact tissues in a saline preparation. In attempts to obtain further evidence for these compounds as intermediates in molting hormone production, cholesterol was not converted to 22,25-dideoxyecdysone or 22-deoxyecdysone when incubated with fat body or prothoracic glands.

Incubated intact prothoracic glands from last-instar *Bombyx* larvae were recently shown (4) to produce a hormonally active compound identified as α-ecdysone. The glands were cultivated at 25°C in a medium containing the protein fraction of hemolymph from diapausing cynthia silkworm pupae, prepared by elution through a Sephadex® G-25[5] column equilibrated with saline and diluted with Wyatt's insect culture medium (without tyrosine). The culture system was provided with an excess of oxygen. The hormone was isolated from the medium and not from the glands, suggesting that the hormone is secreted from the gland immediately after synthesis. The key to success in this system was shown to be the lipoprotein fractions from the hemolymph. Isolated prothoracic glands of the tobacco hornworm have also been reported to produce α-ecdysone (30), consistent with the findings of the Japanese research team. The culture system for the glands from the hornworm included a 10% whole egg ultrafiltrate supplement.

A new metabolite of α-ecdysone, 3-dehydroecdysone, was produced in good yield by a soluble enzyme from light brown *C. vicina* puparia (25). This enzyme system was prepared from the 78,000 X g supernatant from whole puparia homogenized in Tris/HCl buffer, with 0.04% ascorbic acid. The enzyme preparation was purified by ammonium sulfate precipitation, desalted on a Sephadex column, the impurities removed by adsorption on DEAE-Cellulose, rerun on a Sephadex column, and

[5]Mention of a company name or a proprietary product does not necessarily imply endorsement by the U.S. Department of Agriculture.

concentrated by ultrafiltration. The incubations were carried out in 0.1 M phosphate buffer at pH 7.0 and 37°C, with shaking overnight, using unlabeled ecdysone as the substrate. This report contains the most complete description of any cell-free in vitro system for ecdysone metabolism reported in the literature to date.

Recently, a soluble enzyme system from the gut of the early prepupal stage of the tobacco hornworm has been developed that readily produces the 3α-hydroxy epimer of α-ecdysone.[6] Other tissues including fat body, hemolymph, and carcass were devoid of activity. This system also isomerized 20-hydroxyecdysone or 22-deoxyecdysone, but not 22,25-dideoxyecdysone. The incubation was carried out in 0.2 M phosphate buffer (pH 7.2) at 30°C with shaking for 4 hr and included a NADPH generating system. Lesser isomerase activity was present in the larvae and pupae, but none could be demonstrated in the egg or adult. This enzymatic activity may well be responsible for certain of the unknown metabolites reported by other workers.

From the preceding discussion it is readily apparent that due to the diversity of test systems and experimental insects that have been used in the in vitro research reported to date, it would be difficult at this time to either attempt to reach conclusions or to make generalizations based on the available data. Two points, however, are obvious from these investigations: First, the biosynthesis, metabolism, and deactivation of the molting hormones occurs in a number of organs or tissues of insects. Second, the enzyme systems involved in ecdysone metabolism still represent a primary area of challenge and opportunity for investigators interested in molting hormone biochemistry.

CONCLUSIONS

Our knowledge of the intermediary metabolism of the phytosterols and their conversion to cholesterol in insects has greatly increased during the past seven years. A number of the normal intermediates in the biochemical pathways from the plant sterols to cholesterol have been isolated and positively identified. We are now beginning to gain information on the mechanisms involved in the dealkylation of the sterol side chain. Additional interesting differences in the metabolism and utilization of phytosterols have been found to occur among different species of phytophagous insects indicating that considerably more diversification has occurred with regard to sterol metabolism in insects than was formerly believed to exist. The prior placement of insects into two broad categories in regard to their ability to utilize and to metabolize plant sterols—the zoophagous species in one group and the phytophagous and omnivorous species considered together in the second group—no longer appears to be valid. The findings discussed in this chapter suggest further that comparative research on related insects with differing feeding habits such as the phytophagous and predacious species of the family Coccinellidae could well provide some most interesting information that may directly relate biochemical and physiological adaptation to phylogeny and speciation within a related group of insects.

[6]Nigg, H. N. et al. Ecdysone metabolism: Ecdysone dehydrogenase-isomerase. Submitted for publication.

In addition to the isolation and identification of the new insect ecdysone—26-hydroxyecdysone—from an insect embryo, there has been considerable information published on the metabolism and conjugation of ecdysones since the last reviews. These studies have led to postulations that the different ecdysones may function at different stages of insect development and that the various ecdysones, metabolites, and conjugates may also function in the maintenance of molting hormone titer in the various stages of insect development.

Certainly the study of steroid metabolism in insects remains a fertile field for future research. This is particularly true in the case of ecdysone biosynthesis, where the pathway from cholesterol or 7-dehydrocholesterol to the ecdysones is almost completely unknown with the exception of such possible intermediates as 22,25-dideoxyecdysone, 22-deoxyecdysone, or 2,22,25-trideoxyecdysone. The enzyme systems involved in virtually all aspects of steroid metabolism in insects still remain a relatively unexplored but a critically important area of investigation. Finally, it is important to stress the need to obtain as complete an understanding as possible of the utilization and metabolism of steroids in insects since this is perhaps one of the most distinct areas of biochemical and physiological difference between insects and vertebrates. As such, knowledge in this area fulfills a dual function in that it provides information on an area that may be exploited for its practical potential, as well as providing basic information on comparative steroid biochemistry.

Literature Cited

1. Allais, J. P., Alcaide, A., Barbier, M. 1973. Fucosterol-24,28 epoxide and 28-oxo-β-sitosterol as possible intermediates in the conversion of β-sitosterol into cholesterol in the locust, *Locusta migratoria* L. *Experientia* 29:944–45
2. Beck, S. D., Kapadia, G. G. 1957. Insect nutrition and metabolism of sterols. *Science* 126:258–59
3. Cherbas, L., Cherbas, P. 1970. Distribution and metabolism of α-ecdysone in pupae of the silkworm *Antheraea polyphemus*. *Biol. Bull.* 138:115–28
4. Chino, H. et al 1974. Biosynthesis of α-ecdysone by prothoracic glands in vitro. *Science* 183:529–30
5. Chu, H. M., Norris, D. M., Kok, L. T. 1970. Pupation requirement of the beetle, *Xyleborus ferrugineus:* sterols other than cholesterol. *J. Insect Physiol.* 16:1379–87
6. Clayton, R. B., Edwards, A. M. 1963. Conversion of 5α-cholestan-3β-ol to Δ^7-5α-cholesten-3β-ol in cockroaches. *J. Biol. Chem.* 238:1966–72
7. Clever, U., Clever, I., Storbeck, I., Young, N. L. 1973. The apparent requirement of two hormones, α- and β-ecdysone for molting induction in insects. *Develop. Biol.* 31:47–60
8. Earle, N. W., Lambremont, E. N., Burks, M. L., Slatten, B. H., Bennett, A. F. 1967. Conversion of β-sitosterol to cholesterol in the boll weevil and the inhibition of larval development by two azasterols. *J. Econ. Entomol.* 60:291–93
9. Galbraith, M. N., Horn, D. H. S., Middleton, E. J. 1973. Ecdysone biosynthesis in the blowfly *Calliphora stygia*. *Chem. Commun.*, 203–4
10. Galbraith, M. N., Horn, D. H. S., Middleton, E. J., Thomson, J. A. 1970. The biosynthesis of crustecdysone in the blowfly *Calliphora stygia*. *Chem. Commun.* (3):179–80
11. Gilbert, L. I., King, D. S. 1973. Physiology of growth and development: endocrine aspects. In *The Physiology of Insecta,* ed. M. Rockstein, 249–370. New York: Academic
12. Goodnight, K. C., Kircher, H. W. 1971. Metabolism of lathosterol by *Drosophila pachea*. *Lipids* 6:166–69
13. Gorell, T. A., Gilbert, L. I., Tash, J. 1972. The uptake and conversion of α-ecdysone by the pupal tissues of *Hyalophora cecropia*. *Insect Biochem.* 2:91–106
14. Heed, W. B., Kircher, H. W. 1965. Unique sterol in the ecology and nutri-

tion of *Drosophila pachea. Science* 149: 758–61
15. Heinrich, G., Hoffmeister, H. 1970. Bildung von Hormonglykosiden als Inaktivierungsmechanismus bei *Calliphora erythrocephala. Z. Naturforsch.* 25b: 358–61
16. Hoffman, J. A., Koolman, J., Karlson, P., Joly, P. 1974. Molting hormone titer and metabolic fate of injected ecdysone during the fifth larval instar and in adults of *Locusta migratoria. Gen. Comp. Endocrinol.* 22:90–97
17. Horn, D. H. S. 1971. The ecdysones. In *Naturally Occurring Insecticides,* ed. M. Jacobson, D. G. Crosby, 333–459. New York: Dekker. 585 pp.
18. Huber, R., Hoppe, W. 1965. Zur Chemie des Ecdysons, VII. Die Kristall- und Molekulstrukturanalyse des Insektenverpuppungs—hormons Ecdyson mit der automatisierten Faltmolekülmethode. *Chem. Ber.* 98:2403–24
19. Ikekawa, N., Morisaki, M., Ohtaka, H., Chiyoda, Y. 1971. Reaction of fucosterol-24,28-epoxide with boron trifluoride etherate. *Chem. Commun.* 1498
20. Ikekawa, N., Suzuki, M., Kobayashi, M., Tsuda, K. 1966. Studies on the sterol of *Bombyx mori* L. IV. Conversion of the sterol in the silkworm. *Chem. Pharm. Bull.* 14:834–36
21. Kaplanis, J. N., Robbins, W. E., Thompson, M. J., Baumhover, A. H. 1969. Ecdysone analog: Conversion to alpha-ecdysone and 20-hydroxyecdysone by an insect. *Science* 166:1540–41
22. Kaplanis, J. N., Robbins, W. E., Thompson, M. J., Dutky, S. R. 1973. 26-Hydroxyecdysone: New insect molting hormone from the egg of the tobacco hornworm. *Science* 180:307–8
23. Kaplanis, J. N., Thompson, M. J., Yamamoto, R. T., Robbins, W. E., Louloudes, S. J. 1966. Ecdysones from the pupa of the tobacco hornworm, *Manduca sexta* (Johnannson). *Steroids* 8:605–23
24. Karlson, P., Bode, C. 1969. Die Inaktivierung des Ecdysons bei der Schmeissfliege *Calliphora erythrocephala* Meigen. *J. Insect Physiol.* 15: 111–18
25. Karlson, P., Bugany, H., Dopp, H., Hoyer, G.-A. 1972. 3-Dehydroecdyson, ein Stoffwechselprodukt des Ecdysons bei der Schmeissfliege *Calliphora erythrocephala* Meigen. *Hoppe Seyler's Z. Physiol. Chem.* 353:1610–14
26. Karlson, P., Hoffmeister, H., Hummel, H., Hocks, P., Spiteller, G. 1965. Zur chemie des Ecdysons, VI. Reaktionen des Ecdysonmoleküls. *Chem. Ber.* 98:2394–402
27. Karlson, P., Koolman, J. 1973. On the metabolic fate of ecdysone and 3-dehydroecdysone in *Calliphora vicina. Insect Biochem.* 3:409–17
28. King, D. S. 1972. Ecdysone metabolism in insects. *Am. Zool.* 12:343–45
29. King, D. S. 1972. Metabolism of α-ecdysone and possible immediate precursors by insects in vivo and in vitro. *Gen. Comp. Endocrinol. Suppl.* 3: 221–24
30. King, D. S. et al 1974. The secretion of α-ecdysone by the prothoracic glands of *Manduca sexta* in vitro. *Proc. Nat. Acad. Sci. USA* 71:793–96
31. Koolman, J., Hoffman, J. A., Karlson, P. 1973. Sulphate esters as inactivation products of ecdysone in *Locusta migratoria. Hoppe Seyler's Z. Physiol. Chem.* 354:1043–48
32. Louloudes, S. J., Thompson, M. J., Monroe, R. E., Robbins, W. E. 1962. Conversion of cholestanol to Δ^7-cholestenol by the German cockroach. *Biochem. Biophys. Res. Commun.* 8:104–6
33. Miyazaki, H., Ishibashi, M., Mori, C., Ikekawa, N. 1973. Gas phase microanalysis of zooecdysones. *Anal. Chem.* 45:1164–68
34. Morisaki, M., Ohtaka, H., Okubayashi, M., Ikekawa, N. 1972. Fucosterol-24,28-epoxide, as a probable intermediate in the conversion of β-sitosterol to cholesterol in the silkworm. *Chem. Commun.* 1275–76
35. Moriyama, H. et al 1970. On the origin and metabolic fate of α-ecdysone in insects. *Gen. Comp. Endocrinol.* 15:80–87
36. Nakanishi, K., Moriyama, H., Okauchi, T., Fujioka, S., Koreeda, M. 1972. Biosynthesis of α- and β-ecdysones from cholesterol outside the prothoracic gland in *Bombyx mori. Science* 176:51–52
37. Randall, P. J., Lloyd-Jones, J. G., Cook, I. F., Rees, H. H., Goodwin, T. W. 1972. The fate of the C-25 hydrogen of 28-isofucosterol during conversion into cholesterol in the insect *Tenebrio molitor. Chem. Commun.* 1296–98
38. Randall, P. J., Rees, H. H., Goodwin, T. W. 1972. Mechanism of alkylation during sitosterol biosynthesis in *Larix decidua. Chem. Commun.* 1295–96
39. Rees, H. H. 1971. Ecdysones. In *Aspects of Terpenoid Chemistry and Biochemistry,* ed. T. W. Goodwin, 181–222. New York: Academic. 441 pp.
40. Ritter, F. J., Wientjens, W. H. J. M.

1967. Sterol metabolism in insects. *TNO Nieuws* 22:381–92
41. Robbins, W. E., Dutky, R. C., Monroe, R. E., Kaplanis, J. N. 1962. The metabolism of ^{3}H-β-sitosterol by the German cockroach. *Ann. Entomol. Soc. Am.* 55:102–4
42. Robbins, W. E., Kaplanis, J. N., Svoboda, J. A., Thompson, M. J. 1971. Steroid metabolism in insects. *Ann. Rev. Entomol.* 16:53–72
43. Schaefer, C. H., Kaplanis, J. N., Robbins, W. E. 1965. The relationship of the sterols of the Virginia pine sawfly, *Neodiprion pratti* Dyar, to those of two host plants *Pinus virginiana* Mill and *Pinus rigida* Mill. *J. Insect Physiol.* 11: 1013–21
44. Shaaya, E., Karlson, P. 1965. Der Ecdysontiter während der Insektenentwicklung-II. Die postembryonale entwicklung der Schmeissfliege *Calliphora erythrocephala. J. Insect Physiol.* 11: 65–69
45. Svoboda, J. A., Hutchins, R. F. N., Thompson, M. J., Robbins, W. E. 1969. 22-*Trans*-cholesta-5,22,24-trien-3β-ol—An intermediate in the conversion of stigmasterol to cholesterol in the tobacco hornworm, *Manduca sexta* (Johannson). *Steroids* 14:469–76
46. Svoboda, J. A., Robbins, W. E. 1967. Conversion of beta sitosterol to cholesterol blocked in an insect by hypocholesterolemic agents. *Science* 156: 1637–38
47. Svoboda, J. A., Robbins, W. E. 1968. Desmosterol as a common intermediate in the conversion of a number of C_{28} and C_{29} plant sterols to cholesterol by the tobacco hornworm. *Experientia* 24: 1131–32
48. Svoboda, J. A., Robbins, W. E. 1971. The inhibitive effects of azasterols on sterol metabolism and growth and development in insects with special reference to the tobacco hornworm. *Lipids* 6:113–19
49. Svoboda, J. A., Robbins, W. E., Cohen, C. F., Shortino, T. J. 1972. Phytosterol utilization and metabolism in insects: recent studies with *Tribolium confusum.* In *Insect and Mite Nutrition,* ed. J. G. Rodriguez, 505–16. Amsterdam: North-Holland. 702 pp.
50. Svoboda, J. A., Thompson, M. J., Robbins, W. E. 1967. Desmosterol, an intermediate in dealkylation of β-sitosterol in the tobacco hornworm. *Life Sci.* 6:395–404
51. Svoboda, J. A., Thompson, M. J., Robbins, W. E. 1971. Identification of fucosterol as a metabolite and probable intermediate in conversion of β-sitosterol to cholesterol in the tobacco hornworm. *Nature New Biol.* 230:57–58
52. Svoboda, J. A., Thompson, M. J., Robbins, W. E. 1972. 24-Methylenecholesterol: isolation and identification as an intermediate in the conversion of campesterol to cholesterol in the tobacco hornworm. *Lipids* 7:156–58
53. Svoboda, J. A., Womack, M., Thompson, M. J., Robbins, W. E. 1969. Comparative studies on the activity of 3β-hydroxy-Δ^5-norcholenic acid on the Δ^{24}-sterol reductase enzyme(s) in an insect and the rat. *Comp. Biochem. Physiol.* 30:541–49
54. Thompson, M. J., Kaplanis, J. N., Robbins, W. E., Svoboda, J. A. 1973. Metabolism of steroids in insects. *Advan. Lipid Res.* 11:219–65
55. Thompson, M. J., Kaplanis, J. N., Robbins, W. E., Yamamoto, R. T. 1967. 20,26-Dihydroxyecdysone, a new steroid with moulting hormone activity from the tobacco hornworm, *Manduca sexta* (Johannson). *Chem. Commun.* (13):650–53
56. Thompson, M. J., Svoboda, J. A., Kaplanis, J. N., Robbins, W. E. 1972. Metabolic pathways of steroids in insects. *Proc. Roy. Soc. Ser. B* 180:203–21
57. Willig, A., Rees, H. H., Goodwin, T. W. 1971. Biosynthesis of insect moulting hormones in isolated ring glands and whole larvae of *Calliphora. J. Insect Physiol.* 17:2317–26
58. Yang, R. S. H., Wilkinson, C. F. 1972. Enzymic sulphation of p-nitrophenol and steroids by larval gut tissues of the southern army worm (*Prodenia eridania* Cramer). *Biochem. J.* 130: 487–93
59. Young, N. L. 1974. *The in vivo metabolism of ^{3}H-molting hormones during larval development of Chironomus tentans and Calliphora erythrocephala.* Phd Thesis, Purdue Univ., Lafayette, Ind.

RECENT RESEARCH ADVANCES ON THE EUROPEAN CORN BORER IN NORTH AMERICA[1,2,3]

❖6089

T. A. Brindley
Department of Zoology and Entomology, Iowa State University, Ames, Iowa 50010

A. N. Sparks
Agricultural Research Service, U.S. Department of Agriculture, Tifton, Georgia 31794

W. B. Showers
Agricultural Research Service, U.S. Department of Agriculture, Ankeny, Iowa 50021

W. D. Guthrie
Agricultural Research Service, U.S. Department of Agriculture, Ankeny, Iowa 50021

BIOLOGY AND ECOLOGY

Distribution

The recorded distribution of the European corn borer, *Ostrinia nubilalis,* has not changed significantly since Brindley & Dicke's review in 1963 (15), except along the southernmost portion of its range. Annual articles presenting the status of the European corn borer (3) indicate that each year the borer spreads into a few previously uninfested counties within known infested states. Sparks & Young (116) made a survey and found 34 of 35 counties infested in southern Georgia and

[1]The survey of the literature pertaining to this review was concluded in March 1974.

[2]Paper No. J-7841 of the Iowa Agriculture and Home Economics Experiment Station, Ames, Iowa. Project No. 1923. In cooperation with the University of Georgia College of Agriculture Experiment Stations, Coastal Plain Station, Tifton, Georgia.

[3]Mention of a pesticide or proprietary product in this paper does not constitute a recommendation or an endorsement of this product by the USDA or cooperating agencies.

concluded that the borer probably was present in all areas of extensive corn production in Georgia. Light-trap records from Tifton, Georgia, indicate that the borer's seasonal life history is very similar to its life history in South Carolina (31) and in Alabama (32). There are three complete generations each year and a fourth generation completes development in most years. Although official records do not show the presence of the borer in Florida, records from the southern tier of counties in Alabama (H. F. McQueen, personal communication) and in Georgia (116) indicate that corn-growing areas of the Florida panhandle probably are infested. Showers, Reed & Brindley (109) conducted laboratory studies and concluded that the Georgia borer had adapted to the photoperiod-temperature interaction of the region and was capable of producing large numbers of moths for the summer and autumn generations.

Chiang (20) studied the dispersion of the borer in Minnesota and in South Dakota from 1945 to 1970 and suggested that after the initial invasion in 1943, two distinctly different populations could have invaded Minnesota, one in 1952 and one in 1966. Chiang & Hodson (21) concluded that populations in the Waseca, Minnesota, area were kept at relatively low levels by environmental factors, but that with favorable temperatures, the borer populations could return to an economically significant level.

A highly successful cooperative project involving several North Central states has added significantly to available knowledge of the general biology and ecology of the European corn borer (24, 35, 58, 59, 121). Other studies show that the borer can be separated into several biotypes on the basis of differential responses to diapause, survival, and feeding habits when collections from several areas are subjected to common conditions in the field (22, 107, 120) or in the laboratory (107, 117). Morphometric differences involving five characters were used to separate borers from four locations into biotypes in 1967 (73), and then ten morphometric characters of female adults were utilized to further distinguish the biotypes in 1970 (23).

Diapause

The diapause characteristic of the borer has been studied by several researchers. Laboratory studies have shown significant differences in the abilities of diapausing vs nondiapausing larvae to synthesize and incorporate DNA and RNA precusor-type macromolecules (56). Lynch, Brindley & Lewis (85) studied the factors that influence survival of laboratory-reared larvae at subzero temperatures. They measured survival and oxygen consumption and detected no differences in diapause intensity due to exposure of larvae to decreasing photophases or lower temperature during the induction of diapause. After diapause was induced, conditioning to low temperatures and the length of the conditioning were associated with increased survival.

Beck and co-workers (5–8) have conducted numerous laboratory studies concerning the physiology of diapause in the European corn borer. They discovered, named, and further researched a developmental hormone, proctodone, that plays a role in diapause development and prepupal morphogenesis. Proctodone is secreted by the ileal epithelium of the anterior intestine, is photoperiodically sensitive on a rhythmic basis, and is believed to be associated with one of the basic elements of the insect's photoperiodic system.

Sparks, Brindley & Penny (117) and Sparks et al (120) conducted studies with geographical populations of the borer. Egg masses from parental stocks collected in Minnesota, Iowa, and Missouri were used to infest caged corn in Waseca, Minnesota; Ankeny, Iowa; and Portageville, Missouri. The F_1 progeny of these parental stocks exhibited significant differences in survival and diapause characteristics (120). Laboratory and field studies utilized three parental stocks to produce nine F_1 generations for study of the effects of temperature and photoperiod on diapause. These experiments provided evidence that diapause in the borer is controlled by a multigenetic makeup that responds to temperature and photoperiod (117). Further studies with similar techniques by Chiang, Keaster & Reed (22) revealed that the Missouri borer population was more responsive to photoperiod and temperature than was the Minnesota population. Showers, Brindley & Reed (107) conducted field studies at Ankeny, Iowa and Morris, Minnesota in which parental stocks from Minnesota, Alabama, and Maryland were used to produce nine F_1 generations. Again, significant differences in survival and diapause characteristics were shown among the biotypes of the three locations. The degree of expression of these differences is governed by the environment, primarily photoperiod and temperature. W. B. Showers et al (unpublished observations) showed that the effect of the interaction of photoperiod and temperature on the diapause response places the corn borer populations of North America into three ecotypes: northern (univoltine), central (multivoltine), and southern (multivoltine).

Development

A study of the relationship between the borer and constant and variable temperatures resulted in the advancement of theoretical development thresholds of 12.2, 11.1, and 12.8°C for the egg, larval, and pupal stages, respectively (86). Matteson & Decker (86) also found that controlled variable temperatures within the normal range of development had little effect on the durations of the immature stages. In a field study conducted by Blantran de Rozari (10), however, the minimum temperature and fluctuations of temperature usually were the determining factors in the rate of development; these two temperature parameters were responsible for 37–53% of the variation. Jarvis & Brindley (69) summarized 12 years of data from Boone County, Iowa, in which the relationship between temperature accumulations and seasonal development of the borer were correlated. They developed a regression equation that allowed them to predict the date of any desired percentage hatch of first- and second-generation oviposition or moth flight for the Boone County area. Hill & Keith (57) found that above average second-generation populations (225 borers/100 plants) in Nebraska, could be predicted by July 15 of each year on the basis of 1500 or more accumulated degree days. That temperature has a tremendous effect on the borer in nature was documented (106) in the report of a partial third generation produced in 10 of 17 years (1950–1966) in Boone County, Iowa. The partial third generation, however, succumbed to the below-freezing temperatures of autumn. But, the accumulation of 690.5 borer degree days between August 22 and

October 8, 1970, resulted in the development and survival of a third generation and allowed Showers & Reed (106) to speculate that 68% of the Boone County borer population entering the winter of 1970–1971 were third-generation insects.

Mating

The interaction of temperature and mating was first recorded in the laboratory when Sparks (112) observed that copulation was achieved when adults were exposed to a falling temperature in phase with a light-dark 14:10-hr photoperiod. Loughner and Brindley (82–84) made more thorough studies of the effects of photoperiod, thermoperiod, and other environmental factors on mating behavior and success of the borer. Decreasing temperatures (29.4–19.4°C), decreasing light intensities (250–0.1 ft-c), and intermediate to high relative humidities served as exogenous synchronizers for mating. Full moonlight (90) and wind speeds in excess of 15 km/hr were inhibitory. Showers, Reed & Oloumi-Sadeghi (110) discovered that temperature plateaus of at least 2 hr duration per night were essential for activation of the mating response. The plateaus were less of a requisite for the searching behavior of the male than for the pheromone activity of the female.

Pheromones

A chemical stimulus emitted by the female European corn borer was documented by Sparks (113), isolated and confirmed by Klun (74), and identified as *Z*-11-tetradecenyl acetate (*Z*-11-tda) by Klun & Brindley (75). Klun & Robinson (76) reported that the opposite geomterical isomer, *E*-11-tetradecenyl acetate (*E*-11-tda), inhibited attraction of the male European corn borer indigenous to central Iowa. But, Roelofs et al (99) reported that the European corn borer in New York was attracted specifically to the geometrical isomer *E*-11-tda. Later, Klun & Robinson (77) showed elicitation of a sex-attraction response by a combination of geometrical isomers, some of which attracted as many male European corn borers as did relatively pure *Z*-11-tda. Males of other species of Lepidoptera also were attracted to some of the test chemicals. They hypothesized that specific concentrations or blends of chemicals might be important in keeping certain lepidopteran species apart.

In New York, Roelofs et al (99) found that 14-day old *Z*-11-tda was as effective as fresh *E*-11-tda in attracting male European corn borers. Oloumi-Sadeghi (90) reported that the attractancy of a combination of 91.5% *Z* and 8.5% *E*-11-tda decreased as the lure became three nights old. The attractancy of pure *Z*-11-tda increased significantly, however, as it became three nights old. He hypothesized that a change in concentration or isomeric configuration (*Z* to *E*) was taking place. Later Klun et al (78) reported that minor amounts of *E* isomer in the *Z* were necessary for maximum sex attraction by the Iowa strain of European corn borer.

Showers, Reed & Oloumi-Sadeghi (110) found that the male European corn borer in Iowa is attracted to female corn borers of Georgia, Minnesota, and Quebec, as well as females of Iowa, but shows little response to female European corn borers from New York. From diapause response, the female populations from Minnesota and Quebec used in the test by Showers, Reed & Oloumi-Sadeghi (110) represent the northern ecotype (univoltine); those from Iowa, the central ecotype; and those from Georgia, the southern ecotype. Therefore, on the North American continent, the two known pheromone races (predominantly *Z* vs predominantly *E*) do not coincide with, but transcend the diapause races of the European corn borer.

Studies of the European corn borer antennae were undertaken by Cornford, Rowley & Klun (26). They tentatively assigned the roll of pheromone receptor to Type-A sensilla trichodea because the sexual dimorphism favored the male, and secondly because this sensillar type resembled the sex pheromone receptor of *Bombyx mori.*

A great deal of research has been done on olfactory response of the European corn borer to pheromones. But, little is known of their capabilities to perform as a control mechanism. Oloumi-Sadeghi (90) discovered, through a comparison with light traps and pheromone traps, that males respond to synthetic sex attractant only after feral females have mated and initiated egg deposition. He also found that pheromone-baited traps captured a greater proportion of males sooner when adult populations were small. His results suggest that intensive trapping during the spring flight or a small summer flight might reduce subsequent larval populations.

Radiation Sterilization and Juvenile Hormones

Guthrie, Dollinger & Stetson (45) developed techniques and studied spermatogenesis in the borer; they found a haploid chromosome count of 31. They suggested that male sterilization could be obtained by irradiating: (*a*) third- and early fourth-instar larvae to affect only spermatogonia; (*b*) late fourth-instar larvae to affect spermatogonia and early spermatocytes; (*c*) half- to full-grown fifth-instar larvae and early pupae to affect spermatocytes, spermatids, and mature spermatozoa; and (*d*) adults to affect spermatozoa. Chaudhury & Raun (19) completed additional studies of spermatogenesis and testicular development of the borer under a specific temperature regimen. They quantitatively determined the percentages of the testicular components in six developmental stages of germ cells, including spermatogonia, primary spermatocytes, secondary spermatocytes, spermatids, elongating sperm, and mature sperm. They concluded that testicular development increased rapidly during the larval and early pupal stages and that the adult insect contains its full complement of sperm.

Raun et al (97) discovered that severe somatic damage of nondiapausing larvae ensued after irradiation with gamma rays from a cobalt^{60} source, but diapausing larvae were unaffected. Irradiation of diapausing corn borers did not induce gene mutations but affected motility or viability of the sperm. Laboratory studies by Harding (51) and Jackson & Brindley (68) showed that feeding or applying chemosterilants to corn borers sterilized the adults and that just 4% of the eggs deposited were viable. Lewis, Lynch & Berry (81) studied the effect of synthetic juvenile hormones on the corn borer and noted that the compounds exerted a high degree of activity on treated larvae in the laboratory. However, there was no larval response to the JH-like compounds in the field, and the phytotoxicity to corn plants was slight to extreme with granular and liquid formulations, respectively.

Light Traps

Barrett, Day & Hartsock (4) found that when adult European corn borer populations were low, one to five blacklight traps adjacent to sweet corn plots protected the crop from borer damage. There were, however, no significant differences in damaged plants between lighted and unlighted plots when corn borer populations were high (18–51% infestation).

Sound

Pulse rate and amplitude of sound have been used to simulate those emitted by echolocating bats. Belton & Kempster (9) obtained variable results with these devices when they tried to use them to control the corn borer. They attributed the variation to a possible change in moth behavior on wet vs dry nights and suggested that corn plants may act as a sound barrier. Thus, low-flying moths would not be exposed to high levels of sound. Agee (2) studied the tympanic organ of the corn borer and found it to be served by two acoustical cells and by one nonacoustical cell. The organ detected frequencies of 14 and 100 kHz at 90 and 75 db, respectively. More moths showed avoidance response at pulse frequencies of 25 kHz (80–90 db at 1 m). However, Belton & Kempster (9) more than halved the infestation of European corn borer in sweet corn by blanketing the area with a frequency of 50 kHz.

Techniques

Several spin-off projects have resulted in publication of technique papers of possible value to other entomological researchers. Sisson, Brindley & Bancroft (111) reported on a statistical method for combining biological data from European corn borer experiments over years. Sparks & Facto (115) described techniques for use of time-lapse infrared cinematography to study flight, mating, and feeding habits of the borer. Showers, Lewis & Reed (108) reported on a method of identifying released insects that had been reared on wheat germ. Sparks (114) developed a microchamber for replicating photophases in diapause studies. Drecktrah, Knight & Brindley (30) studied the morphology of the internal anatomy of the fifth-instar larvae; Drecktrah & Brindley (29) studied the morphology of the internal reproductive systems of the adults, and Jones (72) did a definitive study on the postembryonic development of the reproductive system of the European corn borer.

CONTROL

Biological

Biological control is defined by Dicke (27) as a balance between a species and biological stress factors that limits the potential population. Research on biological control usually falls into one of five categories, having to do with parasites, predators, pathogens, radiation and genetics, or host plant resistance. The first three categories will be discussed in this section. Radiation and genetics and host plant resistance are discussed elsewhere in this review.

PARASITES In hope of stemming the phenomenal increase of the European corn borer in North America, 24 exotic species of insect parasites were imported during the 1920s, 1930s, and 1940s (15). By 1962, however, just six of these parasite species remained, and a tachinid, *Lydella thompsoni* (=*grisescens*), was considered the most effective and widely established. However, Franklin & Holdaway (36) discovered that the fly was primarily attracted to a specific corn variety (habitat) and, secondarily, to the European corn borer (food host). Sparks, Raun & Carter (118) reported that the fall population of *L. thompsoni* closely followed the fluctuations of the first-generation European corn borer and that in Boone County, Iowa, between 1951–1961, there was a reservoir of 1–29% parasitism by *L. thompsoni* of the fall borer population available to attack the first-generation corn borer during each succeeding spring. During 1958–1964, *L. thompsoni* and *Eriborus terebrons* (= *Horogenes punctorius*), reported by Townes (125), became increasingly prevalent among corn borer populations in Nebraska, and during 1960–1963, these two parasites parasitized 1.6–23% of second-generation corn borers in Ohio (59).

Beginning with the 1963 season, there was a decline in European corn borer and *L. thompsoni* populations throughout the Corn Belt (59, 125). According to Hill et al (58), in 1966 *L. thompsoni* disappeared from Iowa, Minnesota, and Nebraska. Although European corn borer populations in these states attained a level of prosperity during 1968–1971, *L. thompsoni* did not reappear. During a similar resurgence of corn borer populations in Ontario, however, Wressel (125) reported that *L. thompsoni* had reappeared and was maintaining itself at a low level.

Another widely established exotic parasite is the eulophid, *Simpiesis viridula* (15). Thousands of this parasite were released in Quebec from 1931 to 1934 and from 1945 to 1947 (63), but they were not recovered until 1964. Showers & Reed (105) reported that, in Iowa, the female of *S. viridula* rarely searches the bottom portion of the corn plant, but 62–75% of the second-generation European corn borers tunnel into that part of the plant. They concluded that this characteristic behavior of the wasp is a factor that maintains *S. viridula* populations well below host population levels and, therefore, that the wasp would be of minor importance in controlling the European corn borer.

Wressel (125) conducted a census on *L. thompsoni, E. terebrons,* and *S. viridula* in southwestern Ontario from 1948 to 1964 and determined that these parasites are of minor importance in controlling European corn borers. Miller (88) observed the activities of three native parasites, an ichneumonid, *Campoletis* sp., and two tachinids, *Archytas marmoratus* and *Lixophaga* sp., on first-generation European corn borers in Georgia and concluded that parasitism was insignificant.

P. Burbutis and C. Davis (unpublished observations) placed corn borer egg masses parasitized by *Trichogramma nubilalum* on sweet corn in the field and in the greenhouse. They found that parasites released in the lower one third to halfway up the plant were more effective in finding and parasitizing available egg masses. They achieved 5% parasitism by *T. nubilalum* at a time of year when natural parasitism by this species had not been observed.

PREDATORS Dicke & Jarvis (28) reported that predation by *Orius insidiosus* peaked during early larval development in first-generation European corn borers and was a significant factor in control. There was little predation by *O. insidiosus,* however, on early-instar larvae of second-generation corn borers, partly because corn pollination occurred at the same time, and because pollen became the principal food for *O. insidiosus* and the corn borers became incidental food.

In North Dakota, Frye (37) reported that coccinellids and chrysopids are in synchrony with the single-generation corn borer and that when he used a predation index developed by Chiang and Holdaway (Brindley & Dicke, 15) he found that these predators reduced European corn borer populations by ca 83 and 72% during 1966 and 1967, respectively. Carlson & Chiang (18) were able to increase the number of chrysopids and the so-called four-spotted fungus beetle, *Glischrochilus quadrisignatus quadrisignatus,* in experimental corn plots by spraying sucrose solution on the plants. Chrysopid buildup was negated, however, if aphid populations were high. Sparks et al (119) demonstrated that although insect predators play an important part in corn borer population fluctuations at some locations during some years, they cannot be depended upon year after year, or in any given years, to alter significantly the borer population at any specific location.

Several species of birds have been observed preying upon overwintering European corn borers. However, the downey woodpecker, *Dendrocopos pubescens,* was reported to be the most important in Arkansas (124) and North Dakota (38).

PATHOGENS Hudon (62) found two preparations of the bacterium, *Bacillus thuringiensis,* reasonably effective in controlling the European corn borer. He concluded, however, that the bacterium was nothing more than a promising control agent for the corn borer. Raun (93) also reported variable laboratory and field results with strains of the bacterium. Raun, Sutter & Revelo (96) determined that, in the laboratory, the bacterium lost its pathogenicity after 72 hr of ultraviolet irradiation and that mortality of treated corn borer larvae was delayed at temperatures below 32°C. Some of these ecological effects on pathogenicity, however, were decreased by encapsulation of the microbial insecticide (95). McWhorter, Berry & Lewis (87) concluded that the variability between the bacterial varieties *Bacillus thuringiensis* var. *thuringiensis* and *Bacillus thuringiensis* var. *alesti* might be attributed to differences in potency as well as to the environment. They suggested that standardization of bacterial preparations and better means of application were the major problems in obtaining consistent corn borer control. Sutter & Raun (122) determined that the microbial agent did not function as a toxicant in the European corn borer but that *B. thuringiensis* var. *thuringiensis* caused the epithelial cells to slough off into the lumen. It thus exposed areas of the basement membrane to attack by vegetative rods. Death then occurred as the vegetative rods entered the hemocoel and produced a septicemia. Esterline & Zimmack (34) looked at the possibility of using the bacteria *Escherichia coli* and *Serratia marcescens* as microbial agents and concluded from histological results that these bacteria would not be useful in the control of European corn borer.

VanDenburg & Burbutis (123) found that the naturally occurring microsporidian, *Nosema* (=*Perezia*) *pyraustae,* infected 80–85% of the European corn borer population in Delaware during 1959 and 1960. Hill et al (59) reported that the *N. pyraustae* infection of the corn borer population in Iowa combined with the abnormally cool temperatures in the late summer of 1964 were suspected of causing the greatest autumn to postharvest reduction in borer populations recorded in that state. Furthermore, the complete lack of *N. pyraustae* infection in the fall of 1967–1969 partly explained the higher corn borer populations in Cuming County, Nebraska (58). Showers, Brindley & Reed (107) found that *N. pyraustae* infection was most severe within a Minnesota population of European corn borer; the effect on survival was less severe among progeny of Minnesota and Alabama crosses than among progeny of Minnesota and Maryland crosses. The potential use of *N. pyraustae* as a practical control of the European corn borer was considerably enhanced when L. C. Lewis and R. E. Lynch (unpublished observations) developed methods for lyophilizing, vacuum drying, and storing the protozoan in a form that can be formulated into granules or baits for field application.

Laboratory pathogenicity tests of the effects of fungi on corn borer larvae showed that *Beauveria bassiana* and *Metarrhizium anisopliae* produced 93 and 78% mortality, respectively (16). However, *Aspergillus niger,* until recently a chronic problem in laboratory rearing of the corn borer, produced only 1% mortality.

The search for viruses that might affect corn borer populations has been less successful. Raun (94) reported a virus-like disease, and electron microscopy of the diseased fat-body tissues revealed hexagonal particles thought by Adams & Wilcox (1) to be icosahedra, but a pathogen was not isolated.

The survey of research concerning biological control agents for the corn borer (parasites, predators, and pathogens), therefore, points up that there are at least two potential agents that may control the insect, the bacterium, *B. thuringiensis,* and the protozoan, *N. pyraustae.* However, no biological agents are now consistently operative at a sufficient level to control the European corn borer in North America. This was also the conclusion of Chiang & Hodson (21) when they found that there was no consistency or host-density dependence of natural enemies of the corn borer in Minnesota. Further, they concluded that since 1952 corn borer populations have been kept relatively low by climatic and agricultural changes.

Chemical

Interest in the use of chemicals for borer control in recent years has been greatest among the producers of specialty crops, such as hybrid seed, canning, and market garden corn, and peppers. Previously DDT was used extensively. Jackson (65) made 33 comparisons with granular formulations and 13 with spray formulations of other promising insecticides. He found that granular formulations of diazinon, carbaryl, and endrin gave satisfactory control, but that DDT was the most effective of the spray formulations. Harding, Lovely & Dyar (53) concluded that carbaryl, diazinon, and carbofuran gave satisfactory control of the first-generation borer. Carbofuran, EPN [*O*-ethyl *O*-(*p*-nitrophenyl) phenylphosphonothoate], and diazinon were equally as effective in controlling second-generation borers, but carbaryl was significantly less effective. Berry et al (13) found that granular formulations of diazinon, carbaryl, malathion, carbofuran, and EPN gave satisfactory control of the borer. However, persistence of an insecticide depends on the structure of the chemical, as well as on the formulation of the material. Harding et al (55) compared granular, capsular, and ultra-low volume formulations of diazinon. They reported that ultralow volume residues were greatest on leaves and least in the whorl where residual toxicants are the most effective. Residues from the granular and capsular formulations were greatest in the whorl. Residues from all formulations were completely absent seven days after application.

Munson et al (89) reported that when insecticides were applied to corn foliage, 80 to 90% of the granules applied over the row fell to the ground and that the material on the ground then controlled larvae of the western corn rootworm, *Diabrotica virgifera.* Thus the European corn borer and the corn rootworms could be controlled with a single application of insecticide, and this, combined with the control of the two species of insects, would reduce the cost and amount of pesticides used on cropland. Hills, Peters & Berry (60) compared the effectiveness of planting and postplanting treatments for the control of first-generation corn borers and western corn rootworms. Both treatment times were effective in controlling corn rootworms, but the most effective control of borers was achieved with postplanting applications.

Jackson (66, 67) obtained partial control of first-generation borers with American Cyanamid CL 47470 (cyclic propylene *P, P*-diethyl phosphonodithiomidocarbonate) and carbofuran. His work showed that these materials were more effective when applied as a band 6.3 cm to one side of the seed and at the same depth as the seed. Harding (50) reported that granular formulations of CL 47470, CL 47470 + phorate, carbofuran, propoxur, Ortho® 9006 (*O, S*-dimethyl phosphoramidothioate), and dioxacarb were effective in reducing the numbers of borer cavities. Edwards & Berry (33) showed that carbofuran, TD-5032 (hexamethylditin), and CL 47470 applied at time of planting and at a rate of 4.0 lb actual insecticide/acre were effective for borer control 50 days postapplication.

The biology of an insect determines the most effective placement of an insecticide for optimum control. For example, when foam was used by Berry (11) and by Berry, McWhorter & Lovely (12) as an insecticide carrier, these formulations were as effective or, in some instances, more effective than the traditional sprays or granular formulations. Also, reduced rates were as effective as full rates.

Entomologists were also concerned with the possibility that the corn borer might develop resistance to insecticides. Harding & Dyar (52) demonstrated that when borer larvae were exposed to DDT, diazinon, and carbaryl over 12 generations in the laboratory, selection resulted in strains with some resistance to the chemicals. They also found that there was no cross-over of tolerance.

Most recommendations for the control of the European corn borer with insecticides were developed for moderate plant populations arranged in 40-inch rows. In recent years, however, the use of narrow rows and high plant populations has become standard practice. Harding et al (54) determined the effect of these new practices on borer control with DDT. These workers found that neither the establishment nor the control of first- or second-generation borers was significantly affected by row spacing or plant populations. But control of both borer generations tended to decrease as plants were grown closer together.

Insecticide applications for corn borer control are not limited to corn. In some locations, this pest may damage such crops as peppers, potatoes, and snap beans.

Moreover, seed or field corn can tolerate low borer populations without suffering substantial reduction in yield, but vegetable buyers demand high-quality vegetables free from insect damage and contamination. Burbutis et al (17) reported excellent control of the borer on pepper with 1.5 and 2.0 lb DDT/acre; however, it was necessary to maintain a weekly spray schedule from early in July until the first week of September.

Insecticides that show systemic activity have also been used on peppers. Ryder, Burbutis & Kelsey (101) reported that carbofuran and CL 47470 significantly reduced European corn borer infestations in peppers but did not give commercially acceptable control of the borer.

The effects of the corn borer in reducing potato yields are not clear cut; however, in some cases a reduction in yield occurred as a result of the borer feeding on the foliage. Bray (14) reported successful control of the European corn borer with applications of DDT, Kepone® [decachlorooctahydro-1-3-4-methene-2H-cyclobuta(cd) pentalen-2-one], phosphamidon, carbaryl, and endosulfan. Control of the borer population, however, did not result in increases in yield. Hofmaster, Waterfield & Boyd (61) controlled corn borers with soil applications of carbofuran and fensulfothion, but the increases in yield reported by these researchers may have occurred as a result of controlling the Colorado potato beetle, *Leptinotarsa decemlineata,* and the potato tuberworm, *Phthorimaea operculella,* as well as the corn borer.

Varietal Resistance

PLANT BREEDING AND GENETICS Brindley & Dicke (15) reviewed research on host-plant resistance through 1961. Guthrie (40) wrote a comprehensive review of techniques, accomplishments, and the potential of breeding for resistance to European corn borers in corn. Gallun, Starks & Guthrie (39) reviewed the chemical basis of resistance to first-generation borers.

If insects such as the European corn borer have more than one generation each season, the biological relationship between the insect and host plant may not be the same for each generation. In the Corn Belt states, resistance to first-generation borers is actually leaf-feeding resistance, because young larvae of the first generation feed primarily on the spirally rolled leaves in the whorl (43). Resistance to a second-generation infestation is actually resistance to collar- and sheath-feeding (second-generation larvae infest plants when corn is shedding pollen and silks are emerging), because young larvae feed on pollen accumulation at the axils of the leaves and on sheath, collar, and husk tissue. Most of the feeding, however, is on sheath and collar tissue (46, 47). In host-plant resistance research, therefore, the word generation is meaningless; the growth stage of the plant being attacked is important. Researchers should be aware that plant material resistant to insects during the vegetative stage of development may not be resistant during pollen shedding and later stages. For example, inbred lines of corn that are resistant to leaf feeding (first generation) by the European corn borer may be highly susceptible to sheath and collar feeding (second generation). Inbred lines that are resistant to second-generation larvae may not be resistant to first-generation larvae, but some lines may have some resistance to both generations.

Resistance to the first-generation borer has been easy to find (41), but resistance to the second generation has been more difficult to find in corn germ plasm. Although over 600 entries were evaluated for second-generation resistance, only inbred B52 has shown a high level of resistance (48, 92).

During the 1940s and 1950s many corn hybrids, all double crosses, were extremely susceptible to leaf feeding by a first-generation infestation. Today most corn hybrids have a lower level of susceptibility, and many have at least an intermediate degree of resistance (40). In recent years, most farmers have planted single crosses or simulated single crosses instead of double-cross hybrids. In general, single crosses with the following combinations of inbred lines are effective in reducing populations of first-generation borers: Resistant X Resistant, Intermediate X Intermediate, Resistant X Intermediate, or Resistant X Susceptible. Either dominance or incomplete dominance of resistance is necessary if the Resistant X Susceptible combination is to be effective (15).

Hybrids used today differ in degree of tolerance (ability to stand up under a moderate-to-heavy infestion) to a second-generation infestation, but all are susceptible to sheath and collar feeding (W. D. Guthrie, unpublished observations). Breeding for resistance to both generations of the European corn borer is, therefore, of great interest to researchers. Work is underway to evaluate breeding methods for selecting for both first- and second-generation resistance in the same plant population (48).

In most instances, resistance to first-generation European corn borers is conditioned by genes at several loci, and effects are cumulative among loci. Scott, Dicke & Penny (102) used reciprocal translocations and reported that five chromosome arms in CI31A and six arms in B49 carried genes for first-generation resistance. Scott, Hallauer & Dicke (104) determined the type of gene action involved in first-generation resistance by using F_2, F_3, and selfed backcross populations of CI31A(R) X B37(S), plus individual F_2 plants of (CI31A X B37) X CI31A, and individual F_2 plants of (CI31A X B37) X B37; most of the genetic variance for first-generation resistance was of the additive type.

Jennings et al (71) used a generation mean analysis to determine the genetic basis of second-generation resistance. The following nine populations were studied: P_1, P_2, F_1, F_2, F_3, BC_1, BC_2, and selfed progenies of both backcrosses. In four different experiments, B52 was used as the resistant parent (P_1), and B39, L289, Oh43, and WF9 were used as the susceptible parent (P_2). The data indicated no simple genetic basis of resistance and suggested that high resistance to a second-generation infestation may be the cumulative effect of an unknown number of loci. Additive genetic effects were predominant in conditioning resistance, but dominance was significant in all crosses. Scott, Guthrie & Pesho (103) showed that this high resistance of B52 is also transmitted in hybrid combinations. Jennings, Russel & Guthrie (70) used a diallel analysis involving 10 inbred lines and their 45 single crosses to demonstrate further that additive type of gene action conditions resistance to a second-generation infestation.

In early efforts to breed for resistance to first-generation borers, the backcross method was not successful in transferring resistance to susceptible inbred lines. The desired genotypes could not be identified in the segregating generations; when more than two backcrosses were used, the needed level of resistance was lost. The level of resistance could be increased, however, by intermating among resistant plants in progeny of the first or second backcross (100).

Penny, Scott & Guthrie (91) showed that recurrent selection was very effective in increasing the level of resistance to first-generation corn borers in five corn populations. Recurrent selection is essentially a breeding method of concentrating genes for certain desirable characters for which selection is being practiced while maintaining a broad gene base for other characteristics in the population. This procedure allows for the accumulation of desirable genes at numerous loci.

Frequencies of genes that condition resistance in corn to second-generation borers are low in populations of corn in the Corn Belt states (71). Consequently, population improvement is needed to increase the gene frequencies. A recurrent-selection technique is being used for selecting for resistance to both first- and second-generation borers in a ten-line synthetic corn hybrid (48). Ca 500 plants are infested in the whorl stage of development, and resistant plants are self-pollinated. The S_1 progenies are evaluated in replicated trials for resistance to first- and second-generation borers, and the best 10% of the lines are recombined to obtain an improved population.

REARING

In European corn borer resistance investigations, all plots are artificially infested with egg masses produced in the laboratory. Progress would be nil without this technique of artificial infestation (47).

For many years, first-generation egg production was obtained from moths that were collected from large emergence cages that had been filled with infested cornstalks the previous fall (49). Until recently, however, a good source of moths has not been available for second-generation egg production. Moths were obtained primarily by net collecting in patches of grass or weeds near corn fields that had a first-generation infestation or from infested, caged, green, sweet cornstalks (44).

The use of wheat germ marked the advent of the modern era of practical artificial diets for rearing plant-feeding Lepidoptera (25). During the past several years, European corn borer larvae have been reared individually in 3-dram vials on a plug of meridic diet; although over 90% survival is obtained (49, 79), this procedure is too slow to produce the number of moths needed for egg production in biological research. Therefore, during 1965–1973, larvae were reared on a meridic diet in plastic dishes (25.0 cm diam, 8.8 cm deep). However, several disease problems occurred with dish-reared insects. The pathogens of primary importance were bacteria, protozoa, and fungi. This problem was solved in 1969 and 1971 by the use of aureomycin in the diet to suppress bacteria, Fumidil B to suppress protozoa (*Nosema pyraustae,*) and four fungal inhibitors (methyl *p*-hydroxybenzoate, propionic acid, formaldehyde, sorbic acid) to suppress fungi (48, 80, 98).

Data collected during an 8-yr period showed that European corn borers reared continuously on a meridic diet cannot be used for screening inbred lines of corn because leaf-feeding damage is too low to measure resistance (48). For example, one culture at the European Corn Borer Laboratory (Ankeny, Iowa) that has been reared for 108 generations on a meridic diet had lost its virulence to infest corn plants by the thirty-fourth generation (64). Subsequently, crosses between this culture and a native population and backcrosses to each parent showed that the loss in virulence was genetically controlled (additive gene action in the insect) (42; Guthrie, unpublished observations). Cultures reared for 1 to 14 generations on a meridic diet performed as well as a feral population on corn plants (Guthrie, unpublished observations). Therefore, virulence is now maintained in corn borer cultures by dissecting 6000 feral larvae from cornstalks each fall to initiate new cultures (48).

During 1970–1973, ca 1,000,000 egg masses were produced each season for first-generation infestations (ca 25 eggs per mass) and more than 300,000 egg masses were produced for second-generation infestations. This technique has greatly accelerated research on second-generation resistance and on the biology and ecology, biological control, chemical control, and sex pheromones of the European corn borer (48). Also, two commercial seed corn companies are now rearing corn borers on meridic diets to provide large numbers of egg masses for studies of resistance to both first- and second-generation borers (Guthrie, unpublished observations).

With cooperation between federal, state, and private researchers, we are confident that corn hybrids will be developed that are resistant to all generations of the European corn borer. The American farmer will be the beneficiary of this cooperative research effort.

Literature Cited

1. Adams, J. R., Wilcox, T. A. 1965. A viruslike disease condition in the European corn borer, *Ostrinia nubilalis* (Hübner). *J. Invertebr. Pathol.* 7: 265–66
2. Agee, H. R. 1969. Acoustic sensitivity of the European corn borer moth, *Ostrinia nubilalis. Ann. Entomol. Soc. Am.* 62:1364–67
3. Agricultural Research Service. 1968–1973. *USDA Coop. Econ. Insect Rep.*
4. Barrett, J. R., Day, H. O., Hartsock, J. G. 1971. Reduction in insect damage to cucumbers, tomatoes, and sweet corn through use of electric light traps. *J. Econ. Entomol.* 64:1241–49
5. Beck, S. D., Alexander, N. 1964. Hormonal activation of the insect brain. *Science* 143:478–79
6. Beck, S. D., Alexander, N. 1964. Proctodone, an insect developmental hormone. *Biol. Bull.* 126:185–98
7. Beck, S. D., Alexander, N. 1964. Chemically and photoperiodically induced diapause development in the European corn borer, *Ostrinia nubilalis. Biol. Bull.* 126:175–84
8. Beck, S. D., Shane, J. L., Colvin, I. B. 1965. Proctodone production in the European corn borer, *Ostrinia nubilalis. J. Insect Physiol.* 11:297–303
9. Belton, P., Kempster, R. H. 1962. A field test on the use of sound to repel the European corn borer. *Entomol. Exp. Appl.* 5:281–88
10. Blantran de Rozari, M. 1973. *Effect of temperature on the survival and development of the European corn borer, Ostrinia nubilalis (Hubn).* MS thesis. Iowa State Univ., Ames. 58 pp.
11. Berry, E. C. 1972. Alternate methods for the control of European corn borers. *Ill. Custom Spray Operators' Training Sch. Rep.* 24:29–32
12. Berry E. C., McWhorter, G. M., Lovely, W. G. 1972. Foam as a carrier for insecticides and in corn insect control. *Proc. N. Cent. Br. Entomol. Soc. Am.* 27:147
13. Berry, E. C. et al 1972. Further field tests of chemicals for control of the European corn borer. *J. Econ. Entomol.* 65:1113–16

14. Bray, D. F. 1961. European corn borer control in potatoes. *J. Econ. Entomol.* 54:782–84
15. Brindley, T. A., Dicke, F. F. 1963. Significant developments in European corn borer research. *Ann. Rev. Entomol.* 8:155–76
16. Brooks, L. D., Raun, E. S. 1965. Entomogenous fungi from corn insects in Iowa. *J. Invertebr. Pathol.* 7:79–81
17. Burbutis, P., Fieldhouse, D. J., Crossan, D. F., VanDenburg, R. S., Ditman, L. P. 1962. European corn borer, green peach aphid, and cabbage looper control on peppers. *J. Econ. Entomol.* 55:285–88
18. Carlson, R. E. Chiang, H. C. 1973. Reduction of an *Ostrinia nubilalis* population by predatory insects attracted by sucrose sprays. *Entomophaga* 18: 205–11
19. Chaudhury, M. F. B., Raun E. S. 1966. Spermatogenesis and testicular development of the European corn borer, *Ostrinia nubilalis. Ann. Entomol. Soc. Am.* 59:1157–59
20. Chiang, H. C. 1972. Dispersion of the European corn borer in Minnesota and South Dakota, 1945 to 1970. *Environ. Entomol.* 2:157–61
21. Chiang, H. C., Hodson, A. C. 1972. Population fluctuations of the European corn borer, *Ostrinia nubilalis,* at Waseca, Minnesota, 1948–70. *Environ. Entomol.* 1:7–16
22. Chiang, H. C., Keaster, A. J., Reed, G. L. 1968. Differences in ecological responses of three biotypes of *Ostrinia nubilalis* from the North Central United States. *Ann. Entomol. Soc. Am.* 61:140–46
23. Chiang, H. C., Kim, K. C., Brown, B. W. 1970. Morphometric variability related to ecological conditions of three biotypes of *Ostrinia nubilalis* in the North Central United States. *Ann. Entomol. Soc. Am.* 63:1013–16
24. Chiang, H. C. et al 1961. Populations of European corn borer, *Ostrinia nubilalis* (Hbn) in field corn, *Zea mays* (L.). *Mo. Univ. Agr. Exp. Sta. Res. Bull.* 776, 95 pp.
25. Chippendale, G. M. 1972. Composition of meridic diets for rearing plant-feeding lepidopterous larvae. *Proc. N. Cent. Br. Entomol. Soc. Am.* 27:114–21
26. Cornford, M. E., Rowley, W. A., Klun, J. A. 1973. Scanning electron microscopy of antennal sensilla of the European corn borer, *Ostrinia nubilalis. Ann. Entomol. Soc. Am.* 66:1079–88
27. Dicke, F. F. 1972. Philosophy on the biological control of insect pests. *J. Environ. Qual.* 1:249–53
28. Dicke, F. F., Jarvis, J. L. 1962. The habits and seasonal abundance of *Orius insidiosus* (Say) on corn. *J. Kans. Entomol. Soc.* 35:339–44
29. Drecktrah, H. G., Brindley, T. A. 1967. Morphology of the internal reproductive systems of the European corn borer. *Iowa State J. Sci.* 41:467–80
30. Drecktrah, H. G., Knight, K. L., Brindley, T. A. 1966. Morphological investigations of the internal anatomy of the fifth larval instar of the European corn borer. *Iowa State J. Sci.* 40:257–86
31. Durant, J. A. 1969. Seasonal history of the European corn borer at Florence, South Carolina. *J. Econ. Entomol.* 62:1072–75
32. Eden, W. G. 1959. Research on European corn borer in Alabama. *J. Ga. Entomol. Soc.* 5:1–3
33. Edwards, C. R., Berry, E. C. 1972. Evaluation of five systemic insecticides for control of the European corn borer. *J. Econ. Entomol.* 65:1129–32
34. Esterline, A. L., Zimmack, H. L. 1972. Histological effects of *Escherichia coli* and *Serratia marcescens* upon the European corn borer. *J. Econ. Entomol.* 65:283–84
35. Everett, T. R., Chiang, H. C., Hibbs, E. T. 1958. Some factors influencing populations of European corn borer, *Pyrausta nubilalis* (Hbn.) in the north central states. *Minn. Univ. Agr. Exp. Sta. Tech. Bull.* 229, 63 pp.
36. Franklin, R. T., Holdaway, F. G. 1966. A relationship of the plant to parasitism of European corn borer by the tachinid parasite *Lydella grisescens. J. Econ. Entomol.* 59:440–41
37. Frye, R. D. 1972. Evaluation of insect predation on European corn borer in North Dakota. *Environ. Entomol.* 1:535–36
38. Frye, R. D. 1972. Bird predation on the European corn borer. *N. Dak. Farm Res.* 29:28–30
39. Gallun, R. L., Starks, K. J., Guthrie, W. D. 1975. Plant resistance to insects attacking cereals. *Ann. Rev. Entomol.* 20:337–57
40. Guthrie, W. D. 1974. Techniques, accomplishments and future potential of breeding for resistance to European corn borer in corn. In *Biological Control of Plants Insects and Diseases,* ed. F. G. Maxwell, F. A. Harris, 359–80. Jackson, Miss.: Univ. Press. 670 pp.

41. Guthrie, W. D., Dicke, F. F. 1972. Resistance of inbred lines of dent corn to leaf feeding by 1st-brood European corn borers. *Iowa State J. Sci.* 46:339–57
42. Guthrie, W. D., Carter, S. W. 1972. Backcrossing to increase survival of larvae of a laboratory culture of the European corn borer on field corn. *Ann. Entomol. Soc. Am.* 65:108–9
43. Guthrie, W. D., Dicke, F. F., Neiswander, C. R. 1960. Leaf and sheath feeding resistance to the European corn borer in eight inbred lines of dent corn. *Ohio Agr. Exp. Sta. Res. Bull.* 860, 38 pp.
44. Guthrie, W. D., Dicke, F. F., Pesho, G. R. 1965. Utilization of European corn borer egg masses for research programs. *Proc. N. Cent. Br. Entomol. Soc. Am.* 20:48–50
45. Guthrie, W. D., Dollinger E. J., Stetson, J. F. 1965. Chromosome studies of the European corn borer, smartweed borer, and lotus borer. *Ann. Entomol. Soc. Am.* 58:100–5
46. Guthrie, W. D., Huggans, J. L., Chatterji, S. M. 1969. Influence of corn pollen on the survival and development of 2nd-brood larvae of the European corn borer. *Iowa State J. Sci.* 44:185–92
47. Guthrie, W. D., Huggans, J. L., Chatterji, S. M. 1970. Sheath and collar feeding resistance to the second-brood European corn borer in six inbred lines of dent corn. *Iowa State J. Sci.* 44:297–311
48. Guthrie, W. D., Russell, W. A., Jennings, C. W. 1971. Resistance of maize to second-brood European corn borers. *Ann. Corn Sorghum Res. Conf. Proc.* 26:165–79
49. Guthrie, W. D., Raun, E. S., Dicke, F. F., Pesho, G. R., Carter, S. W. 1965. Laboratory production of European corn borer egg masses. *Iowa State J. Sci.* 40:65–83
50. Harding, J. A. 1966. Field studies with crop protection chemicals against the European corn borer. *USDA-ARS Special Rep. V-369.* Ankeny, Iowa. 14 pp.
51. Harding, J. A. 1967. Chemosterilization of male European corn borers by feeding of tepa and apholate to larvae. *J. Econ. Entomol.* 60:1631–32
52. Harding, J. A., Dyar, R. C. 1970. Resistance induced in the European corn borer in the laboratory by exposing successive generations to DDT, diazinon, or carbaryl. *J. Econ. Entomol.* 63:250–53
53. Harding, J. A., Lovely, W. G., Dyar, R. C. 1968. Field tests of chemicals for control of the European corn borer. *J. Econ. Entomol.* 61:1427–30
54. Harding, J. A., Brindley, T. A., Corley, C., Lovely, W. G. 1971. Effect of corn row spacing and of plant populations on establishment and control of the European corn borer. *J. Econ. Entomol.* 64:1524–27
55. Harding, J. A., Corley, C., Beroza, M., Lovely, W. G. 1969. Residues of diazinon on field corn treated with granular, capsular, and ULV formulations for control of the European corn borer. *J. Econ. Entomol.* 62:832–33
56. Hayes, D. K., Reynolds, P. S., McGuire, J. U., Schechter, M. S. 1972. Synthesis of macromolecules in diapausing and nondiapausing larvae of the European corn borer. *J. Econ. Entomol.* 65:676–79
57. Hill, R. E., Keith, D. L. 1971. Corn borers—population fluctuation and control. *Nebr. Farm Ranch Home Quart.* 18:16–20
58. Hill, R. E., Chiang, H. C., Keaster, A. J., Showers, W. B., Reed, G. L. 1973. Seasonal abundance of the European corn borer, *Ostrinia nubilalis* (Hbn.), within the North Central States. *Nebr. Agr. Exp. Sta. Res. Bull.* 255, 82 pp.
59. Hill, R. E. et al 1967. European corn borer, *Ostrinia nubilalis* (Hbn.). Population in field corn, *Zea mays* (L.), in the North Central United States. *Nebr. Agr. Exp. Sta. Res. Bull.* 225, 100 pp.
60. Hills, T. M., Peters, D. C., Berry, E. C. 1972. Timing of insecticide applications to control the western rootworm and the European corn borer. *J. Econ. Entomol.* 65:1697–1700
61. Hofmaster, R. N., Waterfield, R. L., Boyd, J. C. 1967. Insecticides applied to the soil for control of eight species of insects on Irish potatoes in Virginia. *J. Econ. Entomol.* 60:1311–18
62. Hudon, M. 1963. Further field experiments on the use of *Bacillus thuringiensis* and chemical insecticides for the control of the European corn borer, *Ostrinia nubilalis,* on sweet corn in Southwestern Quebec. *J. Econ. Entomol.* 56:804–8
63. Hudon, M. 1965. First recovery of the imported parasite, *Simpiesis viridula,* (Thomson), of the European corn borer, *Ostrinia nubilalis* (Hübner), in Quebec. *Phytoprotection* 46:113–15
64. Huggans, J. L., Guthrie, W. D. 1970. Influence of egg source on the efficacy of European corn borer larvae. *Iowa State J. Sci.* 44:313–53
65. Jackson, R. D. 1963. Insecticide screening tests on European corn borer. *Proc. N. Cent. Br. Entomol. Soc. Am.* 18:73

66. Jackson, R. D. 1965. Chemical control of the European corn borer in field tests in 1963. *USDA-ARS Special Rep. V-321.* Ankeny, Iowa. 15 pp.
67. Jackson, R. D. 1965. Chemical control of the European corn borer in field tests in 1964. *USDA-ARS Special Rep. V-328.* Ankeny, Iowa. 9 pp.
68. Jackson, R. D., Brindley, T. A. 1971. Hempa and metepa as chemosterilants of imagos of the European corn borer. *J. Econ. Entomol.* 64:1065–68
69. Jarvis, J. L., Brindley, T. A. 1965. Predicting moth flight and oviposition of European corn borer by the use of temperature accumulations. *J. Econ. Entomol.* 58:300–2
70. Jennings, C. W., Russell, W. A., Guthrie, W. D. 1974. Genetics of resistance in maize to first- and second-brood European corn borer. *Crop Sci.* 14:In press
71. Jennings, C. W., Russell, W. A., Guthrie, W. D., Grindeland, R. L. 1974. Genetics of resistance in maize to second-brood European corn borer. *Iowa State J. Res.* 48:267–80
72. Jones, J. A. 1973. *Postembryonic development of the reproductive system of the European corn borer,* Ostrinia nubilalis *(Hübner).* PhD dissertation. Iowa State Univ., Ames, Iowa. 167 pp.
73. Kim, K., Chiang, H. C., Brown, B. W. 1967. Morphometric differences among four biotypes of *Ostrinia nubilalis. Ann. Entomol. Soc. Am.* 60:796–801
74. Klun, J. A. 1968. Isolation of a sex pheromone of European corn borer. *J. Econ. Entomol.* 61:484
75. Klun, J. A., Brindley, T. A. 1970. *Cis*-11-tetradecenyl acetate a sex stimulent of European corn borer. *J. Econ. Entomol.* 63:779–80
76. Klun, J. A., Robinson, J. F. 1971. European corn borer moth: Sex attractant and sex attraction inhibitors. *Ann. Entomol. Soc. Am.* 64:1083–86
77. Klun, J. A., Robinson, F. J. 1972. Olfactory discrimination in the European corn borer and several pheromonally analogous moths. *Ann. Entomol. Soc. Am.* 65:1337–40
78. Klun, J. A. et al 1973. Insect pheromones: Minor amount of opposite geometrical isomer critical to sex attraction. *Science* 181:661–63
79. Lewis, L. C., Lynch, R. E. 1969. Rearing the European corn borer, *Ostrinia nubilalis* (Hübner), on diets containing corn leaf and wheat germ. *Iowa State J. Sci.* 44:9–14
80. Lewis, L. C., Lynch, R. E. 1970. Treatment of *Ostrinia nubilalis* larvae with Fumidil B to control infections caused by *Perezia pyraustae. J. Invertebr. Pathol.* 15:43–48
81. Lewis, L. C., Lynch, R. E., Berry, E. C. 1973. Synthetic juvenile hormones: Activity against the European corn borer in the field and the laboratory. *J. Econ. Entomol.* 66:1315–17
82. Loughner, G. E. 1971. Precopulatory behavior and mating success of the European corn borer under controlled conditions. *Iowa State J. Sci.* 46:1–5
83. Loughner, G. E. 1972. Mating behavior of the European corn borer, *Ostrinia nubilalis,* as influenced by photoperiod and thermoperiod. *Ann. Entomol. Soc. Am.* 65:1016–19
84. Loughner, G. E., Brindley, T. A. 1971. Mating success of the European corn borer, *Ostrinia nubilalis,* as influenced by environmental factors. *Ann. Entomol. Soc. Am.* 64:1091–94
85. Lynch, R. E., Brindley, T. A., Lewis, L. C. 1972. Influence of photophase and temperature on survival and oxygen consumption of diapausing European corn borers. *Ann. Entomol. Soc. Am.* 65:433–36
86. Matteson, J. W., Decker, G. C. 1964. Development of the European corn borer at controlled constant and variable temperatures. *J. Econ. Entomol.* 58:344–49
87. McWhorter, G. M., Berry, E. C., Lewis, L. C. 1972. Control of the European corn borer with two varieties of *Bacillus thuringiensis. J. Econ. Entomol.* 65: 1414–17
88. Miller, M. C. 1971. Parasitism of the corn earworm, *Heliothis zea,* and the European corn borer, *Ostrinia nubilalis,* on corn in North Georgia. *J. Ga. Entomol. Soc.* 6:246–49
89. Munson, R. E., Brindley, T. A., Peters, D. C., Lovely, W. G. 1970. Control of both the European corn borer and western rootworm with one application of insecticide. *J. Econ. Entomol.* 63: 385–90
90. Oloumi-Sadeghi, H. 1973. *Development of methods for the prediction of European corn borer flight and egg deposition.* PhD dissertation. Iowa State Univ, Ames. 237 pp.
91. Penny, L. H., Scott, G. E., Guthrie, W. D. 1967. Recurrent selection for European corn borer resistance in maize. *Crop Sci.* 7:407–9
92. Pesho, G. R., Dicke, F. F., Russell, W. A. 1965. Resistance of inbred lines of corn (*Zea mays* L.) to the second brood of the European corn borer, [*Ostrinia*

nubilalis (Hübner)]. *Iowa State J. Sci.* 40:85–98

93. Raun, E. S. 1963. Corn borer control with *Bacillus thuringiensis* Berliner. *Iowa State J. Sci.* 38:141–50
94. Raun, E. S. 1963. A virus-like disease of the European corn borer. *Proc. N. Cent. Br. Entomol. Soc. Am.* 18:21
95. Raun, E. S., Jackson, R. D. 1966. Encapsulation as a technique for formulating microbial and chemical insecticide. *J. Econ. Entomol.* 59:620–22
96. Raun, E. S., Sutter, G. R., Revelo, M. A. 1966. Ecological factors affecting the pathogenicity of *Bacillus thuringiensis* var. *thuringiensis* to the European corn borer and fall armyworm. *J. Invertebr. Pathol.* 8:365–75
97. Raun, E. S., Lewis, L. C., Picken, J. C., Hotchkiss, D. K. 1967. Gamma irradiation of European corn borer larvae. *J. Econ. Entomol.* 60:1724–30
98. Reed, G. L., Showers, W. B., Huggans, J. L., Carter, S. W. 1972. Improved procedure for mass rearing the European corn borer. *J. Econ. Entomol.* 65:1472–76
99. Roelofs, W. L., Carde, R. T., Vartell, R. J., Tierney, P. G. 1972. Sex attraction trapping of the European corn borer in New York. *Environ. Entomol.* 1:606–8
100. Russell, W. A. 1972. A breeder looks at host plant resistance for insects. *Proc. N. Cent. Br. Entomol. Soc. Am.* 27:77–87
101. Ryder, J. C., Burbutis, P. P., Kelsey, L. P. 1969. Systemic insecticides for control of European corn borer and green peach aphid on peppers. *J. Econ. Entomol.* 62:1150–51
102. Scott, G. E., Dicke, F. F., Penny, L. H. 1966. Location of genes conditioning resistance in corn to leaf feeding of the European corn borer. *Crop Sci.* 6:444–46
103. Scott, G. E., Guthrie, W. D., Pesho, G. R. 1967. Effect of second-brood European corn borer infestations on 45 single-cross corn hybrids. *Crop Sci.* 7:229–30
104. Scott, G. E., Hallauer, A. R., Dicke, F. F. 1964. Types of gene action conditioning resistance to European corn borer leaf feeding. *Crop. Sci.* 4:603–4
105. Showers, W. B., Reed, G. L. 1969. Effect of host population levels on incidence of parasitism by a Eulophid wasp. *Proc. N. Cent. Br. Entomol. Soc. Am.* 24:111–14
106. Showers, W. B., Reed, G. L. 1971. Three generations of European corn borer in central Iowa. *Proc. N. Cent. Br. Entomol. Soc. Am.* 26:53–56
107. Showers, W. B., Brindley, T. A., Reed, G. L. 1972. Survival and diapause characteristics of hybrids of three geographical races of the European corn borer. *Ann. Entomol. Soc. Am.* 65:450–57
108. Showers, W. B., Lewis, L. C., Reed, G. L. 1968. A possible marker for European corn borer moths. *J. Econ. Entomol.* 61:1464–65
109. Showers, W. B., Reed, G. L., Brindley, T. A. 1971. An adaptation of the European corn borer in the Gulf South. *Ann. Entomol. Soc. Am.* 64:1369–73
110. Showers, W. B., Reed, G. L., Oloumi-Sadeghi, H. 1974. European corn borer: Attraction of males to synthetic lure and to females of different strains. *Environ. Entomol.* 3:51–58
111. Sisson, D. V., Brindley, T. A., Bancroft, T. A. 1965. Combining biological data from European corn borer experiments over years. *Iowa State J. Sci.* 39:403–5
112. Sparks, A. N. 1963. Preliminary studies of factors influencing mating of the European corn borer. *Proc. N. Cent. Br. Entomol. Soc. Am.* 18:95
113. Sparks, A. N. 1963. The use of infra-red photography to study mating of the European corn borer. *Proc. N. Cent. Br. Entomol. Soc. Am.* 18:96
114. Sparks, A. N. 1966. A microchamber for replicating photophases in diapause studies with the European corn borer. *J. Econ. Entomol.* 59:492–93
115. Sparks, A. N., Facto, L. A. 1966. Mechanics of infrared cinematography in studies with the European corn borer. *J. Econ. Entomol.* 59:420–22
116. Sparks, A. N., Young, J. R. 1971. European corn borer activity in Georgia. *J. Ga. Entomol. Soc.* 6:211–15
117. Sparks, A. N., Brindley, T. A., Penny, N. D. 1966. Laboratory and field studies of F_1 progenies from reciprocal matings of biotypes of the European corn borer. *J. Econ. Entomol.* 59:915–21
118. Sparks, A. N., Raun, E. S., Carter, S. W. 1963. *Lydella grisescens* R. D. populations in corn borers of Boone County, Iowa. *Proc. N. Cent. Br. Entomol. Soc. Am.* 18:20–21
119. Sparks, A. N., Chiang, H. C., Burkhardt, C. C., Fairchild, M. L., Weekman, G. T. 1966. Evaluation of the influence of predation on corn borer populations. *J. Econ. Entomol.* 59:104–7
120. Sparks, A. N., Chiang, H. C., Keaster, A. J., Fairchild, M. L., Brindley, T. A. 1966. Field studies of European corn

borer biotypes in the midwest. *J. Econ. Entomol.* 59:922–28

121. Sparks, A. N., Chiang, H. C., Triplehorn, C. A., Guthrie, W. D., Brindley, T. A. 1967. Some factors influencing populations of the European corn borer, *Ostrinia nubilàlis* (Hübner), in the North Central States: Resistance of corn, time of planting, and weather conditions. Part II. *N. Cent. Reg. Res. Publ. 180 (Iowa State Univ. Agr. Home Econ. Exp. Sta. Res. Bull.)* 559:66–103
122. Sutter, G. R., Raun, E. S. 1967. Histopathology of European corn borer larvae treated with *Bacillus thuringiensis. J. Invertebr. Pathol.* 9:90–103
123. VanDenburg, R. S., Burbutis, P. P. 1962. The host-parasite relationship of the European corn borer, *Ostrinia nubilalis,* and the protozoan, *Perezia pyraustae,* in Delaware. *J. Econ. Entomol.* 55:65–67
124. Wall, M. L., Whitcomb, W. H. 1964. The effect of bird predators on winter survival of the southern and European corn borers in Arkansas. *J. Kans. Entomol. Soc.* 37:187–92
125. Wressel, H. B. 1973. The role of parasites in the control of the European corn borer, *Ostrinia nubilalis,* in Southwestern Ontario. *Can. Entomol.* 105:553–57

RECENT ADVANCES IN OUR KNOWLEDGE OF THE ORDER SIPHONAPTERA

❖6090

Miriam Rothschild
Ashton, Peterborough, PE8 5L2 Great Britain

In 1964 Holland (43) reviewed the evolution, classification, and host relationships of the Siphonaptera, and he concluded "We are now beginning to understand the natural and phyletic relationships of flea groups though many problems remain. But we are a long way from understanding the ecology and physiology of fleas, and from tracing a meaningful history of their host association." In the following pages we have concentrated on those areas not covered by Holland, especially the lesser known aspects of the order, together with a few unpublished observations. Systematics, taxonomy (including descriptions of new taxa and external morphology), parasites of fleas, medical and veterinary literature, and control by insecticides and other methods are deliberately excluded.

THE ORIGIN OF FLEAS

Hinton (42) supported Tillyard (142) in assigning the Siphonaptera a Mecopteran-like ancestor, basing his assumption on larval characters. Recently a number of observations have been made which lend support to the Tillyard/Hinton hypothesis.

Additional Evidence for the Descent from Winged Ancestors

A recent reexamination of the cuticular ridges and musculature of the thorax presents further evidence on this point. It has been shown (11, 84, 114, 115) that resilin, a rubber-like protein, is present in the pleural arch of fleas and its position on top of the pleural ridge indicates its homology with the pleural wing hinge process of flying insects, which is also composed of resilin. Two other features of the thorax of fleas thus become intelligible: (*a*) the fold on the surface of the pleural ridge, along which certain of the epipleural muscle fibers are inserted and which invariably bends anteriorly, can be homologized with the pleural ridge fold of scorpion-flies and dragonflies; and (*b*) the transverse ridge of the metanotum [= notal ridge in Rothschild & Traub (116)] can be homologized with the ridge which in winged insects divides the wing-bearing notum into an anterior scutal region and a posterior scutellar region (136).

A detailed study of the musculature (J. Schlein & M. Rothschild, personal observation) of *Xenopsylla cheopis* reveals that the flea has retained several direct and indirect flight muscles which have secondarily been incorporated in the jumping mechanism. The fact that the flea has become flattened from side to side, and the wing hinge ligament thus displaced from a dorsal to a lateral position, is the principal modification which makes this possible (116). The starter muscle (= tergo-trochanteral depressor) is highly developed in all fleas [including sessile species (115)]. Epipleural muscles (= subalars and basalars) are present which, before the flea jumps, help to compress the resilin in the pleural arch. The presence of the subalars (attached to the posterior rim of the coxa) was correctly recorded by Snodgrass (136) and Lewis (71), but both failed to note the presence of the basalars inserted on the anterior rim of the metacoxa and originating along the thoracic ridges (115, 116). This additional information concerning the sclerotized ridges (including the pleural arch and resilin) and muscles of the thorax strongly supports the theory that fleas are descended from winged ancestors.

The mysterious outgrowths, the so-called wing buds of Sharif (124) of the pupa of ceratophyllid fleas have been reexamined by Poenicke (91) and Rothschild (unpublished observations). Poenicke has confirmed their presence in the pupae of five families of ceratophyllids and their absence in four genera of pulicid fleas (although in the latter he found a lateral fold in the prepupal stage resembling those of the prepupa of *Ceratophyllus*). He concludes that since these structures are derivatives of the pleurae and do not arise in the suture between the pleurum and the tergum, they are unlikely to be relict wing buds. Furthermore their position below the level of the pleural arch (which is placed more dorsally in ceratophyllids than in pulicids) does not support the theory that they represent wing buds.

Additional Specializations Shared with the Mecoptera

Sections of the pharate adult of fleas show that resilin is secreted within the pleural arch (at the juncture of the pleural ridge and notal ridges) in a different manner from that recorded for the locust (1). In the latter species the same cells lining the pleural wing process first secrete cuticle and then change over to the secretion of resilin. In the flea, a finger-like invagination of the pleural arch cavity, composed of tall specialized cells, secretes the resilin, and the cuticular layers are provided by much smaller cells lining the rest of the cavity. Insufficient material of larval *Boreus* was available to prove conclusively that resilin was laid down in the same manner as in the flea (the earliest stages were lacking); nevertheless histological studies of the various layers of the pleural arch wall strongly support the view that the process of resilin secretion is similar in the flea and *Boreus.* Resilin is also present in the wing-hinge ligament of *Panorpa* (84).

Studies on the structure of the sperm of fleas and Mecoptera (4, 32, 105, 135) show that both groups display two unusual features. The flagella spiral round the mitochondria (a rare characteristic found only in sperm of a very few insects), and the inner circle of nine fibrils is lacking. These two features combined strongly support the theory that the two orders are related. Their sperm also possess a thick glycoprotein coat composed of transverse threads.

Other less important but probably significant similarities have been noted: 1. Multiple sex chromosomes are recorded for both superfamilies of fleas (8); *X. cheopis* exhibiting a chromosome complement of $2n = 14A + X_1X_2Y$ (♂) and $14A + X_1X_1X_2X_2$ (♀). This characteristic is also shared with *Boreus* (21). 2. Histological studies have stressed the resemblance between the proventricular spines of fleas and *Boreus* (102, 116). 3. Mecoptera and Siphonaptera display sexual dimorphism in the nerve cord, the males possessing an additional abdominal ganglion. 4. The enervation and structure of the eyes of fleas show they are reduced compound eyes, not displaced dorsal ocelli [see Hopkins & Rothschild, Vol. IV, p. 6 & Plate 12A (46) and Wachmann (163)]. 5. The first and second thoracic link-plates of fleas can be homologized with the two skeletal bridges—in this case still fused with the pleurosternum—of *Merope tuber* (Protomecoptera) which also connect the thoracic segments. The first pronotal bridge encompasses the first thoracic spiracle, which is the characteristic feature of the first thoracic link-plate of fleas, and connects the pronotum to the mesepisternum. The second bridge which forms a hook-like protuberance at the ventro-posterior corner of the mesepimeron and extends to the metepisternum (it is separated from the mesopleuron by a suture and may be flexible), curves round the base of the second thoracic spiracle in a similar manner to the second link-plate in fleas (Schlein & Rothschild, unpublished observations). 6. Snodgrass (136) described the pleuro-coxal articulation between the fleas' fore-coxa and propleura as "an arrangement just the reverse of that which ordinarily hinges the leg to the body." This arrangement of a pivot on the basal foramen of the coxa and a receptive acetabulum on the pleurum also exists in *Panorpa* (Mecoptera) (Schlein and Rothschild, unpublished observations). 7. The presence of a transverse interfurcal muscle (71) in the prothorax has been confirmed (Schlein & Rothschild, unpublished observations).

LIFE CYCLES AND REPRODUCTION

Several interesting and more or less aberrant life cycles differing from the two broad categories associated with ceratophyllid and pulicid fleas (106), which are linked to anatomical and physiological adaptations or modifications of a very pronounced kind, have recently been described.

Uropsylla tasmanica, a parasite of Dasyurids [Warneke in Dunnet (28)], passes the whole life cycle, except the pupal stage, on the body of the host. The eggs are glued onto individual hairs of the pelt and the larvae burrow beneath the skin and live as endoparasites. The larvae of *Hoplopsyllus* live as ectoparasites in the fur of the arctic hare, and occasionally larvae of *Tunga* and *Dasypsyllus* (106) are found as facultative ectoparasites on the body of the host. Larvae of *Spilopsyllus cuniculi* burrow into the hair and feed upon the tissues of baby rabbits which have recently died in the nest (Rothschild, unpublished observation).

Barnes & Radovsky (6) have investigated the life cycle and biology of *Tunga monositus* parasitizing the ears of *Peromyscus* from Baja California. The larvae emerge from the egg with degenerate and nonfunctional mouthparts and no locomotary aids are present. They never feed. There is only one larval molt. Presumably there is sufficient nutrient material in the egg to carry the larva through to the pupal and adult stage. This situation was intuitively foreseen by Pausch & Fraenkel (85) who found that flea larvae did not depend on external sources and on intracellular microorganisms for sterols and certain vitamins, and concluded these substances were supplied via the relatively enormous egg. When the male *T. monositus* hatches from the pupa the testis are open, and fully developed ripe sperm is already present in the epididymis (Rothschild, unpublished observations). It fertilizes the neosomic female (3, 98) and then dies. The gut of both sexes contains pupal meconium at eclosion and the male does not partake of a blood meal, and, in fact, never imbibes nourishment at any stage of the life cycle. Certain curious morphological modifications of this species are considered below.

Another unusual type of life cycle occurs in the subfamily Spilopsyllinae parasitizing the Leporidae. The reproduction of the genera *Spilopsyllus* and *Cediopsylla* is under the control of the hormones circulating in the peripheral blood of the mammalian host (110) [see also Solomon (139)]. Maturation of the female and maximum maturation of the male (111) can only occur if the fleas are primed by imbibing the raised concentration of hormones present in the blood of the pregnant doe rabbit or the newborn young (1–10 days old). In nature, copulation and impregnation of such primed fleas occurs only in the presence of nestlings (75, 110). Estrogens and corticosteroids are the principal hormones stimulating maturation, whereas progestins delay or inhibit maturation and can also initiate resorption of developing oocytes. An airborne kairomone emanating from the newborn young, which is also present in their urine, boosts copulation and reproduction in the nest and partly accounts for the increased fertility of fleas feeding on large litters (110). Experiments have shown that the external application (to the flea) of estrogens and corticosteroids (as well as juvenile hormone analogues) can also stimulate maturation, thereby demonstrating that the mammalian steroid hormones act directly upon the flea, not just indirectly by initiating some change in the host's physiology. The most striking difference between the life cycles of these two species of fleas and other species so far investigated is that maturation is limited to adults feeding on certain types of rabbits, namely the pregnant doe and newborn young. Development does not occur on the healthy buck rabbit, estrous doe, or young rabbit 3–9 weeks of age (i.e. before puberty). Some indication that different age groups of the host provide better breeding conditions than others has already been suggested. Thus *Xenopsylla cheopis* is said to lay more fertile eggs if reared on adult rather than baby mice [although this has been denied by Prasad (92)]. Krampitz (68) found that *Leptopsylla segnis* laid more fertile eggs when it had access to family groups rather than individual male, female, or baby mice as hosts. *X. cheopis, Xenopsylla astia, Nosopsyllus fasciatus,* and *L. segnis* can be reared on hypophysectomized or castrated hosts (68, 95, 112) and are thus independent of high concentrations of mammalian sex hormones. *X. astia,* however, was less fertile when fed on orchidectomized rats than on the controls. Joseph (58) found that the race (or possibly distinct species) of *Ctenocephalides felis,* which in India parasitizes buffaloes and horses, did not breed on castrated cattle. Unfortunately his experimental animals were not in perfect health and until his work is repeated a doubt remains as to whether the lack of hormones or the aftermath of illness was responsible for the fleas' disinclination to pair and breed on castrated hosts.

In addition to these very unusual life cycles several have recently been described which demonstrate a wide variety of specializations connected with the breeding habits of the host. Thus freshly emerged *Echidnophaga gallinacea* show a marked heliotropism which assists them in finding and setting on the wattles of cocks or hens (140). The larval and pupal stages and some adults of *Glaciopsyllus* can withstand freezing in the host's ice-covered nests for nine months of the year (M. D. Murray, personal communication); adult *Neopsylla setosa* is active at freezing temperatures (65); unfed *Ceratophyllus styx* can jump into Sandmartins hovering in front of old nesting sites in the quarry face (7). A large number of scattered references occur in the literature (61, 92, 123) indicating that many fleas which are capable of living on a wide range of mammals or birds are nevertheless more fertile feeding on one species than on another, while in some cases a few acceptable hosts nevertheless prove totally unsuitable for flea reproduction.

Vaughan & Mead-Briggs (160) have shown that the urine of the host attracts the rabbit flea which moves upwind towards the source. The latter has also demonstrated the great ability of this species to reach a host from the substratum in seminatural conditions (74). Humphries (51) has studied the reaction of *C. gallinae* to shadow (in relation to host finding) and Benton & Lee (12) have demonstrated the activation of the cat flea *C. felis* by carbon dioxide and the presence of warm objects. Molyneux (79) has shown that the larvae of *N. fasciatus* solicit fecal blood from parent fleas laying eggs or hiding in the nest. The larvae nip hold of the adult with their large mandibles, often in their pygidial region. The fleas respond by instant defecation of semiliquid blood stored in the rectal ampulla, which is immediately imbibed by the larvae.

Parallel evolution can be expected in habits as well as morphological details, and not infrequently the two are connected (147). Thus the size of the pleural arch (114) is often related to the size of the host. Likewise the antennal suckers of the male, which serve to grasp the female during copulation (110, 113), are lost in *Spilopsyllus cuniculi* and the totally unrelated *Chimaeropsylla potis,* sessile or semisessile species, in which case, apparently, the female does not have to be restrained while pairing. Their absence in *Ancistropsylla* suggests that in this group the female is also attached—possibly by the genal hooks (R. Traub, personal communication) rather than the mouthparts. It is interesting that two related genera of lagomorph fleas, *Hoplopsyllus* and *Spilopsyllus,* have adopted such contrasting specializations in order to ensure larval development—in one case a parasitic larval state, and in the other mammalian hormone dependence (110).

Copulation and Impregnation

It is well known that some fleas, especially bird fleas, copulate and impregnate the females before partaking of the initial blood meal (112). Others require a considerable period feeding either on both or one sex of the host. The rabbit flea in nature only copulates in the presence of the newborn young (75); before copulation can take place both sexes must be primed through the intake of mammalian blood containing a critical concentration of hormones. If fleas are feeding on the pregnant doe a transfer to the nestlings is required before pairing occurs, but introduced directly onto the newborn young it can take place without a transfer. Impregnation is only possible after the terminal plug in the testis is resorbed and sperm has passed into the epididymis. Within the testis migration and mixing of sperm bundles occurs (112). Both heads and tails of developing spermatozoa are at this stage confined in gelatinous caps. If the European rat flea is raised on rats, fleas sometimes emerge from their pupae with the testis fully open, the sperm released from the bundles, and yolk developing in the oocytes. A rise in the ambient temperature is then sufficient to trigger copulation and impregnation (55), but raised on mice, or different strains of rats, the male emerges with the testis still blocked and a blood meal of several hours duration is required before pairing takes place (111). The larval food, therefore, seems to exert a critical influence on development at eclosion.

Humphries (49) has described the mating sequence in *Ceratophyllus gallinae.* He postulates a cuticular hormone which is perceived by the male after palp contact with the female. Suter (140) previously noted that amputation of the male palps inhibited copulation. The female rabbit flea certainly releases the initial mating signal, but the males are seen to raise their claspers in response without being in contact with the females. Coating the antennae with vaseline (110) reduces copulation and impregnation. The fully primed male *S. cuniculi* cannot mate with unprimed or partially primed females (which presumably do not, in this condition, give the signal, even though newborn rabbits are present), whereas only partially primed males can respond to fully primed females. Males which fail to recognize unprimed females as mates, immediately copulate with primed females if these are substituted for them. The time lag or delay between transfer and copulation, which varies according to the degree of priming, suggests a chemical intermediary is involved (110).

Suter (140) noted that in *Echidnophaga gallinacea* egg laying only occurred after pairing, and ceased when the spermatheca was empty, but recommenced following renewed copulation. In the rabbit flea, sterile egg laying can occur before pairing and with an empty spermatheca.

The reproductive organs of the male flea (*Hystrichopsylla, Ctenocephalides, Ceratophyllus*) and their method of functioning were brilliantly elucidated and described by Günther (38). He showed that the penis rods in the species he examined were the only extrusable organ of the flea capable of entering the female. He postulated that at some point the penis transferred sperm to the pumping bulb (125), called by Günther the capsula, and thence to the penis rods. In the rabbit flea, sperm have been observed (105) wound round the notched tip of the longer penis rod "like spaghetti on a fork." The shorter of the two rods lodges in the bursa copulatrix and a slot in its spoon-like terminal expansion guides the longer rod through the aperture of the spermathecal duct like a rope running over a pulley (104).

F. J. Radovsky and associates (personal communication) have pointed out that in the case of *Tunga monositus* it is the sclerotized inner tube which enters the female during copulation and is capable of movement in the vertical plane, thus corroborating the observations of Geigy & Suter (33) for *Tunga penetrans.* Smit (131) and Traub (147) believe this is also true for various Rhopalopsyllids and Pygiopsyllids.

Goncharov (36) has shown how the various diverticulae and invaginations of the vagina help to clamp the aedeagus in position during pairing.

INTERNAL STRUCTURE AND RELATIONSHIP

A matchless paper by Wenk (165) describes the internal anatomy of the head capsule of *Ctenocephalides canis* and a sound description of the general internal anatomy of *Xenopsylla cheopis* is supplied by Wasserburger (164) who gives an excellent description of the spermathecal glands and other glandular structures. A considerable body of information concerning the internal anatomy of fleas has been accumulated during the study of the life cycle of *S. cuniculi* since species of 20 different genera of fleas have been cut in serial section for comparison.

It was found that the internal organs of fleas, like the external anatomy, although uniform in their basic plan, are all variable. For example, in the species examined there are two paired salivary glands of the same basic type, but so far every genus displays salivary glands which differ in some easily observed characteristic, be it size, number of cells, width and sclerotization of ducts, staining properties, etc.

Many of the internal organs undergo a well-defined cycle of development during reproduction (5, 25, 96, 112, 155–159): in particular, the reproductive organs themselves and the accessory glands, gut epithelium, fat body, oocytes, and salivary glands.

Some internal anatomical structures and modifications are nevertheless characteristic of the higher taxonomic groups. Thus: (*a*) Pulicoidea entirely lack branch 63b (136) of the trochanteral depressor muscle, some vestiges of which are present in all fleas so far examined in other families (Schlein and Rothschild, unpublished observation). (*b*) All fleas except the Tunginae possess six rectal pads, whereas in this subfamily they are reduced to two.

The most striking differences occur, however, in the Hystrichopsyllids, supporting K. Jordan's hunch [Hopkins & Rothschild, Vol. III (46) and Smit (133)] that "the Hystrichopsyllids probably form a separate group, equal in importance to the combined families of Leptopsyllidae, Ancistropsyllidae, Ceratophyllidae and Ischnopsyllidae." Examples of these internal structures (which do not all appear in the same genera) are: (*a*) polytrophic ovaries, considered as psuedo-polytrophic, i.e. secondarily modified panoistic ovaries by Hopkins & Rothschild [Vol. IV (Plate 12) (46)] especially well developed in *Stenoponia* in which nurse cells are also present (70); (*b*) a vaginal clamp dividing the vagina into two parts, and in *Hystrichopsylla* having an epithelial lining to the proximal portion of the vagina (116). An epithelial lining to the vagina or part of it does not occur apparently in fleas in any other family.

In addition there are numerous minor diagnostic characters at the specific or generic levels which are found in the internal organs. Thus *T. monositus* and the unnamed sibling species from Utah also parasitizing *Peromyscus* (R. Traub, personal communication) are the only two species of fleas known which are totally lacking proventricular spines, although these vary in number from species to species (80). Giant cells are present lining the rectum in this species (lacking in *T. penetrans*). The oviduct is muscular in species of the *caecata* group, which enables the eggs to be expelled forcibly and flung clear of the body of the host. Jordan (57) pointed out the bristles of the terminal segments were modified in this species for the same purpose. The large spermathecae of *Echidnophaga* and its allies are associated with very large elongated spermatozoa. The salivary glands (as already stated above) are liable to great specific as well as group variation. For example, those of the Tunginae are enormously enlarged, are composed of very many cells, and occupy large areas of the body, whereas in another fixed species, *E. gallinacea,* they are greatly reduced in size (only 4 cells each, but up to 30 cells in *Spilopsyllus cuniculi*). In the reproductive organs of both sexes there are many subtle differences. Owing to the contrasting staining properties of the extrachorion material secreted in the oviduct (109, 116) this structure can exhibit striking peculiarities. In species laying in very dry conditions (for example, *Stenoponia tripectinata* in the Middle East), the eggs are not only very large but may be covered with extrachorion material of great hardness and, presumably, impermeability. There are also considerable histological and morphological differences in the intestine and midgut, linked in some cases to the different types of long- and short-term digestion described by Darskaya and others (19, 23, 158) and the result of different habits and life cycles, but in other cases also reflecting a phylogenetic relationship.

The nerve cord of *Tunga monositus* in the male undergoes a most extraordinary displacement during the pharate adult stage. In the late pupal stages the nerve cords of both sexes occupy the usual ventral position. During development of the enormous aedeagus the terminal half is pushed dorsad in the male until the last abdominal ganglion comes to occupy a position in the tergum—the nerve cord thus forming a symmetrical V-shaped figure instead of a straight line.

Enough has been learned of the internal anatomy of fleas to show that it can provide important evidence for the classification of the higher categories, as well as identification at species or generic level.

FEEDING AND DIGESTION

The oral secretion of the cat flea (*Ctenocephalides felis felis*) was studied by Benjamini et al (10) and Michaeli et al (76, 77). They demonstrated the presence of a multitude of substances such as peptides, amino acids, aromatic compounds, phosphorus, and many fluorescent materials. They showed that the allergenic activity of the secretion is caused by a material of low molecular weight (under 2000) which can combine with proteins or peptides present and thus appear in a high molecular fraction. The free hapten, when deposited in the host's skin, proceeds to combine chemically with collagen, and free amino groups of collagen and other proteins react with the flea hapten to yield a complete antigen.

The presence in the saliva and salivary glands of *Xenopsylla cheopis* of an anticoagulant was demonstrated by Deoras & Prasad (26). They also described lateral movement of the distal end of the lacinae, which has been confirmed by the author.

They found that *X. cheopis* and *X. astia* may adopt either capillary or pool feeding, but prefer the former. Rothschild et al (112) have suggested that pool feeding is probably characteristic of *S. cuniculi.* These authors suggest blood is partly digested outside the body of the flea and noted that on hatching, before taking a blood meal, the fleas swallow their own saliva and digest the remains of the pupal gut content (112).

In an important paper Galun (31) has shown that fleas fed artificially through a membrane gorge twice as much on blood cells as on plasma. Isotonic solutions of sodium salts which have small ions such as iodide, choride, bromide, acetates, and nitrates can be substituted for plasma. Increasing the tonicity of lactose solutions increases uptake by the fleas. Galun postulated that the effect of the osmotic solutions is due to osmoreceptors which are stimulated by shrinkage in hypotonic solutions. The stimulatory effect of blood cells can be produced by solutions of 5×10^{-3} M ATP (adenosine triphosphate) in 0.15 M NaCl. Fleas showed the highest specialization among the insects tested and in their case ATP was the only compound which had phago-stimulatory properties among a wide range of nucleotides tested and which stimulated them to gorge.

Munshi (80) described the microanatomy of the proventriculus of *X. cheopis.* Richards & Richards (102) and Rothschild & Traub (116) showed that the proventricular spines are modified setae, not true spines. The former authors also described a heterogenous hexagonally organized layer in the midgut. This beaded layer consists of two types of short cylinders of greatly different diameters, each type arranged in a mixed hexagonal array (100, 101).

This digestive cycle in *X. cheopis* has been described by Vashchyonok & Solina (158) and they have distinguished three distinct phases, but their interpretation, although accurate, oversimplifies the process somewhat. In the cycle in the rabbit flea, development of the midgut epithelium and the feeding and defecation rates are partially under the control of the sex hormones of the rabbit (112). The external application of hormones or JH (juvenile hormone) stimulates maturation and with it increased development of the gut epithelium and feeding (112). Bibikova (14) describes secretion, absorption, and regeneration cells, which are zoned in the midgut, associated with a merocrine type of secretion. Fleas engorge in 2–10 min, those piercing a small blood vessel feeding faster than the pool feeders. Bibikova states that females imbibe larger amounts of blood than males, 0.18 mg (at the initial feed) as opposed to 0.10 mg. She notes that blood from the normal host is digested faster than from unusual hosts, and that a rise in temperature accelerates digestion. The Russian authors have also studied the digestive process (13), the pH, and the proteolytic enzymes (17). Drinking of water by fleas has been described by Galun (31) and Humphries (47).

ADAPTIVE STRUCTURES AND INTRASPECIFIC EVOLUTION

A brilliant and detailed comparative study by Traub (145–147, 150–152) of the spines and combs of a large number of species has demonstrated their adaptive modifications which are often related to the vestiture, affinities, and habits of their host. The subtleties of convergence are extremely well illustrated. For example, similarities exist between the combs of fleas parasitizing volant hosts such as birds and flying squirrels, or moles as opposed to shrews, and the stout coarse bristles of porcupine fleas, demonstrating that not only the microhabitat of fur and feather is involved but also the life cycle and habits of the host. Smit (132) adds the development of the pretarsal ungues of fleas as an adaptive character correlated with the type of pelt and skin of the host. Humphries (48) has suggested that the distance between the pronotal spines is related to the diameter of individual hairs and density of the host's pelt. Rothschild & Traub (116) have shown conclusively by studies of the pupal stages that the so-called spines in the combs of fleas are modified setae and not true spines. The pseudosetae have not yet been studied in the pupal stages but Jordan (56) drew attention to the fact that if the mesonotum is injured in larval life the pseudosetae may develop into spines similar to those in the pronotal comb.

Peus (86) considers the intraspecific evolution and marginal distribution of various European species, particularly in relation to characters which appear to have no selective value. He concludes that the phenomenon of intraspecific evolution of fleas corresponds in principle to the concept of species or race developed by O. Kleinschmidt. Peus calls attention to the enigmatic character of geographical variation in species such as *Ctenophthalmus.* Smit (127, 129, 134) also discusses intraspecific variation in *Ctenophthalmus agyrtes.*

The size of the pleural arch is clearly adaptive. In fleas which parasitize large hosts, especially those like deer and sheep which have no well-defined nest, the quantity of resilin in the arch is large (for example, *Vermipsylla alakurt, Ancistropsylla nepalensis*) and the fleas' jumping capacity correspondingly great (115). The variety of the cat flea, *Ctenocephalides felis orientis,* which parasitizes horses and buffaloes in India, is a better jumper than that found on the more usual hosts (58), and has more resilin in the pleural arch (Rothschild, unpublished observation). Disseminated throughout the order are related and unrelated groups and single species which have secondarily lost the structure. This is not infrequent where the hosts, such as squirrels and swallows, construct high aerial nests, since speculative jumping in such situations is dangerous. Fleas parasitizing moles show reduced pleural arches and sometimes lack them altogether. Similarly certain fleas inhabiting nests of desert rodents, or snow-bound nests, where it is risky to wander from the security of the breeding area, also lack a pleural arch.

ECOLOGICAL AND FAUNISTIC STUDIES

During the last decade a large number of faunistic surveys and ecological studies, particularly in the USSR, have been made, often as a background to epidemiological investigation. Many of these are primarily studies on the host, e.g. *Dipus sagitta* (121), *Arvicola terrestris* (82), and *Microtus agrestis* (154), but contain considerable information on ectoparasites. Russian authors have also studied joint annual cycles and the close integration of host and parasite (15, 16, 22, 24, 69, 161). At the present time there is an urgent need for analysis and coordination of this widely scattered information, and a comprehensive volume on the ecology and biology of fleas is the primary requirement of the day. Investigations are now carried out on a much larger scale than heretofore. Thus Darskaya et al (25) in their studies of the annual cycle of the gerbil flea, *Xenopsylla conformis,* de-fleaed 4000 gerbils, examined 92 nests and the rakings from 10,000 entrance holes, and dug out 300 burrows; 7000 bats were searched for ectoparasites in Malaysia (9); and 92,694 fleas were collected from 620 hosts in southern California (118–120) and 46,962 from 468 *Citellus beecheyi* hosts. Various important and little-known aspects of flea biology can be discovered by reading through these studies, apart from the more classical data relating to host relationship, distribution, bionomics, and climatic influence. Thus in Peus' papers on the ecology of the fleas of Germany (87, 89, 90) he deals with coexistence and interspecific competition within the nest, the infestation of newborn young, and the transmission of fear reactions from host to flea, among other topics. Hůrka's (52–54) faunistic and biological studies of bat fleas are also important. It would be extremely useful if, for example, such interesting items as changes in host specificity with altitude and temperature limitations (41), breeding during estivation of the host (81), sojourn of fleas in burrows (64, 168), reinfestation of trapped/released bird and mammal hosts (103, 120), competition for feeding sites on the body of the host (93), relation of flea infestations to spacing between rabbits (78), maturation of *Xenopsylla* assessed by the tanning of the spermatheca (35), seasonal variation in numbers of *Xenopsylla cheopis* (148), and host suitability for *Xenopsylla* (39) were sorted out and collected in a readily accessible volume.

In connection with the deliberate spread of the virus of myxomatosis in Australia (30) the first large-scale introduction of a flea vector has been organized. Pupae were originally obtained from the UK (137, 138), cultures were developed, and up to 11,000 fleas were hatched in the laboratory per day). Breeding populations were established by release of adult fleas onto wild rabbits at various points in New South Wales. The flea is spreading slowly, the fastest rate observed being 5–6½ miles per annum, but both infestation rate and flea index can be higher than any reported in nature in the UK. There is evidence that in certain parts of the release areas (166) fleas spend considerable periods free in the burrows of their hosts, probably in response to higher ambient temperatures.

By the use of radioactive isotopes (141) the distribution and feeding habits of *Xenopsylla* parasitizing *Rhombomys* were followed in their complex burrows. In the cold season (December through February) the fleas moved to the deeper passages where they became inactive and fed little if at all. In spring most of the fleas moved to the upper passages, entrance, etc, and the feeding rate was high. The highest feeding level of males occurred in June, and of females, in July.

GEOGRAPHICAL DISTRIBUTION

A resurgence of interest in Wegener's theory of continental drift has stimulated research into the geographical distribution of fleas. Important papers by Traub (147, 149) on the distribution of flea groups on the southern continents support the theory of continental drift and also show how the knowledge of the evolution and distribution of mammals and their ectoparasites can contribute to our understanding of infectious diseases associated with these animals (153). These studies also lead to the redefinition of the zoogeographical subregions, arising from a better knowledge of the effects of altitude on distribution. The affinities of squirrel fleas indicate the past history of the Asiatic Sciuridae (149). Other papers dealing with geographical distribution are Smit (130)—Scandinavia and Finland; George (34)—Great Britain; Haddow et al (40)—Ceratophyllidae; and Lewis (72)—Siphonaptera and host preferences.

CONTRIBUTIONS OF GENERAL INTEREST

Kessel (60) has described the embryology of fleas and Kamala Bai (59) has described the histochemistry of egg yolk in *Xenopsylla.* Vashchyonok (155) studied the histological characteristics of oogenesis in *Echidnophaga oschanini.* Pausch & Fraenkel (85) described the rearing of *X. cheopis* larvae on artificial diet and demonstrated that this species can develop successfully on food of very poor nutritional quality. Margalit & Shulov (73) investigated the effect of temperature on the development of prepupa and pupa of *X. cheopis,* and Shulov & Naor (126) investigated the olfactory responses of the adult of this species. Yinon, Shulov & Margalit (167) examined the hygroreaction of *X. cheopis.* Others have compared oxygen consumption and temperature requirements in different species, and have thereby assessed their activity (18, 65–67). Sterile male fleas of *X. cheopis* were found in the F_4 generation of inbred fleas by Prasad (94). Nelson (83) has recorded bird fleas from bat guano in Lovelock Cave, Nevada, dated (by radio carbon analysis) 4570 ± 110 yr before the present time. Rothschild (107) identified a female *Pulex irritans* from a Viking pit near Dublin, Ireland, dated about the tenth century. Peus (88) described a second species of flea from Baltic amber and considers that the evolution of fleas (as an order) effectively ceased in the Eocene $50–60 \times 10^6$ years ago. Castration by parasites and intersexuality in *Amphipsylla sibirica* is described by Brinck-Lindroth & Smit (20). Goncharov (37) considered basic structure of the abdomen based on the study of castrated males. Prokop'ev (97) studied the different morphological types of cocoons, and Humphries (50) studied the movement of the pupa and pharate adult before eclosion. Řeháček & Fischer (99) have evolved a technique for maintaining primary cell cultures derived from larvae of *Ctenocephalides felis.* Flea larvae are described by Kir'yakova (62, 63) and Vysochkaya & Kir'yakova (162). Rudall & Kenchington (117) examined the cocoon silk of fleas and found it displays certain unusual features. Detinova (27) reviewed the age structure of flea populations. Rothschild et al (114) showed that fleas jump by means of a click mechanism which releases energy stored in resilin in the pleural arch. Smit (128) has produced the first bibliography of fleas of China. New terminology of the aedeagus and vagina are proposed by Smit (131) and Rothschild & Traub (116), and improved photographic techniques for illustration are proposed by Rothschild and Traub (116, 147). Notable monographs are Sakaguti's fleas of Japan (122), Dunnet & Mardon's fleas of Australia (29), Holland's fleas of New Guinea (44) and Australia (45), Tipton & Machado-Allison's fleas of Venezuela (143), and Tipton & Mendez's fleas of Panama (144). In addition, the following types are described: fleas of the Hawaiian Islands (39a); Siphonaptera in South Africa (south of the Sahara) (40a); fleas of Alaska (46a) and of Central Asia and Kazakhstan (54a); Siphonaptera of Egypt (71a); fleas of Afganistan (72a); bird fleas of Germany (86a); fleas of Poland (126a); Siphonaptera of New Zealand (128a), Switzerland (129a), and Mongolia (129b); a review of the genus *Thrassis* (139a); fleas of Rumania (139b) and Hungary (141a); ectoparasites of Israel (141b); and a revision of the *Stivalius robinsoni* group (147). A good general account of flea biology for students is provided by Askew (2) and Rothschild & Clay (108).

Literature Cited

1. Anderson, S. O., Weis-Fogh, T. 1964. Resilin, a rubberlike protein in arthropod cuticle. In *Advances in Insect Physiology*, ed. J. W. L. Beament, J. E. Treherne, V. B. Wigglesworth, 2:1–65. New York: Academic. 364 pp.
2. Askew, R. R. 1971. *Parasitic Insects*. London: Heinemann. 316 pp.
3. Audy, J. R., Radovsky, F. J., Vercammen-Grandjean, P. H. 1972. Neosomy: Radical intrastadial metamorphosis associated with Arthropod symbiosis. *J. Med. Entomol.* 9(6):487–94
4. Baccetti, B. 1972. Insect sperm cells. In *Advances in Insect Physiology*, ed. J. E. Treherne, M. J. Berridge, V. B. Wigglesworth. 9:315–97. New York: Academic. 438 pp.
5. Balashov, Yu. S., Bibikova, V. A., Polunina, O. A., Murzakhemetova, K. M. 1965. Feeding and impairment of the valve function of the pre-stomach in fleas. *Med. Parazitol. Parazit. Bolez.* 34(4):471–76
6. Barnes, A. M., Radovsky, F. J. 1969. A new *Tunga* from the Nearctic region with description of all stages. *J. Med. Entomol.* 6(1):19–36
7. Bates, J. K. 1962. Field studies on the behavior of bird fleas. 1. Behavior of the adults of three species of bird fleas in the wild. *Parasitology* 52:113–32
8. Bayreuther, K., Bräuning, S. 1971. Die Cytogenetik der Flöhe. II. *Xenopsylla cheopis* Rothschild 1903, und *Leptopsylla segnis* Schönherr, 1811. *Chromosoma* 33:19–29
9. Beck, A. J. 1971. A survey of bat ectoparasites in West Malaysia. *J. Med. Entomol.* 8:147–52
10. Benjamini, E., Feingold, B. F., Kartman, L. 1960. Allergy to flea bites. III. The experimental induction of flea bite sensitivity in guinea pigs by exposure to flea bites and by antigen prepared from whole flea extracts of *Ctenocephalides felis felis*. *Exp. Parasitol.* 10:214–22
11. Bennet-Clark, H. C., Lucey, E. C. A. 1967. The jump of the flea: a study of the energetics and a model of the mechanism. *J. Exp. Biol.* 47(1):59–76
12. Benton, A. H., Lee, S. Y. 1965. Sensory reactions of Siphonaptera in relation to host-finding. *Am. Midl. Natur.* 74(1): 119–25
13. Bibikova, V. A. 1963. Experiments in application of microscopy and spectroanalytical method in the blood study of the host in fleas' stomachs. *Mater. Nauch. Konf. Prirodn. Ochag. Profilakt. Chumy. Alma-Ata*, 21–22
14. Bibikova, V. A. 1965. Conditions for the existence of the plague microbe in fleas. *Cesk. Parazitol.* 12:41–46
15. Bibikova, V. A., Gerasimova, N. G. 1967. On biology of *Xenopsylla skrjabini* Joff, 1928. Communication 2. The nutrition of fleas in experimental conditions. *Zool. Zh.* 46(5):730–36
16. Bibikova, V. A., Il'inskaya, V. L., Kaluzhenova, Z. P., Morozova, I. V., Shmuter, M. F. 1963. Contribution to the biology of fleas gen. *Xenopsylla* in the desert Sary-Ishikotrau. *Zool. Zh.* 42(7):1045–51
17. Bibikova, V. A., Murzakhemetova, K. M., Terent'eva, L. I. 1964. Locating pH of the gastro-intestinal tract in fleas. *Trudÿ, V. nauchn. Konf. po priorodnoĭ ochagovosti Bolezneĭ i voprosam parasitologii, respublik Sredneĭ Azii i Kazakhstana*, 229–31
18. Bibikova, V. A., Pavlova, A. E. 1970. Infection influence of fleas' respiration by a pest microbe. *Vop. Prirodn. Ochag. Bol. Alma-Ata*, 155–159
19. Bibikova, V. A., Sosunova, A. N., Morozova, I. V., Fedorova, V. I. 1971. Survival of formed host blood-corpuscles in the gut of Siphonaptera. *Entomol. Obozr.* 50:38–42
20. Brinck-Lindroth, G., Smit, F. G. A. M. 1973. Parasitic nematodes in fleas in northern Scandinavia and notes on intersexuality and castration in *Amphipsylla sibirica*. *Entomol. Scand.* 4(4): 302–22

21. Cooper, K. W. 1951. Compound sex chromosomes with anaphasic precocity in the male Mecopteran *Boreus brumalis* Fitch. *J. Morphol.* 89:37–57
22. Darskaya, N. F. 1963. Observations of *Xenopsylla conformis* Wagn. relating to the mechanism of the autumnal change in the activities of gerbil fleas of this genus. *Probl. Parazitol. Tr. Nauch. Konf. Parazitol. Ukr. SSR 4th,* 338–39
23. Darskaya, N. F. 1964. On the comparative ecology of bird fleas of the genus *Ceratophyllus* Curt. 1832. Ectoparasites: fauna, biology, and practical significance. *Proc. Study Fauna Flora USSR Sect. Zool.* Vol. 4, 39(54): 31–180
24. Darskaya, N. F. 1970. On the study of annual cycles of fleas gen. *Xenopsylla* Roths. 1903. In *Carriers of Especially Dangerous Infections and Their Control,* 108–31. Moscow: State Publishing House
25. Darskaya, N. F., Bakeyev, N. N., Kadatskaya, K. P. 1962. Contribution to the study of the annual cycle of the gerbil flea *Xenopsylla conformis* Wagner in Azerbaydzhan. *Med. Parazitol. Parazit. Bolez.* 31(3):342–46
26. Deoras, P. J., Prasad, R. S. 1967. Feeding mechanism of Indian fleas *X. cheopis* (Roths.) and *X. astia* (Roths.) *Indian J. Med. Res.* 55(10):1041–50
27. Detinova, T. S. 1968. Age structure of insect populations of medical importance. *Ann. Rev. Entomol.* 13:427–50
28. Dunnet, G. M. 1970. Siphonaptera (Fleas). In *The Insects of Australia,* CSIRO, Canberra, 647–55. Melbourne, Aust.: Melbourne Univ. Press. 1029 pp.
29. Dunnet, G. M., Mardon, D. K. 1974. A Monograph of Australian Fleas. *Aust. J. Zool. Suppl. Ser.* 30:In press
30. Fenner, F., Ratcliffe, F. N. 1965. *Myxomatosis.* Cambridge: Cambridge Univ. Press. 379 pp.
31. Galun, R. 1966. Feeding stimulants of the rat flea *Xenopsylla cheopis* Roths. *Life Sci.* 5:1335–42
32. Gassner, G. III, Breland, O. P., Biesele, J. J. 1972. The spermatozoa of the Scorpionfly *Panorpa nuptialis:* a transmission electron microscope study. *Ann. Entomol. Soc. Am.* 65(6):1302–9
33. Geigy, R., Suter, P. 1960. Zur Copulation der Flöhe. *Rev. Suisse Zool.* 67:206–10
34. George, R. S., Ed. 1974. *Provisional Atlas of Insects of the British Isles,* Part 4. Siphonaptera. Abbots Ripton, U.K.: Inst. Terrestrial Ecol., Monks Wood Exp. Sta. In press
35. Gerasimova, N. G. 1971. The rate of maturation in adults of *Xenopsylla skrjabini* and *X. nuttalli. Parazitologiya* 5:137–39
36. Goncharov, A. I. 1966. Contribution to the study of the morphological and functional peculiarities of the modified segments and the separate parts of the aedeagus in some species of the genus *Ctenophthalmus. Osobo Opasnye Infektii na Kavkaze, Stavropol,* 54–56
37. Goncharov, A. I. 1972. On the basic structure of the abdomen in Siphonaptera, based on studies of castrated males. *Parazitologiya* 6:465–68
38. Günther, K. K. 1961. Funktionell-anatomische Untersuchung des männlichen Kopulationsapparates der Flöhe unter besonderer Berücksichtigung seiner postembryonalen Entwicklung. *Deut. Entomol. Z.* 8(3–4):258–349
39. Haas, G. E. 1969. Quantitative relationships between fleas and rodents in a Hawaiian cane field. *Pac. Sci.* 23(1):70–82
39a. Haas, G. E., Wilson, N., Tomich, P. Q. 1972. Ectoparasites of the Hawaiian Islands. I. Siphonaptera. *Contrib. Am. Entomol. Inst.* 8(5):1–76
40. Haddow, J., Traub, R., Rothschild, M. 1975. *The Distribution of the Ceratophyllidae with an Illustrated Key to the Genera.* Cambridge: Cambridge Univ. Press. In preparation
40a. Haeselbarth, E. 1966. A note on the subspecies of *Ctenocephalides felis* in Africa south of the Sahara. *Zool. Anz.* 176(5):357–65
41. Harmsen, R., Jabball, I. 1968. Distribution and host specificity of a number of fleas collected in South and Central Kenya (including the collection of the University College, Nairobi, Mount Kenya Expedition March 1966). *J. E. Afr. Natur. Hist. Soc. Nat. Mus.* 27:157–61
42. Hinton, H. E. 1958. The Phylogeny of the Panorpoid orders. *Ann. Rev. Entomol.* 3:181–206
43. Holland, G. P. 1964. Evolution, classification and host relationships of Siphonaptera. *Ann. Rev. Entomol.* 9:123–46
44. Holland, G. P. 1969. Contribution towards a monograph of the fleas of New Guinea. *Mem. Entomol. Soc. Can.* (61): 1–77
45. Holland, G. P. 1971. Two new species and three new subspecies of *Acanthopsylla* from Australia, with some notes on the genus. *Can. Entomol.* 103(2): 213–34

46. Hopkins, G. H. E., Rothschild, M. 1953–1971. *An Illustrated Catalogue of the Rothschild Collection of Fleas in the British Museum (Natural History)*, Vols. I–V. London: Brit. Mus. (Natur. Hist.)
46a. Hopla, C. E. 1964. Alaskan hematophagous insects, their feeding habits and potential as vectors of pathogenic organisms. Part I. The Siphonaptera of Alaska. *Final Report Between the Arctic Aeromedical Laboratory, US Air Force, Fort Wainright, Alaska and the University of Oklahoma Research Institute, Norman, Oklahoma.* 346 pp.
47. Humphries, D. A. 1966. Drinking of water by fleas. *Entomol. Mon. Mag.* 102:260–62
48. Humphries, D. A. 1966. The function of combs in fleas. *Entomol. Mon. Mag.* 102:232–36
49. Humphries, D. A. 1967. The mating behavior of the hen flea *Ceratophyllus gallinae* (Schrank). *Anim. Behav.* 15(1): 82–90
50. Humphries, D. A. 1967. The behavior of fleas within the cocoon. *Proc. Roy. Entomol. Soc. London Ser. A* 42(4–6): 62–70
51. Humphries, D. A. 1968. The host-finding behavior of the hen flea *Ceratophyllus gallinae* (Schrank). *Parasitology* 58:403–14
52. Hůrka, K. 1963. Bat fleas of Czechoslovakia. Contribution to the Distribution, Morphology, Bionomy, Ecology and Systematics. Part I. Subgenus *Ischnopsyllus* Westw. *Acta Faun. Entomol. Mus. Nat. Prague.* 9:57–120
53. Hůrka, K. 1963. Bat fleas of Czechoslovakia. II. Subgenus *Hexactenopsylla* Oud., genus *Rhinolophopsylla* Oud., Subgenus *Nycteriodopsylla* Oud., Subgenus *Dinycteropsylla* Ioff. *Acta. Univ. Carol. Biol.* 1963. No. 1:1–73
54. Hurka, K. 1969. Systematic, faunal and bionomical notes on the European and Asiatic flea species of the family Ischnopsyllidae. *Acta Univ. Carol. Biol.* 1969:11–26
54a. Ioff, I. G., Mikulin, M. A., Skalon, O. I. 1965. *Handbook for the Identification of the Fleas of Central Asia and Kazakhstan.* Moscow: State Publishing House (In Russian)
55. Iqbal, Q. J., Humphries, D. A. 1970. Temperature as a critical factor in the mating behavior of the rat flea *Nosopsyllus fasciatus* (Bosc.). *Parasitology* 61:375–80
56. Jordan, K. 1950. On characteristics common to all known species of Suctoria and some trends of evolution in this order of insects. *Proc. Int. Congr. Entomol., 8th, Stockholm,* 87–95
57. Jordan, K. 1962. Notes on *Tunga caecigena. Bull. Brit. Mus. Natur. Hist. London Entomol.* 12(7):353–64
58. Joseph, S. A. 1972. *A study of the flea fauna of the Madras State, with special reference to the ecology, bionomics and host relations of the genus Ctenocephalides.* PhD Thesis, Madras University, Madras, India
59. Bai, M. K. 1972. Studies on host-flea relationship. Part I. Histochemistry of egg yolk in *Xenopsylla cheopis* (Roths) and *Xenopsylla astia* (Roths). *Indian J. Med. Res.* 60:752–56
60. Kessel, E. L. 1939. The Embryology of Fleas. *Smithson. Misc. Coll.* 98(3):1–78
61. Kir'yakova, A. N. 1960. The connection of the rates of oviposition in bird fleas to that of the time of the brooding period of their hosts. *Dokl. Akad. Nauk. SSSR* 131(6):1476–77
62. Kir'yakova, A. N. 1968. The external morphology of flea larvae. Family Pulicidae. (Comm. 6). *Parasitologiya* 2: 548–58
63. Kir'yakova, A. N. 1968. A comparative-morphological review of larvae of some genera of fleas. *Entomol. Obozr.* 47(1):71–79
64. Kir'yakova, A. N. 1973. On the life span of fleas in burrows. *Parasitologiya* 7(3):261–63
65. Kondrashkina, K. I., Dudnikova, A. F. 1964. A comparative analysis of the intensity of oxygen consumption by fleas of different species. *Zool. Zh.* 43(12): 1874–76
66. Kondrashkina, K. I., Dudnikova, A. F. 1967. Consumption of oxygen by the fleas of grey rats. B. K. Fenyuk *Grȳzunȳ i ikh Ektoparasitȳ,* 87–91
67. Kondrashkina, K. I., Gerasimova, N. G., Kuraev, I. I., Luk'yanova, A. D. 1969. Characteristics of metabolism in individual groups of zooids *Xenopsylla cheopis,* infected by the pest microbes. *Problemȳ Osobo Opasnȳkh Infektsii* 6:210–12
68. Krampitz, H. E. 1965. Beobachtungen an einer Labortoriumszucht von *Leptopsylla segnis* Schöhnerr. 1811. *Z. Parasitenk.* 26:197–214
69. Kulakova, Z. G. 1964. Nutrition of *Xenopsylla gerbilli caspica* Joff and of certain other fleas. *Ektoparazitȳ* 4: 205–20
70. Kunitskaya, N. T. 1960. On the reproductive organs in female fleas and determination of their physiological age.

Med. Parazitol. Parazit. Bolez. 29(6): 688–701

71. Lewis, R. E. 1961. The thoracic musculature of the Indian rat flea *Xenopsylla cheopis. Ann. Entomol. Soc. Am.* 54(33):387–97

71a. Lewis, R. E. 1967. The fleas of Egypt, an illustrated and annotated key. *J. Parasitol.* 53(4):863–85

72. Lewis, R. E. 1972–1973. Notes on the geographical distribution and host preferences in the order Siphonaptera. *J. Med. Entomol.* 9(6):511–20 (Part I); 10(3):255–60 (Part II)

72a. Lewis, R. E. 1973. Siphonaptera collected during the 1965 Street expedition to Afghanistan. *Fieldiana Zool.* 64:i–xi, 1–161

73. Margalit, J., Shulov, A. S. 1972. Effect of temperature on the development of prepupa and pupa of the rat flea *Xenopsylla cheopis* Rothschild. *J. Med. Entomol.* 9(2):117–25

74. Mead-Briggs, A. R. 1964. Some experiments concerning the interchange of rabbit fleas *Spilopsyllus cuniculi* (Dale) between living hosts. *J. Anim. Ecol.* 33:13–26

75. Mead-Briggs, A. R., Vaughan, J. A. 1969. Some requirements for mating in the rabbit flea *Spilopsyllus cuniculi* (Dale). *J. Exp. Biol.* 51:495–511

76. Michaeli, D., Benjamini, E., de Buren, F. P., Larrivee, D. H., Feingold, B. F. 1965. The role of collagen in the induction of flea bite hypersensitivity. *J. Immunol.* 95:162–70

77. Michaeli, D., Benjamini, E., Miner, R. C., Feingold, B. F. 1966. In vitro studies on the role of collagen in the induction of hypersensitivity to flea bites. *J. Immunol.* 97:402–6

78. Mohr, C. O., Adams, L. 1963. Relation of flea infestations to spacing between cottontail rabbits. *J. Wildl. Manage.* 27(1):71–76

79. Molyneux, D. H. 1967. Feeding behavior of the larval rat flea *Nosopsyllus fasciatus* Bosc. *Nature* 215(5102):779

80. Munshi, D. M. 1960. Micro-anatomy of the proventriculus of the common rat flea *Xenopsylla cheopis* (Rothschild). *J. Parasitol.* 46(3):362–72

81. Myalkovskaya, S. A., Bryukhanova, L. V. 1972. On the Siphonaptera of *Citellus pygmaeus* (Rodentia) during the aestivation of the hosts. *Zool. Zh.* 51:308–10

82. Nazarova, I. V., Tikhvinskaya, M. V. 1971. Siphonaptera of *Arvicola terrestris* L. (Rodentia) in central Povolozh'e (USSR). *Parazitologiya* 5:413–16

83. Nelson, B. C. 1972. Fleas from the archaeological site at Lovelock Cave, Nevada. *J. Med. Entomol.* 9(3):211–18

84. Neville, C., Rothschild, M. 1967. Fleas—insects which fly with their legs. *Proc. Roy. Entomol. Soc. London Ser.* C 32(3):9–10

85. Pausch, R. D., Fraenkel, G. 1966. The nutrition of the larva of the Oriental rat flea *Xenopsylla cheopis* (Rothschild). *Physiol. Zool.* 39(3):202–22

86. Peus, F. 1966. Intraspezifische Evolution und Randverbreitung bei Flöhen. *Zool. Anz.* 177(1):50–83

86a. Peus, F. 1967. Zur Kenntnis der Flöhe Deutschlands. I. Zur Taxonomie der Vogelflöhe. *Deut. Entomol.* 14:81–108

87. Peus, F. 1968. Zur Kenntnis der Flöhe Deutschlands. II. Faunistik und Ökologie der Vogelflöhe *Zool. Jahrb. Syst.* 95:571–633

88. Peus, F. 1968. Uber die beiden Bernstein-Flöhe. *Palaeontol. Z.* 42:62–72

89. Peus, F. 1970. Zur Kenntnis der Flöhe Deutschlands. III. Faunistik und Ökologie der Säugetierflöhe. Insectivora, Lagomorpha, Rodentia. *Zool. Jahrb. Syst.* 97:1–54

90. Peus, F. 1972. Zur Kenntnis der Flöhe Deutschlands (Schluss). IV. Faunistik und Ökologie der Säugertierflöhe. *Zool. Jahrb. Syst.* 99:408–504

91. Poenicke, H-W. 1969. Über die postlarvale Entwicklung von Flöhen, unter besonderer Berücksichtigung der sogenannten "Flügelanlagen". *Z. Morphol. Oekol. Tiere* 65:143–86

92. Prasad, R. S. 1969. Influence of host on fecundity of the Indian rat flea, *Xenopsylla cheopis* (Roths.). *J. Med. Entomol.* 6(4):443–47

93. Prasad, R. S. 1972. Different site selections by the rat fleas *Xenopsylla cheopis* (Roths.) and *X. astia* (Roths.). *Entomol. Mon. Mag.* 108:63–64

94. Prasad, R. S. 1972. Sterile male of the rat-flea *Xenopsylla cheopis* Rothschild. *Trans. Roy. Soc. Trop. Med. Hyg.* 66(6): 945–46

95. Prasad, R. S. 1973. Studies on host-flea relationship. II. Sex hormones of the host and fecundity of rat flea *Xenopsylla astia* (Rothschild) and *Xenopsylla cheopis* (Rothschild). *Indian J. Med. Res.* 61(1):38–44

96. Prokopiyev, V. N. 1958. The method of determining the physiological age of the female of *Oropsylla silantiewi* Wagn. and seasonal changes in age-composition of flea populations. *Izv. Irkutsk. Protivochumn. Inst. Sib. Dal'n. Vost.* 17:91–108

97. Prokopiyev, V. N. 1969. Morphological types of cocoons and their outlet mechanism of imago of fleas. *Mater. VI Nauch. Konf. Protivochumnykh Uchrezhdenii S. Azii i Kazakhstana,* 182–84
98. Radovsky, F. J. 1972. Fixed parasitism in the Siphonaptera. *J. Med. Entomol.* 9(6):602
99. Řeháček, J., Fischer, R. G. 1971. Primary cell cultures derived from larvae of *Ct. felis* (Bouché). *J. Med. Entomol.* 8:66–67
100. Reinhart, C., Schulz, U., Hecker, H., Freyvogel, T. A. 1972. Zur Ultrastruktur des Mitteldermepithels bei Flöhen. *Rev. Suisse Zool.* 79(3):1130–37
101. Richards, P. A., Richards, A. G. 1968. Flea *Ctenophthalmus:* Heterogeneous hexagonally organised layer in the midgut. *Science* 160:423–25
102. Richards, P. A., Richards, A. G. 1969. Acanthae: a new type of cuticular process in the proventriculus of Mecoptera and Siphonaptera. *Zool. Jahrb. Anat.* 86(8):158–76
103. Rothschild, M. 1958. The bird fleas of Fair Isle. *Parasitology* 48(3–4):382–412
104. Rothschild, M. 1964. Transfer of sperm by the penis rods of the male rabbit flea. (Note of an exhibit) *Proc. Roy. Entomol. Soc. London Ser. C.* 29(5):18
105. Rothschild, M. 1965. Fleas. *Sci. Am.* 213(6):44–53
106. Rothschild, M. 1966. Remarks on the life-cycles of fleas. *Proc. Int. Congr. Parasitol., 1st, Rome, 1964,* ed. A. Corradetti, 29–30. Oxford: Pergamon. 624 pp.
107. Rothschild, M. 1973. A specimen of *Pulex irritans* found in a Viking pit in Ireland. (Note of an exhibit). *Proc. Roy. Entomol. Soc. London Ser. C.* 38(7):29
108. Rothschild, M., Clay, T. 1952. *Fleas, Flukes & Cuckoos.* London: Collins. 4th ed. 304 pp.
109. Rothschild, M., Ford, B. 1965. Observations on gravid rabbit fleas *Spilopsyllus cuniculi* (Dale) parasitising the hare (*Lepus europaeus* Pallas), together with further speculations concerning the course of myxomatosis at Ashton, Northants. *Proc. Roy. Entomol. Soc. London Ser. A* 40(7–9):109–17
110. Rothschild, M., Ford, B. 1973. Factors influencing the breeding of the rabbit flea (*Spilopsyllus cuniculi*): a springtime accelerator and a kairomone in nestling rabbit urine, with notes on *Cediopsylla simplex,* another "hormone bound" species. *J. Zool.* 170:87–137
111. Rothschild, M., Ford, B. 1973. Differences in the mating behaviour of the rat flea *Nosopsyllus fasciatus* (Bosc.). *J. Entomol.* (A)47(2):157–59
112. Rothschild, M., Ford, B., Hughes, M. 1970. Maturation of the male rabbit flea (*Spilopsyllus cuniculi*) and the Oriental rat flea (*Xenopsylla cheopis*): some effects of mammalian hormones on development and impregnation. *Trans. Zool. Soc. London* 32:105–88
113. Rothschild, M., Hinton, H. E. 1968. Holding organs on the antennae of male fleas. *Proc. Roy. Entomol. Soc. London Ser. A.* 43(7–9):105–7
114. Rothschild, M., Schlein, Y., Parker, K., Neville, C., Sternberg, S. 1973. The flying leap of the flea. *Sci. Am.* 229(5): 92–100
115. Rothschild, M., Schlein, Y., Parker, K., Sternberg, S. 1972. Jump of the Oriental rat flea *Xenopsylla cheopis* (Roths.). *Nature* 239(5366):45–48
116. Rothschild, M., Traub, R. 1971. A revised glossary of terms used in the taxonomy and morphology of fleas. In: *An Illustrated Catalogue of the Rothschild Collection of Fleas in the British Museum (Natural History)*, ed. G. H. E. Hopkins, M. Rothschild, 5:8–85, plates 1–13. London: Trustees of the Brit. Mus. (Natur. Hist.)
117. Rudall, K. M., Kenchington, W. 1971. Arthropod silks: the problem of fibrous proteins in animal tissues. *Ann. Rev. Entomol.* 16:73–96
118. Ryckman, R. E. 1971. Plague vector studies. Part I. The rate of transfer of fleas among *Citellus* (Rodentia), *Rattus* (Rodentia) and *Sylvilagus* (Lagomorpha) under field conditions in southern California (USA). *J. Med. Entomol.* 8:535–40
119. Ryckman, R. E. 1971. Plague vector studies. Part II. The role of climatic factors in determining seasonal fluctuations of flea species associated with the California ground squirrel (*Citellus beecheyi:* Rodentia). *J. Med. Entomol.* 8:541–49
120. Ryckman, R. E. 1971. Plague vector studies. Part III. The rate deparasitized ground squirrels (*Citellus beecheyi:* Rodentia) are reinfested with fleas under field conditions. *J. Med. Entomol.* 8:668–70
121. Sabilayev, A. S. 1971. The ecology of *Dipus sagitta* in the north-west Kyzylkumy (USSR) (Rodentia). *Zool. Zh.* 50:1553–63
122. Sakaguti, K. 1963. *A Monograph of the Siphonaptera of Japan.* Osaka. 255 pp.

123. Samarina, G. P., Alexeyev, A. N., Shiranovich, P. I. 1968. The study of fertility of the rat fleas (*Xenopsylla cheopis* Rothsch. and *Ceratophyllus fasciatus* Bosc.) under their feeding on different animals. *Zool. Zh.* 47:261–68
124. Sharif, M. 1937. On the internal anatomy of the larva of the rat-flea, *Nosopsyllus fasciatus* (Bosc.). *Phil. Trans. Roy. Soc. Ser. B.* 227(547): 465–538
125. Sharif, M. 1945. On the structure of the so-called "penis" of the Oriental rat flea, *Ctenocephalides felis* subsp. *orientis* (Jordan), and homologies of the external male genitalia in Siphonaptera. *Proc. Nat. Inst. Sci. India* 11(2):80–95
126. Shulov, A., Naor, D. 1964. Experiments on the olfactory responses and host-specificity of the Oriental rat flea (*Xenopsylla cheopis*). *Parasitology* 54: 225–31
126a. Skuratowicz, W. 1967. Pchly-Siphonaptera in Klucze do oznaczania owadów polski. *Pol. Towarzystwo Entomol.* 53:1–141
127. Smit, F. G. A. M. 1963. Species groups in *Ctenophthalmus. Bull. Brit Mus. (Natur. Hist.) London Entomol.* 14:107–52
128. Smit, F. G. A. M. 1965. Siphonaptera collected by Dr. K. Zimmerman in north-east China (with a bibliography of Chinese Siphonaptera). *Ann. Mag. Natur. Hist.* 13(7):473–91
128a. Smit, F. G. A. M. 1965. Siphonaptera of New Zealand. *Trans. Roy. Soc. N.Z.* 7(1):1–50
129. Smit, F. G. A. M. 1966. Distribution of subspecies of the flea *Ctenophthalmus agyrtes* in and around Austria. *Entomol. Ber. Amsterdam* 26:216–21
129a. Smit, F. G. A. M. 1966. Siphonaptera. *Insect. Helv. Cat.* 1:1–107
129b. Smit, F. G. A. M. 1967. Siphonaptera of Mongolia. Results of the Mongolian-German Biological Expeditions since 1962. *Mitt. Zool. Mus. Berlin* 43: 77–115
130. Smit, F. G. A. M. 1969. A catalogue of the Siphonaptera of Finland with distribution maps of all Fennoscandian species. *Ann. Zool. Fenn.* 6:47–86
131. Smit, F. G. A. M. 1970. Siphonaptera. In *Taxonomist's Glossary of Genitalia in Insects,* ed. S. L. Tuxen, 141–56. Copenhagen: Munksgaard. 2nd ed. 359 pp.
132. Smit, F. G. A. M. 1972. On some adaptive structures in Siphonaptera. *Folia Parasitol. Prague* 19(1):5–17
133. Smit, F. G. A. M. 1973. The male of *Stephanopsylla thomasi. Entomol. Ber. Amsterdam* 33(11):215–17
134. Smit, F. G. A. M., Szabó, I. 1967. The distribution of subspecies of *Ctenophthalmus agyrtes* in Hungary. *Ann. Hist. Natur. Mus. Nat. Hungarica* 59:345–52
135. Smith, D. S. 1968. *Insect Cells, Their Structure and Function.* Edinburgh: Oliver & Boyd. 372 pp. (See Plate 106)
136. Snodgrass, R. E. 1946. The skeletal anatomy of fleas. *Smithson. Misc. Collect.* 104(18):1–89
137. Sobey, W. R., Conolly, D. 1971. Myxomatosis: the introduction of the European rabbit flea *Spilopsyllus cuniculi* (Dale) into wild rabbit populations in Australia. *J. Hyg.* 69:331–46
138. Sobey, W. R., Menzies, W. 1969. Myxomatosis: the introduction of the European rabbit flea *Spilopsyllus cuniculi* (Dale) into Australia. *Aust. J. Sci.* 31:404
139. Solomon, G. B. 1969. Host hormones and parasitic infections. *Int. Rev. Trop. Med.* 31:101–58
139a. Stark, H. E. 1970. A revision of the flea genus *Thrassis* Jordan 1933 with observations on ecology and relationship to plague. *Univ. Calif. Publ. Entomol.* 53:1–184
139b. Suciu, M. 1967. Contributions to the study of Siphonaptera in the Socialist Republic of Romania. *Comm. Zool. Bucuresti* 5:75–94
140. Suter, P. R. 1964. Biologie von *Echidnophaga gallinacea* (Westw.) und Vergleich mit andern Verhaltenstypen bei Flöhen. *Acta Trop.* 21:193–238
141. Sviridov, G. G., Morozova, I. V., Kaluzhenova, Z. P., Il'inskaya, V. L. 1963. The use of radioactive isotopes for the study of the ecology of fleas. Part I. Alimentary relations of fleas of the genus *Xenopsylla* with *Rhombomys opimus* Pall. under natural conditions. *Zool. Zh.* 42(4):546–50
141a. Szabo, I. 1962. A hazai Siphonaptera Kutatások Története. Rovartania Kozlemenyek. *Folia Entomol. Hungarica* (*Sera. Nova*). 15(18):327–33
141b. Theodor, O., Costa, M. 1967. *A Survey of the Parasites of Wild Mammals and Birds in Israel. I. Ectoparasites.* Jerusalem: Isr. Acad. Sci. Human. 117 pp.
142. Tillyard, R. J. 1935. The evolution of the Scorpion-flies and their derivatives (order Mecoptera). *Ann. Entomol. Soc. Am.* 28(1):37–45
143. Tipton, V. J., Machado-Allison, C. E. 1972. Fleas of Venezuela. *Brigham Young Univ. Sci. Bull. Biol. Ser.* 17(6): 1–115

144. Tipton, V. J., Méndez, E. 1966. The fleas of Panama. In *Ectoparasites of Panama,* ed. R. L. Wenzel, V. J. Tipton, 289–386. Chicago: Field Mus. Natur. Hist. 862 pp.
145. Traub, R. 1968. *Smitella thambetosa,* n. gen. and n. sp., a remarkable "helmeted" flea from New Guinea with notes on convergent evolution. *J. Med. Entomol.* 5:375–404
146. Traub, R. 1969. *Muesebeckella,* a new genus of flea from New Guinea, with notes on convergent evolution. *Proc. Entomol. Soc. Washington* 71(3): 374–96
147. Traub, R. 1972. Notes on zoogeography, convergent evolution and taxonomy of fleas, based on collections from Gunong Benom and elsewhere in south-east Asia. *Bull. Brit. Mus. Natur. Hist. Zool.* 23(9):201–305 (Part I); 23(10):307–87 (Part II); 23(11):389–450 (Part III)
148. Traub, R. 1972. Notes on fleas and the ecology of plague. *J. Med. Entomol.* 9(6):603
149. Traub, R. 1972. The zoogeography of fleas as supporting the theory of continental drift. *J. Med. Entomol.* 9(6): 584–89
150. Traub, R., Barrera, A. 1966. A new species of *Ctenophthalmus* from Mexico, with notes on the ctenidia of shrew fleas as examples of convergent evolution. *J. Med. Entomol.* 3(2):127–45
151. Traub, R., Dunnet, G. M. 1973. Revision of the Siphonapteran genus *Stephanocircus* Skuse 1893. *Aust. J. Zool.* Suppl. Ser. No. 20:41–128
152. Traub, R., Evans, T. M. 1967. Descriptions of new species of hystrichopsyllid fleas with notes on arched pronotal combs, convergent evolution and zoogeography. *Pac. Insects* 9(4):603–77
153. Traub, R., Wisseman, C. L. 1974. The ecology of chiggerborne rickettsiosis (scrub typhus). *J. Med. Entomol.* In press
154. Ulmanen, I., Myllymäki, A. 1971. Species composition and numbers of fleas in a local population of the field vole, *Microtus agrestis* (L.) (Rodentia). *Ann. Zool. Fenn.* 8:374–84
155. Vashchyonok, V. S. 1966. Histological characteristics of oogenesis in fleas *Echidnophaga oshanini* Wagn. *Zool. Zh.* 45(12):1815–25
156. Vashchyonok, V. S. 1966. Morphological and physiological changes, taking place in the organism of fleas *Echidnophaga oshanini* Wagn. during feeding and reproduction. *Entomol. Obozr.* 45(2):715–27
157. Vashchyonok, V. S. 1967. Gonotrophic relations in fleas *Ctenophyllus consimies* Wagn. *Parazitol. Sbornik* 23:222–35
158. Vashchyonok, V. S., Solina, L. T., 1969. The digestion of fleas *Xenopsylla cheopis* Roths. *Parazitologiya* 3(5):451–60
159. Vashchyonok, V. S., Solina, L. T. 1972. Age-determined changes in the fat tissue of female fleas *Xenopsylla cheopis. Zool. Zh.* 51:79–85
160. Vaughan, J. A., Mead-Briggs, A. R. 1970. Host-finding behavior of the rabbit flea *Spilopsyllus cuniculi* with special reference to the significance of urine as an attractant. *Parasitology* 61:397–409
161. Volyanskiy, Yu. Ye. 1972. Seasonal changes in the numbers of Siphonaptera and in the species composition in nests of *Microtus arvalis* (Rodentia). *Parazitologiya* 6:54–56
162. Vysochkaya, S. O., Kir'yakova, A. N. 1970. Methods of collecting and studying fleas and their larvae. *Methods Parasitol.* Invest. Leningrad 2:5–83
163. Wachmann, E. 1972. Das Auge des Hühnerflohs *Ceratophyllus gallinae* (Schrank). *Z. Morphol. Oekol. Tiere* 73:315–24
164. Wasserburger, H. J. 1961. Beiträge zur Histologie und mikroskopischen Anatomie von *Xenopsylla cheopis* Rothschild. *Deut. Entomol. Z.* 8:373–414
165. Wenk, P. 1953. Der Kopf von *Ctenocephalus canis* (Curt.). *Z. Parasitenk.* 11:593–606
166. Williams, R. T. 1973. Establishment and seasonal variation in abundance of the European rabbit flea, *Spilopsyllus cuniculi* (Dale), on wild rabbits in Australia. *J. Entomol.* (A)48(1):117–27
167. Yinon, U., Shulov, A., Margalit, J. 1967. The hygroreaction of the larvae of the Oriental rat flea *Xenopsylla cheopis* Rothsch. *Parasitology* 57:315–19
168. Zhovtyi, I. F. 1968. A study of the conditions of the dwelling of fleas in the burrows of rodents in Siberia and the Far East. *Izv. Irkutsk. Gos. Nauch. Issled Protiv-Inst. Sibiri Vostoka* 27:212–30

ADAPTATIONS OF ARTHROPODA TO ARID ENVIRONMENTS [1]

◆6091

J. L. Cloudsley-Thompson
Department of Zoology, Birkbeck College, University of London,
London, Great Britain WCIE 7HX

The principal environmental factors in arid regions to which Arthropoda have become adapted are as follows: extremes of temperature and humidity, so that the saturation deficiency of the atmosphere may be very great; excessive insolation and radiant energy; low, irregular rainfall, which tends to be seasonal; long periods of drought; strong winds; a soil surface often of loose sand, hard clay, or rock; and a sparse and usually very specialized covering of vegetation. The ways in which arthropods have become adapted to climatic and environmental extremes of the desert are often merely extensions of their adaptations to terrestrial life itself. They comprise avoidance of unfavorable conditions by behavioral responses; diurnal and seasonal rhythms of activity; diapause; resistance to transpirational desiccation; absorption of moisture through the mouth, anus, or integument; conservation of metabolic water; ability to survive dehydration; adaptations to high and low temperature, ultraviolet light, and food shortage. These adaptations were reviewed in 1964 (40) and the microclimatological aspects were discussed to some extent in *Annual Review of Entomology,* two years earlier (34). Subsequent research has validated the approach then adopted. The present review is intended to supplement these earlier publications and to bring them up to date. Only papers published from 1961 to the present day are cited. Much information concerning the physiology of arthropods in relation to heat and water has appeared in the last 13 years but, in general, only work on species that inhabit arid environments has been quoted here.

During the last decade, it has become increasingly evident that the phylum Arthropoda is polyphyletic (132, 133) and that arthropodization must have occurred at least three times in evolutionary history, resulting in three separate phyla (or subphyla) viz. Crustacea, Uniramia (including 'myriapods' and insects), and Chelicerata. Similarities between woodlice (Isopoda: Oniscoidea) and myriapods,

[1]The survey of literature pertaining to this review was concluded on January 1, 1974.

and between insects and arachnids, must therefore result from convergent evolution. These different groups are discussed separately in the following pages in order to make apparent the extraordinarily close parallels that have evolved independently in response to desert conditions.

Terrestrial arthropods display varying degrees of adaptation to terrestrial life. On an ecological basis, they may be divided into two main groups. The first comprises the crustaceans and myriapods, which are very susceptible to water loss through the integument and are therefore essentially nocturnal and cryptozoic. The second group consists of insects and arachnids which can exploit a wide range of terrestrial habitats, both at night and during the day, because they possess epicuticular layers of lipid which severely restrict transpiration (29, 30). An important difference between the two groups also lies in the nature of their nitrogenous excretory products. Crustacea and myriapods eliminate nitrogen mainly as ammonia; whereas insects and arachnids excrete the insoluble purines uric acid and guanine, respectively (18). Excretion of waste nitrogen in an insoluble form greatly reduces loss of water via the excretory system: the degree of terrestrialness achieved by an animal depends on its ability to conserve water and thus to maintain osmotic homeostasis.

The fact that arthropods are relatively small has several important consequences of which, perhaps, the most important is that evaporative cooling cannot be used for temperature regulation over anything except very short periods (65). An insect the size of a tsetse fly, weighing about 0.02 gm, would have to lose water equal to twice its body weight each hour in order to keep cool in desert conditions (70). Small size, correlated with a large surface area, also prohibits endothermy for any length of time. On the other hand, it permits arthropods to occupy microhabitats, in which climatic conditions are favorable, that are completely inaccessible to larger animals (34).

The adaptations of Arthropoda to arid environments comprise morphological, behavioral, physiological, and ecological modifications which are closely interrelated in nature. For reasons of clarity they will be treated separately in the following pages. The need for conciseness precludes considerations of purely physiological aspects of the various phenomena discussed.

MORPHOLOGICAL ADAPTATIONS

Morphological adaptations to arid environments fall into two broad classes: adaptations for locomotion and burrowing in sand and adaptations that result in reduction of respiratory water loss. In addition, desert arthropods tend either to have cryptic coloration or to be black. The latter color probably has an aposematic function and, in some cases, may represent Müllerian mimicry (40, 70). Temperature differences due to color are of minor importance (69, 70, 79, 169), although this has been disputed (93).

Adaptations for Living in Sand

Some species are adapted for burrowing, others for running over the hot surface; the legs and body form may be adapted accordingly (70, 122). The shapes of the notum and setae are characteristic of psammophilous tenebrionid larvae, but the significance of these modifications is unknown (161). Tenebrionid beetles of the group Stizopina found in the Namib desert are arenicolous, but, where stones are present, they often hide under them during the day. Psammophilous modifications include contraction of the body, hypertrophy of the tactile bristles and armature on the legs, reduction in the number of corneal facets of the eye, enlargement of the metasternum, and so on (123). Adaptations of the scorpion, *Anuroctonus phaeodactylus,* for burrowing include unusually thick and heavy pedipalps, reduced metasoma, and small pectines with few teeth (188).

Adaptations that Reduce Water Loss

Attention has recently been drawn to the many striking morphological similarities between plants and animals of arid environments. These include the presence of setae, hairs, scales, and so on, that create a boundary layer which reduces the flow of heat from the environment, surface-volume relationships, resistance to diffusion through stomata and spiracles, etc (82). Although loss of water through the cuticle can be drastically reduced by physiological means (12, 13, 14, 61, 78), some loss through the respiratory membranes is probably inevitable. Land arthropods without occlusable spiracles or lung-books are therefore at a grave disadvantage in comparison with those that have mechanisms for closing the respiratory apertures. The lung-books of arachnids, like the spiracles of arachnids and insects, are kept shut by special muscles until the amount of carbon dioxide in the respiratory system exceeds about 5%: only then are they relaxed to permit ventilation (100, 159). It appears that unfed female *Hyalomma dromedarii* and *Ornithodoros savignyi* have a metabolically controlled mechanism which does not function in hydrated females (100). In *Haemaphysalis* spp. closing of the spiracles depends on hemolymph pressure and the action of atrial and other body muscles (159). There is some dispute regarding the morphology of the spiracles of ticks (100, 104, 159).

WOODLICE The sowbugs represent a group that probably invaded the land much later than did other terrestrial arthropods. It is now believed that there have been several invasions of the land by oniscoid isopods (177), probably via the littoral zone. Adaptation to terrestrial conditions has occurred mainly through modification of the respiratory organs with the development of pseudotracheae (185), the ability to roll up into a ball, and so on. Conglobation, as this is called, considerably reduces the amount of water lost in transpiration because the respiratory organs, from which most of the loss takes place, are situated on the ventral side of the body (182). Most terrestrial Isopoda can take up moisture from the substratum by means of the uropods and transfer it to the capillary water system, but this ability has been lost in genera such as *Hemilepistus, Armadillo,* and *Venezillo,* which are best adapted to arid conditions. Presumably more water would be lost than gained by such a system. The isopods utilize two distinct respiratory mechanisms: the endopodite type, consisting of a blood sac with a thin membrane always bathed in a film of water, functioning as a gill; and the exopodite type, which consists of pseudotracheae. These are formed as invaginations of the outer wall of the exopodite and blood is separated from the air by a thin membrane. Both types can be present in the same species, but woodlice from arid regions possess only the second. Moreover, these pseudotracheae open into a single aperture, located laterally in a pocket-like depression, although there does not appear to be any mechanism for closing this opening (182).

MYRIAPODS The Indian millipedes *Cingalobolus* and *Aulacobolus* spp. possess epicuticles of lipoprotein in summer months. Although these lack an outer lipid layer, they restrict the permeability of the cuticle. At the commencement of the rainy season, however, when there is no necessity to restrict water loss, the new cuticle that is formed after molting is without an epicuticle (153). Although spiracular structure has not been studied in desert centipedes, the size of the spiracular opening and the degree of development of the lappets in the spiracle cap may be important in controlling the rate of water loss in European species of Geophilomorpha (127).

INSECTS The spiracle of insects (155) and ticks (100, 104, 159) living in dry environments tend to be small and often sunken or hidden (11). It has long been known that the spiracles of xerophilous buprestid beetles are covered with a basketwork of outgrowths which are believed to impede the diffusion of water molecules more than those of oxygen or carbon dioxide (82). A special case is found in desert Tenebrionidae with a subelytral cavity (a morphological adaptation to conditions of extreme aridity) whose major function is to reduce water loss by transpiration, for the spiracles open into it (41). The size of the subelytral air space does not appear to be an adaptive feature, however—large and small cavities being found in both diurnal and nocturnal species of the Namib desert (122). Water-balance studies and field observations on *Eleodes armata* support this finding, but it seems doubtful whether there is much thermal advantage in the arrangement (7, 10).

BEHAVIORAL ADAPTATIONS

These consist mainly of responses to environmental stimuli and activity rhythms, as a result of which arthropods maintain themselves in favorable microhabitats most of the time and, if they do have to leave them to feed or mate, do so when conditions elsewhere are not too adverse.

WOODLICE The behavior of woodlice is clearly correlated with their environment. When tested in temperature gradients, most animals are found most of the time at the lowest temperature available in the ranges 10–20, 13–27, 21–35, and 28–45°C, but the reaction is especially pronounced in species from mesic environments and least marked in *Venezillo arizonicus* from xeric habitats (180). Unlike the others, this species moves less frequently and at a lower speed when temperatures are high. Such behavior may be of survival value in arid regions, where temperatures are lower in its microhabitat under stones, than on the ground outside (181). The responses of terrestrial isopods to humidity are, again, more pronounced in species from humid than from xeric environments. Among species from both the littoral zone and mesic habitats, the main response to light is photonegative, unless the animals are dehydrated or the temperature rises, whereupon the response becomes photopositive (185). Isopods inhabiting grassland or forest (e.g. *Armadillidium* spp.) are more efficient in regulating their water balance and can, therefore, afford photopositive behavior. In the desert, the main pattern is of negative reactions to light (e.g. *Armadillo* spp. and *V. arizonicus*); with the exception of the North African *Hemilepistus reaumuri* which is crepuscular (185).

MYRIAPODS It has long been known that millipedes are mainly nocturnal in habit (29, 30). This has been confirmed by recent experiments and observations in the field (174) on West African species. Whereas *Spirostreptus assiniensis* is consistently photonegative—which relates to its habit of remaining in the dark, moist surroundings during the daytime—*Oxydesmus* sp. and *Habrodesmus falx* are photopositive, moving about in exposed situations in the field, especially in daylight, but only when the temperature and humidity of the air are favorable (172). The elaborate nests which are built primarily for molting also protect the larvae of *Oxydesmus* sp. during the dry season when moisture conditions become unfavorable. The swarming behavior of *H. falx* may also assist survival by reducing water loss (175). Desert centipedes, *Scolopendra* spp., are uniformly nocturnal. During the daytime they secrete themselves under rocks and fallen timber, and in holes in the ground. The rhythm of activity is endogenous and persists in darkness, but is quickly lost in constant light (53).

INSECTS South African Thysanura, *Ctenolepisma longicaudata* and *Machiloides delanyi,* actively avoid surface temperatures of 40–43°C and 33–36°C, respectively. Both species are strongly photonegative. *M. delanyi* also shows a marked avoidance of relative humidities outside the range 70–80%, while *C. longicaudata* is normally indifferent to atmospheric humidity. When desiccated, however, both species react positively to the highest humidity available. Avoidance of high temperatures overrules light and humidity responses and, in normal animals, the negative photoresponse overrides the humidity reactions; on desiccation, the order of precedence is reversed (98). The Egyptian desert grasshopper, *Sphingonotus carinatus,* normally shows a preference for dry air; its preferred temperature lies around 45°C (88, 89). Humidity and temperature receptors of this species and of *Aiolopus thalassinus* (86, 87) probably lie mainly in the antennae (90). The ecology of the desert cockroaches, *Arenivaga* spp., is based on their burrowing behavior. During the day, the insects are found at depths of 20–60 cm in dune sand. At night, females and nymphs ascend to 1–3 cm and males come out on the surface (94). Such behavior is obviously controlled by an endogenous rhythm (29, 30, 45, 48), of which the zeitgeber may be the level of temperature inversion in the soil (70). Rhythmic activity in the day-active grasshopper *Poecilocerus hieroglyphicus* is likewise endogenous, but is synchronized primarily by alternating light and darkness. It is correlated with the daily pattern of feeding, mating, and oviposition (4). This species is photopositive and negatively geotactic—responses that take it to the leaves of its specific host plant, *Calotropis procera* (3). Daily activity of *Diceroprocta apache* is closely dependent on temperature, and the insect is adapted to its environment by orientation behavior which emphasizes heat loss and inhibits heat gain (96). The body temperature of the desert locust is regulated by behavioral means that supplement physiological mechanisms (169); orientation of flying swarms has recently been reviewed (179).

Even in the extreme daytime temperatures of Khartoum, *Musca domestica* is day-active, having an endogenous rhythm which disappears rapidly in constant light (149). Ants, however, are active both at night and during the day. Orientation of *Cataglyphis bicolor* in the Tunisian desert, where there are no efficient land marks, depends on the pattern of polarized light. On moonless nights, ants react anemomenotactically (60). Tiny terrestrial visual clues are also employed over short distances round the nest (186). The blister-beetle, *Epicauta aethiops,* responds negatively to gravity and positively to light, as a result of which it climbs the plants on which it feeds. When disturbed, the insects immediately drop to the ground and take cover. The reactions to light override those to gravity (74), as in the case of *P. hieroglyphicus* (3). Beetles of the family Tenebrionidae form a conspicuous element of the fauna of most desert regions. Some species are day-active, others crepuscular or nocturnal (40). In a comparative study of the responses to light and diurnal rhythms of three species common around Khartoum, *Ocnera hispida* and, to a lesser extent, *Pimelia grandis* exhibited photonegative responses at all temperatures, while *Adesmia antiqua* was photopositive. The intensity of the light reactions of the first two species was not influenced by temperature, but *A. antiqua* was more markedly photopositive at higher temperatures. All species responded to near-lethal temperatures by digging into the sand. Analysis of aktograph records showed that *O. hispida* and *P. grandis* are strictly nocturnal; *A. antiqua* is diurnal (38). At the same time the day-active *A. antiqua* shows preference for a higher temperature than *P. grandis* (72), but both species are hygro-negative unless desiccated, when they become hygro-positive (73). Activity of *Blaps sulcata* in Israel begins at sunset, reaches a peak towards midnight and ceases at dawn, when the photonegative insects aggregate under thick, flat-bottomed stones (114). While most tenebrionid beetles show either diurnal or nocturnal activity, *Lepidochora argentogrisea* is active only shortly after sunset. The endogenous circadian rhythm of activity is synchronized by temperature (124). All species of the genus *Cardiosis,* on the other hand, are day-active, although the rhythms of the different species respond to ambient thermal conditions. On cool and windy days, activity extends through the middle hours, but, on hot and calm days, it becomes bimodal and limited to periods when the body temperature can be maintained above 30°C and at substantially below the lethal limits (92).

ARACHNIDS Desert forms are mainly nocturnal burrowers. Although they do not discriminate between light and darkness, burrowing is characteristic of desiccated whip-scorpions, *Mastigoproctus giganteus.* By using the sensitive front legs, the animals first detect nonhorizontal surfaces and, provided the substrate is moist, dig burrows against and beneath rocks, etc so that further water loss is presumably avoided. Desiccated specimens also respond positively to moist air in a humidity gradient (56). Burrowing is characteristic of scorpions (42) and has been described in *Nebo hierochonticus* (160) and in *Anuroctonus phaeodactylus* (188). In this species the appendages are cleaned by sponge bathing; the telson is wetted in the pre-oral cavity and then dabbed over the body. Aspects of the physiology and behavior of *Leiurus quinquestriatus* have been investigated in relation to the desert environment (1, 33). The scorpion reacts positively to humidity and negatively to light and to temperatures above 39°C (2). Scorpions are nocturnal in their habits and show well-marked endogenous rhythms of activity (1, 29, 30, 40, 42, 50). The locomotory rhythm is reflected in the physiological respiratory rhythms of *Euscorpius carpathicus* and *Euscorpius italicus* (59). In *Centruroides sculpturatus,* on the other hand, the rhythm is represented by a change of location rather than by an outburst of activity and there is no respiratory rhythm. This species remains in bark crevices or under some form of cover during the day, but, soon after sunset, it comes to the surface where it sits motionless, awaiting potential prey (83). The circadian rhythm of *Buthotus minax* is entrained by darkness and rising temperature; it is neither advanced nor retarded by constant darkness and, therefore, does not follow the circadian rule (50). Among the Solifugidae, the camel-spider, *Galeodes granti,* has likewise been shown by aktograph experiments, as well as field observations, to be nocturnal in habit (31). It spends the daytime in a deep burrow, the mouth of which is closed with a plug of dead leaves (32). Comparable behavior is characteristic of tarantulas, *Aphonopelma* spp. (162). Adult giant velvet mites, *Dinothrombium tinctorium,* appear in the Sudanese desert after rain. They are diurnal in habit, but become crepuscular as the soil dries. They are positively phototactic and negatively geotactic in dry sand, but dig burrows where it is damp (36). Similarly, *Dinothrombium pandorae* emerge on the first sunny day after rain in the southern California desert and feed voraciously on alate termites, with whose emergence their own is coordinated. They spend most of their adult lives in vertical burrows (140, 171). Host-finding behavior of parasitic acarines has been reviewed (23).

PHYSIOLOGICAL ADAPTATIONS

Physiological adaptations common to plants and animals of arid environments include tolerance of high temperature, facultative hyperthermia, relatively low cuticular and respiratory transpiration, efficient salt excretion, atmospheric water uptake, conservation of metabolic water, and resistance to desiccation. Although no plant or animal exhibits all these features, the number common to every desert organism reflects the need to maintain optimum temperature and water balance under adverse conditions (82). In general, evaporation cooling is of but slight importance to forms that possess an impervious integument (40, 47).

Adaptations to Heat

Arthropods in diapause show an enhanced resistance to heat. An example is afforded by the eggs of the notostracan *Triops granarius* which survive in dried mud or dust throughout the dry season where they may be exposed to temperatures up to 80°C (51). Laboratory experiments show that they can withstand 98 ± 1°C for over 16 hr (24). Even when not in diapause, desert Arthropoda are usually able to tolerate higher temperatures than can related species from less arid regions (40). Much work has been carried out in the past on the upper lethal temperatures of arthropods but, in earlier studies, acclimation was seldom taken into account (20).

WOODLICE Woodlice show acclimation to high temperatures, both in their ability to withstand lethal temperatures (46, 62) and in their rates of oxygen consumption (46, 63). This is not surprising; most invertebrate poikilotherms probably do so (25).

INSECTS The hygrophilous grasshopper, *Aiolopus thalassinus,* is able to resist rising temperatures up to 45°C without any apparent effect on its activity, and *Sphingonotus carinatus* can withstand temperatures up to 41°C (88). American desert roaches, *Arenivaga investigata,* have an upper lethal temperature in dry air of 48.5°C for exposures of 30 min (70); the beetle *Centrioptera muricata* one of nearly 50°C for 30 min in nearly dry air (8); and *Onymacris plana* from the Namib desert, a lethal temperature of 51°C for exposures of 30 min in saturated air (69). With experiments of short duration, such as these, it is questionable how long the whole body of the animal would have been at the experimental temperature. Experiments carried out over a period of 24 hr are, therefore, of greater significance. The lethal temperatures of Saharan tenebrionids, tested over 24 hr at relative humidities below 5% are as follows: *Adesmia antiqua,* 46°C (73); *Ocnera hispida,* 45°C; and *Pimelia grandis,* 43°C (37). [*A. antiqua* is day-active; *O. hispida* and *P. grandis* are nocturnal (38).] The figure for *Eleodes* spp. from Albuquerque is 39.5–40°C, from Las Cruces it is 41°C (52). Heat death in arthropods cannot be attributed to any single factor (40); in desert species such as *O. hispida,* it is associated with a decrease of pH in the hemolymph and may be primarily related to the accumulation of acid metabolites (37). In the case of *C. muricata,* both uric and lactic acid increase significantly in the hemolymph during heat death (8).

Temperature regulation in arthropods is very largely behavioral (47). The desert sphinx-moth, *Manduca sexta,* however, controls heat loss from the thorax by regulating the amount of blood flow (101), and it has been suggested that desert insects adjust the production and loss of heat more than do insects from other biomes (95). Acclimation to 34°C resulted in lower oxygen consumption in *A. antiqua, O. hispida,* and other desert arthropods than in specimens acclimated to lower temperatures (25).

Before leaving the subjects of heat and lethal temperatures in insects, brief mention should be made of cryptobiosis, a state in which metabolism comes to a reversible standstill (105, 106). For instance, larvae of *Polypedilum vanderplancki* are capable of tolerating desiccation, and some metamorphosed after exposure to 102–104°C for 1 min when dry. Others recovered temporarily after exposure to 106°C for 3 hr, or to 200°C for 5 min. Cryptobiosis is clearly exploited by these insects as a means of surviving in an extreme and unstable environment, for active larvae do not withstand exposure to temperatures over 43°C for more than about 1 hr (103).

ARACHNIDS The lethal temperatures of desert arachnids are even higher than those of insects. For example, the lethal temperature of *Galeodes granti,* for an exposure of 24 hr at a relative humidity below 10%, is 50°C; while for the scorpions *Leiurus quinquestriatus* and *Buthotus minax* the temperatures are 47 and 45°C, respectively (37). The upper lethal temperature of *Vejovis* sp. from the Chihuahuan desert is 41°C (52). In scorpions, lethal temperatures, like water relations (39), appear to be correlated with habitat and distribution. Acclimation has been demonstrated in the respiratory functions of these animals (25) but acclimation to 10 or 30°C had no effect on tarantulas, *Aphonopelma* spp., which showed critical thermal maxima of 43°C in temperatures rising 0.5°C/min (162). Perhaps some acclimation may have occurred during the course of the experiment, however. Duration of survival at high temperature is affected by the temperature of acclimation of *Lycosa carolinensis,* and the spiders survive better at high temperatures when the humidity is also high (139).

Adaptations to Cold

Freezing conditions provide a challenge to life because living organisms are composed predominantly of water which solidifies as ice at temperatures below zero. The effects of cold on arthropods are manifold, but a primary distinction must be made on the basis of whether freezing occurs or not. Considerable research has been carried out on the physiology of survival at low temperatures of arctic insects, but little appears to be known about the effects of cold on tropical Arthropoda. The available information is summarized below.

MYRIAPODS Julid millipedes in New Mexico have been found to possess supercooling points of –6.3 ± 0.38°C and to recover well from freezing (52). The mean supercooling point of large *Scolopendra polymorpha,* conditioned in an insectiary at 27°C, was found to be –3.1 ± 0.48°C, and a single supercooling period was injurious or fatal (53). This suggests that cold winter temperatures must be avoided by burrowing deeply.

INSECTS The resistance of insects to cold has also been extensively studied, but seldom with respect to desert arthropods. Species of *Gryllus* and *Eleodes* from Arizona and New Mexico have been shown to have supercooling points of –7.2 ± 0.42°C and –11.8 ± 0.62°C, respectively, which may be of importance in enabling them to survive low winter temperatures (52). On the other hand, the supercooling points of *O. hispida* and *A. antiqua* from the Sudan are –10.4 ± 0.70°C and –9.0 ± 0.89°C, respectively, and it has been argued that the ability to supercool may be a taxonomic rather than an adaptive feature (49). Hydration caused a significant reduction in the supercooling point of *Schistocerca gregaria,* but not of *Locusta migratoria migratorioides.* Preconditioning significantly lowered the supercooling point of *S. gregaria*; this was not achieved by dehydration, the elimination of feces, an increase in osmolarity, or by the secretion of glycerol. It could, however, have resulted from a change in the nature or location of food already in the alimentary canal (49).

When fully hydrated larvae of *P. vanderplanki* are frozen, they recover temporarily but fail to complete their development (125). Desiccated larvae in a state of cryptobiosis, on the other hand, can survive immersion in liquid air or liquid helium (103).

ARACHNIDS Scorpions, *Vejovis* sp., from Arizona and New Mexico have a supercooling point of –5.8 ± 0.61°C, but do not recover well from freezing (52). Although it could be argued that this comparatively low freezing point is an adaptation to cold winters, *L. quinquestriatus* from Sudan, where temperatures seldom fall below 5.6°C (51), has a supercooling point of –7.5 ± 0.31°C (49). The ability to resist freezing cannot, therefore, have any adaptive significance whatsoever in this species.

Adaptations to Drought

RATES OF TRANSPIRATION Reduction in the rate of transpiration is a universal adaptation of desert arthropods (40).

Woodlice The rate of transpiration from desert woodlice is lower than that from species of more humid regions (67). Since the regulation of body temperature is related to the rate at which water is lost, evaporative thermoregulation does not play a substantial role in controlling the body temperatures of isopods from more xeric habitats. Thus, both *Armadillidium vulgare* and *Venezillo arizonicus* kept a high body temperature and survived well, irrespective of their water loss (182). *Buddelundia albinogriseus* from regions of low rainfall in South Australia provides a link between *V. arizonicus* (Armadillidae) and the Armadillidiidae in this respect. The degree of its adaptation to life on land is intermediate—sufficient for the species to extend into semiarid habitats (183, 184). Figures for water loss by transpiration from the African *Periscyphis jannonei* and the cosmopolitan *Metoponorthus pruinosus,* collected from localities around Khartoum, are almost as low as those for *Hemilepistus reaumuri.* Different populations of the same species of woodlouse probably show transpiration rates which are related to the dryness of their environments, but rapid acclimation in the rate of water loss does not occur (46).

It is well established that the rate of cuticular transpiration in arachnids and insects is insignificant below a critical temperature (13, 65, 78). Although similar claims have been made for woodlice, the temperature effect can usually be explained by a higher rate of diffusion of water molecules. Even such a low rate of evaporation, as is found in *H. reaumuri* (67), can be attributed to calcification of the integument and also, possibly, to the lack of a capillary water system. In *V. arizonicus* there is, therefore, for the first time, evidence in an isopod of an epicuticular layer of wax having a distinct critical temperature between 38 and 40°C (182). The point has often been made that one of two alternative courses was open to the ancestors of the terrestrial arthropods: to evolve an impervious integument and all the physiological adjustments that must accompany it; or, like woodlice, to remain in relatively moist surroundings by means of behavior responses, including nocturnal habits (29, 30, 40). That a desert isopod, which has, presumably, followed the first of these to its furthest limits, should then have been able to evolve a discrete epicuticular wax layer de novo is, consequently, a matter of very considerable surprise; it certainly requires further investigation.

Myriapods Millipedes are generally less well adapted than are centipedes to life in arid places; recent observations from tropical Africa (172, 173) support this view. Transpiration from the North American desert millipede, *Orthoporus ornatus,* is 0.07 ± 0.01%/hr at 25°C in dry air (55), less than one tenth that of the Nigerian *Spirostreptus assiniensis,* which loses an average of 0.57 ± 0.05%/hr in dry air at 24 ± 1.5°C (173). Therefore an order of magnitude separates the transpiration rates of these two species, although they belong to the same family, Spirostreptidae. An unexpected increase occurs in cuticular permeability between 35–40°C in large, but not small, *O. ornatus* (55), but it is not on a scale comparable with that found in arachnids and insects (13, 21, 78). Rates of water loss in dry air have been measured from various species of tropical and temperate centipedes, but transpiration was always directly related to saturation deficiency, and in no case is there evidence of a critical, transitional temperature at which the cuticle shows a sudden increase in permeability. Nevertheless, water is retained more efficiently by *S. polymorpha* from New Mexico than it is by forms from less arid environments (53).

Insects Water regulation in insects has been reviewed (12, 164), and the discussion below is restricted to recent observations on desert species. The permeability of the integument has been the subject of extensive research (13, 61), and it is clear that the rate of transpiration is particularly low in desert species (70, 82). There is evidence, moreover, that epidermal cells of migratory locusts, *L. migratoria migratorioides,* expend energy continuously in regulating the water balance of the cuticle which, to be in equilibrium with the blood, would contain 60% more water than it actually does (190). This may result from active transport (190), or from a cuticular water pump (189). At present, available evidence is insufficient to draw definite conclusions (130). The permeability of locust cuticle in dry air at 30°C is 0.71 mg/cm^2/hr. There is, however, a change in permeability at low humidities (130). The rate of water loss from *Anacridium melanorhodon* exceeds that from *Poecilocerus hieroglyphicus,* the two species surviving in dry air at temperatures above 37°C for 11 and 18 days, respectively (5). Cuticular transpiration, although low in absolute rate, is a greater source of water loss from tenebrionid beetles of the Sonoran desert than is respiratory transpiration. This suggests that spiracular control of water loss is of considerable importance in maintaining water balance. A marked increase in respiratory transpiration occurs at 40°C for the autumn species *Eleodes armata,* and at 42.5°C for the summer species *Centrioptera muricata* and *Cryptoglossa verrucosa.* Cuticular transitional temperatures for the three species are 40°C, 47.5°C, and 50°C, respectively (7). *E. armata* exhibits greater water loss than *C. verrucosa* at all temperatures, but no uptake of atmospheric moisture occurs in either species (10).

Stringent control over water loss by closing the spiracles has been demonstrated in many insects (21, 70). Ventilation of the tracheal system of resting desert locusts, *Schistocerca gregaria,* caused by abdominal pumping, is 42 l/kg/hr (187). If the respiratory gases become 99% saturated, a locust breathing air at 20% relative humidity at 30°C will lose water at a rate of 2.1 mg/2 g locust/hr (91). Adding to this, an estimated cuticular loss of 7 mg/locust/hr gives a total water loss of 9.1 mg/locust/hr, of which respiratory loss would comprise 23% of the total (163). This may not be constant, however, since spiracular opening is subject to physiological control (134–136). In migratory locusts, spiracular loss at 30°C is about half the total loss (130, 131). Proportionately, less water is lost from the spiracles at higher saturation deficiencies, ventilatory movements decrease in rate and amplitude, and there is an increased incidence of discontinuities—all modifications that tend to conserve water (130, 131). The cyclical release of carbon dioxide by diapausing insect larvae and pupae serves a similar function (113). The importance of the elytra of desert Tenebrionidae has already been mentioned; one of the adaptive features of these long-lived insects is the ability to produce secondary sclerotization if the elytra are damaged (22). Desert beetles lose water from four to ten times faster than desert scorpions, when both are expressed as percentages of total weight per unit time (79). If this criterion is used (7), ecologically significant though it may be, it gives no indication about cuticle permeability. If a scorpion weighs a hundred times more than a thysanuran, and both lose an equal percentage of weight in unit time, the rate of loss per unit area will be five times greater in the scorpion (68). *Arenivaga* spp. appear to differ little from *Periplaneta americana* as regards physiological responses to dehydration. There is a reduction in the amount of water in the hemolymph of both and, although the osmotic pressure increases, it is subject to strong osmoregulation (66).

Larvae of *Polypedilum* spp., inhabiting rock depressions in Nigeria, withstand periods of drought and heat in a state of cryptobiosis (103, 106, 137). Larvae of the stratiomyid, *Cyrtopus fastuosus,* on the other hand, survive for only relatively short periods. Although they are in diapause, their metabolism is continued at a low level and depends on a limited rate of entry of oxygen through an almost closed system of spiracles. When a depression fills with water, these larvae become active much more quickly than those of *Polypedilum* spp. which take about 30 min or more to begin feeding. Since the moisture may last for less than 24 hr and since food is scarce, a quick start may be advantageous (137).

Arachnids As in insects, transpiration takes place both through the integument and through the respiratory openings of arachnids. The rates of water loss from different species of African scorpion are related to their distribution in terms of the dryness of their environments (39). Total water loss from the American desert scorpion, *Hadrurus arizonensis,* 0.028% body wt/hr at 30°C in dry air (80), is comparable to 0.030% at 33°C from *Leiurus quinquestriatus* (33) and 0.090% from *Galeodes granti* at 33°C (31), 0.147% from *Eurypelma* sp. at 33°C (43), 0.245% from *E. armata* at 30°C (10), and 0.320% from *Locusta migratoria migratorioides* (130) at 30°C. Transpiration rates of scorpions are thus well below those of other desert arthropods under similar conditions. Increased cuticular permeability after death is interpreted as evidence for the existence of an active water-retaining mechanism (80). In contrast to desert scorpions, the semimontane *Diplocentrus spitzeri* has a fairly high rate of water loss, resembling that of *Buthus occitanus*—a North African species of European origin that burrows deeply beneath rocks (57). Although *G. granti* has a fairly high rate of water loss (0.116 mg/cm^2/hr or 0.090% wt/hr) when compared with scorpions, it does not die until it has lost up to two thirds of its total weight (31). Water loss in dry air is low from male tarantulas, *Aphonopelma* spp., at 30°C, averaging 1.76% (162). This is only 16% of the rate reported for *Dugesiella* (=*Eurypelma*) *californica* at 25°C and 25% relative humidity (102), and 57.5% of the mean rate from *Eurypelma* (=*Aphonopelma*?) sp. at 33°C (43). Cuticular permeability has long been known to be exceptionally low in unfed desert ticks. The transition temperature of the epicuticular waxes is 52°C for *Hyalomma dromedarii,* and 63°C for *Ornithodoros savignyi* (84). Compared with larger arthropods, the flour mite *Acarus siro* has high rates of water loss and uptake when measured as a percentage of its weight, but low rates when measured as a percentage area of its surface (167). Conflicting results have been obtained with regard to transpiration from the whipscorpion *Mastigoproctus giganteus.* Although some large specimens lost water in dry air at different temperatures in proportion to the saturation deficiency of the atmosphere (56), other smaller individuals exhibited a marked increase at approximately 37.5°C, which appeared to be the result of a change in cuticular permeability (6). The difference may be due to variations in size; it is more likely, however, that the cuticles of the larger animals had been abraided by the desert soil (56).

WATER UPTAKE Terrestrial arthropods obtain water via the mouth, the anus (150, 168), or through the integument (12–15, 70). They may also make use of metabolic water, although the increased respiration necessary may prevent a net gain from resulting (187). Eggs of locusts and other insects are known to absorb moisture against an osmotic gradient (19). The mechanism is little understood, but there is no suggestion that desert species are particularly efficient in this respect. Studies on the eggs of Acrididae have shown that the uptake of water is affected by temperature (166) and that development ceases temporarily when the environment is dry (108, 109, 165).

Insects Many insects ingest substantial quantities of preformed water with their food. This is especially true of species that feed on plant juices or blood (21). Some are capable of absorbing water vapor from unsaturated air. This has been demonstrated in the firebrat, *Thermobia domestica* (16, 141, 142); *Ctenolepisma longicaudata* (97); larvae and adult female cockroaches of the genus *Arenivaga* (64); and prepupae of *Xenopsylla cheopis* (119). It usually occurs at high humidities (163), either through the integument (14, 15) or the rectum, since blocking the anus arrests the process in *T. domestica* (142). But this may be due to nervous inhibition (144, 145). *Ctenolepisma terebrans* gains weight by absorption of water vapor at relative humidities above 47.5% until the normal water content of the body has been restored (68). Even in summer the microclimate in which this species lives is such that the relative humidity rises above this level at night. Although the mechanism of water uptake is still obscure, the body volume of *T. domestica* is apparently a crucial factor, and neither desiccation nor rehydration have much effect on concentrations of Na^+ and K^+ in the hemolymph (146) . At one time it seemed possible that the mechanism might be related to the properties of anomalous water, or polywater, but it appears unlikely that this hypothesis can be tenable (75). The process is certainly adaptive but it is surprising that it should not be more widespread among desert arthropods. Even the desert millipede *Orthoporus ornatus* is unable to absorb moisture across the cuticle, although both oral and anal uptake have been demonstrated from moist surfaces (55); nor can tenebrionid beetles do so (10, 40), although the cryptonephridial renal complex is associated with efficient removal of water from the feces (154).

Arachnids Parasitic mites and ticks obtain ample water from their food. Ticks show no spiracular control when engorged (100), while salivary secretion is invoked as a means of eliminating excess moisture by *Boophilus microplus* (170) and probably other Ixodidae. When separated from their hosts, however, desert acarines may have to withstand long periods of starvation under extremely desiccating circumstances, when a low rate of transpiration and the ability to take up moisture from unsaturated air may be of vital importance. The critical equilibrium humidity differs from 45 to 95% in different species (121). In general, however, the osmotic pressures of the hemolymphs of terrestrial arthropods are in equilibrium with air at approximately 99% relative humidity (163).

The species of Arachnida in which active absorption occurs are chiefly acarines, including ticks (17, 85, 117, 120, 126) and mites (112, 115, 116, 118, 167). For example, the critical equilibrium humidities of *H. dromedarii* and *O. savignyi* are about 75%, being below this in larval stages and above it in later instances and adults (84). In the case of *Dermacentor variabilis* (120) and *A. siro* (116), the absence of trachea strongly indicates the integument or anus as the site of uptake. The whipscorpion *M. giganteus* gains water by drinking from a moist substrate, as well as from the blood of its prey (56), as does the scorpion *Centruroides sculpturatus* (81).

ECOLOGICAL AND PHENOLOGICAL ADAPTATIONS

Studies of micrometeorology and energy exchange in desert arthropods have emphasized the importance of burrowing (34). Vertical movements in the burrow provide a wide choice of microenvironments (79), while the microenvironment of rodent burrows likewise influences acarine distribution (192). Phenological mechanisms permit animals to take maximum advantage of short, climatically favorable seasons. Such adaptations permit the avoidance of desert conditions and are associated only indirectly with water relations or thermal stress. At the same time, many arthropods show seasonal acclimatization to heat and cold, and longitudinally separated populations may vary genetically in their thermal responses (47).

WOODLICE Not only do Sudanese woodlice show short-term acclimation to high temperatures, but individuals acclimated in winter for 48 hr to summer temperatures (34 ± 1°C) did not survive as well on damp filter paper at 41.5°C as did specimens tested shortly after being collected in the field. On the other hand, animals acclimated for 48 hr in a cold room in summer survived better at 41.5°C than did animals previously kept in equivalent room temperature (28 ± 1°C) in winter. It seems, therefore, that seasonal differences in the ability to withstand high temperatures may not be related solely to acclimatization (46).

MYRIAPODS Observations on populations of *Rhysida nuda togoensis* and *Ethmostigmus trigonopodus* in northern Nigeria show rapid growth rates and life cycles that are shorter than those of the centipedes of temperate regions. It is suggested that both species are enabled to continue feeding in damp areas, after the rains have ended, by virtue of their mobility, large size, and consequently greater resistance to desiccation. Increasing aridity drives them into crevices in the soil where a relatively rich fauna may persist throughout the dry season. The young appear at the beginning of the rains, thereby having the entire rainy season for feeding (128).

INSECTS The influence of man on insect ecology has already been discussed (156). The adaptations of insects inhabiting the Naryn sands, Transvolga, are concerned with obtaining food, moisture, and an equable temperature (151). The chief factor limiting the distribution of Machilida in South Africa is availability of moisture; Lepismatida are not thus restricted (99). Population peaks of *Gryllus bimaculatus* in Israel are related to the temperature during nymphal development (157). Extremes of heat and cold are avoided by *Onymacris rugatipennis* whose activity is bimodal in summer and unimodal in winter (107). The water economy of *Onymacris* spp. is more efficient than that of *Lepidochora argentogrisea* which inhabits the same vegetationless environment, feeding on wind-blown detritus and drinking droplets that collect on its integument during the fogs it encounters (129). The ecology of Moroccan species of *Timarcha* (Chrysomelidae) is related to diapause, phenology, and circadian rhythms (111). The only character distinguishing Saharan species of ants is the greater impermeability of their cuticles. These insects occupy an important place in the entomology of the desert through their social life and the precision of their behavior (58). The influence of rainfall in arid regions has received much attention. The majority of Coleoptera, lepidopterous larvae, and other insects are found in the desert near Khartoum in August, at the time of the annual rains, and the presence of dead beetles during the dry season suggests that the older generation tends to die off at this time of year, while the young adults and larvae bury themselves deeply in the ground (54). The importance of moisture on insect abundance and diversity during the dry season in the forest of Costa Rica (110) provides an interesting comparison with grazed and ungrazed areas of desert (54). The distribution of terrestrial arthropods on the Erkowit plateau, near Sinkat, Sudan, is likewise to be interpreted in terms of rainfall and vegetation; while, in the Red Sea coastal plain, the thermal advantage of vegetation and rock cover is manifest (35). Fewer tenebrionid beetles were taken in pitfall traps near Phoenix, Arizona during extreme summer conditions and throughout the colder winter months than at other times of year. Increases in the number of beetles collected resulted from general increases in population density during seasonal periods of eclosion (9). In the Russian deserts, phytophagous insects are most numerous at the time of the spring rains (77). The main factors determining the numbers of insects collected in light traps near Bairam-Ali, Turkmenia, are temperature in April and humidity from June onwards. Numbers are also influenced by wind and atmospheric pressure (176). Breeding of grasshoppers in Ghana increases during the rainy season (27) and oviposition of *Nomadacris septemfasciata* is synchronized by the onset of rains (191). Plagues and recessions of the desert locust *Schistocerca gregaria* are dependent upon rainfall (178) which, likewise, influences phase change (152) and sexual maturation (143). Outbreaks of these insects fall into three distinct phases: circulation, followed by multiplication, and then gregarization (158). Maturation is delayed by a dry season diet low in gibberellins and essential oils (71). The onset of the rains, however, coincides with bud-burst and vegetation rich in gibberellin, eugenol, and other monoterpenoids which act as environmental cues and induce maturation and subsequent breeding (26). *Anopheles gambiae* survives the long dry season in the Sudan in a stage of ovarian diapause. When the rains finally come, the mosquitoes are ready to oviposit (147, 148).

ARACHNIDS Deserts are, traditionally, the home of biting and stinging animals, and the great reduction of plant cover makes surface activity hazardous except for the speedy and the well armed. The venoms of desert arthropods can thus be regarded in part as physiological adaptation to the environment (138). An ecological study of the spiders of a California desert community shows a predominence of hunting forms over web-spinners (ratio of 50:3 individuals). Although not diverse, the fauna exploits rather fully the limited resources of the habitat (28).

Attention has been drawn to the lack of integration between the disciplines of physiology and ecology, yet without such integration, ecological observations will be capable only of partial interpretation (20, 21). Detailed, comprehensive studies of all aspects of the temperature and water balance of desert arthropods are necessary. The work reviewed in this article emphasizes the compartmentalization of ideas that takes place and the need for a more synthetic approach. That entomologists are aware of this is clear from the broader attitude of several recent publications (44, 70, 79, 82). Even the formidable problem of identifying the insect fauna from tracks left in desert sand has now been overcome (76). It seems reasonable therefore, to hope that integrated studies will become more numerous in the future.

Literature Cited

1. Abushama, F. T. 1963. Bioclimate, diurnal rhythms and water loss in the scorpion *Leiurus quinquestriatus* (H. & E.). *Entomol. Mon. Mag.* 98:216–24
2. Abushama, F. T. 1964. On the behavior and sensory physiology of the scorpion *Leiurus quinquestriatus* (H. & E.). *Anim. Behav.* 12:140–53
3. Abushama, F. T. 1967. Geotaxis in *Poecilocerus hieroglyphicus. J. Zool.* 153: 453–61
4. Abushama, F. T. 1968. Rhythmic activity of the grasshopper *Poecilocerus hieroglyphicus. Entomol. Exp. Appl.* 11:341–47
5. Abushama, F. T. 1970. Loss of water from the grasshopper *Poecilocerus hieroglyphicus* (Klug), compared with the tree locust *Anacridium melanorhodon* (Walker). *Z. Angew. Entomol.* 66:160–67
6. Ahearn, G. A. 1970. Water balance in the whipscorpion, *Mastigoproctus*

giganteus (Lucas). *Comp. Biochem. Physiol.* 35:339–53
7. Ahearn, G. A. 1970. The control of water loss in desert tenebrionid beetles. *J. Exp. Biol.* 53:573–95
8. Ahearn, G. A. 1970. Changes in hemolymph properties accompanying heat death in the desert tenebrionid beetle *Centrioptera muricata. Comp. Biochem. Physiol.* 33:845–57
9. Ahearn, G. A. 1971. Ecological factors affecting population sampling of desert tenebrionid beetles. *Am. Midl. Natur.* 86:385–406
10. Ahearn, G. A., Hadley, N. F. 1969. The effects of temperature and humidity on water loss in two desert tenebrionid beetles *Eleodes armata* and *Cryptoglossa verrucosa. Comp. Biochem. Physiol.* 30:739–49
11. Barrington, E. J. W. 1967. *Invertebrate Structure and Function.* London: Nelson 549 pp.
12. Barton-Browne, L. B. 1964. Water regulation in insects. *Ann. Rev. Entomol.* 9:63–82
13. Beament, J. W. L. 1961. The water relations of insect cuticle. *Biol. Rev.* 36:281–320
14. Beament, J. W. L. 1964. The active transport and passive movement of water in insects. *Advan. Insect Physiol.* 2:67–129
15. Beament, J. W. L. 1965. The active transport of water: evidence models and mechanisms. *Symp. Soc. Exp. Biol.* 19:273–98
16. Beament, J. W. L., Noble-Nesbitt, J., Watson, J. A. L. 1964. The waterproofing mechanism of arthropods. 111. Cuticular permeability in the firebrat, *Thermobia domestica* (Packard). *J. Exp. Biol.* 41:323–30
17. Belozerov, V. N., Seravin, L. N. 1960. Water balance regulation in *Alectrobius tholozani* at different atmospheric humidity. *Med. Parazitol. Parazit. Bolez.* 3:308–13
18. Berridge, M. J. 1970. Osmoregulation in terrestrial arthropods. *Chemical Zoology,* ed. M. Florkin, B. T. Scheer, 5:287–319. New York: Academic. 460 pp.
19. Browning, T. O. 1969. Permeability to water of the shell of the egg of *Locusta migratoria migratorioides,* with observations in the egg of *Teleogryllus commodus. J. Exp. Biol.* 51:99–105
20. Bursell, E. 1964. Environmental aspects: temperature. *The Physiology of Insecta,* ed. M. Rockstein, 1:283–321. New York: Academic. 640 pp.
21. Bursell, E. 1964. Environmental aspects: humidity. See Ref. 20, pp. 323–61
22. Byzova, J. B. 1960. Secondary integument sclerotization in tenebrionids. *Zool. Zh.* 39:540–45
23. Camin, J. H. 1963. Relations between host-finding behaviour and life histories in ecto-parasitic Acarina. *Advan. Acarol.* 1:411–24
24. Carlisle, D. B. 1968. *Triops* (Entomostraca) eggs killed only by boiling. *Science* 161:279–80
25. Carlisle, D. B., Cloudsley-Thompson, J. L. 1968. Respiratory function and thermal acclimation in tropical invertebrates. *Nature* 218:684–85
26. Carlisle, D. B., Ellis, P. E., Betts, E. 1965. The influence of aromatic shrubs in sexual maturation in the desert locust *Schistocerca gregaria. J. Insect Physiol.* 11:1541–58
27. Chapman, R. F. 1962. The ecology and distribution of grasshoppers in Ghana. *Proc. Zool. Soc. London* 139:1–66
28. Chew, R. M. 1961. Ecology of the spiders of a desert community. *J. NY Entomol. Soc.* 69:5–41
29. Cloudsley-Thompson, J. L. 1961. Adaptive functions of circadian rhythms. *Cold Spring Harbor Symp. Quant. Biol.* 25:345–55
30. Cloudsley-Thompson, J. L. 1961. *Rhythmic Activity in Animal Physiology and Behaviour.* London: Academic. 236 pp.
31. Cloudsley-Thompson, J. L. 1961. Some aspects of the physiology and behaviour of *Galeodes arabs. Entomol. Exp. Appl.* 4:257–63
32. Cloudsley-Thompson, J. L. 1961. Observations on the biology of the 'camel-spider' *Galeodes arabs* C. L. Koch in the Sudan. *Entomol. Mon. Mag.* 97: 145–52
33. Cloudsley-Thompson, J. L. 1961. Observations on the biology of the scorpion *Leiurus quinquestriatus* (H. & E.) in the Sudan. *Entomol. Mon. Mag.* 97:153–55
34. Cloudsley-Thompson, J. L. 1962. Microclimates and the distribution of terrestrial arthropods. *Ann. Rev. Entomol.* 7:199–222
35. Cloudsley-Thompson, J. L. 1962. Bioclimatic observations in the Red Sea hills and coastal plain, a major habitat of the desert locust. *Proc. Roy. Entomol. Soc. London* (A)37:27–34
36. Cloudsley-Thompson, J. L. 1962. Some aspects of the physiology and behaviour of *Dinothrombium. Entomol. Exp. Appl.* 5:69–73

37. Cloudsley-Thompson, J. L. 1962. Lethal temperatures of some desert arthropods and the mechanism of heat death. *Entomol. Exp. Appl.* 5:270–80
38. Cloudsley-Thompson, J. L. 1963. Light responses and diurnal rhythms in desert Tenebrionidae. *Entomol. Exp. Appl.* 6:75–78
39. Cloudsley-Thompson, J. L. 1963. Some aspects of the physiology of *Buthotus minax* with remarks on other African scorpions. *Entomol. Mon. Mag.* 98: 243–46
40. Cloudsley-Thompson, J. L. 1964. Terrestrial animals in dry desert heat: arthropods. *Adapt. Environ.* 4:451–65
41. Cloudsley-Thompson, J. L. 1965. On the function of the sub-elytral cavity in desert Tenebrionidae (Col). *Entomol. Mon. Mag.* 100:148–51
42. Cloudsley-Thompson, J. L. 1965. The scorpion. *Sci. J.* 1(5):35–41
43. Cloudsley-Thompson, J. L. 1967. The water-relations of scorpions and tarantulas from the Sonoran desert. *Entomol. Mon. Mag.* 103:217–20
44. Cloudsley-Thompson, J. L. 1968. The Markhiyat jebels: a desert community. *Desert Biology,* ed. G. W. Brown Jr, 1:1–20. New York: Academic. 635 pp.
45. Cloudsley-Thompson, J. L. 1969. *The Zoology of Tropical Africa.* London: Weidenfeld & Nicolson, 355 pp.
46. Cloudsley-Thompson, J. L. 1969. Acclimation, water and temperature relations of the woodlice *Metoponorthus pruinosus* and *Periscyphis jannonei* in the Sudan. *J. Zool.* 158:267–76
47. Cloudsley-Thompson, J. L. 1970. Terrestrial invertebrates. *Comparative Physiology of Thermoregulation,* ed. G. C. Whittow, 1:15–77. New York: Academic. 333 pp.
48. Cloudsley-Thompson, J. L. 1970. Recent work on the adaptive functions of circadian and seasonal rhythms in animals. *J. Interdiscipl. Cycle Res.* 1:5–19
49. Cloudsley-Thompson, J. L. 1973. Factors influencing the supercooling of tropical arthropods, especially locusts. *J. Natur. Hist.* 7:471–80
50. Cloudsley-Thompson, J. L. 1973. Entrainment of the "circadian clock" in *Buthotus minax. J. Interdiscipl. Cycle Res.* 4:119–23
51. Cloudsley-Thompson, J. L., Chadwick, M. J. 1964. *Life in Deserts.* London: Foulis. 218 pp.
52. Cloudsley-Thompson, J. L., Crawford, C. S. 1970. Lethal temperatures of some arthropods of the southwestern United States. *Entomol. Mon. Mag.* 106:26–29
53. Cloudsley-Thompson, J. L., Crawford, C. S. 1970. Water and temperature relations, and diurnal rhythms of scolopendomorph centipedes. *Entomol. Exp. Appl.* 13:187–93
54. Cloudsley-Thompson, J. L., Idris, B. E. M. 1964. The insect fauna of the desert near Khartoum: seasonal fluctuation and the effect of grazing. *Proc. Roy. Entomol. Soc. London* (A)39:41–46
55. Crawford, C. S. 1972. Water relations in a desert millipede *Orthoporus ornatus* (Girard). *Comp. Biochem. Physiol.* 42A:521–35
56. Crawford, C. S., Cloudsley-Thompson, J. L. 1971. Water relations and desiccation-avoiding behaviour in the vinegaroon *Mastigoproctus giganteus. Entomol. Exp. Appl.* 14:99–106
57. Crawford, C. S., Wooten, R. C. Jr. 1973. Water relations in *Diplocentrus spitzeri* a semimontane scorpion from the southwestern United States. *Physiol. Zool.* 46:218–29
58. Délye, G. 1969. *Recherches sur l'écologie, la physiologie et l'éthologie des fourmis du Sahara.* Thèses: Université d'Aix-Marseille. 155 pp.
59. Dresco-Derouet, L. 1961. La métabolisme respiratoire des scorpions. 1. Existence d'un rythme nycthéméral de la consommation d'oxygène. *Bull. Mus. Natur. Hist. Nat.* 32:553–57
60. Duelli, P. 1972. The relation of astromenotactic and anemomenotactic orientation mechanisms in desert ants, *Cataglyphis bicolor. Information Processing in the Visual Systems of Arthropods,* ed. R. Wehner 281–86. Berlin: Springer. 334 pp.
61. Ebling, W. 1964. The permeability of insect cuticle. *The Physiology of Insecta,* ed. M. Rockstein, 3:507–56. New York: Academic. 692 pp.
62. Edney, E. B. 1964. Acclimation to temperature in terrestrial isopods. I. Lethal temperature. *Physiol. Zool.* 37:364–77
63. Edney, E. B. 1964. Acclimation to temperature in terrestrial isopods. II. Heart rate and standard metabolic rate. *Physiol. Zool.* 37:378–94
64. Edney, E. B. 1966. Absorption of water vapour from unsaturated air by *Arenivaga* sp. *Comp. Biochem. Physiol.* 19:387–408
65. Edney, E. B. 1967. Water balance in desert arthropods. *Science* 156:1059–66
66. Edney, E. B. 1968. The effect of water loss on the haemolymph of *Arenivaga* sp. and *Periplanata americana. Comp. Biochem. Physiol.* 25:149–58

67. Edney, E. B. 1968. Transition from water to land in isopod crustaceans *Am. Zool.* 8:309–26
68. Edney, E. B. 1971. Some aspects of water balance in tenebrionid beetles and a thysanuran from the Namib Desert of southern Africa. *Physiol. Zool.* 44: 61–76
69. Edney, E. B. 1971. The body temperature of tenebrionid beetles in the Namib desert of southern Africa. *J. Exp. Biol.* 55:253–72
70. Edney, E. B. 1974. Desert arthropods. *Desert Biology,* ed. G. W. Brown Jr, 2:311–83. New York: Academic. 601 pp.
71. Ellis, P. E., Carlisle, D. B. 1965. Desert locusts: sexual maturation delayed by feeding on senescent vegetation. *Science* 149:546–47
72. El Rayah, E. A. 1970. Some reactions of two desert beetles *Adesmia antiqua* and *Pimelia grandis* to temperature. *Entomol. Exp. Appl.* 13:286–92
73. El Rayah, E. A. 1970. Humidity responses of two desert beetles, *Adesmia antiqua* and *Pimelia grandis. Entomol. Exp. Appl.* 13:438–47
74. El Rayah, E. A. 1973. On the gravity reactions and sexual behaviour of *Epicauta aethiops. Rev. Zool. Bot. Afr.* 87:402–12
75. Everett, D. H., Haynes, J. M., McElroy, P. J. 1971. The story of anomalous water. *Sci. Progr., Oxford* 59:279–308
76. Fiori, G., Mellini, E., Crovetti, A. 1966. Brevi considerazioni sulle orme lasciate sulla sabbia da alcuni insetti subdeserticoli e deserticoli. *Studi Sassaresi* (3) 14:1–23
77. Ghilarov, M. S. 1964. The main directions in insect adaptation to the life in the desert. *Zool. Zh.* 43:443–54
78. Hackman, R. H. 1971. The integument of Arthropoda. *Chemical Zoology,* ed. M. Florkin, B. T. Scheer, 6:1–62. New York: Academic. 484 pp.
79. Hadley, N. F. 1970. Micrometeorology and energy exchange in two desert arthropods. *Ecology* 51:434–44
80. Hadley, N. F. 1970. Water relations of the desert scorpion, *Hadrurus arizonensis. J. Exp. Biol.* 53:547–58
81. Hadley, N. F. 1971. Water uptake by drinking in the scorpion, *Centruroides sculpturatus. Southwest. Natur.* 15:495–505
82. Hadley, N. F. 1972. Desert species and adaptation. *Am. Sci.* 60:338–47
83. Hadley, N. F., Hill, R. D. 1969. Oxygen consumption of the scorpion *Centruroides sculpturatus. Comp. Biochem. Physiol.* 29:217–26
84. Hafez, M., El-Ziady, S., Hefnawy, T. 1970. Biochemical and physiological studies of certain ticks. Cuticular permeability of *Hyalomma (H.) dromedarii* Koch (Ixodidae) and *Ornithodorus (O.) savignyi* (Audouin). *J. Parasitol.* 56:154–68
85. Hafez, M., El-Ziady, S., Hefnawy, T. 1970. Biochemical and physiological studies of certain ticks. Uptake of water vapour by the different developmental stages of *Hyalomma (H.) dromedarii* Koch and *Ornithodoros (O.) savignyi* (Audouin). *J. Parasitol.* 56:354–61
86. Hafez, M., Ibrahim, M. M. 1963. Field and laboratory studies on the behaviour of the grasshopper *Aiolopus thalassinus* F. toward humidity. *Bull. Soc. Entomol. Egypte* 47:75–96
87. Hafez, M., Ibrahim, M. M. 1963. The temperature reactions of *Aiolopus thalassinus* F. *Bull Soc. Entomol. Egypte* 47:105–16
88. Hafez, M., Ibrahim, M. M. 1964. On the ecology and biology of the desert grasshopper *Sphingonotus carinatus* Sauss., in Egypt. *Bull. Soc. Entomol. Egypte* 48:193–217
89. Hafez, M., Ibrahim, M. M. 1964. Studies on the behaviour of the desert grasshopper, *Sphingonotus carinatus* Sauss., toward humidity and temperature. *Bull. Soc. Entomol. Egypte* 48:229–43
90. Hafez, M., Ibrahim, M. M. 1964. The possible receptors of humidity and temperature in two Egyptian grasshoppers *Aiolopus thalassinus* F. and *Sphingonotus carinatus* Sauss. *Bull. Soc. Entomol. Egypte* 48:245–57
91. Hamilton, A. G. 1964. Occurrence of periodic or continuous discharge of carbon dioxide by male locusts (*Schistocerca gregaria* Forskål). *Proc. Roy. Soc. Ser. B* 160:373–95
92. Hamilton, W. J. III 1971. Competition and thermoregulatory behaviour of the Namib desert tenebrionid beetle genus *Cardiosis. Ecology* 52:810–22
93. Hamilton, W. J. III 1973. *Life's Color Code.* New York: McGraw-Hill. 238 pp.
94. Hawke, S. D., Farley, R. D. 1973. Ecology and behavior of the desert cockroach, *Arenivaga* sp. *Oecologia* 11:263–79
95. Heath, J. E., Hanegan, J. L., Wilkin, P. J., Heath, M. S. 1971. Adaptation and thermal responses of insects. *Am. Zool.* 11:147–58

96. Heath, J. E., Wilkin, P. J. 1970. Temperature responses of the desert cicada *Diceroprocta apache*. *Physiol. Zool.* 43:145–54
97. Heeg, J. 1967. Studies on Thysanura. I. The water economy of *Machiloides delanyi* Wygodzinsky and *Ctenolepisma longicaudata* Escherich. *Zool. Afr.* 3:21–41
98. Heeg, J. 1967. Studies on Thysanura. II. Orientation reactions of *Machiloides delanyi* Wygodzinsky and *Ctenolepisma longicaudata* Escherich to temperature, light and atmospheric humidity. *Zool. Afr.* 3:43–57
99. Heeg, J. 1969. Studies on Thysanura. III. Some factors affecting the distribution of South African Thysanura. *Zool. Afr.* 4:135–43
100. Hefnawy, T. 1970. Biochemical and physiological studies of certain ticks (Ixodidea). Water loss from the spiracles of *Hyalomma (H.) dromedarii* Koch and *Ornithodoros (O.) savignyi* (Audouin). *J. Parasitol.* 56:362–66
101. Heinrich, B. 1971. Temperature regulation of the sphinx moth, *Manduca sexta.* I. Flight energetics and body temperature during free and tethered flight. II. Regulation of heat loss by control of blood circulation. *J. Exp. Biol.* 54:153–66
102. Herreid, C. F. 1969. Water loss of crabs from different habitats. *Comp. Biochem. Physiol.* 28:829–35
103. Hinton, H. E. 1960. Cryptobiosis in the larva of *Polypedilum vanderplanki* Hint. *J. Insect Physiol.* 5:286–300
104. Hinton, H. E. 1967. The structure of the spiracles of the cattle tick *Boophilus microplus. Aust. J. Zool.* 15:941–45
105. Hinton, H. E. 1968. Reversible suspension of metabolism and the origin of life. *Proc. Roy. Soc. Ser. B* 171:43–57
106. Hinton, H. E. 1971. Reversible suspension of metabolism *C. R. Conf. Int. Phys. Théor. Biol., 2e, Versailles, 1969,* 69–89
107. Holm, E., Edney, E. B. 1973. Daily activity of Namib desert arthropods in relation to climate. *Ecology* 54:45–56
108. Hunter-Jones, P. 1964. Egg development in the desert locust (*Schistocera gregaria* Forsk.) in relation to the availability of water. *Proc. Roy. Entomol. Soc. London* (A)39:25–33
109. Hunter-Jones, P., Lambert, J. G. 1961. Egg development of *Humbe tenuicornis* Schaum in relation to availability of water. *Proc. Roy. Entomol. Soc. London* (A)36:75–80
110. Janzen, D. H., Schoener, T. W. 1968. Differences in insect abundance and diversity between wetter and drier sites during a tropical dry season. *Ecology* 49:96–110
111. Jolivet, P. 1965. Notes sur l'écologie des *Timarcha* marocaines. *Bull. Soc. Sci. Natur. Phys. Maroc.* 1965:159–90
112. Kanungo, K. 1965. Oxygen uptake in relation to water balance of a mite (*Echinolaelaps echidninus*) in unsaturated air. *J. Insect Physiol.* 11:557–68
113. Kanwisher, J. W. 1966. Tracheal gas dynamics in pupae of the cecropia silkworm. *Biol. Bull.* 130:96–105
114. Kaufmann, T. 1966. Observations on some factors which influence aggregated by *Blaps sulcata* in Israel. *Ann. Entomol. Soc. Am.* 59:660–64
115. Knülle, W. 1962. Die Abhängigkeit der Luftfeuchte—Reaktionen der Mehlmilbe (*Acarus siro* L.) vom Wassergehalt des Körpers. *Z. Vergl. Physiol.* 45:233–46
116. Knülle, W. 1965. Die Sorption und Transpiration des Wasserdampfes bei der Mehlmilbe (*Acarus siro* L.) *Z. Vergl. Physiol.* 49:586–604
117. Knülle, W. 1966. Equilibrium humidities and survival of some tick larvae. *J. Med. Entomol.* 2:335–38
118. Knülle, W. 1967. Significance of fluctuating humidities and frequency of blood meals on the survival of the spiny rat mite, *Echinolaelaps echidninus* (Berlese) *J. Med. Entomol.* 4:322–25
119. Knülle, W. 1967. Physiological properties and biological implications of the water vapour sorption mechanisms, in larvae of the oriental rat flea, *Xenopsylla cheopis* (Roths.). *J. Insect Physiol.* 13:333–57
120. Knülle, W., Devine, T. 1972. Evidence for active and passive components of sorption of atmospheric water vapour by larvae of the tick *Dermacentor varibilis. J. Insect Physiol.* 18:1653–64
121. Knülle, W., Spadafora, R. R. 1969. Water vapour sorption and humidity relationships in *Liposcelis. J. Stored Prod. Res.* 5:49–55
122. Koch, C. 1961. Some aspects of abundant life in the vegetationless sand of the Namib desert dunes. Positive psammotropism in Tenebrionid beetles. *J. Southwest. Afr. Sci. Soc.* 15:8–34, 76–92
123. Koch, C. 1963. The Tenebrionidae of southern Africa. xxix. *Luebbertia plana* gen. et spec. nov., with a dichotomic analysis of Stizopina. *Sci. Pap. Namib Desert Res. Sta.* No. 18:1–87

124. Kühnelt, G. 1969. On the biology and temperature accommodation of *Lepidochora argentogrisea* Koch. *Sci. Pap. Namib Desert. Res. Sta.* No. 51:121–28
125. Leader, J. P. 1962. Tolerance to freezing of hydrated and partially hydrated larvae of *Polypedilum. J. Insect Physiol.* 8:155–63
126. Lees, A. D. 1964. The effect of ageing and locomotor activity on the water transport mechanism of ticks. (*Proc. 1st Int. Congr. Acarology*), *Acarologia,* 6:315–23
127. Lewis, J. G. E. 1963. On the spiracle structure and resistance to desiccation of four species of geophilomorph centipede. *Entomol. Exp. Appl.* 6:89–94
128. Lewis, J. G. E. 1972. The life histories and distribution of the centipedes *Rhysida nuda togoensis* and *Ethmostigmus trigonopodus* in Nigeria. *J. Zool.* 167:399–414
129. Louw, G. N., Hamilton, W. J. III. 1972. Physiological and behavioral ecology of the ultra-psammophilous Namib Desert tenebrionid, *Lepidochora argentogrisea. Madoqua* (2)1(54–62):87–95
130. Loveridge, J. P. 1968. The control of water loss in *Locusta migratoria migratoriodes* R. & F. I. Cuticular water loss. *J. Exp. Biol.* 49:1–13
131. Loveridge, J. P. 1968. The control of water loss in *Locusta migratoria migratorioides* R. & F. II. Water loss through the spiracles *J. Exp. Biol.* 49:15–29
132. Manton, S. M. 1964. Mandibular mechanisms and the evolution of arthropods. *Phil. Trans. Roy. Soc. London Ser. B.* 247:1–183
133. Manton, S. M. 1973. Arthropod phylogeny—a modern synthesis. *J. Zool.* 171:111–30
134. Miller, P. L. 1960. Respiration in the desert locust. I. The control of ventilation. *J. Exp. Biol.* 37:224–36
135. Miller, P. L. 1960. Respiration in the desert locust. II. The control of the spiracles. *J. Exp. Biol.* 37:237–63
136. Miller, P. L. 1960. Respiration in the desert locust. III. Ventilation and the spiracles during flight. *J. Exp. Biol.* 37:264–78
137. Miller, P. L. 1970. On the occurrence and some characteristics of *Cyrtopus fastuosus* Bigot and *Polypedilum* sp. from temporary habitats in Western Nigeria. *Entomol. Mon. Mag.* 105:233–38
138. Minton, S. A. Jr. 1968. Venoms of desert animals. *Desert Biology,* ed. G. W. Brown Jr., 1:487–516. New York: Academic. 635 pp.
139. Moeur, J. E., Eriksen, C. H. 1972. Metabolic responses to temperature of a desert spider, *Lycosa (Pardosa) carolinensis. Physiol. Zool.* 45:290–301
140. Newell, I. M., Tevis, L. Jr. 1960. *Angelothrombium pandorae* n.g. n. sp., and notes on the biology of the giant red velvet mites. *Ann. Entomol. Soc. Am.* 53:293–304
141. Noble-Nesbitt, J. 1969. Water balance in the firebrat, *Thermobia domestica* (Packard). Exchange of water with the atmosphere. *J. Exp. Biol.* 50:745–69
142. Noble-Nesbitt, J. 1970. Water balance in the firebrat *Thermobia domestica* (Packard). The site of uptake of water from the atmosphere. *J. Exp. Biol.* 52:193–200
143. Norris, M. J. 1964. Environmental control of sexual maturation in insects. *Symp. Roy. Entomol. Soc. London* 2:56–65
144. Okasha, A. Y. K. 1971. Water relations in an insect, *Thermobia domestica.* I. Water uptake from sub-saturated atmosphere as a means of volume regulation. *J. Exp. Biol.* 55:435–48
145. Okasha, A. Y. K. 1973. Water relations in an insect, *Thermobia domestica.* II. Relationships between water content, water uptake from sub-saturated atmospheres and water loss. *J. Exp. Biol.* 57:285–96
146. Okasha, A. Y. K. 1973. Water relations of an insect, *Thermobia domestica.* III. Effects of desiccation and rehydration on the haemolymph. *J. Exp. Biol.* 58:385–400
147. Omer, S. M., Cloudsley-Thompson, J. L. 1968. Dry season biology of *Anopheles gambiae* Giles in the Sudan. *Nature* 217:879–80
148. Omer, S. M., Cloudsley-Thompson, J. L. 1970. Survival of female *Anopheles gambiae* Giles through a 9-month dry season in Sudan. *Bull. WHO* 42:319–30
149. Parker, A. H. 1972. Studies on the diurnal rhythms of the housefly, *Musca domestica* L., in a dry tropical environment. *Acta Trop.* 19:97–119
150. Phillips, J. E. 1964. Rectal absorption in the desert locust *Schistocerca gregaria* Forskal, I. Water. II. Sodium, potassium and chloride. III. The nature of the excretory process. *J. Exp. Biol.* 41:15–80
151. Rafes, P. M. 1960. The life forms of insects inhabiting the Naryn Sands of the semidesert Transvolga region. *Entomol. Obozr.* 38:19–31
152. Rainey, R. C. 1962. Some effects of environmental factors on movements and

phase-change of locust populations in the field. *Colloq. Int. Cent. Nat. Rech. Sci.* No. 114:175–99

153. Rajulu, G. S., Krishanan, G. 1968. The epicuticle of millipedes belonging to the genera *Cingabololus* and *Aulacobolus* with special reference to seasonal variations. *Z. Naturforsch. B* 23:845–51
154. Ramsay, J. A. 1964. The rectal complex of the mealworm *Tenebrio molitor* L. *Phil. Trans. Roy. Soc. London Ser. B* 248:279–314
155. Ritcher, P. O. 1969. Spiracles of adult Scarabaeoidea and their phylogenetic significance. 1. The abdominal spiracles. *Ann. Entomol. Soc. Am.* 62:869–80
156. Rivnay, E. 1964. The influence of men on insect ecology in arid zones. *Ann. Rev. Entomol.* 9:41–62
157. Rivnay, E., Ziv, M. 1963. A contribution to the biology of *Gryllus bimaculatus* Deg. in Israel. *Bull. Entomol. Res.* 54:37–43
158. Roffey, J., Popov, G. 1968. Environment and behavioral process in a desert locust outbreak. *Nature* 219:446–50
159. Roshdy, M. A., Hefnawy, T. 1973. The functional morphology of *Haemaphysalis* spiracles. *Parasitenk.* 42:1–10
160. Rosin, R., Shulov, A. 1963. Studies on the scorpion *Nebo hierochonticus. Proc. Zool. Soc. London* 140:547–75
161. Schultze, L. 1969. The Tenebrionidae of southern Africa Part xlii: Description of the early stages of *Carchares macer* Pascoe and *Herpiscius sommeri* Solier with a discussion of some phylogenetic aspects arising from the incongruities of adult and larval systematics. *Sci. Pap. Namib Desert Res. Sta.* No. 53:139–49
162. Seymour, R. S., Vinegar, A. 1973. Thermal relations, water loss and oxygen consumption of a North American tarantula. *Comp. Biochem. Physiol.* 44A: 83–96
163. Shaw, J., Stobbart, R. H. 1963. Osmotic and ionic regulation in insects. *Advan. Insect Physiol.* 1:315–99
164. Shaw, J., Stobbart, R. H. 1972. The water balance and osmoregulatory physiology of the desert locust (*Schistocerca gregaria*) and other desert and xeric arthropods. *Symp. Zool. Soc. London* No. 31:15–38
165. Shulov, A. S. 1970. The development of eggs of the red locust *Nomadacris septemfasciata* (Serv.), and African migratory locust *Locusta migratoria migratorioides* (R. and F.), and its interruption under particular conditions of humidity. *Anti-Locust Bull.* No. 48: 1–22
166. Shulov, A., Pener, M. P. 1963. Studies on the development of eggs of the desert locust (*Schistocerca gregaria* Forskål) and its interruption under particular conditions of humidity. *Anti-Locust Bull.* No. 41:1–59
167. Solomon, M. E. 1966. Moisture gains, losses and equilibria of flour mites, *Acarus siro* L., in comparison with larger arthropods. *Entomol. Exp. Appl.* 9:25–41
168. Stobbart, R. H. 1968. Ion movements and water transport in the rectum of the locust *Schistocerca gregaria. J. Insect Physiol.* 14:269–75
169. Stower, W. J., Griffiths, J. E. 1966. The body temperature of the desert locust (*Schistocerca gregaria*). *Entomol. Exp. Appl.* 9:127–78
170. Tatchell, R. J. 1967. Salivary secretion in the cattle tick as a means of water elimination. *Nature* 213:940–41
171. Tevis, L. Jr., Newell, I. M. 1962. Studies on the biology and seasonal cycle of the giant red velvet mite *Dinothrombium pandorae. Ecology* 43:497–505
172. Toye, S. A. 1966. Studies on the locomotory activity of three species of Nigerian millipedes: *Spirostreptus assiniensis, Oxydesmus* sp., and *Habrodesmus falx. Entomol. Exp. Appl.* 9: 369–77
173. Toye, S. A. 1966. The effect of desiccation on the behaviour of three species of Nigerian millipedes: *Spirostreptus assiniensis, Oxydesmus* sp., and *Habrodesmus falx. Entomol. Exp. Appl.* 9: 378–84
174. Toye, S. A. 1966. The reactions of three species of Nigerian millipedes (*Spirostreptus assiniensis, Oxydesmus* sp. and *Habrodesmus falx*) to light, humidity and temperature. *Entomol. Exp. Appl.* 9:468–84
175. Toye, S. A. 1967. Observations on the biology of three species of Nigerian millipedes. *J. Zool.* 152:67–78
176. Tshernyskev, W. B., Bogush, P. P. 1973. Influence of weather on flight of insects toward light in Central Asia. *Zool. Zh.* 52:700–8
177. Vandel, A. 1964. De l'emploi des appareils respiratoires pour l'établissement d'une classification rationelle des isopodes terrestres (Oniscoidea). *Bull. Soc. Zool. Fr.* 89:730–36
178. Waloff, Z. 1962. Flight activity of different phases of the desert locust in relation to plague dynamics. *Colloq. Int. Cent. Nat. Rech. Sci.* No. 114:201–16
179. Waloff, Z. 1972. Orientation of flying locusts, *Schistocerca gregaria* (Forsk.),

in migrating swarms. *Bull. Entomol. Res.* 62:1–72

180. Warburg, M. R. 1964. The response of isopods towards temperature, humidity and light. *Anim. Behav.* 12:175–86
181. Warburg, M. R. 1965. The microclimate in the habitats of two isopod species in southern Arizona. *Am. Midl. Natur.* 73:363–75
182. Warburg, M. R. 1965. Water relation and internal body temperature of isopods from mesic and xeric habitats. *Physiol. Zool.* 38:99–109
183. Warburg, M. R. 1965. The evaporative water loss of three isopods from semiarid habitats in South Australia. *Crustaceana* 9:302–8
184. Warburg, M. R. 1968. Simultaneous measurement of body temperature and water loss in isopods. *Crustaceana* 14:39–44
185. Warburg, M. R. 1968. Behavioural adaptations of terrestrial isopods. *Am. Zool.* 8:545–59
186. Wehner, R., Flatt, I. 1972. The visual orientation of desert ants, *Cataglyphis bicolor,* by means of terrestrial clues. *Information Processing in the Visual Systems of Arthropods,* ed. R. Wehner, 295–302. Berlin: Springer. 334 pp.
187. Weis-Fogh, T. 1967. Respiration and tracheal ventilation in locusts and other flying insects. *J. Exp. Biol.* 47:561–87
188. Williams, S. C. 1966. Burrowing activities of the scorpion *Anuroctonus phaeodactylus* (Wood). *Proc. Calif. Acad. Sci.* (4)34:419–28
189. Winston, P. W. 1967. New evidence for a cuticular water pump in insects. *Nature* 214:383–84
190. Winston, P. W., Beament, J. W. L. 1969. An active reduction of water level in insect cuticle. *J. Exp. Biol.* 50:541–46
191. Woodrow, D. F. 1965. Observations on the red locust (*Nomadacris septemfasciata* Serv.) in the Rukwa Valley, Tanganyika, during its breeding season. *J. Anim. Ecol.* 34:187–200
192. Yunker, C. E., Guirgis, S. S. 1969. Studies of rodent burrows and their ectoparasites in the Egyptian desert. I. Environment and microenvironment; some factors influencing acarine distribution. *J. Egypt. Pub. Health Assoc.* 44:498–542

RESPONSES OF ARTHROPOD NATURAL ENEMIES TO INSECTICIDES[1]

❖6092

B. A. Croft and A. W. A. Brown
Pesticide Research Center and Department of Entomology, Michigan State University, East Lansing, Michigan 48824

During the past quarter of a century, the study of insecticides and their side effects has evolved into a highly technical and precise science. The knowledge obtained is highly relevant to the emphasis now laid by applied entomologists on pest management and the worldwide efforts to use insecticides more judiciously or to supplant them where possible with less disruptive control measures. Implicit in the pest management concept is the maximum use of natural enemies of arthropod pests (diseases, predators, parasites), supplemented with selective insecticides when necessary. The present emphasis on integration of chemical and biological methods reflects the consensus that insecticides will continue as part of most pest management programs until alternative methods sufficient to achieve the desired control are found.

The responses of arthropod natural enemies to the insecticides employed against the pests which they attack were first observed qualitatively in the field, rather than measured quantitatively. The most apparent one, natural enemy mortality, was often associated with pest outbreaks, which many authors from the late 1930s to the 1950s (165, 171, 185, 186, 219) attributed to the disruptive effects of insecticides on the balance between pest populations and those of their natural enemies. These conclusions led to the concept of integrated control or pest management, and prompted studies of insecticidal selectivity vis-à-vis natural enemies. The study of natural enemy responses to insecticides, however, has not been comparable in volume to the detailed toxicological investigations, in the laboratory as well as the field, which have been directed to the responses of pests and later to the development of resistance. Contributing to this differential emphasis and effort have been: 1. the preferential attention given to control of direct competitors rather than to conservation of benefactors that were sometimes recognized, 2. the assumption that natural

[1]Survey of the literature pertaining to this review was concluded in December 1973.

enemies respond to insecticides in the same way as pests and thus their study was unnecessary, 3. the disproportionately greater monetary resources made available for studying the responses of pests as compared to those of predators and parasites, 4. the greater difficulty of rearing or culturing predators and parasites in the large numbers required for detailed experimentation, and 5. the lack of standardized toxicological test methods for natural enemies similar to those developed for pest evaluations.

In recent years, the need for more study of the specific responses of predators and parasites to insecticides has been emphasized in reviews by Croft (50), Georghiou (85), and Newsom (164). Each author addressed himself to the major question—do pests and natural enemies respond similarly? Although there is abundant evidence that both target pests and nontarget natural enemy species are readily killed by insecticides, the question remains whether there may not be essential differences between the responses of these two groups at the biochemical, organismal, or population levels. Applied pest control literature, for example, contains implications that natural enemies characteristically are susceptible to broad-spectrum insecticides while it is only the pests which develop resistances. On the other hand, Newsom (163, 164) and Ferguson (77) point out that the existing experimental evidence does not justify the assumption of any physiological differences between the responses of pests and of natural enemies to insecticides. The main reason for questioning whether these two groups respond differently is the fact that more than 224 cases of insecticide resistance are known for pests (34), but less than 10 have been reported for natural enemies (50, 85).

The present review concentrates on collecting the available data on the direct toxicity of insecticides to natural enemies in terms of LD_{50} and other log-dosage mortality expressions, and on the indirect effects arising from secondary poisoning through the prey or pest species. After reviewing in detail the nine known instances of developed resistance in arthropod natural enemies, it considers the possible reasons for their paucity as compared to the situation in pest species. Finally, using tetranychid mites and their phytoseiid predators as an example, it discusses how a full understanding of the responses of arthropod enemies to insecticides may be turned to advantage in pest management.

SUSCEPTIBILITY TO DIRECT TOXIC DOSES

The mortality caused directly by contact of a natural enemy with a toxicant has been abundantly documented in terms of the reductions in their numbers, or in the degree of parasitism or predation which followed insecticide applications in the field. Such data are helpful in evaluating the general effects of an insecticide, but it is usually impossible to distinguish between the direct toxic effects and indirect effects caused by destruction of the prey, hosts, competitors, or alternate food sources of the natural enemy population.

A more precise and quantitative assessment of direct acute toxicity is the dosage-mortality relationship. The literature surveyed for this review yielded toxicity assessments of this exact type in only 6 papers published before 1958, 17 between 1958 and 1966, and 42 between 1967 and 1973. Most of these determinations were for natural enemies of pests of heavily treated crops such as cotton, deciduous fruits, and citrus. More were for predators than for parasites, probably reflecting the greater ease with which predators could be identified and collected in sufficient quantity for test evaluation.

Figures for median lethal doses, deposits, or concentrations (LC_{50} or LD_{50}) from 20 investigations with 10 species of coccinellids (Table 1) show an order of toxicity for the common insecticides similar to that which had been deduced without exact LD_{50} or LC_{50} figures (20, 46) for yet 5 other lady-bird species (*Cryptolaemus montrouzieri, Stethorus picipes, Rodolia cardinalis, Lindorus lophanthae,* and *Stethorus punctum*). Taking the results as a whole, the insecticides arrange themselves into 5 classes of descending toxicity, the 5 insecticides in each class also being written in the order of descending toxicity: 1. parathion, parathionmethyl, malathion, azinphosmethyl, carbaryl; 2. mevinphos, phosphamidon, diazinon, dimethoate, ethion; 3. demeton, demetonmethyl, carbophenothion, trichlorfon, thiometon; 4. lindane, toxaphene, endrin, DDT, endosulfan; 5. chlorobenzilate, schradan, binapacryl, tetradifon, dicofol. The figures obtained for LD_{50} or LC_{50} levels for 4 other coleopteran species, 5 hemipterans, 1 chrysopid, and 3 acarines (Table 2) showed a roughly similar ranking of insecticides for their lethality to these natural enemies.

With respect to parasites, the results obtained for adults of 17 species of Hymenoptera and 1 tachinid (Table 3) show the organophosphorus (OP) compounds to have approximately the same ranking of toxicity as for predators, but they reveal that the parasites differ from predators, being especially susceptible to the cyclodiene compounds and lindane. The order of descending toxicity of the common insecticides to parasites is as follows: 1. endrin, dieldrin, aldrin, lindane, DDT; 2. parathion, parathionmethyl, malathion, carbaryl, azinphosmethyl, toxaphene, phosphamidon; 3. carbophenothion, trichlorfon, demetonmethyl, endosulfan, thiometon.

Comparisons Between Predators, Parasites, and Pest Species

From the literature, 13 instances can be found of predator-prey complexes where the LD_{50} or LC_{50} values have been determined for predators and prey by the same methods (Tables 4 and 5). When the figure for each predator species is compared with that for its prey, the predator shows the greater tolerance, provided the prey has not developed resistance. Of the 77 predator-prey comparisons that exclude resistant prey strains, there are 63 cases where the predator is more tolerant as against 14 where the prey is less susceptible, 59 to 5 where insects are the prey and 4 to 9 where mites are the prey. With coccinellid predators the score is 31 to 3 with insect prey and 2 to 6 with mite prey.

Although the data summarized in Tables 4 and 5 indicate that predators are generally more tolerant to insecticides than are their prey, it should be noted that as many as 34 of 77 of the predators represented in these comparisons were coccinellids, which are well known as being tolerant to a variety of insecticides (7, 10, 22, 117). Moreover, in 4 of the 13 predator-prey complexes assessed, the strains of prey species were of pristine susceptibility, whereas predators were taken from localitites which had been under insecticide pressure for some time or were populations actually suspected of having some tolerance. Furthermore, many of the compounds with which they were tested had been chosen as being likely to be less toxic to the predators than to their prey. Therefore the above toxicological assessments, which constitute only a minor portion of the published data on this topic, may give a distorted comparison between predator and prey susceptibility, and one would therefore be erroneous in generalizing these proportions to field populations.

Where parasites have been compared with their hosts for susceptibility (Table 6), it is found that the situation is the reverse of that found with predators, there being 11 out of 15 instances where the parasite is the more susceptible. If hyperparasites and dipterans are excluded, the score for hymenopterous primary parasites becomes 3–10 in favor of the more tolerant hosts, and if isopropyl-parathion is excluded it is only 1–10. Although the relative proportion of cases favoring a greater tolerance among predators than their prey are strikingly greater than those for parasite to host comparisons, this generalization should be further qualified. It is unlikely that LD_{50} and LC_{50} comparisons for randomly selected field populations of predators and their prey would be as excessively in favor of the natural enemy as indicated by the data presented in Tables 4 and 5. On the other hand, it is likely that a more representative survey would still show the same trend. Indeed, it has proved easier to find selective insecticides which favor predators in preference to their prey than to discover compounds which favor parasites in relation to their host.

To compare parasites with predators directly and irrespective of the pest, there are 77 cases where LC_{50} or LD_{50} values may be found in the data summarized in Tables 1–3 in which both groups are simultaneously assessed by the same methods. They show that for 66 cases in which the predator is more tolerant, there are 11 cases where the parasite is less susceptible. If the data in the 2 publications of Lingren and co-workers (135, 136) are included, the score becomes 90 to 11. The considerably larger volume of published work on the effect of single dose levels on either group, which is not cited in this review, gives substantially the same picture. That is, in most cases predators are more tolerant of insecticides than are the parasites which occur in the same habitat and attack the same pests.

Since one or more stages of a natural enemy must search out prey or hosts, it might be expected that predators and adult parasites would pick up greater amounts of toxicant and thus suffer greater mortality from residual deposits than would the more sedentary pests occupying a similar habitat. There is little evidence published to date indicating that the behavior of these natural enemies confers a greater exposure to insecticides. Where the biological activity of residues has been assessed over a period of time after application, the percentage mortalities of predators (17, 48, 57, 94, 153) and adult parasites (17, 39, 207, 208) often yield a decay curve which resembles closely that of the chemical deposit itself.

Insecticides may also exert a direct lethal effect on predators and adult parasites by acting as stomach poisons, being taken up in the nectar, pollen, honeydews, or even in free water, contaminated in the process of their application to the ecosystem. Bartlett (23) used honey baits to assess 61 common pesticides for their acceptance and toxicity to 2 representative hymenopterous parasites and 2 coccinellid predators. He found that most of them were gustatory repellents, but that despite this the

Table 1 Susceptibility of coccinellid predators to insecticides, ranked in ascending order of LD_{50} or LC_{50}

Species	Stage	Insecticides	Method (Ref.)
Adonia variegata	A[a]	malathion, trichlorfon, thiometon	D (245)
	A	dimethoate, demetonmethyl, vamidothion	D (83)
Coccinella 11-*punctata*	L, A	parathion, azinphosmethyl, carbaryl, trichlorfon, lindane, endrin, DDT, toxaphene	D (94)
Coccinella 7-*punctata*	L	malathion, parathion, diazinon, tepp, demeton	D (178)
	A	malathion, trichlorfon, thiometon	D (245)
	A	mevinphos, phorate, phosphamidon, carbaryl, carbophenothion, trichlorfon, chlorobenzilate	D (205)
	A	dimethoate, demetonmethyl, thiometon, vamidothion	D (83)
Coleomegilla maculata	A	endrin, toxaphene, DDT	T (13)
Cycloneda limbifer	A	malathion, fenitrothion, trichlorfon	R (245)
Hippodamia convergens	L	parathion, malathion, mevinphos, demeton, carbophenothion, schradan, nicotine	R (19)
	A	parathionmethyl, parathion, DDT, heptachlor, endrin, dieldrin	R (37)
	A	dicrotophos, parathionmethyl, phosphamidon, demeton, trichlorfon	T (135)
	A	parathion, malathion	R (93)
	A	azinphosmethyl, parathion, carbaryl, diazinon, DDT, endosulfan, binapacryl	T (153)
	A	parathionmethyl, carbaryl, malathion, toxaphene	R (243)
Hippodamia 5-*signata*	A	malathion, parathion, mevinphos, demeton, carbophenothion, schradan, nicotine	R (19)
Pharoscymnus numidicus	L, A	dimethoate, phosphamidon, malathion, ethion	R (117)
Stethorus pauperculus	A	mevinphos, malathion, parathion, azinphosmethyl, dimethoate, endosulfan, DDT, dicofol	D (113)
	A	parathion, phosphamidon, diazinon, chlorobenzilate	R (182)
Stethorus punctillum	A	parathion, diazinon, malathion, demeton, chlorobenzilate, tetradifon, dicofol	R (33)

[a]A = Adults, L = Larvae, D = Direct spray, R = Exposure to residues, and T = Topical application.

Table 2 Susceptibility of coleopteran, hemipteran, chrysopid, and acarine predators to insecticides, ranked in ascending order of LD_{50} or LC_{50}

Species	Stage	Insecticides	Method (Ref.)
Feronia melanaria (Carab.)	A[a]	azinphosmethyl, dieldrin, diazinon, thionazin	S (159)
Pseudophonus rufipes (Carab.)	A	parathion, dieldrin, carbaryl, DDT, toxaphene	R (236)
Pterostichus vulgaris (Carab.)	A	phorate, thionazin, disulfoton, menazon	S[c] (48)
Paederus alfierii (Staphylin.)	A	parathion, azinphosmethyl, lindane, endrin, dieldrin, DDT, toxaphene	D (94)
Nabis americoferus (Nab.)	A	parathionmethyl, dicrotophos, trichlorfon, phosphamidon, demeton	T (135)
	A	parathion, malathion	R (93)
Anthororis confusus[b] (Anthocor.)	N	dimethoate, phorate, menazon	D (72)
Geocoris punctipes (Anthocor.)	A	parathionmethyl, dicrotophos, phosphamidon, trichlorfon, demeton	T (135)
Orius insidiosus (Anthocor.)	A	parathion, parathionmethyl, malathion, aldrin, toxaphene, dieldrin, DDT	R (37)
	A	parathionmethyl, dicrotophos, trichlorfon, phosphamidon, demeton	R (135)
Chrysopa carnea (Chrysop.)	L	malathion, thiometon, trichlorfon	D (245)
	L	phosphamidon, dicrotophos, parathionmethyl, demeton, trichlorfon	T (135)
	L	parathionmethyl, carbaryl, toxaphene, malathion	R (243)
	A	parathionmethyl, carbaryl, malathion, toxaphene	R (243)
Typhlodromus occidentalis (Phytos.)	F	fundal, azinphosmethyl, gardona, carbaryl, plictran, dicofol, omite	D (54)
	(F)	fundal, plictran, carbaryl, gardona, dicofol, omite, azinphosmethyl	D (54)
	(F)	carbaryl, diazinon, gardona, phosalone, azinphosmethyl, tepp	R (58)
Amblyseius fallacis (Phytos.)	F	gardona, azinphosmethyl, imidan, parathion, plictran, omite	I (198)
	F	azinphosmethyl, parathion, carbaryl	R (157)
	(F)	carbaryl, phosalone, diazinon, gardona, tepp, azinphosmethyl, phosmet	R (58)
Macrocheles muscaedomesticae (Mac.)	F	naled, diazinon, fenthion, dimethoate, trichlorfon, dimetilan, lindane, DDT	S (14)

[a] A = Adults, L = Larvae, N = Nymphs, F = Females, (F) = Strain made resistant by azinphosmethyl pressure in field, D = Direct spray, R = Exposure to residues, T = Topical application, I = Immersion, and S = Mixed with soil or substrate.

[b] Same result for *A. nemorum* also.

[c] LT_{50} value.

Table 3 Susceptibility of adult parasites to insecticides, ranked in ascending order of LD_{50} or LC_{50}

Parasite	Insecticides	Method	Ref.
Voria ruralis (Tach.)	parathionmethyl, toxaphene, malathion, carbaryl	R[a]	(243)
Apanteles marginiventris (Bracon.)	parathionmethyl, monocrotophos, azinphosmethyl, trichlorfon, chlordimeform	T	(136)
Bracon mellitor (Bracon.)	carbaryl, DDT, parathionmethyl	T	(2)
Chelonus blackburni (Bracon.)	parathionmethyl, toxaphene, malathion, carbaryl	R	(243)
Lysiphlebus fabarum (Bracon.)	malathion, thiometon	R	(245)
Meteorus leviventris (Bracon.)	parathionmethyl, carbaryl, toxaphene, malathion	R	(243)
Microbracon brevicornis (Bracon.)	parathion, phosphamidon, malathion, carbophenothion, trichlorfon, carbaryl, endosulfan, DDT	R	(204)
Opius oophilus (Bracon.)	endrin, dieldrin, aldrin, lindane, parathion, chlordane, DDT	T	(224)
Opius persulcatus (Bracon.)	lindane, parathion, parathionmethyl, tepp, rotenone, pyrethrins, nicotine	D	(148)
Praon abjectum (Bracon.)	malathion, thiometon, trichlorfon	R	(245)
Campoletis sonorensis (Ichn.)	parathionmethyl, carbaryl, malathion, toxaphene	R	(243)
Campoletis perdistinctus (Ichn.)	parathionmethyl, parathion, malathion, azinphosmethyl, carbaryl, demeton, trichlorfon	T	(136)
	monocrotophos, aldicarb, disulfoton	T	(39)
Brachymeria intermedia (Chalc.)	parathionmethyl, carbaryl, malathion, toxaphene	R	(243)
Pauridia perégrina (Encyrt.)	mercaptothion, parathion, fenthion, carbaryl, trichlorfon, azinphosmethyl, DDT, dicofol	R	(208)
Aphytis holoxanthus (Aphelin.)	dimethoate, dioxathion, binapacryl, chlorobenzilate, dicofol, tetradifon	R[b]	(200)
Mormoniella vitripennis (Pterom.)	carbaryl, endosulfan, binapacryl, oxythioquinox, isolan, chlorbenside, dicofol	R	(8)
Trichospilus pupivora (Euloph.)	phosphamidon, dimethoate, parathion, carbaryl, malathion, endosulfan, trichlorfon	R[b]	(41)
Trichogramma evanescens (Trich.)	fenthion, carbaryl, demetonmethyl (to prepupae inside host egg)	D	(125)

[a] R = Exposure to residues, D = Direct spray, and T = Topical application.
[b] LT_{50} value.

Table 4 LC_{50} and LD_{50} levels for various complexes of pests and their predators

Complex tested	Ecosystem	Method Units	Pest Predator	Compound	LC_{50} or LD_{50} level:[a] *Pest*, Predator	Reference
1	Cotton	Direct spray	*Aphis gossypii* (A)[b]	Endosulfan	*0.011*, 0.274	(205)
		% concentration	*Coccinella 7-punctata* (A)	Trichlorfon	*0.088*, 0.016	
				Phorate	*0.00036*, 0.00098	
				Mevinphos	*0.00005*, 0.00053	
				Carbaryl	*0.0006*, 0.0037	
2	Cotton	Direct spray	*Aphis craccivora* (A)	Dimetonmethyl	*6.6*, 295.1, 83.2	(83)
		mg/1	*Coccinella 7-punctata* (A)	Dimethoate	*7.4*, 25.6, 7.9	
			Adonia variegata (A)	Thiometon	*59;* 1549, —	
				Vamidothion	*76;* 3236; 209	
3	Cotton	Topical	*Heliothis virescens* (L)	Trichlorfon	*77; 1.6;* 20000	(36)
		μg/g	*Lygus hesperus* (A)	Dichlorvos	*63, 0.8,* 200	
			Chrysopa carnea (L)			
4	Alfalfa	Residues	*Therioaphis maculata* (A)	Nicotine	*0.22*, 748, 9	(19)
		lbs/100 gals	*Hippodamia 5-signata* (A)	Pyrethrins	*0.22*, 0.30, 70.6	
			Hippodamia convergens (L)	Lindane	*0.06*, 0.75, 0.30	
				Parathion	*0.006*, 0.036, 0.010	
				Demeton	*0.0023*, 0.21, 0.45	
				Mevinphos	*0.0012*, 0.0037, 0.0028	
5	Beets	Direct spray	*Aphis fabae* (A)	Malathion	*3.0, 2.6,* 8.0, 3.2, 0.25, 1.2	(245)
		mg/m^2	*Aphis craccivora* (A)	Thiometon	*7.5, 5.7,* 450, —, 110, 60	
			Coccinella 7-punctata (A)	Trichlorfon	*45, 50,* 40, —, 70, 2000	
			Coccinella 5-punctata (A)			
			Adonia variegata (A)			
			Chrysopa carnea (A)			

Table 4 (Continued)

Complex tested	Ecosystem	Method Units	Pest Predator	Compound	LC_{50} or LD_{50} level:[a] *Pest*, Predator	Reference
6	Castor bean	Residue on leaves	*Eutetranychus banksi* (A)	Chlorobenzilate	*0.14*, 0.5	(182)
		% concentration	*Stethorus pauperculus* (A)	Parathion	*0.00021*, 0.00011	
				Diazinon	*0.0010*, 0.0037	
				Phosphamidon	*0.0052*, 0.0025	
7	Cereals	Direct spray	*Macrosiphum avenae* (A)	Malathion	*0.0013*, 0.73, 0.046	(93)
		$\mu g/cm^2$	*Hippodamia convergens* (A)	Parathion	*0.0002*, 0.13, 0.0051	
			Nabis americoferus (A)			
8	Water	p.p.b.	*Culex quinquefasciatus* (L)	Chlorpyrifos	*1.0*, 35.2, 2.1, 5.4	(193)
			Notonecta undulata (A)			
			Laccophilus fasciatus (A)			
			Chaoborus punctipennis (L)			
9	Manure	p.p.b.	*Musca domestica* (L)	Fenthion	*0.87*, 1.3	(14)
			Macrocheles muscaedomesticae (A)	Diazinon	*0.58*, 1.0	
				Naled	*0.43*, 0.44	
				Trichlorfon	*0.98*, 7.7	
				Dimethoate	*0.05*, 2.6	
				Dimetilan	*1.9*, 24	

[a] LC_{50} (or LD_{50}) values listed in horizontal sequence pertain to the pest(s) and their predator(s) listed vertically in column 4 for each complex.

[b] (A) = Adults and (L) = Larvae.

Table 5 LC_{50} levels for complexes of pests and their predators in which both S and R strains exist

Complex tested	Ecosystem	Method Units	Pest Predator	Compound	LC_{50} level:[a] *Pest*, predator	Reference
1	Rice	Immersion	S[b]—*Nephotettix cincticeps* (A)	Malathion	*0.006, 0.074,* 0.050, 0.217	(114)
		% concentration	R —*Nephotettix cincticeps* (A)	Diazinon	*0.027, 0.062,* 0.020, 0.322	
			Lycosa pseudoannulata (A)	Fenitrothion	*0.046, 0.632,* 0.178, 0.361	
			Oedothorax insecticeps (A)	MIPC	*0.005, 0.017,* 0.015, 0.153	
				BMPC	*0.009, 0.018,* 0.025, 0.435	
2	Greenhouse crops	Direct spray	S —*Tetranychus urticae* (F)	Parathion	*5, 600,* 0.02, 2	(33)
		mg/1 concentration	R —*Tetranychus urticae* (F)	Malathion	*75, 900,* 0.33, 3	
			Stethorus punctillum (A)	Carbophenothion	*4, 12,* 4, 15	
			Typhlodromus longipilus (A)	Diazinon	*20, 30,* 0.1, 1	
				Demeton	*12, 48,* 0.5, 20	
3	Apple	Slide-dip	R —*Tetranychus urticae* (F)	Azinphosmethyl	*6.2, 4.2,* 0.07, 5.8	(198)
		lbs/100 gals	R —*Panonychus ulmi* (F)	Parathion	---, *11,* 0.26, 15.3	
			S —*Amblyseius fallacis* (F)	Phosmet	*10.9, 6.4,* 0.23, 10.6	
			R—*Amblyseius fallacis* (F)	Omite	*7.2, 0.28,* 24.0, 50.8	
				Plictran	*0.54, 0.20,* 21.9, 2.9	
4	Apple	Slide-dip	R —*Tetranychus urticae* (F)	.Dinocap	*0.80, 0.08,* 0.51	(199)
		lbs/100 gals	R —*Panonychus ulmi* (F)	Dikar	*24.1, 0.80,* 7.8	
			R—*Amblyseius fallacis* (F)			

[a] LC_{50} values listed in horizontal sequence pertain to the pest(s) and their predator(s) listed vertically in column 4 for each complex.
[b] S = Susceptible strain, (A) = Adults, R = Resistant strain, and (F) = Females.

Table 6 LD_{50} levels for various complexes of pests and their parasites

Complex tested	Ecosystem	Method Units	Pest Parasite	Compound	LD_{50} level:[a] *Pest*, parasite	Reference
1	Citrus	Topical application μg/gm	*Dacus dorsalis* (L)[b] *Opius oophilus* (L)	Aldrin Lindane Parathion	*156.6*, 10.8 *42.3*, 20.7 *11.0*, 25.2	(224)
2	Citrus	Topical application μg/gm	*Dacus dorsalis* (A) *Opius persulcatus* (A) *Opius longicaudatus* (A)	Parathion Isopropyl-parathion	*1.2*, 2.0, 1.2 *3.5*, 100, 100	(149)
3	Citrus	Direct spray gm/100 ml	*Dacus dorsalis* (A) *Opius persulcatus* (A)	Lindane Tepp Schradan	*0.0032*, 0.0017 *0.013*, 0.0045 *1.2*, 0.1	(148)
4	Beets	Impregnated paper mg/cm^2	*Aphis fabae* (A) *Praon abjectum* (A) *Lysiphlebus fabarum* (A)	Malathion Thiometon	*22*, 11, 0.7 *45*, 55, 6.8	(245)
5	Oak	Direct spray oz/acre	*Phryganidea californica* (L) *Itoplectis behrensii* (A) *Dibrachys cavus* (A)[c]	Zectran	*0.45*, 0.12, 0.11	(195)
6	Hardwoods	Topical application μg/insect	*Porthetria dispar* (A) *Apanteles melanoscelus* (A) *Brachymeria intermedia* (A) *Exorista rossa* (A)[d]	Carbaryl	*0.26*, 0.03, 0.03, 1.0	(230)

[a] LD_{50} values listed in horizontal sequence pertain to the pest(s) and their parasite(s) listed vertically in column 4 for each complex.
[b] (L) = Larvae and (A) = Adults.
[c] Hyperparasite.
[d] Tachinid.

mortality was unexpectedly rapid, sometimes occurring immediately after tasting and rejection. Although the most toxic pesticides were those most poorly accepted, their repellency afforded no or little protection to these natural enemies, the merest taste being fatal. When the parasite *Campoletis perdistinctus* was confined and then force-fed on honey-water solutions of systemic insecticides such as monocrotophos, aldicarb, and disulfoton, an appreciable mortality resulted, much more than the insignificant mortality observed when they were caged on cotton plants which had received stem or soil applications of these compounds. Cate et al (39) therefore concluded that stem applications of monocrotophos and soil applications of aldicarb to cotton were virtually harmless to adult *C. perdistinctus* in the field, although they had found that nectars collected from treated plants killed a considerable proportion of these parasites when they were force-fed on them in the laboratory.

Environmental and Physiological Factors Affecting Susceptibility

The influence of environmental and physiological factors on the direct toxicity of insecticides to natural enemies has received but a fraction of the attention given to their influence on pest susceptibility. Among the environmental factors, Critchley (48) found that the toxicity of thionazin to predaceous carabids in soils was increased by higher temperature and moisture and by lower soil pH, compaction, and illumination—factors which had the same effect on pest arthropods occurring in the same habitat. He also found that smaller, younger, and inadequately nourished beetles were the more susceptible. Similarly, starvation of the adult parasite *Habrobracon juglandis* decreased the oviposition and increased the egg mortality on exposure of the adults to the alkylating agent ethyl methane sulfonate (98), and starved adults of *Trichogramma evanescens* were almost 10 times more susceptible to metasystox than those fed on sugar syrup (126). Adults of the predator *Coleomegilla maculata* (13) and of the parasite *Aphelinus mali* (207) were considerably more tolerant to insecticides when in the diapause condition.

Among coccinellids, the predaceous *Coleomegilla maculata* closely resembles the plant pest *Epilachna varivestris* in being naturally tolerant to DDT. The figures obtained by Atallah & Nettles (12) for the delayed cuticular penetration of DDT and its rapid detoxication to DDE in *Coleomegilla* are remarkably similar to those obtained by Sternburg & Kearns (218) for *Epilachna.* In assessments by Brattsten & Metcalf (32) of the detoxificative abilities of a great variety of insects, based on the extent to which the synergist piperonyl butoxide increased the toxicity of carbaryl, the coccinellid predator *Hippodamia convergens* showed a synergistic ratio similar to that of the phytophagous *Epilachna.* Moreover, the ratios that they found in a variety of hymenopterous parasites had a range not dissimilar to those for a variety of pest caterpillars. Larvae of *Chrysopa carnea* were much less susceptible than the tobacco budworm to trichlorfon, the difference being in a much slower cuticular penetration (36). The egg of the predator *Anthocoris nemorum,* which is laid inside the leaf tissue of plants liable to be infested by aphids, is covered by an unsclerotized chorion. Systemic insecticides such as phorate, menazon, and dimethoate kill the embryo, but not until it is fully developed, indicating that the vulnerable site is the cholinesterase which finally appears as the nervous system develops (71). With dimethoate, trichlorfon, demetonmethyl, or vamidothion used as cotton insecticides, the predator *Coccinella 7-punctata* suffers the same reductions in cholinesterase and aliesterase activity as does its aphid prey (83).

Although there is little evidence to indicate any physiological difference between arthropod natural enemies and pest arthropods in their responses to insecticides, attention should be paid to the hypothesis made by Gordon (89) several years ago, that "the extraordinarily high and generalized tolerance of the larval feeding stages of relatively polyphagous holometabolous insects to contact insecticides is probably the result of selection for endurance of prolonged and varied biochemical stresses associated with a diversity of their natural food plants." Krieger et al (128) have recently confirmed that among lepidopterous larvae the activity of aldrin epoxidase in the midgut tissues was higher in oligophagous than monophagous species and higher yet in the polyphagous species. They considered their results to indicate that the enzyme activities had been adjusted by natural selection to detoxify plant toxins, which may include alkaloids, rotenoids, cyanides, etc—in fact a great variety of natural insecticides. Dyte (68) cites the example of the presence of nicotine resistance in three lepidopterous pests of tobacco, namely *Manduca sexta, Heliothis virescens,* and *Trichoplusia ni.* By contrast, the hymenopterous honey bee is characterized by an unusual susceptibility to carbaryl, which Gilbert (86) suggested is due to low detoxicative ability; this is confirmed by the finding of a low synergistic ratio with piperonyl butoxide for this species (32).

The hypothesis that the evolutionary conditioning by natural plant toxicants has given pests a greater opportunity to have developed preadaptations, needs to be carefully assessed experimentally. Another factor which may be important is the dependence of natural enemies on the survival of sufficient numbers of their hosts or prey to provide them with food; hence the susceptibility of populations occurring in heavily treated environments at any point in time would depend on that of the pest at the lower trophic level. The hypothesis described in the previous paragraph, when applied to the dynamic processes which occur in nature, would imply that the natural enemy may be able to express its capacity to adapt to a natural plant toxicant only after the pest had exploited its genetic plasticity in resisting the toxicant. Also in relation to conventional synthetic insecticides, it must be remembered that susceptibility assessments on pests and their natural enemies are made on samples taken from populations which almost always have already had some exposure to insecticides in the field. It is therefore possible that many of the cases reported from the field where a natural enemy has appeared to be much more susceptible than a pest, or where an insecticide dosage which was sublethal to a host species taken from the field was toxic to its internal parasites (226, 229), are due to the fact that the pest has begun to develop tolerances or resistances while the natural enemy has not yet been able to do so. As is discussed in more detail in a later section, these two points are borne out in the examples of sequential development of resistance first among pests and then later among their natural enemies to a natural plant toxicant (229) and to several synthetic organic insecticides (50).

SUSCEPTIBILITY OF DEVELOPMENTAL STAGES

There is a considerable body of published data on the insecticide susceptibility of the successive life stages of a variety of natural enemy species. In most cases, these evaluations were made to determine when predators and parasites would be in their least susceptible stages, so that sprays could be applied with minimal effect on them. Early reviews (22, 186, 233) on integrated control gave considerable emphasis to this type of research. Bartlett (22) has generalized that it is in the adult stages that predators and parasites are the most susceptible to insecticides, while eggs are the least affected. He generalized that holometabolous larvae were intermediately susceptible, but heterometabolous nymphs were more prone to be poisoned than adult forms.

Since the review of Bartlett (22), further research has tended to confirm his observations and several preliminary studies have indicated possible reasons why they are so. Wilkinson et al (243) and Lingren & Ridgway (135) reported that larvae of *C. carnea* were considerably more tolerant to a variety of insecticides than were adults, a contributing factor in the case of trichlorfon being the slower penetration rate and the reduced detoxication to dichlorvos in the larva (36). Among parasites, the adult stage has proved to be the most susceptible (74, 75, 134, 136), often being the only stage to live an exposed life. Larvae of *H. convergens* were found to be less susceptible than adults to malathion and parathion. Hamilton & Kieckhefer (93) suggested that the difference could be mainly due to the greater glycogen and fat-body reserves accumulated by these larvae than by the adults, which readily metabolize these materials during metamorphosis and reproductive activities, and that detoxication in fat body tissues could account for the greater tolerance of the larvae. This postulation is similar to that of Hoffman & Grosch (98), who proposed a detoxicative function for the fat body of *Habrobracon juglandis* after assessing its activity in giving this parasite protection against the effect of ethyl methane sulfonate.

INDIRECT EFFECTS OF INSECTICIDES

Pesticides can affect predators and parasites indirectly through their influence on the pest species which constitute their prey or hosts, either by eliminating them as a source of food or by leaving them as sources of secondary poisoning. Numerous reviews (22, 186, 216, 233) have already covered the subject of pest resurgence, resulting from the sequence of pest elimination, starvation among the remaining natural enemies, and pest reinvasion before the natural enemies are reestablished. The present review discusses the indirect effects of pesticides due to secondary poisoning and their accumalation in food chains.

Predator-Prey Relationships

Initial studies on this topic were focused principally on determining whether selectivity could be obtained with broad-spectrum insecticides by exploiting the differences between predators and their prey in their behavior and feeding habits. Early advocates of integrated control (22, 186, 216, 233) stressed the usefulness of compounds that acted as systemic insecticides and thus could be presented to plant-feeding pests in a way that would not affect the natural enemies appreciably. Bartlett (22), reviewing the limited published data at the time, concluded that "destruction of some predators by feeding upon hosts poisoned with systemics . . . has not appeared to be of critical importance in the destruction of natural enemies. . . ." While many experiments have demonstrated the selectivity of administering toxicants systemically (7, 25, 73, 150, 184, 189, 190, 191, 217, 245), in other more recent studies significant mortalities of predators have been encountered either from direct toxicity and fumigant action (25, 27, 42, 72, 73, 184, 245) or from their ingestion of prey species which had taken up the insecticide (7, 15, 25, 27, 72, 109, 145, 245). Because of the differences observed between one insecticide and another, many workers have concluded that they should be evaluated individually to determine their influence on specific natural enemies.

Although predatory species can be killed by secondary poisoning from contaminated prey, little is known about the fate of a toxicant inside the body of the prey and the factors which determine its toxicity to a predator. A safe assumption might be that it would be analogous to the uptake of systemic insecticides from the treated plant by a phytophagous pest. The most detailed study is that of Kiritani & Kawahara (123) made on the fate and effect of BHC passing through a food chain from irrigated soil to rice plants to green rice leafhoppers (*Nephotettix cincticeps*) and finally to a predaceous spider (*Lycosa pseudoannulata*). They found that the leafhoppers came to contain 3 times as much BHC as the rice plants after they had fed on them for 2 days. The original 13% of gamma isomer in the BHC applied was found to increase as it was stepwise accumulated successively in plants and then in the leafhoppers, in which the proportion of this isomer reached 35%. This preferential concentration of gamma-BHC in the leafhopper was particularly significant since the *Lycosa* spider is much more susceptible to this isomer than its leafhopper prey.

The nutritional condition of the prey as it affects the fertility or appetite of a predator may reflect a similar food-chain effect. European red mites reared on apple leaves with a high nitrogen content allow the predator *Amblyseius potentillae* to lay as many eggs with the consumption of fewer prey (234). The increase in numbers of coccinellid and anthocorid predators in cotton fields heavily treated with fertilizers (5) may be attributable not only to their bollworm prey being more abundant but also to these larvae being more nutritious. Similarly, the probable factors governing the effect of an insecticide on a predator through the food chain would be related to the behavior and feeding habits of the prey (amount and rate of uptake from the plant), the fate of the toxicant inside the prey (localization, concentration, and metabolism of the toxicant), the feeding habits of the predator (numbers of prey taken), and the metabolic and detoxicative abilities of the predator.

Biological magnification in food chains, which has been thoroughly studied in ecosystems culminating in vertebrate predators such as raptorial birds, has not been studied to any extent in arthropod natural enemies. In the case of the *Lycosa* spider cited above, there was no biological concentration of BHC at the predator level, but only in plants and leafhoppers (123). Residue levels of chlorinated hydrocarbons that were higher than normal have been reported in two species of carabid beetles and several aquatic predators (70), and in adult *Coleomegilla maculata* (12), but they were not excessively high. When dragonfly nymphs were exposed to sublethal rates of DDT in water (242), or mantids were fed dieldrin-treated adult *Drosophila* (237), no significant tendencies to accumulate toxicants were observed. Newson (164) suggested that the lack of interest in studying residue accumulation in arthropod natural enemies may be due to its relative insignificance as compared to the much greater influence of insecticides from direct contact. It is evident that, if food-chain poisoning were added to the effect of direct exposure, the two factors would almost entirely account for the toxic effect of insecticides on predator populations in the field. Residue accumulations probably are of little importance in conferring toxic effects because the predator is so close to the primary consumer level and in many cases shares the same habitat as its prey.

Parasite-Host Relationships

As noted in several previous sections, the parasite-host relationship can dramatically influence the toxicity of an insecticide to a parasite. The interaction of nicotine produced by the tobacco plant with the tomato hornworm and its principal parasite *Apanteles congregatus* is probably the best example of this. While the pest has acquired a resistance to this natural toxicant, immature stages of the parasite are appreciably affected by the chemical through the physiology of the budworms, on certain tobacco varieties (87, 229). For adult parasites which feed on the host when they oviposit, or for larval stages which feed externally on an exposed host, the discussion presented on food-chain toxicity would apply. For parasites which feed internally within their hosts or those protected by other host-associated structures (e.g. scale parasites), the toxicity of the insecticide is mediated through the physiology of the host to the natural enemy, or is determined by its ability to penetrate the host-associated structure or host cuticle and to reach the feeding parasite. For internal parasites, the effect of the insecticide, in many cases, depends on the response of the host. If it kills the host, this may be sufficient to ensure the death of the parasite. If the host survives, the effect of the insecticide is mediated by the ability of the host's physiology to detoxify it or not, or even in some cases to convert it to more toxic metabolites.

From the applied standpoint, it has long been recognized that since internal parasites are often protected from direct exposure by proper timing of insecticidal application, these beneficial arthropods can be selectively conserved (82, 90). Host-parasite relationships involving several species of natural enemies have been frequently investigated, and several recent publications (4, 24, 61, 115, 134, 136, 195) demonstrate how they may be exploited by proper timing. The most common method is to treat with a nonpersistent insecticide when the immature stages of the parasite are protected, at a time late enough to avoid killing the host in its early stages of development (otherwise the parasite would not have the proper requisites to complete its life cycle), but early enough to avoid the end of the developmental period (otherwise the parasite may be killed when it emerges from its host or cocoon). It is usually found that application of a short-residual toxicant to bracket the middle portion of the parasite's development period is the least disruptive.

It has been reported that internal parasites become more susceptible to poisoning as they approach pupation. With *Apanteles congregatus,* applications of nicotine to parasitized tomato hornworms were found to be fatal to the parasite larvae inside if made within 24 hours of their exit to spin cocoons. Since applications made before that period did not kill the *Apanteles,* Thurston & Fox (229) postulated that these parasite larvae underwent some physiological or morphological change which made them more susceptible as they approached pupation. Adult parasites, being the stage most susceptible to insecticides, are especially likely to be killed in the process of emerging from their host (22, 115, 126, 136, 168). This is probably due partly to their contacting the insecticide residues present on the surface of the host or their own cocoon, and partly to their being more susceptible before the adult cuticle has hardened. Once emerged, the adult would tend to contact many insecticide-contaminated surfaces in its search for suitable hosts.

That there may be differences between chemicals in their effect on parasite and host was observed by Tamashiro & Sherman (224), who applied insecticides topically to larvae of the oriental fruit fly *Dacus dorsalis,* unparasitized or parasitized by the primary parasite *Opius oophilus.* With aldrin, dieldrin, endrin, chlordane, and lindane, to which the parasite was more susceptible than the host larvae that it inhabited, dosages could be found that allowed the emergence of the fruit flies from unparasitized puparia, but caused the parasites to die in the pupal stage within parasitized host puparia. With parathion, however, to which the parasite was more tolerant than the host, dosages could be found at which unparasitized larvae formed elongated puparia from which fruit flies would not emerge, but allowed the occasional emergence of a parasite from such abnormal puparia. The cyclodiene insecticides and lindane also caused latent toxicity in that there would be subsequent mortality in the adult stage resulting from the treatment of larvae; this was exhibited by both the *Dacus* host and the *Opius* parasite, but was not manifest in parathion applications to either species. An opposite difference between the organochlorine and organophosphorus compounds was reported by Evenhuis (74) for their effect on larvae of *Aphelinus mali* within the mummified bodies of the aphid, *Eriosoma lanigerum.* Whereas DDT, BHC, lindane, and endosulfan had little effect, parathion and malathion were very toxic to the late pupal stages of the parasite. Kot & Plewka (126) concluded that the tolerance of *Trichogramma evanescens* inside the eggs of *Sitotroga* was related to the inability of insecticides to penetrate the chorion of the host egg and therefore was more a function of the toxicology of the host than of the parasite. Additional observations on parasite-host-pesticide interactions have been described in several publications (19, 27, 88, 115, 126, 127, 136, 154, 209, 215). In each case the effect of individual insecticides on host and parasite was relatively specific to the compound employed and the species treated.

It was noted by Bartlett (18) that internal parasites caused their hosts to stop feeding, and he hypothesized that this behavioral influence could decrease the likelihood of the parasitized pest ingesting sufficient amounts of toxicant to kill itself or the internal parasite. There are examples of parasitized hosts, for example hornworms containing *Apanteles* (228), being more difficult to kill than unparasitized ones, which would have a favorable effect on the proportion of parasites to pests. But there probably are more examples where parasitism has no effect on host susceptibility (61). The third case, where insecticides kill parasitized hosts more readily than unparasitized ones, for example salt marsh caterpillars parasitized by tachinids (226), would be highly undesirable, especially if the natural enemy was in the early larval stage. Indeed, the percentage parasitism of the spruce budworm by the primary parasite *Apanteles fumiferanae* increased considerably one week after treatment of Canadian forests with DDT (138). The increase was attributed to the following events: 1. a few parasite larvae survived which were within cocoons between the time of treatment and of sampling; 2. budworms already moribund or barely active because they contained nearly full-grown *Apanteles* larvae did not take up sufficient DDT from the residues to poison the parasites; and 3. some *Apanteles* were able to emerge from budworms which were knocked down by DDT and unable to return to their feeding sites in the trees.

An additional way in which the selective application of insecticides might be useful is to control those species that are competitors, predators, and hyperparasites of primary parasites. The use of insecticides for this type of arthropod management has been studied in a few cases, although the practice may not always be desirable because these species may be important in maintaining ecosystem stability. DeBach & Bartlett (62) reviewed examples of several ant species which fed on honeydews of homopteran pests and thus disturbed the oviposition activities of certain chalcidoid parasites. So active were these competitors that in some cases they prevented effective biological control from occurring. Inserra (110) found that parasitism of the Florida red scale by *Aphelinus melinus* was substantially increased by controlling the Argentine ant with lindane, and concluded that the ant must be eradicated or greatly suppressed if the introduced parasite was to be effective. Since hyperparasites may also have an unfavorable influence on primary parasitism, attempts have been made to identify insecticides that have a favorable activity. For example, the carbamate Zectran applied for California oakworm control was only slightly more toxic to the hyperparasite *Dibrachys cavus* than it was to the primary parasite *Itoplectis behrensii* (195). On the other hand, at least five aphicides have been reported to be more toxic to the adult hyperparasite *Alloxysta curvicornis* than to the adult primary parasite *Diaretiella rapae* (240).

Responses to Sublethal Levels of Insecticides

The effects of sublethal doses of insecticides, either at a level which causes no mortality in the population or at a toxic level which leaves some survivors, have been thoroughly reviewed by Moriarty (156) but almost exclusively for pest arthropods. The limited amount of data for predators and parasites, as collected below, mainly concerns the more obvious effects on developmental rate, reproductive rate, and longevity, with a few cases of behavioral responses observed in the laboratory and field. Latent toxicity, a property already described for the cyclodiene derivatives and for lindane, affects the internal braconid parasite *O. oophilus* when these insecticides are topically applied to the larvae of its host, and may be regarded in this context as a greatly reduced longevity potential (224). Among predators, slower larval and pupal developmental rates, following sublethal doses applied to the larvae, were reported for 3 species of coccinellids treated with the herbicide 2,4-D amine (3) and for *Chrysopa rufilabris* treated with azinphosmethyl, ethion, carbophenothion, or carbaryl (131). Reduced longevity was reported for *Coleomegilla maculata* when the adults had been topically treated with endrin or toxaphene (13) and for *Eublemma amabilis* and *Holcocera pulverea,* lepidopteran predators of the lac insect, exposed in the adult stage to deposits of aldrin or dieldrin (142). These treatments also reduced their fecundity almost to the point of no oviposition at all. Among parasites, exposure of the aphelinid wasp *Aphytis holoxanthus* to sublethal deposits of dimethoate, endosulfan, dioxathion, or sulfur (200) resulted in greatly decreased fecundity. An additional effect of sulfur, observed in the encyrtid *Metaphycus helvolus,* was to render the parasite permanently unable to recognize its coccid hosts (80). In investigating the effects of deposits of 3 insecticides and 2 fungicides on *Encarsia formosa,* the principal parasite of the greenhouse whitefly, Irving & Wyatt (111) found that they all decreased the amount of oviposition, including the fungicide pirimicarb, which could have the opposite effect under certain circumstances.

In contrast, DDT when applied to pupae of *Bracon hebetor* results in a significant increase in the oviposition rates of the emerged females (92). In *Coleomegilla* also, DDT applied to the adults increased the level of oviposition by at least 60%, although the F_1 generation hatching from the eggs showed a 25% decrease in survival (13). In the field, adults of *Chrysopa californica* have been found to lay more eggs on DDT-treated trees than on untreated trees, independent of the amount of prey available (81).

It has been observed that persisting low-level residues of insecticide treatments have been stimulatory to populations of pest insects or mites. This has been suggested by Luckey (137) to be partially explained by the phenomenon termed hormoligosis, namely the stimulatory effects of harmful agents when applied at certain levels in the subharmful or stress range. Indeed, he found that most of the 14 organochlorine and organophosphorus insecticides that he added to the diet of the house cricket, *Acheta domestica,* were stimulatory in terms of increasing the bodyweights attained when concentrations were at some point between one tenth and one hundredth of the median lethal dose. In this context it is relevant to note that Plapp & Casida (176) have found that DDT and dieldrin added to the diet of the house fly stimulates the microsomal oxidase activity and results in greater detoxication of a range of different insecticides. It is possible that the increased oviposition induced by DDT may be a hormoligotic effect resulting from increased enzyme activity, if not simply due to a nervous response from the stimulatory effect of light doses on the oviposition reflexes. With respect to organophosphorus insecticides, hormoligosis-inducing concentrations probably would seldom be present in the field if at all, since the bulk of the evidence for the predators and parasites in the post-spray period indicates an inhibitory effect on reproduction and development. The question as to whether hormoligotic effects may operate among natural enemies needs to be clarified further.

The activity of carabid beetles is greatly increased by the soil insecticides aldrin, dieldrin, DDT, and certain organophosphorus compounds (e.g. thonazin) (45, 48, 63, 69). With *Bembidion lampros,* this behavioral response results in greater catches in pitfall traps (45). *Harpalus aeneus,* when exposed to sublethal levels of insecticides, responds by releasing large amounts of a chemical (probably formic acid), which the experiments of Critchley (48) indicate might be sufficient to cause self-annihilation in the confines of its burrow. Adult *H. rufipes,* while exposed to sublethal deposits of DDT, showed a reduction in feeding rate which Dempster (63) considered may prevent this predator from controlling *Pieris rapae* for some time after spray treatments.

Gustatory repellency of an insecticide could produce a behavioral response in a natural enemy that would convert a lethal dose into a sublethal one. The syrphid *Xanthogramma aegyptium,* for example, stops feeding on phosphamidon-poisoned aphids before a lethal dose is ingested (15). But among 60 pesticides tested in honey bait by Bartlett (23) for the responses they induced in the coccinellids *Lindorus* and *Cryptolaemus,* the aphelinid *Aphytis,* and the encyrtid parasite *Metaphycus,* those which caused a strong gustatory repellency were so insecticidal that the rejection of the tasted bait afforded no protection.

With respect to genetic effects of sublethal levels of insecticides, the evidence from braconid parasites is not different from that obtained from other insect species. Topical applications of DDT (92) or heptachlor (91) to virgin females of *B. hebetor* were found to induce no lethal mutations as judged by the percentage hatch of the male embryos that they produced. This result parallels the nonmutagenicity of DDT or lindane already reported for *Drosophila melanogaster* (173). On the other hand, the alkylating compound apholate proved to be as good a chemosterilant for *B. hebetor* (232) as for pest insects, and chromosome aberrations were produced in females of the parasite *Habrobracon juglandis* topically treated with the alkylating agent ethyl methane sulfonate (98).

Responses to Pathogens and JH Compounds

The diseases of arthropod pests have recently been exploited to control them with some success. They are usually applied by the same methods as used for insecticides, and like insecticides they must be assessed for their side-effects on natural enemies. By direct uptake they can cause mortality of the free-living predators and adult parasites, and the exposed larvae of external parasites. With respect to internal parasites, Tamashiro (223) distinguished the following types of effects caused by insect pathogens: 1. they may directly affect the parasite through the host, 2. they may cause premature death of the host and indirectly cause death of the parasite, 3. they may alter the physiology of the host so that it no longer is nutritionally suitable for the development of the parasite, and 4. they may infect and alter the host sufficiently so the adult parasite will not oviposit in the infected host. To date, little is known about the last three modes of action, but the effects of these agents on free-living predators and internal parasites have been assessed for the major disease groups including the pathogenic bacteria, viruses, and protozoa.

The effect of *Bacillus thuringiensis* (BT) spores on a wide range of predators and parasites has been tested. Deposits of the BT product Thuricide® HPC had no effect on adults of *Brachymeria, Campoletis, Chelonus, Meteorus,* or *Voria* parasites, nor on predaceous *Chrysopa* larvae or adult *Hippodamia* (243). Field treatments of Thuricide HPC, Dipel® wettable, and Biotrol® XK wettable formulations had no effect on populations of *Telenomus alsophilae,* an egg parasite of the elm spanworm (115). BT spores were found not only to be without effect on *Nemeritis canescens,* but also to be distributed by the parasite among the host populations of the Indian meal moth (129). Slight reductions in parasite populations were observed when the product Enterobacterin-3® was applied against the diamond-back moth on cabbage in the USSR (4) or when BT spores were applied against satin moth populations occurring on poplars in Belgium which were parasitized by *Micropalpus retroflexa, Apanteles solitarius,* and other species (161). A significant reduction in emergence of the parasites *Solenotus intermedius* and *Halticoptera aenea* from pupae of *Liromyza* leafminers was observed when the formulation Biotrol® 5B was applied to tomatoes in California (210). When the more potent formulation HD-1 was applied to lettuce in Arizona, Vail et al (231) found that it reduced the percentage parasitization of the cabbage looper and alfalfa looper, which prompted them to discuss in some detail the factors which must be considered in interpreting data from field evaluations.

Among the baculoviruses now being developed for control of plant-feeding caterpillars, the nuclear polyhedrosis of the satin moth was found to have little effect on its parasites beyond delaying their emergence (161). On the other hand, a granulosis virus of the cabbageworm caused considerable mortality of the parasite *Apanteles glomeratus* (118). The granulosis virus of the armyworm was found by Kaya & Tanada (116) to produce a toxin which killed a high percentage of the parasite *Apanteles militaris;* toxins from other viruses produced in other caterpillars were also lethal to *A. militaris* and to another parasite, *Phanerotoma flavotestacea.* Application of nuclear polyhedrosis virus of the cabbage looper and alfalfa looper had the effect of reducing the parasitization of these caterpillars; however, since the overall control was good, Vail et al (231) considered this to be a case of competition between the virus and the parasites.

Among the protozoa, the microsporidan *Nosema polyvora* infecting cabbageworms was found to reduce the survival, size, and diapause ability of *A. glomeratus* parasitizing them (112). Moreover this species of parasite is known to be capable of transmitting protozoan diseases from one host to another (225). Although the neogregarine *Mattesia grandis* infects the larvae and adults of the parasite *Bracon mellitor* after they have fed on infected boll weevils, this parasite does not transmit the protozoan among larvae of the boll weevil even if it is itself infected (146).

With respect to the effect of juvenile hormone (JH) mimics on parasites, Bracken & Nair (31) were the first to observe that farnesol, fed to the adult ichneumonid *Exoristes comstocki* in a sucrose solution, stimulated yolk deposition and egg production. They suggested that host feeding may provide the parasite with a hormonal stimulus for ovarian development as well as the nutrients for egg production. Neal et al (160) found that treatment of the diapause alfalfa weevils with 10,11-epoxy-farnesenic acid, methyl ester, (a C_{16} juvenile hormone analog) stimulated the braconid parasite *Microctonus aethiops* within this host to emerge prematurely. They suggested that advantage could be taken of this response as a practical method for extracting parasites from diapausing or sexually immature weevils. Treatment of parasitized overwintering larvae of *Chilo suppressalis* by dipping them in several plant-derived JH mimics (phytoecdysones) causes accelerated pupation of the parasite *Chelonus munakatae* (206). Among hymenopterous predators, *Polistes* wasps showed significant changes in ovarian development and diapause relationships when topically treated with a synthetic mixture described as 8 geometric isomers of juvenile hormone (26). On the other hand, 4 JH chemicals applied to tobacco budworm larvae delayed development of the parasitoid *Microplitis croceipes* within them, and reduced the percentage of pupation and emergence (29). When topically applied to larvae of tobacco budworm predators, several JH compounds severely affected the completion of metamorphosis by *Chrysopa carnea,* while causing no preimaginal mortalities. Conversely, the percentage effect on metamorphosis of *Nabis roseipennis* was only moderate, while preimaginal abnormalities were common (28). Among the dytiscid larvae predaceous on mosquito larvae, those of *Thermonectes* were significantly reduced in numbers by application of the JH mimic Monsanto 585 (214), while those of *Laccophilus* remained unaffected by the JH mimic ZR-515 (Altosid, methoprene) applied in a slow-release formulation (152).

To date, both selective and nonselective properties have been demonstrated for arthropod diseases and JH compounds. Even an agent as intrinsically specific as an insect pheromone can be attractive to the natural enemies of a pest insect as well as to its mate (183, 201). Although these agents do have some side effects on beneficial species, they appear to be considerably more selective than conventional insecticides and it is likely that future research will demonstrate how they may be utilized in pest management.

RESISTANCE IN NATURAL ENEMY SPECIES

When developed resistance to insecticides first became a problem among pests in the 1950s, the question whether predators and parasites could do likewise was frequently posed (174, 213, 241). Because resistance among these natural enemies had not been observed in the field, the subject was not pursued to any extent until recent publications by Croft (50), Georghiou (85), and Newsom (164) reintroduced the question. While many authors have recognized the potential value of resistance occurring in a natural enemy population (50, 97, 157, 203, 244), few studies to see if it were present have been undertaken. Probably this is one reason why the list of natural enemies that have developed resistance (Table 7) includes only 9 species, 5 of which are phytoseiid mites; of the remainder, 2 are braconid parasites in which resistant strains have been produced by laboratory selection and 2 are resistant species of predatory insects which were selected in nature.

Pielou and his co-workers (172, 174. 175, 241) were the first to study the possibilities for selecting a resistant natural enemy, when in 1949 they commenced submit-

Table 7 Predatory mites and insect natural enemies which have developed resistance to insecticides

Taxonomic group	Species	Insecticide	Resistance ratio	Reference
Acarina: Phytoseiidae	*Typhlodromus occidentalis*	Azinphosmethyl	101	(54)
			104	(6)
	Typhlodromus caudiglans	DDT	—[b]	(97)
	Typhlodromus pyri	Azinphosmethyl	10	(104)
	Amblyseius fallacis	Azinphosmethyl	100	(157)
			83	(198)
			117	(6)
			944	(56)
		Carbaryl	25, 77	(56)
		DDT	—[b]	(211)
		Gardona	25	(198)
			28	(56)
		Methoxychlor	—[b]	(211)
		Parathion	103	(157)
			59	(198)
		Phosmet	46	(198)
	Amblyseius hibisci	Parathion	—[b]	(119)
Coleoptera: Coccinellidae	*Coleomegilla maculata*	DDT	6	(13)
		Parathionmethyl	10	—[c]
			35	(40)
Diptera: Anthomyiidae	*Ophyra leucostoma*	DDT	—[b]	(84)
		Dieldrin	—[b]	(84)
Hymenoptera: Braconidae	*Macrocentrus ancylivorus*[a]	DDT	4.4	(175)
			12	(194)
	Bracon mellitor[a]	DDT - toxaphene	8	(2)
		DDT	4	(2)
		Carbaryl	4	(2)
		Parathionmethyl	4	(2)

[a]Resistance induced in the laboratory; all others have developed resistance in the field.
[b]Not determined.
[c]R. Mohamed, personal communication.

ting *Macrocentrus ancylivorus,* a parasite of the oriental fruit moth, to laboratory selection with DDT, their intent being to develop a resistant strain for release in peach orchards to be treated with this insecticide. After 9 months' selection a 4.4-fold increase in DDT resistance was attained, which was regarded as encouraging. Continuation of this selection produced a maximum resistance ratio of 12 times the normal by the F_{19} generation (194). Subsequently, despite the continuation of selection pressure on both sexes until the F_{29} and on females only until the F_{71}, the resistance levels gradually declined. At this point, after 6 years of work involving the selection of a total of 3 million individual parents, selection was discontinued at the F_{72}, whereupon resistance regressed back to almost the original level by the F_{85} generation. The second major attempt to develop resistant strains was made with *Bracon mellitor,* a parasite of the boll weevil, by selection of lines with DDT, toxaphene, parathionmethyl, carbaryl, and a DDT-toxaphene mixture for at least 5 generations (2). The highest resistance ratio achieved was to the DDT-toxaphene mixture, but was less than 10, while only 4-fold increases were obtained with the other 4 compounds.

Resistance in field populations of predators and parasites does not betray its presence as it does in pest species because these natural enemies seldom become more abundant upon first developing the resistance. Whereas pest species have an assured food supply, the numbers of natural enemies depend on the abundance of hosts or prey, which is usually lacking after the insecticide application. As with pests, however, certain situations confuse the picture with regard to possible resistance in predators and parasites. There are cases where natural enemies have extremely high migratory capacities and rapidly reinvade a habitat following an insecticidal treatment (30, 35, 64, 76, 143), and cases where a significant portion of the predator or parasite population is protected within an unsprayed portion of the habitat or within their host (24, 138, 143, 179, 220, 227). There also is the problem of confusing resistance with the inherent tolerances which are the normal characteristics of a variety of predator and parasite species (7, 13, 20, 21, 30, 63, 97, 135, 144, 162, 224, 243).

The longest-known and best-documented cases of resistance among arthropod natural enemies are those for the phytoseiid mites *Typhlodromus occidentalis* and *Amblyseius fallacis.* Both species are specialized predators of tetranychid mites on several crops including decidous tree fruits. As early as 1953, Huffaker & Kenneth (107) published data indicating that populations of *T. occidentalis* preying upon the cyclamen mite in California strawberry fields was considerably more tolerant to parathion than those of its congener *Typhlodromus reticulatus* which also was present. In 1958, a similar persistence of *T. occidentalis* was noted in apple orchards in the interior of British Columbia, but it was as yet uncertain whether this was due to a natural tolerance of this species or whether it had developed a resistance to parathion (155). Again, it was observed in Washington apple orchards in 1961 that *T. occidentalis* populations were surviving sprays applied for codling moth control which included parathion as well as azinphosmethyl, phosmet, and DDT (103).

In 1970, Croft & Jeppson (54) compared populations of *T. occidentalis* from central Washington and from Utah with presumably susceptible strains from southern California for their susceptibility to azinphosmethyl. They found that mites from Washington had a 101-fold higher LC_{50} and those from Utah an 88-fold higher LC_{50} to azinphosmethyl, along with moderately developed tolerances to Gardona® (5x), Omite® (2x), and oxythioquinox (2x) among the mites from Washington. The strains from southern California, which had no known history of exposure, were as tolerant to parathion as those from Washington and Utah. Since these initial studies, the approximately 100-fold resistance ratio between Washington and California strains for azinphosmethyl has been further substantiated (6), and cross resistances have been reported to a variety of OP compounds including phosmet (58), Gardona (54, 58), tepp (58, 79), and phosalone (141, 180, 181). If the distribution of the cited cases where resistance has been proved by laboratory test (L) and where populations have been reported to survive heavy OP treatments in the field (F) are mapped (Figure 1), it may be seen that the places where *T. occidentalis* has developed OP resistance on several crops in irrigated areas are widely distributed throughout the arid regions of western North America.

In almost parallel fashion, insecticide resistance has developed in populations of *A. fallacis* throughout midwestern and eastern North America. The first indication was published by Oatman & Legner (166), who observed that *A. fallacis* was the most abundant mite predator present in sprayed apple orchards of Door County, Wisconsin. Smith et al (211) later reported a tolerance or resistance to methoxychlor and DDT in *A. fallacis* populations occurring in Maryland greenhouses. In the 1960s, both in Missouri (177) and New Jersey apple orchards (221, 222), it was noted that populations of *A. fallacis* readily tolerated treatments with certain OP compounds and it was suggested that resistance had developed. Motoyama et al (157) tested the susceptibility levels of North Carolina populations in comparison with normal strains and found that the resistance was real; subsequently Croft & Nelson (57) found that most Michigan populations from apple orchards were resistant to azinphosmethyl, and one of them was significantly tolerant to diazinon. Cross resistances were found to involve such OP compounds as phosmet, demeton, dimethoate, leptophos, phosphamidon, Gardona, and Supracide® (57). However, in contrast to *T. occidentalis*, the OP resistance in *A. fallacis* does not give cross resistance to tepp and phosalone (58).

A study of the possible resistance mechanisms made in vivo by Motoyama et al (158), revealed that an OP-resistant strain of *A. fallacis* from North Carolina degraded azinphosmethyl faster than a susceptible strain, and that it suffered less inhibition of its cholinesterase activity. Since no difference was detectable between S and R strains in bimolecular rate constants, the resistance mechanism appeared not to be associated with a modified cholinesterase, but rather to a higher nonspecific esterase activity. In addition to OP resistance, resistance to carbaryl was found in *A. fallacis* populations collected from an Indiana apple orchard where this carbamate insecticide had been applied for nine consecutive years. From this material Croft & Meyer (56) developed a strain simultaneously resistant to OP compounds and carbaryl by means of laboratory selections and hybrid crosses.

A survey of 22 orchards in Michigan revealed azinphosmethyl resistance in *A. fallacis* to be widespread throughout the state, occurring wherever this insecticide had been intensively applied for several years (57). Indeed, Croft & Meyer (56) had found that experimental treatment of an apple orchard with 5–7 annual applications for a period of 4 years increased the LC_{50} level of *A. fallacis* mites by more than 300 times. When the distribution of resistant populations as indicated by laboratory (L) and field evaluations (F) is mapped (Figure 2), it is seen that OP resistance among *A. fallacis* populations is widespread throughout the midwest and eastern

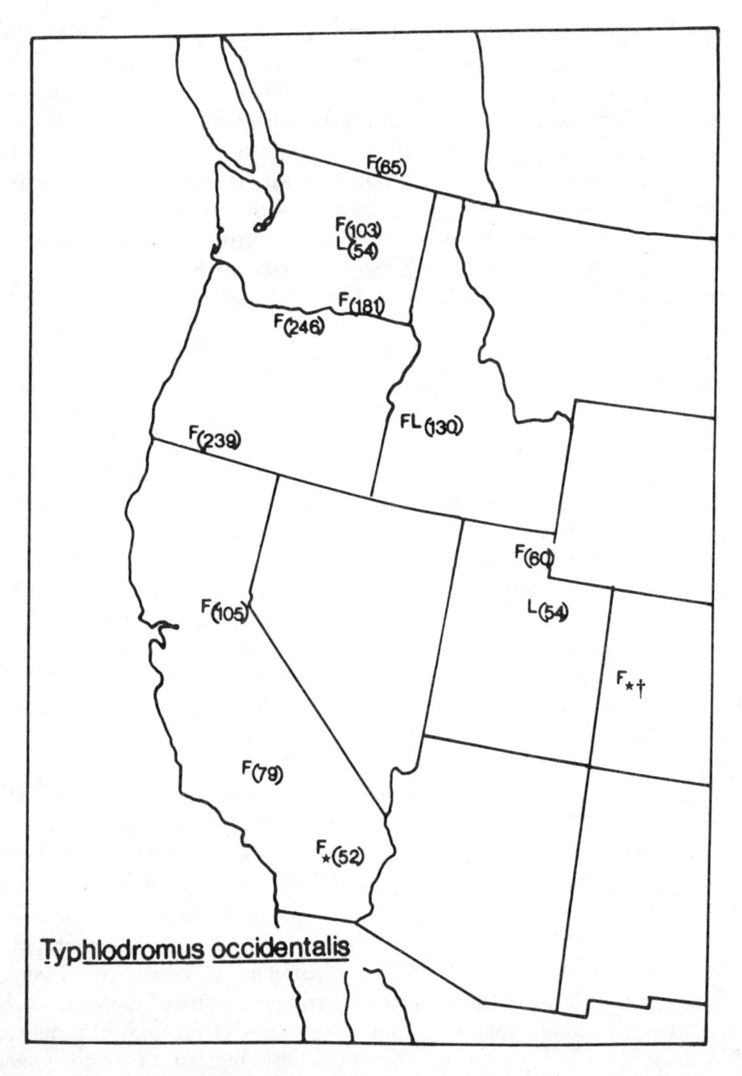

Figure 1 Distribution of OP-resistant populations of *Typhlodromus occidentalis* in western North America. F = resistance detected by field experience, L = resistance assessed by laboratory test, ( ) = subscript figures in brackets refer to literature citations, ★ = established by field release experiments, and † = J. A. Quist, personal communication.

United States. Moreover, it is probably present in apple orchards of eastern Canada where this species occurs (96, 133, 169, 170).

Lower levels of resistance in more restricted areas have been reported for three other species of phytoseiid mites. A population of *Typhlodromus caudiglans* in an Ontario peach orchard was reported resistant to DDT (97). There are also reports of populations of *Amblyseius hibisci* resistant to parathion in several citrus groves of the San Joaquin valley of California (119) and of *Typhlodromus pyri* to azinphos-methyl in an apple orchard in New Zealand (104). In none of these three cases was the resistance level sufficiently high to ensure the immunity that resistant *T. occidentalis* and *A. fallacis* populations now enjoy from OP insecticides applied at the normal field rates.

There are two examples where predatory insects have been found to have developed insecticide resistance in the field. Georghiou (84) in California detected in 1967

Figure 2 Distribution of OP-resistant populations of *Amblyseius fallacis* in midwestern and eastern North America. F = resistance detected by field experience, L = resistance assessed by laboratory test, ( ) = subscript figures in brackets refer to literature cited, ** = resistance to carbaryl, vide (56), and * = C. M. Watve, personal communication.

an incipient resistance to dieldrin and DDT in the black garbage fly, *Ophyra leucostoma,* which is known to be a predator of the house fly. Atallah & Newsom (13) had reported a six-fold resistance to DDT in samples of the coccinellid *Coleomegilla maculata* collected in Louisiana in 1966. More recently, samples of *C. maculata* collected from cotton fields in Louisiana (R. Mohamed, personal communication) and Mississippi (40) have shown, respectively, a 10-fold and a 35-fold resistance to parathionmethyl, which has been applied on this crop for almost two decades.

FACTORS GOVERNING THE DEVELOPMENT OF RESISTANCE IN NATURAL ENEMIES

Of the factors influencing the development of resistance, such as length of life cycle, type of reproduction, sex ratio, vagility, dispersal behavior, degree of isolation, and proportion of population selected, very much less is known for natural enemies than for pest species. Predators and parasites also may have additional features peculiar to their ecological mode of life which make them respond differently from pests to the selective action of insecticides.

At the outset, it is a question whether the markedly greater frequency of resistance reported among pests than among predators and parasites is simply due to the fact that these natural enemies have been inadequately sampled and tested. Newsom (164) has stated that "observational evidence from field experiments strongly suggests that resistance has probably evolved to most of the widely used insecticides in many other predaceous insects and mites." On the other hand, Georghiou (85) concludes that "there is little doubt that the development of resistance in natural enemies is severely hindered and lags far behind that of their prey or host species."

This question is difficult to answer in the present state of knowledge. While the rapidly increasing interest in biological control during the past decade has resulted in many evaluations of the toxicity of insecticides to a variety of natural enemies, few susceptibility tests comparing different populations have been made upon them. Had these arthropods been pests and had resistance developed, it would seem unlikely that widespread resistance would not be detected. However, in addition to the complication arising from predators that are naturally tolerant or capable of rapidly reinvading from untreated sites, there is a unique detection problem which makes the discovery of resistant strains more difficult for natural enemies than for pests. If a crop has a low economic threshold for a pest which also is the prey or host of a resistant natural enemy and intensive chemical control measures are necessary to maintain it below these levels, then a resistant natural enemy could only be abundant when the pests were also abundant. In this infrequent event, a subsequent treatment would reduce the pest to subthreshold levels, with the result that the resistant natural enemy would appear to be susceptible because its numbers had been severely reduced by lack of hosts. Unless the resistant natural enemy could find alternate prey or host species in the sprayed habitat, it would never be abundant and the resistance would probably go undetected.

If the statement of Newsom (164) is correct and the situation described above is a significant factor, then the systematic testing of samples of natural enemies in heavily treated ecosystems should detect more cases of resistance, provided of course that susceptible strains can be obtained for comparison from untreated areas. It may be useful in documenting the responses to chemicals in the future to establish baseline data before the compound is released or widely used, as is the current practice for pest species. To do this, standardized susceptibility test methods specifically for predators and parasites must be developed and established.

If the paucity of known resistant natural enemies as compared to the abundance of known resistant pests is a true reflection of the real situation, then these two groups must inherently differ in their potential for adapting to these toxicants. However, the hypothesis that plant-feeding pests can adapt faster because of a higher genetic variability resulting from their greater exposure to plant toxicants lacks sufficient evidence. On the other hand, there is a considerable amount of evidence that pests and natural enemies are similar physiologically, and therefore we should look to differences in exposure, selection pressure, or population dynamics to account for the differential frequency in resistance responses among these two groups.

Concerning differential exposure, it would seem evident that natural enemies often are less intensively selected than pests simply because toxicants, being primarily directed at the pests, affect the natural enemies only insofar as they occupy the same habitat as their prey or host. It is the exception rather than the rule that pests and their natural enemies occupy the same habitat throughout their life cycles, and even then differences in their biology, behavior, and feeding habits ensure a certain differential in the exposure of either group to the toxicants. On the other hand, there are many reports that high mortality of natural enemies often occurs, and that most treated habitats exert a significant selection pressure on the predator and parasite populations. Where the development of multiple resistance by the pest has dictated the use of a variety of insecticide types, the natural enemies might be even more intensively selected where they had not already developed such resistances.

Among the predators which have acquired resistance, it is the phytoseiid mites that, with respect to resistance levels attained, number of species involved, and variety of chemical resistance types, have shown the greatest potential of any natural enemy group studied to date. This propensity is undoubtedly related to their biological characteristics, and to their association with spider mites which also are notorious for developing resistance to pesticides (95). All five species of phytoseiid mites with resistant populations (Table 7) are specialized predators of spider mites (147). Both *Typhlodromus occidentalis* and *Amblyseius fallacis* multiply no less rapidly than their prey, have many generations per season, are continuously exposed in all life stages in the same habitat as their prey, have limited powers of dispersal which would reduce interbreeding with wild unexposed types, can persist under relatively low prey densities, and can be maintained in the absence of prey by alternative foods including pollen, honeydews, and plant nectars. Such characteristics ensure that these predators are exposed to a selection process similar to that experienced by their prey, and that the resistant genotypes thus selected would find adequate food for their survival and multiplication after the acaricide treatment. Since most plants can sustain considerable foliar damage from spider mites before any significant economic damage is sustained, these predaceous mites more often find prey levels that are sufficient to allow for their multiplication after the insecticidal treatment.

Although there is only one reported case of resistance among coccinellid predators, namely a 35-fold increase in the LD_{50} of parathionmethyl shown by *Coleomegilla maculata* in Mississippi (40), it appears likely that this family of natural enemies also has the capacity to develop highly resistant populations. If allowances are made for its longer life cycle, it is possible that the resistance developed in the coccinellid beetle in as few generations as in the phytoseiid mites.

Probably the most significant element in relation to the action of insecticide application on natural enemies is related to the additional stress imposed by the virtual elimination of their food source (50, 85, 164). Although the toxicant may exert a similar physiological stress on both pest and natural enemy, the subsequent events which govern survival and multiplication of the remaining and possibly resistant genotypes are considerably different for these two groups. Almost always, pests have an unlimited food supply available to the selected populations that survive treatment. But for natural enemies, the more effective the treatment on the pest species, the scarcer becomes the prey or hosts on which the predators or parasites depend. Since at some minimum threshold of prey or host density the natural enemy will be unable to find sufficient prey for their development or hosts for their oviposition, the net result is either starvation of the surviving genotype or their dispersal to areas where their resistance potential is diluted by interbreeding with wild untreated types. On the other hand, the natural enemy could persist if it is polyphagous or can turn to an alternative food supply, and it would be among these species that resistance would be most likely to develop first.

It seems probable, considering the dependence of a natural enemy on an adequate food supply, that resistance would develop sequentially, first in the pest or host and subsequently in the natural enemy population. Although the common practice of discontinuing the particular insecticide involved in a case of decisive pest resistance would set a limit to this process in the natural enemy, this hypothesis leads to the conclusion that there must be cases where natural enemies have been given the opportunity to be further selected once the pest which they predated or parasitized had developed resistance. The OP resistance of those tetranychid mites which are the principal prey of the phytoseiid mites which have developed it, namely *Tetranychus mcdanieli, Tetranychus pacificus, Tetranychus urticae, Panonychus ulmi,* and *Panonychus citri,* developed at some period between 1947 and 1959, depending on the part of the world involved (95). With the continued application of OP compounds such as azinphosmethyl to control the entire complex of fruit pests, the resistance developing in the predators reached the point of being detected and was reported by the following dates: 1969–1970 for *Typhlodromus occidentalis* (54, 103); 1968–1970 for *Amblyseius fallacis* (157, 221); and 1972 for *Typhlodromus pyri* (104). Resistance to DDT, applied in peach orchards to control the oriental fruit moth, also occurred first in the European red mite and subsequently in the predator *Typhlodromus caudiglans* (97).

A similar sequence may be discerned with the coccinellid *Coleomegilla,* predaceous on aphids and on noctuid larvae including the cotton bollworm and the tobacco budworm. With the progressive introduction of parathionmethyl to control cotton pests starting around 1955, both of these species of *Heliothis* have been reported to have developed OP-resistant populations by 1970 (34). Nevertheless parathionmethyl has continued to be applied on cotton since the boll weevil is still susceptible to it, and these treatments have evidently allowed the coccinellid predator to be further selected to the point where a significant resistance level has been attained.

With parasites, the record of reported cases (Table 7) would indicate that they do not have as great a potential for developing resistance as predators. In laboratory selection experiments parasites did not show the ability to produce highly resistant strains which have been well documented for pest species when selected in the laboratory (85). Colonies of the boll weevil parasite *Bracon mellitor* derived from field collections from three different parts of Mississippi developed only a 4–8-fold tolerance to several cotton insecticides (2). A colony of *Macrocentrus ancylivorus* (174, 194, 241), which like *Trichogramma evanescens* (125) showed a flat d-m regression line indicative of considerable heterogeneity, attained only a 12-fold resistance to DDT after 19 generations of selection. Furthermore, resistance actually decreased despite continued selection for a further 52 generations, and when selection pressure was removed susceptibility reverted almost to its original level in 13 generations.

Evidence to refute this generalization has been reported by Spiller (213) who noted that there was circumstantial proof that *Aphelinus mali,* a parasite of the woolly apple aphid, developed resistance to DDT and lindane in New Zealand. Chesnut & Cross (43) have also reviewed the intensity of parasitism of the boll weevil in the United States over the past 30 years and interpreted the increase of average parasitism rates by *Bracon hebetor* from an original 4% to a final 13% to indicate a developed resistance to cotton insecticides. Another example is found in *Apanteles congregatus.* Its host, *Manduca sexta,* has characteristically become tolerant to all varieties of tobacco plants, but populations feeding on dark tobacco, a variety with a high nicotine content, were not successfully parasitized by *A. congregatus* during the early years of the present century (87). In the 1970s, Thurston & Fox (229) found that sublethal doses of nicotine topically applied to parasitized hornworm larvae failed to assure the mortality of the parasite larvae within (although some mortality occurred when the parasite emerged). By comparing their results with the earlier observations, they concluded that *A. congregatus* in a sequential pattern similar to those observed for predator mites and *Coleomegilla* has been in the process of adapting physiologically to nicotine over the past 70 years after the hornworm had completed a similar adaptation.

There are at least three factors which may limit the ability of parasites to develop resistance more than that of predators. First, predators are generally more polyphagous than parasites; thus while the surviving plus-variants of predators could more readily turn to alternate prey or other foods after the decimation of their prey by insecticidal treatment, the survivors among the monophagous parasites would probably have nowhere else to turn after the elimination of their specific hosts. Second, predators more often share the same habitat with their prey at all stages of their life cycle and are exposed externally. By contrast parasites usually are not associated with their host at all stages and many are internally protected from exposure to insecticides. Finally, since the survival of a resistant genotype of an endoparasitic species is dependent upon the survival of its specific host, the chances of its survival are a product of the frequencies of resistant genotypes in the host species as well as in the parasite species. Thus if the frequency was 1 in 10,000 in the parasite and 1 in 100 in the host, the chance of the resistant parasite genotype surviving would be 1 in 1,000,000. With a predator, the survival of a resistant individual is only dependent on its frequency of occurrence and its ability to find an individual of any prey species that survives the treatment. Thus it may be concluded that just as parasites are generally more susceptible to insecticides than predators (as has been concluded above), so also they are less able to develop tolerance or resistance, as compared to the predators occurring in the same habitats. In this sence, it may be worthwhile to consider, with Huffaker (106), that parasites in developing their specialized mode of life have sacrificed enough genetic diversity to handicap them in adapting to new stresses such as insecticides. While Georghiou (85) conceded that it was possible that a highly specialized parasite has a lower genetic plasticity than an omnivorous insect, he pointed out that it would be difficult to prove.

APPLICATION OF DATA TO PRINCIPLES AND PRACTICE

The question whether natural enemy complexes exhibit the same responses to insecticide treatments as are commonly observed among pests, giving rise to the phenomena of species displacement, trophic level simplification, and pest resurgence, has recently been posed by Newsom (164). He also raised the corollary question whether it would be possible to augment the populations of predatory species by using insecticides to relieve the pressures from their natural enemies (secondary parasites or predators and competitors) reasoning that it should be possible to do by design for predators what has been done accidentally for pest species. In other words, could natural enemies be managed more efficiently with insecticides if we understood more precisely how they are influenced by these chemicals?

The complex of predators that attack tetranychid mites in deciduous fruit orchards will now be taken as an example, and in particular those systems where the phytoseiid mites *Typhlodromus occidentalis* and *Amblyseius fallacis* are the principal species, since these predators have been the subject of the most detailed research towards management by insecticide manipulation. The relevant characteristics of these systems, as described in three recent reviews (108, 147, 235), are as follows: 1. In unsprayed habitats throughout the world, spider mite pests are normally regulated by predators at low equilibrium levels. 2. Their predatory complexes may differ species-wise from region to region but will usually include members of the Acarina, Araneida, Coleoptera, Neuroptera, Hemiptera, Thysanoptera, and Diptera. 3. The life cycles and population dynamics of these predators and pest mites are relatively well understood, as well as the secondary predation relationships between predators and their natural enemies. 4. Individual natural enemy species have been characterized with respect to their control efficiencies, namely their ability to control the pest as quickly as possible and/or to regulate the pest at as low a density as possible without themselves dying from starvation. Thus the groups composing the predator complex can be ranked in the following (approximate) order of descending efficiency: (*a*) phytoseiid mites, Coccinellidae (*Stethorus*); (*b*) Araneida, Thripidae, stigmaeid and other predaceous mites; (*c*) hemipteran predators (Anthocoridae, Miridae, Lygaeidae, Nabidae); (*d*) Neuroptera (Chrysopidae, Hemerobiidae, Coniopterygidae); (*e*) dipterous predators (Cecidomyidae, Syrphidae). 5. Spider mites soon develop resistances to most of the insecticides applied to control other fruit pests (e.g. codling moth) as well as to specific acaricides, while most natural enemies continue to be killed by sprays; consequently spider mites become severe insecticide-induced pests. 6. An immense body of data has been obtained in the last 25 years on the responses of the spider mites and their natural enemies to a wide variety of insecticides and acaricides.

There now exists a sufficiently detailed understanding of the selective action of insecticides to allow for the development of effective spider-mite control programs based on the integration of chemicals and biological agents. The data available for the systems where *Typhlodromus occidentalis* and *Amblyseius fallacis* are the principal predators may indicate the potential for management in other systems of pest and natural enemy, as their responses to insecticides become better understood. In answer to the first question posed by Newsom (164), certain individual species of phytoseiid mites have in fact responded to insecticide treatments by becoming abundant, and the predator complex has constricted with respect to species, at least within the constraints imposed by the lower trophic levels of their spider-mite prey and the original plant host. Downing & Moilliet (66, 67) observed that *T. occidentalis* rapidly increased in density in British Columbia apple orchards which had been intensively treated with OP insecticides for the previous two years. *Typhlodromus caudiglans* and *Typhlodromus pyri,* characteristic predators of the low-density populations in unsprayed orchards, declined to almost an undetectable level along with several less abundant phytoseiid species. Evidently the insecticide regime had caused species displacement of those phytoseiids that were competing with *T. occidentalis,* which then rapidly multiplied in response to the increased populations of spider mites that had been released from the more diverse and effective predatory complex. If sprays were withheld for the subsequent two years, the resident complex of phytoseiid mites reappeared in the orchard and *T. occidentalis* was relegated to its original insignificant role. In other areas of western North America, *T. occidentalis* is the species most commonly associated with intensively sprayed orchards, despite wide variations in the complex of predatory species associated with unsprayed orchards in these areas (49, 55, 120, 132, 246). In many cases, intensive insecticide applications result in similar patterns of ecosystem simplification by displacement of certain predatory species, followed by the resurgence of a single or relatively few species accompanied by an increase in the pest. An almost identical response has been observed for the predator *A. fallacis* in central and eastern North American apple orchards wherever these relationships have been extensively studied (124, 177, 212).

Thus the experience with orchard mites conclusively demonstrates that just as insecticides simplify the composition of phytophagous pest populations and favor those species most tolerant or able to adapt, so also insecticides induce certain predatory species to become abundant as other species which are competitors or secondary predators become eliminated. In other ecosystems and natural enemy complexes, most of which have been less intensively studied, similar responses have been reported. The literature yields selected examples of each type of response, not only species displacement (30, 48, 179, 238), but also species resurgence (1, 30, 38, 47, 63, 88, 238) and ecosystem simplification (16, 38, 48, 63, 193). The ecosystems involved include a variety of habitat types (e.g. annual crops, coniferous forests, aquatic systems), and almost all the major taxonomic groupings of natural enemies.

Passing to the corollary question raised by Newsom (164), it is not unreasonable to assume, since most agricultural systems are artificially produced and maintained, that the natural enemy complexes could similarly be manipulated to take advantage of the most efficient combinations of predators and parasites. While it is preferable to establish a natural enemy complex consisting of a diverse assemblage of species, the combination in which maximum diversity is maintained for stability purposes may not be the most useful one for relevant biological control. On short-term annual crops, for example, where the disruptive cultural practices followed serve to encourage pest instability, a single or restricted natural enemy complex which rapidly responds and overexploits the available prey or hosts may be the most desirable (109, 167).

While management of natural enemy complexes, in view of the dearth of selective insecticides and our limited knowledge of the biology and toxicology of predators and parasites, would seem somewhat premature for general application, nevertheless this possibility has been quite extensively investigated for the natural enemies of tetranychid mites. The data obtained for systems where the phytoseiid mites *T. occidentalis* and *A. fallacis* are the principal predatory elements will now be reviewed under the headings of insecticide and acaricide selectivity, natural enemy manipulations, the use of resistant natural enemies, and the dynamics of long-term selection. The danger in making generalizations for other complexes of pests and their natural enemies is acknowledged.

Insecticide and Acaricide Selectivity

The initial step in developing an integrated pest control program is to assess the pesticides applied against pest arthropods for their effect on the natural enemies. For predators and tetranychid mites, this type of study has greatly increased during the past decade and has reached the point where almost all the pesticides commonly applied to fruit crops have been tested for their effect on such important species as *T. occidentalis* and especially *A. fallacis* (Table 8). At present, several research groups in the United States test new experimental chemicals as much for their influence on these predators as for their effectiveness against the spider mites. Data of this type are a prerequisite for the development and maintenance of integrated programs of mite control, and this has been recognized and has probably received more attention than any other topic connected with management of natural enemies by chemical means.

Table 8 Concentration levels at which pesticides are negligibly, moderately, or highly toxic to *Amblyseius fallacis*

Compound	% concentration	Method	References
		Negligibly toxic	
Aramite	0.009-.400	F, R[a]	44, 192, 211
Azinphosmethyl[b]	.030-.060	F, S, R	57, 59, 221, 222
Benomyl	.023	S, R	57
Carbaryl[b]	.030	R	56
Carzol	.015-.018	F	101
Chlorfenson	.060	R	211
Chlorobenzilate	.030	R	211
DDT[b]	.060	R	211
Dicofol	.044	F	151, 211
Dieldrin	.030-.060	F, R	151, 211, 222
Dikar	.048-.096	F	59, 151
Dimethoate[b]	.034	S, R	57, 59
Dinocap	.008	F	59
Dithianon	.045	F, S, R	57, 151
Dodine	.029-.156	F, S, R	57, 59, 101, 151, 211
Endosulfan	.030	F, S, R	57, 59
Ferbam	.200	F	44
Folpet	.120	R	211
Glyodin	.150-.200	F	44, 192, 211
Maneb	.168	R	211
Menazon	.009	F	202
Metiram	.192	F	59, 101
Omite	.026-.045	F, S, R	57, 59, 101, 151, 197
Oxythioquinox	.008	F	151
Parathion[b]	.275	S	198
Phosmet[b]	.030-.360	F, S, R	57, 59, 101, 151, 221
Phosphamidon[b]	.008	S, R	57, 151
Plictran	.005-.015	F, S, R	57, 101
Tetradifon	.060	F, R	211
Zineb	.156	F	211
		Moderately toxic	
Aramite	0.027	F	44
Benomyl	.015-.023	F	59
Binapacryl	.030-.060	F	151
Carbaryl[b]	.060-.300	S, R	56
Carzol	.060	F	57
Chlorbenside	.024-.048	F, R	44, 202, 211
Chlorfenson	.045	F	44
Chlorobenzilate	.003-.030	F, R	44
Chloropropylate	.030	F	59, 101, 151
Diazinon	.030-.060	F, S, R	44, 57, 59, 151, 211, 221
Dichlone	.023-.030	F	44
Dicofol	.021-.052	F	59, 101
Dieldrin	.060-.500	F, R	44, 192
Dikar	.096-.192	F, S, R	57, 59, 101, 151
Dimethoate[b]	.030-.068	F, S, R	57, 59, 151
Dinocap	015-.023	F, S, R	44, 57, 59, 151

Table 8 (Continued)

Compound	% concentration	Method	References
	Moderately toxic (Continued)		
Endosulfan	0.060	F, S, R	57, 59, 211
Ferbam	.140	R	211
Gardona[b]	.030–.045	F, S, R	57, 59, 211
Heptachlor	.030–.100	R	191, 211
Lime-Sulfur	1.8 – 2.0	F	44, 59
Lindane	.060–.170	R, F	44, 211
Methoxychlor	.060–.180	F, R	44, 211
Milbex	.038	F	202
Oxythioquinox	.015–.047	F, S, R	57, 59, 151, 197, 202
Phosphamidon[b]	.015–.030	F, S, R	57, 59, 151, 202
Plictran	.015–.030	F, S, R	57, 59, 101, 151
Sulfur	.15 –.30	F	97, 151
Tetradifon	.090–.090	F	101, 202
Zineb	.156	F	44
	Highly toxic		
Azinphosmethyl	0.025–.030	F, S	151, 197, 198, 202, 211
Carbaryl	.030–.120	F, S, R	56–59, 97, 101, 151, 211, 222
Carbofuran	.002–.090	F, R	56, 151, 222
Carbophenothion	.006–.030	F	202, 221
Chlorfenson	.090	F	44
DDT	.003–.120	F, R	44, 97, 192, 222
Demeton	.010–.030	F, S, R	44, 57, 59, 151, 211
Diazinon	.090	F	44
Dicofol	.037	F	202
Dimethoate	.028–.100	F, R	202, 211
Dinobuton	.060	F	151
Dioxathion	.060	F	221
EPN	.032–.048	F, R	44, 211
Ethion	.030–.036	F, R	211, 221
Fundal	.030–.060	F, S, R	57, 59, 101
Galecron	.060	F	59, 151
Gardona	.030–.062	S, R	57, 59, 197
Heptachlor	.500	R	192
Leptophos	.045	S, R	57
Lovozal	.036	F	151
Lindane	.020	R	192
Malathion	.001–.060	F, R	44, 151, 192, 202, 211
Mesurol	.015	F	151
Monocrotophos	.015	F	151
Parathion	.001–.036	F, S, R	44, 97, 197, 211
Phosalone	.034–.060	F, S, R	57, 58, 59, 151
Phosmet	.050–.060	F, S, R	197, 202
Phosphamidon	.030	S, R	57, 59
Schradan	.054	F	44
Sulfur	.30 –.60	F	44, 57, 101

[a]F = Field application, S = Slide-dip method, and R = Exposure to residue.
[b]Pesticide tested on a strain of *Amblyseius* resistant to that pesticide.

Insecticide selectivity to benefit a natural enemy, as defined by Ripper (186–188), can be attained either by a physiological or by an ecological mode. There are several chemicals which are physiologically selective by virtue of being more toxic for spider mites than for the phytoseiid predators *T. occidentalis* and *A. fallacis,* as may be seen in a tabulation of the LC_{50} levels for *A. fallacis* and two spider mite pests for the compounds Omite, Plictran®, dinocap, and Dikar® in complexes 3 and 4 of Table 5. With the exception of the greater toxicity of dinocap and Dikar to *A. fallacis* as compared to *Tetranychus urticae,* these acaricides or acaricide-funigicides provide for selective mortality of pest mites in situations where the ratios of *A. fallacis* to spider mites are not sufficiently high to ensure effective biological control.

Selectivity in the ecological mode can be gained in a variety of ways; in the case of *Amblyseius fallacis* and its prey (57, 59), it is based on correct timing of applications of pesticides chosen for their toxicity and persistence (Table 9) in relation to the habits of this predator in commercial apple orchards. For example, most of the adult female *A. fallacis* overwinter in the debris and ground cover near the bases of deciduous fruit trees. During the early season (April-June) the predators increase by feeding on the phytophagous mites that they find in the ground cover, and only after mid-June and into July do they migrate into the fruit trees and control spider-mite populations. After controlling the pest mites in the trees, they return to feed on the ground-cover mites or to overwinter. It has been substantiated that sprays applied to fruit trees between April and mid-June containing compounds known to be moderately toxic to *A. fallacis* do not have an appreciable effect on the predator populations even though some of the spray drifts onto the ground cover (Table 9). During the period of predator-prey interaction in the tree, only compounds of slight or negligible acaricidal activity are applied. After control has been attained and the predators have dispersed back to the ground cover, compounds of moderate toxicity to mites can again be employed.

Another example of ecological selectivity is the practice of applying carbaryl for fruit thinning only on the outer tree foliage in early springs. When followed in the orchards of western North America, it has the effect of conserving the populations of *T. occidentalis,* which are mostly confined to the inner tree regions during the spring (103, 140).

Manipulations of a Natural Enemy Complex

Preliminary to any attempts at manipulation by insecticide treatments, the food-web relationships of the natural enemy complex, e.g. those species centering on the phytoseiid *A. fallacis* which feed on spider mites and on each other in apple orchards (Figure 3), must be thoroughly understood. Conventional broad-spectrum insecticides applied at rates recommended for deciduous fruit pest control are almost always toxic to all macropredators and certain micropredators of the complex (100, 102, 151). When the phytoseiid present as the principal predator develops resistance to OP compounds as *A. fallacis* has done, only this species and the naturally OP-tolerant stigmaeid mites *Agistemus fleschneri* and *Zetzellia mali* are able to persist after OP compounds are applied (59, 102, 151). At this point, by modifying spray programs, the composition of the natural-enemy complex can be altered and

Table 9 Insecticide timing for ecological selectivity to *Amblyseius fallacis* in Michigan apple orchards (vide 59)

Compound	Formulation	Rate/ 100 gals	Rate/ acre	Useable to June 1 or 1st cover	Useable to June 10 or 2nd cover	Useable June 10 to harvest	Useable after predator-prey interaction has occurred
Phosphamidon	8 EC	1/2 pt	2 pts	X			X
Phosphamidon	8 EC	1/4 pt	1 pt	X	X		X
Phosphamidon	8 EC	1/8 pt	1/2 pt	X	X	X	X
Demeton	6 EC	1/3 pt	1 1/3 pt	X	X		X
Demeton	6 EC	1/4 pt	1 pt	X			
Dimethoate	2.7 EC	1 1/2 pts	6 pts	X	X		X
Dimethoate	2.7 EC	1 pt	4 pts	X	X	X	X
Dimethoate	25 WP	2 lbs	8 lbs	X	X		X
Dimethoate	25 WP	1 lb	4 lbs	X	X	X	X
Azinphosmethyl	50 WP	1/2 lb	2 lbs	X	X	X	X
Phosmet	50 WP	2 lbs	4 lbs	X	X	X	X
Gardona	75 WP	2/3 lb	2 2/3 lbs	X[a]			
Endosulfan	50 WP	1 lb	4 lbs	X	X		X
Endosulfan	50 WP	3/4 lb	3 lbs	X	X	X	X
Carbaryl	50 W	2 lbs	8 lbs	X[a]			
Diazinon	50 W	1 lb	4 lbs	X	X		X
Phosalone	3 EC	1 pt	4 pts	X[a]			

[a] Use as early in the season as possible.

structured in several different ways. First, either the stigmaeid mites or *A. fallacis*, or both, can be eliminated and manipulated by application of certain selective acaricides (57, 162). In Michigan apple orchards, preliminary studies (B. A. Croft, unpublished data) have indicated that selective chemical control of stigmaeid mites may be desirable and even necessary if *A. fallacis* is to exert its full potential for biological control. Second, if the dosages of certain of the broad-spectrum insecticides are reduced, *Stethorus* can be conserved and maintained as an addition to the two mite predators or by different chemical combinations can be maintained as the sole principal predator (9–11). Third, if the conventional insecticides are applied at very low rates, or if they are replaced by ryania, lead arsenate, or *Bacillus thuringiensis* spore preparations, the entire predator complex can be conserved (99, 139). In these cases, however, the spider-mite control attained is not accompanied by effective control of other fruit pests.

Insecticide manipulations of the types discussed above, when performed experimentally, can be useful tools for assessing the components of a natural enemy complex. The role of individual species can be separately evaluated, the subtleties of interspecific competition can be characterized, and controlled interactions between species at various density levels can be artificially maintained by selective chemical use. In using these techniques, one must be aware of the indirect effects that insecticides may have when administered in sublethal doses on pest mites (108) and their natural enemies (as discussed in a previous section). The use of these insecticides as investigational tools also provides a means of obtaining the biological

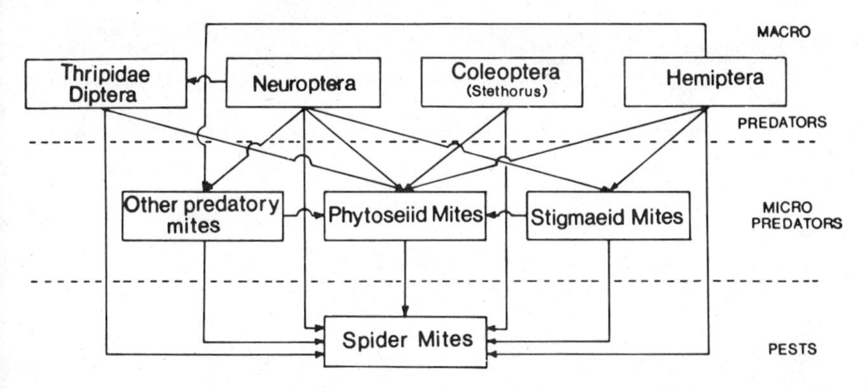

Figure 3 Food-web relationships between spider mites and their principal arthropod predators.

and chemical impact data necessary for modeling the interactions of spider mites and their natural enemies (51). Since the maintenance of the entire predator complex is impossible under field conditions, it is necessary to determine what manipulation of natural enemies provides the most effective management of the mite pests. Among predator mites, for example, the presence of stigmaeids and phytoseiids at the same time may not be the most favorable situation. Although it has been suggested that the stigmaeid mites might be conserved in early spring, since they may provide appreciable biological control of spider mites on the trees at a time when *A. fallacis* is as yet on the ground (R. P. Holdsworth, personal communications), they might well be selectively suppressed later on, in order to allow the more effective phytoseiids which have now ascended the apple trees to assert their far greater potential for control. Through combining an understanding of the habits of the natural enemy components with a knowledge of their responses to the insecticides available, rather refined manipulations are indeed possible and more advantageous levels of pest control may be achieved.

Resistant Natural Enemies in Pest Management

The occurrence of insecticide resistance in natural enemies as effective as *Typhlodromus occidentalis* and *Amblyseius fallacis* greatly improves the potential for integrated chemical and biological pest control. This was proved by an experiment in a southern California apple orchard with two releases of OP-resistant *T. occidentalis,* for comparison with a release of susceptible *T. occidentalis.* Here each of the resistant populations, from Washington (A) and Utah (B), proved to control McDaniel spider mites early in the season before the pests could exceed 10 or 20/leaf (Figure 4). The release of susceptible predators at the same density proved unable to provide control until spider mites attained 40/leaf (Figure 4C). In other trees, where no releases were made and the only source of predators was immigration

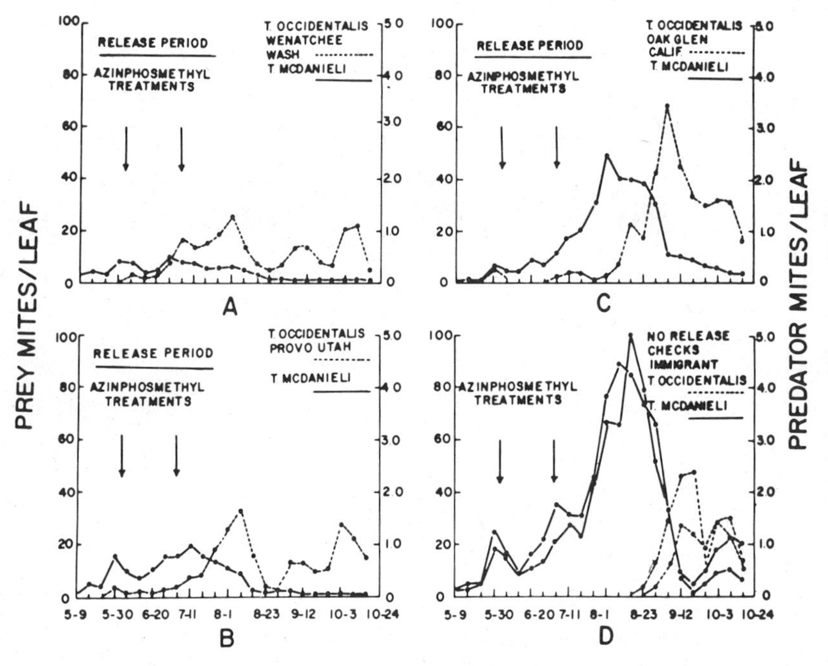

Figure 4 The numerical abundance of *Typhlodromus occidentalis* and its prey *Tetranychus mcdanieli,* following release of OP-resistant strains (A and B) and a susceptible strain (C) of this predator for comparison with a check without releases (D), vide (53).

(Figure 4D), the pest density attained more than 100/leaf (52). This greater effectiveness of resistant predator strains has also been exploited among naturally occurring populations. In Washington state, *T. occidentalis* mites which are resistant to OP insecticides are carefully monitored in apple orchards where they provide effective biological control of pest mites (103, 105). Before 1965, chemical control of the McDaniel spider mite was very costly since it had developed resistance to many acaricides. A selective spray program was subsequently developed and recommended which did not kill the resistant predatory mite but still produced control of all the major insect pests. This combination of selective chemicals and resistant predators provided cheaper and more permanent control of spider mites than that attained with pesticides alone (105). Similar integrated programs for mite control on apples in British Columbia (65), Utah (60), California (52, 53), Colorado (J. A. Quist, personal communication), and Oregon (246); pears in Oregon (239); plums in Washington (E. W. Anthon, personal communication); peaches (105) and grapes (78, 79, 121) in California; and hops (180, 181) in Washington have utilized the biological control of spider mites made possible by insecticide-resistant populations of *T. occidentalis.* In New Jersey (221, 222), North Carolina (196), Illinois (151), Ohio (101), and Michigan (59), similar programs with OP-resistant *A. fallacis* when adopted by fruit growers allow substantial reductions to be made in acaricide use while obtaining effective control of the European red mite and two-spotted spider mite.

An additional manipulation which has a possible practical application is to detect insecticide-resistant natural enemies where they first appear and then distribute them into regions where they have not developed or do not occur. Its feasibility was demonstrated by the experimental introduction of azinphosmethyl-resistant strains of *T. occidentalis* from Washington and Utah into a southern California apple orchard (52, 53). Whereas an indigenous susceptible population could not survive sufficiently to exert any control in a treated environment, the introduced resistant strains interbred with the native mites and established a hybrid resistant population which provided good control of the McDaniel spider mite. Subsequent samples taken in 1972 have further verified that the resistance character has been permanently established, and it probably will continue indefinitely if insecticide selection is continued (B. A. Croft, unpublished data). Since this initial study, introductions of OP-resistant strains of *T. occidentalis* have been made into orchards in Idaho (130), Colorado (J. A. Quist, personal communication), Australia (J. L. Readshaw, unpublished observations), and the Netherlands (122), and of *A. fallacis* into British Columbia, Washington, and New Zealand (B. A. Croft, unpublished records). Although it is too early to evaluate the successes of these releases, Kirby (122) has suggested that more attempts to establish these resistant populations should be made in the future.

Although the OP-resistance features available in *T. occidentalis* and *A. fallacis* populations are sufficient to allow phytoseiid survival where certain OP insecticides such as azinphosmethyl and phosmet are applied to control insect pests such as the codling moth, plum curculio, apple maggot, and red-banded leafroller, they would be useless should any one of these pests develop OP resistance and dictate a substitution of a new group of compounds. It therefore would be better if the phytoseiid employed were multiresistant, and towards this end the laboratory development of a hybrid strain of *A. fallacis* developed from crosses of OP-resistant and carbaryl-resistant populations collected from the field (56) is a constructive measure which fully warrants the steps being taken to establish it in the field. Artificially developed multiresistant strains would anticipate what it may take natural populations years to develop, if at all.

Long-Term Dynamics of Insecticide Resistance

A detailed understanding of the responses of arthropod natural enemies to insecticides could eventually lead to prognosis and management of predators and parasites, and of their prey or hosts, even while they are in the process of developing adaptations to the multiplicity of insecticides now available for application. Past experience has shown that when economic thresholds for pest density were considered to be "no pests present" and applications were made solely with that objective, they have given rise to a resistance picture in which the benefits of its presence in the natural enemy do not begin to counterbalance the detriment resulting from its intensity and frequency of occurrence in the pest population. It may be possible, however, to modify this trend in the future by setting more realistic thresholds for pest density and by applying insecticides less frequently and more carefully. In the apple orchards of Washington state, for example, higher thresholds for certain indirect pest arthropods including spider mites were established in 1965, and since that date lesser amounts and fewer applications of broad-spectrum insecticides and acaricides have been applied (103, 105). There has been no case of a deciduous fruit pest acquiring significant resistance during the nine-year period since this modified pest management program was adopted (S. C. Hoyt, personal communication), in marked contrast to the previous ten-year period. Moreover, in the four years from 1965 to 1969 the natural enemy *Typhlodromus occidentalis* rapidly developed low-level tolerances to a variety of insecticides (54), in addition to intensifying its OP-resistance. It would be interesting to see whether these tolerance and resistance levels have further increased during the last 5 years.

From this recent experience, two conclusions may be tentatively drawn which could have widespread implications for future pest management on a variety of crops. First, it would seem likely that the addition of a selection pressure from the natural enemies thus preserved, taken in conjunction with the reduction in the insecticide pressure, would delay the development of resistance in the pest. Second, the survival of sufficient prey or host species ensured by a program with a more relaxed economic threshold would allow the natural enemy to multiply and respond by becoming resistant to a wider spectrum of insecticides. It would therefore follow that although economic thresholds for pests will continue to be largely dictated on the grounds of acceptable damage, some consideration should be given to ensuring that sufficient numbers of prey or hosts remain to serve as food for the resistant genotypes of natural enemies surviving the applications. Those natural enemies which are capable of finding prey or hosts at low density levels may develop resistance at the same time as their prey. For other natural enemies, however, the limitations of a low economic pest threshold in relation to their searching capacity may militate against pest-control applications that allow such simultaneous adaptations to occur. There is a need for the entomologist to follow the selection process in natural enemies as well as in the pests by periodically assessing the insecticide-susceptibility levels of both these elements of the ecosystem. The greater overall understanding of selection in both groups thus obtained could result in an optimization of spray practices leading to a greater balance in the adaptation of both groups to insecticides and ultimately to a greater balance in the ecosystem; thus avoiding our past error of destroying natural enemies and inducing resistance in the pest.

Acknowledgments

This work was supported in part by NSF grant No. GB-34718 and USDA-ARS, ERS 12-14-100-11,200(33) entitled "Principles, Strategies and Tactics of Pest Population, Regulation and Control in Pome and Stone Fruit Ecosystems." Thanks are expressed to M. J. Nakashima and L. D. Newsom for their help in preparing this manuscript. Published as article 6799 of Michigan State University Agricultural Experiment Station, Michigan State University, East Lansing.

Literature Cited

1. Abdel-Salam, F. 1967. Uber die Wirkung von Phosphosaureestern auf einige Arthropoden innerhalb der Apfelbaum-Biozönose in Abhängigkeit von ihrer Dichte. *Z. Angew. Zool.* 54: 233–83
2. Adams, C. H., Cross, W. H. 1967. Insecticide resistance in *Bracon mellitor*, a parasite of the boll weevil. *J. Econ. Entomol.* 60:1016–20
3. Adams, J. B. 1960. Effects of spraying 2,4-D amine on coccinellid larvae *Can. J. Zool.* 38:285–88
4. Adashkevich, B. P. 1966. Effect of chemical and microbiological treatments upon parasites of *Plutella maculipennis. Zool. Zh.* 45:1040–46
5. Adkisson, P. L. 1958. The influence of fertilizer applications on populations of *Heliothis zea* (Boddie) and certain insect predators. *J. Econ. Entomol.* 51: 757–59
6. Ahlstrom, K. R., Rock, G. C. 1973. Comparative studies on *Neoseiulus fallacis* and *Metaseiulus occidentalis* for azinphosmethyl toxicity and effects of prey and pollen on growth. *Ann. Entomol. Soc. Am.* 66:1109–13
7. Ahmed, M. K. 1955. Comparative effect of systox and schradan on some predators of aphids in Egypt. *J. Econ. Entomol.* 48:530–32
8. Ankersmit, G. W., Locher, J. T., Velthuis, H. H. W., Zwart, K. W. R. 1962. Effect of insecticides, acaricides and fungicides on *Mormoniella vitripennis. Entomophaga* 7:251–55
9. Asquith, D. 1972. Initiating integrated pest management in apple orchards. *Proc. Mass. Fruit Grow. Assoc.* 78:24–35
10. Asquith, D., Colburn, R. 1971. Integrated pest management in Pennsylvania apple orchards. *Bull. Entomol. Soc. Am.* 17:89–91
11. Asquith, D., Hull, L. A. 1973. *Stethorus punctum* and pest-population responses to pesticide treatment on apple trees. *J. Econ. Entomol.* 66:1197–1203
12. Atallah, Y. H., Nettles, W. C. 1966. DDT-metabolism and excretion in *Coleomegilla maculata* DeGeer. *J. Econ. Entomol.* 59:560–64
13. Atallah, Y. H., Newsom, L. D. 1966. The effect of DDT, toxaphene and endrin on the reproductive and survival potentials of *Coleomegilla maculata. J. Econ. Entomol.* 59:1181–87
14. Axtell, R. C. 1966. Comparative toxicities of insecticides to house fly larvae and *Macrocheles muscaedomesticae*, a mite predator of the house fly. *J. Econ. Entomol.* 59:1128–30
15. Azab, A. K., Tawfik, M. F. S., Fahmy, H. S. M., Awadallah, K. T. 1971. Effect of some insecticides on the larvae of the aphidophagous syrphid, *Xanthogramma aegyptium* Wied. *Bull. Entomol. Soc. Egypt, Econ. Ser.* 5:37–45
16. Barrett, G. W. 1968. The effects of an acute insecticide stress on a semienclosed grassland ecosystem. *Ecology* 49:1019–35
17. Bartlett, B. R. 1953. Retentive toxicity of field-weathered insecticide residues to entomophagous insects associated with citrus pests in California. *J. Econ. Entomol.* 46:565–69
18. Bartlett, B. R. 1956. Natural predators: can selective insecticides help to preserve biotic control? *Agr. Chem.* 11:42–44, 107
19. Bartlett, B. R. 1958. Laboratory studies on selective aphicides favoring natural enemies of the spotted alfalfa aphid. *J. Econ. Entomol.* 51:374–78
20. Bartlett, B. R. 1963. The contact toxicity of some pesticide residues to hymenopterous parasites and coccinellid predators. *J. Econ. Entomol.* 56:694–98
21. Bartlett, B. R. 1964. Toxicity of some pesticides to eggs, larvae and adults of the green lacewing, *Chrysopa carnea. J. Econ. Entomol.* 57:366–69
22. Bartlett, B. R. 1964. Integration of chemical and biological control. In *Biological Control of Insect Pests and Weeds*, ed. P. DeBach, 489–514. New York: Reinhold. 844 pp.
23. Bartlett, B. R. 1966. Toxicity and acceptance of some pesticides fed to parasitic Hymenoptera and predatory coccinellids. *J. Econ. Entomol.* 59:1142–49
24. Bess, H. A. 1964. Populations of the leaf-miner, *Leucoptera meyricki* Ghesa. and its parasite in sprayed and unsprayed coffee in Kenya. *Bull. Entomol. Res.* 55:59–82
25. Binns, E. S. 1971. The toxicity of some soil-applied systemic insecticides to *Aphis gossypii* and *Phytoseiulus persimilis* on cucumbers. *Ann. Appl. Biol.* 67:211–22
26. Bohm, M. K. 1972. Effects of environment and juvenile hormone on ovaries of the wasp *Polistes metricus. J. Insect Physiol.* 18:1875–83
27. Bonnemaison, L. 1962. Toxicité de divers insecticides de contact ou endothérapiques vis-a-vis des prédateurs et

parasites des pucerons. *Phytiat. Phytopharm.* 11:67–84
28. Boone, C. O., Hammond, A. M. 1974. A comparative study of the effects of JH chemicals on certain beneficial and pest insects. *Environ. Entomol.* In press
29. Boone, C. O., Hammond, A. M. 1974. Juvenile hormones: Adverse effects on the growth and development of the hymenopterous parasite, *Microplitis croceipes* (Cresson). *Environ. Entomol.* In press
30. Boyce, H. R., Dustan, G. G. 1955. Parasitism of twig-infesting larvae of the oriental fruit moth *Grapholitha molesta* in Ontario, 1939–1953. *Rep. Entomol. Soc. Ontario.* 84:48–55
31. Bracken, G. K., Nair, K. K. 1967. Stimulation of yolk deposition in an ichneumonid parasitoid by feeding synthetic juvenile hormone. *Nature* 216: 483–84
32. Brattsten, L. B., Metcalf, R. L. 1970. The synergistic ratio of carbaryl with piperonyl butoxide as an indicator of the distribution of multifunction oxidases in the Insecta. *J. Econ. Entomol.* 63:101–4
33. Bravenboer, L. 1959. Die empfindlichkeit von *Tetranychus urticae* und ihren naturlichen Feinden *Typhlodromus longipilus* and *Stethorus punctillum* gegen Insektizide, Akarizide und Fungizide. *IV Int. Congr. Plant. Prot. 1957* 1: 937–38
34. Brown, A. W. A. 1972. Pest resistance to pesticides. In *Insecticides in the Environment,* ed. R. White-Stevens, I:457–552. New York: Marcel Dekker
35. Brunson, M. H. 1960. Effect of parathion on parasites of oriental fruit moth cocoons and *Trichogramma minutum* in peach orchards. *J. Econ. Entomol.* 53:304–6
36. Bull, D. L., Ridgway, R. L. 1969. Metabolism of trichlorfon in animals and plants. *J. Agr. Food Chem.* 17:837–41
37. Burke, H. R. 1959. Toxicity of several insecticides to two species of beneficial insects on cotton. *J. Econ. Entomol.* 52:616–18
38. Carolin, V. M., Coulter, W. K. 1971. Trends of western spruce budworm and associated insects in Pacific northwest forests sprayed with DDT. *J. Econ. Entomol.* 64:291–97
39. Cate, J. R., Ridgway, R. L., Lingren, P. D. 1972. Effects of systemic insecticides applied to cotton on adults of an ichneumonid parasite, *Campoletis perdistinctus. J. Econ. Entomol.* 65:484–88
40. Chambers, H. W. 1973. Comparative tolerance of selected beneficial insects to methyl parathion. Communication to *Ann. Meet. Entomol. Soc. Am., November 28,* p. 68
41. Chandrika, S., Nair, M. R. G. K. 1968. Relative toxicity of insecticides to adults of *Trichospilus pupivora* (Ferriere). *Agr. Res. J. Kerala* 6:33–36
42. Cherry, E. T., Pless, C. D. 1969. Bioassay of leaves from tobacco grown on soil treated with certain systemic insecticides. *J. Econ. Entomol.* 62:1313–16
43. Chesnut, T. L., Cross, W. H. 1971. Arthropod parasites of the boll weevil, *Anthonomus grandis:* 2. Comparisons of their importance in the United States over a period of thirty-eight years. *Ann. Entomol. Soc. Am.* 64:549–57
44. Clancy, D. W., McAlister, H. J. 1958. Effects of spray practices on apple mites and their predators in West Virginia. In *Proc. Int. Congr. Entomol., 10th, Montreal 1956* 4:597–601
45. Coaker, T. H. 1966. The effect of soil insecticides on the predators and parasites of the cabbage root fly (*Erioischia brassicae*) and on the subsequent damage caused by the pest. *Ann. Appl. Biol.* 57:397–407
46. Colburn, R., Asquith, D. 1971. Tolerance of the stages of *Stethorus punctum* to selected insecticides and miticides. *J. Econ. Entomol.* 64:1072–74
47. Cone, W. W. 1963. Insecticides as a factor in population fluctuations of mites on alfalfa. *Wash. Agr. Exp. Sta. Tech. Bull.* 41:1–55
48. Critchley, B. R. 1972. A laboratory study of the effects of some soil-applied organophosphorus pesticides on Carabidae (Coleoptera). *Bull. Entomol. Res.* 62:229–42
49. Croft, B. A. 1970. *Comparative studies on four strains of* Typhlodromus occidentalis *Nesbitt. I. Description of orchard mite relationships at Oak Glen, California.* PhD thesis. Univ. California, Riverside. 91 pp.
50. Croft, B. A. 1972. Resistant natural enemies in pest management systems. *Span* 15:19–21
51. Croft, B. A. 1974. Deciduous fruits. In *Introduction to Pest Management,* ed. R. L. Metcalf, W. H., Luckman. New York: Wiley. In press
52. Croft, B. A., Barnes, M. M. 1971. Comparative studies on four strains of *Typhlodromus occidentalis.* III. Evaluations of releases of insecticide resistant strains into an apple orchard ecosystem. *J. Econ. Entomol.* 64:845–50

53. Croft, B. A., Barnes, M. M. 1972. Comparative studies on four strains of *Typhlodromus occidentalis.* VI. Persistence of insecticide resistant strains in an apple orchard ecosystem. *J. Econ. Entomol.* 65:211–16
54. Croft, B. A., Jeppson, L. R. 1970. Comparative studies on four strains of *Typhlodromus occidentalis.* II. Laboratory toxicity of ten compounds common to apple pest control. *J. Econ. Entomol.* 63:1528–31
55. Croft, B. A., Jorgensen, C. D. 1969. Life history of *Typhlodromus mcgregori. Ann. Entomol. Soc. Am.* 62:1261–67
56. Croft, B. A., Meyer, R. H. 1973. Carbamate and organophosphorus resistance patterns in populations of *Amblyseius fallacis. Environ. Entomol.* 2:691–95
57. Croft, B. A., Nelson, E. E. 1972. Toxicity of apple orchard pesticides to Michigan populations of *Amblyseius fallacis. Environ. Entomol.* 1:576–79
58. Croft, B. A., Stewart, P. G. 1973. Toxicity of one carbamate and six organophosphorus insecticides to O-P resistant strains of *Typhlodromus occidentalis* and *Amblyseius fallacis. Environ. Entomol.* 2:486–88
59. Croft, B. A., Thompson, W. W., Howitt, A. J., Jones, A. L., Putnam, A. R. 1974. Integrated control of plant feeding mites in Michigan apple orchards. *Res. Rep. Mich. State Univ.* In press
60. Davis, D. W. 1970. Integrated control of apple pests in Utah. *Utah Sci.* 31:43–45, 48
61. Davis, D. W. 1970. Insecticidal control of the alfalfa weevil in northern Utah and some resulting effects on the weevil parasite *Bathyplectes curculionis. J. Econ. Entomol.* 63:119–25
62. DeBach, P., Bartlett, B. 1964. Methods of colonization, recovery and evaluation. In *Biological Control of Insect Pests and Weeds,* ed. P. DeBach, 402–26. New York: Reinhold. 844 pp.
63. Dempster, J. P. 1968. The sublethal effect of DDT on the rate of feeding by the ground beetle *Harpalus rufipes. Entomol. Exp. Appl.* 11:51–54
64. Dominick, C. B. 1964. Parasitism of the tobacco flea beetle in tobacco fields sprayed with endrin. *J. Econ. Entomol.* 57:1004–5
65. Downing, R. S., Arrand, J. C. 1968. Integrated control of orchard mites in British Columbia. *Brit. Columbia Dep. Agr. Entomol. Br. Publ.* 7 pp.
66. Downing, R. S., Moilliet, T. K. 1971. Occurrence of phytoseiid mites in apple orchards in south central British Columbia. *J. Entomol. Soc. Brit. Columbia* 68:33–36
67. Downing, R. S., Moilliet, T. K. 1972. Replacement of *Typhlodromus occidentalis* by *T. caudiglans* and *T. pyri* after cessation of spray on apple trees. *Can. Entomol.* 104:937–40
68. Dyte, C. E. 1967. Possible new approach to the chemical control of plant feeding insects. *Nature* 216:298
69. Edwards, C. A., Lofty, J. R., Stafford, C. J. 1970. Effects of pesticides on predatory beetles. *Rep. Rothamsted Exp. Sta. 1969* Pt. I:246
70. El Sayed, E. I., Graves, J. B., Bonner, F. L. 1967. Chlorinated hydrocarbon insecticide residues in selected insects and birds found in association with cotton fields. *Agr. Food Chem.* 15:1014–17
71. Elliott, W. M. 1966. *The action of systemic insecticides on aphids and their anthocorid predators.* PhD thesis. Univ. London. 184 pp.
72. Elliott, W. M. 1970. The action of some systemic aphicides on the nymphs of *Anthocoris nemorum* and *A. confusum. Ann. Appl. Biol.* 66:313–21
73. Elliott, W. M., Way, M. J. 1968. The action of some systemic aphicides on the eggs of *Anthocoris nemorum* and *A. confusum. Ann. Appl. Biol.* 62:215–26
74. Evenhuis, H. H. 1959. Effect van insekticiden op de bloedluisparasiet *Aphelinus mali. Meded. Di Thuinb.* 22:306–11
75. Fan, S. H., Ho, K. K. 1971. A preliminary study on the life history, rearing method of *Apanteles plutellae* and the effects of different insecticides to it. *Chung-hua Chih Wu Pao Hu Hsueh Hui Plant Prot. Bull.* 13:156–61
76. Fenton, F. A. 1959. The effect of several insecticides on the total arthropod population in alfalfa. *J. Econ. Entomol.* 52:428–32
77. Ferguson, D. E. 1969. The compatible existence of non-target species to pesticides. *Bull. Entomol. Soc. Am.* 15: 363–66
78. Flaherty, D. W., Huffaker, C. B. 1970. Biological control of Pacific mites and Willamette mites in San Joaquin valley vineyards. I. Role of *Metaseiulus occidentalis. Hilgardia* 40:267–308
79. Flaherty, D. W., Lynn, C., Jensen, F., Hoy, M. 1972. Correcting imbalances—spider mite populations in southern San Joaquin vineyards. *Calif. Agr.* 26:10–12
80. Flanders, S. E. 1943. The susceptibility of parasitic Hymenoptera to sulfur. *J. Econ. Entomol.* 36:469

81. Fleschner, C. A., Scriven, G. T. 1957. Effect of soil-type and DDT on ovipositional responses of *Chrysopa californica* (Coq.) on lemon trees. *J. Econ. Entomol.* 50:221–22
82. Froggatt, W. W. 1905. The effects of a fumigation with hydrocyanic gas upon ladybird beetle larvae and other parasites. *Agr. Gaz. N.S.W.* 16:1088–89
83. Gargav, V. P. 1968. *A study of selectivity in the action of organophosphorus compounds on aphids that damage cotton and their coccinellid predators.* Summ. Thesis Tashkent Sol'-Khoz. Inst., 16 pp.
84. Georghiou, G. P. 1967. Differential susceptibility and resistance to insecticides of coexisting populations of *Musca domestica, Fannia canicularis, F. femoralis,* and *Ophyra leucostoma. J. Econ. Entomol.* 60:1338–44
85. Georghiou, G. P. 1972. The evolution of resistance to pesticides. *Ann. Rev. Ecol. Syst.* 3:133–68
86. Gilbert, M. 1971. Enzymes: can they protect honey bees from pesticides? *Am. Bee J.* 111:471
87. Gilmore, J. U. 1938. Notes on *Apanteles congregatus* (Say) a parasite of tobacco hornworms. *J. Econ. Entomol.* 31: 712–15
88. Godfrey, G. L., Root, R. B. 1968. Emergence of parasites associated with the cabbage aphid during a chemical-control program. *J. Econ. Entomol.* 61:1762–63
89. Gordon, H. T. 1961. Nutritional factors in insect resistance to chemicals. *Ann. Rev. Entomol.* 6:27–54
90. Green, E. E. 1917. Note on the immunity of chalcid parasites to hydrocyanic acid gas. *Ann. Appl. Biol.* 4:90
91. Grosch, D. S. 1970. Reproductive performance of a braconid after heptachlor poisoning. *J. Econ. Entomol.* 63: 1348–49
92. Grosch, D. S., Valcovic, L. R. 1967. Chlorinated hydrocarbon insecticides are not mutagenic in *Bracon hebetor* tests. *J. Econ. Entomol.* 60:1177–79
93. Hamilton, E. W., Kieckhefer, R. W. 1969. Toxicity of malathion and parathion to predators of the English grain aphid. *J. Econ. Entomol.* 62:1190–92
94. Hassanein, M. H., Khalil, F. 1968. Effect of insecticides on predators of the cotton leaf worm. *Bull. Entomol. Soc. Egypt* 2:247–64
95. Helle, W. 1965. Resistance in Acarina: Mites. *Advan. Acarol.* 3:71–93
96. Herbert, H. J. 1952. Progress report on predaceous mite investigations in Nova Scotia. *Ann. Rep. Entomol. Soc. Ontario, 83rd,* pp. 27–29
97. Herne, D. H. C., Putman, W. L. 1966. Toxicity of some pesticides to predaceous arthropods in Ontario peach orchards. *Can. Entomol.* 98:936–42
98. Hoffman, A. C., Grosch, D. S. 1971. The effects of ethyl methane sulfonate on the fecundity and fertility of *Bracon (Habrobracon)* females. 1. The influences of route of entry and physiological state. *Pestic. Biochem. Physiol.* 1:319–26
99. Holdsworth, R. P. 1970. Codling moth control as part of an integrated program in Ohio. *J. Econ. Entomol.* 63:894–97
100. Holdsworth, R. P. 1971. Major predators of the European red mite on apple in Ohio. *Ohio Agr. Res. Develop. Center Res. Circ.* 192:17 pp.
101. Holdsworth, R. P. 1974. Integrated control of apple pest insects in Ohio. In: 1974 Commercial fruit spray recommendations for Ohio. *Bull. Ohio State Univ. Extension Serv.* 506:20–26
102. Horsburg, R. L. Asquith, D. 1968. Initial survey of arthropod predators of the European red mite in south-central Pennsylvania. *J. Econ. Entomol.* 61: 1752–54
103. Hoyt, S. C. 1969. Integrated chemical control of insects and biological control of mites on apples in Washington. *J. Econ. Entomol.* 62:74–86
104. Hoyt, S. C. 1972. Resistance to azinphosmethyl of *Typhlodromus pyri* from New Zealand. *N.Z. J. Sci.* 15:16–21
105. Hoyt, S. C., Caltagirone, L. E. 1971. The developing programs of integrated control of pests of apples in Washington and peaches in California. In *Biological Control,* ed. C. B. Huffaker, 395–421. New York: Plenum
106. Huffaker, C. B. 1971. The ecology of pesticide interference with insect populations. In *Agricultural Chemicals—Harmony or Discord for Food, People and the Environment,* ed. J. E. Swift, 92–107. *Univ. Calif. Div. Agr. Sci.,* 151 pp.
107. Huffaker, C. B., Kennett, C. E. 1953. Differential tolerance to parathion in two *Typhlodromus* predatory on cyclamen mite. *J. Econ. Entomol.* 46:707–8
108. Huffaker, C. B., Van de Vrie, M., McMurtry, J. A. 1970. Ecology of tetranychid mites and their natural enemies: tetranychid populations and their possible control by predators: an evaluation. *Hilgardia* 40:391–458
109. Hussey, N. W., Parr, W. J., Gould, H. J. 1965. Observations on the control

of *Tetranychus urticae* Koch on cucumbers by the predatory mite *Phytoseiulus riegeli* Dosse. *Entomol. Exp. Appl.* 8:271–81

110. Inserra, S. 1969. L' *Iridomyrmex humilis* Mayr un temibile nemico dell' *Aphytis melinus* DeBach. *Entomologica* 5:79–84
111. Irving, S. N., Wyatt, I. J. 1973. Effects of sublethal doses of pesticides on the oviposition behavior of *Encarsia formosa. Ann. Appl. Biol.* 75:57–62
112. Issi, I. V., Maslennikova, V. A. 1964. Effects of microsporidiosis on the diapause and survival of the parasite *Apanteles glomeratus* L. and the cabbage white butterfly *Pieris brassicae* L. *Entomol. Obozr.* 43:112–17
113. Jotwani, M. G., Sarup, P., Pradhan, S. 1960. Effect of some important insecticides on the predator *Stethorus pauperculus* Weise. *Indian J. Entomol.* 22: 272–76
114. Kawahara, S., Kiritani, K., Sasaba, T. 1971. The selective activity of rice-pest insecticides against the green rice leafhopper and spiders. *Bochu Kagaku* 36:121–28
115. Kaya, H. K., Dunbar, D. M. 1972. Effect of *Bacillus thuringiensis* and carbaryl on an elm spanworm egg parasite *Telenomus alsophilae. J. Econ. Entomol.* 65:1132–34
116. Kaya, H. K., Tanada, Y. 1972. Response of *Apanteles militaris* to a toxin produced in a granulosis-virus-infected host. *H. Invertebr. Pathol.* 19:1–17
117. Kehat, M., Swirski, E. 1964. Chemical control of the date palm scale *Parlatoria blanchardi* and the effect of some insecticides on the lady beetle *Pharoscymnus* aff. *numidicus* Pic. *Israel J. Agr. Res.* 14:101–10
118. Kelsey, J. M. 1964. Interaction of virus and insect parasites of *Pieris rapae.* L. *Proc. Int. Congr. Entomol., 11th, Vienna, 1960* 2:790–96
119. Kennett, C. E. 1970. Resistance to parathion in the phytoseiid mite *Amblyseius hibisci. J. Econ. Entomol.* 63: 1999–2000
120. Kinn, D. N., Doutt, R. L. 1972. Initial survey of arthropod fauna in north coast vineyards of California. *Environ. Entomol.* 1:508–13
121. Kinn, D. N., Doutt, R. L. 1972. Natural control of spider mites on wine grape varieties in northern California. *Environ. Entomol.* 1:513–18
122. Kirby, A. M. A. 1973. Progress in the control of orchard pests by integrated methods. *Commonw. Bur. Hort. Plant Crops Abstr.* 43:1–16, 57–65
123. Kiritani, K., Kawahara, S. 1973. Food-chain toxicity of granular formulations of insecticides to a predator, *Lycosa pseudoannulata,* of *Nephotettix cincticeps. Bochu Kagaku* 38:69–75
124. Knisley, C. B., Swift, F. C. 1972. Qualitative study of mite fauna associated with apple foliage in New Jersey. *J. Econ. Entomol.* 65:445–48
125. Kot, J. 1970. The phenomenon of partial resistance to insecticides in some arthropods. *Ekol. Pol.* 18:351–59
126. Kot, J., Plewka, T. 1970. The influence of Metasystox on different stages of the development of *Trichogramma evanescens. Deut. Akad. Landwirtschaftwiss. Tagungsber.* 110:185–92
127. Kowalska, T., Pruszynski, S., Szczepanska, K. 1971. Toxic action of the fenitrothion preparations on the natural enemies of the cabbage aphid *Brevicoryne brassicae. Biol. Inst. Ochrony Roslin, Zeszyt* 48:163–74
128. Krieger, R. I., Feeny, P. P., Wilkinson, C. F. 1971. Detoxification enzymes in the guts of caterpillars: An evolutionary answer to plant defenses? *Science* 172:579–81
129. Kurstak, E. 1965. Microbial infections with *B. thuringiensis* in *Anagasta kuehniella* in the presence of the parasite *Nemeritis canescens. Proc. Int. Congr. Entomol., 12th, London, 1964,* p. 738
130. Larson, H. 1970. Raising resistant predator mites. *Idaho State Hort. Soc. Trans., 76th,* pp. 32–37
131. Lawrence, P. O., Kerr, S. H., Whitcomb, W. H. 1973. *Chrysopa rufilabris:* effect of selected pesticides on duration of third larval stadium, pupal stage and adult survival. *Environ. Entomol.* 2: 477–80
132. Lee, M. S., Davis, D. W. 1968. Life history and behavior of the predatory mite *Typhlodromus occidentalis* in Utah. *Ann. Entomol. Soc. Am.* 61:251–55
133. LeRoux, E. J. 1961. Effects of "modified" and "commercial" spray programs on the fauna of apple orchards in Quebec. *Ann. Entomol. Soc. Quebec* 6:87–121
134. Lingappa, S. S., Starks, K. J., Eikenbary, R. D. 1972. Insecticidal effect on *Lysiphlebus testaceipes,* a parasite of the greenbug, at all three developmental stages. *Environ. Entomol.* 1:520–21
135. Lingren, P. D., Ridgway, R. L. 1967. Toxicity of five insecticides to several insect predators. *J. Econ. Entomol.* 60:1639–41

136. Lingren, P. D., Wolfenbarger, D. A., Nosky, J. B., Diaz, M. 1972. Response of *Campoletis perdistinctus* and *Apanteles marginiventris* to insecticides. *J. Econ. Entomol.* 65:1295–99
137. Luckey, T. D. 1968. Insect hormoligosis. *J. Econ. Entomol.* 61:7–12
138. Macdonald, D. R., Webb, F. E. 1963. Insecticides and the spruce budworm. In: Spruce Budworm Dynamics, ed. R. F. Morris. *Mem. Entomol. Soc. Can.* 31:288–310
139. MacPhee, A. W., Sanford, K. H. 1961. The influence of spray programs on the fauna of apple orchards in Nova Scotia. XII. Second Supplement to VII. Effects on beneficial arthropods. *Can. Entomol.* 93:671–73
140. Madsen, H. F. 1968. Integrated control of deciduous tree fruit pests. *World Crops* 20:21–23
141. Madsen, H. F. 1970. Insecticides for codling moth control and their effect on other insects and mites of apple in British Columbia. *J. Econ. Entomol.* 63: 1521–23
142. Majumdar, N., Bhattacharya, A. 1968. Effect of insecticidal sprays on the longevity and fecundity of *Eublemma amabilis* and *Holcocera pulverea*. *Indian J. Entomol.* 30:239–40
143. Manser, P. D., Bennett, F. D. 1962. Possible effects of the application of malathion on the small moth borer *Diatraea saccharalis* (F.) and its parasite *Lixophaga diatraeae* (Tns.) in Jamaica. *Bull. Entomol. Res.* 53:75–82
144. Mardzhanyan, G. M., Ust'yan, A. K. 1965. Integrated control of the green peach aphid *Myzodes persicae* Sulz., on tobacco. *Entomol. Obozr.* 44:441–48
145. McClanahan, R. J. 1967. Food-chain toxicity of systemic acaricides to predaceous mites. *Nature* 215:1001
146. McLaughlin, R. E., Adams, C. H. 1966. Infection of *Bracon mellitor* by *Mattesia grandis*. *Ann. Entomol. Soc. Am.* 59:800–2
147. McMurtry, J. A., Huffaker, C. B., Van de Vrie, M. 1970. Ecology of tetranychid mites and their natural enemies: a review I. Tetranychid enemies: their biological characters and the impact of spray practices. *Hilgardia* 40:331–90
148. Messenger, P. S. 1951. Report for the period January-March, Fruit Fly Investigations Project. *Div. Biol. Contr. Univ. Calif.*, pp. 70–80
149. Metcalf, R. L. 1964. Selective toxicity of pesticides. *World Rev. Pest Contr.* 3:28–43
150. Metcalf, R. L., Reynolds, H. T., Fukuto, T. R. 1967. Carbamate insecticides: where are they headed? *Farm Chem.* 130:45, 48–49, 52, 54
151. Meyer, R. H. 1974. Management of phytophagous and predatory mites in Illinois orchards. *Environ. Entomol.* 3:333–40
152. Miura, T., Takahashi, R. M. 1973. Insect developmental inhibitors. 3. Effects on non-target aquatic organisms. *J. Econ. Entomol.* 66:917–22
153. Moffitt, H. R., Anthon, E. W., Smith, L. O. 1972. Toxicity of several commonly used orchard pesticides to adult *Hippodamia convergens*. *Environ. Entomol.* 1:20–23
154. Monaco, R. 1969. The action taken against *Dacus oleae* by *Opius concolor* distributed in Apulia in the Gargano olive groves, and by indigenous parasites in the same habitat. *Entomologica* 5:139–91
155. Morgan, C. V. G., Anderson, N. H. 1958. Notes on parathion-resistant strains of two phytophagous mites and a predaceous mite in British Columbia. *Can. Entomol.* 90:92–97
156. Moriarty, F. 1969. The sublethal effects of synthetic insecticides on insects. *Biol. Rev.* 44:321–57
157. Motoyama, N., Rock, G. C., Dauterman, W. C. 1970. Organophosphorus resistance in an apple orchard population of *Typhlodromus (Amblyseius) fallacis*. *J. Econ. Entomol.* 63:1439–42
158. Motoyama, N., Rock, G. C., Dauterman, W. C. 1971. Studies on the mechanism of azinphosmethyl resistance in the predaceous mite *Neoseiulus (T.) fallacis*. *Pestic. Biochem. Physiol.* 1:205–15
159. Mowat, D. J., Coaker, T. H. 1967. The toxicity of some soil insecticides to carabid predators of the cabbage root fly (*Erioischia brassicae*). *Ann. Appl. Biol.* 59:349–54
160. Neal, J. W., Bickley, W. E., Blickenstaff, C. C. 1970. Recovery of the parasite *Microctonus aethiops* from the alfalfa weevil after hormonal treatment. *J. Econ. Entomol.* 63:681–82
161. Nef, L. 1972. Influence of chemical and microbial treatments on a population of *Stilpnotia (=Leucoma) salicis* and on its parasites. *Z. Angew. Entomol.* 69: 357–67
162. Nelson, E. E., Croft, B. A., Howitt, A. J., Jones, A. L. 1973. Toxicity of apple orchard pesticides to *Agistemus fleschneri*. *Environ. Entomol.* 2:219–22
163. Newsom, L. D. 1967. Consequences of

insecticide use on nontarget organisms. *Ann. Rev. Entomol.* 12:257–86
164. Newsom, L. D. 1974. Predator insecticide relationships. *Entomophaga Mem. Ser. 7,* 88 pp.
165. Nicholson, A. J. 1939. Indirect effects of spray practices on pest populations. *Verh. Int. Kongr. Entomol., 7th, Berlin* 4:3022–28
166. Oatman, E. R., Legner, E. F. 1962. Integrated control of apple insects and mite pests in Wisconsin. *Proc. N. Cent. Br. Entomol. Soc. Am.* 17:110–15
167. Oatman, E. R., McMurtry, J. A., Voth, V. 1968. Suppression of the two-spotted spider mite on strawberry with mass releases of *Phytoseiulus persimilis. J. Econ. Entomol.* 61:1517–21
168. Obrtel, R. 1961. Effect of two insecticides on *Aphidius ervi* Hal., an internal parasite of *Acyrthosiphon onobrychis* (Boyer). *Zool. Listy Fol. Zool.* 10:1–8
169. Paradis, E. R. 1955. Cycle évolutif de tetranyque à deux points, *Tetranychus bimaculatus* Harvey dans le sud-ouest du Quebec. *Natur. Can.* 82:5–30
170. Parent, B. 1958. Efficacité des prédateurs sur la tetranyque rouge de pommier, *Panonychus ulmi* (Koch) dans les vergers du Quebec. *Ann. Entomol. Soc. Quebec* 4:62–69
171. Pickett, A. D. 1949. A critique on insect chemical control measures. *Can. Entomol.* 81:67–76
172. Pielou, D. P. 1950. Selection for DDT tolerance in *Macrocentrus ancylivorus. Ann. Rep. Entomol. Soc. Ontario, 81st,* pp. 44–45
173. Pielou, D. P. 1952. The non-mutagenic effect of p,p'-DDT and gammahexachlorocyclohexane in *Drosophila melanogaster* Meig. *Can. J. Zool.* 30: 375–77
174. Pielou, D. P., Glasser, R. F. 1951. Selection for DDT tolerance in a beneficial parasite, *Macrocentrus ancylivorus* 1. Some survival characteristics and the DDT resistance of the original laboratory strain. *Can. J. Zool* 29:90–101
175. Pielou, D. P., Glasser, R. F. 1952. Selection for DDT resistance in a beneficial insect. *Science* 115:117–18
176. Plapp, F. W. Jr., Casida, J. 1970. Induction by DDT and dieldrin of insecticide metabolism by house fly enzymes. *J. Econ. Entomol.* 63:1091–92
177. Poe, S. L., Enns, W. R. 1969. Predaceous mites associated with Missouri orchards. *Trans. Mo. Acad. Sci.* 3:69–82
178. Pradhan, S., Jotwani, M. G., Sarup, P. 1959. Effect of some important insecticides on *Coccinella septempunctata,* a predator of mustard aphid (*Lipaphis erysimi* Kalt.) *Indian Oilseeds J.* 3:121–24
179. Price, P. W. 1972. Immediate and long-term effects of insecticide application on parasitoids in jackpine stands in Quebec. *Can. Entomol.* 104:263–70
180. Pruszynski, S., Cone, W. W. 1972. Relationship between *Phytoseiulus persimilis* and other enemies of the two-spotted spider mite on hops. *Environ. Entomol.* 1:431–33
181. Pruszynski, S., Cone, W. W. 1973. Biological observations of *Typhlodromus occidentalis* on hops. *Ann. Entomol. Soc. Am.* 66:47–51
182. Rai, B. K., Kanta, S., Lal, R. 1964. Relative toxicity of some pesticides to *Eutetranychus banksi* (McGregor). *Indian Oilseeds J.* 8:360–63
183. Rice, R. E. 1969. Response of some predators and parasites of *Ips confusus* (Lec.) to olfactory attractants. *Contrib. Boyce Thompson Inst.* 24:189–94
184. Ridgway, R. L., Lingren, P. D., Cowan, C. B., Davis, J. W. 1967. Populations of arthropod predators and *Heliothis* spp. after application of systemic insecticides to cotton. *J. Econ. Entomol.* 60: 1012–16
185. Ripper, W. E. 1944. Biological control as a supplement to chemical control of insect pests. *Nature* 153:448–52
186. Ripper, W. E. 1956. Effects of pesticides on balance of arthropod populations. *Ann. Rev. Entomol.* 1:403–38
187. Ripper, W. E. 1958. Experiments on the integration of biological and chemical control of insect pests. *Proc. Int. Congr. Entomol., 10th* 3:479–93
188. Ripper, W. E. 1960. Selective insect control and its application to the resistance problem. *Misc. Pub. Entomol. Soc. Am.* 2:153–56
189. Ripper, W. E., Greenslade, R. M., Hartley, G. S. 1950. A new systemic insecticide, bis (bis dimethylaminophosphorus) anhydride. *Bull. Entomol. Res.* 40:481–501
190. Ripper, W. E., Greenslade, R. M., Hartley, G. S. 1951. Selective insecticides and biological control. *J. Econ. Entomol.* 44:448–59
191. Ripper, W. E., Greenslade, R. M., Lickerish, L. A. 1949. Combined chemical and biological control of insects by means of a systemic insecticide. *Nature* 163:787–89
192. Ristich, S. S. 1956. Toxicity of pesticides to *Typhlodromus fallacis* (Gar.) *J. Econ. Entomol.* 49:511–15

193. Roberts, D. R., Roberts, L. W., Miller, T. A., Nelson, L. L., Young, W. W. 1973. Effects of a polyethylene formulation of chlorpyrifos on non-target populations naturally infesting artificial field pools. *Mosquito News* 33:165–72
194. Robertson, J. G. 1957. Changes in resistance to DDT in *Macrocentrus ancylivorus* Rohw. *Can. J. Zool* 35: 629–33
195. Robertson, J. L. 1972. Toxicity of Zectran aerosol to the California oakworm, a primary parasite, and a hyperparasite. *Environ. Entomol.* 1:115–17
196. Rock, G. C. 1972. Integrated control of mites. *Agr. Exp. Ser., Fruit Insect Note, A-1 North Carolina State Univ. Publ.*, 18 pp.
197. Rock, G. C., Yeargan, D. R. 1970. Relative toxicity of Plictran to the European red mite, the two-spotted spider mite and the predaceous mite *Neoseiulus (Typhlodromus) fallacis. Down Earth* 26:1–4
198. Rock, G. C., Yeargan, D. R. 1971. Relative toxicity of pesticides to organophosphorus-resistant orchard populations of *Neoseiulus fallacis* and its prey. *J. Econ. Entomol.* 64:350–52
199. Rock, G. C., Yeargan, D. R. 1972. Laboratory studies on toxicity of dinocap to *Neoseiulus fallacis* and its prey. *J. Econ. Entomol.* 65:932–33
200. Rosen, D. 1967. Effect of commercial pesticides on the fecundity and survival of *Aphytis holoxanthus. Israel J. Agr. Res.* 17:47–52
201. Rudinsky, J. A., Novak, V., Svihra, P. 1971. Attraction of the bark beetle *Ips typographus* L. to terpenes and a male-produced pheromone. *Z. Angew. Entomol.* 67:179–88
202. Sanford, K. H. 1967. The influence of spray programs on the fauna of apple orchards in Nova Scotia XVII. Effect on some predaceous mites. *Can. Entomol.* 99:197–201
203. Sanger, A. M. H. 1958. The effects of aldrin, dieldrin and endrin on insect parasites and predators. *Span* 1:16–20
204. Saradamma, K., Nair, M.R.G.K. 1968. Relative toxicity of insecticides to adults of *Microbracon brevicornis* (Wesmael). *Agr. Res. J. Kerala* 6:98–100
205. Sarup, P., Jotwani, M. G., Singh, D. S. 1965. Further studies on the effect of some important insecticides on *Coccinella septempunctata* Linn. *Indian J. Entomol.* 27:72–76
206. Sato, Y. 1968. Insecticidal action of phytoecdysones. *Appl. Entomol. Zool.* 3:155–62
207. Schneider, H. 1958. Untersuchungen über den Einfluss neuzeitlicher Insektizide und Fungizide auf die Blutlauszehrwespe (*Aphelinus mali* Hald.) *Z. Angew. Entomol.* 43:173–96
208. Searle, C. M. S. 1965. The susceptibility of *Pauridia peregrina* Timb. to some pesticide formulations. *J. Entomol. Soc. S. Africa* 27:239–49
209. Shorey, H. H. 1963. Differential toxicity of insecticides to the cabbage aphid and two associated entomophagous insect species. *J. Econ. Entomol.* 56: 844–47
210. Shorey, H. H., Hall, I. M. 1963. Toxicity of chemicals and microbial insecticides to pest and beneficial insects on poled tomatoes. *J. Econ. Entomol.* 56:813–17
211. Smith, F. F., Henneberry, T. J., Boswell, A. L. 1963. The pesticide tolerance of *Typhlodromus fallacis* and *Phytoseiulus persimilis* with some observations on the predatory efficiency of *P. persimilis. J. Econ. Entomol.* 56:274–78
212. Specht, H. B. 1968. Phytoseiid in the New Jersey apple orchard environment with descriptions of spermathecae and three new species. *Can. Entomol.* 100:673–92
213. Spiller, D. 1958. Resistance of insects to insecticides. *Entomologist* 2:1–18
214. Steelman, C. D., Schilling, P. E. 1972. Effects of a juvenile hormone mimic on *Psorophora confinnis* and non-target aquatic insects. *Mosquito News* 32: 350–54
215. Stern, V. M. 1963. The effect of various insecticides on *Trichogramma semifumatum* and certain predators in southern California. *J. Econ. Entomol.* 56:348–50
216. Stern, V. M., Smith, R. F., Van den Bosch, R., Hagen, K. S. 1959. The integrated control concept. *Hilgardia* 29: 81–101
217. Stern, V. M., Van den Bosch, R., Born, D. 1958. New control for alfalfa aphid. *Calif. Agr.* 12:4–5
218. Sternburg, J., Kearns, C. W. 1952. Metabolic fate of DDT when applied to certain naturally tolerant insects. *J. Econ. Entomol.* 45:497–505
219. Strickland, E. H. 1948. Could the widespread use of DDT be a disaster? *Entomol. News* 61:85
220. Struble, G. R. 1965. Effect of aerial sprays on parasites of the lodgepole needle miner. *J. Econ. Entomol.* 58:226–27
221. Swift, F. C. 1968. Population densities of the European red mite and the predaceous mite *Typhlodromus (A.) fallacis*

on apple foliage following treatment with various insecticides. *J. Econ. Entomol.* 61:1489–91

222. Swift, F. C. 1970. Predation of *Typhlodromus (A.) fallacis* on the European red mite as measured by the insecticidal check method. *J. Econ. Entomol* 63:1617–18
223. Tamashiro, M. 1968. Effect of insect pathogens on some biological control agents in Hawaii. *Proc. Joint US-Japan Seminar on Microbial Control of Insect Pests, Fukuoka,* pp. 147–53
224. Tamashiro, M., Sherman, M. 1955. Direct and latent toxicity of insecticides to oriental fruit fly larvae and their internal parasites. *J. Econ. Entomol* 48: 75–79
225. Tanada, Y. 1955. Field observations on a microsporidian parasite of *Pieris rapae* (L.) and *Apanteles glomeratus* (L.). *Proc. Hawaii. Entomol. Soc.* 15:609–16
226. Taylor, E. A. 1954. Parasitization of the salt marsh caterpillar in Arizona. *J. Econ. Entomol.* 47:525–30
227. Telford, A. D. 1961. Lodgepole needle miner parasites: biological control and insecticides. *J. Econ. Entomol.* 54: 347–55
228. Thurston, R. 1960. Effect of insecticidal sprays on hornworms and on *Apanteles congregatus. J. Econ. Entomol.* 53:976
229. Thurston, R., Fox, P. M. 1972. Inhibition by nicotine of emergence of *Apanteles congregatus* from its host, the tobacco hornworm. *Ann Entomol. Soc. Am.* 65:547–50
230. Tomlin, A. D., Forgash, A. J. 1972. Toxicity of insecticides to gypsy moth larvae. *J. Econ. Entomol.* 65:953–59
231. Vail, P. V., Soo Hoo, C. F., Seay, R. S., Killinen, R. G., Wolf, W. W. 1972. Microbial control of lepidopterous pests of fall lettuce in Arizona and effects of chemicals and microbial pesticides on parasitoids. *Environ. Entomol.* 1:780–85
232. Valcovi, L. R., Grosch, D. S. 1968. Apholate induced sterility in *Bracon hebetor* Say. *J. Econ. Entomol.* 61: 1514–17
233. Van den Bosch, R., Stern, V. M. 1962. The integration of chemical and biological control of arthropod pests. *Ann. Rev. Entomol.* 7:367–86
234. Van de Vrie, M., Boersma, A. 1970. The influence of the predatory mite *Typhlodromus (A.) potentillae* (Garman) on the development of *Panonychus ulmi* (Koch) on apple grown under various nitrogen conditions. *Entomophaga* 15: 291–304
235. Van de Vrie, M., McMurtry, J. A., Huffaker, C. B. 1972. Ecology of tetranychid mites and their natural enemies. III. Biology, ecology and pest status and host-plant relations of tetranychids. *Hilgardia* 41:343–432
236. Van Dinther, J. B. M. 1963. Residual effect of a number of insecticides on adults of the carabid *Pseudophonus rufipes* (Dej.). *Entomophaga* 8:43–48
237. Van Halteren, P. 1971. Preliminary investigations on dieldrin accumulation in an insect food chain. *J. Econ. Entomol.* 64:1055–56
238. Washino, R. K., Ahmed, W., Linn, J. D., Whitesell, K. G. 1972. Rice field mosquito control studies with low volume Dursban sprays in Colusa county, California. IV. Effects upon aquatic nontarget organisms. *Mosquito News* 32:531–37
239. Westigard, P. H. 1971. Integrated control of spider mites on pear. *J. Econ. Entomol.* 64:496–501
240. Wiackowski, S. K., Herman, E. 1968. Laboratory investigations on the effect of insecticides on adults of primary and secondary aphid parasites. *Pol. Pismo Entomol.* 38:593–600
241. Wilkes, A., Pielou, D. P., Glasser, R. F. 1951. Selection for DDT tolerance in a beneficial insect. In: Conference on Insecticide Resistance and Insect Physiology. *Nat. Acad. Sci. Publ.* No. 219, pp. 78–81
242. Wilkes, F. G., Weiss, C. M. 1971. Accumulation of DDT by dragonfly nymph, *Tetragoneuria. Trans. Am. Fish Soc.* 100:222–36
243. Wilkinson, J. D., Biever, K. D., Ignoffo, C. M. 1974. Contact toxicity of some chemical and biological pesticides to several insect parasitoids and predators. *Environ. Entomol.* In press
244. Winteringham, F. P. W. 1959. Pest resistance in the context of integrated control. *FAO Symp. Integrated Pest Control, Rome 1956* 1:25–32
245. Zeleny, J. 1965. The effect of insecticides (Fosfotion, Intration, Soldep) on some predators and parasites of aphids (*Aphis craccivora* Koch., *Aphis fabae* Scop.). *Rozpravy ČSAV* 75:3–73
246. Zwick, R. W. 1972. Studies on the integrated control of spider mites on apples in Oregon's Hood River Valley. *Environ. Entomol.* 1:169–76

PLANT RESISTANCE TO INSECTS ATTACKING CEREALS[1]

❖6093

R. L. Gallun
Agricultural Research Service, U.S. Department of Agriculture, W. Lafayette, Indiana 47907

K. J. Starks
Agricultural Research Service, U.S. Department of Agriculture, Stillwater, Oklahoma 74074

W. D. Guthrie
Agricultural Research Service, U.S. Department of Agriculture, Ankeny, Iowa 50021

The development of cereal crops that resist specific insect pests was the goal of entomologists and plant breeders as early as 1782 when "Underhill" wheat was found to resist attacks of the Hessian fly, *Mayetiola destructor,* in New York (44). However, the development of resistance in insects to insecticides and public awareness of problems relating to the use of insecticides have caused increased interest in the possibility of developing plants resistant to insects.

Although the use of resistant crop plants is now being recognized as a method of controlling insects, research in this field has actually been in progress for many years. R. H. Painter of Kansas State University probably did more than any other person to make agricultural scientists and administrators cognizant that this method is an ideal way to protect the crop against insect pests. In his book (103), he reviewed more than 2100 papers that were published on plant resistance to insects prior to 1949. At about the same time, Vavilov (141) published his review of papers between 1910 and 1933 on the subject of varietal resistance to insects attacking fruit, vegetables, cotton, and cereals; 76 papers are cited, the majority from foreign periodicals. Painter (104) subsquently updated his original list by reviewing 178 articles and discussing 52 insect-crop relationships for 20 crops. He listed insect-resistant varieties of crop plants grown in the United States at the time and the insect pests and crops being studied for resistance. Also, he discussed the nature of resistance, the inheritance of resistance, and the importance of biotypes.

[1]In cooperation with the Purdue University Agricultural Experiment Station, Journal Paper No. *5302;* Oklahoma Agricultural Experiment Station, Journal Article No. *J–2762;* and the Iowa Agriculture and Home Economics Experiment Station, Journal Paper No. *J–7730;* respectively.

Since Painter's last review, numerous others have been made. For example, Beck (8) made an extensive review of investigations into the nature of resistance of plants to insects for the period 1956–1963. He cited 179 publications and emphasized resistance to oviposition, resistance to feeding, and biophysical and biochemical resistance to the survival of insects. Chesnokov (20) published a lengthy review of Russian research on plant resistance to insects in which he listed 83 articles that were reviewed mostly for techniques used in studying pest resistance in agricultural crops. More recently, Maxwell et al (90) published a comprehensive review of host-plant resistance studies conducted during the period 1958–1972 in which they cite work done in field crops, horticulture crops, forest trees, tobacco, and sugar beets. Also, some problems associated with the development of resistant varieties and the impact of resistant varieties on specific insect populations were discussed. In addition, the inheritance of resistance, the nature of resistance, and physiological races were considered to a limited degree. Approximately 1400 papers were examined; 555 were cited. Thus this review is extremely comprehensive and serves as a reference for pertinent literature on the subject of plant resistance to insects. Other generalized reviews have been made by Horber (71) and Pathak (109).

In the present article, we limit our review to papers dealing with plant resistance in cereal crops. No general reviews have been devoted to resistance in cereals, though tremendous amounts of research have been concerned with cereals, with the glaring exception of millets (85). However, reviews of resistance to specific cereal insect pests are available (58, 135, 143). For example, resistance to insect pests of rice was partly reviewed by Pathak (107). Most research relating to resistance and rice has been done with stem borers, *Chilo* spp. (18, 74, 97, 99, 110), but impressive work has also been done with a planthopper, *Nilaparavata lugens* (4, 22); a leafhopper, *Nephotettix* sp. (19, 22); a gall midge, *Pachydiplosis oryzae* (96, 101); and a weevil, *Sitophilus zeamais* (125). (These references prove that plant resistance transcends language barriers because some of the nations involved are India, Taiwan, the Philippines, Japan, Thailand, and the United States.) We also slight research concerning resistance of cereals to stored-grain pests, though we are aware that exciting work is being done in nations such as Russia (131) and the United States (5, 116, 129).

In general, publications since 1957 on cereal insects are reviewed to show the accomplishments and to point out the problems that have arisen in developing insect-resistant varieties since the earlier work was discussed. We thus exclude such early research as that dealing with resistance to chinch bugs, *Blissus leucopterus,* in sorghum because it was so highly successful that there have been no recent publications. In all, we deal with more than 500 articles relating to resistance in cereal crops to more than 50 insect pests, all published during the last 15 years in more than 190 periodicals.

COOPERATION

The ultimate goal of any research having to do with plant resistance to insects should be the development and continued release of resistant and otherwise useful varieties. This goal is accomplished through the cooperative efforts of members of a team that each contribute a share of the research effort (104). Key members are the plant breeder, the entomologist, and the pathologist; because the genetic diversity of the crop must be preserved, constant vigilance must be exercised against new hazards, and the resources of resistant germplasm must be expanded and refined to combat new pests. A lack of cooperation among workers representing several disciplines, regardless of the justifications, can reduce long-term benefits. For example, CSH-1 hybrid sorghum gave highly significant better quality and higher yield in India, but its susceptibility to the so-called sorghum shoot fly, *Atherigona varia* var *soccata,* has somewhat limited its acceptance (139). Conversely, entomologists sometimes report resistance but fail to follow through with the cooperative efforts necessary if insect-resistant cultivars with adequate agronomical characteristics are to be released.

REARING

When one is breeding for insect resistance, enough insects must be in one place at one time before new plant introductions, or the breeder's plant material that is segregating for resistance, can be evaluated. The entomologist must therefore rely on natural field populations or rear the insects in the laboratory. However, natural infestations have to be increased by a technique of successive plantings similar to the technique used to increase *Atherigona varia* in test plantings (137). A list (2) exists of the strain type, name or number, and place and date of origin of 283 species of insects, mites, ticks, and spiders currently being reared in 77 laboratories of the Agricultural Research Service. Insects from these colonies are available on request to researchers wishing to establish their own research populations.

The resistance to many cereal insects can be studied with the native host plant in a greenhouse insectary or in a growth chamber, for example: the cereal leaf beetle, *Oulema melanopus* (23, 24); the English grain aphid, *Macrosiphum avenae* (119, 120); the greenbug, *Schizaphis graminum* (15, 31); the Hessian fly, *M. destructor* (14); the oat bird-cherry aphid, *Rhopalosiphum padi* (72, 124); and the wheat midge, *Sitodiplosis mosellana* (6). These insect pests have short life cycles, so large populations can be reared on seedling plants. However, other insect pests of cereals such as the wheat stem sawfly, *Cephus cinctus* (69, 87); European corn borer, *Pyrausta nubilalis* (21, 115); corn earworm, *Heliothis zea* (59, 146, 156); Asiatic rice borer, *Chilo suppressalis* (108); sorghum midge, *Contarinia sorghicola* (34, 76); the southwestern corn borer, *Zeadiatrea grandissella* (32); and the northern corn rootworm, *Diabrotica longicornis* (150), require large plants for evaluation. Laboratory or greenhouse facilities, therefore, are normally not adequate. Nevertheless, both field and laboratory means of producing borers are used with the European corn borer (61).

The need for large numbers of certain insects for resistance or other studies has led to the development of artificial diets. With these diets, the insect can be reared off the plant. Artificial diets have been used to rear the greenbug (26); wheat stem sawfly (77); sugar cane borer, *Diatraea saccharalis* (142); western corn rootworm, *Diabrotica virgifera* (56, 57); European corn borer (7); pink stem borer, *Sesamia inferens* (17); sorghum stem borer *Chilo zonellus* (30); and fall armyworm, *Spodopt-*

era frugiperda (91). Indeed, most lepidopteran pests of cereal crops can be reared on variations of the diet developed by Vanderzant et al (140). However, some standard diets must be supplemented with chemicals or other materials to satisfy the nutritional requirements of the particular insect being studied. For instance, the addition of zinc, iron, and manganese to the diet increased the reproduction and rate of growth of greenbugs (26). Also, lyophilized corn kernels or corn leaves added to an agar-base diet for corn earworm larvae increased their weight and resulted in the retention of more food than when these ingredients were not added (136). Even the age of the plant material used in diets can make a significant difference in the weight gain and retention of food by larvae (151), and cellulose, dust, or paraffin added to a synthetic diet for the western corn rootworm increased survival (55). In addition, the occurrence of protozoan bacterial and fungal infections must be considered when artificial diets are used (61, 85). Antibiotics added to a diet for European corn borer at a rate as low as 200 ppm significantly reduced the level of infection of the insects with *Nosema pyraustae* (formerly *Perezia pyraustae*) (86).

Finally, when large numbers of insects are reared on synthetic diets for successive generations, inbreeding can occur. For instance, European corn borers grown on a meridic diet for 34–54 generations and then placed on susceptible corn plants had low larval establishment and survival compared with wild-type borers placed on the same plants (73). Also, leaf-feeding ratings and the number of leaf lesions were less for the diet-reared borers than for the wild-type borers. Nevertheless, survival equal to that of the wild-type borers could be recovered by one backcross to a wild-type parent, so diet-reared insects could be used for evaluation. In contrast, Mayo (91) found that continuous rearing of fall armyworms had no effect on larval establishment, survival on host plants, or feeding damage to sorghum. However, the number of generations of continuous rearing was only about half that for the corn borer, so inbreeding for a longer time could have some adverse effect.

The techniques used to rear insects on artificial diets vary according to the species. For the wheat stem sawfly, drinking straws have been packed with a synthetic diet and plugged with a damp cotton wool plug (77), and test tubes, small bore glass tubing, or grooved plastic trays have also been used (92). For the sugar cane borer, Public Health Service sputum cups (95) or Erlenmeyer flasks containing sugar cane tops or an artificial sugar cane diet (105) have been used. For the western corn rootworm, chunks of European corn borer diet covered with cellulose, dust, or paraffin, or buried in the soil, were satisfactory (56). For corn earworm larvae (10) and European corn borer larvae (60), vials or plastic cups containing diets have been used. Also, large numbers of European corn borers are reared on a meridic diet in plastic dishes, 10 inches in diameter and 3½ inches deep (61).

When insects are reared on the natural host plant, pots or greenhouse flats are usually planted with the crop required; then the insects are placed on the seedlings and are covered with a tent or cage to confine them to the plants. This sort of rearing is being done with the greenbug (31) and with the Hessian fly (13). However, when plastics are used for the confining cage, care must be taken to use the right kind: Chada (16) found that cellulose acetate and vinyl plastic were toxic to greenbugs.

Regardless of whether insects are reared on artificial diets or host plants, techniques must be developed for infesting the test plants. The stage of insect to be used, the number of insects per plant, and the growth stage of the plant should be considered.

METHODS OF EVALUATION

In any program for developing resistant varieties, there must be a way to evaluate the plants for resistance to insects. However, the techniques must be as simple as possible, efficient, and hightly accurate. Most publications dealing with resistance in plants to specific insects have a section describing the methods used. However, Dahms (28) reviewed techniques for obtaining cultivars for testing controlled insect infestations and the criteria used to evaluate resistance. In his review, he refers to Chesnokov's book, *Methods of Investigating Plant Resistance to Pests* (20). This publication, translated into English from Russian, outlines methods of investigating plant resistance to various types of pests. Chesnokov obtained most of his information while studying samples from the collection of the All Union Institute of Plant Cultivation. He describes methods of estimating the number of pests on crop varieties, of evaluating the degree of damage suffered, and of estimating decreases in yield in individual varieties with varying degrees of damage, especially for the frit fly, *Oscinella frit,* and the Hessian fly. Every student of host-plant resistance should have this publication in his library. Another publication of value to students of resistance in crops to insects is *Crop Loss Assessment Methods* (38) published by the Food and Agricultural Organization of the United Nations. The primary objective of this publication is to provide plant-protection workers with the general principles that should guide them as they plant and conduct field experiments to measure crop losses. However, techniques and apparatus commonly used to assess population densities of harmful organisms are discussed, as are general methods that can be used to estimate crop damage and special methods that have been developed specifically to assess populations of specific organisms.

As noted, evaluation for resistance can be done in the field, laboratory, or greenhouse. (Tests in insectaries and growth chambers are considered laboratory tests.) In field evaluations, either natural or artificial infestations of insects are used, but the results are subject to environmental interactions between plant and insect (6) that may be difficult to control. Thus, critical assessment of resistance may be hindered by fluctuations in the population density of the insect and the effects of the environment on both the insect and the plant. However, in a field test, the response of the plant is to the native field population of the insect instead of to inbred lines or to the more homozygous populations used in the laboratory. As noted in the section on rearing, laboratory populations may lose their capability to infest a plant that would be infested by native strains, and the diet-fed European corn borer was indeed unable to establish on corn plants (61, 73). As a result, in investigations of resistance to the European corn borer, the egg masses that are produced in the laboratory for artificial infestation of corn are obtained by starting a new borer colony each spring from a native population (62).

Laboratory and field-grown plants also may differ. Roberts (123) described differences in percentage of stem cutting by the wheat stem sawfly when susceptible plants were grown in the greenhouse and field. Those grown in the greenhouse during the summer or winter had more infested stems than those grown under irrigation in the field. On the other hand, some insect-plant relationships show very little difference in effects when the plants are evaluated in the greenhouse and the laboratory. Hsu & Robinson (72) found no significant difference in field and greenhouse tests of the resistance of barley varieties to the oat bird-cherry aphid.

The methods or techniques used to select for resistance are as diversified as the insect-plant relationships. For example, Chada (15) developed an insectary technique for rapid testing of varieties for resistance to the greenbug. The test plants were grown in trays covered with plastic cages, subjected to uniform infestations for 10–14 days, and then rated according to the amount of leaf area damaged. Later, this technique was modified by Wood (153); the insects were not caged and were allowed to move more freely from plant to plant to simulate field conditions. Plants destroyed within two weeks were rated susceptible; others were rated tolerant.

McNeal et al (93) used kernel weight from wheat stems infested by the wheat stem sawfly as an index to determine the weight loss of kernels from tunneled stems regardless of whether or not they had been cut. Holmes et al (70) found visual estimates to be accurate enough to rank varieties cut by the wheat stem sawfly in their proper order according to percentage of stems cut. The length of time an insect remains on barley plants in the field (100) has also been used to determine resistance to the oat bird-cherry aphid and the English grain aphid.

Larval weight gain is another method often used to determine resistance of a particular crop plant to an insect. Wilson & Shade (149) weighed adult cereal leaf beetles after they were confined to different species of graminae to determine the preferred hosts of this insect. Gallun et al (48) used the number of eggs and larvae per plant as an index of resistance. Later, Schillinger (127) modified this technique to include larval growth in the laboratory.

Number of immature corn leaf aphids, *Rhopalosiphum maidis,* remaining on corn plants grown in the greenhouse after a certain period was used to evaluate inbred lines for resistance. Dishner & Everly (33) compared data reported by previous workers and showed a high degree of concurrence between greenhouse data and data collected in the field. Robinson (124), however, showed that the fecundity of the oat bird-cherry aphid, commonly used as a measurement of antibiosis, was a valid measure only when the environment was strictly controlled. He therefore made several suggestions concerning ways to reduce variability in aphid fecundity: use aphids of the same clone, use pureline seed, sow all seeds at the same time, cage the aphids on the plant, and ensure that temperature and light conditions are similar for all tests. The foregoing examples of methods of evaluating for resistance are just a few of the many that can be used.

GERMPLASM BANKS

Germplasm banks of various plant species are located throughout the world and are a source of plant material to examine in a program to develop insect resistant cultivars. The plant materials provide the genetic diversity required when improving crops by adding insect resistance. Even though indigenous germplasm affords more rapid progress in obtaining adaptability, exotic material should not be ignored. Indeed, resistance may be found in material from areas where the insect pest has never been recorded. The importance of these genetic resources is discussed by Creech & Reitz (25). They describe the origin of the different germplasm; how it is evaluated, increased, and maintained; how it is used; and where it is stored. They state that the United States is a major contributor to the wealth of conserved germplasm. The working stocks of the Agricultural Research Service of the U.S. Department of Agriculture alone are estimated to be about 120,000 items. Other collections are held in state experiment stations, Soil Conservation Service, public institutions, and industry.

The Plant Production and Protection Division, Crop Ecology, and Genetic Resources Unit of FAO publishes lists of plant genetic resources available in countries throughout the world. Almost two million samples in approximately 1500 collections have been located, but the lists are not regarded as complete. For cereals, they have published lists of wheat and wheat relatives, maize, sorghum, barley, oats, rye, and rice. Specifically, 125 collections from 45 countries are listed for wheat and wheat relatives (39). The 3 institutions listed as having the largest collections are: Small Grains Collection, Agricultural Research Service, U.S. Department of Agriculture, Beltsville, Maryland, USA (21,400 samples); the Vavilov Institute of Plant Industry, Leningrad, USSR (21,219 samples); and the Australian Wheat Collection, Tamworth, N.S.W. Australia (18,000 samples).

A total of 98 collections are listed for barley, oats, and rye in 34 countries (43). The 3 largest collections are: the Swedish Seed Association, Svalov, Sweden (10,000 samples); the Research Institute of Cereals, Kramery, Havlivkovo, Czechoslovakia (6000 samples); and the Agricultural Experiment Station, University of South Dakota, Brookings, South Dakota, USA (5075 samples).

In 44 countries 105 maize collections are reported (42). The 3 largest collections are found in the Department of Agronomy, University of Illinois, Urbana, Illinois, USA (75,400 samples); CIMMYT, Apartado Postal 6–64, Mexico (12,000 samples); and the Instituto Columbiana Agropecuorio, Bogota, Columbia (6800 samples).

Some 63 sorghum collections are listed for 33 countries (40). The largest are: the Rockefeller Foundation, New Delhi, India (9000 samples); the East African Agricultural Research Organization, Muguga, Kenya (9000 samples); and the National Seed Storage Laboratory, Agricultural Research Service, U.S. Department of Agriculture, Fort Collins, Colorado, USA (7261 samples).

For rice, 90 collections are listed in 36 countries (41) with the 3 largest collections being: the Central Agricultural Station, Non Repos, B.C. Guyana (20,000 samples); the National Seed Storage Laboratory, Agricultural Research Service, U.S. Department of Agriculture, Fort Collins, Colorado, USA (16,340 samples); and the International Rice Research Institute, Manilla, Philippines (11,900 samples).

GENETICS AND BREEDING FOR RESISTANCE

Knowledge of the genetics of resistance of a plant to a specific insect pest is helpful in breeding a resistant cultivar, and a number of cereal-insect interrelationships have been studied from this point of view. One example is the investigation of the resistance in several rice cultivars to a planthopper, *Nilaparavata lugens,* and also a leafhopper, *Nephotettix impicticeps* (3). The resistance of 'Mudgo', 'Manavara Co22', and 'Dalwa Sannam MJU15' to the planthopper was found to be controlled by a single dominant gene that seems to be allelomorphic; and another cultivar, 'Karsamba Red ASD7', possesses a single recessive gene for planthopper resistance that is either allelic or closely linked to the locus that conditions resistance in the other 3 cultivars. Resistance to the leafhopper in the cultivars 'Pankhari 203', 'ASE7', and 'IR8' is also controlled by single genes that are nonallelic and dominant (3). Likewise, a monogenic type inheritance was shown in 37-day-old corn plants to western corn rootworm adults, this resistance is recessive (132). Resistance in sorghum to shoot fly attack was found to be quite complex, but additive gene action contributed most to the variation (138). However, in ovipositional studies by Blum (9), the size of the shoot fly population was found to have a pronounced effect on the rates of oviposition. Thus, nonpreference was partially dominant with low populations, but susceptibility (preference) was dominant with high populations. In contrast, dominance and/or specific combining ability was relatively unimportant in some maize inbreds resistant to the fall armyworm, but general combining ability was highly significant. Therefore, selection for resistance among the lines and their progeny depends on the accumulation of additive gene effects, and recurrent selection based on the performance of selfed progeny is suggested for developing a resistant plant population (147). Also, genetic analysis of another lepidopteran pest, *Chilo zonellus,* was reported by Rana & Murty (118).

The genetic studies of resistance in corn to the European corn borer have ranged from segregation of F_2 and backcross populations to reciprocal translocation studies, marker lines, and mutable loci systems. A single dominant gene pair was found to condition resistance in one study (112), and three gene pairs were identified in another (113). Translocation studies showed that the genes conditioning resistance in the inbred CI 31A are located on the short arms of chromosomes 1, 2, and 4 and on the long arms of chromosomes 4 and 6 (130).

Studies to determine resistance in sweet corn to the corn earworm have produced inconsistent results. Some studies of F_1 hybrids showed resistance to be a dominant trait; others with different inbreds showed it to be recessive (144); some others showed that general combining effects account for part of the genetic variability (146). Chromosomes associated with resistance in six sweet corn inbreds to the corn earworm were identified by Widstrom & Wiseman (148) as 4 and 5 for inbred 245, 3 for inbred 20, and 8 for inbred La2W. Those implicated less strongly were 4 and

8 for inbred 20, 1 and 3 for inbred 81–1, and 6 for inbred 322. A series of 18 waxy-marked chromosomes and 9 reciprocal translocations was used for this study.

Resistance to the greenbug has been studied in wheat, barley, oats, and sorghum. In wheat, hybrids of resistant and susceptible varieties indicated that a single recessive gene pair was responsible for resistance (27). But in later tests of crosses between the wheat varieties 'Concho' X 'Dickinson Sel 28A', there seemed to be an absence of dominance of the factor or factors conditioning the expression of resistance, and segregation of the populations suggested a mode of inheritance more complex than monogenic inheritance (117). In barley, resistance in the cultivars 'Omugi' and 'Dobaku' was determined to be controlled by a single dominant gene (53, 55, 134). In oats, resistance to the greenbug in the variety 'Russian 77' was conditioned by a single gene pair (54). In sorghum, resistance to this pest was controlled by single incompletely dominant factors (63, 145).

Resistance to the wheat stem sawfly is generally attributed to the solidness of the wheat stem, which acts as a deterrent to egg laying and larval development. Thus, most genetic and cytogenetic research over the years has had to do with the inheritance of resistance and the location of the chromosomes responsible for stem solidness. Wallace & McNeal (143) published a review of this research to 1962 in which they cite over 40 papers, most published before 1957. However, they thereby nearly covered the sum total of research on the subject as little has been published since. In their review, Wallace and McNeal cite and discuss genetic and cytological studies of F_1 segregation families from crosses between diploids and tetraploids, tetraploids X tetraploids, tetraploids X hexaploids, and hexaploids X hexaploids. From this research, stem solidness is essentially conditioned by one or more dominant, recessive, or complementary genes, depending on the parents used, the cross, and the ploidy involved. However, this work was all reported before 1957 and, hence, is not reviewed here.

Studies of the chromosomes responsible for stem solidness and hollowness and the responsible genome have also been quite extensive and mostly published before 1957, though some research has been reported since then. Essentially, the D genome is known to tend to produce hollow stems and the A and B genomes to produce solidness. For example, Larson (81, 82) found that in 'Chinese Spring', chromosomes 2A, 2D, 6D, and 7D carry genes for hollow stem and chromosome 4B has a gene for solid stem. However, Larson & MacDonald (83) found that in the monosomic lines of S-615, chromosomes 3B, 3D, 5A, 5B, and 5D carry genes for solid stem, and chromosomes 2D, 6D, and 7D have genes for hollow stem. They also found (84) that chromosome 3B in 'Rescue' wheat has a very important gene for stem solidness, that 6A and 6D seem to make the culms more hollow, and that 5A makes the stem solid, especially in lower internodes.

Likewise, most of the studies of the genetics of resistance to the Hessian fly were conducted before 1957 and were included in an excellent review by Allan et al (1), and by Gallun & Reitz (51). Resistance to the Hessian fly in a wheat variety is usually conditioned by a single gene (H_1 H_2, H_3, H_5, and H_6). Resistance is dominant or incompletely dominant, but the Marquillo resistance is more complex, and in PI 94587, a 14-chromosome wheat, there seem to be five or more genes that condition resistance (12). Also, two duplicate factor genes, H_7 H_8, have been associated with resistance in 'Seneca' to Race E of the Hessian fly (111).

Few genetic studies have been made of resistance in corn to the corn leaf aphid. However, Everly (37) studied the establishment and development of the corn leaf aphid on inbred and single-cross dent corn and came to the conclusion that resistance in corn is composed of many genes with varying degrees of dominance and additivity. Also, a high degree of resistance to the cereal leaf beetle has been found in wheat plants with hairy leaves (48, 128). Adult beetles will not oviposit on leaves with dense mats of hair, even when they are confined to the plant in cages. Genetic studies of segregating plant populations from crosses between high pubescent and glabrous-type parent showed that the gene action responsible for resistance in the form of hairs was mainly additive with partial dominance for pubescent density (122). J. A. Webster (personal communication) compiled an annotated bibliography of the association of plant hairs and insect resistance; he lists one reference for sorghum; two each for corn, oats, and rice; and 14 for wheat.

Resistance in barley to the cereal leaf beetle seems to be recessive. The mechanism is nonpreference by the larvae for feeding and differential egg laying (64), but transgressive inheritance was found in crosses between two lines, indicating that there is a possibility of obtaining higher resistance.

Pathak (109) listed inheritance patterns in rice for three insect pests. Inheritance for resistance to the striped borer, *Chilo suppressalis,* was polygenic; for the planthopper, *Nilaparavata lugens,* there were both dominant and recessive genes; and for the leafhopper, *Nephotettix impicticeps,* there was a single dominant gene.

Different methods of transferring resistance into good agronomic types have been used and are being used in cereal crops, but the method least used is probably selection from a heavily infested field and subsequent increase in seed from this plant for further use. Such selection is more adapted to legume crops and crops that can be propagated vegetatively. Hybridization is undoubtedly the technique most used (28). It usually involves improving a number of characters simultaneously while retaining other acceptable characters.

For self-pollinated crops such as wheat, oats, and barley, breeding for resistance to insects has usually been accomplished by backcrossing or by a modification of the pedigree system whereby detailed studies of individual plants are made during the segregating generations (28). Backcrossing can improve an acceptable variety by adding one or more characters, such as insect resistance and disease resistance, because it involves making crosses between the susceptible parent and a parent having resistance and then crossing the progeny of the cross back to the recurring parent. Selection is made among the backcross progeny for the derived character, and these plants are backcrossed again to the original parent. (Evaluation for resistance can be made at every generation, but it is not necessary because every plant is either homozygous or heterozygous for the resistant character.) At the end of several generations of backcrossing, the acceptable types are allowed to self-pollinate, and resistant selections are made. Several varieties resistant to the Hessian fly have been developed by this method or a modification. For example, 'Dual' wheat, the first certified variety resistant to the Hessian fly released in Indiana, was the product of 32 years of breeding and research during which desirable characters from 8 parent cultivars were combined (11). 'Arthur', another soft red winter wheat, was selected in 1960 from the F_3 generation of a Purdue cross made in 1957 (126). The detailed parentage of this variety is as follows: 'Minhardi' X 'Wabash' 5 X 'Fultz Sel.' CI 11512 X 'Hungarian' 2 X W38; 3 X 'Wabash' 4 X 'Fairfield' 6 X 'Redcoat' sib X 'Wisconsin' CI 12633 7 X 'Vigo' 4 X 'Trumbull' 2 X 'Hope' X 'Hussar' 3 X 'Fulhio' X 'Purkof^3' 5 X 'Kenya Farmer'.

For cross-pollinated crops such as corn, recurrent selection is the method of developing superior insect-resistant lines. In general, the technique involves self-pollination in some heterozygous populations, outcrossing to some appropriate tester, interpollinating superior extracted lines followed by one or more cycles until the objective is reached, evaluation of the test cross progenies, and, finally, utilization of superior test crosses of the selfed seed. The procedure was used by Penny et al (114) to increase the level of resistance to European corn borers in some lines. Two cycles were sufficient to shift resistance to a higher level, and three cycles produced essentially borer-resistant cultivars.

INSECT BIOTYPES

Genetic diversity exists within an insect species, as well as within a plant species. Therefore, when an insect population is subjected to extreme selection pressure in the form of a resistant crop variety, those variants within the population that are able to survive may interbreed to form populations of a new biotype or race; the resistant plant may then become susceptible. When antibiosis is the mechanism of resistance, such selection pressure is most successful if it results in the death of most of the insects. [Races, strains, or biotypes seldom if ever develop when tolerance or nonpreference is the main mechanism of resistance inherent in the plant (52).]

Biotypes or races arise most frequently among aphids, and this is understandable because of the parthenogenetic reproduction of these insects. With this type of reproduction and the relatively short life cycle, new biotypes can become abundant within one or two growing seasons. In the case of cereal insects, biotypes or races are known within the greenbug, the corn leaf aphid, the rice weevil, and the Hessian fly. Therefore, in breeding for resistance to these insect pests, one must consider the problem of biotypes.

The greenbug is recorded as having three biotypes (A, B, and C) that differ in their ability to live on and attack varieties of wheat, barley, and sorghum. For example, two wheat lines, 'Dickinson Sel 28A' and CI 9058, are resistant to the original field population (biotype A), but in 1961, biotype B developed and could kill Sel 28A and CI 9058 (154). More recently, biotype C was identified as being able to infest and survive on Piper sudangrass, though biotypes A and C cannot (155). Indeed, the reports and records indicate that biotype C built up into widely distributed virulent populations in one year. Such an increase is hard to comprehend, but figures by Dahms (29) on greenbug reproduction show it to be possible: one alate female could have approximately four million offspring 50 days after the first nymph was born.

Another pest of sorghum, the corn leaf aphid, has four or more biotypes (133). With this insect species, biotypes are usually distinguished by survival capacity and fecundity on host plants, though damage to white 'Martin' sorghum and 'Spartan' barley has also been used (106). There are no instances recorded in which the biotypes are active in the field.

Probably the most comprehensive investigation of insect biotypes in cereals has been concerned with races of Hessian flies. The first were described by Painter et al (102), but we now have eight. The identifying characteristics are ability or inability to survive on and stunt wheat plants having different genes for resistance (47, 51, 66). The distinction is based solely on plant and insect reaction, there being no morphological or other traits that distinguish the races.

Table 1 lists the eight races of Hessian fly and the genes in the wheat plant that condition resistance to a particular race (52). The genes in the wheat plant that condition resistance are dominant or incompletely dominant. However, the genes in the insect that condition virulence to a specific wheat are recessive, and the avirulent genes are dominant (67). Also, dominant genes for resistance in the wheat plant are specific for comparable recessive genes in the insect for virulence. A wheat plant, therefore, remains resistant as long as the Hessian fly does not have a set of recessive genes for virulence that can overcome the dominant genes for resistance in the plant. However, a wheat plant may have any number of dominant genes for resistance, each working independently of one another, and any one pair may provide resistance as long as the insect does not have a pair of recessive genes specific for virulence to that resistance. This situation substantiates the gene-for-gene theory to the effect that for every gene in a plant which conditions resistance, there may be a comparable gene in the insect which can overcome that resistance (52). These particular findings occurred as a result of genetic studies by Gallun and Hatchett (49, 50, 67) that confirmed previous cytological research by Metcalfe (94) that chromosome elimination occurs during spermatogenesis of the Hessian fly.

The discovery of the dominance of avirulence in the Hessian fly led to the suggestion that a Hessian fly race avirulent to specific wheats could be released in an area where these wheats are growing and infested with new virulent races or biotypes. The resulting progeny of the interacial matings will then die (65). For example, the Great Plains (GP) race cannot live on the majority of wheats in the

Table 1 Genes in wheat that condition resistance to races of Hessian fly

Hessian fly races	Genes in wheat
GP	H_1 H_2, H_3, H_5, H_6, H_7 H_8
A	H_3, H_5, H_6
B	H_5, H_6
C	H_3, H_5
D	H_5
E	H_1 H_2, H_5, H_6, H_7 H_8
F	H_1 H_2, H_3, H_5, H_7 H_8
G	H_1 H_2, H_5, H_7 H_8

eastern United States, and it carries dominant genes for avirulence. Therefore, progeny of matings between Great Plains flies will die on these wheats, progenies of matings between the Great Plains and native flies will die, and progeny of any matings between native flies of the same race will survive. Then if large numbers of the Great Plains race are released on wheats in the east, the Hessian fly could be eradicated. In fact, laboratory studies and tests in small-scale research plots have substantiated that the Hessian fly can be eradicated in three generations in such an area by releasing 19 Great Plains flies to every 1 native fly (45, 46).

Obviously, the races of Hessian flies are important in breeding resistant wheats. They can be used to identify different sources of resistance in the world collections; they can distinguish between wheats having different genes for resistance; and they can be used to determine whether combinations of genes occur in a single plant (52). Such knowledge about the interrelationship that occurs between plant and insect assists entomologists and plant breeders to combat insect pests of crops and at the same time to maintain genetic diversity in the crop.

NATURE OF RESISTANCE

The cause of resistance to insect pests in plants has been the objective of many research studies over the years, though such knowledge has only rarely been a major influence in the development of resistant varieties. For example, attractants, suppressants, feeding deterrents, inhibitors, amino acids, sugars, fatty acids, and growth hormones have been attributed to resistance in plants to specific insect pests. Beck (8) was concerned with the fundamental concepts of plant resistance in his review of biochemical and biophysical resistance to insect survival. Maxwell (89) emphasized the nutritional relationships related to antibiosis and preference in his review in which he cites papers dealing with the sugar content of plants and the amino acids and mineral requirements of insects in relation to their host plant. Because of the many papers published on the nature of resistance in plants to insects, we refer the reader to these reviews and consider here the situation in regard to only one insect, the European corn borer; to only one plant, corn; and to the single biochemical isolated and identified as being responsible for resitance in corn to the corn borer.

Klun & Brindley (78) found a correlation between the concentration of 6-methoxy-2-benzoxazolinone (MBOA) in the dried whorl tissue of 11 inbred lines of corn and resistance to leaf feeding by the first-brood European corn borer. Then Klun et al (79) identified the biochemical entity responsible for resistance in corn to the borer as 2,4-dihydroxy-7-methoxy-2H-1,4-benzoxazin-3-(4H)-one (DIMBOA), a compound formed by enzymatic hydrolysis of a glucoside precursor that undergoes slow chemical decomposition to MBOA. Finally, Klun et al (80) showed that the correlation between the concentration of DIMBOA in plant tissue and level of resistance to leaf feeding by the first-brood borer was highly significant for 11 inbreds ($r = 0.89$) and for 55 single crosses ($r = 0.74$). Therefore, DIMBOA is a chemical factor in the resistance of corn to leaf feeding (first-brood) by the European corn borer. However, DIMBOA may not be the only chemical involved because 2 of the 11 inbred lines (W22, which is intermediate for resistance to leaf feeding, and

R101, which is completely susceptible) had relatively high concentrations of DIMBOA without being noticeably resistant.

Nevertheless, the results were such that Klun et al (80) suggested the possibility that an objective chemical analysis for DIMBOA in the whorl portion of the maize plant could serve as an indicator of resistance to leaf feeding by the first-brood borers. This technique would eliminate the rating of plants for resistance by visual evidence of larval feeding damage. However, for a chemical procedure to be useful in selecting for corn-borer resistance in segregating populations of corn, the method must be fast and efficient. At present, the chemical evaluation is much too slow to be practical: it would require one person 1 year to determine the amount of DIMBOA in only 5000 samples of corn, while one person with a tape recorder can rate 25,000 plants for resistance in 10 days. Furthermore, the concentration of DIMBOA in the plant does not correlate with resistance to sheath feeding (second-brood), and inbred lines resistant to the first-brood are not necessarily resistant to the second-brood. We do not mean to imply that research of this nature should be discontinued. If methods of analysis involving biochemical techniques are more efficient than methods involving the actual insect pest, then by all means they should be used, but unless they are, the man with the recorder and the insect will have to do the selecting for resistance. Meanwhile, offshoots of the research may provide leads to other types of research and may open the door to the solution of problems that are not associated with insect resistance in plants.

RESISTANT VARIETIES

The ultimate goal of any program involving plant resistance to pest insects is the development and release of a resistant cultivar, whether synthetic or inbred. Thus, it is surprising that so little information is published concerning resistant varieties per se. Reports of varietal registration in *Crop Science* and miscellaneous state agricultural experiment station bulletins are almost the only sources, but one must scrutinize every sentence of these publications to find a mention of insect resistance. Probably the only complete report concerning varieties resistant to a specific pest insect is a publication by Gallun & Reitz (51). The authors discuss the genetics of resistance in 28 varieties of wheat to the 7 races of the Hessian fly, describe each variety, and report the estimated acreage planted to each variety by wheat class and by state during 1969. However, the acreages cited (8½ million planted to 24 resistant varieties in 34 states) were preliminary estimates based on the national wheat-variety survey for 1969 conducted by the USDA and cooperating state officials (121).

Nevertheless, some varieties resistant to the wheat stem sawfly are discussed in the monograph of Wallace & McNeal (143). Three varieties, 'Rescue', 'Chinook', and 'Cypress', are spring wheats that resulted from wheat-breeding programs in Canada over the last 10 years. However, in 1969, 7 sawfly-resistant varieties were grown on an estimated 1½ million acres (121) in the United States.

A similar estimate of acreages planted to corn that is resistant to the European corn borer is practically impossible because of the closed pedigrees of corn produced by commercial seed companies. Therefore, Sprague (98) used figures for corn inbreds produced by public agencies to determine that publicly developed lines were used in producing 40 to 50% of the total seed requirements in 1969. Since he considered that any trends established for public lines would also apply to the corn lines used in commercial hybrids, he then used little arithmetic to establish that the European corn borer resistant inbreds were grown in the United States in 1969 on approximately 21½ million acres.

In East Africa, more than 70% of sorghum varieties 'Serena' and 'Namatare' showed good recovery from (tolerance to) shoot fly attack (35) and yielded normally. However, seedling resistance gave inconsistent results. Sorghum germplasm resistant to the sorghum midge has been also registered. For example, Wiseman et al (152) registered SGIRL-MR-1, the product of seven years of selection, which exhibits the nonpreference mechanism of resistance.

Varieties of wheat resistant to the greenbug have yet to be released, but several varieties of barley that are resistant to this insect are available (36, 53, 75, 134). The acreages in the United States planted to these varieties have not been reported, but they probably could be obtained by using the statistical lists of acreages seeded to barleys during the year.

'IR20', a multiresistant rice variety, has varying degrees of resistance to the striped borer and to the green leafhopper as well as resistance to specific diseases. At present, this variety is the most widely grown improved plant variety in Bangladesh, the Philippines, and Vietnam and is reportedly grown on a total area of approximately five million acres (M. D. Pathak, personal communication).

VALUE OF RESISTANT VARIETIES

Some varieties of crops that have been developed for resistance to insects have resulted in savings to agriculture of millions of dollars annually. However, estimates of crop losses due to insects are generally educated guesses based on actual yields, the elimination of the cost of insecticidal treatments that are not incurred, and the value of the varieties as improvements over older varieties.

Thus the loss estimates of wheat to the Hessian fly are still based on the results of a 1943 test of yield by Hill et al (68) in which the losses were estimated to range from 0.04 bushel per acre when only 1% of the productive culms were infested. These estimates were determined for now extinct varieties and do not apply to the newer, more productive varieties. Nevertheless, the advantage resistant varieties have over susceptible varieties should be at least one dollar per acre. Then, the annual savings to agriculture achieved by planting varieties resistant to the Hessian fly and wheat stem sawfly must be approximately ten million dollars at a minimum. Also, the rice variety, IR20, must give annual savings of upwards of five million dollars.

When the cost of the research required to develop varieties resistant to the Hessian fly, the wheat stem sawfly, the European corn borer, and the spotted alfalfa aphid is compared with the actual dollars saved, it works out to a ratio of approximately 1 to 300 as reported by Luginbill (88) and in *Integrated Pest Management,* a study prepared by the Council on Environmental Quality in November 1972 and published by the Superintendent of Documents, US Government Printing Office, Washington, D.C. The latter report states: "Despite the lengthy development time and the costs of developing resistant strains, the economic rewards are great. The total cost of research conducted by federal and state agencies and private companies to develop resistant varieties for the Hessian fly, wheat stem sawfly, European corn borer, and spotted alfalfa aphid, was about $9.3 million. But the annual savings in reduced losses to the farmer is estimated at $308 million. The net monetary value of the research is about $3 billion over a ten-year period, or a return for each research dollar invested of approximately $300 in reduced crop losses." In addition, the report notes an indirect value in that the total insect population is so reduced that damage to susceptible varieties is minimized.

Literature Cited

1. Allan, R. E., Heyne, E. G., Jones, E. T., Johnston, C. O. 1959. Genetic analyses of ten sources of Hessian fly resistance, their interrelationships and association with leaf rust reaction in wheat. *Kans. Agr. Exp. Sta. Tech. Bull.* 104. 51 pp.
2. Agricultural Research Service, Biological Investigations, Pesticides Chemicals Research Branch, Entomology Research Division. 1969. Colonies of insects, mites, ticks, spiders, and insect cell lines maintained in laboratories of the Entomology Research and Market Quality Research Divisions of the Agricultureal Research Service. *US Dept. Agr.* 88 pp.
3. Athwal, D. S., Pathak, M. D., Bacalangco, E. H., Pura, C. D. 1971. Genetics of resistance to brown planthoppers and green leafhoppers in *Oryza sativa* L. *Crop Sci.* 11:747–50
4. Bae, S. H. 1966. *Studies on some aspects of the life history and habits of the brown planthopper, Nilaparavata lugens (Stol.).* M.S. thesis, Univ. of the Philippines, Los Banos, Laguna
5. Bains, S. S., Kaur, B., Athwal, A. S. 1971. Relative varietal resistance in some new wheats to attack of *Trogoderma granarium* Everts. *Bull Grain Technol.* 9:197–202
6. Basedow, T. 1972. Relations between host plants and phenology of gall midges *Contarinia tritici* (Kirby) and *Sitodiplois mosellana* (Gehin). *Z. Angew. Entomol.* 71:359–67
7. Beck, S. D., Smissman, E. E. 1960. The European corn borer, *Pyrausta nubilalis,* and its principal host plant. VIII. Laboratory evaluation of host resistance to larval growth and survival. *Ann. Entomol. Soc. Am.* 53:755–62
8. Beck, S. D. 1965. Resistance of plants to insects. *Ann. Rev. Entomol.* 10:207–32
9. Blum, A. 1969. Oviposition preference by the sorghum shootfly (*Atherigona varia soccata*) in progenies of susceptible X resistant sorghum crosses. *Crop Sci.* 9:695–96
10. Burton, R. L., Harrell, E. A., Cox, H. C., Hare, W. W. 1966. Devices to facilitate rearing of lepidopterous larvae [*Heliothis zea* (Boddie)]. *J. Econ. Entomol.* 59:594–96
11. Caldwell, R. M., Cartwright, W. B., Compton, L. E. 1959. Dual, a Hessian fly resistant soft red winter wheat. *Purdue Univ. Agr. Exp. Sta. Mimeo ID–44.* 1 p.
12. Caldwell, R. M., Gallun, R. L., Compton, L. E. 1966. Genetics and expression of resistance to Hessian fly, *Phytophaga destructor* (Say). *Proc. Int. Wheat Genet. Symp., 2nd, Lund 1963, Hereditas, Suppl.* 2:462–63
13. Cartwright, W. B., LaHue, D. W. 1944. Testing wheats in the greenhouse for Hessian fly resistance. *J. Econ. Entomol.* 37:385–87
14. Cartwright, W. B., Caldwell, R. M., Compton, L. E. 1959. Response of resistant and susceptible wheats to Hessianfly attack. *Agron. J.* 51:529–31
15. Chada, H. L. 1959. Insectary technique for testing the resistance of small grains to the greenbug. *J. Econ. Entomol.* 52:276–79
16. Chada, H. L. 1962. Toxicity of cellulose acetate and vinyl plastic cages to barley plants and greenbugs. *J. Econ. Entomol.* 55:970–72
17. Chatterji, S. M., Sharma, G. C., Sidiqui, K. H., Panwar, V. P. S., Young, W. R. 1969. Laboratory rearing of the pink stem borer, *Sesamia inferens* Walker,

on artificial diets. *Indian J. Entomol.* 31:75–77

18. Chatterji, S. M. et al 1971. Relative susceptibility of some promising exotic maize material to *Chilo zonellus* Swinhoe under artificial infestations. *Indian J. Entomol.* 33:209–13
19. Cheng, C. H., Pathak, M. D. 1972. Resistance to *Nephotettix virescens* in rice varieties. *J. Econ. Entomol.* 65:1148–53
20. Chesnokov, P. G. 1962. Methods of investigating plant resistance to pests. *Nat. Tech. Inform. Serv., US Dep. Commerce.* 107 pp.
21. Chiang, H. C., Holdaway, F. G. 1960. Relative effectiveness of resistance of field corn to the European corn borer, *Pyrausta nubilalis,* in crop protection and in population control. *J. Econ. Entomol.* 53:918–24
22. Chou, W. D., Cheng, C. H. 1971. Field reactions of rice varieties screened for their resistance to *Nilaparavata lugens* and *Nephotettix cincticeps* in insectary. *Taiwan Agr. Res. J.* 20:68–75
23. Connin, R. V., Cobb, D. L., Gomulinski, M. S., Arnsman, J. C. 1966. Plaster of Paris as an aid in rearing insects pupating in the soil. *J. Econ. Entomol.* 59:1530
24. Connin, R. V., Cobb, D. L., Arnsman, J. C., Lawson, G. 1968. Mass rearing of the cereal leaf beetle, *Oulema melanopus. US Dep. Agr. ARS* 33–125. 11 pp.
25. Creech, J. L., Reitz, L. P. 1971. Plant germ plasm now and for tomorrow. *Advan. Agron.* 23:1–49
26. Cress, D. C., Chada, H. L. 1971. Development of a synthetic diet for the greenbug, *Schizaphis graminum.* 3. Response of greenbug biotypes A and B to the same diet medium. *Ann. Entomol. Soc. Am.* 64:1245–47
27. Curtis, B. C., Schlehuber, A. M., Wood, E. A. 1960. Genetics of greenbug resistance in two strains of common wheat. *Agron. J.* 52:599–602
28. Dahms, R. G. 1972. Techniques in the evaluation and development of host-plant resistance. *J. Environ. Qual.* 1:254–59
29. Dahms, R. G. 1972. The role of host plant resistance in integrated insect control. In *Control of Sorghum Shoot Fly,* 152–67. New Delhi: Oxford & IBH
30. Dang, K., Anand, M., Jotwani, M. G. 1970. A simple improved diet for mass rearing of sorghum stem borer, *Chilo zonellus* Swinhoe. *Indian J. Entomol.* 32:130–33
31. Daniels, N. E., Porter, K. B. 1958. Greenbug resistance studies in winter wheat. *J. Econ. Entomol.* 51:702–4
32. Davis, F. M., Scott, G. E., Henderson, C. A. 1973. Southwestern corn borer: preliminary screening of corn genotypes for resistance. *J. Econ. Entomol.* 66:503–6
33. Dishner, G. H., Everly, R. T. 1961. Greenhouse studies on the resistance of corn and barley varieties to survival of the corn leaf aphid. *Proc. Indiana Acad. Sci.* 71:138–41
34. Doering, G. W., Randolph, N. M. 1960. Field methods to determine the infestation of the sorghum webworm and damage by the sorghum midge in grain sorghum. *J. Econ. Entomol.* 53:749–50
35. Doggett, H., Starks, K. J., Eberhart, S. A. 1970. Breeding for resitance to the sorghum shoot fly. *Crop Sci.* 10:528–31
36. Edwards, L. H., Smith, E. L., Pass, H., Wood, E. A. 1970. Registration of Kerr barley. *Crop Sci.* 10:725
37. Everly, R. T. 1967. Establishment and development of corn leaf aphid populations on inbred and single cross dent corn. *Proc. N. Cent. Br. Entomol. Soc. Am.* 22:80–84
38. FAO. 1970. Crop loss assessment methods. *FAO Manual, Evaluation and Prevention of Losses by Pests, Diseases, and Weeds.* 87 pp.
39. FAO. 1970. World list of collections—wheat and wheat relatives. *Plant Introd. Newslett.* 24:18–31
40. FAO. 1971. World list of collections—sorghum. *Plant Genet. Resour. Newslett.* 25:30–36
41. FAO. 1972. World list of germplasm collections—rice. *Plant Genet. Resour. Newslett.* 26:27–35
42. FAO. 1972. World list of germplasm collections—maize. *Plant Genet. Resour. Newslett.* 27:22–32
43. FAO. 1972. World list of barley, oats, and rye collections. *Plant Genet. Resour. Newslett.* 28:24–33
44. Fitch, A. 1847. The Hessian fly, its history, character, transformations, and habits. *Trans. NY State Agr. Soc.* 6:12–13
45. Foster, J. E., Gallun, R. L. 1972. Populations of the eastern races of the Hessian fly controlled by release of the dominant avirulent Great Plains race. *Ann. Entomol. Soc. Am.* 65:750–54
46. Foster, J. E., Gallun, R. L. 1973. Controlling Race B Hessian flies in field cages with the Great Plains race. *Ann. Entomol. Soc. Am.* 66:567–70
47. Gallun, R. L., Deay, H. O., Cartwright, W. B. 1961. Four races of Hessian fly

selected and developed from an Indiana population. *Purdue Univ. Res. Bull.* 732. 7 pp.
48. Gallun, R. L., Ruppel, R., Everson, E. H. 1966. Resistance of small grains to the cereal leaf beetle. *J. Econ. Entomol.* 59:827–29
49. Gallun, R. L., Hatchett, J. H. 1968. Interrelationship between races of Hessian fly, *Mayetiola destructor* (Say), and resistance in wheat. *Proc. Int. Wheat Genet. Symp., 3rd, Aust. Acad. Sci. Canberra,* 258–62
50. Gallun, R. L., Hatchett, J. H. 1969. Genetic evidence of elimination of chromosomes in the Hessian fly. *Ann. Entomol. Soc. Am.* 62:1095–1101
51. Gallun, R. L., Reitz, L. P. 1971. Wheat cultivars resistant to races of Hessian fly. *US Dep. Agr. Prod. Res. Rep.* 134. 16 pp.
52. Gallun, R. L. 1972. Genetic interrelationships between host plants and insects. *J. Environ. Qual.* 1:259–65
53. Gardenhire, J. H., Chada, H. L. 1961. Inheritance of greenbug resistance in barley. *Crop Sci.* 1:349–52
54. Gardenhire, J. H. 1964. Inheritance of resistance in oats. *Crop Sci.* 4:443
55. Gardenhire, J. H. 1965. Inheritance and linkage studies on greenbug resistance in barley (*Hordeum vulgare* L.). *Crop Sci.* 5:28–29
56. George, B. W. 1963. Experiments on egg hatching and the use of artificial diets for the western corn rootworm. *Proc. N.Cent. Br. Entomol. Soc. Am.* 18:94–95
57. George, B. W., Ortman, E. E. 1965. Rearing the western corn rootworm in the laboratory. *J. Econ. Entomol.* 58:375–77
58. Guthrie, W. D., Dicke, F. F., Neiswander, C. R. 1960. Leaf and sheath feeding resistance to the European corn borer in eight inbred lines of dent corn. *Ohio Agr. Exp. Sta. Res. Bull.* 860. 38 pp.
59. Guthrie, W. D., Walter, E. V. 1961. Corn earworm and European corn borer resistance in sweet corn inbred lines. *J. Econ. Entomol.* 54:1248–50
60. Guthrie, W. D., Raun, E. S., Dicke, F. F., Pesho, G. R., Carter, S. W. 1965. Laboratory production of European corn borer egg masses. *Iowa State J. Sci.* 40:65–83
61. Guthrie, W. D., Carter, S. W. 1972. Backcrossing to increase survival of larvae of a laboratory culture of the European corn borer on field corn. *Ann. Entomol. Soc. Am.* 65:108–9
62. Guthrie, W. D., Russell, W. A., Jennings, C. W. 1971. Resistance of maize to second-brood European corn borers. *Proc. Ann. Corn Sorghum Res. Conf.* 26:165–79
63. Hackerott, H. L., Harvey, T. L., Ross, W. M. 1969. Greenbug resistance in sorghums. *Corp Sci.* 9:656–58
64. Hahn, S. K. 1968. Resistance of barley (*Hordeum vulgare* L. Emend. Lam.) to cereal leaf beetle (*Oulema melanopus* L.). *Crop Sci.* 8:461–64
65. Hatchett, J. H., Gallun, R. L. 1967. Genetic control of the Hessian fly. *Proc. N. Cent. Br. Entomol. Soc. Am.* 22:100–1
66. Hatchett, J. H. 1969. Race E, sixth race of the Hessian fly, *Mayetiola destructor,* discovered in Georgia wheat fields. *Ann. Entomol. Soc. Am.* 62:677–78
67. Hatchett, J. H., Gallun, R. L. 1970. Genetics of the ability of the Hessian fly, *Mayetiola destructor,* to survive on wheats having different genes for resistance. *Ann. Entomol. Soc. Am.* 63: 1400–7
68. Hill, C. C., Udine, U. J., Pinckney, J. S. 1943. A method of estimating reduction in yield of wheat caused by Hessian fly infestation. *US Dep. Agr. Cir.* 663. 10 pp.
69. Holmes, N. D., Peterson, L. K. 1957. Effect of continuous rearing in Rescue wheat on survival of the wheat stem sawfly, *Cephus cinctus* Nort. *Can. Entomol.* 89:363–64
70. Holmes, N. D., McKenzie, H., Peterson, L. K., Grant, M. N. 1968. Accuracy of visual estimates of damage by the wheat stem sawfly. *J. Econ. Entomol.* 61:679–84
71. Horber, E. 1972. Plant resistance to insects. *Agr. Sci. Rev. Coop. State Res. Serv., US Dep. Agr.* 10. 11 pp.
72. Hsu, S. J., Robinson, A. G. 1962. Resistance of barley varieties to the aphid *Rhopalosiphum padi* (L.). *Can. J. Plant Sci.* 42:257–61
73. Huggans, J. L., Guthrie, W. D. 1970. Influence of egg source on the efficacy of European corn borer larvae. *Iowa State J. Sci.* 44:313–53
74. Israel, P. 1967. Varietal resistance to rice stem borer in India. *Proc. Symp. Int. Rice Res. Inst., 1964. On the Major Insect Pests of Rice,* 391–403
75. Jackson, B. R., Schlehuber, A. M., Oswalt, R. M., Wood, E. A. Jr., Peck, R. A. 1964. Will winter barley. *Okla. Exp. Sta. Bull.* B–631. 16 pp.
76. Jotwani, M. G., Singh, S. P., Chaudhari, S. 1971. Relative susceptibility of some sorghum lines to midge damage.

Indian Agr. Res. Inst. Final Tech.Rep., 123–30

77. Kasting, R., McGinnis, A. J. 1958. Note on a method of artificially sustaining larvae of the wheat stem sawfly, *Cephus cinctus* Nort. *Can. Entomol.* 90:63–64
78. Klun, J. A., Brindley, T. A. 1966. Role of 6-methoxybenzoxazolinone in inbred resistance of host plant (maize) to first brood larvae of European corn borer. *J. Econ. Entomol.* 59:711–18
79. Klun, J. A., Tipton, C. L., Brindley, T. A. 1967. 2,4-Dihydroxy-7-methoxy-1,4-benzoxazin-3-one (DIMBOA), an active agent in the resistance of maize to the European corn borer. *J. Econ. Entomol.* 60:1529–33
80. Klun, J. A., Guthrie, W. D., Hallauer, A. R., Russell, W. A. 1970. Genetic nature of concentration of 2,4-dihydroxy-7-methoxy-2H-1,4-benzoxazine-3(4H)-one and resistance to the European corn borer in a diallel set of eleven maize inbreds. *Crop Sci.* 10:87–90
81. Larson, R. I. 1957. Cytogenetics of solid stem in common wheat: monosomic F_2 analysis of the variety S–615. *Wheat Inform. Serv.* 6:2–3
82. Larson, R. I. 1957. Cytogenetics of solid stem in common wheat. I. Monosomic F_2 analysis of the variety S–615. *Can. J. Bot.* 37:135–56
83. Larson, R. I., MacDonald, M. D. 1959. Cytogenetics of solid stem in common wheat. II. Stem solidness of monosomic lines of the variety S–615. *Can. J. Bot.* 37:365–78
84. Larson, R. I., MacDonald, M. D. 1962. Cytogenetics of solid stem in common wheat. IV. Aneuploid lines of the variety Rescue. *Can. J. Genet. Cytol.* 4:97–104
85. Leuck, D. B. 1972. Induced fall armyworm resistance in pearl-millett. *J. Econ. Entomol.* 65:1608–11
86. Lewis, L. C., Lynch, R. E. 1970. Treatment of *Ostrinia nubilalis* larvae with fumidil B to control infections caused by *Perezia pyraustae*. *J. Invertebr. Pathol.* 15:43–48
87. Luginbill, P. Jr. 1958. Influence of seeding density and row spacing on the resistance of spring wheats to the wheat stem sawfly. *J. Econ. Entomol.* 51: 804–8
88. Luginbill, P. Jr. 1969. Developing resistant plants—the ideal method of controlling insects. *US Dep. Agr. Prod. Res. Rep.* 111. 14 pp.
89. Maxwell, F. G. 1972. Host plant resistance to insects—nutritional and pest management relationships. *Insect and Mite Nutrition*, ed. J. G. Rodriguez, 599–609. Amsterdam: North-Holland. 702 pp.
90. Maxwell, F. G., Jenkins, J. N., Parrott, W. L. 1972. Resistance of plants to insects. *Advan. Agron.* 24:187–265
91. Mayo, Z. B. Jr. 1972. Damage to sorghum in the greenhouse by fall armyworms reared on artificial diet for different lengths of time. *J. Econ. Entomol.* 65:927–28
92. McGinnis, A. J., Kasting, R. 1962. Method of rearing larvae of the wheat stem sawfly, *Cephus cinctus* Nort., under artificial conditions. *Can. Entomol.* 94:573–74
93. McNeal, F. H., Berg, M. A., Luginbill, P. Jr. 1955. Wheat stem sawfly damage in four spring wheat varieties as influenced by date of seeding. *Agron. J.* 47:522–25
94. Metcalfe, M. E. 1935. The germ cell cycle in *Phytophaga destructor* (Say). *Quart. J. Microsc. Sci.* 77:585–604
95. Miskemen, G. W. 1965. Nonaseptic laboratory rearing of the sugarcane borer, *Diatraea saccharalis*. *Ann. Entomol. Soc. Am.* 58:820–23
96. Modder, W. W. D., Alagoda, A. 1972. A comparison of susceptibility of rice varieties, IR–8 and Warangel 1263, to attack by gall midge, *Pachydiplosis oryzae* (Wood-Mason). *Bull. Entomol. Res.* 61:745–53
97. Munakata, K., Okamoto, D. 1967. Varietal resistance to rice stem borers in Japan. In: Major Insects Pests of Rice Plant. *Proc. Symp. Int. Rice Res. Inst., 1964,* 419–20
98. National Academy of Science. 1972. *Genetic Vulnerability of Major Crops.* Washington, DC. 300 pp.
99. Oliver, B. F., Gifford, J. R., Trahan, G. B. 1972. Differential infestation of rice lines by the rice stalk borer. *J. Econ. Entomol.* 65:711–13
100. Orlab, K. B. 1961. Host plant preference for cereal aphids in the field in relation to the ecology of barley yellow dwarf virus. *Entomol. Exp. Appl.* 4: 62–72
101. Ou, S. H., Kanjanasoon, P. 1961. A note on gall midge resistant rice variety in Thailand. *Int. Rice Commission Newslett.* 10
102. Painter, R. H., Salmon, S. C., Parker, J. H. 1931. Resistance of varieties of winter wheat to Hessian fly, *Phytophaga destructor* (Say). *Kans. State Agr. Exp. Sta. Tech. Bull.* 27. 58 pp.

103. Painter, R. H. 1951. *Insect Resistance in Crop Plants.* New York: Macmillian. 520 pp.
104. Painter, R. H. 1958. Resistance of plants to insects. *Ann. Rev. Entomol.* 3:267–90
105. Pan, Y. P., Long, W. H. 1961. Diets for rearing the sugarcane borer. *J. Econ. Entomol.* 54:257–61
106. Pathak, M. D., Painter, R. H. 1958. Effect of the feeding of the four biotypes of corn leaf aphid, *Rhopalosiphum maidis* (Fitch), on susceptible white martin sorghum and spartan barley plants. *J. Kans. Entomol. Soc.* 31:93–100
107. Pathak, M. D. 1968. Ecology of common insect pests of rice. *Ann. Rev. Entomol.* 13:257–94
108. Pathak, M. D. 1969. Stemborer and leafhopper—planthopper resistance in rice varieties. *Entomol. Exp. Appl.* 12: 789–800
109. Pathak, M. D. 1970. Genetics of plants in pest management. In *Concepts of Pest Management,* ed. R. L. Rabb, F. E. Guthrie, 138–57. Raleigh, N.C.: North Carolina State Univ. Press. 242 pp.
110. Pathak, M. D., Andres, F., Galacgac, N., Raros, R. 1971. Resistance of rice varieties to striped rice borers. *Int. Rice Res. Inst. Tech. Bull., Los Banos,* 11. 69 pp.
111. Patterson, F. L., Gallun, R. L. 1973. Inheritance of resistance of Seneca wheat to Race E of Hessian fly. *Proc. Int. Wheat Genet. Symp., 4th, Columbia, Mo.,* 445–49
112. Penny, L. H., Dicke, F. F. 1957. A single gene-pair controlling segregation for European corn borer resistance. *Agron. J.* 49:193–96
113. Penny, L. H., Dicke, F. F. 1966. Inheritance of resistance in corn to leaf feeding of the European corn borer. *Agron. J.* 48:200–3
114. Penny, L. H., Scott, G. E., Guthrie, W. D. 1967. Recurrent selection for European corn borer resistance in maize. *Crop Sci.* 7:407–9
115. Pesho, G. R., Dicke, F. F., Russell, W. A. 1965. Resistance of inbred lines of corn (*Zea mays* L.) to the second brood of the European corn borer (*Ostrinia nubilalis* Hubner). *Iowa State J. Sci.* 40:85–98
116. Peters, L. L., Fairchild, M. L., Zuber, M. S. 1972. Effect of corn endosperm containing different levels of amylose on Angoumois grain moth biology. 3. Interrelationship of amylose levels and moisture content of diets. *J. Econ. Entomol.* 65:1168–69
117. Porter, K. B., Daniels, N. E. 1963. Inheritance and heritability of greenbug resistance in a common wheat cross. *Crop Sci.* 3:116–18
118. Rana, B. S., Murty, B. R. 1971. Genetic analysis of resistance to stem borer in sorghum. *Indian J. Genet. Plant Breed.* 31:521–29
119. Rautapaa, J. 1966. The effect of the English grain aphid, *Macrosiphum avenae* (F.), on the yield and quality of wheat. *Ann. Agr. Fenn.* 5:334–41
120. Rautapaa, J. 1970. Preference of cereal aphids for various cereal varieties and species of Gramineae, Juncaceae, and Cyperaceae. *Ann. Agr. Fenn.* 9:267–77
121. Reitz, L. P., Lebsock, K. L. 1972. Distribution of the varieties and classes of wheat in the United States in 1969. *US Dept. Agr. Statist. Bull* 475. 70 pp.
122. Ringlund, K., Everson, E. H. 1968. Leaf pubescence in common wheat, *Triticum aestivum* L., and resistance to the cereal leaf beetle, *Oulema melanopus. Crop Sci.* 8:705–10
123. Roberts, D. W. A. 1957. Sawfly resistance in wheat. II. Differences between wheat grown in the greenhouse and on irrigated land. *Can. J. Plant Sci.* 37: 292–99
124. Robinson, A. G. 1964. Variability of resistance of barley varieties to the aphid *Rhopalosiphum padi* (L.) in different environments. *Proc. Int. Congr. Entomol., 12th, London,* 533
125. Rossetto, C. J., Painter, H., Wilbur, D. A. 1971. Resistance of varieties of rough rice to *Sitophilus zeamais* Motschulsky. *Int. Congr. Entomol., 13th, Moscow, 1968, Tr.* 2i:381–82
126. Schafer, J. F. et al 1968. Arthur soft red winter wheat, a breakthrough to a new yield level. *Purdue Univ. Agr. Exp. Sta. Res. Progr. Rep.* 335. 4 pp.
127. Schillinger, J. A. 1966. Larval growth as a method of screening *Triticum* sp. for resistance to the cereal leaf beetle. *J. Econ. Entomol.* 59:1163–66
128. Schillinger, J. A., Gallun, R. L. 1968. Leaf pubescence of wheat as a deterrent to the cereal leaf beetle, *Oulema melanopus. Ann. Entomol. Soc. Am.* 61:900–3
129. Schoonhoven, A. V., Horber, E. 1972. Development of maize weevil on kernels of opaque-2 and floury-2, nearly isogenic corn inbred lines. *Crop Sci.* 12:862–63
130. Scott, G. E., Dicke, F. F., Pesho, G. R. 1966. Location of genes conditioning resistance in corn to leaf feeding of the

European corn borer (*Ostrinia nubilalis*). *Crop Sci.* 6:444–46

131. Shmaraev, G. E. 1971. Results of studying resistance to pests of a food corn collection. *Tr. Prikl. Bot. Genet. Selek.* 43:109–20
132. Sifuentes, J. A., Painter, R. H. 1964. Inheritance of resistance to western corn rootworm adults in field corn. *J. Econ. Entomol.* 57:475–77
133. Singh, S. R., Painter, R. H. 1965. Reaction of four biotypes of corn leaf aphid, *Rhopalosiphum maidis* (Fitch), to differences in host plant nutrition. *Proc. Int. Congr. Entomol., 12th, London,* 543
134. Smith, O. D., Schlehuber, A. M., Curtis, B. C. 1962. Inheritance studies of greenbug resistance in four varieties of winter barley. *Crop Sci.* 2:489–91
135. Sprague, G. F., Dahms, R. G. 1972. Development of crop resistance to insects. *J. Environ. Qual.* 1:28–34
136. Starks, K. J., Bowman, M. C., McMillian, W. W. 1967. Resistance in corn to the corn earworm, *Heliothis zea,* and the fall armyworm, *Spodoptera frugiperda.* III. Use of plant parts of inbred corn lines by the larvae. *Ann. Entomol. Soc. Am* 60:873–74
137. Starks, K. J. 1970. Increasing infestations of the sorghum shoot fly in experimental plots. *J. Econ. Entomol.* 63: 1715–16
138. Starks, K. J., Eberhart, S. A., Doggett, H. 1970. Recovery from shoot fly attack in a sorghum diallel. *Crop Sci.* 10:9–21
139. Thobbi, V. V. et al 1968. Insect control studies on the new hybrid sorghum CSH-1 in India. *Indian J. Entomol.* 30:45–57
140. Vanderzant, E. S., Richardson, C. S., Fort, S. W. 1962. Rearing the bollworm on artificial diet. *J. Econ. Entomol.* 55:140
141. Vavilov, N. I. 1950. The origin, variation, immunity and breeding of cultivated plants. *Chron. Bot.* 13:147
142. Walker, D. W., Alemany, A., Quintana, V., Padovani, F., Hagan, K. S. 1966. Improved xenic diets for rearing the sugar cane borer in Puerto Rico. *J. Econ. Entomol.* 59:1–4
143. Wallace, L. E., McNeal, F. H. 1966. Stem sawflies of economic importance in grain crops in the United States. *US Dep. Agr. Tech. Bull.* 1350. 50 pp.
144. Walter, E. V. 1962. Sources of earworm resistance in sweet corn. *Proc. Am. Soc. Hort. Sci.* 80:485–87
145. Weibel, D. E., Starks, K. J., Wood, E. A. Jr., Morrison, R. D. 1972. Sorghum cultivars and progenies rated for resistance to greenbugs. *Crop Sci.* 12:334–36
146. Widstrom, N. W., Hamm, J. J. 1969. Combining abilities and relative dominance among maize inbreds for resistance to earworm injury. *Crop Sci.* 9:216–19
147. Widstrom, N. W., Wiseman, B. R., McMillian, W. W. 1972. Resistance among some maize inbreds and single crosses to fall armyworm injury. *Crop Sci.* 12:290–92
148. Widstrom, N. W., Wiseman, B. R. 1973. Locating major genes for resistance to the corn earworm in maize inbreds. *J. Hered.* 64:83–86
149. Wilson, M. C., Shade, R. E. 1964. The influence of various Gramineae on weight gains of postdiapause adults of the cereal leaf beetle, *Oulema melanopa. Ann. Entomol. Soc. Am.* 57:659–61
150. Wilson, R. L., Peters, D. C. 1973. Plant introductions of *Zea mays* as sources of corn rootworm tolerance. *J. Econ. Entomol.* 66:101–4
151. Wiseman, B. R., McMillian, W. W., Bowman, M. C. 1970. Retention of laboratory diets containing corn kernels or leaves of different ages by larvae of the corn earworm and the fall armyworm. *J. Econ. Entomol.* 63:731–32
152. Wiseman, B. R., McMillian, W. W., Widstrom, N. W. 1973. Registration of SGIRL-MR-1 sorghum germplasm. *Crop Sci.* 13:398
153. Wood, E. A. Jr. 1961. Description and results of a new greenhouse technique for evaluating tolerance of small grains to the greenbug. *J. Econ. Entomol.* 54:303–5
154. Wood, E. A. Jr. 1961. Biological studies of a new greenbug biotype. *J. Econ. Entomol.* 54:1171–73
155. Wood, E. A. Jr., Chada, H. L., Saxton, P. N. 1969. Reaction of small grains and grain sorghum to three greenbug biotypes. *Okla. State Univ. Agr. Res. Progr. Rep.* 618. 7 pp.
156. Zuber, M. S. et al 1971. Evaluation of 10 generations of mass selection for corn earworm resistance. *Crop Sci.* 11:16–18

BRAIN STRUCTURE AND BEHAVIOR IN INSECTS

❖6094

P. E. Howse
Department of Biology, University of Southampton, Southampton S09 5NH, Great Britain

INTRODUCTION

In spite of fierce arguments over the supposed intelligence of insects, most biologists prior to the present century were loath to admit that insects possess brains. Indeed, one criterion of Linnaeus' definition of the Insecta was the lack of a brain. The behavior of decapitated insects made this easy to believe. The great eighteenth century naturalist, Buffon, wrote: "the hippobosca, equina, or horsefly, will live, run, nay, even copulate, after being deprived of its head."

The brain is unnecessary for the execution of many of the stereotyped behavior patterns of insects. Decapitated *Drosophila* species live for several days and engage in preening, flying, walking, and copulation (93). The latter, admittedly, occurs only if the insect is virtually raped, as the headless females reject courting males. *Gryllus campestris,* when headless, can be induced to sing several of its song patterns when the neck connectives are stimulated (74), male mantids can complete a sequence of mating behavior (82), and debrained pupae of *Hyalophora cecropia* will undergo a sequence of behavior enabling them to escape from the cocoon (103). It was first shown by Horridge (40) that headless cockroaches are capable of learning leg position responses. Chen et al (11) later found that intact insects learned such a response faster than headless ones, and the learning persisted for much longer. Long-term memory is apparently established in the head and thoracic ganglia, but will persist in the latter after removal of the head. Circadian locomotory rhythms in cockroaches are lost on decapitation (35).

In some of the examples quoted above, the brain is acting as a trigger for a sequence of behavior—in *H. cecropia,* for example, implantation of a brain into a decerebrate pupa will initiate pre-eclosion and eclosion behavior, a sequence in which sensory feedback is of minor, if any, importance. This behavior, like the singing of crickets (46), is therefore centrally programmed.

A consequence of central programming of behavior is that an insect may be unable to adapt to a change in its sensory field and its behavior then tends to recycle. Examples may be found commonly in descriptions of the behavior of hunting wasps. Fabre (19) described the behavior of the wasp *Pelopaeus,* which provisions its nest with spiders and lays an egg on the first spider it obtains. Fabre removed the first spider with its egg. The wasp then brought a replacement, but did not lay another egg, and continued for two days to bring spiders as fast as Fabre removed them. Peckham & Peckham (77) described how a species of *Pompilus* wasp, after paralyzing a spider, hung it on a plant while it opened its nest hole. They replaced the spider with a similar immobile one, but the wasp did not accept this and flew off to return with another spider which it put down again while it dug a new nest hole adjacent to the existing one. Many similar observations are found in the writings of these and other biologists who studied hunting wasps. They illustrate a remarkable inflexibility of behavior and suggest that links in a sequence of activities are provided only by very limited features of a stimulus situation. This has been confirmed in general terms by many more recent workers such as Tinbergen (on *Philanthus*) (101), Steiner (on *Liris nigra*) (95), and Baerends (on *Ammophila*) (4).

While insect behavior often appears to consist of the release of stored programs there is also much important evidence for one action having aftereffects on the threshold or nature of performance of a subsequent action. Examples of this are common, and are found in the behavior of aphids, in which flight and settling have reciprocal effects upon one another (48); in the settling movements of hemileucine moths, which are quantitatively related to preceding flight activity (6); in the stinging activity of ammophiline wasps, which gradually changes as a result of sensory feedback from the caterpillar (28); and in the dance language of honeybees (23), in which flight distance and direction determine the characters of the dance.

Part of the hallmark of insect behavior is an association of very inflexible behavior patterns with a propensity for learning that is often startling. Thus dragonflies and damselflies will attempt to fly upside down if illuminated from below (22), but are capable of recognizing and defending a territory with a complexity of visual features (38). Honeybees, blinded in one eye, will allow their behavior to be dictated by a light beam, but are capable of learning foraging routes with respect to visual or olfactory features (22, 23) and can be conditioned in a matter of seconds to show a feeding response in association with a certain scent, color, visual pattern, or time of day (60).

To summarize, evidence from behavioral studies suggests that the insect brain is essentially a selection apparatus acting upon motor mechanisms in the nerve cord. It may effect the likelihood of transitions between activities and link them in such a way that one has aftereffects upon the other. Behavioral thresholds also undergo large circadian changes, for which the brain may be responsible. Prodigious learning abilities are shown by many of the aculeate Hymenoptera and would appear to demand a large proportion of brain tissue.

ANATOMY OF THE BRAIN

Reference to detailed anatomical descriptions of the insect brain will be found in Holmgren (39), Bullock & Horridge (9), and in other works that are listed below. The purpose of this review is to examine the structure and function of the mushroom bodies (corpora pedunculata) and the central body which are the most conspicuous neuropile masses of the brain. The optic and antennal lobes will be discussed only in relation to these bodies.

Dujardin (18) was the first to describe the insect brain in any detail. He drew attention to the paired mushroom-shaped bodies and noted that their calyces were especially large in "intelligent" insects such as ants and bees. Their convoluted appearance in whole mounts led him to suggest that they were equivalent to the vertebrate cerebral cortex and hence the seat of intelligence. A number of workers then began comparative studies, among them Flögel (20) who found that Hemiptera had very small mushroom bodies and rudimentary calyces, while wasps, bees, and ants had the largest mushroom bodies with deep double calyces. Other insects, including, Lepidoptera, *Blatta, Forficula, Dytiscus, Aeshna,* and *Tabanus* formed an intermediate group. Such variations did not exist in the size of the central body. Von Alten (2) made a more detailed survey of the brains of Hymenoptera. He found the corpora pedunculata poorly developed in sawflies; better developed in cynipids, urocerids, ichneumonids, and braconids; and largest in social and subsocial bees and wasps. The dimensions of the mushroom bodies of bees increased in a series from *Osmia* through *Chalcidoma* and *Megachile* to *Anthidium.* The best developed mushroom bodies were in *Bombus* species, where they were somewhat larger than in *Apis mellifera.* No very marked differences were found between parasitic bees, such as *Nomada* and *Psythirus,* and social and subsocial bees. Von Alten's comparisons, however, were made on dimensions of brain components in only one plane and did not take into account the three-dimensional structure.

Caste differences are found in the brains of social insects. Forel (21) found that the brains and mushroom bodies of queen ants were usually smaller than those of the workers, and that the brains of the males were atrophied. He therefore supported the view of Dujardin that the mushroom bodies were the cerebrum of insects. An alternative theory, that they were connected with vision, was refuted by findings of both Forel (21) and Rabl-Rückhard (78) that blind ants had well-developed mushroom bodies.

Jonescu (47) found the mushroom bodies of the honeybee, *A. mellifera,* to be largest in the workers (Table 1) and smallest in the drones. This trend was not paralleled by the optic lobe, which was about twice as large in drones as in workers and queens, and so the supposition that the mushroom bodies were concerned with vision was further refuted. More recent measurements by Lucht-Bertram (62), however, show that the apparent differences in the mushroom bodies virtually disappear if the differential development of the optic lobe is taken into account. Evidently the caste differences are not great: all that can be said is that the optic lobe is relatively much larger in drones and the mushroom bodies are significantly larger in workers than in other castes.

Caste differences in the brain structure of termites were claimed by Thompson (100), who examined *Reticulitermes flavipes,* but her findings were not confirmed by subsequent investigators. In a study of many ant species, Pandazis (75) found that the mushroom bodies were generally largest in workers, and smallest in males. Although he surveyed ants with many quite different habits of life, no clear-cut correlations emerged between behavior and structure of the mushroom bodies apart

Table 1 Size of the brain components of *Apis mellifera*

Part of brain	Queen	Worker	Drone
Optic lobes[a]	33.9	31.4	67.6
Corpora pedunculata[a]	9.2	13.5	5.6
Corpora pedunculata as % of whole brain[a]	14.39	21.37	7.76
Corpora pedunculata as % of brain minus optic lobes[b]	28.35	39.05	29.31
Central body as % of brain minus optic lobes[b]	0.54	0.51	0.52

[a]From Dujardin (18).
[b]From Lucht-Bertram (62).

from those among castes. For example, size of optic lobes could not be correlated with size of mushroom bodies. It appears that a given taxonomic group has the same basic endowment of brain structure which is as much a speciality as any basic morphological features which the species have in common. This general rule is borne out also by studies on nonsocial insects. Thus Goossen (30) found relatively few differences among the mushroom bodies of a number of coleopteran species, and Groth (31) similarly found little variation in a study of 29 species of Diptera. Both these workers found that the most evident relationship was between body size and mushroom body size. Goossen made his comparisons on width measurements of the brain in suitable transverse sections, which effectively obscured any quantitative differences that might exist between coleopteran and hymenopteran brain structures.

Comparative studies of insects from a number of different orders were made by Hanström (32–34), whose results are more valuable than many others because they include volume measurements. These findings broadly confirm those of Flögel (20). Correlations of the size of the mushroom bodies with that of any one sensory center were difficult to make, and Hanström was forced to conclude that their size was determined by a number of factors including connections with all the main sensory centers, and that the mushroom bodies had undergone a separate development in the pterygote orders he examined. The Apidae and Formicidae were found to have the best developed mushroom bodies, which were seen not only as association centers for sensory inflow, but as centers controlling the complex instinctive activities of these insects. Hanström found the central body to be relatively constant in volume in insects of different orders.

Howse (42) and Howse & Williams (43) in a survey of the brains of over 30 insect species also found the central body to be remarkably constant in structure even among insects with widely differing life styles. A direct correlation was established between the size of the species, judged by head width, and the volume of the central body. In a survey of 29 species of Diptera, Groth (31) found a similar relationship, although this was nonlinear.

It was argued that the volume of the central body formed a convenient standard against which to compare the size of the corpora pedunculata in a given insect. The mushroom to central body ratio was found to be very high in social and subsocial Hymenoptera and in Isoptera, but low in all nonsocial insects (including *Pieris, Blaberus, Tenthredo, Mantis, Chrysops*). This index of mushroom body size is therefore proportional to behavioral complexity, as far as this can be usefully assessed.

VARIATIONS IN THE FORM OF THE MUSHROOM BODIES

Relative differences in the shape of the mushroom bodies have been catalogued by Flögel (20) and Bullock & Horridge (9). Both these accounts contain inaccuracies and misleading comparisons. Although there are still many gaps in our knowledge, it is possible to divide insects into about six main groups on the basis of the gross structure of the mushroom bodies.

Hymenoptera

The social and subsocial aculeate Hymenoptera have mushroom bodies with deep double calyces. The calyx wall in bees is divisible into three main regions (42) or possibly four (98). The basal region, which receives input from the antennal lobe, has been called the basal ring (42, 109) and has a marked glomerular structure. Most of the side wall is formed by tissue containing numerous branched endings of intrinsic cells (i.e. neurones with endings confined to the mushroom bodies) and has been called the collar (42). A distal lip is also clearly distinguishable. In ants, the lip is much the largest element of the calyx wall, while in bees, social wasps, and hunting wasps the lip and basal ring are roughly equal in size and the collar is the largest part (42). Hymenoptera which do not have well-developed learning abilities (woodwasps, sawflies) have a calyx wall which appears entirely glomerular and undifferentiated.

Fibers from the calyx wall enter the stalk (pedunculus) in an orderly arrangement and form three concentric cylinders of chromophilic tissue, which appear as rings in cross section. These cylinders of tissue split and change their orientation reappearing in the α lobe (*rückläufiger Steil*) and β lobe (*Balken*) as transverse bands seen in cross section. The rearrangement of fibers has been described in somewhat differing accounts by Goll (29), Strausfeld (98), and Howse (42). Nevertheless, it is plain that the spatial separation of elements in the calyces is preserved in the stalk and lobes. There are about three main chromophilic bands in the lobes, and a number of subsidiary ones. The bands differ in their thickness and the degree to which they stain.

Dictyoptera and Isoptera

The cockroaches (*Blatta, Periplaneta, Blaberus*) have two well-developed calyces to each mushroom body, but the calyx wall is entirely glomerular and probably homologous with the basal ring of the hymenopteran calyx. The stalk and lobes are of smaller diameter and much longer than those of Hymenoptera: the α lobe has a slightly sigmoid shape, turning dorsally to end close to the calyces. In transverse section, the lobes show 10–12 chromophilic bands which are all more or less similarly spaced and equal in thickness.

The mushroom bodies of termites are similar to those of cockroaches, but the calyces are relatively much smaller, and the lobes are relatively much longer. The α lobe is similar to that of cockroaches, but the β lobe, instead of ending near the mid-line below the central body as it does in cockroaches and many other insects, bends under the central body and continues dorsally in the median line to end level with the calyces (33). This formation of the β lobe is peculiar to termites. The banding pattern of the lobes resembles that of cockroaches. The mushroom bodies of primitive termites (e.g. *Zootermopsis*) resemble most closely those of cockroaches, but those of the higher termites (e.g. *Macrotermes, Apicotermes*) are relatively smaller with long thin lobes (43).

Lepidoptera

The Lepidoptera studied by Pearson (76) have a double calyx of uniform glomerular structure on each side. The lobe system is complicated by the presence of up to four additional lobes, the largest of which, the Y lobe, arises close to the origin of the α lobe and is connected by a tract with the ipsilateral calyces. The Y lobe was not found in the butterfly, *Pieris,* but may be peculiar to Lepidoptera. Banding patterns are not found in the lobes which are subdivided into a number of fiber bundles within which there are probably extensive synaptic connections.

Orthoptera

Acheta domesticus has a distinct double calyx on each side. The anterior calyx of each mushroom body gives rise to the central region of the stalk (91) and the posterior calyx gives rise to fibers which join the central region peripherally. Three zones are seen in a transverse section of the pedunculus, a dark outer ring, a dark central axis, and an intervening light zone (87). These zones can be traced into the α and β lobes. The calyx of the locust *Schistocerca gregaria* is essentially a single cup in which the posterior calyx is represented by a distinct lobe (112). In the stalk, a dark zone arises from the posterior lobe and a light zone from the anterior lobe. The α lobe is short and thinner than the β lobe, and divided into five bundles almost circular in cross section.

Other Orders

The other orders, in our present state of knowledge, can be grouped together because the mushroom bodies are poorly developed and are not known to show any outstanding peculiarities. Some Coleoptera, such as *Cetonia* and *Melonontha,* have well-developed double calyces (30), but in others, such as *Geotrupes, Carabus,* and *Dytiscus,* the calyces are little more than club-like expansions of the stalk. They are also barely evident in Diptera (31) and contain few cell bodies. The stalk and α lobe, however, have a concentric structure like those of *Acheta* (31, 42). Mushroom bodies are even less distinguishable in Odonata and Hemiptera.

General Comparisons

There appears to be a correlation between the feeding and foraging habits of the insects and the degree to which the mushroom bodies are developed. The Hemiptera have simple feeding strategies and form one end of a spectrum with social Hymenoptera at the other. The syrphid Diptera and Odonata are of interest in having small mushroom bodies compensated for by very large optic lobes. Although these insects are often territorial, it may be that their visual memory resides in the optic lobes. In the social and subsocial Hymenoptera, it would appear to lie in the collar, which is a large calyx element peculiar to bees and wasps, while olfactory memory may lie in the lip.

Neuroanatomy of the Mushroom Bodies

Kenyon (49) using the Golgi selective silver staining technique, found that the globuli cells of the calyx of the bee were monopolar neurones. One branch of each neurone arborizes in the wall of the calyx, and a colateral runs into the stalk and divides giving one branch which runs to the end of the α lobe and another which runs to the end of the β lobe. These neurones form the great majority of mushroom body tissues and have been called Kenyon cells (43).

Goll (29) has found four kinds of Kenyon cells in the brains of *Formica rufa* and *Formica pratensis.* These differ according to the locations of their cell bodies and the region of the calyx wall which they innervate. Type I neurones have cell bodies at the bottom of the calyx and innervate the basal ring region, type II neurones have cell bodies filling the remainder of the cup and innervate the collar and lip regions, type III neurones have cell bodies peripheral to the lip which they innervate, and type IV neurones have cell bodies on the outside of the collar and basal ring which they innervate. In Lepidoptera, Pearson (76) found only two types of Kenyon cells, one with many spiny branches in the calyx and the other with bunched claw-like endings covering a narrow cylindrical field. Schürmann (91) found at least three types in *Acheta domesticus,* all similar to those found in *Sphinx.* Spiny cells are present with numerous arborizations in the posterior calyx, and two kinds of bunched cells, those with short branches mainly in the anterior calyx and those with long branches distributed throughout many regions of the calyx. The latter are the most numerous type. The Kenyon cells of social Hymenoptera differ from those of other insects in having very large arborizations in the calyx (43, 76). This applies especially to the cells supplying the collar region in bees and wasps, and it has been suggested that these arborizations form a substrate for memory storage (43).

It has been suggested that nervous transmission occurs between adjacent Kenyon cells. Landholt (58) found apparent glial windows between cell bodies in the calyces of *Formica lugubris.* The stalk of *Acheta* contains about 55,000 fibers of which about 90% belong to intrinsic neurones (91). In the brain of the honeybee, there are many synaptic contacts between these parallel running fibers in the stalk (88). Bundles of fibers are separated by glial processes in the bee, and also in *Periplaneta* (26), *Acheta* (89, 91), and *Sphinx* (76). In the bee, the fibers and fiber bundles run parallel in the lobes, but in *Acheta* and *Sphinx* they frequently twist over one another.

The fine structure of the calycal glomeruli has been studied by a number of authors (57, 63, 94, 102). In *F. rufa* and *Camponotus ligniperda* there are large presynaptic end-knobs with many postsynaptic feet converging on them (57). Selective silver staining shows that the main input to the calyces of *Sphinx* is from the antennal tract, and the fibers of this have knob-like terminals. It is believed that these form the center of glomeruli and fit the claw-like terminals of bunched intrinsic cells (76). Accessory cells (76), connecting the calyces and Y lobe of Lepidoptera, also have claw-like terminals which form glomeruli. Many types of extrinsic cells are found with endings in the stalk and in the lobes, but the location of their perikarya and other branches are, in all cases, very uncertain. The endings in the lobes commonly show extensive arborization over a wide field, but in some neurones these arborizations are largely confined to one plane (91). In *Formica* Goll (29) described a fiber with endings extending across the α lobe in a narrow transverse zone.

Most earlier workers described tracts connecting the mushroom bodies with the central body, and with other neuropile masses. Detailed examination of selectively stained preparations, however, has failed to show such direct connections between the mushroom body lobes and the central body (76, 91, 112).

LOCATION OF NEUROTRANSMITTERS

Acetylcholine

Van der Kloot (107) demonstrated the presence of acetylcholine esterase in *Hyalophora cecropia.* This enzyme declines in quantity in the brain tissue at the beginning of pupation, and at the same time the brain becomes electrically inexcitable. At the end of diapause the cholinesterase level rises and the brain resumes excitability. Cymborowski et al (16) showed a circadian rhythm of fluctuation of cholinesterase activity in the brain of *Acheta domesticus* which correlated with variations in locomotor activity.

Landolt & Sandri (59) suggested that vesicles of 300–600 Å in diameter found in the calyces of *F. lugubris* contained acetylcholine, and Steiger (94) demonstrated the existence of presynaptic endings in the calyces of *F. rufa* and *C. ligniperda* which were characterized by many such vesicles and mitochondria. In addition, presumed presynaptic vesicles of 1000–1500 Å in diameter were present, and tight junctions occurred between the synaptic endings. Steiger considered that the former might be containers for catecholamines. However, most studies have failed to show catecholamines in the calyces or perikarya cells of the insect brain, although Schürmann & Klemm (92) found evidence for small quantities of indolamines in the calyces of *A. domesticus.*

Frontali et al (27) found cholinesterase activity in parts of the central body, in the protocerebral bridge, lobula, and medulla. There was a smaller amount present in the calyces and antennal glomeruli, and none in the calycal globuli cells or in the α and β lobes. This general picture has been confirmed by the work of Hess (37) who noted that the distributions of cholinesterase and catecholamines in the cockroach brain were largely complementary. A tentative suggestion has been made that the antennal glomeruli and the calyces from an interrelated system. They both contain acetylcholine esterase and are linked by the olfactorio-globularis tract which also contains the esterase. The ocellar nerve and the protocerebral bridge are connected (according to some authors) and both contain cholinesterase-positive fibers (26).

The existence of acetylcholine esterase does not indicate with certainty that cholinergic neurones are present, but the application of cholinergic blocking agents such as atropine and scopolamine to the brain of ants rendered them less aggressive and decreased the amplitude of electrical (EEG) discharges in the brain (54). Also, application of acetylcholine to the calyces gave rise to strong electrical activity.

Monoamines

The distribution of monoamines in the insect brain has been studied quite extensively. The brain and subesophageal ganglion of *Periplaneta americana* have been shown to contain about 2.5 μg of dopamine and 0.4 μg of noradrenaline per gram of fresh tissue (12). 5-Hydroxytryptamine (5-HT) has been found in the head of the potter wasp, *Sceliphron,* (0.47 μg/g fresh tissue) and of the honeybee, *Apis mellifera,* (0.07–0.16 μg/g fresh tissue) (111). There is also some evidence for indolamines in the brain of *Acheta* and for 5-hydroxytryptophan which is a precursor of 5-HT.

In the cockroach, the catecholamines are localized mainly in the central body and α and β lobes (24) and also in the fiber network of the antennal lobe (37). The calyx and the stalk have no catecholamines, and in the α and β lobes the catecholamines are confined to alternate bands. Frontali and her co-workers have interpreted these as transverse bands, but they may correspond to those running in the axis of the lobes (see above) which can appear transverse according to the plane of section. Catecholamines are also especially dense in the actual body and in the α and β lobes of certain Trichoptera (50, 51). But in *Schistocerca gregaria* (52) and *A. domesticus* (92) smaller quantities were also found in the stalk and calyces. Dopamine and noradrenaline appear to be the most common monoamines in the cockroach (25) and locust (52). Dopamine has been localized in the α and β lobes of Trichoptera (53). 5-HT has been found in perikarya of the optic lobe of the locust and also in the pars intercerebralis (52), where it may co-exist with catecholamines. The monoamines in the stalk and lobes of the cricket mushroom bodies have not yet been identified, but the calyces are thought to contain indolamines (92).

Reserpine, which removes stores of catecholamines and 5-HT, has a tranquilizing effect on houseflies (36). In crickets, it was found to reduce locomotory activity considerably and no circadian rhythm of locomotion was detectable (13). In ants (*Formica rufa*) it also inhibited locomotor activity without causing any evident disturbance in muscular coordination (56); phototaxis was markedly suppressed and outbursts of aggressive behavior sometimes occurred. Cholinergic blocking agents such as atropine and scopolamine decreased aggression of ants towards the beetle *Geotrupes* and reduced the amplitude of brain potential recorded extracellularly (54, 55). Application of 5-HT and 5-HTP increased intraspecific aggression and increased spontaneous electrical activity of the optic lobe and the whole brain (55) of *F. rufa.* Steiner & Pieri (96) report that microelectrophoretic application of dopamine to the β lobe of *Formica* produces a strong inhibitory effect.

Some progress has been made with the localization of monamines within neurones. Mancini & Frontali (64) found that catecholamines in the β lobe of the cockroach were probably confined to the fibers which ended in small electron-dense vesicles (around 320 Å diameter). Such vesicles appeared in tissue which was incubated in a medium containing noradrenaline or α-methyl-noradrenaline, which will replace dopamine. Hence it may be that the β lobes contain only dopamine, which would suggest that they are functionally different from other parts of the mushroom bodies.

Frontali et al (26) considered that the fluorescent bands of catecholamines in the cockroach lobes correspond with extrinsic neurones and their branches, although these are relatively free of vesicles. Schürmann & Klemm (92), on the basis of their studies on *Acheta,* express a contrary view, that the fluorescence results from accumulation of transmitters in presynaptic fibers and that the extrinsic fibers are postsynaptic. The evidence suggests that the Kenyon cells are cholinergic, although they do not appear to have transmitter substances in their perikarya.

In summary there is evidence (none of it conclusive) that presynaptic fibers in the calyces are cholinergic or sometimes aminergic. The Kenyon cells have catecholamines in their projections through the stalk and in the α and β lobes. It is possible that there are different proportions of monoamines in the two lobes, reflecting a functional differentiation. The central body contains, in different parts, both cholinergic and aminergic fibers, which suggests that it influences more than one brain region (the same is true of the optic lobes). Few workers have paid attention to the protocerebral bridge, but Klemm & Axelsson (52) report that this contains indolamines in the locust.

ELECTROPHYSIOLOGICAL RECORDING

While the electrical responses of the optic lobes have been studied extensively, there are few studies on other parts of the brain, and these few are difficult to relate to brain anatomy. Blest & Collett (8) found neurones in the medial protocerebral lobes of various moths which responded to visual stimulation. Some neurones connected the two optic lobes, others connected the optic and medial protocerebral lobes. Others with a long latency were found in the medial lobes which presumably were postsynaptic to those coming from the optic lobes. Also in moths, Collett (10) found neurones with large binocular receptive fields which connected the lobula with the medial protocerebrum. Neurones were also found which arise in the medial protocerebral lobes and project to the medulla or lobula, or to the ventral nerve cord.

An area in the ventrolateral protocerebrum of noctuid moths responds to acoustic stimulation of the tympanal organ (83). The neurones respond with phasic or tonic spike sequences, and some show lapses in excitability lasting from a few seconds to an hour. A light stimulus inhibits the responsiveness of some neurones. Adam (1) found an auditory center in the locust in the same general area, at the borders of the proto- and deutocerebrum. Some of the neurones here responded selectively to acoustic stimulation of certain pulse duration, pulse repetition frequencies, and frequency range. Horridge (41) also found acoustically sensitive units in the optic lobes and in the lateral protocerebrum of the locust, many of which also responded to light, but he failed to find such neurones in the mushroom bodies. Interneurones sensitive to visual stimulation were found by Mimura et al (69) in the fly, *Boettcherisia peregrina.* Phasic responses to the onset of illumination were found in the medial protocerebrum, near the central body. Tonic responses and phasic responses to occlusion of light were found in the deuto- and tritocerebrum and the subesophageal ganglion. The response of the tonic units could be modulated by illuminating the ocelli, and convergence of ocellar and antennal input was found to occur in extensive regions of the brain (70). The ocelli here appear to have the function of regulating brain activity.

Perhaps the largest cell body in the locust brain belongs to an interneurone known as the Descending Contralateral Movement Detector (DCMD), the properties of which have been reviewed by Rowell (85). This is especially sensitive to small contrasting moving objects. It shows a response decrement which is specific to stimulation of a given retinal area and which undergoes long-term changes and can show spontaneous recovery. In the aroused active animal there is no response decrement, but antennal cleaning is more frequent when DCMD responsiveness is low (86). This, together with other evidence, suggests that its activity is modulated by a general arousal system driven by many sensory inputs. The neurone soma is laterally placed on the posterior protocerebrum (72). One branch runs towards the optic lobe on the same side, but apparently does not reach it. The other branch crosses over to the other side of the brain and runs through the ventral nerve cord, terminating in the metathoracic ganglion. It mediates rapid escape responses, including jumping.

Responses recorded from the mushroom bodies of the cockroach on antennal stimulation have a characteristic long latency, varying from 40–70 msec (65). The responses in the homolateral calyx are large compound waves which spread to the α and β lobes (66). Concurrent smaller waves appear to inhibit spontaneous activity in the mushroom bodies. Vowles (110) also recorded spike discharges from the mushroom bodies of the honeybee, which again occurred with a very long latency (40 msec to 2–3 min) after stimulation of sense organs.

These electrophysiological results indicate that much integrative activity occurs in the undifferentiated neuropile subsequent to integration in the optic and antennal lobes. The latencies of response in the mushroom bodies are too long for these to be implicated in startle responses; in the cockroach, for example, an escape response to antennal stimulation occurs after about 10 msec (65).

STIMULATION EXPERIMENTS

There have been remarkably few brain stimulation experiments on insects. Rowell (84) elicited locomotion and feeding movements by electrical stimulation of the locust brain. Locomotion occurred after a latency of 10 sec or more on stimulation of the upper protocerebrum, while arousal was much faster in the lower cerebrum. It was suggested that the slow arousal was mediated by stimulation of sensory input to the calyces. Foraging and feeding behavior could be elicited from almost anywhere within the brain, if the locust was sufficiently hungry, by a process of disinhibition. Vowles (110) obtained similar results from the honeybee. Stimulation of the mushroom bodies gave rise to locomotion, cleaning, feeding, and aggressive behavior after a long latency of between 2 and 90 sec.

The stimulation experiments carried out on crickets by Huber and his co-workers are of considerable importance for our present understanding of insect brain function. In the region of the central complex he obtained increases or decreases in the ventilatory rhythm. In the region of the mushroom bodies such changes were coupled with increases or decreases in locomotion (44). Stimulation of the calycal region inhibited singing, while the rivalry song and calling song could be obtained by stimulation of the stalk and lobes. Abnormal songs were obtained on stimulation in and around the central body. On the basis of this evidence it was suggested that song types are selected in the mushroom bodies and the appropriate neural patterning generated in the central body. Subsequent work by Otto (74), however, showed that the various song patterns could be elicited by stimulation of the neck connectives of headless crickets and that the normal machinery of the song was in the thoracic ganglia. A typical song pattern could be obtained by stimulation of various brain regions or of the neck connectives with currents of abnormally high or abnormally low (unphysiological) intensity. From many regions in and around the mushroom bodies stimulation at a single locus elicited two or three song patterns. The courtship and rivalry songs required strong stimulation, and therefore presumably the excitation of fiber systems over a relatively wide field around the stimulating electrode. Inhibitory loci, found in and around the central body and calyces, control general activity levels; locomotory movements, ventilatory movements, and certain other responses are inhibited as well as singing. This points to the existence of a unified inhibitory mechanism. The brain commands to the song centers in the thorax were found to be rather unspecific with the same fibers activating any of the three song patterns.

Otto (73) was also able to induce complex natural sequences of behavior by stimulation in the region of the mushroom bodies and central optic tracts of freely moving crickets. This included foraging, feeding (if the insect was hungry), digging, singing, and territorial behavior. The different behavior patterns could be elicited from the same locus with the same stimulus intensity, although aggressive behavior sometimes demanded stronger stimulation.

The results of brain stimulation experiments, therefore, support the view that the mushroom bodies function as switchgear, selecting the appropriate behavior patterns and controlling their sequential appearance. The evidence is also not inconsistent with the view that the central complex is an arousal system.

LESION AND TRANSPLANTATION EXPERIMENTS

Early studies on gross brain lesions and their effects upon posture and locomotion have been reviewed by Bullock & Horridge (9) and Ten Cate (99). Bethe (7), in a study of several species, found that behavioral thresholds were generally lowered after removal of the brain and concluded that the brain contained inhibitory centers which affected muscle tonus and reflex activity. Unilateral lesions gave rise to circus movements towards the intact side. Loeb (61) and others believed that the changes in tonus were due to the direct effects of asymmetrical sensory stimulation, but Alverdes (3), among others, showed that destruction of the compound eye on one side had different effects upon the stance of mayfly larvae from destruction of one side of the brain. Roeder (81) showed that removal of the protocerebrum on one side of the mantis brain resulted in increased locomotion and continual rotation towards the intact side. Removal of both protocerebral lobes predisposed the insect to long periods of continuous locomotion which occurred when the animal was stimulated. Median division of the lobes resulted in reduced and hesitant locomotion and an increased tendency to track moving objects without the fatigue effects shown by normal mantids. Roeder thus conceived that the two halves of the protocerebrum were mutually inhibitory and both inhibited the activity of thoracic centers controlling locomotion. They also inhibited the subesophageal ganglion; when the latter was removed the insects became quite inactive.

The model of Roeder's was subsequently modified by Huber (45) as a result of his experiments on *Gomphocerus rufus.* For several hours after removal of one calyx the insects showed no obvious abnormality of behavior apart from increased locomotion. Removal of both calyces increased locomotion to an even greater extent but also abolished singing and mating behavior. Destruction of the central body alone also prevented singing and mating but led to a marked reduction in locomotory activity. It was therefore suggested that the mushroom bodies inhibit the central body, which otherwise excites the subesophageal ganglion and thoracic centers. Huber thus identified the inhibitory centers as the mushroom bodies and the central body as an excitatory center additional to the subesophageal ganglion.

Lesion experiments of Drescher (17) on the brain of *Periplaneta americana* failed to confirm the general applicability of Huber's model. Damage to the central body by median section of the brain or removal of one half resulted in hyperphagia, autophagy of antennae and legs, a reduction in the threshold of escape responses, and loss of cataleptic abilities. When the lesion was asymmetric, marked circus movements resulted. Similar experiments on stick insects (97) also showed that surgical removal of the central body prevented catalepsy from occuring. Drescher concluded that the central body was inhibitory. Removal of one calyx induced no apparent changes in behavior, a finding that has also been made with honeybees (68), but removal of both calyces resulted in an initial increase in locomotion in the cockroach followed by a reduction of activity. The insects had increased thresholds to sensory stimulation but reacted very strongly when these were overcome. Optomotor reactions could still be elicited, but, unlike Huber's grasshoppers, the insects could still mate without mushroom bodies.

Van der Kloot & Williams (108) found that the mushroom bodies were essential for the coordination of cocoon spinning in *Hyalophora cecropia.* Radiocautery of the antennal lobes, optic lobes, and undifferentiated neuropile, and exact medial transection of the brain had no effect upon spinning, but transections to one side of the midline or slight damage to the mushroom bodies resulted in marked aberrations of cocoon spinning. With a cut close to the midline, the insect spun two silk layers, one above the other. With the cell bodies of one mushroom body destroyed or with a β lobe cauterized, only one flat layer was spun, but two layers were spun if only an α lobe was cauterized; the β lobe thus appears to control production of the two silk layers.

Menzel et al (67) have implicated the α lobe in short-term learning in the honeybee. By applying a cold needle to parts of the brain they established that interactions between the ipsilateral antennal lobe, calyx, and α lobe were most important in storage of short-term memory, but there was also transference of the memory to the contralateral mushroom body.

Mid-saggittal sections of the brain of *Schistocerca gregaria* produced no abnormalities of behavior (42), but lesions to the calyces resulted in abnormalities of stance and locomotion, suggesting that reflexes normally in competition were being released simultaneously. Median section of the brain can be achieved without damage to the mushroom bodies in locusts, but this is not possible in honeybees. Split-brain bees were found to be unable to fly, and about one third were unable to walk. Nearly all bees with damaged calyces showed disinhibition of competitive reflexes. For example, the same insect at one time could show protrusion of the sting, extension of the proboscis, antennal preening, and walking. Lesions to the central complex resulting from splitting the brain had, on the whole, different consequences: some honeybees became highly unresponsive, others became hyperactive and stung themselves to death.

These results suggest that the mushroom bodies inhibit the expression of behavior patterns and in this way probably control their occurrence on the basis of the input they receive. The central complex may function to modulate total responsiveness and has indeed been implicated in the control of circadian activity rhythms. Roberts (79) observed hyperactivity and disturbances in locomotory rhythms in cockroaches following median section of the brain, and has found that microelectrode lesions in the central body often produce arhythmicity (80). Nishiitsutsuji-Uwo & Pittendrigh (71), however, found that mid-sagittal cuts in the cockroach brain did not affect locomotory rhythms but cuts that were off center caused arhythmia. This could be a result of damage to the lateral neurosecretory cells of the pars intercerebralis (71), but might be a result of damage to the central body (80). Tyshchenko (106), however, has recently localized electrical circadian rhythms in cricket and cockroach brains to the region containing lateral neurosecretory cells. Lesions in other regions, including the central complex, do not destroy such rhythms. Circadian rhythms in accumulation and release of neurosecretion of pars intercerebralis neurones of crickets and similar rhythms of protein synthesis have been demonstrated by Cymborowski and his co-workers (13, 15). Release of neurosecretion coincides with inhibition of locomotion. Administration of reserpine abolishes rhythmicity but reduces locomotor activity (13), and, therefore, may be achieving this effect by removing catecholamines which are localized in the central complex and lobes of the mushroom bodies and stimulate locomotory activity.

A third part of the circadian clock, additional to the pars intercerebralis neurosecretory cells and the central complex which are likely components, is the optic lobe. Nishiitsutsuji-Uwo & Pittendrigh (71) first demonstrated that removal of both optic lobes produced arhythmia, and Roberts (80) later showed by lesion experiments that the lobula and medulla were essential elements of the circadian clock controlling locomotion.

Parabiosis experiments (14, 35) indicate that there is a hormonal link or a blood-borne factor of some other kind in the control of locomotion rhythms. This is supported by experiments of Ball & Chaudhury (5) in which the optic lobes, or brain and optic lobes, were transplanted to the abdomen of a blinded cockroach. They then generated a circadian rhythm of locomotion in phase with an imposed cycle of illumination. This experiment has been criticized on methodological grounds by Roberts (80), but transplantation experiments have been strikingly successful in the search for the clock controlling eclosion in saturniid moths. Truman & Riddiford (105) showed that implantation of a brain into the abdomen of a debrained pupa could induce emergence at the normal time. Exchange of brains between pupae of *H. cecropia* and *Antheraea pernyi* resulted in an exchange of emergence times. Subsequent experiments (104) showed that medial section of a pupal brain or removal of the optic lobes did not change the emergence time. Implantation of the central part of the brain (containing the medial neurosecretory cells) or of the lateral lobes (containing the lateral neurosecretory cells) produced no timing mechanism. The medial part of the brain, intact with one lateral lobe, is essential for transference of a rhythm. The lateral neurosecretory cells produce an eclosion hormone (103).

THEORIES OF BRAIN FUNCTION

The evidence considered here indicates that the mushroom bodies in most insects play little role in visual integration. The social and subsocial Hymenoptera are a major exception to this, and there is some support for Horridge's (9) view that the calyx is a circular projection of the environment. It is plain that a great deal of integration is carried out in the optic and antennal lobes, and it is likely that only some of the output of these centers is passed to the mushroom bodies. In particular, rapid evasive responses are probably not mediated by the mushroom bodies.

Theories put forward in recent years (42, 44, 90, 110) propose that the mushroom bodies evaluate sensory input, which is fed into the calyces, and select appropriate behavior patterns. Vowles (109, 110) established by electrophysiological recording in the honeybee that tracts connected the optic and antennal lobes with the calyx, and with the α lobe. He suggested the β lobe was a source of premotor output. He proposed that association of sensory input occurred in the calyx, and was then fed along Kenyon cell fibers and distributed equally to the α and β lobes. If this information was inappropriate to the insect, a corrective signal was fed back into the calyx in a neuronal arc from the α lobe to the appropriate sensory lobe and from there to the calyx (Figure 1). Further anatomical evidence for the connections that Vowles claimed has not, however, been forthcoming (see above). The results of Menzel et al (67), showing that chilling of the lobe interferes with short-term olfactory learning, however, suggests that the lobe has connections with the antennal lobe which must play some part in information store. This supports Vowles' hypothesis of connections between the α lobe and sensory centers as far as short-term memory is concerned, but this may be a special feature of the brains of social

Hymenoptera. It cannot fully account for the functions of the α lobes, which are well developed in insects such as termites in which rapid learning processes are not known.

Huber's early model of brain action in the cricket also presented the mushroom bodies as an apparatus for selecting behavior patterns. His hypothesis that the central body formulated the neural pattern for the songs had to be rejected when it was found that headless crickets could sing (74) and that a typical song could be elicited from many parts of the brain. Further, direct connections have not been found between the mushroom bodies and the central complex in silver-stained preparations (see above).

An alternative model has recently been put forward (42) which takes into account the neuronal architecture of the mushroom bodies and uses this to explain how these bodies could control sequences of behavior. It is supposed that sensory information is fed into the calyx and initiates certain excitation patterns in the calyx walls. If an impulse pattern corresponds with a pre-existing pattern of sensitivity in intrinsic

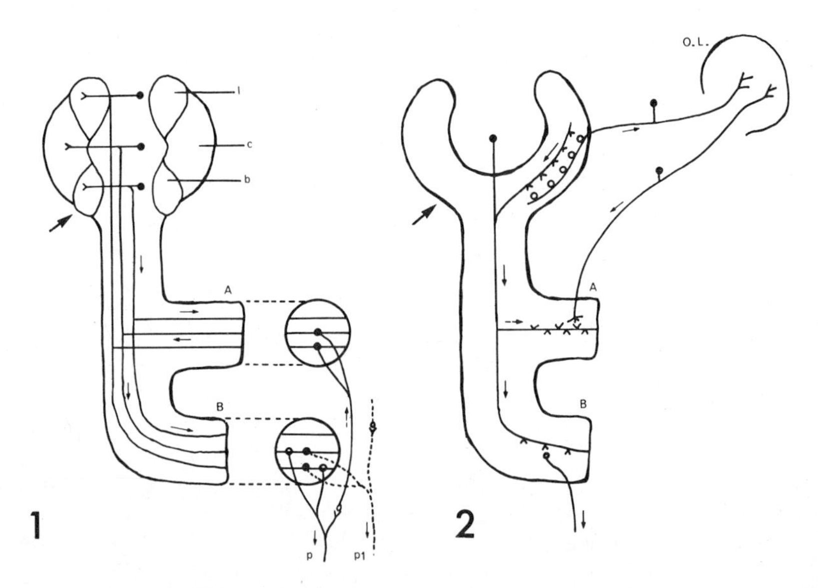

Figure 1 Hypothetical models of mushroom body function: 1, after Howse (42); 2, after Vowles (110). In 1, sensory input to the calyx excites Kenyon cells, and the excitation is carried to bands of the α and β lobes. A premotor neuron (p) is excited by a particular spatial pattern of Kenyon cell activity. A colateral branch of p activates different Kenyon cell endings in the α lobe, thereby also exciting corresponding endings in the β lobe and activating a second premotor cell (p_1). In this way sequences of behavior can be produced. l = lip of calyx, c = collar, b = basal ring, A = α lobe, B = β lobe. In 2, input from sensory centers such as the optic lobe (O.L.) also activates premotor neurones with endings in the α lobe. A feedback loop exists between the α lobe and the sensory centers whereby corrective signals are fed back into the calyx if inappropriate information is fed to the lower motor centers.

mushroom body neurones, the pattern will be transmitted to the lobes, with the spatial separation preserved. Extrinsic premotor neurones which have endings placed to tap the excitation will then be excited and set an appropriate lower motor center into operation. Pre-existing patterns of sensitivity in the calyx may either be determined genetically, or some of them (as in the lip and collar regions of social Hymenoptera) may be determined as a result of experience and may represent a memory store. Sequences of behavior may be determined either by appropriate changes in sensory input, or by feedback between the lobes or both. Figure 1 shows a hypothetical model of the way in which feedback between the lobes could directly excite or presensitize another group of Kenyon cell fibers which would lead to another action in a sequence.

This model has the advantage of explaining the significance of the banding patterns in the lobes of Hymenoptera, Isoptera, and cockroaches. Electrotonic spread of excitation may occur in the bands which are rich in transmitter substances and contain innumerable finely branched endings. Therefore the need for point to point connections of extrinsic and intrinsic fibers would be greatly reduced. Mushroom bodies with long lobes would make possible long sequences of behavior: both are found in Isoptera. Whether the α lobe and/or the β lobes connect to the motor centers is not known and makes little difference to this theory.

It has further been suggested (42) that the central complex is concerned with the regulation of levels of responsiveness, including threshold variations expressed as circadian rhythms. This is consistent with interpretations of neuroanatomy, distribution of transmitter substances, and the results of lesion and stimulation experiments already discussed. Indeed, it is difficult to envisage an alternative function for the central complex. But the recent work of Tyshchenko (106) appears to contradict this interpretation: punctate lesions of the lateral neurosecretory cells of the pars intercerebralis of crickets destroyed circadian rhythmicities, but lesions of the central complex failed to do so. This does not, however, exclude the possibility that the neurosecretory cells are controlled in some way by the central complex but can free run alone.

Very recently, Schürmann (90) has also proposed that the mushroom bodies are concerned with control of behavior sequences. He suggests that temporal sequences of activities are determined largely by interactions between intrinsic fibers. Spatial separation of excitation or inhibition by synaptic interactions then ensures a temporal sequence of activation or inhibition of postsynaptic extrinsic fibers with endings in the lobes.

Literature Cited

1. Adam, L. 1969. Neurophysiologie des Hörens und Bioakustik einer Feldheuschrecke (*Locusta migratoria*). *Z. Vergl. Physiol.* 63:227–89
2. von Alten, H. 1910. Zur Phylogenie des Hymenopterengehirns. *Jena. Z. Naturwiss.* 46:511–90
3. Alverdes, F. 1925. Körperstellung und Lokomotion bei Insekten nach Eingriffen am Gehirn. *Biol. Zentralbl.* 45:353–64
4. Baerends, G. P. 1941. Fortpflanzungsverhalten und Orientierung der Grabwespe *Ammophila campestris* Jur. *Tijdschr. Entomol.* 84:68–275
5. Ball, H. J., Chaudhury, M. F. B. 1973. Photic entrainment of circadian rhythms by illumination of implanted brain tissues in the cockroach *Blaberus craniifer*. *J. Insect Physiol.* 19:823–30
6. Bastock, M., Blest, A. D. 1958. An analysis of behaviour sequences in *Automeris aurantiaca* Weym. *Behaviour* 12:243–84
7. Bethe, A. 1897. Vergleichende Untersuchungen über die Funktionen des Centralnervensystems der Arthropoden. *Pfluegers Arch. Gesamte Physiol. Menschen Tiere* 68:449–545
8. Blest, A. D., Collett, T. 1965. Microelectrode studies of the medial protocerebrum of some Lepidoptera. I. Response to simple, binocular visual stimulation. *J. Insect Physiol.* 11:1079–1103
9. Bullock, T. H., Horridge, G. A. 1965. *Structure and Function in the Nervous System of Invertebrates*, Vol. II. San Francisco: Freeman
10. Collett, T. 1971. Connections between wide-field monocular and binocular movement detectors in the brain of a hawkmoth. *Z. Vergl. Physiol.* 75:1–31
11. Chen, W. Y., Aranda, L. C., Luco, J. V. 1970. Learning and long- and short-term memory in cockroaches. *Anim. Behav.* 18:725–32
12. Colhoun, E. H. 1964. Aspects of biologically active substances in insects with particular reference to the cockroach, *Periplaneta americana.* In *Comparative Neurochemistry,* ed. D. Richter. Oxford: Pergamon
13. Cymborowski, B. 1973. Control of the circadian rhythm of locomotor activity in the house cricket. *J. Insect. Physiol.* 19:1423–40
14. Cymborowski, B., Brady, J. 1972. Insect circadian rhythms transmitted by parabiosis—a re-examination. *Nature New Biol.* 236:221–22
15. Cymborowski, B., Dutkowski, A. 1969. Circadian changes in RNA synthesis in the neurosecretory cells of brain and subesophageal ganglion of the house cricket (*Acheta domesticus* L.). *J. Insect Physiol.* 15:1187–97
16. Cymborowski, B., Skangiel-Kramska, J., Dutkowski, A. 1970. Circadian changes of acetylcholinesterase activity in the brain of the house crickets (*Acheta domesticus* L.) *Comp. Biochem. Physiol.* 32:367–70
17. Drescher, W. 1960. Regenerationsversuche am Gehirne von *Periplaneta americana* unter Berücksichtigung von Verhaltensänderung und Neurosekretion. *Z. Morphol. Oekol. Tiere* 48:576–649
18. Dujardin, F. 1850. Mémoire sur le système nerveux des insectes. *Ann. Sci. Natur. Zool.* 14:195–206
19. Fabre, J. H. 1880. *Souvenirs Entomologiques.* Paris: Delagrave
20. Flögel, J. H. L. 1878. Über den einheitlichen Bau des Gehirns in den verschiedenen Insekten Ordnungen. *Z. Wiss. Zool.* (Suppl.) 30:556–92
21. Forel, A. 1874. Les fourmis de la Suisse. *Nouv. Mem. Soc. Helv. Sci. Natur.* 26:1–200
22. Fraenkel, G., Gunn, D. L. 1940. *The Orientation of Animals.* Oxford: Clarendon
23. von Frisch, K. 1967. *The Dance Language and Orientation of Bees.* Cambridge, Mass.: Belknap Press–Harvard. 566 pp.
24. Frontali, N. 1968. Histochemical localization of catecholamines in the brain of normal and drug-treated cockroaches. *J. Insect Physiol.* 14:881–86
25. Frontali, N., Häggendal, J. 1969. Noradrenaline and dopamine content in the brain of the cockroach *Periplaneta americana. Brain Res.* 14:540–42
26. Frontali, N., Mancini, G. 1970. Studies on the neuronal organization of cockroach corpora pedunculata. *J. Insect Physiol.* 16:2293–2301
27. Frontali, N., Piazza, R., Scopelliti, R. 1971. Localization of acetylcholinesterase in the brain of *Periplaneta americana. J. Insect Physiol.* 17:1833–42
28. Gervet, J., Fulcraud, J. 1970. Le thème de pîque dans la paralysation de sa proie par l'ammophile *Podalonia hirsuta* Scolopi. *Z. Tierpsychol.* 27:82–97

29. Goll, W. 1967. Strukturnuntersuchungen am Gehirn von *Formica. Z. Morphol Oekol. Tiere* 59:143–210
30. Goossen, M. 1949. Untersuchungen an Gehirnen verschieden grosser, jeweils verwandter Coleopteren- und Hymenopterenarten. *Zool. Jahrb. (Abt. Allg. Zool. Physiol.*) 62:1–64
31. Groth, U. 1971. Vergleichende Untersuchungen über die Topographie und Histologie des Gehirns der Dipteren. *Zool. Jahrb. Anat.* 88:203–319
32. Hanström, B. 1928. *Vergleichende Anatomie des Nervensystems der wirbellosen Tiere.* Berlin: Springer
33. Hanström, B. 1930. Über das Gehirn von *Termopsis nevadensis* u. *Phyllium pulchrifolium* nebst Beiträgen zur Phylogenie der Corpora Pedunculata der Arthropoden. *Z. Morphol. Oekol. Tiere* 19:732–73
34. Hanström, B. 1940. Inkretorische Organe, Sinnesorgane und Nervensystem des Kopfes einiger niederer Insektenordnungen. *Kgl. Sv. Vetenskapsakad Handl. Ser. 3B* 18:4–266
35. Harker, J. E. 1961. Diurnal rhythms. *Ann. Rev. Entomol.* 6:131–46
36. Hays, S. B. 1965. Some effects of reserpine, a tranquillizer, on the housefly. *J. Econ. Entomol.* 58:782–83
37. Hess, A. 1972. Histochemical localization of cholinesterase in the brain of the cockroach (*Periplaneta americana*). *Brain Res.* 46:287–95
38. Heymer, A. 1973. Verhaltensstudien an Prachtlibellen *Z. Tierpsychol.* Suppl. 11:1–100
39. Holmgren, N. 1916. Zur vergleichenden Anatomie des Gehirns von Polychaeten, Onychophoren, Xiphosuren, Arachniden, Crustaceen, Myriapoden und Insekten. *Kgl. Sv. Vetenskapsakad. Handl.* 56:1–299
40. Horridge, G. A. 1962. Learning of leg position by the ventral nerve cord in headless insects. *Proc. Roy. Soc. London B* 157:33–52
41. Horridge, G. A. 1964. Multimodal units of locust optic lobes. *Nature* 204:499–500
42. Howse, P. E. 1974. Design and function in the insect brain. In *Experimental Analysis of Insect Behaviour,* ed. L. Barton-Browne. Berlin: Springer
43. Howse, P. E., Williams, J. L. D. 1969. The brains of social insects in relation to behaviour. *Proc. VI Congr. IUSSI, Bern,* 59–64
44. Huber, F. 1960. Untersuchungen über die Funktion des Zentralnervensystems und insbesondere des Gehirns bei der Fortbewegung und Lauterzeugung der Grillen. *Z. Vergl. Physiol.* 44:60–132
45. Huber, F. 1965. Neural integration (Central nervous system). In *The Physiology of Insecta,* ed. M. Rockstein, II. London: Academic
46. Huber, F. 1967. Central control of movements and behaviour in invertebrates. In *Invertebrate Central Nervous Systems,* ed. C. A. G. Wiersma, 333–54. Chicago: Chicago Univ. Press
47. Jonescu, C. 1909. Vergleichende Untersuchungen über das Gehirn der Honigbiene. *Jena. Z. Naturwiss.* 45:111–80
48. Kennedy, J. S. 1965. Co-ordination of successive activities in an aphid. Reciprocal effects of settling on flight. *J. Exp. Biol.* 43:489–501
49. Kenyon, C. F. 1896. The brain of the bee. A preliminary contribution to the morphology of the nervous system of arthropoda. *J. Comp. Neurol.* 6:133–210
50. Klemm, N. 1968. Monoaminhaltige Strukturen im Zentralnervensystem der Trichoptera. Teil I. *Z. Zellforsch. Mikrosk. Anat.* 92:487–502
51. Klemm, N. 1971. Teil II. *Z. Zellforsch. Mikrosk. Anat.* 117:537–58
52. Klemm, N., Axelsson, S. 1972. Determination of dopamine, noradrenaline and 5-hydroxytryptamine in the brain of the desert locust, *Schistocerca gregaria* Forsk. *Brain Res.* 57:289–98
53. Klemm, N., Bjöklund, A. 1971. Identification of dopamine and noradrenaline in nervous structures of the insect brain. *Brain Res.* 26:459–64
54. Kostowski, W. 1968. A note on the effects of some cholinergic and anticholinergic drugs on the aggressive behaviour and spontaneous electrical activity of the central nervous system in the ant, *Formica rufa. J. Pharm. Pharmacol.* 20:381–84
55. Kostowski, W., Beck, J., Meszaros, J. 1965. Drugs affecting the behaviour and spontaneous bioelectrical activity of the nervous system in the ant, *Formica rufa. J. Pharm. Pharmacol.* 17:253–55
56. Kostowski, W., Tarchalska, B. 1972. The effects of some drugs affecting brain 5-HT on the aggressive behaviour and spontaneous electrical activity of the central nervous system of the ant, *Formica rufa. Brain Res.* 38:143–49
57. Lamparter, H. E., Steiger, U., Sandri, C., Akert, K. 1969. Zum Feinbau der Synapsen im Zentralnervensystem der Insekten. *Z. Zellforsch. Mikrosk. Anat.* 99:435–42

58. Landolt, A. M. 1965. Electronenmikroskopische Untersuchungen an der Perikaryenschicht der Corpora pedunculata der Waldameise (*Formica lugubris* Zett) mit besonderer Berücksichtigung der Neuron-Glia-Beziehung. *Z. Zellforsch. Mikrosk. Anat.* 66:701–36
59. Landolt, A. M., Sandri, C. 1966. Cholinergische Synapsen im Oberschlundganglion der Waldameise (*Formica lugubris* Zett.) *Z. Zellforsch. Mikrosk. Anat.* 69:246–59
60. Lindauer, M. 1970. Lernen und Gedächtnis, Versuche an der Honigbiene. *Naturwissenschaften* 57:463–67
61. Loeb, J. 1918. *Forced movements, Tropisms and Animal Conduct.* Philadelphia: Lippincott
62. Lucht-Bertram, E. 1962. Das postembryonale Wachstum von Hirnteilen bei *Apis mellifica* L. und *Myrmeleon europeus* L. *Z. Morphol. Oekol. Tiere* 50:543–75
63. Mancini, G., Frontali, N. 1967. Fine structure of the mushroom body neuropile of the brain of the roach, *Periplaneta americana. Z. Zellforsch. Mikrosk. Anat.* 83:334–43
64. Mancini, G., Frontali, N. 1970. On the ultrastructural localization of catecholamines in the β-lobes (corpora pedunculata) of *Periplaneta americana. Z. Zellforsch. Mikrosk. Anat.* 103:341–50
65. Maynard, D. M. 1956. Electrical activity in the cockroach cerebrum. *Nature* 177:529–30
66. Maynard, D. M. 1967. Organization of central ganglia. In *Invertebrate Nervous Systems,* ed. C. A. G. Wiersma, 231–55. Chicago: Univ. Chicago Press
67. Menzel, R., Erber, J., Masuhr, T. 1974. Learning and memory in the honeybee. In *Experimental Analysis of Insect Behaviour,* ed. L. Barton-Browne. Berlin: Springer
68. Micheli, L. 1941. Di alcuni ricerche sperimentali sulla fisiologia del sistema nervoso degli imenottari. *Atti. Soc. Ital. Sci. Natur. Mus. Civ. Stor. Natur. Milano* 80:193–98
69. Mimura, K., Tateda, H., Morita, H., Kuwabara, M. 1969. Regulation of insect brain excitability by ocellus. *Z. Vergl. Physiol.* 62:382–94
70. Mimura, K., Tateda, H., Morita, H., Kuwabara, M. 1970. Convergence of antennal and ocellar inputs in the insect brain. *Z. Vergl. Physiol.* 68:301–10
71. Nishiitsutsuji-Uwo, J., Pittendrigh, C. S. 1968. Central nervous system control of circadian rhythmicity in the cockroach. III. The optic lobes, locus of the driving oscillation. *Z. Vergl. Physiol.* 58:14–46
72. O'Shea, M., Rowell, C. H. F., Williams, J. L. D. 1974. The anatomy of a locust visual interneurone; the descending contralateral movement detector. *J. Exp. Biol.* 60:1–12
73. Otto, D. 1969. Hirnreizinduzierte komplexe Verhaltensfolgen bei Grillen *Zool. Anz.* 33(Suppl.):472–77
74. Otto, D. 1971. Untersuchungen zur zentralnervösen Kontrolle der Lauterzeugung von Grillen. *Z. Vergl. Physiol.* 74:227–71
75. Pandazis, G. 1930. Über die relative Ausbildung der Gehirnzentren bei biologisch verschiedenen Ameisenarten. *Z. Morphol. Oekol. Tiere* 18:114–69
76. Pearson, L. 1971. The corpora pedunculata of *Sphinx Ligustri* L. and other Lepidoptera: an anatomical study. *Phil. Trans. Roy. Soc.* 259:477–516
77. Peckham, G. W., Peckham, E. G. 1898. *On the Instincts and Habits of Solitary Wasps.* Wisconsin Geol. and Natur. Hist. Surv., Madison
78. Rabl-Rückhard, T. 1875. Studien über Insektengehirne. I. Das Gehirn der Ameise. *Arch. Anat. Physiol. Wiss. Med.* 42:480–99
79. Roberts, S. K. de F. 1966. Circadian activity rhythms in cockroaches. III. The role of endocrine and neural factors. *J. Cell. Comp. Physiol.* 67:473–86
80. Roberts, S. K. 1974. Circadian rhythms in cockroaches. Effects of optic lesions. *J. Comp. Physiol.* 88:21–30
81. Roeder, K. D. 1937. The control of tonus and locomotor activity in the praying mantis (*Mantis religiosa* L.). *J. Exp. Zool.* 76:353–74
82. Roeder, K. D. 1963. *Nerve Cells and Insect Behaviour.* Cambridge, Mass.: Harvard Univ. Press
83. Roeder, K. D, 1969. Acoustic interneurons in the brain of noctuid moths. *J. Insect Physiol.* 15:825–38
84. Rowell, C. H. F. 1963. A method for chronically implanting stimulating electrodes into the brains of locusts, and some results of stimulation. *J. Exp. Biol.* 40:271–84
85. Rowell, C. H. F. 1971. The Orthopteran descending movement detector (D.M.D.) neurones: a characterisation and review. *Z. Vergl. Physiol.* 71:167–94
86. Rowell, C. H. F. 1971. Antennal cleaning, arousal, and visual interneurone re-

sponsiveness in a locust. *J. Exp. Biol.* 55:749–61
87. Schürmann, F. W. 1970. Über die Struktur der Pilzkörper des Insektengehirns. I. Synapsen im Pedunculus. *Z. Zellforsch. Mikrosk. Anat.* 103:365–81
88. Schürmann, F. W. 1971. Synaptic contacts of association fibres in the brain of the bee. *Brain Res.* 26:169–76
89. Schürmann, F. W. 1972. Über die Struktur der Pilzkörper des Insektenhirns. II. Synaptische Schaltungen im Alpha-Lobus des Heimchens *Acheta domesticus* L. *Z. Zellforsch. Mikrosk. Anat.* 127:240–57
90. Schürmann, F. W. 1974. Bemerkungen zur Funktion der Corpora pedunculata im Gehirn der Insekten aus morphologischer Sicht. *Exp. Brain Res.* 20:406–32
91. Schürmann, F. W. 1973. Über die Struktur der Pilzkörper des Insektenhirns. III. Die Anatomie der Nervenfasern in den Corpora pedunculata bei *Acheta domesticus.* L. Eine Golgi-Studie. *Z. Zellforsch. Mikrosk. Anat.* 145:247–85
92. Schürmann, F. W., Klemm, N. 1973. Zur Monoaminverteilung in den Corpora pedunculata des Gehirns von *Acheta domesticus* L. *Z. Zellforsch. Mikrosk. Anat.* 136:393–414
93. Spieth, H. T. 1966. Drosophilid mating behaviour: the behaviour of decapitated females. *Anim. Behav.* 14:226–35
94. Steiger, U. 1967. Über den Feinbau des Neuropils im Corpus pedunculatum der Waldameise. Elektronenoptische Untersuchungen. *Z. Zellforsch. Mikrosk. Anat.* 81:511–36
95. Steiner, A. 1962. Etude du comportement prédateur d'un Hyménoptère Sphégien, *Liris nigra* V. d. L. (=*Notogonia pompiliformis* *Pz.*). *Ann. Sci. Natur. Zool. Ser. 12* 4:1–126
96. Steiner, F. A., Pieri, L. 1969. Comparative microelectrophoretic studies of invertebrate and vertebrate neurons. *Progr. Brain Res.* 31:191–99
97. Steiniger F. 1933. Die Erscheinung der Katalepsie bei Stabheuschrecke und Wasserläufe. *Z. Morphol. Oekol. Tiere* 26:591–708
98. Strausfeld, N. J. 1970. Variations and invariants of cell arrangements in the nervous systems of insects. (A review of neuronal arrangements in the visual system and corpora pedunculata). *Verh. Zool. Ges. Köln* 64:97–108
99. Ten Cate, J. 1931. Physiologie der Gangliensysteme der Wirbellosen. *Ergeb. Physiol.* 33:137–336
100. Thompson, C. B. 1916. The brain and frontal gland of the castes of the white ant *Leucotermes flavipes. J. Comp. Neurol.* 23:515–72
101. Tinbergen, N. 1972. *The Animal in its World. Vol. I. Field Studies.* London: Allen & Unwin
102. Trujillo-Cenoz, O., Melamed, J. 1962. Electron-microscope observations on the calyces of the insect brain. *J. Ultrastruct. Res.* 7:389–98
103. Truman, J. W. 1971. The physiology of insect ecdysis. I. The eclosion behaviour of saturniid moths and its hormonal release. *J. Exp. Biol.* 54:805–14
104. Truman, J. W. 1972. Physiology of insect rhythms. II. The silkmoth brain as the location of the biological clock controlling eclosion. *Z. Vergl. Physiol.* 81:99–114
105. Truman, J. W., Riddiford, L. M. 1970. Neuroendocrine control of ecdysis in silkmoths. *Science* 167:1624–26
106. Tyshchenko, V. P. 1973. The role of the nervous cells in circadian rhythm regulation in Insecta. In *Neurobiology of Invertebrates,* ed. J. Salanki. Budapest: Akadémiai kiadó
107. Van der Kloot, W. G. 1955. The control of neurosecretion and diapause by physiological changes in the brain of the *Cecropia* silkworm. *Biol. Bull.* 109: 276–94
108. Van der Kloot, W. G., Williams, C. M. 1953. Cocoon construction by the *Cecropia* silkworm. *Behaviour* 5:141–74
109. Vowles, D. M. 1955. The structure and connexions of the corpora pedunculata in bees and ants. *Quart. J. Microsc. Sci.* 96:239–55
110. Vowles, D. M. 1964. Models in the insect brain. In *Neural Theory and Modeling,* ed. E. Reiss. Stanford, Calif.: Stanford Univ. Press
111. Welsh, J. H., Moorhead, M. 1960. The quantitative distribution of 5-hydroxytryptamine in invertebrates, especially in their nervous systems. *J. Neurochem.* 6:146–69
112. Williams, J. L. D. 1972. *Some observations on the neuronal organisation of the supra-oesophageal ganglion in* Schistocerca gregaria *Forskal with particular reference to the central complex.* PhD thesis, Univ. of Wales, Cardiff

STRUCTURE OF CUTICULAR MECHANORECEPTORS OF ARTHROPODS

◆6095

Susan B. McIver
Department of Parasitology, School of Hygiene, University of Toronto,
Toronto M5S IAI, Canada

Mechanoreception is the perception of mechanical distortion of the body caused by the external stimuli of touch and air- or water-borne vibrations, or due to the internal forces generated by activities of the muscles. Mechanical stimuli are involved in more behavioral activities than any other type of external stimulus. Among these are locomotion, posture, feeding, orientation, oviposition, and hearing. Arthropods with their rigid exoskeleton are particularly well suited for the detection of mechanical stimuli and have developed a number of specialized structures for this purpose. The sensilla for mechanoreception may be classified according to their function, such as auditory, tactile, stretch, and position receptors, or according to their structure. One structural type may have more than one function, e.g. hair sensilla, although usually tactile receptors, may also perceive air-borne vibrations.

Two general groups of mechanoreceptors are known. Type I are those sensilla in which the dendrite of the bipolar neuron is associated with the cuticle or its invagination. Type II (commonly called stretch receptors) are sensilla with multipolar neurons which are associated with the walls of the alimentary canal, inner surface of the body wall, muscles, and connective tissues, but not the cuticle. Type I may be subdivided into (*a*) those sensilla which, although associated with the inner aspect of the cuticle, lack an external cuticular component, the scolopidial or chordotonal sensilla, and (*b*) those with an external cuticular portion, the cuticular mechanoreceptors, which are the subject of this review.

For additional general information the reader is referred to Bullock & Horridge (15), Dethier (32), Hoffman (65), Sinoir (113), Slifer (114), Snodgrass (121, 122), and Schwartzkopff (110) on mechanoreceptors in general; to Howse (69) for chordotonal organs; to Osborne (91) for stretch receptors; and to Finlayson (35), Hoffman (66), and Pringle (96) for proprioceptors.

Numerous light microscope studies have determined the types and basic structure of cuticular mechanoreceptors. More recent application of electron microscopic techniques has revealed the structure in more detail. Electrophysiological and ultrastructural studies together are helping to provide the basis for the eventual understanding of sensory transduction, the conversion of the mechanical energy of the stimulus to the electrical energy of the nerve impulse. Better understanding of the structure and function of cuticular mechanoreceptors will facilitate more critical analysis of the roles they play in behavior. This review emphasizes the structure of cuticular mechanoreceptors, relying especially on the literature of the last ten years. Limited space has necessitated critical selection of the papers cited, so that many early and some later works could not be included.

STRUCTURE

A sensillum is derived from one epidermal mother cell which divides to form the trichogen, tormogen, neurilemma, and nerve cell (e.g. 27, 134). The fully differentiated sensillum consists of (*a*) cuticular components, (*b*) sensory neuron (s), and (*c*) sheath cells. Sensillar structure will be discussed in this order.

Cuticular Components

HAIR SENSILLA Hair sensilla are the most abundant, widespread, and extensively investigated type of cuticular mechanoreceptor. The external part is either hair-shaped or clearly a derivative of a hair, such as a scale, filament, or peg. The hair-shaped types are commonly called sensilla chaetica, Bohm's bristles, or sensilla trichodea. Trichobothria belong to this group also, but because of unusual features are herein considered separately. Hair sensilla may function solely as mechanoreceptors or as chemoreceptors as well.

Mechanoreceptors In insects these structures typically bear no pores or openings in the hair wall and are innervated by one neuron. The hair is usually drawn to a sharp tip and exteriorly may bear cuticular sculpturings such as grooves or spicules. The hair is attached to the socket by an articulating membrane, which may consist of the rubber-like protein resilin (125). The walls of the socket are cuticular and may bear inward projecting ribs or diaphragms. These projections as well as the height and diameter of the socket restrict the movement of the hair.

The dendrite is attached by the cuticular sheath (112) either to the center or to one side of the base of the hair (Figure 1). The cuticular sheath may continue beyond the end of the dendrite and insert into the wall of the hair at various distances beyond the base (e.g. 43, 51, 89, 114, 115). In some hemimetabolous insects the extended cuticular sheath forms the inner wall of the ecdysial canal which is located in the hair wall above the base (52) (Figure 2).

During stimulation, the sensitive tip of the dendrite apparently must be held firmly in position. Invaginations of the cuticular sheath, which could prevent a downward movement of the dendrite have been observed (51, 114), as have auxilliary cuticular levers which, when the hair is moved, transmit the mechanical force, to the tip of the dendrite (43) (Figure 3).

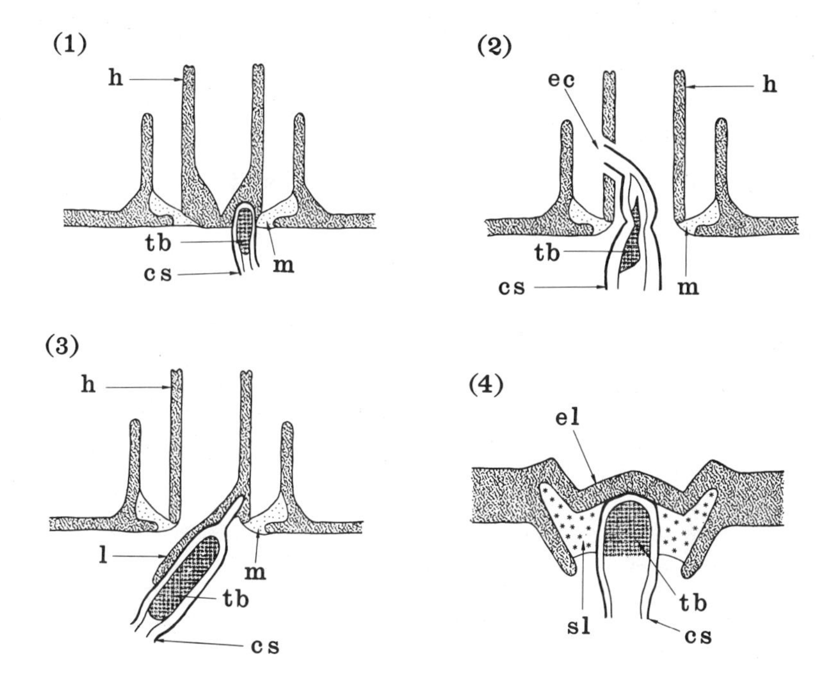

Figures 1–3 Stylized diagrams of examples of types of relationships between the tip of the dendrite and socket in hair sensilla (adapted from references 82, 62, and 43, respectively).

Figure 4 Campaniform sensillum. cs = cuticular sheath, ec = ecdysial canal, el = exocuticular layer, h = hair, l = lever, m = articulating membrane, sl = spongy layer, tb = tubular body.

Scales or sensilla squamiformia are widespread in insects and usually are not innervated. In Lepidoptera, however, certain scales on the wings, abdomen (15), and antennae (109) are innervated.

In arachnids hairs which function solely as mechanoreceptors are in general similar to those in insects. In spiders and ticks, however, multiple rather than single innervation appears to be characteristic. Hairs of spiders have three neurons (37, 40) and those of ticks two (26). In some hairs of ticks the dendrite ends at the base while the cuticular sheath continues distally to fuse with the wall of the ecdysial canal (26). In spiders, Foelix (37) reported that the tip of the dendrite fastens to the articulating membrane. Later, with Chu-Wang (40), he described a three part connection of the dendrite to the hair and membrane. Presumed strictly mechanosensory hairs occur on mites (14).

In 1851 Leydig (76) first described setae in crustaceans, but not until 1973 (123) was information from fine structure studies available. Antennal hairs of copepods are not set into a socket, but rather are constricted basally. Such constrictions presumably render some degree of flexibility. Each hair is innervated by bipolar neurons which do not extend into the setal lumen and which have basal body-like structures. Surrounding the basal body region is a complex of microtubules which may connect with bands of microtubules in the setal lumen. Possibly the microtubules mechanically connect the neurons with the moving tip of the hair and function as a piezoelectric crystal (123).

Trichobothria (Fadenhaare) were first described by Dahl (30) in spiders and since then have been found in both the mandibulate and chelicerate arthropods. Usually the hair is long and filamentous, but slender short hairs arising from typical sockets occur (40), as do hairs expanded into bulbs (62). Trichobothria are sensitive to small variations in air currents (e.g. 55, 56, 68, 88). The length of the hair is related to the frequencies perceived (33, 57).

The socket has the shape of a deep cup or flask with cuticular projections on the inner surface. A large socket orifice permits the hair a wide angle of deflection. The hair attaches to a membrane which forms the base of the socket. Below the membrane there may be one (21, 117) or more (68) chambers, within which the tip of the dendrite(s) is located.

Single innervations of trichobothria have been reported in insects (33, 50, 52, 88) and scorpions (68). Spiders have multiple innervations with two to four neurons (21, 55, 56) and Symphyla have 16 (62). The dendrite(s) may attach to one side of the hair base (68) or to the membrane that forms the floor of the cup-shaped depression of the socket (117) or may end between cuticular pegs in the articulating membrane (62) or in an ecdysial canal (50, 52). In spiders a sturdy cuticular sheath (helmet) attached to the base of the hair covers the dendritic tips (21, 22).

Mechano- and chemoreceptors In insects thick-walled hair sensilla, which function as contact chemoreceptors, frequently contain a mechanosensitive neuron. Although most common on the tarsi, mouthparts, and antennae, these sensilla may occur widely on the body surface (116). Each of the interpseudotracheal papillae, which are reduced hairs on the labella of blow flies and flesh flies, is innervated by a mechanosensitive neuron, as well as by chemosensitive neurons (92, 130).

Typically the hair has a blunt tip perforated by a single large pore. The chemosensitive dendrites extend the length of the hair to the pore, while the mechanosensitive dendrite attaches to one side of the hair base (e.g. 1, 31, 60, 73, 92). In insects there is usually one mechanosensitive neuron per sensillum, but two have been found associated with a contact chemoreceptor on the antennae of an ant (34). An unusual situation occurs in the short bristles on the cerci of crickets in which the tubular body, a characteristic feature of mechanoreceptors and the presumed site of transduction, is located at the hair base within one of the chemosensitive dendrites (108). Blaney & Chapman (12) reported a tubular body-like structure in one of the dendrites, which has no specialized attachment at the hair base, of the terminal sensilla on the maxillary palps of a grasshopper. Possibly this chemosensitive dendrite functions as a mechanoreceptor, as in crickets.

In the contact chemoreceptors of spiders and ticks two mechanosensitive dendrites are typically found (26, 38, 39, 41, 61, 100). In some cases in ticks the dendrites are attached by the cuticular sheath to the articulating membrane (38, 39), while in others the cuticular sheath extends beyond the dendritic tip at the base of the hair to end in an ecdysial canal (26). In spiders the cuticular sheath does not terminate in the articulating membrane as in a purely mechanoreceptor sensillum. In some spiders fibers connect one side of the sheath to the hair base and enveloping cells provide an indirect attachment to the cuticle (41), while in others no attachment mechanism is found (61).

Thick-walled hairs with numerous pores occur in mites. These sensilla, which are presumably olfactory receptors, also have a neuron displaying features characteristic of mechanoreceptors (14).

CAMPANIFORM SENSILLA Berlese (10) first used the term campaniform to describe sensilla previously known as dome, bell, and cupola-shaped structures. Campaniform sensilla, structures homologous to hair sensilla (e. g. 10, 74, 106, 122) and peculiar to insects, are proprioceptors responding to strains in the exoskeleton (93, 94). These sensilla occur widely on the body of adult insects, frequently being concentrated near joints or on structures subject to cuticular distortion, such as legs and halteres. Larvae of some species also have campaniform sensilla.

In a vertical section a campaniform sensillum generally appears as a small cuticular cap surrounded by a ring of raised cuticle. Characteristically one neuron innervates the sensillum (Figure 4). Usually the dendrite attaches to the center of the cap, but eccentric attachment is known (109). Variation in size and structure of the cuticular portions is common and relates to the specific characteristics of the stimulus perceived at the particular location of the sensillum (e.g. 42, 45, 119). The cap may be circular or oval, raised or sunken, and flat, arched, or formed into a small peg. Cuticular hinges may attach the cap to the body cuticle. In hemimetabolous insects the cap may have a pore which is the ecdysial canal (107). In holometabolous insects at least one instance is known of a pore occurring in the cap (83). The ring of raised cuticle surrounding the cap may be incomplete (13).

Light microscope studies originally revealed differences in the structure of the cap (112, 121). Recent electron microscopic investigations (20, 28, 52, 63, 83, 86, 105, 119, 124, 125, 131) have more clearly illustrated these differences. The cap usually consists of at least two layers (Figure 4): a modified exocuticular layer on the outer surface and an inner spongy layer, which may be the rubber-like protein, resilin (125). Additional subdivisions of the inner layer have been reported (119). Hawke el al (63) described a campaniform sensillum in which the spongy layer is lacking. Sometimes a rim of sturdy cuticle occurs below the spongy layer (20, 52, 105, 131).

Campaniform and hair sensilla may be closely associated to form a functional unit. Sihler (112) reported the proximity of campaniform sensilla to the sockets of trichobothria on the cerci of crickets. Also on crickets, Gnatzy & Schmidt (52) found the number of campaniform sensilla to be directly proportional to the diameter of the trichobothrium; the smallest have none and the largest have five, distributed asymmetrically around the socket. Presumably, distortion of the cuticle in the socket region would stimulate the campaniform sensilla and provide information different from that perceived by the trichobothrium. In tactile spines on cockroach legs a different type of association occurs in which a single campaniform sensillum is located in the wall of the hair at the junction with the articulating membrane (19). The tactile spine itself is not innervated, but movements of the hair stimulate the campaniform sensillum which functions as the sensory element.

SLIT ORGANS In the exoskeleton of most members of the Chelicerata are trough-shaped structures known as slit organs which were first reported from spiders by Bertkau (11). The slits may occur singly or in groups. In the latter case they are referred to as compound slit organs or lyriform organs. In the hunting spider, compound slit organs are restricted to the appendages, usually being located near the joints, while single slits occur on all body parts and mostly at some distance from the joints (8).

Pringle (95) demonstrated that slit organs in the appendages of a scorpion and an amblypygid are sensitive to strains in the cuticle and considered them to be analogous to the campaniform sensilla of insects. Sensitivity to air-borne vibrations has been shown for slit organs on the legs of spiders (3, 132). In the hunting spider, kinesthetic orientation, i. e. an animal's ability to direct its locomotion by making use of information about its own previous movement sequences, is dependent upon the compound slit organs in the legs (9, 111).

Externally a slit organ appears as a trough in the cuticle, the center region being deeper than the ends (4, 70, 81, 102). The slits themselves are enclosed in cavities in the cuticle. In compound organs, each slit is separated by cuticular lamellae. In the spider, *Cupiennius salei,* the enclosed cavity is divided into two chambers by a membrane similar in appearance to the epicuticle (4). The outer chamber lying in the exocuticle is bounded on top by a thin layer of cuticle. Basally the inner chamber opens as a bell into the meso- and endocuticle. Each slit is innervated by two neurons, the dendrite of one terminating close to the membrane separating the two chambers, and that of the second extending through the separating membrane and outer chamber to form a finger-like projection. In the metatarsi of the common house spider, *Achaearanea tepidariorum,* one neuron attaches to the cuticular wall of the slit (102). Kaston (70) reported one neuron per slit in a variety of spider species studied with the light microscope. In general the single and compound slit organs on the same and different spider species appear to have a similar structure (4, 70).

The cuticular portions of slit organs play an important role in stimulus conductions (5–7). Features such as size, shape, and composition are well adapted to transform even small amounts of mechanical energy applied to the hard cuticle into deformation of the slit and to focus the resulting forces on the tip of the dendrite. Deformation of the slit is greatest in the middle region and with stress applied at right angles to the long axis. Degree of deformation increases with the length of the slit (5).

PECTINES Behind the posterior pair of legs of scorpions are paired organs, the pectines, which bear a series of stout cuticular teeth. On each tooth is a field of blunt pegs. Each unhinged tooth is set into a cuticular depression (17, 18). Within the base of the peg is a cylindrical membrane, possibly cuticular in nature, pleated into eight inwardly projecting radial folds. To each pleat is attached a dendrite. Carthy (18) considered that bending of the peg would lead to deformation of the internal cylindrical membrane, stretching or relaxing the attached dendrites. The radial arrangement of the pleats would make it possible to detect deformation in various directions and would provide an anatomical explanation for the electrophysiological finding (67) that movement in two directions may be separately sensed.

When moving, scorpions lower the pectines in such a manner that the fields of pegs are in contact with the substrate. Carthy (18) proposed that the pegs function in the selection of substrate with a particular size of sand grain.

UNUSUAL STRUCTURES As knowledge of cuticular mechanoreceptors expands, structures are being found which do not conveniently fit into presently conceived categories. Such findings are to be expected and in time will probably necessitate a restructuring of the present classification system. Examples of these unusual structures follow. On the mouthparts of beetle larvae typical mechanoreceptor neurons innervate (*a*) sensilla tigella (stalks), and (*b*) structures with conical bases distally divided into cones and crescents (29). Spheroid sensilla, dense bulbs on slender stalks, occur in close association with trichobothria on the cerci of certain cockroaches and may function as gravity detectors (101). A broad spectrum of campaniform-like sensilla exists. Soft-bodied larvae bear several of these, including pedunculate forms (36, 137). Variation on the contact chemoreceptor theme is provided by spots, pits, papillae, and unclassified sensilla on house fly larvae (23, 24) and domes on wasp ovipositors (63). All of these have a pore to the exterior and are innervated by neurons with features characteristic of chemo- and mechanoreceptors.

Sensory Neurons

The sensory neuron is composed of an axon, cell body, and dendrite which is divided into inner (proximal, Sinnesfortsatz) and outer (distal, Sinnescilium) segments by the ciliary region. Except for the characteristic outer dendritic segment, the neurons of cuticular mechanoreceptors and chemoreceptors (116) have a similar structure. The cell body has a large spherical nucleus, relatively pale cytoplasm, probably due to a high water content (86), and the usual complement of cell organelles. Proximally, the cell body gives rise to the axon which extends uninterrupted to the central nervous system. Microtubules and mitochondria are the most frequently reported organelles in the cytoplasm of the axon. Microfibers and multivesicular bodies occur also (60), as well as organelles common to axons in general. Distally from the cell body, the dendrite extends to the cuticular portion of the sensillum. Within the inner segment the usual assortment of organelles, as well as glycogen (20, 24), occurs, whereas the outer segment contains only microtubules, filaments, vesicles, and vacuoles. In the distal portion of the inner segment there is usually a concentration of mitochondria, which in the dendrite of a slit organ is associated with a pronounced swelling (4).

The dendrite typically narrows at the ciliary region and distally expands to a diameter not greater than that of the inner segment. However, in some campaniform sensilla a bulbous dilation occurs in this region (20, 131), and, in a few hair sensilla, an expanded area with lobe-like extensions has been found (99, 114). The outer segment is encased in a cuticular sheath which extends down to the ciliary region.

A region assuming the structure of a cilium was first described in insects by Gray and Pumphrey (58) in the hearing organ of the locust. The ciliary region is usually composed of nine sets of peripherally located doublet tubules, but lacks the central pair characteristic of motile cilia, with the resulting ciliary formula of $9 \times 2 + 0$. In the soft ticks, Argasidae, the ciliary configuration of $11 \times 2 + 0$ is apparently characteristic (38, 39). As in the ciliary region of chemoreceptors (116), central tubules are occasionally observed (20, 124). The outer member of each peripheral doublet lacks the arms found in motile cilia. Distally the ciliary microtubules extend into the outer dendritic segment while basally they connect with a centriole-like structure, the basal body, which has nine sets of peripheral triplet tubules. Arising from the basal body are periodically banded rootlets which extend basally into the inner dendritic segment. In most papers two basal bodies have been reported, although several workers have found only one. When two basal bodies are present, rootlets have been observed extending from each. Chu-Wang & Axtell (24) found ciliary tubules as well as rootlets originating from both basal bodies.

In the constricted ciliary region just distal to the basal body, two unusual modifications of the dendrite have been reported. One is an aggregate of cords (fibers?) termed connecting body in a campaniform sensillum (131). The other is a fibrillar body which consists of an accumulation of fibers through which the microtubules pass (24, 25, 119).

The characteristic feature of cuticular mechanoreceptors is an accumulation of microtubules in the distal region of the outer segment. The importance of the microtubules was emphasized when Thurm (125–127) reported from work on hair and campaniform sensilla of honey bees that a structure he called the tubular body most likely was the site of sensory transduction. The tubular bodies studied by Thurm occur at the distal end of the outer segment and consist of 50–100 tubules lying parallel to one another in an electron dense material. Subsequently, greater complexity and structural variation in tubular bodies have been demonstrated in numerous arthropod species.

In some tubular bodies filaments as well as microtubules occur (24–26, 39, 60, 119). The electron dense material in several tubular bodies has a striated appearance due to its arrangement in layers perpendicular to the axis of the microtubules (2, 37, 40, 43, 52, 83, 119). The number of microtubules varies from a few dozen (97) to about 1000 (40, 90). Sometimes a group of microtubules encloses a space containing a filament. Arrangement of the microtubules occurs in all degrees of complexity from the simple, consisting of a single row of peripheral microtubules (97), to the complicated (40, 43, 86, 119). Variation in arrangement of microtubules within the same tubular body is known (43, 63, 119); for example, in the antennal hair sensilla of the red cotton bug the peripheral tubules are 100–200 Å apart and occur in palisades whereas the central ones are 300 Å apart and are hexagonally arranged (43).

The peripheral microtubules and plasma membrane of the dendrite have been reported to be connected by dense patches (20) or filaments (40, 43). Between the plasma membrane and the cuticular sheath trivial desmosomal connections (131) (hemidesmosomes?) and large granules (4, 21, 40, 63) may occur. Possibly the granules serve as attachment points between the dendrite and the sheath (21, 63).

The wide spectrum of structural variation of the distal end of the dendrite has led to problems in nomenclature. Some authors, due to differences in the amount of electron dense material and number and configuration of microtubules, have considered the term tubular body, as defined by Thurm (125), to be inappropriate. Subsequently Foelix & Chu-Wang (40) proposed that tubular body be redefined as "any dense arrangement of microtubules which is restricted to a dendritic terminal, regardless of the amount of electron dense material." This new definition takes into account the variation in the amount of electron dense material, but not that of the number of microtubules. The recent demonstration by Rice and associates (97), that a dendrite with only a few microtubules is a mechanoreceptor, suggests that the concept of tubular body needs to be expanded further to include both the wide variation in number and configuration of microtubules and the amount of electron dense material. The concept of tubular body as proposed by Foelix & Chu-Wang (40) and expanded here would allow for a more uniform nomenclature.

Although the emphasis here is on structure, a few comments on the major ideas pertaining to the mode of sensory transduction should aid in a better appreciation of the functional morphology of these sensilla. The main point under consideration by workers in this field is the role of the microtubules within the tubular body. Thurm (125–127) considered the neuron to be excited by compression of the tubular body. Moran & Varela (85) demonstrated by chemical disassembly that the microtubules within the tubular body are essential for transduction. Two hypotheses for transduction, based on the compression concept, have been proposed: the "stress-activated muscle working backwards" (119) and the "strain-activated mechanochemical engine in reverse" (86). Another line of thought suggests that the microtubules act as a cytoskeleton (104), across which the receptor membrane is stretched causing change in shape of pores in the membrane with resulting increase in ionic conductance and depolarization (97). The membrane stretch theory of Rice et al (97) was developed for a dendrite with only a single row of microtubules, whereas the tubular bodies studied by Thurm (125–127) are of a more complex nature.

Lewis (75) suggests that the response, adaption, and recovery of cuticular mechanoreceptor neurons depend not only on the transducer membrane but also on the viscous and elastic properties of the contents of the dendrite in the stressed region. This suggestion could well be an important clue in understanding the variation in structure, which presumably controls the viscous and elastic properties, of the tubular body.

Sheath Cells

Surrounding the neuron are two or more specialized epidermal cells which are known collectively as sheath cells (Hüllzellen). In sensilla with two sheath cells, the names trichogen and tormogen are commonly used. When three cells occur, they are either numbered or referred to as neurilemma, trichogen, and tormogen cells.

In campaniform sensilla a variety of terms such as enveloping and accessory supporting cells have been used. Interestingly in ticks two sheath cells envelope mechanoreceptor neurons, whereas three surround chemoreceptors within the same sensillum (26, 39). These variations in terms and numbers make it difficult to establish the identity of corresponding sheath cells in various sensilla. Schmidt (106) presented evidence from electron microscopic studies that the sheath cells of hair and campaniform sensilla as well as scolopidia are homologous. More such developmental studies at the fine structure level are needed to determine not only the identity, but also the function of the sheath cells.

In development the trichogen cell forms the hair or corresponding structure while the tormogen cell builds the socket region. Upon completion of development, the cells withdraw leaving a fluid-filled extracellular space, the receptor-lymph cavity (89), beneath the cuticle. Zacharuk (138) and Slifer (116) reported that the trichogen cell secretes the cuticular sheath as well as the hair. Schmidt & Gnatzy (107) considered the internal enveloping cell to form the cuticular sheath and the middle enveloping cell the hair.

The sheath cells contain the common array of cytoplasmic organelles. Lysosome-like bodies have also been found (2). Within the region nearest the dendrite of the inner cell may occur numerous microtubules running parallel to the axis of the dendrite (4, 38, 83, 86, 119, 124). Various types of linkages, including regular, septate, and macula adherens desmosomes and septate junctions, connect the sheath cells to each other, to the dendrite, and to the surrounding epidermal cells.

Surfaces of the sheath cells contacting the extracellular space bear numerous lamellae or microvilli. Occasionally authors incorrectly use the term microvilli to describe structures which are in fact lamellae. The microvilli show indications of secretory activity and in all probability produce and regulate the content of the fluid in the extracellular space. Evidence for the secretory role of the microvilli is (*a*) the occurrence of numerous mitochondria located in dilated regions (119) or at the bases (82) of the microvilli, (*b*) the presence near the plasma membrane of the microvilli of small particles which may have a role in transmembrane ion transport (119), and (*c*) the existence of vesicles and vacuoles in the microvilli (2, 20)

The hypothesis for the production of the receptor potential proposed by Thurm (128, 129) necessitates the participation of both the neuron and the sheath cells and may explain the significance of some of the previously mentioned features of the sheath cells. Thurm (128) found that the fluid in the extracellular space is positively charged and the interior of the neuron and sheath cells negatively charged with regard to the neutral hemolymph. This situation is presumably brought about by an electrogenic ion pump which is located in the microvilli of the sheath cells and which secretes potassium ions into the extracellular fluid. The septate junctions between the neuron and the sheath cells and the sheath cells themselves probably prevent communication, which would result in a short circuit between the extracellular fluid and the hemolymph. The mechanical stimulus controls the permeability of the dendritic membrane to potassium.

In addition to the possible involvement in the electrical activities of the dendrite (128, 129) the fluid in the extracellular space might have a role in the nutritive and excretory functions of the outer segment of the dendrite as this region is largely devoid of cell organelles. Also in certain mechanoreceptors the fluid could possibly transmit the stimulus to the tubular body by hydraulic compression (20, 37, 102).

MOLTING

During the molting the fate of various portions of the sensillum, especially the dendrite with possible loss of function, is of particular interest. Shedding of the cuticular sheath has been reported for several types of mechanoreceptors (e. g. 118, 137, 138). There is clear evidence that the tips of the dendrites of hair and campaniform sensilla are lost at ecdysis (53, 54, 84, 107). It is known from light microscopy (e.g. 134) that the distal portion of the dendrite remains connected to the old cuticle during apolysis, presumably permitting functional continuity until an advanced stage in molting. This has been verified by electron microscopic studies (53, 54, 107) which show that at least in the campaniform and hair sensilla of crickets, the tubular body stays in contact with the old cuticle during apolysis by an extension of the dendrite and a new tubular body is formed below the old one in the same dendrite. The dendritic extension containing the old tubular body leaves the new cuticle via the ecdysial canal (Figure 2). In hair sensilla the ecdysial canal is located above or near the base of the hair and in campaniform sensilla in the middle of the cap. When the tubular body is located within one of the chemosensitive dendrites in a contact chemoreceptor, it is shed through the apical pore along with the distal portion of that dendrite (108).

Hair and campaniform sensilla are probably receptive through the old tubular body until the time of ecdysis and thereafter through the new one (53, 54, 107). In slit organs of spiders the dendrite remains within the old cuticle until ecdysis and only a gradual loss of sensitivity occurs (133).

MECHANORECEPTORS AND BEHAVIOR

Dethier's book (32) provides a review of the literature and analysis of the roles played by mechanoreceptors in behavior. Limited space permits mention of only a few of the papers since that time. Nevertheless, these few examples provide some insight into the types of behavioral activities influenced or governed by cuticular mechanoreceptors.

Most attention has been focused on the participation of various mechanoreceptors, including hair and campaniform sensilla, in the initiation, control, and maintenance of flight. In locusts, the in-flight function of the sensilla on the antennae (46–48), wings (44, 72), legs (71), and head (16, 47, 48, 59, 120) has been thoroughly analyzed. In blow flies the entire antenna can function as an air-current sense organ involved in flight control through perception by various sensilla of the differential deflection of the antennal segments (49, 103). Sidewind deviation during flight in honey bees is controlled at least partly by hair sensilla on the compound eyes (87).

Mechanoreceptors play an important role in the life of the social Hymenoptera. Hair plates located between body regions function as gravity receptors enabling bees and ants to find a goal using gravity for orientation (77–79). In addition cervical hair plates of honey bees are necessary in order to produce properly oriented honey combs and cells. Bees lacking these plates cannot build a comb. Hair sensilla at the tips of the antennae are used for controlling the thickness of the walls of the cells and for checking the smoothness (80).

Silkworm moths lay their eggs in compact, mono-layered clusters. On the anal papillae of the female are numerous hair sensilla which by detection of unevenness of the oviposition site, regulate the proper positioning of the eggs within the cluster. The sensory information from these hairs acts as the triggering signals for the action of the abdominal muscles and the genital organs (135, 136).

Mechanoreceptors have been shown to be particularly important in feeding by completely hematophagous insects. The mouthparts of adult tsetse flies have a preponderance of mechanoreceptors over chemoreceptors, which is thought to relate to the limited number of chemical phagostimulants encountered and the importance of physical stimuli to feeding (98). In contrast omnivorous blow flies have an abundance of chemoreceptors (64).

CONCLUDING REMARKS

Among arthropod cuticular mechanoreceptors the ultrastructure of only the hair and campaniform sensilla of insects and the hair sensilla and slit organs of arachnids are known in significant detail. Future endeavor should (*a*) elucidate the structure of the sensilla in all arthropod groups, (*b*) determine the variability of sensillar structure within a particular group and establish how this variability relates to function, and (*c*) apply better microscopic techniques as they become available to determine more exactly the ultrastructure of the various sensilla. Only careful fine structure work will establish the morphological basis necessary for understanding sensillar function, which in turn will facilitate a better appreciation of the roles played by cuticular mechanoreceptors in behavior.

NOTE ADDED IN PROOF Chapman et al (Chapman, K. M. et al 1973. Form and role of deformation in excitation of an insect mechanoreceptor. *Nature* 244:453–54) present evidence that in campaniform sensilla longitudinal compression due to the indentation of the dome, rather than stretching, pinching, or bending, of the dendrite is the physiological means of excitation during proprioception.

The reader's attention is drawn to the recent papers by U. Thurm, "Basics of the generation of receptor potentials in epidermal mechanoreceptors in insects," and by J. Küppers, "Measurements on the ionic milieu of the receptor terminal in mechanoreceptive sensilla in insects," published in *Mechanoreception* (Symposium Bochum, October 14–18, 1973), Abhandlungen der Rheinisch—Westfälischen Akademie der Wissenschaften, Westdeutscher Verlag, Opladen, 1974.

Acknowledgments

I thank Drs. K. S. Boo, A. M. Fallis, R. S. Freeman, A. Hudson, E. H. Slifer, and K. A. Wright for their constructive criticisms of the manuscript. My work has been supported by the Medical Research Council (MA 2909) and the Defence Research Board (6801–50).

Literature Cited

1. Adams, J. R., Holbert, P. E., Forgash, A. J. 1965. Electron microscopy of the contact chemoreceptors of the stable fly, *Stomoxys calcitrans. Ann. Entomol. Soc. Am.* 58:909–17
2. Altner, H., Ernst, K. D., Karuhize, G. 1970. Untersuchungen am Postantennalorgan der Collembolen. I. Die Feinstruktur der postantennalen Sinnesborste von *Sminthurus fuscus* (L.). *Z. Zellforsch. Mikrosk. Anat.* 111:263–85
3. Barth, F. G. 1967. Ein einzelnes Spaltsinnesorgan auf dem Spinnentarsus: seine Erregung in Abhängigkeit von den Parametern des Luftschallreiges. *Z. Vergl. Physiol.* 55:407–49
4. Barth, F. G. 1971. Der sensorische Apparat der Spaltsinnesorgane (*Cupiennius salei* Key., Araneae). *Z. Zellforsch. Mikrosk. Anat.* 112:212–46
5. Barth, F. G. 1972. Die Physiologie der Spaltsinnesorgane I. Modellversuche zur Rolle des cuticularen Spaltes beim Reiztransport. *J. Comp. Physiol.* 78: 315–36
6. Barth, F. G. 1972. Die Physiologie der Spaltsinnesorgane II. Funktionelle Morphologie eines Mechanoreceptors. *J. Comp. Physiol.* 81:159–86
7. Barth, F. G. 1973. Bauprinzipien und adäquater Reiz bei einem Mechanoreceptor. *Verh. Deut. Zool. Ges.* 66: 25–30
8. Barth, F. G., Libera, W. 1970. Ein Atlas der Spaltsinnesorgane von *Cupiennius salei* Keys. Chelicerata. *Z. Morphol. Tiere* 68:343–69
9. Barth, F. G., Seyfarth, E. A. 1971. Slit sense organs and kinesthetic orientation. *Z. Vergl. Physiol.* 74:326–28
10. Berlese, A. 1909. *Gli insetti. I. Embriologia e morfologia.* Milan: Societa Editrice Libraria
11. Bertkau, P. 1878. Versuch einer natürlichen Anordnung der Spinnen nebst Bemerkungen zu einzelnen Gattungen. *Arch. Naturgesch.* 44:354
12. Blaney, W. M., Chapman, R. F. 1969. The fine structure of the terminal sensilla on the maxillary palps of *Schistocerca gregaria* (Forskal). *Z. Zellforsch. Mikrosk. Anat.* 99:74–97
13. Blaney, W. M., Chapman, R. F. 1969. The anatomy and histology of the maxillary palp of *Schistocerca gregaria. J. Zool.* 157:509–35
14. Bostanian, N. J., Morrison, F. O. 1973. Morphology and ultrastructure of sense organs in the twospotted spider mite. *Ann. Entomol. Soc. Am.* 66:379–83
15. Bullock, T. H., Horridge, G. A. 1965. *Structure and Function in the Nervous Systems of Invertebrates,* 2:1005–62. San Francisco: Freeman. 1719 pp.
16. Camhi, J. M. 1969. Locust wind receptors III. Contribution to flight initiation and lift control. *J. Exp. Biol.* 50:363–73
17. Carthy, J. D. 1966. Fine structure and function of the sensory pegs on the scorpion pectine. *Experientia* 22:89
18. Carthy, J. D. 1968. The pectine of scorpions. *Symp. Zool. Soc. London* 23: 251–61
19. Chapman, K. M. 1965. Campaniform sensilla on the tactile spines of the legs of the cockroach. *J. Exp. Biol.* 42:191–203
20. Chevalier, R. L. 1969. The fine structure of campaniform sensilla on the halteres of *Drosophila melanogaster. J. Morphol.* 128:443–64
21. Christian, U. 1971. Zur Feinstruktur der Trichobothrien der Winkelspinne *Tegenaria derhami* (Scopoli). *Cytobiologie* 4:172–85
22. Christian, U. 1973. Trichobothrien, ein Mechanorezeptor bei Spinnen. Elektronenmikroskopische Befunde bei der Winkelspinne *Tegenaria derhami.* (Scopoli). *Verh. Deut. Zool. Ges.* 66: 31–36
23. Chu, I-Wu, Axtell, R. C. 1971. Fine structure of the dorsal organ of the house fly larva, *Musca domestica* L. *Z. Zellforsch. Mikrosk. Anat.* 117:17–34
24. Chu-Wang, I-Wu, Axtell, R. C. 1972. Fine structure of the terminal organ of the house fly larva, *Musca domestica* L. *Z. Zellforsch. Mikrosk. Anat.* 127:287–305
25. Chu-Wang, I-Wu, Axtell, R. C. 1972. Fine structure of the ventral organ of the house fly larva, *Musca domestica* L. *Z. Zellforsch. Mikrosk. Anat.* 130:489–95
26. Chu-Wang, I-Wu, Axtell, R. C. 1973. Comparative fine structure of the claw sensillum of a soft tick, *Argas (Persicargas) arboreus* Kaiser, Hoogstraal, and Kohls, and a hard tick, *Amblyomma americanum* (L.). *J. Parasitol.* 59: 545–55
27. Clever, U. 1958. Untersuchungen zur Zelldifferenzierung und Musterbildung der Sinnesorgane und des Nervensystems in Wachsmottenflügel. *Z. Morphol. Tiere* 47:201–48
28. Corbiere-Tichane, G. 1971. Ultrastructure de l'équipement sensoriel de la mandible chez la larve du *Speophyes*

lucidulus Delar. (Coléoptère cavernicole de la sous-famille des Bathysciinae). *Z. Zellforsch. Mikrosk. Anat.* 112:129–38
29. Corbiere-Tichane, G. 1971. Ultrastructure du système sensoriel de la maxille chez la larve du Coléoptère cavernicole *Speophyes lucidulus* Delar. *J. Ultrastruct. Res.* 36:318–41
30. Dahl, F. 1883. Über die Hörhaare bei den Arachnoiden. *Zool. Anz.* 6:267–70
31. Dethier, V. G. 1955. The physiology and histology of the contact chemoreceptors of the blowfly. *Quart. Rev. Biol.* 30:348–71
32. Dethier, V. G. 1963. *The Physiology of Insect Senses.* New York: Wiley. 266 pp.
33. Draslar, K. 1973. Functional properties of trichobothria in the bug *Pyrrhocoris apterus* (L.). *J. Comp. Physiol.* 84: 175–84
34. Dumpert, K. 1972. Bau und Verteilung der Sensillen auf der Antennengeissel von *Lasius fulginosus* (Latr.). *Z. Morphol. Tiere* 73:95–116
35. Finlayson, L. H. 1968. Proprioceptors in the invertebrates. *Symp. Zool. Soc. London* 23:217–49
36. Finlayson, L. H. 1972. Chemoreceptors, cuticular mechanoreceptors, and peripheral multiterminal neurones in the larva of the tsetse fly (*Glossina*). *J. Insect Physiol.* 18:2265–75
37. Foelix, R. F. 1970. Structure and function of tarsal sensilla in the spider *Araneus diadematus. J. Exp. Zool.* 175:99–124
38. Foelix, R. F., Axtell, R. C. 1971. Fine structure of tarsal sensilla in the tick *Amblyomma americanum.* (L.). *Z. Zellforsch. Mikrosk. Anat.* 114:22–37
39. Foelix, R. F., Chu-Wang, I-Wu. 1972. Fine structural analysis of palpal receptors in the tick *Amblyomma americanum* (L.). *Z. Zellforsch. Mikrosk. Anat.* 129:548–60
40. Foelix, R. F., Chu-Wang, I-Wu. 1973. The morphology of spider sensilla I. Mechanoreceptors. *Tissue Cell* 5: 451–60
41. Foelix, R. F., Chu-Wang, I-Wu. 1973. The morphology of spider sensilla II. Chemoreceptors. *Tissue Cell* 5:461–78
42. Fraenkel, G., Pringle, J. W. S. 1938. Halteres of flies as gyroscopic organs of equilibrium. *Nature* 141:919–20
43. Gaffal, K. P., Hansen, K. 1972. Mechanoreceptive Strukturen der antennalen Haarsensillen der Baumwollwanze *Dysdercus intermedius* Dist. *Z. Zellforsch. Mikrosk. Anat.* 132:78–94
44. Gettrup, E. 1965. Sensory mechanisms in locomotion: the campaniform sensilla of the insect wing and their function during flight. *Cold Spring Harbor Symp. Quant. Biol.* 30:615–22
45. Gettrup, E. 1973. Stimulus transmission in cuticular mechanoreceptors. *Naturwissenschaften* 60:52–53
46. Gewecke, M. 1972. Bewegungsmechanismus und Gelenkrezeptoren der Antennen von *Locusta migratoria* L. *Z. Morphol. Tiere* 71:128–49
47. Gewecke, M. 1972. Antennen und Stirn-Scheitelhaare von *Locusta migratoris* L. als Luftströmungs-Sinnesorgane bei der Flugsteuerung. *J. Comp. Physiol.* 80:57–94
48. Gewecke, M. 1972. Die Regelung der Fluggeschwindigkeit bei Heuschrecken und ihre Bedeutung für die Wanderflüge. *Verh. Deut. Zool. Ges.* 65:247–50
49. Gewecke, M., Schlegel, P. 1970. Die Schwingungen der Antenne und ihre Bedeutung für die Flugsteuerung bei *Calliphora erythrocephala. Z. Vergl. Physiol.* 67:325–62
50. Gnatzy, W. 1973. Die Feinstruktur der Fadenhaare auf den Cerci von *Periplaneta americana* L. *Verh. Deut. Zool. Ges.* 66:37–42
51. Gnatzy, W., Rupprecht, R. 1972. Die Bauchblase von *Nemurella picteti* Klapalek. *Z. Morphol. Tiere* 73:325–42
52. Gnatzy, W., Schmidt, K. 1971. Die Feinstruktur der Sinneshaare auf den Cerci von *Gryllus bimaculatus* Deg. I. Faden-und Keulenhaare. *Z. Zellforsch. Mikrosk. Anat.* 122:190–209
53. Gnatzy, W., Schmidt, K. 1972. Die Feinstruktur der Sinneshaare auf den Cerci von *Gryllus bimaculatus* Deg. IV. Die Häutung der kurzen Borstenhaare. *Z. Zellforsch. Mikrosk. Anat.* 126: 223–39
54. Gnatzy, W., Schmidt, K. 1972. Die Feinstruktur der Sinneshaare auf den Cerci von *Gryllus bimaculatus* Deg. V. Die Häutung der langen Borstenhaare an der Cercusbasis. *J. Microsc.* 14: 75–84
55. Görner, P. 1965. Mehrfach innerviete Mechanorezeptoren (Trichobothrien) bei Spinnen. *Naturwissenschaften* 52: 437
56. Görner, P. 1965. A proposed transducing mechanism for a multiply-innervated mechanoreceptor (Trichobothrium) in spiders. *Cold Spring Harbor Symp. Quant. Biol.* 30:69–73
57. Görner, P., Andrews, P. 1969. Trichobothrien, ein Ferntastsinnesorgan bei

Webespinnen. *Z. Vergl. Physiol.* 64: 301–17

58. Gray, E. G., Pumphrey, R. J. 1958. Ultrastructure of the insect ear. *Nature* 181:618
59. Guthrie, D. M. 1966. The function and fine structure of the cephalic airflow receptor in *Schistocerca gregaria. J. Cell Sci.* 1:463–70
60. Hansen, K., Heuman, H. G. 1971. Die Feinstruktur der tarsalen Schmeckhaare der Fliege *Phormia terraenovae* Rob.-Desv. *Z. Zellforsch. Mikrosk. Anat.* 117:419–22
61. Harris, D. J., Mill, P. J. 1973. The ultrastructure of chemoreceptor sensilla in *Ciniflo. Tissue Cell* 5:679–89
62. Haupt, J. 1970. Beitrag zur Kenntnis der Sinnesorgane von Symphylen. I. Elektronenmikroskopische Untersuchung des Trichobothriums von *Scutigerella immaculata* Newport. *Z. Zellforsch. Mikrosk. Anat.* 110:588–99
63. Hawke, S. D., Farley, R. D., Greany, P. D. 1973. The fine structure of sense organs in the ovipositor of the parasitic wasp, *Orgilus lepidus* Musebeck. *Tissue Cell* 5:171–84
64. Hodgson, E. S. 1968. Taste receptors in arthropods. *Symp. Zool. Soc. London* 23:269–77
65. Hoffman, C. 1963. Vergleichende Physiologie der mechanischen Sinne. *Fortschr. Zool.* 16:269–332
66. Hoffman, C. 1964. Bau und Vorkommen von proprioceptiven Sinnesorganen bei den Arthropoden. *Ergeb. Biol.* 27:1–38
67. Hoffman, C. 1964. Zur Funktion der kammförmigen Organe von Skorpionen. *Naturwissenschaften* 7:172
68. Hoffman, C. 1967. Bau und Funktion der Trichobothrien von *Euscorpius carpathicus* L. *Z. Vergl. Physiol.* 54:290–352
69. Howse, P. E. 1968. The fine structure and functional organization of chordotonal organs. *Symp. Zool. Soc. London* 23:167–93
70. Kaston, B. J. 1935. The slit sense organs of spiders. *J. Morphol.* 58:189–209
71. Knyazeva, N. I. 1969. The locust leg receptors, participating in the reflex of initiation of flight (*Locusta migratoria* L.). *Trans. All-Union Entomol. Soc.* 53:132–47. In Russian
72. Knyazeva, N. I. 1970. Receptors of the wing apparatus regulating the flight of the migratory locust, *Locusta migratoria* L. *Entomol. Rev.* 49:311–17 (Engl. translation)
73. Larsen, J. R., Dethier, V. G. 1963. The fine structure of the labellar and antennal chemoreceptors of the blowfly, *Phormia regina. Proc. Int. Congr. Zool., 16th, 1962* 3:81–83
74. Lees, A. D. 1942. Homology of the campaniform organs on the wings of *Drosophila melanogaster. Nature* 150: 375
75. Lewis, C. T. 1970. Structure and function in some external receptors. *Symp. Roy. Entomol. Soc. London* 5:59–76
76. Leydig, F. 1851. Über *Artemia salina* und *Branchipus stagnalis. Z. Wiss. Zool.* 3:280–307
77. Markl, H. 1964. Geomenotaktische Fehlorientierung bei *Formia polyctena. Z. Vergl. Physiol.* 48:552–86
78. Markl, H. 1966. Schwerkraftdressuren an Honigbienen I. Die Geomenotaktische Fehlorientierung. *Z. Vergl. Physiol.* 53:328–52
79. Markl, H. 1966. Schwerkraftdressuren an Honigbienen II. Die Rolle der Schwererezeptorischen Borstenfelder verschiedener Gelenke für die Schwerekompassorientierung. *Z. Vergl. Physiol.* 53:353–71
80. Martin, H., Lindauer, M. 1966. Sinnesphysiologische Leistungen beim Wabenbau der Honigbiene. *Z. Vergl. Physiol.* 53:372–404
81. McIndoo, N. E. 1911. The lyriform organs and tactile hairs of araneids. *Proc. Acad. Natur. Sci. Philadelphia* 63:375–418
82. McIver, S. B. 1972. Fine structure of the sensilla chaetica on the antennae of *Aedes aegypti. Ann. Entomol. Soc. Am.* 65:1390–97
83. Moeck, H. A. 1968. Electron microscopic studies of antennal sensilla in the ambrosia beetle *Trypodendron lineatum* (Olivier). *Can. J. Zool.* 46:521–56
84. Moran, D. T. 1971. Loss of the sensory process of an insect receptor at ecdysis. *Nature* 234:476–77
85. Moran, D. T., Varela, F. G. 1971. Microtubules and sensory transduction. *Proc. Nat. Acad. Sci. USA* 68:757–60
86. Moran, D. T., Chapman, K. M., Ellis, R. A. 1971. The fine structure of cockroach campaniform sensilla. *J. Cell Biol.* 48:155–73
87. Neese, V. 1965. Zur Funktion der Augenborsten bei der Honigbiene. *Z. Vergl. Physiol.* 49:543–85
88. Nicklaus, R. 1965. Die Erregung einzelner Fadenhaare von *Periplaneta americana* in Abhängigkeit von der Grösse und Richtung der Auslenkung. *Z. Vergl. Physiol.* 50:331–62

89. Nicklaus, R., Lundquist, P. G., Wersäll, J. 1967. Elektronmikroskopie am sensorischen Apparat der Fadenhaare auf den Cerci der Schabe *Periplaneta americana*. *Z. Vergl. Physiol.* 56:412–15
90. Nicklaus, R., Lundquist, P. G., Wersäll, J. 1968. Die Übertragung des Reizes auf den distalen Fortsatz der Sinneszelle bei den Fadenhaaren von *Periplaneta americana*. *Verh. Deut. Zool. Ges., Zool. Anz. Suppl.* 31:578–84
91. Osborne, M. P. 1970. Structure and function of neuromuscular junctions and stretch receptors. *Symp. Roy. Entomol. Soc. London* 5:77–100
92. Peters, W., Richter, S. 1963. Morphological investigation on the sense organs of the labella of the blowfly, *Calliphora erythrocephala* Mg. *Proc. Int. Congr. Zool., 16th, 1962* 3:89–92
93. Pringle, J. W. S. 1938. Proprioception in insects I. A new type of mechanical receptor from the palps of the cockroach. *J. Exp. Biol.* 15:101–13
94. Pringle, J. W. S. 1938. Proprioception in insects II. The action of the campaniform sensilla on the legs. *J. Exp. Biol.* 15:114–31
95. Pringle, J. W. S. 1955. The function of the lyriform organs of arachnids. *J. Exp. Biol.* 22:270–78
96. Pringle, J. W. S. 1963. The proprioceptive background to mechanisms of orientation. *Ergeb. Biol.* 26:1–11
97. Rice, M. J., Galun, R., Finlayson, L. H. 1973. Mechanotransduction in insect neurones. *Nature New Biol.* 241:286–88
98. Rice, M. J., Galun, R., Margalit, J. 1973. Mouthpart sensilla of the tsetse fly and their function. II: Labial sensilla. *Ann. Trop. Med. Parasitol.* 67:101–8
99. Richter, S. 1964. Die Feinstruktur des für die Mechanorezeption wichtigen Bereichs der Stellungshaare auf dem Prosternum von *Calliphora erythrocephala* Mg. *Z. Morphol. Tiere* 54:202–18
100. Roshdy, M. A., Foelix, R. F., Axtell, R. C. 1972. The subgenus *Persicargas*. 16. Fine structure of Haller's Organ and associated tarsal setae of adult *A.* (*P.*) *arboreus* Kaiser, Hoogstraal, and Kohls. *J. Parasitol.* 58:805–16
101. Roth, L. M., Slifer, E. H. 1973. Spheroid sense organs on the cerci of polyphagid cockroaches. *Int. J. Insect Morphol. Embryol.* 2:13–24
102. Salpeter, M. M., Walcott, C. 1960. An electron microscope study of a vibration receptor in the spider. *Exp. Neurol.* 2:232–50
103. Schlegel, P. 1970. Die Leistungen eines Gelenkrezeptors der Antenne von *Calliphora* für die Perzeption von Luftströmungen. Elektrophysiologische Untersuchungen. *Z. Vergl. Physiol.* 66: 45–77
104. Schmidt, K. 1969. Die Feinbau der stiftführenden Sinnesorgane im Pedicellus der Florfliege, *Chrysopa* Leach. *Z. Zellforsch. Mikrosk. Anat.* 99:357–88
105. Schmidt, K. 1969. Die campaniformen Sensillen im Pedicellus der Florfliege (*Chrysopa,* Planipennia). *Z. Zellforsch. Mikrosk. Anat.* 96:478–89
106. Schmidt, K. 1973. Vergleichende morphologische Untersuchungen an Mechanorezeptoren der Insekten. *Verh. Deut. Zool. Ges.* 66:15–25
107. Schmidt, K., Gnatzy, W. 1971. Die Feinstruktur der Sinneshaare auf den Cerci von *Gryllus bimaculatus* Deg. II. Die Häutung der Faden-und Keulenhaare. *Z. Zellforsch. Mikrosk. Anat.* 122:210–26
108. Schmidt, K., Gnatzy, W. 1972. Die Feinstruktur der Sinneshaare auf den Cerci von *Gryllus bimaculatus* Deg. III. Die kurzen Borstenhaare. *Z. Zellforsch. Mikrosk. Anat.* 126:206–22
109. Schneider, D., Kaissling, K. E. 1957. Der Bau der Antenne des Seidenspinners *Bombyx mori* L. II. Sensillen cuticulare Bildungen und innerer Bau. *Zool. Jahrb. Abt. Anat. Ontog. Tiere* 76:223–50
110. Schwartzkopff, J. 1964. Mechanoreception. In *The Physiology of Insecta,* ed. M. Rockstein, 1:509–61. New York: Academic. 640 pp.
111. Seyfarth, E. A., Barth, F. G. 1972. Compound slit sense organs on the spider leg: mechanoreceptors involved in kinesthetic orientation. *J. Comp. Physiol.* 78:176–91
112. Sihler, H. 1924. Die Sinnesorgane an der Cerci der Insekten. *Zool. Jahrb. Anat.* 45:519–80
113. Sinoir, Y. 1969. L'ultrastructure des organes sensoriels des insectes. *Ann. Zool. Ecol. Anim.* 1:339–56
114. Slifer, E. H. 1961. The fine structure of insect sense organs. *Int. Rev. Cytol.* 11:125–59
115. Slifer, E. H. 1968. Sense organs on the antennal flagellum of a giant cockroach, *Gromphadorhina portentosa,* and a comparison with those of several other species. *J. Morphol.* 126:19–30
116. Slifer, E. H. 1970. The structure of arthropod chemoreceptors. *Ann. Rev. Entomol.* 15:121–42

117. Slifer, E. H., Sekhon, S. S. 1970. Sense organs of a thysanuran, *Ctenolepisma lineata pilifera,* with special reference to those on the antennal flagellum. *J. Morphol.* 132:1–26
118. Slifer, E. H., Prestage, J. J., Beams, H. W. 1959. The chemoreceptors and other sense organs on the antennal flagellum of the grasshopper. *J. Morphol.* 105:145–91
119. Smith, D. S. 1969. The fine structure of haltere sensilla in the blowfly, *Calliphora erythrocephala* (Meig.), with scanning electron microscopic observations on the haltere surface. *Tissue Cell* 1:443–84
120. Smola, U. 1970. Untersuchung zur Topographie, Mechanik und Strömungsmechanik der Sinneshaare auf dem Kopf der Wanderheuschrecke *Locusta migratoria. Z. Vergl. Physiol.* 67:382–402
121. Snodgrass, R. E. 1926. The morphology of insect sense organs and the sensory nervous system. *Smithson. Misc. Collect.* 77:1–80
122. Snodgrass, R. E. 1935. *Principles of Insect Morphology.* New York: McGraw-Hill. 667 pp.
123. Strickler, J. R., Bal, A. K. 1973. Setae of the first antennae of the copepod *Cyclops scutifer* (Sars): Their structure and importance. *Proc. Nat. Acad. Sci. USA* 70:2656–59
124. Stuart, A. M., Satir, P. 1968. Morphological and functional aspects of an insect epidermal gland. *J. Cell Biol.* 36:527–49
125. Thurm, U. 1964. Mechanoreceptors in the cuticle of the honeybee. Fine structure and stimulus mechanism. *Science* 145:1063–65
126. Thurm, U. 1964. Das Rezeptorpotential einzelner mechanorezeptorischer Zellen von Bienen. *Z. Vergl. Physiol.* 48:131–56
127. Thurm, U. 1965. An insect mechanoreceptor. Part I: Fine structure and adequate stimulus. *Cold Spring Harbor Symp. Quant. Biol.* 30:75–82
128. Thurm, U. 1970. Untersuchungen zur funktionellen Organisation sensorischer Zellverbände. *Verh. Deut. Zool. Ges.* 64:79–88
129. Thurm, U. 1972. The generation of receptor potentials in epithelial receptors. In *Olfaction and Taste IV,* ed. D. Schneider, 95–101. Stuttgart: Wissenschaftliche Verlagsgesellschaft MBH. 400 pp.
130. Tominaga, Y., Kabuta, H., Kuwabara, M. 1969. The fine structure of the interpseudotracheal papillae of the fleshfly. *Annot. Zool. Jap.* 42:91–104
131. Uga, S., Kuwabara, M. 1967. The fine structure of the campaniform sensillum on the haltere of the fleshfly, *Boettcherisca peregrina. J. Electronmicrosc.* 16:304–12
132. Walcott, C. 1969. A spider's vibration receptor: its anatomy and physiology. *Am. Zool.* 9:133–44
133. Walcott, C., Salpeter, M. M. 1966. The effect of molting upon the vibration receptor of the spider (*Achaearanea tepidariorum*). *J. Morphol.* 119:383–92
134. Wigglesworth, V. B. 1953. The origin of sensory neurones in an insect, *Rhodnius prolixus. Quart. J. Microsc. Sci.* 94:93–112
135. Yamaoka, K., Hirao, T. 1973. Releasing signals of oviposition behaviour in *Bombyx mori. J. Insect Physiol.* 19:2215–23
136. Yamaoka, K., Hoshino, M., Hirao, T. 1971. Role of sensory hairs on the anal papillae in oviposition behaviour of *Bombyx mori. J. Insect Physiol.* 17:897–911
137. Zacharuk, R. Y. 1962. Sense organs of the head of larvae of some Elateridae: Their distribution, structure and innervation. *J. Morphol.* 111:1–34
138. Zacharuk, R. Y. 1962. Exuvial sheaths of sensory neurones in the larva of *Ctenicera destructor* (Brown). *J. Morphol.* 111:35–47

THE BRAZILIAN BEE PROBLEM[1]

❖6096

Charles D. Michener
Department of Entomology and Department of Systematics and Ecology, The University of Kansas, Lawrence, Kansas 66045

There have been repeated and sometimes lurid references to the introduction of African honey bees into South America, to the spread of such bees or their hybrid descendants (Africanized bees, i.e. the Brazilian honey bee), and to the influence of such bees on apiculture, pollination, animal husbandry, and public health in South America and, prospectively, in North America. Much of what is known about these bees in South America appears in the Final Report of the Committee on the African Honey Bee (27) and in two subsequently annotated versions of the same report (31). One function of the present review is to place some of the content of these reports in more accessible form. Another is to review the numerous more recent contributions on the subject, largely from Brazilian sources (including references 13, and 14; proof sheets of only a few articles in reference 13 were available when reference 27 was prepared). Emphasis is placed on the history, present status, and principal attributes of these bees. An account of their possible impact in North America is given in the first report mentioned (27).

Some of the questions about these bees have so far been answered from reports of beekeepers, rather than by controlled experiments or observations. Because of variability of the population over a vast area, the results of experiments even where they have been carried out may not be applicable everywhere. Sources of information are therefore documented below. The code B indicates information from beekeepers, including anecdotal information from scientists. SP refers to observations or experiments made on the population of bees of the central part of the state of São Paulo (Piricicaba, Rio Claro, Ribeirão Preto, etc), where most of the experimental work has been done; in effect, it means under the conditions of central São Paulo.

[1]Contribution number 1551 from the Department of Entomology, The University of Kansas, Lawrence, Kansas, 66045.

HISTORY AND CURRENT STATUS

Origin and Spread

Before its spread by man, *Apis mellifera* ranged from northern Europe to the Cape of Good Hope and eastward into western Asia. It varies geographically, as indicated by the recognition of various subspecies (41, 52, 70), some of them regarded as species by Maa (57). European subspecies were long ago carried to many parts of the New World; the history of their introduction into Brazil is given by Nogueira-Neto (63). In tropical American countries they were not especially successful. The widespread subspecies of subsaharan Africa, *Apis mellifera adansonii,* however, is very abundant both in equatorial and warm temperate southern Africa and is an excellent honey producer (52, 72, 73). It is often aggressive and appears to quickly eliminate all colonies of European bees taken into its range. Writing in Tanzania before the Brazilian bee problem was recognized, Smith (72) said: "Viciousness is the first characteristic which beekeepers and others encounter. . . . The colonies always appear to be alerted, ever ready to defend the hive, and on occasion the whole colony goes berserk and stings every living thing in sight. A beekeeper can usually smoke a colony and get it under control while he collects the crop. But if he appears near the hive a second time within several weeks the bees are on him." Portugal-Araújo (66, 68) indicates similar problems in Angola on the opposite side of Africa. Other authors (1) speak of it as gentle; obviously there is great variation among colonies and populations of African bees and indeed other subspecific names have been given to certain high montane and coastal populations (74), as well as to the form occupying a limited area around the Cape of Good Hope.

A. m. adansonii was introduced into Brazil in 1956 in order to develop a tropical- or subtropical-adapted strain having improved honey productivity. Queens that reached Brazil were mostly from near Pretoria, South Africa (over 1500 m altitude, 26° S; therefore, an area with cool winters), although one was from Tabora, Tanzania. Colonies showing desirable characteristics were to be selected from among hybrids resulting from crosses of European and African bees. However, in 1957, 26 queens mated in Africa escaped with swarms. Thus *A. m. adansonii* was introduced into South America near Rio Claro, São Paulo, Brazil. (For details of introduction see 27, 31, 40, 43, 44, 67, 77.) The progeny of 26 queens and the perhaps 200 males with which they had mated introduced genetic materials that rapidly spread over an area already approaching that of the contiguous states of the United States, a biologically remarkable performance.

The spread is shown in Figure 1. It is impossible to know how much of the dispersal has been natural and how much due to movement of queens or colonies by beekeepers. No doubt both factors have been involved. The African bee is regularly migratory in certain areas (72) and it is not surprising that Brazilian bees show signs of similar abilities and have spread rapidly by swarming and by absconding of colonies. At least for a time the rate of spread was in the vicinity of 320 km per year (27, 31). The name Brazilian honey bee has been applied to the strain

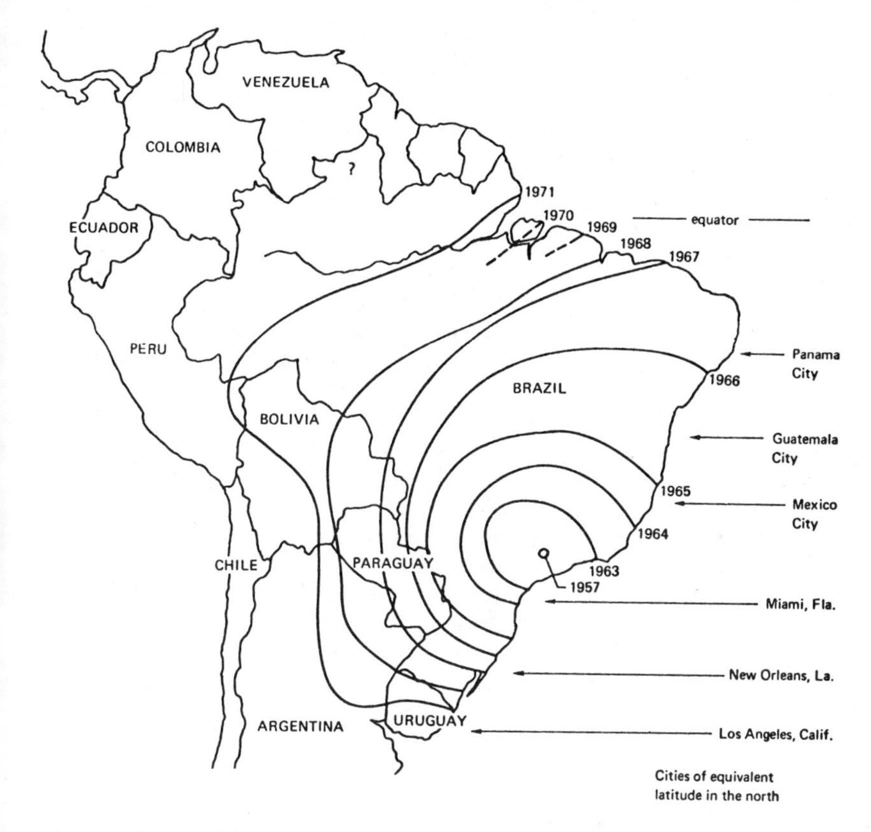

Figure 1 Spread of the Brazilian honey bee (27). ? = area of a possible introduction into Roraima Territory, Brazil. Southward and westward distribution appears to have been slight since 1971. Further northward spread is unverified.

spreading in South America (27), the attributes of which are in many ways similar to those of *A. m. adansonii* in spite of interbreeding with the European bees. European bees have largely vanished from the area where Brazilian bees have been for several years (27, 31).

There are two ways in which such an introduced race might spread. One is by either aggression against the established race or by winning in the competition for environmental resources, such as food or nest sites. This could happen with little interbreeding. The other is by gene flow into the established population. Both have probably occurred in this instance. There are reports of colonies entering apiaries, clustering on or under a hive, and ultimately invading, replacing the queen of the hive. Such aggressive replacement appears to have been common in some areas (B;

coastal São Paulo), rare in others (only one hive in a thousand invaded, SP, 31). It is verified by the finding of dead Brazilian queens in front of Italian hives protected by queen excluders at their entrances (16); these queens were not able to get in when their workers did and died outside. Some of the spread of African genes, however, has doubtless been by means of gene flow. Most European queens returned to their colonies after mating flights and produced workers having at least some African attributes, indicating that they had mated with drones of the Brazilian population (B; 27, 31).

Individual colonies of Brazilian bees (SP) may have a higher life expectancy than do those of Italian bees. A probability of survival of Brazilian colonies four times longer than for Italian colonies has been indicated, and attributed to characteristics of the workers, not of queens (31). The basis for this statement is not clear, but for 10 colonies of Italians and 10 colonies of Brazilians (SP), survival was as follows: Italians—after 6 months seven colonies survived but after 12 and 15 months, only one (which had a hybrid queen) survived; Africans—after 6 months six colonies survived, after 12 months five survived, and after 15 months four survived (51). These 20 colonies were initially of about equal strength, 1500–2000 bees, and were left in boxes essentially undisturbed, subject to periods of food shortage, attacks of ants, and the other vicissitudes experienced by wild small colonies. If verified by more data, these findings may explain part of the success of the Brazilian bee. The longer survival of Brazilian colonies suggests that they may be better at replacing queens than are European colonies under Brazilian conditions.

Queen replacement (SP) is apparently more frequently necessary than in temperate climates; significant racial differences were not noted. Five normally mated Brazilian queens in hive colonies survived a mean of 8.4 months; 6 normally mated Italian X Brazilian queens survived for a mean of 7.4 months. Eighty artificially inseminated queens, half of them Italians and half Brazilians, survived an average of only 4.2 months (51).

The reproductive behavior of males seems similar to that of European males in such matters as time of day and durations and numbers of flights (28). An experiment appears to show partial reproductive isolation that diminishes intersubspecific matings when equal numbers of reproductives of both Italian and Brazilian (SP) races are present (48, 49). (It must be noted that if there were Brazilian drones of unknown origin in the mating area, so that their number exceeded that of Italian drones, interpretations of this experiment would be in some ways different.) Twenty virgin Brazilian queens were inseminated by Brazilian drones in 58.5% of the inseminations, and 21 virgin Italian queens by Italian drones in 64.8% of the matings. These results are based on identification of the race (or hybridity) of 200 worker progeny of each young queen. Recognition must have been based on morphology or color and is therefore questionable but in this limited context as distinguished from variable populations from wide areas, the recognition of the races and their hybrids was presumably accurate. In this experiment the Brazilian queens mated, on the average, with about 7.5 males while Italian queens mated with about 5.3 males each. The latter number is little over half that reported for temperate regions. It would be interesting to know if Italian queens, under SP conditions, mate relatively few times even when undisturbed or if Brazilian drones somehow interfere with mating by the Italian bees.

The number of sperm cells produced by males of Italians is somewhat less (mean 5,463,000; N = 14) than by Brazilians (mean 7,123,000; N = 14) (SP, 48). F_1 hybrids contain a mean of only 3,205,750 (N = 12)—a distinctly reduced number further suggesting slight reproductive isolation—and F_2 hybrids showed great variability, as might be expected (see reference 48 for statistical details).

An outstanding feature of both African and Brazilian bees is the large number of colonies found per unit area. The high frequency of interracial matings that was partially responsible for the demise of European bees in Brazil must be largely due to the enormous feral populations attained by Brazilian bees. There are no data on abundance of feral European bees in Brazil prior to 1957. However, my general observations (1955–1956) indicate that in southern Brazil (Paraná, São Paulo) they were about as scarce as they are in woods in the United States. This means that although workers were common on flowers, it was difficult to find nests or to attract a swarm to an empty box. In tropical forested parts of Brazil, Panama, Costa Rica, etc, in areas not yet reached by Brazilian bees, comparable observations indicate that European bees are far less common than in the United States. One can often collect flower-visiting insects for long periods without seeing an *Apis.* The Brazilian bee changed this; the only quantitative data are from savanna (cerrado) areas in the states of Goias and Mato Grosso where Kerr (46) found an average of 107.5 colonies of feral *Apis* per km^2. Nests are common not only in the traditional places but in holes in the ground, in termite mounds, etc. Reportedly, Brazilian bees will occupy smaller cavities than those utilized by European races (B; 27, 31). On highly attractive flowers in Paraná and São Paulo, J. S. Moure (personal communication) reports up to 15 workers per m^2 in areas without beekeeping. A result of the large feral populations of Brazilian bees is that virgin queens from apiaries encounter large numbers of Brazilian drones with which they can mate. Of course another result of the large feral population is intense competition for food resources with the hive bees of whatever race (35).

Variation

A frustrating aspect of the literature on Brazilian honey bees has been the divergence of opinions and information presented. The contradictory statements are probably mostly justified for different times and areas. Great genetic variability is to be expected in a population of hybrid origin. Moreover, differential selection is to be expected—the distribution encompasses well over 30° of latitude from treeless temperate grasslands to the xeric scrub of northeastern Brazil and the equatorial rain forest of the Amazon basin.

Subsequent sections contain references to differences between northern and southern populations of Brazilian bees. No one has experimentally moved bees from one area to another to verify that these differences are genetic rather than the direct effects of environmental and colonial conditions (weather, nectar flow, colony strength, etc). Yet it seems extremely probable that most of the differences recorded do result from differing amounts of hybridization with European bees and differing subsequent selection, and so have genetic bases. In the states of São Paulo and even Minas Gerais and the more southern states of Paraná, Santa Catarina, and Rio Grande do Sul, there were rather numerous beekeepers and some feral European bees. Hence there were many European bees with which invading Brazilian bees could mate. To the north and especially in the far north and northeast, beekeepers were much less numerous and feral European bees were rare. One would, therefore, expect the Brazilian bee of the northern states to be more like its African ancestors than is the southern population. This initial probable difference has been augmented by several factors, as follows.

In view of the effect of the Brazilian bee on apiculture and because of the public outcry against it, W. E. Kerr produced and supplied to beekeepers queens of European bees (mostly Italians) with the objective of diminishing the undesirable features of Brazilian bees. In 1963–1964 200 Italian queens mated primarily with Italian drones were distributed; in 1965–1972 23,000 virgin Italian (including a few Caucasian) queens were distributed (31). W. E. Kerr is convinced that this program played a significant role in improving the characteristics of the bees over a wide area, in spite of reports that some beekeepers killed the queens (especially those mated to Italian drones) because of their belief that the European attributes included lower honey production (27, 31). Since the European queens went to beekeepers, they further augmented the European attributes of the population in southern Brazil.

Some beekeepers destroy or requeen their most aggressive colonies and put frames of drone comb into their gentlest and most productive colonies, as recommended as early as 1964 (12, 30). Kerr (45, 47) provides practical recommendations for selection toward mildness. Cosenza (22), advocating continuation of such programs of improvement by beekeepers, also recommends further introduction of European genes but Weise, Kerr, and many other Brazilian apiculturalists do not, for fear of reducing the high productivity.

These practices, plus perhaps natural selection, appear to be leading to the development of a reasonably tractable southern Brazilian population, well adapted to its subtropical and warm temperate environment. Some beekeepers work without gloves, a few even without shoes, veils, or shirts, to demonstrate the gentleness of their bees (27, 31). In short, the suggestion in the title of Kerr's (42) paper "The Solution is to Create a New Race" is being realized (80), albeit belatedly and after considerable grief. Those who have seen only this form often feel that the situation does not warrant costly measures to prevent further spread. Comparison of the present situation in southern Brazil with the strong reaction against the Brazilian bees in the early years of their invasion of these same areas (10–12, 62) and currently in Argentina, Bolivia, Paraguay, and Uruguay (4, 5, 15, 26, 34–39, 65) seems to verify improvement of the bee population, although familiarization of beekeepers with attributes of a race of bees new to them is doubtless part of the story (27, 31, 75).

Meanwhile, to the north and especially northeast, where there are few beekeepers, selection may have favored those that flew farthest into areas with almost no competing *Apis,* and thus largely escaped the European influence. In any case, the population in arid states like Pernambuco, Rio Grande do Norte, and Ceará is highly aggressive, quite different from that in the south (27, 31). The appearance of people who sell honey from wild colonies (31), in an area where *Apis* honey was previously little known, shows that the species has become common in the bush. Destruction of wild colonies in the process of taking honey is probable and selection favoring aggressiveness is therefore likely at present. The same thing may be occurring in other areas; for example certain Paraguayan Indians formerly little interested in honey are now producing much more, mostly from wild colonies (79). Exportation of large amounts of honey now (first major year, 1972; 31) from northeast Brazil (Recife) suggests that apiaries there too are productive and evolution like that in the south may result. The most noteworthy attribute of the northeastern populations (at least in 1971; 27) is frightening and is described below in the section on Stinging.

Northward Dispersal

By 1971 Brazilian honey bees had reached Belém, the Ilha de Marajó, and across the Amazon, Macapá, humid areas all near the mouth of the Amazon (27, 31). African bees occur in wet tropical forests (73; Michener, personal observations in Cameroon), although probably less commonly than in savanna. Whether or not the Brazilian bee reached the Amazon under its own power or was carried there, it seemed likely that it would continue without delay across the rest of the Amazon forest. This may now be happening, although recent reports indicate that it is not common on flowering plants near Belém (31). It may be that all but the best located colonies are destroyed during wet seasons and that progress through the humid tropics will be slower than anticipated in 1971. Nesting sites in the Amazon forest are reported to be restricted to large dead dry trees (33). Unverified press reports of the Brazilian bee as far north as the Orinoco River in Venezuela in 1973 suggest, on the other hand, that the Amazon forest did not delay it greatly.

The attributes of the bees which will presumably finally pass through (or around) the forest and continue on their way toward North America cannot be predicted. A single report from Porto Velho, Rondônia (B), farther west in the Amazon area, compares the bees favorably with those of southern Brazil (31). At least there are not many people in the forest whose activities might cause selection for aggressiveness.

For the future, one must presume that the Brazilian bee will migrate through Central to North America, to latitudes where the climate is similar to that of the southernmost areas now occupied in Uruguay and Argentina. But what its attributes will then be cannot be predicted. Current Brazilian conditions show that it can be modified; changes in it will doubtless occur as it moves northward. It could be as intractable as the present population of northeastern Brazil, or it could have attributes like the southern Brazilian population. Much will depend on the stocks of *Apis* which it encounters and on their abundance. A barrier of abundant gentle bees (27) could presumably modify it as it moves northward.

An important factor in the future spread of Brazilian bees is their overwintering ability. There are no records of winter cluster formation. Thermoregulation is reported to be less efficient for African bees than for European ones (25), although no direct comparisons were made; the frequency of subterranian nests in Africa may be a result of the uniform temperature of such sites. In cold weather Brazilian bees are reported to abscond or to continue activity in the nest (B; 27); in the latter case, stores are quickly used up. Long cold spells are therefore commonly fatal, leading to high autumn mortality on the southern Brazilian plateau and presumably explaining the failure of African and Brazilian bees to become established when introduced into central or eastern Europe (81) or the United States. [Stort reports that a beekeeper in South Africa has sent (illegally) various African queens to the US in recent years (31). Presumably their colonies would thrive only in the southernmost states.] Winter mortality would be a serious problem for southern queen or package bee producers marketing in regions of colder climates, no matter what other features the population might have.

CHARACTERISTICS OF THE BRAZILIAN HONEY BEE

Some attributes of the Brazilian bee that conspicuously relate to its history and spread are discussed above. Other features are summarized below. No one feature will always distinguish Brazilian or even African bees from European bees. Means differ, and the combination of characteristics is distinctive. Differences all are quantitative, i.e. of degree and not of kind. Every seemingly peculiar feature of Brazilian bee behavior has its counterpart, perhaps seen only rarely, in European bees.

Color, Size, and Morphology

Color is highly variable but most workers have yellow abdominal bands. J. S. Moure (personal communication) reports about 15% of workers to be black in areas in Paraná and Santa Catarina. As in Africa, a single colony commonly has both black and partly yellow workers (27). Kerr (30) considers chestnut bronze abdominal terga to be characteristic of drones (SP) but other color patterns (with yellow) are also found in northern Brazilian drones (27) and in Africa (66, 67).

In size Brazilian bees average slightly smaller than European races. Virgin queens of Italian and Caucasian bees weigh 208.8 ± 25.39 mg SD (N = 111) and 207.7 ± 21.07 mg SD (N = 132), respectively, whereas equivalent figures for Brazilian queens (SP) are 199.3 ± 25.59 mg SD (N = 110) (30). Numerous measurements of various structures of workers verify the smaller size of Brazilian bees (23, 69). Of possible importance in relation to feeding and pollination is the observation that glossal length of Brazilian bees averages 3.87 mm, compared to 4.15 mm for Caucasian bees and 4.02 mm for hybrids (23).

Size of worker cells is correlated with bee size. For Brazilian bees, measurements across 25 adjacent cells (not made on foundation) averaged 12.6 cm (range 12.1–13.6, N = 55 colonies) (27). There was no significant difference between northern and southern regions. For European bees at Guelph, Ontario, comparable figures were 13.5 cm (range 12.9–14.3, N = 41 colonies). Measurements of comb from small numbers of colonies by others (e.g. 23) give similar results.

Kerr and his associates emphasize that while none of the newly introduced African bees would use standard comb foundation, Brazilian apiary bees, especially in southern Brazil, now do so. Even there wild swarms may not use standard foundation (B; 27, 31). Selection for European genes including larger-sized bees is presumably responsible for the change.

Maa (57) gave some morphological characters for the recognition of *Apis mellifera adansonii.* No one of them seems reliable, probably especially for the Brazilian bee. However, H. V. Daly (not published), using numerous measurements of each bee and multivariate analysis, has demonstrated the possibility of distinguishing Brazilian from European forms by simultaneous consideration of the various features. Size was the most important factor in his study but after its removal appreciable variation remained and aided in making distinctions. The greatest need in tracing the spread of the Brazilian honey bee is ability to identify it on the basis of few specimens. Daly's methods offer the best hope for doing so. The situation is not simple, however, as illustrated by the following comment from W. E. Kerr (personal communication): "In Colombia and Venezuela there is a hybrid between *A. m. mellifera* and *A. m. ligustica* that has many genetic segregants that resemble the African bee, both in external morphology and in temper." He first saw this form in 1952, four years before introduction of African bees to Brazil. This form evidently lacks the ability found in Brazilian bees to form large feral populations, to spread, and to replace other European bees.

Gonçalves (29) and Stort (78) have analyzed the inheritance of various morphological features in relation to aggressiveness in crosses and backcrosses between Italian and Brazilian (SP) bees. Characters which within one group of crosses or another showed significant correlations with one or more components of aggressiveness included diameter of median ocellus, tergal and sternal measurements, and mesoscutal measurements.

Isozymes

Mestriner (59) found a difference in allelic frequencies between Brazilian bees (SP) and Italian bees, on the basis of 68 colonies of the former and 7 of the latter, a total of 300 workers analyzed. At what he called the P_3 locus, both races were variable, the Brazilian bee with the allele P_3S at a frequency of .995, P_3F at .006. Comparable figures for Italian bees were P_3S .531, P_3F .469. Thus the Brazilian bees were almost monomorphic at this locus. Other loci examined showed no racial differences.

Individual and Colonial Development

The developmental period of workers of Brazilian bees is reported to be about a day shorter than for European bees (51, 80). Although similarly short development may occur in European races (27), a mean difference in developmental period seems to exist for workers (but not for reproductives) reared in the same region.

Colony growth of Brazilian bees is reported to be rapid (B; 27, 31). That these reports are justified is suggested by the large numbers of eggs laid in 3 colonies of Brazilian bees (SP), as compared to 3 colonies of Italian bees and 3 of hybrids. Over a 12-month period, the totals for the 3 colonies of each type were 104,520; 58,164; and 55,390 eggs, respectively (51).

General Activity

Brazilian and African bees seem more excitable and active than European bees, although not as much as *Apis cerana* (as I have seen it in Java and Malaysia). Flight around flowers is quicker and more nervous than that of European bees (27, 31). Brazilian bees often fly right into the hive entrance rather than alighting and walking in. Núñez reports that Brazilian bees make decisions more rapidly than European bees (SP; in 31).

Waggle dances often indicate the directions to food sources as little as 20 m away (SP; 27, 31) (or 33 m in African bees, in Tanzania; 71). These distances are much shorter than for European bees.

The general high activity of Brazilian bees is evident the moment a hive is opened. When a comb is lifted the bees run on it, often form hanging festoons, and many of them may take flight. These features are most evident and very troublesome in northern Brazil, less obvious in southern Brazil, and scarcely apparent in the gentlest (most Europeanized?) colonies (27, 31).

Movement of hives in migratory bee culture is sometimes said to lead to balling and killing of queens (27; Santa Catarina), but in São Paulo (31, 50) and with African bees in South Africa (32, 60) no such problem is reported.

Foraging

Brazilian bees are reported to start work earlier and finish later than European bees, often working on cool days and almost until full darkness in the evening (B; 27, 31, 44). However, analysis of data on temperatures and times of day of flights of some 885,000 bees (SP; 51) does not strongly verify this statement. There were no appreciable differences in hour of starting and Brazilian bees were recorded flying only 12 min later in the evening than Italian bees. Temperatures required for activity also scarcely differed for Brazilian and Italian colonies. However, there was a marked difference in the time of maximum flight activity. Maximal activity for Italian bees was from 8:30 to 10:30 A.M., with decreasing flight activity during the afternoon. Brazilian bees as well as hybrids, however, showed maximal activity in the afternoon. If one wishes to work hives when the maximum number of bees are afield, one works Italian colonies in the morning, Brazilian colonies in the afternoon (B, SP; 51). These results were obtained over a period of 13 months and hence are unlikely to result from a flower favored by the Brazilian colonies that produces nectar or pollen largely in the afternoon. Since most flowers are richest in nectar in the morning, the behavior of the Brazilian bees is unexpected.

At calibrated syrup sources (50% sucrose), compared to Italian bees a Brazilian bee (SP) makes briefer visits and carries less to the hive per trip (64). For high flow rates this difference could result merely from the smaller crop capacity, 54.6 ± 7.34 mg ($N = 42$) for Brazilians, 65.1 ± 9.26 ($N = 50$) for Italians (2), but at low flow rates the difference is augmented. At a rate of 0.70 μl/min, Brazilians carry 67% less per trip than Italians. Below that rate motivation for collecting by Brazilian bees disappears while Italians continue working at least to 0.35 μl/min. Because Brazilian bees take less time to take food, they return more often to their colonies. This should lead to more communication and to larger recruitable populations at any time. Thus Brazilian bees may be adapted to large nectar sources and to quickly recruiting many bees to exploit them (64).

Production figures for typical Brazilian municipios show sharp reductions in number of hives and total honey crop during 3 or 4 years after arrival of Brazilian bees (27). Comparisons of hives of different races in the same apiary consistently show the excellent production of Brazilian bees compared to European races at the same times and places (40). Cosenza (19, 21) compared production of 6 colonies of Caucasian bees, 22 colonies of Brazilian bees, and 22 colonies of Brazilian X Caucasian hybrids during one month in Minas Gerais. All colonies were about the same size. The Caucasian bees produced nil; the Brazilians and hybrids produced equally, averaging about 9 and 5.5 kg of honey each in two different apiaries. As another example, Portugal-Araújo (67) reports average honey production of 8.8 kg, 19.2 kg, and 35.5 kg per hive for 10 hives each of black European, Italian, and Brazilian bees, respectively, during a nectar flow of about 65 days (SP). Presumably the 30 hives were in the same apiary and similar in strength. There are many reports of excellent annual productivity, e.g. 80 colonies (SP) that produced an average of 68 kg of honey each in 1969; in the first four months of 1970 some of the same colonies had gathered over 100 kg of honey and one, over 200 kg (31, 51). Or again, in Paraná, 180 kg of honey per hive are recorded annually (75). Annual production figures should be higher than in cool temperate climates where cold often prevents foraging.

Because of excellent productivity, most Brazilian beekeepers, particularly in the south, favor the Brazilian bee, although some (but not others) say that temperate area colonies use so much honey to keep alive in winter that the advantage of this race disappears (B; 27, 31).

Most of the high production noted by beekeepers is with fewer colonies in the apiaries than formerly, when European bees were used. Thus there may be more forage per colony than was customary for the European bees. However, the large competing feral populations should have the opposite effect; some beekeepers destroy feral colonies before moving hives into an area. Simultaneous controlled tests of productivity should be repeated. Such tests must be done in such a way as to guard against robbing, for Brazilian bees are excellent robbers (27, 31) and for this test must be prevented from foraging in the hives of their competitors. A laboratory assessment of hoarding might prove informative (53).

Brazilian bees are highly variable in propolization but some collect even more propolis than Caucasian bees, making working of hives difficult and slow (27).

Absconding and Swarming

Brazilian bees do not die in the hive for lack of food and water; they abscond instead and migrate until they either find a suitable site or die. Small colonies, such as baby nuclei, often abscond when the queen makes her nuptial flight (80); thus it is difficult to use small nuclei for mating queens. Larger colonies do not give trouble by absconding if food, water, and space are available (B; 27, 31).

The large number of feral colonies, probably often absconding or swarming because of inadequate space resulting from the rapid colony growth so that a colony quickly outgrows a small cavity, leads to frequent observation of swarms. (Absconding colonies look just like reproductive swarms.) Cosenza (18, 19) found that 79% of 31 swarms in Minas Gerais occurred during a time of food scarcity, suggesting that they were migratory colonies; 52% of these 31 swarms contained more than one queen, probably as a result of fusion of smaller swarms as is reported in Africa (72, 73) for migrating colonies (but not reproductive swarms). Only 5.7% of 175 swarms observed in northeastern Brazil (state of Ceará) in all months of the year contained more than one queen (51). Migrating colonies with up to 14 queens have been reported (B; 27, 31) and they are sometimes seen to divide into small monogynous colonies. If fission does not occur, all but one of the queens are killed after the bees enter a nest (31; based on 8 polygynous swarms, SP). The same thing evidently occurs in Africa (1, 72), although multiple laying queens are also reported (68). Queens in swarms are usually mated; all those in Cosenza's 31 swarms had sperm cells in their spermathecae.

Swarms are often seen flying, sometimes high (30 m), sometimes across rivers or cities (B; 27, 31). One was reported as landing on a boat 10 km from land in the Amazon estuary. Swarms are often seemingly attracted from the bush to apiaries, where they may invade apiary colonies as indicated above. Swarms are less commonly seen now in southern Brazil than in the years shortly after the Brazilian bee arrived there (B).

Stinging

The best known characteristic of Brazilian bees is their aggressiveness (sense of 56). Individual stings of Brazilian bees produce effects comparable to stings from other honey bees; no chemical differences between venoms of Brazilian and Italian bees are known (58). But Brazilian bees, especially in the northern states, differ dramatically from other races in their sensitivity to disturbance, their ability to communicate alarm within and between colonies, and their capacity to respond quickly by massive and persistent attack on intruders (27, 31). The mechanisms are not clear but could involve release of greater quantities of alarm pheromones; different behavior in pheromone release, e.g. release at the hive entrances instead of in the air; enhanced responsiveness to alarm pheromones; increased use of visual or other signals around the victim; or some combination of the above factors. Enhanced responsiveness to the alarm pheromone is suggested by Stort's (31) finding of a positive correlation in Brazilian bees between the number of olfactory sensilla on the worker antennae and both numbers of stings and distance that bees follow a stung object (SP). Among Italian bees no such relationship was found. A study of European bees showed both higher responsiveness to the alarm pheromone and higher production of it in an aggressive line than in a gentle line (3).

The slightest disturbance at or near the hive entrance in northern populations of Brazilian bees can set off a chain reaction that explodes within seconds; whole apiaries may go out of control. Hundreds of bees become airborne and pursue and sting any animals or people within perhaps 100 m of the apiary (27, 31). The length of time required for recovery from such a disturbance by Brazilian bees (SP) differs greatly from that of Italian bees. According to Stort (77), the former require, on the average, 28.2 min to return to the normal quiet state when there is an enemy standing 3 m away (N = 45 tests, 9 colonies); Italian bees require 2.9 min (N = 25, 5 colonies). F_1 hybrids were intermediate but more like the Italians, although highly variable, with a mean of 9.0 min (N = 15, 3 colonies).

In southern Brazil, colonies rarely go out of control; only rarely in that area do bees from an undisturbed colony fly (2–3 m) and strike the veil of an observer. This is in contrast to experience in the northern states, where bees often attack persons even before they enter apiaries. Some observers report that behavior like that which occurs in the north was prevalent in the south early in the spread of Brazilian bees.

Stort (76, 77) provides some documentation for the racial differences in aggressiveness at hive entrances. Using two colonies of Brazilian bees (SP) and four colonies of Italian bees, he jiggled small leather balls in the air for 60 sec at varying distances from the hive entrances. At 30 cm from the entrance, Brazilian bees stung the ball a mean of 14.0 times (5 trials). The comparable figure for Italian bees was 2.2. With greater distances from the hive entrance of course the number of stings diminishes regularly; it will suffice to note that at 150 cm from the entrance Brazilian bees stung a mean of 1.2 times, Italian bees 0.2. Cosenza (17, 20) reports on similar tests made on five colonies each of Caucasian, Brazilian, and hybrid bees in Minas Gerais. The balls were jiggled 30 cm from the hive entrances and the time to the first sting and the number of stings in the 60 seconds after the first sting were recorded. Mean times in seconds until the first sting were 229, 14, and 89, respectively, for the three racial groups; mean numbers of stings were 1.4, 34.9, and 10.5. The differences were highly significant; the hybrids were intermediate between the Caucasian and Brazilian bees in each case.

Disturbed Brazilian bees pursue a person or animal that has been stung for distances far greater than is usual for European bees. For the distance followed stung leather balls carried away from the hive by an observer, Stort (77) obtained a mean of 160.2 m ± 40.36 (N = 40 tests, 8 colonies) for Brazilian bees (SP), 21.5 ± 11.67 for Italians (N = 20, 4 colonies), and 38.8 ± 25.12 (N = 15, 3 colonies) for the hybrids.

A particularly dangerous attribute of Brazilian bees is their aggressiveness after slight jarring or vibration of the hive. For example, a colony in Rio Grande do Sul judged to be only moderately aggressive when tested using a modification of Stort's method, was extremely aggressive when similarly tested after the hive had been inadvertently slightly jarred (27, 31). The test-leather (2.5 X 2.5 cm) received 92 stings (it was jiggled for only 5 sec) and the bees followed the leather (and the person carrying it) for over a kilometer!

Within probably every race of bees there is much variation in aggressiveness between colonies. Brazilian bees are said to vary particularly from day to day, depending upon previous and present conditions (B). Some persons report colonies particularly aggressive during a nectar flow. Others report them particularly aggressive when there is little food. Some report the bees more aggressive in hot weather, others in cool weather. Some in the tropics reported greater aggressiveness in humid weather. All agreed on one point: aggressiveness is variable, and a colony that is hard to handle one day may be relatively gentle another day. Some data on aggressiveness at different seasons are available (17).

Readiness to sting must be determined by a number of different and independently inherited responses, e.g. amounts of alarm pheromones produced and responses to alarm pheromones, vibrations, movements, breath of an intruder, etc. Inheritance of aggressiveness is therefore complex and polygenic. Stort (76) indicated the existence of eleven genes controlling aggressive behavior responsible for the difference between Brazilian (SP) and Italian bees. Kerr (31, footnote 32), because of correlations between some of Stort's characteristics, attributed the effects to eight genes.

In addition to the problems around hives or nests of feral Brazilian bees, swarms and absconding colonies constitute a perhaps more severe stinging hazard. Possibly only because of their abundance, swarms land on people and probably on animals from time to time (27, 31). Even a gentle swarm, landing on a person unfamiliar with bees, would probably lead him to attempt to brush them off, inciting stings, in turn probably followed by massive stinging in response to the release of alarm pheromones. There are persistent reports that swarms sometimes attack and sting without provocation, and Smith (73) indicates that in Africa absconding colonies, particularly those which have undergone fusion so that they have several queens, may attack if disturbed, unlike true swarms. There are many reports of animals and even people being killed by stings of Brazilian bees. Some deaths, especially those involving allergic persons, could have occurred had the African bees never been introduced into Brazil. Others resulted from rash efforts to destroy hives in the hysteria arising from overpublicity about the dangers. But there appear to be a substantial number of cases in which animals and men have been severely or fatally stung because of the abundance and special behavioral characteristics of Brazilian bees.

Changes in Brazilian Beekeeping Caused by the Brazilian Bee

Before 1957 bees were often kept in or near settlements, especially in southern Brazil and countries to the south. There was no difficulty in obtaining apiary sites. In the tropics, there was little interest in honey bees.

After the spread of the Brazilian bee, the situation changed. Most beekeepers had to move their apiaries away from settlements (7, 39, 80). Beekeeping as a hobby almost disappeared and many beekeepers went out of business because they did not want to work with such vicious bees or because they could not operate apiaries on farms along with livestock. Others moved their bees to isolated spots where they could not disturb animals. Although in many areas beekeepers became unpopular, commercial beekeepers especially in southern Brazil mostly feel that the Brazilian bee is superior to any race they have had before. Even in the north this attitude may be developing (for Bahia, see 55). Beekeepers value the large crops of honey that the Brazilian bee produces. In the Recife area in the north, beekeeping has shifted from native stingless bees to honey bees.

Changes in bee management resulting from use of the Brazilian bee in Brazil include *(a)* dispersal of hives into scattered small apiaries (75), *(b)* wide dispersal of hives within apiaries and facing them in different directions to reduce interactions among bees of nearby colonies (6), *(c)* use of large smokers (8, 54) and other special equipment (9, 80). In Bolivia the gas N_2O at night has been recommended to anesthetize the bees while hives are being worked (37, 38, 39). In southern Brazil some beekeepers make large fires on one side of the apiary to blanket it with smoke before working hives.

In southernmost Brazil, in a climate to which the Italian bee is probably well adapted, serious efforts are being made to maintain populations of Italian bees in spite of the feral Brazilian bees (16). This is only possible with queen excluders at the hive entrances to prevent queens in swarms of Brazilian bees from entering. In Argentina and Bolivia also, acceptance of Brazilian bees is far less complete than in the honey-producing states of Brazil. The mutual adaptation of bees and beekeepers is most complete in the more central part of the range where the Brazilian bee has existed for five or more years.

Diseases do not seem particularly prevalent in Brazilian bees and seem to present no special management problems; diseases vary from place to place and reports as to susceptibility relative to European bees vary. Brazilian bees in Minas Gerais remove dead brood from comb more effectively than Caucasian bees and hence may be more resistant to larval diseases; hybrids are intermediate (24). *Acarapis woodi,* an acarine parasite not found in North America, occurs in *Apis* in Brazil (61) and is likely to spread with the Brazilian bee.

Acknowledgments

For reading the manuscript and offering useful suggestions, I thank Drs. H. V. Daly, W. E. Kerr, M. D. Levin, W. C. Rothenbuhler, and O. R. Taylor.

Literature Cited

1. Bavaresco, F. A. Apicultura africana. See Ref. 13, pp. 314–21
2. Beig, D., Pizani, J., Kerr, W. E. 1972. Capacidade estomacal de duas subespécies de *Apis mellifera. Cienc. Cult. São Paulo* 24:464–68
3. Boch, R., Rothenbuhler, W. C. 1974. Defensive behavior and alarm pheromone production in honeybees. *J. Apic. Res.* In press
4. Boggino, P. A. 1969. Consideraciones a la abeja africana en Paraguay. *Gac. Colmenar* 31:137–38
5. Boggino, P. A. 1971. My experiences with African bees in Paraguay. *Am. Bee J.* 111:232
6. Both, A. Nova distribuição de colméias em apiários com abelhas africanizadas. See Ref. 13, pp. 88–90
7. Both, A. Nova situação para apiários com abelhas africanizadas. See Ref. 13, pp. 91–92
8. Both, A. Fumigador, fumaça e sua aplicação correta em colméias com abelhas africanizadas. See Ref. 13, pp. 111–15
9. Both, A. Indumentaria e técnica para trabalhar com abelhas africanizadas agressivas. See Ref. 13, pp. 122–25
10. Caldas-Filho, C. F. 1965–1966. As abelhas africanas e suas híbridas. *Zootecnia* [São Paulo] 13:39–42, 47–51
11. Caldas-Filho, C. F. 1967. As abelhas africanas e suas híbridas conquistam o Brasil. *Geografica* [São Paulo] 16:19–23
12. Caldas-Filho, C. F., Silva, R. M. B. da. 1964. Notas preliminares sobre a *Apis mellifera adansonii. Zootecnia* [São Paulo] 11(2): 9–18
13. Congresso Brasileiro de Apicultura, I. 1972. [Florianópolis, 1970] 332 + 8 pp.
14. Congresso Brasileiro de Apicultura, 2. 1973. [Sete Lagoas, 1972] 299 pp.
15. Cornejo, L. G. 1971. Management technology of the "African" bee. *Am. Bee J.* 111:262–63
16. Cornejo, L. G., de Santis, L., Vidal Sarmiento, J. A., Muller, V. A. 1973. Results of work for Italianization of an Africanized zone with *Apis mellifera adansonii* in Rio Grande do Sul State (Brazil). *Apiacta* 8:117–20; Cornejo, L. G., Muller, V. A. 1970. Informe previo sobre un intento de italianización de una zona africanizado con *Apis mellif-*

era adansonii, en el estado de Rio Grande do Sul (Brasil), *V. Congr. Apicola Prov. Buenos Aires* [Junin], 5 pp.

17. Cosenza, G. W. Estudo comparativo da agressividade da abelha africana, da abelha caucasiana e de suas híbridas. See Ref. 13, pp. 125–28
18. Cosenza, G. W. Estudo dos enxames de migração de abelhas africanas. See Ref. 13, pp. 128–29
19. Cosenza, G. W. 1972. Comportamento e produtividade da abelha africana e de sus híbridas. *Ser. Pesquisa/Extensão* [Sete Lagoas, MG] No. 19:1–8
20. Cosenza, G. W. 1972. Comparação entre a agressividade da abelha africana, da abelha caucasiana, e de suas híbridas. *Rev. Brasil. Entomol.* 16(3):13–15
21. Cosenza, G. W. Comparação entre a produtividade da abelha africana, da abelha caucasiana e suas híbridas. See Ref. 14, pp. 50–52
22. Cosenza, G. W. Melhoramento de abelhas por meio da hibridação e seleção. See Ref. 14, pp. 133–35
23. Cosenza, G. W., Batista, J. S. Morfometria da *Apis mellifera adansonii* (Abelha africanizada), da *Apis mellifera caucasiana* (abelha *caucasiana*) e suas híbridas. See Ref. 14, pp. 53–56
24. Cosenza, G. W., Silva, T. 1972. Comparação entre a capacidade de limpeza de favos da abelha africana, da abelha caucasiana e de suas híbridas. *Cienc. Cult. São Paulo* 24:1153–58
25. Darchen, R. 1973. La thermoregulation et l'écologie de quelques espèces d'abeilles sociale d'Afrique (Apidae, Trigonini et *Apis mellifica* var. *adansonii*). *Apidologie* 4:341–70
26. DeSantis, L., Cornejo, L. G. 1968. La abeja africana *(Apis adansonii)* en America del Sur. *Rev. Fac. Agron. La Plata* 44:17–45
27. *Final report, Committee on the African Honey Bee.* 1972. Washington, DC: Nat. Res. Counc. Nat. Acad. Sci. 94 pp.
28. Garofalo, C. A. 1972. Comportamento à maturidade sexual de zangões de *Apis mellifera adansonii.* In *Homenagem a Warwick E. Kerr,* 177–85. Rio Claro, Brazil
29. Gonçalves, L. S. 1970. *Análise genética do cruzamento entre* Apis mellifera ligustica *and* Apis mellifera adansonii. Tese de Doutoramento, Fac. Med. de Ribeirão Preto. 142 pp.
30. Gonçalves, L. S., Kerr, W. E. Noções sôbre genética e melhoramento em abelhas. See Ref. 13, pp. 8–36
31. Gonçalves, L. S., Kerr, W. E., Chaud Netto, J., Stort, A. C. (transl.). 1973. *Relatório final do grupo de estudo americano sobre as abelhas africanas.* Ribeirão Preto, Brazil: Fac. Med. de Ribeirão Preto. 56 pp.; *Relatório final do grupo americano sobre a abelha africana.* See Ref. 14, pp. 211–68
32. Guy, R. D. 1972. How does the Brazilian bee compare to South African bees? *S. Afr. Bee J.* 44:9
33. Juliano, J. C. Abelha africana no Territorio de Rondônia. See Ref. 14, p. 170
34. Katznelson, M. 1967. Algo mas sobre la indeseable abeja africana. *Gac. Colmenar* 29:212, 214–215
35. Katznelson, M. 1971. The African bee in Argentina. *Am. Bee J.* 111:47
36. Katznelson, M., Naveiro, J., Fuhr, A. 1969. Las abejas africanas en Argentina. *Gac. Colmenar* 31:182, 184, 186, 188
37. Kempff Mercado, N. 1969. Anaesthesia in the management of African bees. *Bee World* 50:49–50
38. Kempff Mercado, N. 1971. Contribucíon a la solución del problema de las abejas africanas. *Gac. Colmenar* 33: 87–92
39. Kempff Mercado, N. 1973. The African bees—contribution to their knowledge. *Apiacta* 8:121–26. Also: Abejas africanas—contribución a su conocimiento. *Publ. Ministero de Agricultura y Ganaderia* [Santa Cruz, Bolivia]. 22 pp.
40. Kerr, W. E. 1957. Introdução de abelhas africanas no Brasil. *Brasil Apicola* 3:211–13
41. Kerr, W. E. 1960. Espécies e raças de abelhas. *Coopercotia* 18(127):29–34
42. Kerr, W. E. 1966. A solução e criar uma raça nova. *Guia Rural* 1966/67:20–22. Also: 1968, *O Apicultor* [Porto Alegre] 1:7–10
43. Kerr, W. E. 1967. The history of the introduction of African bees in Brazil. *Apiculture in Western Australia,* 2:53–55; *S. Afr. Bee J.* 39:3–5
44. Kerr, W. E. 1969. Aspects of the evolution of social bees. *Evol. Biol.* 3:119–75
45. Kerr, W. E. 1969. Melhoramento e genética de abelha. In *Melhoramento e Genetica,* Chapter 14. São Paulo: Univ. São Paulo and Edições Melhoramento
46. Kerr, W. E. 1971. Contribuição à ecogenetica da algumas espécies de abelhas. *Cienc. Cult. São Paulo* 23(Suppl.):89–90
47. Kerr, W. E. 1972. Melhoramento em abelhas. In *Manual de Apicultura,* ed. J. M. F. Camargo, 96–115. São Paulo: Agron. Ceres

48. Kerr, W. E., Barbieri, M. R., Bueno, D. Reprodução da abelha africana, italiana e seus híbridas. See Ref. 13, pp. 130–35
49. Kerr, W. E., Bueno, D. 1970. Natural crossing between *Apis mellifera adansonii* and *Apis mellifera ligustica. Evolution* 24:145–55
50. Kerr, W. E., Fava, J. F. de M. Contribuição para a apicultura migratoria racional no estado de São Paulo. See Ref. 13, pp. 80–87
51. Kerr, W. E., Gonçalves, L. S., Blotta, L. F., Maciel, H. B. Biologia comparada entre as abelhas italianas *(Apis mellifera ligustica)* africana *(Apis mellifera adansonii)* e suas híbridas. See Ref. 13, pp. 151–85
52. Kerr, W. E., Portugal-Araújo, V. de 1958. Raças de abelhas de Africa. *Garcia de Orta* [Lisbon] 6:53–59
53. Kulinčevič, J. M., Rothenbuhler, W. C. 1973. Laboratory and field measurements of hoarding behavior of the honeybee. *J. Apic. Res.* 12:179–82
54. Lavigne, G. L. Fumigador, luvas, coleta de colônias de abelhas instaladas em locais de difícil acesso, transferência e pulverizaçoes. See Ref. 13, pp. 109–10
55. Lavigne, G. L. Sobre a presenca das abelhas africanas *(Apis mellifera adansonii)* na Bahia, Brasil. See Ref. 14, pp. 126–29
56. Lecomte, J. 1961. Le comportement agressif des ouvrières d'*Apis mellifica* L. *Ann. Abeille* 4:165–270
57. Maa, T. 1953. An inquiry into the systematics of the tribus Apidini or honeybees. *Treubia* 21:525–640
58. Mello, M. L. S. 1970. A qualitative analysis of the proteins in venoms from *Apis mellifera* (including *Apis mellifera adansonii*) and *Bombus atratus. J. Apic. Res.* 9:113–20
59. Mestriner, M. A. 1970. *Polimorfismo protéico em sistema haplodiplóide* (Apis mellifera *Linné 1758*). Tese de Doutoramento, Fac. Med. de Ribeirão Preto. 70 pp.; Mestriner, M. A., Contel, E. P. B. 1972. The *P-3* and *EST* loci in the honeybee *Apis mellifera. Genetics* 72:733–38
60. Mountain, P. 1973. Honey production in South Africa. *Apiacta* 8:141–44
61. Nascimento, C. B., Mello, R. P. de, Santos, M. W. dos, Nascimento, R. V. do, Souza, D. J. de. 1971. Ocorrência de acariose em *Apis mellifera* do Brasil. *Pesq. Agropec. Bras., Sér. Veterinária* 6:57–60
62. Nogueira-Neto, P. 1964. The spread of a fierce African bee in Brazil. *Bee World* 45:119–212
63. Nogueira-Neto, P. Notas sôbre a historia da apicultura brasileira. See Ref. 47, pp. 17–32
64. Núñez, J. A. 1973. Estudo cuantitativo del comportamiento de *Apis mellifera ligustica* Spinola, y *Apis mellifera adansonii* Latreille: Factores energeticos e informacionales condicionantes y estrategia del trabaja recolector. *Apiacta* 8:151–54
65. Oksman, M. 1970. Algunas reflexiones sobre el problema de las abejas "africanas." *Gac. Colmenar* 32:98–100, 102, 104–6
66. Portugal-Araújo, V. de 1956. Notas bionomicas sobre *Apis mellifera adansonii* Latr. *Dusenia* 7:91–102
67. Portugal-Araújo, V. de 1971. The central African bee in South America. *Bee World* 52:116–21
68. Portugal-Araújo, V. de 1973. *Projecto de desenvolvemento apícola para o planalto central.* Missão de Extensão Rural de Angola. 69 pp.
69. Rinaldi, A. J. M., Pailhé, L. A., Popolizio, E. R. Indices alares, tarsales y glosales (en 3 razas de abejas). See Ref. 13, pp. 228–33; 1971. *Publ. Univ. Nac. Tucumán, Fac. Agron. Zootec.* no. 1058:1–10, Misc. no. 41:1–10
70. Ruttner, F. 1968. Les races d'abeilles. *Traité de biologie de l'abeille,* ed. R. Chauvin, 1:27–44. Paris: Masson et Cie
71. Smith, F. G. 1958. Communication and foraging range of African bees compared with that of European and Asian bees. *Bee World* 39:249–52
72. Smith, F. G. 1958. Beekeeping observations in Tanganyika 1949–1957. *Bee World* 39:29–36
73. Smith, F. G. 1960. Beekeeping in the tropics. London: Longmans. 265 pp.
74. Smith, F. G. 1961. The races of honeybees in Africa. *Bee World* 42:255–60
75. Sommer, P. G. Observações em apiarios no estado do Paraná, colhidas no periodo de 1942–1972. See Ref. 14, pp. 130–32
76. Stort, A. C. G. 1970. *Genética da agressividade em abelhas.* Tese de Doutoramento, Fac. Med. de Ribeirão Preto. 166 pp.
77. Stort, A. C. Metodologia para o estudo da genetica de aggressividade de *Apis mellifera.* See Ref. 13, pp. 36–50
78. Stort, A. C. Relações entre caracters do comportamento agressivo e caracteres morfológicos de abelhas do gênero *Apis.* See Ref. 28, pp. 275–83
79. Townsend, G. F. 1973. The African bee *(Apis mellifera adansonii). Apiacta* 8:105–6

80. Wiese, H. 1970. Abelhas africanas, suas caracteristicas e tecnologia de manejo. *Proj. Apicultura, Ministerio de Agricult. Florianopolis,* pp. 1–15; See Ref. 13, pp. 95–108

81. Woyke, J. 1973. Experiences with *Apis mellifera adansonii* in Brazil and in Poland. *Apiacta* 8:115–16

INSECT GROWTH REGULATORS WITH JUVENILE HORMONE ACTIVITY[1]

❖6097

G. B. Staal
Research Laboratory, Zoecon Corporation, Palo Alto, California 94304

The term insect growth regulator (IGR) was designed to describe a new class of bio-rational compounds. Through greater selectivity of action these compounds appear to fit the requirements for third generation pesticides, such as the absence of undesirable effects on man, wildlife, and the environment and compatibility with modern insect pest management principles. Although the majority of new compounds in this class are the result of research on insect juvenile hormones, the general term IGR would also accommodate compounds with a different mode of action. The scope of this review, however, is restricted to those IGRs that produce only or primarily juvenile hormone-type responses. The nomenclature of the chemicals and mixtures with JH activity has given rise to much confusion, and terms such as synthetic JH have occasionally been misused. The terms juvenoid, JH mimic, JH analog (JHA), etc all describe the general class of compounds fairly well. In this review we will use the term juvenile hormone (JH) for the natural hormones, IGR when compounds are used with an intention to control pests, and on occasion JHA for less defined situations. IGR, JH, and JHA may therefore occasionally be used for the same compound.

Less than ten years have elapsed since the chemical structure of the first insect juvenile hormone (JH I) was elucidated in 1967. This was followed by the positive identification of two closely related structures (JH II and JH III, see Table 1). Their presence in various combinations was subsequently proven in various insects, primarily Lepidoptera (82). Schmialek in 1961 (180) was the first to elucidate the JH active isoprenoids farnesol and farnesal in excrement of *Tenebrio molitor.* The low specific activity of these structures made it unlikely, however, that these were actual hormones. It is remarkable that Bowers, Thompson & Uebel (18) synthesized the compound that later became known as JH III as early as 1965, recognized its biological activity, and suggested that it had characteristics of a hormone. Industrial and federal involvement started a great flood of chemical analogs of the juvenile

[1]Contribution No. 21 from the Research Laboratory of Zoecon Corporation.

hormones following the identification of JH I, but several compounds with JH activity were known earlier. Some were even tried in small-scale experiments to evaluate insect control potential. The existence of the juvenile hormone and its source, the corpora allata, was recognized in the early 1930s. Since then a great many studies have been devoted to its role in insect development and a surprising diversity of functions has become known. Yet, the details of the mode of action, the intracellular receptors, the mode of transport in the insect body, the mechanism of the regulation of the corpus allatum secretion, and many other aspects are even now not fully known. Nor has it become clear what the reasons are for the selectivity of several compound types for insect orders or even families. However, certain empirical conclusions can be drawn from long lists of structures compiled with scarcely more than empirical rationale. A discussion of structure-activity relationships is outside the scope of this review. The reader is referred to other reviews, such as those by Slama (193, 194), Jacobson et al (77), Bowers (15), and Pallos & Menn (138).

What makes a juvenile hormone or synthetic analog into a successful pest control agent? The reason cannot be very different from the rationale behind other pest control agents: It must be able to reduce insect populations to below the level of economic damage at a cost below the value of the damage prevented, or to reduce insects of medical or nuisance importance with the funds available. Insect growth regulators do seem to lack most, if not all, of the undesirable properties of current control agents, but the demonstration of actual efficacy is complicated and time consuming. The practical realization of IGRs as operational items will depend on several additional socioeconomic considerations.

Commercial development of any compound requires reasonable certainty that there will be a satisfactory return on investments. The size of the initial investment required, caused to a considerable extent by government imposed regulatory requirements, necessitates the protection of proprietary rights for a given compound structure. Only the easing of government requirements or the provision of financial inducements can change this situation (47).

This review will consider those aspects of the mode of action of structures with JH activity that have a relation to control. For the more theoretical aspects the reader is referred to other reviews such as those for biochemical aspects (254), morphogenetic action (236, 242), reproduction (29, 50), bioassay (57, 194, 207), physiological aspects (54, 130, 238), and behavior (163, 166, 216).

MODE OF ACTION OF INSECT GROWTH REGULATORS

Morphogenetic Effects

Abnormal morphogenesis of the integument is generally irreversible and is the first and most readily observed aspect of IGR action upon insects. Most species will respond to IGR treatment by producing extra larval, nymphal, or pupal forms which may range from giant, but almost perfect forms, to a scale of intermediates between immature and adult forms. Perfect extra larvae or nymphs consume additional food and therefore increase damage, but the extra damage always has a limit, even when several extra stages are formed (185). Less perfect intermediates often succumb earlier because ecdysis is impaired. Generally, the period of greatest sensitivity for metamorphic inhibition is the last larval or nymphal instar (and the pupa in Holometabola), but exceptions occur (penultimate instar in Homoptera). The extent and character of the response depends on the species, the time of the application, the dose, the mode of administration, and the type of compound. Since the several formative processes in an insect are not all sensitive at exactly the same moment, a longer duration of exposure yields a more complete inhibition. In addition, there is good evidence that a superseding early channeling mechanism exists in lepidopterous larvae that determines whether or not metamorphosis will occur. Thus, only an above-threshold dose of JH, applied before a critical moment, can program the formation of perfect superlarvae, while later or lower dosages can only partially inhibit the previously programmed metamorphic processes (186, 187). Holometabola seem to possess a greater ability to compensate for superlarvae through a more extensive redifferentiation and a varying ability to postpone pupation temporarily when exposed to JH than is the case among Hemimetabola.

Metamorphic defects can also result in indirect mortality through the impairment of sensory functions, behavior, feeding, etc. The inhibition of metamorphosis of internal organs or functions may be causally related to otherwise unexplained mortality as a result of IGR application. It is not known whether these inhibitions are intrinsically more reversible than are those of the exoskeleton (53, 83, 84, 147, 182, 184, 188).

Effects on Embryonic Development

The interference of exogenous IGRs with embryonic development has not been prognosticated by prior endocrinological experience. Its mechanism of action is still obscure. There is no doubt, however, that a true growth effect is involved since death is not immediate as would be the case with toxicants such as organophosphates, and since the period of sensitivity is restricted to the early stages of embryonic development. While the final result is nonemergence, only detailed observations can reveal effects which differ from those produced by conventional toxicants. The most potent inhibitors of metamorphosis are usually very potent inhibitors of embryonic development as well, but there is no strict correlation between these two properties in the range of currently known IGR compounds.

Many IGRs are far more potent than current insecticides against eggs, but only when applied before the blastokinesis phase. Hydroprene (see Table 1), for instance, has an LD_{50} of 0.00003 μg per egg in *Samia cynthia,* whereas no insecticide tested showed a value lower than 0.01 μg (G. B. Staal, unpublished observations).

Disruption of embryogenesis by IGRs has been demonstrated in Lepidoptera (161, 162, 167), Thysanura (108, 169), Orthoptera (131, 132, 197), Coleoptera (224), Hemiptera (106, 107), and Homoptera (121). Diptera appear to be much less or not at all sensitive in this respect. Unfortunately, practical applicability is restricted, since only very young eggs are sensitive. Substrate applications would be practical but these fail for reasons not fully known. Either the proteinaceous glue barrier or the apical location of the micropyle may be responsible. However, the vapors of IGRs are known to have effect and could be practical in enclosed situations (223).

Applications after blastokinesis often have resulted in delayed effects in further ontogenesis, e.g. the occurrence of extra larval or nymphal instars similar to those obtained by contemporaneous JH applications in the last larval instar (76, 103, 159–162). This programming effect is explained by Riddiford (160) as the result of interference with sequential determination of ontogenesis during embryogenesis. The effect is probably on the level of JH secretion rather than on the level of the receptor tissues (165, 243). This aspect has not been investigated for control purposes. Riddiford (162) states that the dose required to affect programming is three to ten times higher than that required to prevent hatching.

Effects on Reproduction

The embryonic effects produced by IGRs on direct topical application to eggs can, in principle, be obtained by application to gravid females, but unfortunately the dose required often appears to be too high to be practical (9). Hopes were raised by early experiments on *Pyrrhocoris apterus,* where complete sterilization of females required only 1 μg of compound D (104). This amount can even be repeatedly transferred to females by a treated male. This stimulated interest in the so-called venereal approach which superficially resembles male sterility techniques (103, 105). Since JH fulfills an essential gonadotropic function in most insects (the exception is the group of species that mature their eggs prior to adult emergence), it follows that sterilizing doses must be larger than the endogenous levels necessary to ensure gonadotropic function. Low IGR doses may result in slightly increased fecundity for the same reason (19). Yet, the doses required for female sterilization often seem to compare very favorably with those of alkylating chemosterilants, which require 2000 to 20,000 times higher doses and possess other unwanted side effects prohibiting their use under field conditions (91).

Complete inhibition of reproduction was shown to result from low doses of IGRs applied during the differentiation of follicular epithelial cells prior to adult ecdysis in a wide range of species (101, 102, 109, 168).

Male sterilization, although sometimes observed (213), does not appear to be a very obvious IGR effect. However, Deseö (45) found a reduction of fecundity through the treatment of male codling moths with 1 μg of farnesyl methyl ether.

External and internal morphogenetic effects can influence mating and other reproductive functions directly or indirectly. A well-known example is the inhibition of male genital rotation in Diptera, leading to inability to mate. Sometimes only a pronounced shortening of adult longevity seems responsible for reduced fecundity.

Effects on Diapause

The variation among diapausing stages in different species and the variety of induction and recovery mechanisms explain why the effects of exogenous JH on diapause are difficult to generalize.

EGG DIAPAUSE JH is not known to influence this type of diapause. It usually occurs in more or less mature embryos which have developed beyond the blastokinesis stage.

LARVAL DIAPAUSE Convincing evidence is available that this phenomenon is, if not induced, then at least characterized by high endogenous JH levels (35, 256). Consequently, exogenous JH has no effect. In several nondiapausing species, particularly Coleoptera, exogenous JH may delay pupation, a situation that resembles spontaneous larval diapause. The degree of delay induced seems to be related to final mortality, but exact data are not available.

PUPAL DIAPAUSE Although JH apparently has no direct function in this phenomenon, injected JH can break pupal diapause and simultaneously affect metamorphosis. This effect is referred to as ecdysiotropic or prothoracotropic and is thought to result from stimulation of the prothoracic gland to produce molting hormone (56, 89). The phenomenon is probably not of practical significance since exogenous applications would require too high a dose.

ADULT REPRODUCTIVE DIAPAUSE Adult reproductive diapause may be wholly or primarily induced by JH deficiency and can therefore easily be disrupted by exogenous JH application and active IGRs (10, 16, 37, 46, 51, 67, 75, 85, 126, 140). The reversibility of this disruption is questionable (181), and some experiments (Staal, unpublished observations) indicate that even a low IGR dose can result in mortality without a return to diapause in males as well as females. The number of eggs produced appears to be quantitatively related to the dose applied, indicating a difference with natural diapause termination. The significance of the phenomenon is in the stimulation of out-of-season reproduction, thus decreasing successful subsequent reproduction even if the adults should return to the diapause condition. In the induction of diapause, a different mechanism may be involved that in itself need not be JH sensitive.

Effects on Polymorphism

Virtually all caste, morph, form, and phase determinations in insects appear to be under some form of endocrine control. Often JH exerts a function in these processes that is fully independent of its function in metamorphosis and reproduction. The possible consequences for insect control will be discussed in later chapters.

Effects on Behavior

A profusion of specialized behavioral aspects may be influenced directly or indirectly by JH. Disturbances of these relationships may vary from being inconsequential or reversible to being totally disruptive. Specific studies under field conditions will be required to show their significance for practical control.

THE STRUCTURAL TYPES OF INSECT GROWTH REGULATORS

Even a limited review of all active structures, whether obtained by synthesis or by plant or animal extraction, is not within the scope of this review. Compounds like the paper factor (juvabione, methylester of todomatuic acid; 17) and the peptide analogs (149, 257) are interesting curiosities, but because of their selectivity for Pyrrhocoridae they are not of practical importance. Farnesyl methyl ether was one of the earliest compounds with JH activity that was used intensively in research. Although many bona fide JH effects could be faithfully induced with this compound, no practical utility developed because of its low specific activity on most insects and its chemical instability. Other compounds with activities too low to be of more than academic interest are dodecyl methyl ether and various compounds known for their insecticide synergism like sesoxane, piperonylbutoxide, etc (14).

As will be discussed elsewhere, relatively little biochemical rationale is available for predicting the selective activity for a given structure. Most active structures have been synthesized according to a multitude of different ideas that have thus far resisted integration into a comprehensive theory. Whereas insect activity certainly depends on features such as chain length, geometric configuration, functional groups, polarity, etc, extra requirements for field stability, etc may require structural modifications that could be antagonistic to the activity principles. The identification of other natural insect juvenile hormones, if they exist, could give new impetus to the search for better compounds, although the hormones themselves can at best only serve as a lead for synthesis. For several insect groups such as Orthoptera, we do not have compounds with a morphogenetic activity high enough to be of practical value. This may be due either to intrinsic physiological peculiarities or to our failure to find the suitable IGR structures.

In Table 1 only specific compounds and representatives of structure types that have been used in insect control experiments in the laboratory or in the field are listed.

A. Natural juvenile hormones: JH I is usually the most active of the three known juvenile hormones, particularly in Lepidoptera, but JH II and JH III approach its activity in several insects. Pure JHs have not been widely used in field experiments, with one major exception (6), because of their cost of synthesis and field instability. A crude isomeric mixture containing ca 10% of JH I (ttc) has been commercially available. The natural JHs are methyl esters, whereas synthetic ethyl esters usually exhibit higher activity.

B. A large group of terpenoid structural analogs of the JHs that was never used extensively in field experiments but that has played a significant role in the development of improved IGRs. Active modifications differ in saturation of positions 6,7 and 10,11, and may have different substituents at C-7 and C-11. Ester functions may be replaced by thio-esters, amides, or keto-functions. These often potentiate activity for specific insects.

C. This reaction mixture, at one time confusingly named synthetic juvenile hormone, contains a variety of chlorinated farnesoate esters in addition to unchlorinated compounds which makes the mixtures slightly less specific than the pure major components (D and E). On most economic pests the activity is insufficient and the undefined and irreproducible composition of the mixture is not reconcilable with serious work.

D. Esters (and amides) of this structural type are quite specific for certain groups of Hemiptera and are generally very poorly active on Holometabola. Most work with

Table 1 Representative structures of IGR types

Code	Structure	Description	Other Names	References
A1	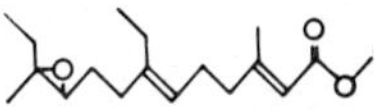	Natural juvenile hormone and its ethyl ester analog	JH I	170, 171
2	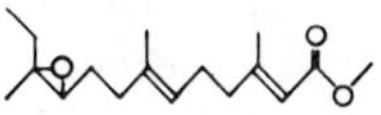	Natural juvenile hormone and its ethyl ester analog	JH II	114
3		Natural juvenile hormone and its ethyl ester analog	JH III	82, 18
B	 O–R	Analogs of 3,7,11-trimethyl-2,6,10-dodecatrienoic acid	Derivatives of farnesol and farnesoic acid	several
C	–	Mixture of hydrochlorinated farnesoates, containing B, D, and E	Law et al mixture	92
D	 Cl Cl	Ethyl or methyl 7,11-dichloro-3,7,11-trimethyl-2-dodecenoate	–	172
E	 Cl	Ethyl or methyl 11-chloro-3,7,11-trimethyl-2,6-dodecadienoate	–	80
F	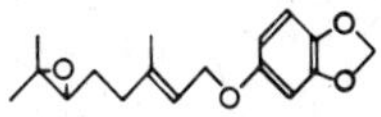	6,7-Epoxygeranyl 3,4-methylenedioxyphenyl ethers with variations in the geranyl chain	Hoffman LaRoche RO 20 3600	22
G	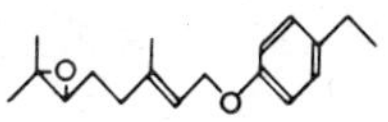	6,7-Epoxygeranyl 4-ethylphenyl ethers including various other ring substitutions	Stauffer R 20458	138, 139
H1	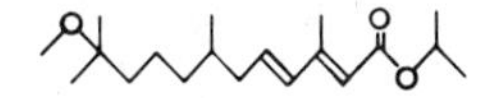	Isopropyl (2*E*,4*E*)-11-methoxy-3,7,11-trimethyl-2,4-dodecadienoate	Methoprene, Altosid® IGR Zoecon ZR-515	64
2	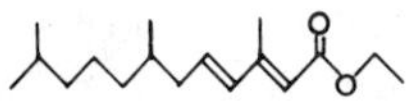	Ethyl (2*E*,4*E*)-3,7,11-trimethyl-2,4-dodecadienoate	Hydroprene, ALTOZAR® IGR Zoecon ZR-512	64
3	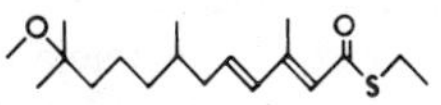 S	*S*-Ethyl (2*E*,4*E*)-11-methoxy-3,7,11-trimethyl-2,4-dodecadienethioate	Triprene Zoecon ZR-619	64
4	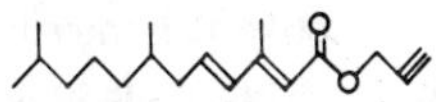	2-Propynyl (2*E*,4*E*)-3,7,11-trimethyl-2,4-dodecadienoate	Kinoprene Zoecon ZR-777	121, 208

these compounds has been done in Czechoslovakia. It was also included in field studies by Bagley & Bauernfeind (6).

E. This monochlorinated compound has a higher activity on Holometabola and a lower one on Pyrrhocoridae than compound D. Reported field activity was often better than could be expected from laboratory research, which could perhaps be related to a greater field stability than that found in many other compounds.

F. Many variations of this structure have shown good activity on Holometabola and Hemimetabola alike. Most active are the 3,7-diethyl analogs, but the dimethyl analog is easier to obtain. This compound, also included in many field studies (6), may lack residual activity under field conditions.

G. The advantages of this type of compound are found in more economic synthesis, a greater stability, and a slightly narrower selectivity than that of the compounds of type F. A variety of field trials have been reported. Degradation pathways in animals and in the environment have been studied in relation to registration requirements (58).

H. Structures of type A, B, F, and G require a terminal epoxide function for good activity but this is detrimental to field stability. Structures of type H show higher biological activity without epoxides at C-10,11 (64). Their field activity, specificity, degradation, toxicology, etc have been intensively studied, particularly for compound H_1 (150, 248), which is currently the only IGR with an experimental label approved by the Environmental Protection Agency in the United States (for control of floodwater mosquitoes) (260).

Geometrical Isomers of JH

Dahm, Röller & Trost (44) showed that the naturally occurring isomer of JH I had the configuration of *trans*-2, *trans*-6, *cis*-10. Earlier it had been found that the *trans*-2, *trans*-6 isomer of farnesol was by far the most active isomer (255).

Schwieter-Peyer (183) tested all eight isomers of JH I (racemic mixtures) and found the natural hormone (ttc) to be the most active on the morphogenesis in three different insects. It was futhermore concluded that the *trans*-2 and the *trans*-6 bonds were the most essential for high activity, and the *cis*-10 isomer was almost inactive. Surprisingly enough the tcc isomer showed highest sterilizing activity in *Dysdercus*, which the author attributed to a slower degradation rate. The C-10 enantiomerism, on the other hand, did not seem to affect the activity in compound A_3 (211), although the natural JH I (A_1) is not a racemic mixture (112, 113). Loew & Johnson (97) synthesized fairly pure enantiomers of JH I and found the + enantiomer to be at least six to nine times more biologically active than the – enantiomer. Specimens of 100% enantiomeric purity have not been produced as far as known, thus an accurate evaluation must wait. The chemical synthesis of JH analogs can often use methods directed at high stereo specificity. The economics of commercial synthesis usually do not allow for complete isomeric purity, but in the case of 2,4-dienoates, a production scale purity of greater than 90% of the active tt isomer is routinely achieved. However, it has been found that in sunlight some formulations of methoprene easily rearrange to a 50–50 mixture of *cis*-2 and *trans*-2 isomers, thus reducing the efficiency somewhat under exposed conditions.

PRACTICAL RESULTS OBTAINED WITH INSECT GROWTH REGULATORS

Diptera

CULICIDAE (MOSQUITOES) In this group IGR development has made the most progress due to the early discovery of sufficient intrinsic activity under laboratory conditions, the realization that adults are the only harmful stage, the short larval life cycle, and the remarkable degree of synchronization in development found in periodically flooded habitats. Pressures of increasing resistance and concomitant increases in environmental hazards with current materials have also contributed to the rapid development.

As early as 1964 Lewallen (94) indicated the specific effect and the subsequent mortality produced by 1 ppm of farnesol to fourth instar *Culex* larvae, but noted that this dose came close to that producing immediate toxicity—clearly not a typical JH effect. Similar results were obtained by Spielman & Williams (203) in 1966 with compound C on *Aedes.* The sensitive period was clearly identified as the later part of the last larval instar, the pupae being almost completely insensitive. These authors noted that the genitalia of males that emerged following exposure to only partly effective doses failed to undergo complete rotation. This process normally occurs after adult ecdysis in mosquitoes (before ecdysis in cyclorrhaphous Diptera). This rotation is essential for mating.

Spielman & Skaff (201) analyzed the sensitive periods and the nature of the effects in detail in *Culex* as well as in *Aedes.* They noted that the main effect was rather unspecific: nonemergence of the adult from the pupal case. In marginally effective doses, the adults often started to emerge but failed to extricate themselves completely from the pupal cuticle. Even with newer, more potent compounds the effects are essentially the same, in that no clear-cut inhibitory effects on the external metamorphosis can be recognized on the adults that fail to emerge. The mechanism of the process of inhibition of emergence is unknown. Therefore, the recognition of nonemergence as a JH effect is based on circumstantial evidence. Compounds such as the natural JHs (A) and active analogs that produce clear-cut JH effects in other insects usually evoke the nonemergence syndrome in mosquitoes while inactive chemical analogs do not (Staal, unpublished observations). However, some chemicals (e.g. Monsanto 585) may produce the nonemergence syndrome at high doses without producing JH effects in other insect species and without being direct toxicants (174, 209). Whereas concentrations of approximately 10 ppm of compound C inhibited adult development, higher concentrations applied to earlier stages inhibited pupation (200, 231). This, however, may also have been due to the presence of toxic compounds in this mixture of uncertain composition.

In what may have been the first large-scale trial with an IGR on mosquito larvae, Spielman (200) noted instability of the preparation under field conditions and its lack of effectiveness against larvae that were not fully synchronized. The author later provided evidence that perhaps the larvae themselves, before the period of maximum sensitivity, are one of the important factors in the degradation of compound C (202) and suggested that combinations with insecticide synergists could improve this situation. His prediction that more potent mosquito control IGRs would soon be synthesized came true and we are at present in a better position to evaluate the practical efficacy of these compounds.

Other compounds that are active against mosquito larvae are compounds of type F (79, 96, 179), type G (21), and many analogs more closely related to the natural hormones, culminating in the rather selective 2,4-dienoate methoprene (H_1) with higher morphogenetic activity on Diptera than most other known IGRs (64, 178,

179). The LD_{50} for this compound under laboratory conditions is 0.1 ppb when applied on late last instar larvae of *Aedes aegypti* and ca 1 ppb on larvae of *Culex pipiens.* Some species selectivity, up to two orders of magnitude, was registered for several compounds (79). It is likely that *Culex* spp. are intrinsically less sensitive to IGRs than, for instance, *Aedes* spp. We should consider also that the duration and the timing of the sensitive periods are not the same for all species according to Schaefer & Wilder (179), and that differences in life style may account for a different exposure to compound residues partitioned out in the water boundary layers or on food particles. Under field conditions the microbial degradation capacity of different water biotopes is also vastly different.

Methoprene (as most other IGRs) is unstable in sunlight, under high water temperatures, and in water with high microbial populations (176–179). Economical mosquito larval control, therefore, only became a reality after the development of protected, slow release formulations such as solid polymers (48) and microcapsules (e.g. Altosid ® "SR 10"; 60, 179). This brought the dose required to provide good control of floodwater mosquitoes down to 1/60 lb per acre (A. I.). A full quantitative analysis of the success factors in this formulation has not been published, but it is known that the microcapsules are readily ingested by the larvae and it is likely that the slow release of methoprene counteracts microbial degradation, sunlight exposure, and unfavorable surface partitioning in the water (176). (The actual solubility of methoprene in water is ca 1.5 ppm.) This must result in a better exposure of asynchronous larval populations.

Methoprene, at present, is the only material sufficiently active for consideration as a practical control agent against the California floodwater mosquito *Aedes nigromaculus,* a species that has become very resistant to most insecticides (24, 177, 178). The SR 10 slow release formulation of methoprene did not always perform well against *Culex* in sewage water for reasons which are as yet undetermined (38). In the course of its commercial development methoprene has been shown to be almost without effect on nontarget organisms in aquatic environments, with perhaps the exception of a few other aquatic Diptera (118). At present the material has an experimental registration for use against floodwater mosquitoes in the United States. Broadening of this label to other species in other environments is to be expected. The near- and long-term vertebrate toxicology performed so far justifies the claim that this IGR is one of the safest materials ever used for insect control (260).

The effects of IGRs on stages other than late larvae appear as rather insignificant compared with the effects on adult emergence. Eggs of *A. aegypti* as well as *C. pipiens* are insensitive even to methoprene (Staal, unpublished observations).

Effects on reproduction of *Aedes* have been obtained by Patterson by topical application of a JHA to engorged adults in doses of 0.25 to 0.5 μg, (141, 142) and by surface residues of JH II (143). The reversibility of this observed temporary and incomplete reduction in fertility indicates that this facet may not be of practical consequence. Bransby-Williams (21) and Wheeler & Thebault (231) failed to produce significant reduction of fertility in *Culex* by spraying 10% solutions of compound G on adults or by applying compound C to adults and eggs.

MUSCIDAE (HOUSE, STABLE, AND OTHER FLIES) The inhibitory activity of IGRs on adult emergence can be demonstrated easily through topical application to wandering larvae and untanned puparia, as well as by treatment of food and pupariation media of muscid larvae (3, 11, 100, 204). As in mosquitoes, extra larval instars cannot be produced by JH alone [molting hormones can produce supernumerary larval cuticle by initiating molting before redifferentiation takes place (259)]. The sensitive period is clearly confined to the above mentioned stages in all cases, but if windows were excised in the puparium wall it could be proven that the actual sensitive period is longer (204). The inhibitory effect of exogenous JH on emergence is very similar to that on mosquitoes, but in this case bristles on the adult abdomen may show inhibition in their morphogenesis as a quantitative dose-related effect in conjunction with effects on internal structures (11). Although the emergence of adults is generally correlated as a dose-response effect with the degree of bristle inhibition, there are indications that this need not always be so. For one of the most active IGRs, methoprene, the nonemergence activity is extremely high, but the corresponding morphogenetic inhibition activity is not outstanding. It is too early, however, to conclude that nonemergence is not a consequence of some form of morphogenetic inhibition. In other experiments the true JH-mimicking properties of this compound in house fly gonadotropic activity have been proven beyond doubt (2). Pronounced ovicidal effects with IGRs on topical application have not been observed for Muscidae, and applications to adults have generally been correspondingly disappointing. Matolín notes, however, that deferred programming effects may occur after treatment of eggs, resulting in abnormalities occurring during the larval-pupal transformation (107).

The general selectivity of different analogs is very similar in flies and mosquitoes, and specifically the same three types of compounds mentioned for mosquitoes have shown laboratory activity against various fly species (78, 246, 247, 251, 252).

Incorporation of methoprene into the rearing medium at 10 ppm showed effective control against several strains of *Musca domestica,* but against different insecticide-resistant strains concentrations of up to 250 ppm were required (78). Incorporation in the medium is not necessarily the most effective route of administration, since the larvae are most susceptible when they have ceased feeding, and fly-breeding medium usually teems with microbial life. Fly larvae generally tend to leave the larval substrate only when this is very moist. In this case, treating surrounding, drier, pupariation substrates appears to be more efficient.

Even more sensitive to IGRs than *M. domestica* are the horn fly (*Haematobia irritans*), the face fly (*Musca autumnalis*), and the stable fly, (*Stomoxys calcitrans*). But, there are substantial relative differences in susceptibility towards the different types of compounds or variations within one type of compound for these species (36, 63, 250)

The results of field tests have been fairly poor when stable and house fly substrates, such as feedlot litter or poultry manure, were superficially sprayed with IGRs (27, 249), presumably because penetration and distribution through these media is poor. This is of particular importance when the larvae pupariate in these media instead of around it, for in the latter case the chance of contact with a treated zone increases.

The same authors obtained excellent stable fly control in natural substrates, such as eelgrass and other beach debris, in Florida. The poor results obtained by topical treatments of manure have inspired studies on oral administration of these compounds to cattle and poultry. This feed through method is feasible for far more toxic insecticides (116). If the IGR survives the digestive tract, mixing through the manure can be assumed to be homogenous. This has proven to be a successful procedure (63, 117), the practical realization of which will depend primarily on considerations of cost and safety of the animals and of the safety of the animal products for human consumption. A dose of 0.7 mg of methoprene per day per cow provided full control of horn flies, while 100 mg per day per cow was required to control stable flies. The residual effect of this treatment was surprisingly good, in that flies could not develop in manure collected up to eight days after the last feeding of the IGR. Since the early experiments were all done with simple emulsifiable formulations, it is likely that suitable formulations could improve the efficacy of the passage through the alimentary tract and in the manure.

Finding an economical and reliable way of continuous administration to cattle is a complicated assignment. Incorporation in water or food is not always practical; mineral licking blocks may not be frequented equally by all individuals in a herd.

The quantity of unformulated methoprene required for fly control in the manure of poultry is far higher, being greater than 25 ppm in the food (65), but may be reduced through proper formulation. Fly control in poultry houses is of interest to the poultry farmer; flies are an indirect nuisance to him and to surrounding communities rather than to the poultry itself. The producers usually spend a lot of money and effort in fly control, through the use of chemicals, frequent clean-ups, the construction of enclosed houses, and manure-treating facilities.

Phytophagous flies have not been the subject of much IGR research, although several species are of regional importance. Skuhravý & Sehnal (191) investigated IGR treatment of gall midges (*Dasineura,* Cecidomyidae) and found compounds F and H to be the most active. Van de Veire & De Loof (217) found that compounds of type A and G had ovicidal activity on eggs of *Hylemyia brassicae* (Anthomyiidae) at fairly high application rates and had nonemergence activity on young pupae as well. This gives some promise for control purposes.

SIMULIIDAE (BLACK FLIES) Under laboratory conditions several species of *Simulium* were found to be affected by rather high concentrations of IGRs in water (approximately 1 ppm of compound G and of methoprene, 41). Control of these species under field conditions in running water could prove to be very difficult.

CHIRONOMIDAE (MIDGES) Norland (127) found larvae of two midge species fully susceptible to methoprene at 0.05 ppm in outdoor experiments. Mulla et al (119), in an extensive study on several midge species, found methoprene considerably more active than compounds F and G. Protected, slow release formulations of methoprene prevented adult emergence in different simulated and actual field experiments at doses below 0.1 ppm or 0.3 lb per acre. Some cross resistance in an organophosphate-resistant strain was observed.

As is the case with mosquitoes, the direct effects on adult reproduction and eggs seem insufficient to consider use of IGRs for control purposes.

We may conclude that the use of IGRs, with their low vertebrate and nontarget toxicity, offers good prospects for the control of Diptera, particularly those associated with cattle or poultry operations and in the aquatic environment.

Lepidoptera

The sensitive periods are very clearly restricted to the last larval instar and the beginning of the pupal stage. Effects obtained by treating earlier are probably related to persistence of the compound in some form in or on the substrate. Intermediates between larvae and pupae have been obtained rather generally. They are usually nonviable and result from treatments made after the first one or two days in the last larval instar, up to the initiation of spinning (pre-pupae). Treatments early in the last larval instar may produce seemingly normal extra larval or supernumerary instars that may continue to feed and thus cause some increased damage. These supernumerary larvae rarely produce normal pupae and adults, presumably because they are usually not completely larval on more detailed inspection. Although it appears that most of the last larval instar is sensitive to IGR application, ideally the best procedure would be to treat late in the instar or, if this cannot be achieved, to avoid overdosing.

The lepidopteran pupa usually has an IGR sensitivity that is approximately equal to or higher than that of the larva, but this is recognized only when the application takes place during the first few days of pupal development. Adult development may be temporarily suspended due to a pupal diapause, which, if it occurs, usually precedes adult development. Some cases are on record in which larval treatment resulted in the production of normal pupae which failed to produce adults, presumably because the induction period of adult development in these cases was considerably more sensitive or the application was made after the sensitive period in the larva had been passed.

NOCTUIDAE (OWLET MOTHS) This family probably includes the greatest number of lepidopteran pest species. The larvae may feed on leaves, in seedpods, fruits and buds, on roots, etc and are therefore not always continuously accessible to chemical applications. However, they often migrate to pupation sites exactly at a time of maximum sensitivity to IGR treatment. Benskin & Vinson (8) state that *Heliothis virescens* late larvae and eggs were the most sensitive stages (towards compounds F and JH II). The effective doses (micrograms per specimen for both eggs and larvae) appear to be excessive. Newer compounds will hopefully be more efficient. Most important is their observation that a dose can be reduced by a factor of ten when the application is repeated at daily intervals. This strongly supports the case for possible success of slow release formulations. Bagley & Bauernfeind (6) report that field treatments with JH I and compounds E and F at one to two pounds per acre yielded no immediate positive control of *H. virescens* larvae. However, they also observed prolonged larval periods, supernumerary instars, and failure to pupate in a fraction of the populations. This suggests that longer term population suppression may be feasible if a satisfactory economical delivery system can be developed.

Laboratory results indicate that some compounds belonging to class H, when continuously present in the food medium, are up to 1000 times more active than compounds in the classes A, B, F, and G (Staal, unpublished obervations). Continuous feeding produces reproducible results in which the timing of the application is eliminated as a complicating factor. The relevance of the procedure to field efficacy is not established, however. Although *H. virescens* larvae often spend most of their lives inside cotton bolls, they do migrate from boll to boll and to pupation sites in soil and litter. They are very sensitive during migration to pupation sites and IGRs cannot produce increased larval damage if applied at this late stage. For other genera, such as *Spodoptera* and *Chorizagrotis,* only laboratory assays have been described (6, 77).

Guerra et al (60a) assayed a variety of compounds on *H. virescens* and concluded that none of these offered immediate prospects for control because of their failure to act before metamorphosis. Although the same argument may apply for effects on reproduction, their finding that compound G at ten μg per larva or one lb per acre foliar spray resulted in 75 and 95% chemosterilant activity, respectively, deserves further consideration, particularly since more active IGRs appear to be available for this insect.

The high degree of acquired insecticide resistance in some Noctuidae, their economic damage, and large area culture conditions can make this family an important test case for the viability of the IGR concept on phytophagous insects in the future.

TORTRICIDAE (LEAFROLLERS) These are rather uniform in life style. The larvae usually feed and pupate in leaves and the eggs are deposited in shingle fashion in batches in close contact to leaf surfaces. In some economically important species several generations per year can occur. Although the damage per larva may be restricted, several successive generations may lead to economic damage in crops while several years of epidemic levels in forests may kill valuable trees. Outram (135, 136) found that JH I had high activity on pupae of *Choristoneura* sp. (spruce budworm). For pupae less than 24 hours old, the LD_{50} was approximately 0.0003 μg. Pupae two days old required far higher doses to obtain similar effective dose effects. Older pupae did not respond. Parallel results were noted on the reproduction of surviving adults, but it was not clear whether reduced fecundity was the result of failure to mate or of other factors. Richmond (158) applied JH III and compounds of type C and F and found that members of the last class were the most active (application two to four days before pupation). Retnakaran (157) used high doses of methoprene (H_1) in laboratory tests on the same species and produced extra larval instars which did not develop into adults. A control dose can be essentially lower as has already been argued. The same authors reported ovicidal effects from several compound types but effective topical doses were high for the compounds used and do suggest a low sensitivity of eggs in this species (155).

Novák & Sehnal (129) performed field spray trials on another species, *Tortrix viridana* (oak leafroller), with compound E and obtained reasonable, but somewhat variable control (82% to 98% efficiency) with a 0.5% spray emulsion when applied to populations of older larvae. The conditions of this test were not completely natural and the compound perhaps not the best choice but the principle was clearly demonstrated. Abdallah (1) found that larvae of *Adoxophyes orana* were sensitive to IGRs, particularly those of type F.

OLETHREUTIDAE Gelbič & Sehnal (52) investigated a variety of compounds of different types on *Laspeyresia pomonella* (codling moth) for morphogenetic and reproduction effects and found JH III and compounds of type E and H most effective on 4 hr old eggs and against last instar larvae, but higher doses per unit weight were required than for most other lepidopteran species. Newly emerged adult females could be sterilized by doses in the order of 10–50 μg. None of these treatments translates directly into applications in the field. Success in the field may depend primarily on the possibility of residual ovicidal effects, and perhaps on exposure of prepupae as they migrate to pupation sites. The last generation overwinters as larvae, the type of diapause not likely to be sensitive to IGRs. Direct treatment of adults as well as topical treatment to eggs during the limited sensitivity periods seems unrealistic.

ARCTIIDAE (TIGER MOTHS) Varjas & Sehnal (218) obtained promising results on *Hyphantria cunea* (fall webworm) by spraying compound E in 0.0005 to 0.5% emulsions corresponding to 0.0075 to 7.5 lb of active ingredient per acre. The higher doses produced supernumerary larvae, the lower ones many unviable intermediates. They evaluated the efficiency of the 0.5 and 0.05% spray, respectively, as 80 and 60% (calculated as the percentage of specimens not reaching the pupal stage). The actual efficiency can be considered to be much higher because many survivors were morphogenetically abnormal. The authors also produced figures indicating that even the lowest dose caused a substantial reduction in the number of emerging adults. They note that the hymenopteran and dipteran parasites present in the population were not noticeably affected.

PYRALIDAE The larvae of many species in this group are classified as borers and spend most of their lives inside plant parts. They often also hibernate inside the host plant as larvae, before pupation in the spring (see diapause section). The metamorphosis to the pupal stage must almost certainly be preceded by a short period of JH sensitivity but the pupae are virtually inaccessible to chemical treatments. Lewis, Lynch & Berry (95) tested several JHs and compounds of types F and G on artificial rearing medium for *Ostrinia nubilalis* (European corn borer) and also tested some of these under field conditions. In the latter case the evaluation technique was obviously not adapted to long range JH effects on the insect population. Intolerable phytotoxicity was observed for some F type compounds at 4 lb per acre.

Other species are known as important pests of stored foods. Mixing IGRs through stored food products, such as whole grain or processed products, has proven to be an efficient control method for populations of *Plodia interpunctella* (Indian meal moth) and *Cadra cautella* (almond moth), preventing further population explosion at concentrations of 5 ppm or less for type H compounds (210). These doses compare favorably with those of insecticides in current use for this purpose. Strong & Diekman also noted that the incidence of webbing, an important nuisance factor in milling machinery, was reduced in IGR-exposed larval populations. The development of IGRs for this purpose will depend not only on toxicological and economical considerations; large-scale experiments must prove that economic damage can be prevented even under conditions of some reinfestation pressure, because damage by first generation larvae is not curtailed. Concurrent prevention of large-scale reinfestation through hygienic, chemical, or physical measures is probably essential. The stability of IGRs in cereal products under the usual conditions of low humidity, temperature, and absence of sunlight is good, and the development of special protective formulations may be required.

LYMANTRIIDAE (TUSSOCK MOTHS) Although early work has suggested that farnesyl methyl ether affects the reproduction of *Malacosoma californicum* when topically applied (229), the high doses involved (310 μg per gram body weight or 55 μg per larva) do not indicate practical utility for this compound.

Novák & Sehnal (128) investigated compound E and hydroprene (H_2) on *Euproctis chrysorrhoea* in laboratory and field tests. They found that the effective doses were relatively low (0.025 μg per gram body weight), approximately 200 times lower than the dose required to produce similar effects in *Galleria mellonella.* Field experiments indicate that 0.001% sprays of hydroprene will provide adequate control of synchronized populations. This concentration is lower than that required with currently used pesticides (0.01 to 0.5%). Success, however, depends on accurate spray timing since the persistence of standard emulsifiable formulations of the compounds on foliage seems to be very short. Overdoses applied at the very beginning of the last larval instar were observed to result in giant pupae. This is probably an indication of short IGR persistence. Avoidance of overdoses or a better timing would prevent this. Tachinid parasites were not noticeably affected in these trials.

YPONOMEUTIDAE Novák & Sehnal (128) also investigated the effect on *Yponomeuta malinella* in which the larvae are colonial in webs. A 0.4% emulsion spray of compound E produced good control and appeared to penetrate the silk nests sufficiently, without unduly harming hymenopteran parasites.

SPHINGIDAE In concurrent field trials, the effects of an application of one to two pounds per acre of JH I and compounds D and F seemed more pronounced on *Manduca sexta* (tobacco hornworm) (6) than on *Heliothis virescens.* The treatments produced supernumerary instars, indicating that lower effective doses could yield control. Pigmentation abnormalities were also noted prior to metamorphosis, but these did not appear to affect viability. Crop damage was not reduced in these trials, as was to be expected.

COLEOPHORIDAE (CASEBEARERS) In a trial by Skuhravý (190), 0.1% sprays of hydroprene (H_2) (ca ½ lb per acre) resulted in a 96% reduction in the field population of *Coleophora laricella* (larch casebearer). A significant decline in the population in the small test plots the following year was attributed to the treatment, and the authors suggested that in larger treatment areas the result would have been even better.

GEOMETRIDAE (LOOPERS) What appears to be one of the most sophisticated field trials with an IGR was performed by Jobin & Retnakaran (81) in 1973 on Anticosta Island (Canada) on *Lambdina fiscellaria* (hemlock looper). Methoprene (H_I) was applied by airplane at rates of 1/64, 1/16, and 3/16 lb per acre on forest plots in which the majority of larvae were in the third and fourth larval instar (the species has four larval instars). The mentioned rates gave 97, 94, and 80% reduction in pupae, which were separated by insecticide-treated forest. The final analysis of long-term effects in this experiment is still underway.

In its totality, the field work with IGRs against Lepidoptera is too scanty to draw definite conclusions. Effects such as prevention and breaking of diapause, reduction of fecundity, residual ovicidal effects, programming, etc, were demonstrated in laboratory experiments but have not received enough attention in the field. The specific activity of many IGRs appears to be sufficiently high. Many of the published studies have not used the currently available, more potent materials or even formulations developed specifically for foliar applications. The feasibility of the concept of population control rather than immediate elimination of damage certainly deserves to be evaluated more extensively.

Coleoptera

Beetles do not appear to be as uniform a group in their response to juvenile hormones as are Lepidoptera. In the response to different analogs, more specificity at the family level is seen than between families of Lepidoptera (210, 224). Add this to the fact that certain species have unlimited flexibility in the number of extra larval instars in the continued presence of juvenile hormone (or adverse external conditions), e.g. *Tenebrio molitor* and *Trogoderma granarium,* while others, such as *Dermestes maculatus* or *Leptinotarsa decemlineata,* respond only by delaying pupation (growing only slightly larger than normal in the process). Morphogenetic, ovicidal, diapause, and sterility effects of JH have been noted. So far, however, little effort has been made to undertake field experiments.

TENEBRIONIDAE (DARKLING BEETLES) Although *T. molitor* is probably the most widely used standard bioassay insect for JH analogs (173, 194), it does not represent a large group of major pest insects. The well-defined grades in the inhibition response obtained when homogenous batches of young pupae are treated with different doses of IGRs have allowed the standardization of biological potency evaluations between different laboratories. The response involves retention of the peculiar abdominal "gin traps" and patches of unpigmented (pupal) cuticle on the ventral side of the abdomen (173). As in most laboratory assays, scoring does not relate directly to the survival of the affected individuals and therefore does not lead to direct conclusions as to effective field dose rates. Two species, *Tribolium confusum* and *Tribolium castaneum,* are economic pests. The effect of relatively low doses incorporated in the rearing medium for *Tribolium* is well established. Strong & Diekman (210) obtained 100% control for both species in media in which 5 to 10 ppm of type H compounds were incorporated. In these test series adults were added to treated media and the effects were then measured in terms of the metamorphosis of the offspring. One must surmise that the morphogenetic activity is dominating over ovicidal activity against eggs deposited in the medium. Early results obtained with compounds such as compound D (12, 214) seem hardly relevant in the light of more recent work.

CHRYSOMELIDAE (LEAF BEETLES) This family contains many species which damage crops and several are crop pests of major importance. The morphogenetic inhibition obtainable is strikingly similar to that noted for cyclorraphous Diptera. Extra larval instars and larval-pupal intermediates are not known to be produced by IGRs but some abnormalities in the adult, such as crumpled wings, defective pupal genitalia, and modified pigmentation patterns on the adult abdomen can easily be interpreted as juvenile hormone effects. Head and mouth parts are not involved, possibly because their determination may be independent of regulation by juvenile hormones. The compound eyes, however, may show an unpigmented (pupal) zone effect similar to that in lepidopteran adults. Last instar larvae of *Leptinotarsa texana* treated with IGRs show long delays in pupation, but even when they finally pupate, the larvae often fail to bury themselves. If they submerge at all after such a delay, death often occurs. The pupae, however, even if they are formed after long delays, always appear to be morphologically normal (Staal, unpublished observations). Reproductive diapause in adults of *Oulema melanopus* (cereal leaf beetle) and *L. decemlineata* can be terminated by relatively high doses of exogenous JH III (37, 126, 181). This has not been explored for control purposes. The subterranean larvae and pupae of *Diabrotica* species do not seem to be prime targets for IGRs due to their life style and relative insensitivity in all stages (6; Staal, unpublished observations).

DERMESTIDAE In this family, *Trogoderma* species, particularly *Trogoderma granarium,* have received the most attention. *Trogoderma inclusum* could be completely controlled by 5 ppm of type H compounds in the food medium (210). The pupae of *T. granarium* are particularly sensitive to type F compounds. Metwally & Sehnal described a response to 0.000005 μg per pupa that they considered to be equivalent to 50% mortality (109–111). Intermediates between larvae and pupae were only obtained by treatments at a very specific time late in the last larval instar. Earlier treatments yielded up to six extra larval molts without appreciable increase in size and probably with only minimal extra food consumption. Superlarvae did occasionally develop into superadults but this would require termination of exposure to juvenile hormone. Karnavar (86) found that the diapause ("quiescent" stage) in larvae of *T. granarium* could be maintained by type F compounds in a dose-related way without producing direct mortality in analogy to the effect on lepidopteran larvae in diapause. Larval-larval molting in this species may continue while the diapause state is maintained (120). The evidence presented, however, does not differentiate this effect from direct delays in pupation produced by exogenous JH in many other holometabolous insects. Metwally and collaborators (109, 111) found that picogram amounts of E, F, and H type compounds applied to young pupae produced recognizable defects in the ovarian development resulting in sterility [typical chemosterilants such as hempa or tepa require amounts three to four orders of magnitude higher to obtain full sterility (91)]. IGR compounds differ considerably in their detailed effects on the ovaries. Vapors of IGRs could repress population growth in *T. granarium.* This is probably due to their effects on reproduction. High doses applied to older pupae, although not resulting in a direct reduction in reproductive capability, reduced the longevity in the ensuing adults from 14–16 days to 4–6 days.

Several stored food beetles belonging to the Anobiidae (*Lasioderma, Stegobium*), Cucujidae (*Oryzaephylus*), and Bostrichidae (*Rhyzopertha*) seem sufficiently responsive to IGRs for practical control (6, 12, 210, 214). Walker & Bowers mention ovicidal effects on *Lasioderma* with low doses of vapors of type F compounds but made no comparison with morphogenetic effects (223).

CURCULIONIDAE (WEEVILS) All studies with weevils show that they are relatively insensitive to IGRs. *Sitophilus* spp., for instance, require a concentration in the medium ten times higher (50 ppm) than do most other beetles (22, 210) to obtain full control. This cannot be due entirely to the hidden life style of the larvae and pupae in the grain kernels because other species with exposed stages such as *Hypera* spp. (alfalfa weevils) are equally insensitive (6; Staal, unpublished observations). The reasons for this relative insensitivity are unknown. Remarkably, even substantial overdoses of the most active compounds produce only minimal metamorphic defects, of which uninflated and crumpled wings and defective ecdysis are the most obvious. It may be that the appropriate type of compound has not been identified, that the integument is a greater barrier than in other insects, or that weevils metabolize IGRs more readily. However, the adult reproductive diapause does respond positively to exogenous JH III in alfalfa weevils (16). Neal & Hower (123) found, however, that summer diapause induction could not be prevented in field experiments on *Hypera postica* with doses up to 4 lb per acre of several IGRs of types A, B, F, and G.

A special problem is presented by the fact that many species have unexposed, often subterranean larvae in addition to being monovoltine. Ovicidal effects obtained on *Pissodes strobi* (white pine weevil) indicate that the relatively high doses required of compounds G & H_4 and the fact that eggs are inaccessible in the oviposition substrate make this approach impractical (156).

BRUCHIDAE (SEED BEETLES) Bruchidae seem equally insensitive to IGRs (6, 12, 110, 214).

COCCINELLIDAE (LADY BEETLES) Foliar applications of 0.01 to 0.05% sprays of compound type F gave acceptable control of *Epilachna varivestis* (Mexican bean beetle) through morphogenetic and ovicidal effects (222). The authors noted the relative inflexibility of this approach in terms of treatment timing and suggested the use of various combination treatments.

Many species of this family are essential beneficials in the natural suppression of aphid populations and it therefore appears important to select IGR compounds for aphid control that are minimally deleterious for these beneficial insects. Bull et al (25) found that the dose required to control aphids was below the danger level for Coccinellidae. Only field observations can give final proof, however. Perhaps the differences in synchronization of the sensitive stages of pests and predators will give sufficient protection under field conditions. The reproductive diapause of aphidophagous Coccinellidae can be broken with IGRs (67), but this may not be of great significance since the hibernation sites are usually outside the crop areas.

It can be concluded that the most immediate practical application of IGRs in Coleoptera is probably the stored food beetle complex in which, with the possible exception of the Curculionidae and Bruchidae, good activities have been observed which justify extension of the investigations into practical situations. The remarkable sterilization effects obtained by Metwally et al (109) in *Trogoderma* should also be investigated in other taxons. It is probably less of an exception than a case of careful observation.

Hemiptera (Homoptera)

This order is extremely heterogenous in life-style, development, and polymorphism. The endocrine mechanisms involved in polymorphism and in the regulation of metamorphosis and reproduction are virtually unstudied in this group of insects. Their small size makes them unattractive subjects for experimental endocrinology. Indirect observations from exogenous applications of IGRs have begun to provide some insight, however.

The damage by Homoptera to plants is as diversified as is their morphology. Damage results from extraction of nutrients, injection of toxic salivary secretions, and/or transmission of viruses and other diseases. When disease transmission is a factor, even low populations of the vector may be intolerable. Populations of endemic Homoptera are usually well controlled by parasites and predators and economic damage seems to prevail primarily when the biological controls are suppressed by broad-spectrum insecticides, by abnormal growing conditions such as those in greenhouses, or in early season crops of cereals when predators and parasites have not yet become efficient due to differences in temperature optima, etc.

Some positive aspects for IGR control of Homoptera are: economic damage often requires a population buildup through several generations with a short life cycle and most life stages are usually well exposed.

APHIDAE (APHIDS) The endocrine mechanism of form determination is poorly understood, partly because the induction periods are often separated from final morphogenetic expression by a considerable time lag that may span more than one generation (66, 93). External stimuli like crowding, inadequate nutrition, and other environmental factors induce switch mechanisms that may or may not involve juvenile and other hormones. It now appears that prior form determinations can be affected or even reversed by exogenous hormones, if not in one embryo then at least in a percentage of all embryos from one female parent (233). The effects are further complicated by short-term effects on metamorphosis. It is also certain that species can differ in induction mechanisms, thresholds, and sensitive periods, which may lead to contradictory conclusions. It is possible to draw some analogy between the programming effects obtained by Riddiford (160) and the influences on form determination of aphids. Particularly the common parthenogenetic reproduction of viviparous females offers excellent opportunity to study phenomena of this kind. The consequences of changes in form determination for insect control are probably as diversified as the phenomenon itself and deserve further examination in detailed field experiments for a few well chosen target pests.

Because of a shorter time lag between induction and effect, the inhibition of metamorphosis is a more easily understood phenomenon. It can be inhibited in all forms (7, 62, 93, 121, 220) and generally the third (penultimate) nymphal instar is the most sensitive. Treatments up to this moment can lead to the production of giant supernumeraries that may molt again but never become viable reproductives. Lower doses and treatments in the last nymphal instar produce intermediates, some of which may also try unsuccessfully to molt again. Parthenogenetic, viviparous nymphal/adult intermediates are known to be able to produce some, at least initially viable, embryos (7, 121) whose fate has not been followed adequately. Some authors conclude that alate nymphs are more sensitive to exogenous JH than are apterous nymphs (212). Besides the fact that inhibition of wing metamorphosis can be more lethal because of mechanical molting difficulties and that supernumeraries in aptera are far more difficult to recognize than those in alatae, there is little evidence to support this conclusion. Topical treatment and incorporation in artificial medium (93) work very well, but, even more important, the effects of residual foliar treatments are reportedly excellent (7, 62, 121). Tamaki (212) studied the possible prevention of host alternation in *Myzus persicae* (green peach aphid) with compounds G and H_2 through the inhibition of alatae on the winter host. Nassar, Staal & Armanious (121) have analyzed the effect of single sprays on population development on the synchronized batches of instars of *Schizaphis graminum* on caged wheat plants under laboratory conditions and thereafter followed the population growth over a two week period. The results indicated that these synchronized groups of aphids could be effectively controlled with spray concentrations as low as 0.0001% in 10 to 12 days (compound H_4), except on the last nymphal instar and adults. For these stages even the highest dose used (0.1%) produced an initial increase in population that later stabilized and decreased to the starting level in approximately 15 days as a consequence of effects on metamorphosis in the next generation. In the controls a rampant population explosion presented a very different picture. While this study provided better insight in the ultimate potential, the complications of asynchronous populations, short-lived residuals, and reimmigration on new growth explain why the initial outdoor field results on small plots have generally been poor.

IGR effects on reproduction can result from several different causes that can be subdivided into direct and indirect sterilization. Indirect sterilization of parthenogenetic forms can be directly traced to incomplete metamorphosis of the genital pore. Thus, supernumerary nymphs, in which the growth of embryos has already started immediately after birth and continues into adult life, finally succumb from embryo congestion. Many more embryos can be counted in these specimens than would ever be present in a normal adult (7, 62, 121). Direct sterility is present when applications to morphogenetically normal adults result in the death of embryos before, during, or immediately after birth. Sometimes embryos are seen in the maternal abdomen that have prematurely undergone their embryonic ecdysis (121).

Direct toxicity of IGRs has been reported in aphids more frequently than in other insect groups (7, 90, 121). There is little doubt that IGRs could never compete successfully with insecticides with 100 to 1000 times greater direct toxicity if this were the only effect. Compound H_4 is different from other IGRs in its relatively high direct toxicity (probably connected with the propynyl toxophore), and direct sterility effects, particularly on adults, but it possesses high JH activity on other stages as well (121).

Form determination is influenced by several environmental stress factors, such as deficient nutrition and crowding, creating opportunities for misinterpretation of IGR effects. Repellency and chemical or mechanical irritation may reduce feeding and increase social contact. Both are factors known to influence form determination. Particularly the effects of relatively inactive or impure materials should therefore be suspect (232, 234). As a typical example, daily rubbing with farnesol, as exercised by one author (220), cannot be considered as a valid procedure unless it can be proven that other chemicals applied in the same way do not produce the effect described. White (234) provides evidence that prenatal JH exposure stimulates wingbud regression in *Brevicoryne brassicae* (cabbage aphid), but the question whether IGRs do or do not stimulate production of apterous forms is only partially answered. Some unpublished observations indicate that highly potent compounds may stimulate alatae production in *Myzus persicae* and at the same time inhibit the metamorphosis of the alatae. Other studies on *Aphis fabae* have indicated that the gynoparae (normally producing oviparous offspring) can be induced to switch completely to the production of viviparous parthenogenetic offspring by IGR treatment (S. G. Nassar, personal communication).

Effects on programming have been noted in offspring produced by adults treated with sterilizing doses before reproduction stops. They may develop into supernumeraries even after taking the precaution of removing them from the contaminated substrate immediately after birth (121). This effect is very similar to that of treatment of late embryonic stages in other insect groups as described by Riddiford (160).

Field studies on aphids have not convincingly produced complete aphid control thus far (6). Many unpublished small plot experiments conducted by Zoecon investigators and collaborators in 1973 have exposed the shortcomings of the small plot approach for evaluation of long-term control coupled with insufficient field stability of the IGRs used (type H compounds). On the other hand, simulated practical tests in greenhouses have generally been very successful and usually yielded full and long lasting control of several species of aphids (74). The continuous presence of hymenopteran parasites and several predators indicated that the IGR method is ecologically sound.

Although area wide population control of aphids seems to be a goal of the distant future, improved formulation may bring control encompassing fairly large acreages into the realm of practical possibility, as it already is for protected closed systems such as greenhouses. A crucial point for control in outdoor situations will be the relative safety of IGRs for beneficial species.

PSEUDOCOCCIDAE (MEALYBUGS) Staal, Nassar & Martin (208) reported laboratory results with this group. Their results indicated that, perhaps because of the less pronounced external metamorphosis in the citrus mealybug (*Pseudococcus citri*), the main effects of IGR treatments were direct mortality and the repression of parthenogenetic reproduction. It is noteworthy that compound H_4 was more toxic to adults than to nymphs. A spray concentration of 0.01% produced good control on batches of various synchronized stages under these artificial conditions. Positive field results have not been reported thus far.

DIASPIDIDAE (ARMORED SCALES) Practical tests with high rates of JH I and compounds D and F failed to produce results in the field on *Unaspis citri* (citrus snow scale) (6). Laboratory experiments on another armored scale, *Aonidiella aurantii* (red scale), were evaluated as more promising. Excellent results were obtained by Nassar, Staal & Martin (122) on *Hemiberlesia lataniae* (latania scale) in laboratory experiments. Virtually all stages showed effects including direct mortality, inhibition of and abnormalities in scale formation, inhibition of differentiation from crawler to nymph and nymph to adult, and production of nonviable offspring following treatment with compound H_3. Some of these effects were obtained with other compounds of type H as well, at rates varying between 0.001 and 0.1% applied as sprays. Experiments aimed at breaking adult diapause have not been described in the literature. This aspect deserves more attention since overwintering scales are usually well exposed.

COCCIDAE (SOFT SCALE) Promising laboratory effects were obtained with a 0.01% spray of compound D on *Coccus hesperidum* (brown soft scale) (6). This scale has a rather lengthy development and for this reason IGR effects become apparent only a long time after treatment. Additionally, a lack of synchronization in development tends to make this group a difficult target for IGRs (J. W. Martin & S. G. Nassar, unpublished observations).

CICADELLIDAE (LEAFHOPPERS) JH III has proven to be active in breaking summer diapause of *Draeculacephala crassicornis* by topical, substrate, and vapor treatments. Since eggs produced after diapause hibernate, the effect on their development was not followed (85). Foliar sprays of compounds of type F and G were later found to be active on the metamorphosis of the nymphs, as well as the reproductive diapause of the females. For the latter application the dose rate required did not appear to be very promising. Compound H_2 rated lower than F and G (154). Bagley & Bauernfeind (6) mention that the diapause disruption in this species could not be obtained in the field with JH I. A more persistent IGR or formulation is required.

DELPHACIDAE Observations on laboratory populations of *Peregrinus maidis* (corn planthopper) indicate that inhibition of metamorphosis can be obtained with IGRs at a fairly high rate (ID_{50} 0.04%) of compound H_4 (S. G. Nassar, personal communication). Adults again proved to be more sensitive than nymphs to the toxic action of compound H_4. A peculiar pigmentation resulted from treatments with any JH active IGR, as the normally cream-colored specimens developed a very pronounced black melanization in the cuticle. As with pigmentation effects in other insects (e.g. Acrididae), this effect was not associated with metamorphosis or mortality factors and could disappear in subsequent molts. One of the sensitive metamorphosis characters proved to be the female ovipositor. Females with shortened ovipositor as the only visable sign of inhibited metamorphosis produced eggs that were only half inserted in the plant tissue. These eggs did not hatch successfully.

ALEYRODIDAE (WHITEFLIES) Whitefly problems are prevalent in field crops, ornamentals, and in greenhouse cultivation. Rapidly increasing resistance has made most chemical control agents obsolete; the best now remaining are pyrethroids (227). Pyrethroids, however, lack activity on eggs and frequent application is essential. Recent unpublished observations (Staal et al) in experimental greenhouses indicate that excellent and long lasting whitefly control can be achieved with IGRs, particularly with compound H_4, due to the effects on immature stages, and particularly the eggs. Although young eggs are sensitive to many other IGRs, compound H_4 excells as a result of its effect throughout the entire span of egg development. Other laboratory results (189) indicate that full control of asynchronous populations can be achieved with two 0.1% sprays, three days apart, of compounds H_1 and H_2, as well as H_4. Malathion at 0.2% did give adequate control after eight days, but the populations resurged after 29 days even following 0.4% sprays. The IGR treatments, however, did not show any population resurgence after 29 days. Since population control with IGRs is not immediate, it can be speculated that initially combined treatments with pyrethroids and IGRs could provide both immediate and long lasting control by virtue of their complementary activity. Population densities below the damage level would only require IGR treatment, however.

PSYLLIDAE (PSYLLIDS) Westigard (230) obtained promising results with hydroprene and methoprene on nymphal and adult populations of pear psylla in the laboratory and under field conditions. Methoprene, at 0.4 lb per 100 gallons, prevented reproduction of treated nymphal populations with negligible effects on beneficial predators and mite pest species.

Hemiptera (Heteroptera)

The early discovery of the natural compounds juvabione and dehydrojuvabione, in paper products derived from balsam fir pulpwood (28, 31, 195, 196, 241), has lead to unrealized hopes that additional and useful IGRs with great specificity would be found. The fact that juvabione occurs primarily in the wood and not in the living tissues of trees (206) makes speculations on defense functions towards insects in ancient or modern times rather irrelevant. Few materials, if any, have been found that are so limited in activity as juvabione in the Pyrrhocoridae. The mixture used by Law et al (C) possesses even higher activity on this group but is less specific, as is obvious from many references in other sections of this review. However, its main component, compound D, is, like other 7,11-disubstituted farnesoate esters, fairly specific for Pyrrhocoridae and Lygaeidae. The most recent addition to this list of highly specific compounds active only on Pyrrhocoridae are certain polypeptides with unusually high activity (88, 149, 257). Compounds of this type have proven to be very effective in spray applications and have shown excellent residual life on plant surfaces (5). Only for this group of IGRs has limited systemic translocation in cotton plants been demonstrated after uptake by the roots (4).

PYRRHOCORIDAE The rather unselective natural hormones show a measurable activity on this family. This makes the explanation of the sensitivity to several very specific compounds very difficult. It may not be related to the existence of a different *Pyrrhocoris* juvenile hormone. *Pyrrhocoris apterus* is of no economic significance; no field work has been reported on this species. None of the compounds selective for Pyrrhocoridae are very active on other families of Hemiptera that are of greater economic significance. Although no compounds with control potential have resulted directly from the work on Pyrrhocoridae, several of the effects demonstrated can be considered exemplary of what can be achieved with IGRs in other pest species once sufficiently active compounds become available.

Morphogenetic activity is maximal when applied to last nymphal instars, but applications to earlier instars, although requiring higher doses, produce similar effects. In Hemimetabola the carry-over effect generally seems higher than in Holometabola. Direct ovicidal effects can be produced by all morphogenetically active compounds. Sensitivity extends past half the embryonic development time. Most interesting was the observation that ovicidal effects can occur through treatment of the females with compound D and also through transfer of compound from treated males. Masner, Sláma & Landa proved conclusively that compound transfer to the females can take place during mating (103). *Pyrrhocoris* also yielded the first example of ovarian dysfunction through injection of submicrogram amounts of compound D (102).

Dysdercus spp. (cotton stainers) have been the subject of several studies (20, 28, 39, 43, 68, 175) that have indicated good possibilities for control, particularly with compound D. Several species cause damage in tropical cotton culture, but with the exception of central Africa, the importance of *Dysdercus* spp. is usually insufficient to justify specific control measures. Most studies ended short of actual field trials. Homberger, Benz & Thommen (68) obtained effective sterility through topical treatment of males and females with 10 μg of compound D per specimen and noted an increased reproduction rate associated with reduced longevity, a rather common IGR effect for morphogenetically marginal doses observed in other insects as well. Details of the effects on embryonic development have been worked out by Riddiford (162) and Matolín (106, 107).

LYGAEIDAE One representative, *Oncopeltus fasciatus* (large milkweed bug), is frequently used in laboratory bioassays. Compound types D and F have shown highest activity in this group so far (14, 77).

MIRIDAE (PLANT BUGS) Worldwide, several species of this family are important local crop pests. Screening operations on *Lygus hesperus* (lygus bug) have not indicated very high activity for any type of compound as yet, and have certainly not shown much correlation of sensitivity with *Pyrrhocoris.* Perhaps most active are the compounds of types F and G (6, 25), but field activity seems to fall short even for these. New compounds have probably not been investigated extensively in this family.

REDUVIIDAE (ASSASSIN BUGS) Laboratory assays by Wigglesworth (239) have indicated highest morphogenetic activity for JH I in *Rhodnius prolixus.* Patterson (144) found type G compounds active on *Triatoma* and *Panstrongylus* spp. in South America, vectors of the debilitating Chagas disease. Work is going on to screen IGRs on some of the species involved. The application of IGRs as a residual treatment in paint work in human dwellings may be feasible.

PENTATOMIDAE (STINKBUGS) Laboratory testing by Kontev et al (88) and Sláma (193) on at least three species have indicated that satisfactory morphogenetic activity can be found in several para-substituted phenylether (G) compounds. Kontev et al performed field tests with a few type G compounds on the monovoltine *Eurygaster integriceps* in wheat fields in the Balkan area after carefully monitoring the population composition to determine the most appropriate spray times. The ID_{50} laboratory values of the IGRs used were not very different from the LD_{50}s of current insecticides, but the field results were not fully satisfactory because of an obvious lack of field stability. In treatments at 1–5 lb per acre a significant proportion of morphogenetic intermediates was found. This indicates that increased stability might improve the degree of long range control and reduce the dose required (26).

The extreme heterogeneity of compound selectivity for the major families of Heteroptera is puzzling. It must be related primarily with receptor activity, since it has been found that selective compounds can penetrate and circulate at fairly high levels in the blood of insensitive species. The prospects for practical control look promising only if compounds of sufficient activity and stability become available.

Isoptera

Despite the complexity of social organization in termites, the agreement of scientists on the effects of exogenous JH is general. All workers noted an increased tendency to presoldier differentiation leading to an increased number of soldiers or intercastes on prolonged exposure to JH and to IGRs of type C, E, G, and H. In the laboratory experiments, the compounds were usually applied to filter paper substrate. Because only minor amounts were consumed as food, the effects were often partly attributed to the action of vapors (71, 72, 98, 115, 225). The termites used were *Zootermopsis* spp. (Kalotermitidae) and *Reticulitermes* spp. (Rhinotermitidae), which showed only minor differences in the sensitivity. Hrdý (72) and Hrdý & Křeček (73) found compounds E and hydroprene (H_2) to be among the most active of the many JHAs screened.

Wanyonyi (225) and Wanyonyi & Lüscher (226) noticed that while high doses of IGRs stimulated soldier development, medium doses stimulated regressive development and the production of intermediates, and low doses repressed the formation of alatae and replacement reproductives (with a repressing effect on molting). These differential effects may illustrate some of the differentiation possibilities to which every larva is periodically susceptible. Lüscher (99) and Meyer & Lüscher (115) have collected evidence that exogenous JH present in the anal secretion of the reproductives could play a role in caste differentiation of larvae to soldiers. Lüscher (99) also found that larvae in the presence of soldiers showed an opposite tendency and therefore postulated the presence of an "anti-JH" emitted by the soldiers, or an alternative function as JH "sink."

The sensitivity window for caste effects is well established and restricted to the middle of the intermolt periods during larval development (226). The minimum exposure was found to be nine days (71, 72), which points to a mode of action that differs from disruption of metamorphosis.

The natural JH titer of reproductives (115) is high and it is perhaps for this reason that effects of JH on reproduction are difficult to obtain. More remarkable is that direct effects on metamorphosis have not been described. This may be because intermediates are destroyed by their fellow workers as soon as they occur. It is also possible that termites, habituated to exogenous JH, have become relatively insensitive to this external influence on metamorphosis and respond only with caste effects or status quo molting.

It is conceivable that impregnation of wood with IGRs could give long range control of termites, primarily through shifts in caste development, but no reports on actual field trials have been published. Studies of this nature will require considerable time, longer than the time-consuming tests with buried impregnated planchets for insecticidal activity. The chances for sufficient stability of IGR compounds absorbed in wood substrates seem good, if one keeps the astounding residual survival of the paper factor in wood and paper products in mind.

Dictyoptera

BLATTELLIDAE (COCKROACHES) Screening of compounds on cockroaches has been very limited. The drawn-out life cycle makes the chances of achieving acceptable success of IGR applications remote when compared with available short-term chemical control applications. The possibility of combining agents that cause short-term mortality with long-term IGR control for residual treatments should not be discounted. Waheed, Fisk & Collins (221) found compound G active on the metamorphosis of *Blattella germanica* (German cockroach). Cruickshank (40) found that amide analogs of JH III absorbed on cornstarch (100 ppm) caused premature death of adults of the same species and rejection of immature egg cases by females. This dose compared favorably with that required for current insecticides. No reports on large-scale tests are available. Whether indeed the amides described by Cruickshank are the most active analogs remains to be established.

Most remarkable is the finding by Hangartner & Masner (61) that JH II and compound G prevented ecdysis in nymphs of *B. germanica* after prolonged exposure or topical treatment. The dose used, 10 μg topical per specimen or 100 μg per cm^2 on filter paper substrate, is 100 times as high as the dose required to inhibit metamorphosis. The ecdysis-inhibiting effect could be produced in any nymphal instar, although it was less pronounced in the last instar nymph. The mechanism of action is still unknown.

Because residual treatments against cockroaches are often applied in control schemes in combination with treatments giving short-term effects, further research on the activity of IGRs against cockroaches seems justified.

Orthoptera

ACRIDIDAE (GRASSHOPPERS) Only nymphs of some locust species have been subjected to comparative screening test with a number of acyclic terpenes (124, 125). Nymphs of many locust and grasshopper species can respond to exogenous JH with a pigmentation change to green along with the simultaneous disappearance of brown and black cuticular pigments. This color change is a normal occurrence during phase transformation in locusts as a response to changes in population density (87, 205). JH appears to override density stimuli. The change to green can occur in any nymphal stage, is reversible, and is not directly related to mortality factors or metamorphosis. The dose necessary to inhibit metamorphosis is not very different from that for provoking a color change in the following instar but is generally high for all compounds tested so far. JH application at the proper moment (early last nymphal instar) can lead to a combination of inhibition of metamorphosis and green pigmentation. Locusts are known to be relatively insensitive to insecticides as well, which may be partly due to their relatively large size. It has not been investigated sufficiently whether the behavior of the gregarious phase changes simultaneously with the pigmentation after JH application under field conditions. If this is the case, interesting consequences may apply to the migratory behavior that plays an essential role in the outbreak epidemiology.

Hymenoptera

FORMICIDAE (ANTS) The IGR literature on this family is very limited. Field work has been reported only for the fire ant (*Solenopsis* sp.) (42, 215). As in other groups of social insects such as the termites, the roles of JH are thought to include functions in the caste determination. The enclosed nest structures and behavioral aspects of the social structure make these insects a very different but worthwhile target.

From laboratory experiments a great variety of effects have been reported, as the result of either topical application, treatment of nest sections, or exposure of baits in the foraging territory (134). These effects include direct morphogenetic effects (particularly on the pupal-adult transformation); effects on brood care; fecundity, fertility, and mortality of queens; caste determination; social disruption resulting in fights and mortality, etc. Compounds of type G, F, and H have failed so far to give the desired degree of control in the field, presumably due to an inadequate field delivery technique. Any combination of the effects cited might suffice to give long-term control, if an appropriate delivery system can be developed. It is generally agreed that the best route of chemical treatment is the offering of a treated bait as is presently done for the insecticide Mirex®. Cupp & O'Neal (42) estimated that a bait containing 3 ppm of methoprene (H_1) would provide control, whereas Troisi & Riddiford (215) found that a bait containing 10 to 100 ppm of methoprene or hydroprene was required to prevent metamorphosis and that lower doses induced a shift to the production of alate males. The variety of behavioral aspects involved (repellency and attraction of baits, trophallaxis, etc) add several new and difficult parameters to this research. With the possible exception of the fire ant, a few domestic species, and a few species that protect agricultural pests, most ants could be considered as beneficial. The chance of harming them accidentally through the use of IGRs can be considered as minimal.

HALICTIDAE AND MEGACHILIDAE (BEES) Topical applications of JH I were found to be active on the pupal-adult metamorphosis of representatives of these families but not on the larval-pupal transformation, by Hsiao & Hsiao (75). Prepupal diapause was not affected by JH I.

VESPIDAE (WASPS) The breaking of female reproductive diapause could be obtained by topical application of compound A (13). No field work has been reported.

APIDAE (HONEY BEES) Bees have been screened primarily in the laboratory. Wirtz (244) and Wirtz & Beetsma (245) found that JH III shifted the caste determination of 3½-day-old larvae in worker cells to queen-like adults and noticed that the effect became less pronounced when larvae older or younger than 3½ days were treated. Affected young larvae were usually removed by the workers. The authors did not describe effects on metamorphosis from larva to pupa or pupa to adult. Žďárek & Haragsim (258) confirmed the effects on caste determination by administration before the cessation of larval feeding and observed inhibition of adult metamorphosis by application to late larvae. They screened several compounds for inhibitory effects on pupae and found compounds of the types C, D, E, F, and H only moderately active, compared with certain acyclic and cyclic ethers. This may indicate that little hazard may have to be expected from field applications, but this waits for final confirmation. Recently, Hrdý (72) administered compounds E and hydroprene (H_2) in food in very high doses (up to 15.5 g per colony!) to foraging bees. Although adult workers were not affected, the larval brood was destroyed and some queens were killed. Similar, but probably more transient, effects were noticed when honeybees were confined in a greenhouse and forced to forage on IGR-treated flowering radish plants.

PARASITIC HYMENOPTERA Wright & Spates (253) found that topical treatments on pupae of the stable fly with doses of compounds of type F and G that would prevent the development of stable fly adults did not affect the metamorphosis and reproduction of the parasites *Muscidivorax raptor* and *Spalangia endius.* Outram (137) studied the survival of hymenopteran parasites in larvae of *Choristoneura fumiferana* treated with H_1 under laboratory conditions. As could be expected, the treatments did reduce the survival of parasites (from 85 to 31% in the highest dose), but only a few of the hosts survived (0.5 compared to 56%). This indicates an excellent potential for this IGR in comparison with broad-spectrum compounds that are often more active on the parasites than on the host, a conclusion that differs from that of Outram. The absence of the IGR effects on adult beneficial hymenopteran parasites, the selectivity, and the limited sensitivity periods make it unlikely that their populations would undergo dramatic reduction, even if some parasites would perish together with their affected hosts.

The evidence obtained thus far on Hymenoptera is insufficient for broad generalizations. It is conceivable that unwanted social species could be controlled by specific formulations of certain IGRs that workers will carry to the queen and the larvae. This could at best provide very long-term control. Detrimental effects on beneficial Hymenoptera do not appear to provide cause for undue concern.

Mallophaga

Chewing lice, pests of domestic animals, received considerable attention in several publications by Chamberlain, Hopkins, and collaborators (32–34, 69, 70). They screened a great variety of compounds in several ways on *Bovicola limbatus* (Angoragoat biting louse) and concluded that the last nymphal (third) instar, was the sensitive stage and that 50 ppm of certain types of compounds, including JH I and JH II, in the diet would provide full control by preventing metamorphosis and reproduction. A concentration of hydroprene (H_2) as low as 5 ppm was fully effective. This means that 0.1 mg of this compound, administered periodically, per goat should suffice to give control (32, 33). Vapor treatments produced ovicidal effects at doses as low as 1 μg per vial, but deferred programming effects subsequent to this treatment could not be detected (33, 69).

Anoplura

Vinson & Williams (219) studied the effects of the type C mixture on *Pediculus humanus humanus* (human body louse) and observed effects on embryogenesis and metamorphosis, which led to high mortality when populations were exposed to impregnated wool fabric. Some of the lice which survived a high dose developed into giant supernumeraries, however. Bagley & Bauernfeind (6) mention ovicidal effects particularly with compound F on *P. humanus* but stated that the absence of immediate effects may preclude their use for public health purposes.

Other Orders of Arthropods

Although most major insect pest species belong to the orders mentioned above, several other orders (e.g. Siphonaptera, Thysanura, Thysanoptera) include species of economic importance. No work directed at control of these species with IGRs has been described. Other orders (e.g. Neuroptera, Odonata, etc) include species that are considered directly beneficial, but IGR work has been equally scarce. Neuropteran predators were found to be relatively insensitive to IGRs at field dose rates (25).

It should also be mentioned that IGRs have not been reported as being active against Acarina (mites and ticks) at reasonable dose rates. This group of arthropods may rely on mechanisms for the regulation of metamorphosis and reproduction different from those of the insects. It is also conceivable that the insect-active hormones are not the appropriate chemical structure for this taxon.

METABOLISM, SYNERGISM, AND RESISTANCE DEVELOPMENT IN INSECT GROWTH REGULATORS

The questions of where, when, and how JH and its analogs are broken down by the insect, whether the insect can develop resistance against IGRs, and whether the action of IGRs can be synergized by metabolic enzyme inhibitors are intimately related. Insects probably depend on breakdown mechanisms at certain moments during the life cycle for the regulation of hormone titers and the possible cleanup of unwanted residues (228, 235). This breakdown probably uses existing enzymes, but the metabolism of several xenobiotics, such as various types of insecticides, also seems dependent on the same classes of enzymes. Therefore it should be no surprise that certain insecticide resistant strains of insects show a cross resistance towards IGRs (8, 30, 49, 148). The importance of cross resistance under field conditions and the possibility for spontaneous selection towards increasing resistance remains to be proven. The fact is that methoprene is the only compound that can still effectively control multi-resistant floodwater mosquito strains in California (24).

It has been speculated repeatedly that insecticide synergists could enhance the activity of IGRs through their effect on the detoxifying enzymes. Several papers do point to synergistic effects of compounds of this type (sesamex, piperonyl butoxide, TOCP, etc) (14, 152, 198) but the effects were not very dramatic and were sometimes even antagonistic (8). This may indicate that the mixed function microsomal oxidases that are usually inhibited by these synergists, perhaps play less of a role than anticipated in IGR metabolism. It is remarkable that several of these synergists, structurally unrelated to JH, exhibit a weak JH activity of their own. However, a full-scale juvenilizing response has never been obtained with any of these compounds (152). The effects of the above synergists may indeed be due to a synergistic or protective function towards a small remaining JH titer (even after allatectomy!), as suggested by Slade & Wilkinson (192) and Brooks (23), rather than to intrinsic activity. One cannot accept their conclusions, however, that all analogs would act primarily in this way. Although truly active analogs may always protect endogenous JH competitively, little doubt exists as to their full capability to replace JH in the gonadotropic sense in allatectomized animals (in a perfect dose response relationship), as well as in morphogenetic assays (156). It should be noted that several weakly active JH analogs were found to slightly synergize more active compounds (153, 164, 199). We believe that these effects can be fully explained by competition for the same metabolic enzymes.

PERSISTENCE AND OTHER QUANTITATIVE PARAMETERS

The half-life of JH I has been shown to be no longer than six to ten hours when injected in insects in pure form, but considerably longer when injected diluted in oil (55, 57, 187). This concurs with the observation that the morphogenetic effect of a given quantity of JH does increase up to a certain point with increasing dilution in oil (when injected). This dilution paradox uncovers an important principle that also explains why the same total amount of JH can be far more effective when administered in a number of small doses (8, 145, 187), continuously (55), or when incorporated in a slow release vehicle, than when given in a single dose. The importance of this cannot be overrated as it makes slow release formulations uniquely suitable for IGRs. There is ample evidence suggesting that flash saturation of JH receptors is far less effective than a drawn out exposure. Sehnal & Schneiderman (187) found in implantation experiments with corpora allata in last instar larvae of *Galleria* that the most complete effects were obtained with a prolonged retention in the recipient because a complex of different morphogenetic phenomena becomes successively affected. Slow release formulations of IGRs give not only a better effect per individual but additionally per group of insects when the synchronization is less than ideal. It is not farfetched to consider the insect cuticle as a slow release sink in itself. Wigglesworth gives evidence that the thinner the cuticle on which the JH is applied the more rapid the penetration and metabolism and the less significant the effect (240). Cuticles are partly resorbed and partly shed during a molt and this may be responsible for the loss of carryover when IGRs are applied topically to an insect in an instar preceding the sensitive one. Notwithstanding the cuticular sink, dilution in a nonvolatile solvent such as oil has on occasion been found to potentiate the activity of certain JH type compounds following topical application (240). Consistent with these views is the finding that emulsified JH is less efficient than pure JH when injected because it is even more exposed to degradation (59, 237). The ecdysiotropic activity of JH analogs measured by Krishnakumaran & Schneiderman (89), however, was not impaired by this procedure in contrast to the JH activity which may indicate an essentially different receptor situation.

Some of the most obvious factors that might help determine the resultant activity of a given compound on topical application, can be summarized as follows: (*a*) receptor fit (steric requirements and functional groups); (*b*) speed of cuticular penetration (neither extreme being desirable); (*c*) resistance against different types of breakdown enzymes; (*d*) protective binding by carrier proteins in the hemolymph; (*e*) protection of existing endogenous JH pools through enzyme competition but counteracted by enzyme induction; (*f*) species selectivity and the ability of the insect to postpone certain sensitive processes in the presence of JH; (*g*) feedback on the endocrine and other regulatory mechanisms; and (*h*) the sensitive stage of the insect. The coexistence of such a multitude of quantitative factors is probably the best reason for the lack of a rational approach to the design of more effective IGRs.

Environmental stability comes into play when compounds are applied on and in food and other substrates frequented by insects. Natural juvenile hormones are quite unstable under ultraviolet light and the persistence outdoors is therefore very short

(146). Synthetic analogs are probably somewhat better in this aspect, and can provide some slow release from sustained surface film contact. Stability still needs much improvement. In this respect specific formulations offer the most immediate promise. Because of the limited periods of susceptibility protection against degradation is far more crucial for IGRs than for current insecticides. However, no degree of persistence can protect new growth on plants from new pest immigrants (with the exception of a number of highly toxic systemic insecticides). It therefore follows that increasing the residual life of foliar sprays also has its limit of usefulness and that repetitive spraying cannot be dispensed with. Reimmigration will play less of a role when larger areas are treated.

CONCLUSION

Except for a direct toxic effect at higher doses on Homoptera, direct toxic effects are generally absent and all other effects are indirect. Although virtually all insect control at present is based on toxic chemicals with immediate effect, the shortcomings of these become increasingly apparent. Other methods are proposed or already used, such as genetic manipulation of populations, sterile male release programs, pheromone techniques, etc. that are a departure from the fast kill methods of the past decades. While nothing can be gained from a chemical method in which the slow mode of action is the only departure from the current practice, there are good indications that IGR methods are longer lasting in their effect, partly due to the intrinsic action but also because they do not eliminate the cooperation of beneficial parasites and predators. However, with a slower activity, the demonstration of efficacy becomes more tedious. Typically, monovoltine pests would only show next generation effects and migration would easily obscure all effects on insect populations in small plots. While the economical and psychological barriers that must be overcome before critical experimentation in large plots can be done are very real, the implementation for actual pest control will require even greater sociological change.

A very useful interim technique may be the initial application of combinations of effective direct toxicants with IGRs. This would satisfy psychological and economical requirements, while the biological effects can be predicted to be at least complementary. Subsequently, populations could be maintained at a minimum level with IGRs alone.

The requirements that a successful modern pesticide must satisfy should include commercial aspects as well. These requirements can be summarized as follows: 1. pest control efficiency on major pest species or group of species; 2. reasonable specificity towards the pest species, avoiding damage to predators and parasites, so that the compounds fit into integrated pest management schedules; 3. low toxicity to man, domestic animals, and wildlife in any form; 4. reasonable cost of application compared with current pest control practices—a novel chemical, however, may command a higher price when it has obvious advantages over the old one; and 5. industrial propriety of the compound structure—without this protection, the capital investments necessary to develop and market the compound cannot be made in the contemporary socioeconomic climate. Several of these factors (e.g. 1 and 4) are interrelated.

How do the IGR compounds rate in these respects? The bottleneck undoubtedly lies in the proof of economic efficacy and the acceptance of control practices which require relatively long periods to achieve effect. Never before has field research been such a crucial factor in the development of new methods of pest control. This research will require increasing sophistication as to the application of knowledge of population dynamics and the utilization of unfamiliar evaluation methods. Many IGRs appear to offer a very acceptable compromise in selectivity between the less desirable extremes of narrow specificity and broad-spectrum activity, which makes them acceptable in modern pest management techniques.

Acknowledgments

I am indebted to G. F. Ludvik, J. B. Siddall, and C. A. Henrick for their scrutiny of the draft and many helpful suggestions which have greatly contributed to this manuscript. I also thank many colleagues who allowed me to cite their manuscripts or unpublished work.

Literature Cited

1. Abdallah, M. D. 1972. Juvenile hormone morphogenetic activity of sesquiterpenoids and non-sesquiterpenoids in last stage larvae of *Adoxophyes orana. Entomol. Exp. Appl.* 15:411–16
2. Adams, T. S. 1974. The role of juvenile hormone in housefly ovarian follicle morphogenesis. *J. Insect Physiol.* 20: 263–76
3. Ashburner, M. 1970. Effects of juvenile hormone on adult differentiation of *Drosophila melanogaster. Nature* 227: 187–89
4. Babu, T. H., Sláma, K. 1971. Systemic activity of a juvenile hormone analog. *Science* 175:78–79
5. Babu, T. H., Sláma, K. 1971. Effectiveness of ethyl pivaloyl-l-alanyl-p-aminobenzoate—A juvenile hormone analogue against red cotton bug, *Dysdercus cingulatus. Indian J. Entomol.* 33:212–15
6. Bagley, R. W., Bauernfeind, J. C. 1972. Field experiences with juvenile hormone mimics. In *Insect Juvenile Hormones,* ed. J. J. Menn, M. Beroza, 113–51. New York: Academic
7. Benskin, J., Perron, J. M. 1973. Effects of an insect growth regulator with high juvenile hormone activity on the apterous form of the potato aphid, *Macrosiphum euphorbiae. Can. Entomol.* 105:619–22
8. Benskin, J., Vinson, S. B. 1973. Factors affecting juvenile hormone analogue activity in the tobacco budworm. *J. Econ. Entomol.* 66:15–20
9. Benz, G. 1971. Failure to demonstrate sterilans effect of juvenile hormone mimetics in *Pieris brassicae* and *Galleria mellonella. Experientia* 27:581–82
10. Benz, G. 1972. Juvenile hormone breaks ovarian diapause in two nymphalide butterflies. *Experientia* 68:1507
11. Bhaskaran, G. 1972. Inhibition of imaginal differentiation in *Sarcophaga bullata* by juvenile hormone. *J. Exp. Zool.* 182:127–41
12. Bhatnagar-Thomas, P. L. 1973. Control of insect pests of stored grains using a juvenile hormone analogue. *J. Econ. Entomol.* 66:277–78
13. Bohm, M. K. 1972. Effects of environment and juvenile hormone on ovaries of the wasp, *Polistes metricus. J. Insect Physiol.* 18:1875–83
14. Bowers, W. S. 1968. Juvenile hormone: Activity of natural and synthetic synergists. *Science* 161:895–97
15. Bowers, W. S. 1971. Juvenile hormones. In *Naturally Occurring Insecticides,* ed. M. Jacobson, D. G. Crosby, 307–32. New York: Dekker
16. Bowers, W. S., Blickenstaff, C. C. 1966. Hormonal termination of diapause in the alfalfa weevil. *Science* 154:1673–74

17. Bowers, W. S., Fales, H. M., Thompson, M. J., Uebel, E. C. 1966. Juvenile hormone: Identification of an active compound from balsam fir. *Science* 154:1020–21
18. Bowers, W. S., Thompson, M. J., Uebel, E. C. 1965. Juvenile and gonadotropic activity of 10,11-epoxyfarnesenic acid methyl ester. *Life Sci.* 4:2323–31
19. Bracken, G. K., Nair, K. K. 1967. Stimulation of yolk deposition in an ichneumonid parasitoid by feeding synthetic juvenile hormone. *Nature* 216:483–84
20. Bransby-Williams, W. R. 1971. Juvenile hormone activity of ethylfarnesoate dihydrochloride with the cotton stainer *Dysdercus cardinalis* Gerst. *Bull. Entomol. Res.* 61:41–47
21. Bransby-Williams, W. R. 1972. Some effects of the juvenile hormone analogue R-20458 on *Culex pipiens fatigans* mosquitoes. *East Afr. Med. J.* 49:509–12
22. Bransby-Williams, W. R. 1972. Activity of two juvenile hormone analogues with *Heliothis armigera* (Hübner), *Sitophilus zeamais* Motshulsky and *Ephestia cautella* (Walker). *East Afr. Agr. Forest. J.* 38:170–74
23. Brooks, G. T. 1973. Insect epoxide hydrase inhibition by juvenile hormone analogues and metabolic inhibitors. *Nature* 245:382–84
24. Brown, A. W. A. 1973. *Insect Resistance. A Continuing Problem Requiring New Chemical Control Agents.* Pap. Presented at Entomol. Soc. Am. Ann. Meet., Dallas, Texas, 1973
25. Bull, D. L. et al 1973. Effects of synthetic juvenile hormone analogues on certain injurious and beneficial arthropods associated with cotton. *J. Econ. Entomol.* 66:623–26
26. Burov, V. N., Drabkina, A. A., Sazonov, A. P., Tsizin, Y. S. 1970. A preliminary evaluation of synthetic analogues of a juvenile hormone as an insecticide for *Eurygaster integriceps. Zool. Zh.* 49:1802–9
27. Campbell, J. B., Wright, J. E. 1973. Efficacy of a juvenile hormone (Stauffer's R-20458) for control of stable flies under field conditions. *Proc. N. Cent. Br. Entomol. Soc. Am.* 28:197
28. Carayon J., Thouvenin, M. 1966. Emploi d'une substance mimétique de l'hormone juvénile pour la lutte contre les *Dysdercus,* hémiptères nuisibles au cotonnier. *Extr. Proc. Verb. Acad. Agr. Fr.* 1966:340–46
29. Cassier, P. 1967. La reproduction des insectes et la régulation de l'activité des corps allates. *Ann. Biol.* 6:595–670
30. Cerf, D. C., Georghiou, G. P. 1972. Evidence of cross-resistance to a juvenile hormone analogue in some insecticide-resistant houseflies. *Nature* 239:401–2
31. Černý, V., Dolejš, L., Lábler, L., Šorm, F., Sláma, K. 1967. Dehydrojuvabione, a new compound with juvenile hormone activity from balsam fir. *Collect. Czech. Chem. Commun.* 32:3926–31
32. Chamberlain, W. F., Hopkins, D. E. 1970. Morphological and physiological changes in *Bovicola limbata* reared on diet containing synthetic juvenile hormone. *Ann. Entomol. Soc. Am.* 63: 1363–65
33. Chamberlain, W. F., Hopkins, D. E., Gingrich, A. R. 1973. Cattle biting louse: Evaluation of compounds for juvenile hormone activity. *J. Econ. Entomol.* 66:127–30
34. Chamberlain, W. F., Hopkins, D. E., Schwarz, M. 1973. Cattle biting louse: Evaluation of compounds for juvenile hormone activity. 2. Hormonal and toxic effects. *J. Econ. Entomol.* 66: 703–6
35. Chippendale, G. M., Yin, C. M. 1973. Endocrine activity retained in diapause insect larvae. *Nature* 246:511–13
36. Chung-Der Huang, Treece, R. E. 1973. Effects of juvenile hormone mimics on the face fly. *Proc. N. Cent. Br. Entomol. Soc. Am.* 28:196
37. Connin, R. V., Jantz, O. K., Bowers, W. S. 1967. Termination of diapause in the cereal leaf beetle by hormones. *J. Econ. Entomol.* 60:1752–53
38. Coombes, L. E., Meisch, M. V. 1973. Control of mosquito larvae with an insect growth regulator in various water types. *Arkansas Farm Res.* 22:4
39. Critchley, B. R., Campion, D. G. 1971. Effects of a juvenile hormone analogue on growth and reproduction in the cotton stainer *Dysdercus fasciatus* Say. *Bull. Entomol. Res.* 61:49–53
40. Cruickshank, P. A. 1971. Some juvenile hormone analogs—A critical appraisal. *Mitt. Schweiz. Entomol. Ges.* 44:98–133
41. Cumming, J. E., McKague, B. 1973. Preliminary studies of effects of juvenile hormone analogues on adult emergence of black flies. *Can. Entomol.* 105: 509–11
42. Cupp, E. W., O'Neal, J. 1973. The morphogenetic effects of two juvenile hormone analogues on larvae of imported fire ants. *Environ. Entomol.* 2:191–94

43. Cuzin-Roudy, J. 1969. Etude comparée, chez *Dysdercus discolor* Walker, de la croissance normale et de la croissance en présence du "paper factor." *Arch. Zool. Exp. Gen.* 109:3–23
44. Dahm, K. H., Röller, H., Trost, B. M. 1968. The juvenile hormone. IV. Stereochemistry of juvenile hormone and biological activity of some of its isomers and related compounds. *Life Sci.* 7: 129–37
45. Deseö, K. V. 1972. The role of farnesylmethyl-ether applied on the male influencing the oviposition of the female codling moth (*Laspreyresia pomonella* L.). *Acta Phytopathol. Acad. Hung.* 7:257–66
46. De Wilde, J. 1970. Hormones and Insect Diapause. *Mem. Soc. Endocrinol.* 18:487–514
47. Djerassi, C., Shih-Coleman, C., Diekman, J. 1974. Operational and policy aspects of future insect control methods. *Science* In press
48. Dunn, R. L., Strong, F. E. 1973. Control of catch-basin mosquitos using Zoecon ZR 515 formulated in a slow release polymer—A preliminary report. *Mosquito News* 33:110–11
49. Dyte, C. E. 1972. Resistance to synthetic juvenile hormone in a strain of the flour beetle, *Tribolium castaneum. Nature* 238:48–49
50. Engelmann, F. 1968. Endocrine control of reproduction in insects. *Ann. Rev. Entomol.* 13:1–26
51. Fockler, C. E., Borden, J. H. 1973. Mating activity and ovariole development of *Trypodendron lineatum:* Effect of a juvenile hormone analogue. *Ann. Entomol. Soc. Am.* 66:509–12
52. Gelbič, I., Sehnal, F. 1973. Effects of juvenile hormone mimics on the codling moth *Cydia pomonella* (L.). *Bull. Entomol. Res.* 63:7–16
53. Gilbert, L. I. 1962. Maintenance of the prothoracic gland by the juvenile hormone in insects. *Nature* 193:1205–7
54. Gilbert, L. I., King, D. S. 1973. Physiology of growth and development: Endocrine aspects. In *The Physiology of Insecta,* Vol. 1. New York: Academic
55. Gilbert, L. I., Schneiderman, H. A. 1958. The inactivation of juvenile hormone extracts by pupal silkworms. *Anat. Rec.* 131:557–58
56. Gilbert, L. I., Schneiderman, H. A. 1959. Prothoracic gland stimulation by juvenile hormone extracts of insects. *Nature* 184:171–73
57. Gilbert, L. I., Schneiderman, H. A. 1960. The development of a bioassay for the juvenile hormone of insects. *Trans. Am. Microsc. Soc.* 79:38–67
58. Gill, S. S., Hammock, B. D., Yamamoto, I., Casida, J. E. 1972. Preliminary chromatographic studies on the metabolites and photodecomposition products of the juvenoid 1-(4'-ethylphenoxy)-6,7-epoxy-3,7-dimethyl-2-octene. In *Insect Juvenile Hormones,* ed. J. J. Menn, M. Beroza, 177–89. New York: Academic
59. Grassmann, A., Geyer, A., Herda, G., Schmialek, P. 1968. Die Inaktivierung von Juvenilhormonwirksamen Substanzen in Insekten. *Acta. Entomol. Bohemoslov.* 64:92–99
60. Graves, T. M., Senior, L. J. 1973. Microencapsulated insect growth regulator (ALTOSID™ SR-10) for mosquito control. *Abstr. 57th Ann. Meet. Pac. Br. Entomol. Soc. Am.* 36–37
60a. Guerra, A. A., Wolfenbarger, D. A., Garcia, R. D. 1973. Activity of juvenile hormone analogues against the tobacco budworm. *J. Econ. Entomol.* 66:833–35
61. Hangartner, W., Masner, P. 1973. Juvenile hormone: Inhibition of ecdysis in larvae of the German cockroach, *Blattella germanica. Experientia* 29: 1358–59
62. Hangarter, W., Peyer, B., Meier, W. 1971. Effects of a juvenile hormone analogue on the apterous form of the bean aphid, *Aphis fabae* Scop. *Meded Fac. Landbouwwetensch. Gent.* 36:866–73
63. Harris, R. L., Frazer, E. D., Younger, R. L. 1973. Horn flies, stable flies, and house flies: Development in feces of bovines treated orally with juvenile hormone analogues. *J. Econ. Entomol.* 66:1099–1102
64. Henrick, C. A., Staal, G. B., Siddall, J. B. 1973. Alkyl 3,7,11-trimethyl-2,4-dodecadienoates, a new class of potent insect growth regulators with juvenile hormone activity. *Agr. Food. Chem.* 21:354–59
65. Herald, F., Knapp, F. W. 1973. Juvenile hormone mimics fed to poultry for house fly control. *Proc. N. Cent. Br. Entomol. Soc. Am.* 28:197
66. Hille Ris Lambers, D. 1966. Polymorphism in Aphididae. *Ann. Rev. Entomol.* 11:47–78
67. Hodek, I., Ruzicka, Z., Sehnal, F. 1973. Termination of diapause by juvenoids in two species of ladybirds. *Experientia* 29:1146–47

68. Homberger, E., Benz, G., Thommen, H. 1971. Investigations with dichlorofarnesenic acid ethyl ester on the cotton stainer *Dysdercus cingulatus* Fabr. *Mitt. Schweiz. Entomol. Ges.* 44:193–95
69. Hopkins, D. E., Chamberlain, W. F. 1972. Susceptibility of the stages of the cattle biting louse to Juveth, an insect juvenile hormone analog. *J. Wash. Acad. Sci.* 62:258–60
70. Hopkins, D. E., Chamberlain, W. F., Wright, J. E. 1970. Morphological and physiological changes in *Bovicola limbata* treated topically with a juvenile hormone analogue. *Ann. Entomol. Soc. Am.* 63:1361–63
71. Hrdý, I. 1972. Der Einfluss von zwei Juvenilhormonanalogen auf die Differenzierung der Soldaten bei *Reticulitermes lucifugus santonensis* Feyt. *Z. Angew. Entomol.* 72:129–34
72. Hrdý, I. 1973. Effect of juvenoids on termites and honeybees. *Proc. 7th Int. Congr. Union Study Social Insects, London,* 158–61
73. Hrdý, I., Křeček, J. 1972. Development of superfluous soldiers induced by juvenile hormone analogues in the termite, *Reticulitermes lucifugus santonensis. Insectes Soc.* 19:105–9
74. Hrdý, I., Zelený, J. 1973. Effects of juvenoids on the population density of *Phorodon humuli* in a cage experiment. *Acta. Entomol. Bohemoslov.* 70:386–89
75. Hsiao, C., Hsiao, T. H. 1969. Insect hormones: Their effects on diapause and development of Hymenoptera. *Life Sci.* 8:767–74
76. Hunt, L. M., Shappirio, D. G. 1973. Larval development following juvenile hormone analogue treatment of eggs of the lesser milkweed bug, *Lygaeus kalmii. J. Insect Physiol.* 19:2129–34
77. Jacobson, M. et al 1972. Juvenile hormone activity of a variety of structural types against several insect species. In *Insect Juvenile Hormones,* ed. J. J. Menn, M. Beroza, 249–302. New York: Academic
78. Jakob, W. L. 1973. Insect development inhibitors: Tests with house fly larvae. *J. Econ. Entomol.* 66:819–20
79. Jakob, W. L., Schoof, H. F. 1971. Studies with juvenile hormone-type compounds against mosquito larvae. *Mosquito News* 31:540–43
80. Jarolim, V., Hejno, K., Sehnal, F., Šorm, F. 1969. Natural and synthetic materials with insect hormone activity 8. Juvenile activity of the farnesane-type compounds on *Galleria mellonella. Life Sci.* 8:831–41
81. Jobin, L., Retnakaran, A. 1974. Symposium contribution to the E.S.A. workshop on insect growth regulators, Las Vegas, U.S.A., 1974.
82. Judy, K. J. et al 1973. Isolation, structure, and absolute configuration of a new natural insect juvenile hormone from *Manduca sexta. Proc. Nat. Acad. Sci. USA* 70:1509–13
83. Judy, K. J., Gilbert, L. I. 1969. Morphology of the alimentary canal during the metamorphosis of *Hyalophora cecropia. Ann. Entomol. Soc. Am.* 62:1438–46
84. Judy, K. J., Gilbert, L. I. 1970. Histology of the alimentary canal during the metamorphosis of *Hyalophora cecropia* (L.). *J. Morphol.* 131:277–300
85. Kamm, J. A., Swenson, K. G. 1972. Termination of diapause in *Draeculacephala crassicornis* with synthetic juvenile hormone. *J. Econ. Entomol.* 65:364–67
86. Karnavar, G. K. 1973. Effects of synthetic juvenile hormone on diapause and metamorphosis of a stored grain pest, *Trogoderma granarium. Indian J. Exp. Biol.* 11:138–40
87. Kennedy, J. S. 1961. Continuous polymorphism in locusts. *Symp. Insect Polymorphism Roy. Entomol. Soc. London* 1961:65–74
88. Kontev, C. A., Žďárek, J., Slàma, K., Sehnal, F., Romaňuk, M. 1973. Laboratory and field effectiveness of selected juvenoids on the wheat pest *Eurygaster integriceps. Acta. Entomol. Bohemoslov.* 70:377–85
89. Krishnakumaran, A., Schneiderman, H. A. 1965. Prothoracotropic activity of compounds that mimic juvenile hormone. *J. Insect Physiol.* 11:1517–32
90. Kuhr, R. J., Cleere, J. S. 1973. Toxic effects of synthetic juvenile hormones on several aphid species. *J. Econ. Entomol.* 66:1019–22
91. Landa, V. 1972. Analogues of the juvenile hormone as chemosterilants. *Abstr. 14th Int. Congr. Entomol, Canberra* 1972:232
92. Law, J. H., Yuan, C., Williams, C. M. 1966. Synthesis of a material with high juvenile hormone activity. *Proc. Nat. Acad. Sci USA* 55:576–78
93. Lees, A. D. 1969. The control of polymorphism in aphids. In *Advances in Insect Physiology,* ed. J. W. L. Beament, J. E. Treherne, V. B. Wigglesworth, 3:207–77. New York: Academic

94. Lewallen, L. L. 1964. Effects of farnesol and Ziram on mosquito larvae. *Mosquito News* 24:43–45
95. Lewis, L. C., Lynch, R. E., Berry, E. C. 1973. Synthetic juvenile hormones: Activity versus the European corn borer in the field and the laboratory. *J. Econ. Entomol.* 66:1315–17
96. Lewis, L. F., Christenson, D. M. 1972. Studies with a juvenile hormone analogue for the control of *Culex pipiens quinquefasciatus* Say. *Proc. 40th Ann. Conf. Calif. Mosquito Contr. Assoc.* 1972:49–50
97. Loew, P., Johnson, W. S. 1971. Synthesis of optically active form of C-18 cecropia juvenile hormone. *J. Am. Chem. Soc.* 93:3765–66
98. Lüscher, M. 1972. Environmental control of juvenile hormone (JH) secretion and caste differentiation in termites. *Gen. Comp. Endocrinol. Suppl.* 3: 509–14
99. Lüscher, M. 1973. The influence of the composition of experimental groups on caste development in *Zootermopsis* (Isoptera). *Proc. 7th Int. Congr. Union Study Social Insects, London.* 1973: 253–56
100. Madhavan, K. 1973. Morphogenetic effects of juvenile hormone and juvenile hormone mimics on adult development of *Drosophila. J. Insect Physiol.* 19: 441–53
101. Masner, P. 1968. The inductors of differentiation of prefollicular tissue and the follicular epithelium in ovarioles of *Pyrrhocoris apterus. J. Embryol. Exp. Morphol.* 20:1–13
102. Masner, P. 1969. The effect of substances with juvenile hormone activity on morphogenesis and function of gonads in *Pyrrhocoris apterus. Acta. Entomol. Bohemoslov.* 66:81–86
103. Masner, P., Sláma, K., Landa, V. 1968. Sexually spread insect sterility induced by the analogues of juvenile hormone. *Nature* 219:395–96
104. Masner, P., Sláma, K., Landa, V. 1968. Natural and synthetic materials with insect hormone activity. IV. Specific female sterility effects produced by a juvenile hormone analogue. *J. Embryol. Exp. Morphol.* 20:25–31
105. Masner, P., Sláma, K., Ždárek, J., Landa, V. 1970. Natural and synthetic materials with insect hormone activity. X. A method of sexually spread insect sterility. *J. Econ. Entomol.* 63:706–10
106. Matolín, S. 1970. Effects of a juvenile hormone analog on embryogenesis in *Pyrrhocoris apterus* L. *Acta Entomol. Bohemoslov.* 67:9–12
107. Matolín, S. 1971. Effets d'un analogue de l'hormone juvenile sur les embryons de trois ordres d'insectes. *Arch. Zool. Exp. Gen.* 112:505–9
108. Matolín, S., Rohdendorf, E. B. 1972. Effects of FME on the embryogenesis of *Lepisma inquilinus* (= *Thermobia domestica*). *Acta. Entomol. Bohemoslov.* 69:1–6
109. Metwally, M. M., Landa, V. 1972. Sterilization of the khapra beetle, *Trogoderma granarium* Everts, with juvenile hormone analogues. *Z. Angew. Entomol.* 72:97–109
110. Metwally, M. M., Sehnal, F. 1973. Effects of juvenile hormone analogues on the metamorphosis of beetles *Trogoderma granarium* and *Caryedon gonagra. Biol. Bull.* 144:368–82
111. Metwally, M. M., Sehnal, F., Landa, V. 1972. Reduction of fecundity and control of the khapra beetle by juvenile hormone mimics. *J. Econ. Entomol.* 65:1603–5
112. Meyer, A. S. 1972. Cecropia juvenile hormone: Harbinger of a new age in pest control. In *Insect Juvenile Hormones,* ed. J. J. Menn, M. Beroza, 317–35. New York: Academic
113. Meyer, A. S., Hanzmann, E. 1970. The optical activity of the cecropia juvenile hormones. *Biochem. Biophys. Res. Commun.* 41:891–93
114. Meyer, A. S., Schneiderman, H. A., Hanzmann, E. 1968. The two juvenile hormones from the cecropia silk moth. *Proc. Nat. Acad. Sci. USA* 60:853
115. Meyer, D., Lüscher, M. 1973. Juvenile hormone activity in the haemolymph and the anal secretion of the queen of *Macrotermes subhyalinus* (Rambur). *Proc. 7th Int. Congr. Union Study Social Insects, London* 1973:268–72
116. Miller, R. W. 1970. Larvicides for fly control—a review. *Bull Entomol. Soc. Am.* 16:154–57
117. Miller, R. W., Uebel, E. C. 1973. Juvenile hormone mimics as feed additives for control of the face fly and house fly. *J. Econ. Entomol.* 67:69–70
118. Miura, T., Takahashi, R. M. 1973. Insect developmental inhibitors. 3. Effects on nontarget aquatic organisms. *J. Econ. Entomol.* 66:917–22
119. Mulla, M. S., Norland, R. L., Ikeshoji, T., Kramer, W. L. 1974. Insect growth regulators for the control of aquatic midges. *J. Econ. Entomol.* 67:165–70

120. Nair, K. S. S. 1974. Studies of the diapause of *Trogoderma granarium:* Effects of juvenile hormone analogues on growth and metamorphosis. *J. Insect Physiol.* 20:231–44
121. Nassar, S. G., Staal, G. B., Armanious, N. I. 1973. Effects and control potential of insect growth regulators with juvenile hormone activity on the greenbug. *J. Econ. Entomol.* 66:847–50
122. Nassar, S. G., Staal, G. B., Martin, J. W. 1972. The biological effects of juvenile hormone analogs on the development of latania scale, *Hemiberlesis lataniae. 56th Ann. Meet. Pac. Br. Entomol. Soc. Am.*
123. Neal, J. W. Jr., Hower, A. A. Jr. 1974. Small plot tests with synthetic juvenile hormones on alfalfa weevil in Maryland and Pennsylvania. *J. Econ. Entomol.* 67:93–96
124. Nemec, V. 1971. Effects de certains analogues de l'hormone juvénile sur deux espèces de criquets; *Locusta migratoria migratorioides* (R. et L.) et *Schistocerca gregaria* (Forsk). *Arch. Zool. Exp. Gen.* 112:511–17
125. Nemec, V., Jarolím, V., Hejno, K., Šorm, F. 1970. Natural and synthetic materials with insect hormone activity. 7. Juvenile activity of the farnesane-type compounds on *Locusta migratoria* L. and *Schistocerca gregaria* (Forsk). *Life Sci.* 9:821–31
126. Nilles, G. P., Zabik, M. J., Connin, R. V., Schuetz, R. D. 1973. Synthesis of bioactive compounds: Juvenile hormone mimetics affecting insect diapause. *J. Agr. Food Chem.* 21:342–47
127. Norland, R. L. 1973. Comparison of juvenile hormone analogue formulations against chironomid midges under semifield conditions. *Proc. 41st Ann. Conf. Calif. Mosquito Contr. Assoc.* 1973:118
128. Novák, K., Sehnal, F. 1973. Action of juvenile hormone analogues on *Euproctis chrysorrhoea* and *Yponomeuta malinella* under field conditions. *Acta. Entomol. Bohemoslov.* 70:20–29
129. Novák, V., Sehnal, F. 1973. Effects of a juvenoid applied under field conditions to the green oak leaf roller, *Tortrix viridana* L. *Z. Angew. Entomol.* 73: 312–18
130. Novák, V. J. A. 1966. *Insect Hormones.* London: Methuen. 478 pp.
131. Novák, V. J. A. 1969. Morphogenetic analysis of the effects of juvenile hormone analogs and other morphogenetically active substances on embryos of *Schistocerca gregaria* (Forskål). *J. Embryol. Exp. Morphol.* 21:1–21
132. Novák, V. J. A. 1972. Juvenilized insect embryos and the principle of animal morphogenesis. *Endocrinol. Exp.* 6: 251–68
133. Ohtaki, T., Takeuchi, S., Mori, K. 1971. Juvenile hormone and synthetic analogues: Effects on larval moult of silkworm, *Bombyx mori. Jap. J. Med. Sci. Biol.* 24:251–55
134. O'Neal, J. 1973. *Juvenile Hormone Screening.* Pap. Presented at the Imported Fire Ant Res. Conf., New Orleans, LA. 1973.
135. Outram, I. 1972. Effects of synthetic juvenile hormone on adult emergence and reproduction of the female spruce budworm, *Choristoneura fumiferana. Can. Entomol.* 104:271–73
136. Outram, I. 1973. Synthetic juvenile hormone: Effect on pupae of the spruce budworm. *J. Econ. Entomol.* 66: 1033–35
137. Outram, I. 1974. Influence of juvenile hormone on the development of some spruce budworm parasitoids. *Environ. Entomol.* 3:361–63
138. Pallos, F. M., Menn, J. J. 1972. Synthesis and activity of juvenile hormone analogs. In *Insect Juvenile Hormones,* ed. J. J. Menn, M. Beroza, 303–16. New York: Academic
139. Pallos, F. M., Menn, J. J., Letchworth, P. E., Miaullis, J. B. 1971. Synthetic mimics of insect juvenile hormone. *Nature* 232:486–87
140. Pan, M. L., Wyatt, G. R. 1971. Juvenile hormone induces vitellogenin synthesis in the monarch butterfly. *Science* 174:503–5
141. Patterson, J. W. 1971. Critical sensitivity of the ovary of *Aedes aegypti* adults to sterilization by juvenile hormone mimics. *Nature New Biol.* 233:176–77
142. Patterson, J. W. 1971. Some effects of juvenile hormone mimics on *Aedes aegypti* and *Rhodnius prolixus. Trans. Roy. Soc. Trop. Med. Hyg.* 65:435
143. Patterson, J. W. 1972. Reduction in fertility induced in *Aedes aegypti* by larval contact with a juvenile hormone mimic. *Trans. Roy. Soc. Trop. Med. Hyg.* 66:21
144. Patterson, J. W. 1973. Prospects of using juvenile hormone mimics in the control of triatomine bugs. *Trans. Roy. Soc. Trop. Med. Hyg.* 67:306
145. Patterson, J. W. 1973. The effect of the persistence of juvenile hormone mimics

on their activity on *Rhodnius prolixus*. *J. Insect Physiol.* 19:1631–37
146. Pawson, B. A., Scheidl, F., Vane, F. 1972. Environmental stability of juvenile hormone mimicking agents. In *Insect Juvenile Hormones,* ed. J. J. Menn, M. Beroza, 191–214. New York: Academic
147. Pipa, R. L. 1963. Ventral nerve cord shortening; a metamorphic process in *Galleria mellonella* L. *Biol. Bull.* 124:293–302
148. Plapp, F. W. Jr., Vinson, S. B. 1973. Juvenile hormone analogs: Toxicity and cross-resistance in the housefly. *Pestic. Biochem. Physiol.* 3:131–36
149. Poduška, K., Šorm, F., Sláma, K. 1971. 9. Structure-juvenile activity relationship in simple peptides. *Z. Naturforsch. B* 26:719–22
150. Quistad, G. B., Staiger, L. E., Schooley, D. 1974. Environmental degradation of the insect growth regulator methoprene (isopropyl (2E, 4E)-11-methoxy-3,7,11-trimethyl-2,4-dodecadienoate). I. metabolism by alfalfa and rice. *J. Agr. Food. Chem.* 22:582
151. Reddy, G., Krishnakumaran, A. 1973. Effects of diet on response to juvenile hormone in *Galleria mellonella* larvae. *Experientia* 29:621–22
152. Redfern, R. E., McGovern, T. P., Beroza, M. 1969. Juvenile hormone activity of Sesamex and related compounds in tests on the yellow mealworm. *J. Econ. Entomol.* 63:540–45
153. Redfern, R. E., Mills, G. D. Jr., Sonnet, P. E. 1972. Aziridines as potentiators of juvenile hormone activity in tests on the yellow mealworm and the large milkweed bug. *J. Econ. Entomol.* 65:1605–7
154. Reissig, W. H., Kamm, J. A. 1974. Effects of foliage sprays of juvenile hormone analogues on development of *Draeculacephala crassicornis. J. Econ. Entomol.* 67:181–83
155. Retnakaran, A. 1970. Blocking of embryonic development in the spruce budworm, *Choristoneura fumiferana* by some compounds with juvenile hormone activity. *Can. Entomol.* 102: 1592–96
156. Retnakaran, A., 1973. Ovicidal effects in the white pine weevil, *Pissodes strobi,* of a synthetic analogue of juvenile hormone. *Can. Entomol.* 105:591–94
157. Retnakaran, A. 1973. Hormonal induction of supernumerary instars in the spruce budworm, *Choristoneura fumiferana. Can. Entomol.* 105:459–61
158. Richmond, C. E. 1972. Juvenile hormone analogues tested on larvae of western spruce budworm. *J. Econ. Entomol.* 65:950–53
159. Riddiford, L. M. 1970. Prevention of metamorphosis by exposure of insect eggs to juvenile hormone analogs. *Science* 167:287–88
160. Riddiford, L. M. 1970. Effects of juvenile hormone on the programming of postembryonic development in eggs of the silkworm, *Hyalophora cecropia. Develop. Biol.* 22:249–63
161. Riddiford, L. M. 1971. Juvenile hormone and insect embryogenesis. *Mitt. Schweiz. Entomol. Ges.* 44:177–86
162. Riddiford, L. M. 1972. Juvenile hormone and insect embryonic development: Its potential role as an ovicide. In *Insect Juvenile Hormones,* ed. J. J. Menn, M. Beroza, 95–111. New York: Academic
163. Riddiford, L. M. 1974. The role of hormones in the reproductive behavior of the female wild silkmoths. In *Experimental Analysis of Insect Behavior,* ed. L. B. Browne, 278–85. New York: Springer
164. Riddiford, L. M., Ajami, A. M., Corey, E. J., Yamamoto, H., Anderson, J. E. 1971. Synthetic imino analogs of cecropia juvenile hormones as potentiators of juvenile hormone activity. *J. Am. Chem. Soc.* 93:1815–16
165. Riddiford, L. M., Truman, J. W. 1972. Delayed effects of juvenile hormone on insect metamorphosis are mediated by the corpus allatum. *Nature* 237:458
166. Riddiford, L. M., Truman, J. W. 1974. Hormones and insect behavior. In *Experimental Analysis of Insect Behavior,* ed. L. B. Browne, 286–96. New York: Springer
167. Riddiford, L. M., Williams, C. M. 1967. The effects of juvenile hormone analogues on the embryonic development of silkworms. *Proc. Nat. Acad. Sci. USA* 57:595–601
168. Rohdendorf, E. B., Sehnal, F. 1972. The induction of ovarian dysfunctions in *Thermobia domestica* by the cecropia juvenile hormones. *Experientia* 28: 1099–1101
169. Rohdendorf, E. B., Sehnal, F. 1973. Inhibition of reproduction and embryogenesis in the firebrat, *Thermobia domestica,* by juvenile hormone analogues. *J. Insect Physiol.* 19:37–56
170. Röller, H., Dahm, K. H. 1968. The chemistry and biology of juvenile hormone. *Recent Progr. Horm. Res.* 24: 651–79

171. Röller, H., Dahm, K. H., Sweeley, C. C., Trost, B. M. 1967. The structure of the juvenile hormone. *Angew. Chem. Int. Ed. Engl.* 6:179–80
172. Romaňuk, M., Sláma, K., Šorm, F. 1967. Constitution of a compound with a pronounced juvenile hormone activity. *Proc. Nat. Acad. Sci. USA* 57: 349–52
173. Rose, M., Westermann, J., Trautmann, H., Schmialek, P., Klauske, J. 1968. I. Juvenilhormonwirkungen bei *Tenebrio molitor* L. in Abhängigkeit von der Konzentration der hormonalen Substanz. *Z. Naturforsch. B* 23:1245–48
174. Sacher, R. M. 1971. A mosquito larvicide with favorable environmental properties. *Mosquito News* 31:513–16
175. Saxena, K. N., Williams, C. M. 1966. 'Paper factor' as an inhibitor of the metamorphosis of the red cotton bug, *Dysdercus koenigii* F. *Nature* 210: 441–42
176. Schaefer, C. H., Dupras, E. F. Jr. 1973. Insect developmental inhibitors. 4. Persistence of ZR-515 in water. *J. Econ. Entomol.* 66:923–25
177. Schaefer, C. H., Dupras, E. F. Jr., Wilder, W. H. 1973. Pond tests with ZR-515: Biological and chemical residues. *Proc. 41st Ann. Conf. Calif. Mosquito Contr. Assoc.* 1973:137–38
178. Schaefer, C. H., Wilder, W. H. 1972. Insect developmental inhibitors: A practical evaluation as mosquito control agents. *J. Econ. Entomol.* 65: 1066–71
179. Schaefer, C. H., Wilder, W. H. 1973. Insect developmental inhibitors. 2. Effects on target mosquito species. *J. Econ. Entomol.* 66:913–16
180. Schmialek, P. 1961. Die Identifizierung zweier in Tenebriokot und in Hefe vorkommender Substanzen mit Juvenilhormonwirkung. *Z. Naturforsch. B* 16: 461–64
181. Schooneveld, H. 1972. Effects of juvenile hormone on the endocrine system during break of diapause in the Colorado beetle. *J. Endocrinol.* 57:55–56
182. Schwartz, J. L. 1971. Inhibition of nerve cord metamorphosis in the western spruce budworm *Choristoneura occidentalis* Freeman by juvenile hormone analogs. *Gen. Comp. Endocrinol.* 17:293–99
183. Schwieter-Peyer, B. 1973. Comparison of the effect of the geometrical isomers of juvenile hormone on *Blattella germanica, Dysdercus cingulatus,* and *Sitophilus granarius. Insect Biochem.* 3:275–82
184. Sehnal, F. 1968. Influence of the corpus allatum on the development of internal organs in *Galleria mellonella* L. *J. Insect Physiol.* 14:73–85
185. Sehnal, F. 1971. Juvenile hormone action and insect growth rate. *Endocrinol. Exp.* 5:29–33
186. Sehnal, F., Meyer, A. S. 1968. Larval-pupal transformation: Control by juvenile hormone. *Science* 159:981–83
187. Sehnal, F., Schneiderman, H. A. 1973. Action of the corpora allata and of juvenilizing substances on the larval-pupal transformation of *Galleria mellonella* L. *Acta. Entomol. Bohemoslov.* 70:289–302
188. Singh, Y. N., Srivastava, U. S. 1973. Anatomical changes in the nervous system of the castor silk moth, *Philosamia ricini* Hutt. during metamorphosis. *Int. J. Insect Morphol. Embryol.* 2:169–75
189. Sipcam Research Center, Milan, Italy. 1973. *Report on Results of 1973 Tests with Insect Growth Regulators and Pherocons of Zoecon Corp.*
190. Skuhravý, V. 1973. Field control of the larch case-bearer moth, *Coleophora laricella,* with a juvenoid. *Acta. Entomol. Bohemoslov.* 70:313–23
191. Skuhravý, V., Sehnal, F. 1973. Inhibition of metamorphosis by juvenoids in the gall midge *Dasyneura laricis. Z. Angew. Entomol.* 74:217–20
192. Slade, M., Wilkinson, C. F. 1973. Juvenile hormone analogs: A possible case of mistaken identity? *Science* 181:672–74
193. Sláma, K. 1971. Insect juvenile hormone analogues. *Ann. Rev. Biochem.* 40:1079–1102
194. Sláma, K., Romanuk, M., Sorm, F. 1974. *Insect Hormones and Bioanalogues.* Wien: Springer. 477 pp.
195. Sláma, K., Williams, C. M. 1965. Juvenile hormone activity for the bug *Pyrrhocoris apterus. Proc. Nat. Acad. Sci. USA* 54:411–14
196. Sláma, K., Williams, C. M. 1966. 'Paper factor' as an inhibitor of embryonic development of the European bug, *Pyrrhocoris apterus. Nature* 210:329–30
197. Socha, R., Gelbič, I. 1973. Stages in the differentiation of ovarian development in *Dixippus morosus* as revealed by the effects of juvenoids. *Acta. Entomol. Bohemoslov.* 70:303–12
198. Solomon, K. R., Bowlus, S. B., Metcalf, R. L., Katzenellenbogen, J. A. 1973. The effect of piperonyl butoxide and triorthocresyl phosphate on the activity of

juvenile hormone mimics and their sulphur isosteres in *Tenebrio molitor* L. and *Oncopeltus fasciatus* (Dallas). *Life Sci.* 13:733–42

199. Sonnet, P. E., Mills, G. D. Jr., Redfern, R. E. 1973. Synergism of synthetic mixed isomers of juvenile hormone. *J. Econ. Entomol.* 66:793–94
200. Spielman, A. 1970. Synthetic juvenile hormones as larvicides for mosquitoes. *Ind. Trop. Health* 7:67–69
201. Spielman, A., Skaff, V. 1967. Inhibition of metamorphosis and of ecdysis in mosquitoes. *J. Insect Physiol.* 13: 1087–95
202. Spielman, A., St. Onge, E. 1974. Stability of exogenous juvenile hormone: Effect of larval mosquitoes. *Environ. Entomol.* 3:259–61
203. Spielman, A., Williams, C. M. 1966. Lethal effects of synthetic juvenile hormone on larvae of the yellow fever mosquito, *Aedes aegypti. Science* 154: 1043–44
204. Srivastava, U. S., Gilbert, L. I. 1968. Juvenile hormone: Effects on a higher dipteran. *Science* 161:61–62
205. Staal, G. B. 1961. Studies on the physiology of phase induction in *Locusta migratoria migratorioides* R. & F. *Publ. Fonds Land. Export Bur.* 40:1–25
206. Staal, G. B. 1967. Plants as a source of insect hormones. *Proc. Kon. Ned. Akad. Wetensch.* 70C:409–18
207. Staal, G. B. 1972. Biological activity and bioassay of juvenile hormone analogs. In *Insect Juvenile Hormones,* ed. J. J. Menn, M. Beroza, 69–94. New York: Academic
208. Staal, G. B., Nassar, S. G., Martin, J. W. 1973. Control of the citrus mealybug with insect growth regulators with juvenile hormone activity. *J. Econ. Entomol.* 66:851–53
209. Steelman, C. D., Schilling, P. E. 1972. Effects of a juvenile hormone mimic on *Psorophora confinnis* (Lynch-Arribalzaga) and non-target aquatic insects. *Mosquito News* 32:350–54
210. Strong, R. G., Diekman, J. 1973. Comparative effectiveness of fifteen insect growth regulators against several pests of stored products. *J. Econ. Entomol.* 66:1167–73
211. Suzuki, Y., Imai, K., Marumo, S., Mitsui, T. 1972. Juvenile hormone activity of S-(–)-, and R-(+)-10,11-epoxy-farnesic acid methyl ester and its alcohol analogue. *Agr. Biol. Chem.* 36: 1849–50
212. Tamaki, G. 1973. Insect developmental inhibitors: Effect of reduction and delay caused by juvenile hormone mimics on the production of winged migrants of *Myzus persicae* on peach trees. *Can. Entomol.* 105:761–65
213. Tang, P. M., Tseng, H. C. 1971. A preliminary observation on the mechanism of sterilization of the brown plant hopper (*Nilaparvata lugens* Stal.) treated with chemosterilants and synthetic juvenile hormones. *Satca Rev.* 16:1–7
214. Thomas, J. P., Bhatnagar-Thomas, P. L. 1968. Use of a juvenile hormone analogue as insecticide for pests of stored grain. *Nature* 219:949
215. Troisi, S. J., Riddiford, L. M. 1974. Juvenile hormone effects on metamorphosis and reproduction of the fire ant, *Solenopsis invicta. Environ. Entomol.* 3:112–16
216. Truman, J. W., Riddiford, L. M. 1974. Hormonal mechanisms underlying insect behavior. *Advan. Insect Physiol.* 10:297–348
217. Van de Veire, M., De Loof, A. 1973. Effects of synthetic cecropia juvenile hormone and of a mimic substance, applied to different stages of the cabbage maggot, *Hylemyia brassicae. Entomol. Exp. Appl.* 16:491–98
218. Varjas, L., Sehnal, F. 1973. Use of a juvenile hormone analogue against the fall webworm, *Hyphantria cunea. Entomol. Exp. Appl.* 16:115–22
219. Vinson, J. W., Williams, C. M. 1967. Lethal effects of synthetic juvenile hormone on the human body louse. *Proc. Nat. Acad. Sci. USA* 58:294–97
220. Von Dehn, M. 1963. Hemmung der Flügelbildung durch Farnesol bei der schwarzen Bohnenlaus, *Doralis fabae* Scop. *Naturwissenschaften* 50:578–79
221. Waheed, A., Fisk, F. W., Collins, W. J. 1973. Effects of a juvenile hormone analog on the development and metamorphosis of the German cockroach, *Blattella germanica. Proc. N. Cent. Br. Entomol. Soc. Am.* 28:166
222. Walker, W. F. 1973. Mexican bean beetle: Compounds with juvenile hormone activity (juvegens) as potential control agents. *J. Econ. Entomol.* 66:30–33
223. Walker, W. F., Bowers, W. S. 1970. Synthetic juvenile hormones as potential coleopteran ovicides. *J. Econ. Entomol.* 63:1231–33
224. Walker, W. F., Bowers, W. S. 1973. Comparative juvenile hormone activity of some terpenoid ethers and esters on

selected Coleoptera. *J. Agr. Food Chem.* 21:145–48

225. Wanyonyi, K. 1974. The influence of the juvenile hormone analogue ZR-512 (Zoecon) on caste development in *Zootermopsis nevadensis* (Hagen). *Insectes Soc.* 21:35–44
226. Wanyonyi, K., Lüscher, M. 1973. The action of juvenile hormone analogues on caste development in *Zootermopsis. Proc. 7th Int. Congr. Union Study Social Insects, London* 1973:392–95
227. Webb, R. E., Smith, F. F., Boswell, A. L., Fields, E. S., Waters, R. M. 1974. Insecticidal control of the greenhouse whitefly on greenhouse ornamental and vegetable plants. *J. Econ. Entomol.* 67:114–18
228. Weirich, G., Wren, J., Siddall, J. B. 1973. Developmental changes of the juvenile hormone esterase activity in haemolymph of the tobacco hornworm, *Manduca sexta. Insect Biochem.* 3:397–407
229. Wellington, W. G. 1969. Effects of three hormonal mimics on mortality, metamorphosis and reproduction of the western tent caterpillar, *Malacosoma californicum ploviale. Can. Entomol.* 101:1163–72
230. Westigard, P. H. 1974. Control of the pear psylla with insect growth regulators and preliminary effects on some non-target species. *Environ. Entomol.* 3:256–58
231. Wheeler, C. M., Thebault, M. 1971. Efficacy of synthetic substances with hormonal activity against IV instar larvae of *Culex pipiens quinquefasciatus* Say in an outdoor insectary. *Mosquito News* 31:170–75
232. White, D. F. 1968. Postnatal treatment of the cabbage aphid with a synthetic juvenile hormone. *J. Insect Physiol.* 14:901–12
233. White, D. F. 1971. Corpus allatum activity associated with development of wingbuds in cabbage aphid embryos and larvae. *J. Insect Physiol.* 17:761–73
234. White, D. F., Lamb, K. P. 1968. Effect of a synthetic juvenile hormone on adult cabbage aphids and their progeny. *J. Insect Physiol.* 14:395–402
235. Whitmore, D., Whitmore, E., Gilbert, L. I. 1972. Juvenile hormone induction of esterases: A mechanism for the regulation of juvenile hormone titer. *Proc. Nat. Acad. Sci. USA* 69:1592–95
236. Whitten, J. 1968. Metamorphic changes in insects. In *Metamorphosis,* ed. W. Etkin, L. I. Gilbert, 43–105. New York: Appleton
237. Wigglesworth, V. B. 1963. The juvenile hormone effect of farnesol and some related compounds: Quantitative experiments. *J. Insect Physiol.* 9:105–19
238. Wigglesworth, V. B. 1964. The hormonal regulation of growth and reproduction in insects. In *Advances in Insect Physiology,* ed J. W. L. Beament, J. E. Treherne, V. B. Wigglesworth, 2:248–336. New York: Academic
239. Wigglesworth, V. B. 1969. Chemical structure and juvenile hormone activity: Comparative tests on *Rhodnius prolixus. J. Insect Physiol.* 15:73–94
240. Wigglesworth, V. B. 1973. Assay on *Rhodnius* for juvenile hormone activity. *J. Insect Physiol.* 19:205–11
241. Williams, C. M., Sláma, K. 1966. The juvenile hormone. VI. Effects of the "paper factor" on growth and metamorphosis of the bug, *Pyrrhocoris apterus. Biol. Bull.* 130:247–53
242. Willis, J. H. 1974. Morphogenetic action of insect hormones. *Ann. Rev. Entomol.* 19:97–115
243. Willis, J. H., Lawrence, P. A. 1970. Deferred action of juvenile hormone. *Nature* 225:81–83
244. Wirtz, P. 1973. Differentiation in the honeybee larva: A histological, electron-microscopical and physiological study of caste induction in *Apis mellifera mellifera* L. *Meded. Landbouwhogesch. Wageningen* 73:1–155
245. Wirtz, P., Beetsma, J. 1972. Induction of caste differentiation in the honeybee (*Apis mellifera*) by juvenile hormone. *Entomol. Exp. Appl.* 15:517–20
246. Wright, J. E. 1970. Hormones for control of livestock arthropods. Development of an assay to select candidate compounds with juvenile hormone activity in the stable fly. *J. Econ. Entomol.* 63:878–83
247. Wright, J. E. 1972. Hormones for control of livestock arthropods. Effectiveness of three juvenile hormone analogues for control of stable flies. *J. Econ. Entomol.* 65:1361–64
248. Wright, J. E., Bowman, M. C. 1973. Determination of the juvenile hormone-active compound Altosid® and its stability in stable fly medium. *J. Econ. Entomol.* 66:707–9
249. Wright, J. E., Campbell, J. B., Hester, P. 1973. Hormones for control of livestock arthropods: Evaluation of two juvenile hormone analogues applied to breeding materials in small plot tests in

Nebraska and Florida for control of the stable fly. *Environ. Entomol.* 2:69–72

250. Wright, J. E., McGovern, T. P., Sarmiento, R., Beroza, M. 1974. Juvenile hormone activity of substituted aryl 3,7-dimethyl-6-octenyl ethers in the stable fly and house fly. *J. Insect Physiol.* 20:423–27
251. Wright, J. E., Spates, G. E. 1971. Biological evaluation of juvenile hormone compounds against pupae of the stable fly. *J. Agr. Food Chem.* 19:289–90
252. Wright, J. E., Spates, G. E. 1972. Laboratory evaluation of compounds to determine juvenile hormone activity against the stable fly. *J. Econ. Entomol.* 65:1346–49
253. Wright, J. E., Spates, G. E. 1972. A new approach in integrated control: Insect juvenile hormone plus a hymenopteran parasite against the stable fly. *Science* 178:1292–93
254. Wyatt, G. R. 1972. Insect hormones. In *Biochemical Actions of Hormones,* ed. G. Litwack, 386–490. New York: Academic
255. Yamamoto, R. T., Jacobson, M. 1962. Juvenile hormone activity of isomers of farnesol. *Nature* 196:908–9
256. Yin, C. M., Chippendale, G. M. 1973. Juvenile hormone regulation of the larval diapause of the southwestern corn borer *Diatraea grandiosella. J. Insect Physiol.* 19:2403–20
257. Zaoral, M., Sláma, K. 1970. Peptides with juvenile hormone activity. *Science* 170:92–93
258. Ždárek, J., Haragsim, O. 1974. Action of juvenoids on metamorphosis of the honey-bee *Apis mellifera. J. Insect Physiol.* 20:209–21
259. Ždárek, J., Sláma, K. 1972. Supernumerary larval instars in cyclorrhaphous Diptera. *Biol. Bull.* 142:350–57
260. Zoecon Technical Bulletin Altosid® IGR, Toxicological Properties, 1974

GENETICAL METHODS OF PEST CONTROL[1]

❖6098

M. J. Whitten and G. G. Foster
Division of Entomology, CSIRO, Canberra, 2601 Australia

A proportion of zygotes in all animal and plant populations carry genetic factors which render them unable to survive the range of environmental conditions they are likely to encounter during development. Lethal factors may arise from point mutations at loci essential for normal development, from some gross genetic imbalance caused by misadventure during the mitotic or meiotic process, or by zygosis of incompatible genomes. Lethals at individual loci are usually inherited as recessives; otherwise they are excluded immediately from the population. Recessive lethality probably accounts for the bulk of genetic death in natural populations. However, there are genetic conditions that give rise to genic or chromosomal imbalances which can be maintained in a population even though these imbalances act as dominant lethals. Lethality generated in this manner may be tolerated by a species and is sometimes exploited to advantage, as in maintaining heterozygosity during a switch from outbreeding to inbreeding in certain plants (30), or in the evolution of isolating mechanisms during speciation (75).

It is important at the outset to distinguish between the elimination of zygotes which cannot survive under any realistic conditions, and the elimination of zygotes incapable of competing successfully with superior genotypes encountered in their sphere of influence, but which would otherwise survive in the absence of such competition. We are concerned here only with unconditional lethality and its relationship to the biotic potential of a population.

The genetic load caused by lethality has to be accommodated by a species if its participation in the evolutionary process is to continue (71). Just how large this load is in natural populations and what the factors are which cause it are questions that have preoccupied population geneticists over the past few decades (35, 70). These studies have suggested ways of contriving lethality by utilizing natural phenomena, or of inducing lethality by employing agents such as irradiation and mutagenic chemicals. They have led to consideration of the possibility of deliberately manipu-

[1]The survey of the literature pertaining to this review was concluded in February 1974.

lating the level of genetic death in natural populations with a view to reducing the abundance of pests. Although an approach to pest control along these lines was outlined in 1940 by a Russian geneticist, Serebrovsky (61), the first significant application of these concepts was proposed by an entomologist, E. F. Knipling. In spite of general skepticism by geneticists (34), Knipling's ideas were applied by his colleagues with spectacular success to the suppression of the screwworm fly, *Cochliomyia hominivorax* (5). In this article we review the developments in the sterile insect release method (SIRM) subsequent to Proverbs' comprehensive treatment in Volume 14 of *Annual Review of Entomology* (49). However, the injection of high genetic loads into natural populations by SIRM or by sterilization of a portion of the field population (32), represents only one aspect of a whole new field of pest control. It is theoretically possible to alter the genetic composition of field populations or to generate high levels of genetic death by means of a single release of fertile rather than sterile insects. Therefore, we devote some space to describing these more ambitious goals of genetic control and discussing their prospects.

THE CURRENT STATUS OF THE STERILE INSECT RELEASE METHOD

In his review of SIRM, Proverbs stated that a need existed for another major success besides the screwworm project, to ensure continued support for similar programs. Six years later this statement is still true; the screwworm project remains without parallel. On a smaller scale, the eradication of the melon fly, *Dacus cucurbitae,* from the Mariana Islands has been substantiated along with the removal of the oriental fruit fly, *Dacus dorsalis,* from the same region by a combination of male annihilation and SIRM (9). The continuous release of irradiation-sterilized Mexican fruit fly, *Anastrepha ludens,* in Baja California is claimed to be responsible for the exclusion of gravid females from southern California (62), but we have not seen figures to substantiate this assertion. In addition to these few instances where SIRM has proved effective, a number of recent projects on diverse insect pests have demonstrated the biological feasibility of SIRM. After some initial failures of SIRM on mosquito control (54), successful reduction of field populations of *Culex fatigans* (= *C. pipiens quinquefasciatus*) (47) and *Anopheles albimanus* (37) has now been demonstrated. However, the failure to reduce population size of *C. fatigans* in villages near New Delhi, India, despite an intensive program (46), emphasizes the difficulty in implementing SIRM under field conditions in developing countries. Populations of the Mediterranean fruit fly, *Ceratitis capitata,* have been suppressed in Spain (39), Italy (16), and Nicaragua (57). Important examples of sustained suppression exist for the codling moth, *Laspeyresia pomonella* (50), and the cotton boll weevil, *Anthonomus grandis* (8).

In the remainder of this section, rather than recount the details of each SIRM project in progress (reports on these can be found elsewhere; see references 3, 5, 9, 23, 26, 29, 32, 34), we shall limit discussion to three case histories, to illustrate some of the biological, economic, and political facts of life which together define the current status of SIRM.

Southwestern Screwworm Eradication Program

Following eradication of the screwworm, *C. hominivorax*, from Curacao in 1954 and from Florida in 1959, the release of sterile screwworms in the southwestern USA was begun in 1962. By 1964–1965, the screwworm problem had virtually ceased to exist in the southwest (5), with only minor reinfestations until 1972. In 1972, the confirmed number of screwworm cases in the southwestern USA exceeded 90,000, compared with fewer than 500 cases in each of the three preceding years (43). In 1973, control of the screwworm appeared to have been reestablished during the spring and summer months, but more than 8000 cases in southern Texas alone were confirmed during October and November (E. C. Corristan, personal communication). Neither the cause of the breakdown nor its long-term significance has yet been determined (6).

Environmental factors which may have contributed to the present problems include unusually favorable weather conditions for screwworm activity during 1972 and 1973; an outbreak of the Gulf Coast tick, *Amblyomma maculatum*, in 1973; and, ironically, modifications in ranching practices which were made possible by the virtual eradication of the screwworm (6). Other possible factors are 1. loss of field adaptation of released flies through prolonged laboratory colonization, and 2. selection in the wild population for flies which do not mate with the released insects (6, 63). Alternative 1 seems less likely, since new screwworm breeding stocks have been established from field collections several times in recent years (6). Alternative 2, which was predicted by Smith & von Borstel (64), could result either from displacement of the original *C. hominivorax* strain in the release area by a migrant strain exhibiting some level of assortative mating, or from genetic change in the field populations enabling them to detect laboratory-reared flies regardless of genotype. If a migrant strain is responsible, the problem could have a simple solution, but the latter possibility could prove more difficult to remedy. It would require identification and modification of the critical feature(s) of the mass-rearing operation to remove its discriminatory effects. Although a genetic response leading to resistance to autocidal control has yet to be proven, the breakdown indicates a need for caution in claiming that this kind of resistance cannot develop (79). It also illustrates the dangers of passing control of SIRM projects, no matter how successful, to professional groups who do not appreciate the dynamic nature of biological material.

Some of the underlying causes of the screwworm project's present difficulties may be traced to its initial success in virtually eliminating the pest from the USA. Because this was achieved with an overkill strategy, requiring only a minimal knowledge of the screwworm's population biology, there appeared to be no need to conduct a sustained program of basic ecological research, other than as an aid to the operation of the release program. Consequently, research on population dynamics, dispersal, and behavior of *Cochliomyia hominivorax* in the field has lagged far behind progress in production and handling techniques (26). While we cannot dispute the elegance of a system which can be so successfully applied with scant knowledge of the target species' biology, with hindsight it is possible to see where some of the present problems might have been avoided if ecological research had received stronger support. Moreover, these problems are aggravated by a continuing struggle for adequate funds even though the project has saved the livestock industry in excess of one billion dollars since 1962 (6). It serves to highlight the difficulty less developed countries may experience in obtaining sustained financial support for continuing programs whose very success could lesson awareness that a threat still exists.

The role of the screwworm project as a model program for other SIRM projects, its virtues, and its deficiencies have been discussed elsewhere (79). If one can extend the overkill strategy to other pests without prior investigation of their population biology, then the technique is clearly powerful. However, it has become apparent that the list of suitable species is very short, and few countries have the resources to implement this sort of strategy. Ecologists now associated with SIRM need little convincing of the importance of thorough long-term ecological research before eradication or suppression attempts are initiated (3, 12, 26, 29, 36, 37, 41, 52, 79). However, by continuing to propose short-term evaluation studies, some entomologists have helped create an atmosphere in which many have come to expect practical results with SIRM during time intervals in which other methods of control rarely make significant progress.

The Sterile Codling Moth Release Project in British Columbia

In a series of field releases between 1962 and 1972 Proverbs and co-workers have convincingly established the biological feasibility of SIRM for controlling the codling moth, *Laspeyresia pomonella,* in the main fruit-growing areas of British Columbia (50). Success was demonstrated using both sexes or males only, using moths reared on apples or artificial diet, and using ground or aerial release methods. It has been established that a release ratio of 40 fully sterile to 1 wild insect is sufficient to cause a rapid downward trend in the wild population. Ecological studies have provided means of estimating the size of overwintering populations and predicting the date and rate of adult emergence in the springtime, enabling the adjustment of release time and numbers to obtain the greatest suppression. Unfortunately, even after insecticide treatments in the year before release to reduce the numbers of overwintering moths, the cost of SIRM control of codling moth (by aerial release) has been estimated to be double the present cost of chemical control of an equivalent area (51). If this technique were applied to the eradication of codling moth from the major pome fruit-growing areas of British Columbia (approximately 11,100 hectares) over a period of ten years, the average cost would be less than present costs of control, even allowing for small reinfestations (53). However, fruit growers in the region have not accepted this proposal and are unlikely to do so until the operational costs of SIRM are comparable with those of chemical control (51, 53). The alternatives now open include: 1. to mark time until chemical control is impracticable either because of legislation or insecticide resistance, or 2. to find means of lessening the costs of SIRM.

It is the second alternative which interests us. Approximately 24% of the total cost could be saved by substituting ground for aerial distribution of sterilized insects (53), but this still would compare unfavorably with chemical control. If a practical way of killing females at an early stage of development could be devised, rearing costs would be lowered, and probably distribution costs also, since evidence exists that sterile males disperse more rapidly and further when released without sterile females (51). Another possible means of reducing the cost of SIRM control of codling moth may be to lower the irradiation dose administered to males. Proverbs (50) reported that males treated with 25–30 krad gamma rays, were considerably more competitive than fully sterile males. Although egg hatch was 10–15% when treated males mated with unirradiated females, the progeny of such matings were so unfit that they neither were able to cause fruit damage nor to survive under field conditions. The survival of these progeny under laboratory conditions led researchers to believe erroneously that these individuals would also survive in the field, and that a higher dose of irradiation was therefore warranted. If further work confirms these findings, release ratios could probably be lowered, permitting a greater area to be controlled by release of a given number of moths. (Meanwhile, seeking anonymity, this species has assumed the various aliases of *Cydia pomonella, Carpocapsa pomonella,* and *Laspeyresia pomonella.* To avoid confusion it is known to applied entomologists as codling moth. With a capricious sweep of the pen taxonomists can achieve what applied entomologists often fail to do in a lifetime.)

The Boll Weevil and SIRM

The cotton boll weevil, *Anthonomous grandis*, is estimated to cost the cotton industry in the USA between $200 and $300 million annually (10). Because of these costs, and the fear of resistance developing to organophosphorous insecticides, the cotton industry has pressed the US government to attempt eradication of this pest. In response, millions of dollars were invested in boll weevil research from the late 1950s onwards, resulting eventually in a mammoth plan for eradication of the weevil from the US. It involves intensive insecticide application, field-by-field destruction of cotton plants after harvest, sex-attractant traps, and SIRM (8). The cost estimates range from $665 to $1,500 million over a six- to ten-year period. Undeniably, if the campaign were successful, large benefits would immediately accrue, not only financial ones to the cotton growers, but less tangible ones brought about by a sharply reduced need for insecticides. Roughly one third of all agricultural insecticides currently used in the US are aimed at the boll weevil (10). However, there is considerable dispute among entomologists as to whether eradication can be achieved. A pilot trial in 1971 gave results that were encouraging to the proponents of eradication, but at the same time revealed operational problems that promise to become astronomical in a cotton belt-wide program (8). Many entomologists feel that assessment of alternative biological control measures, improvements in mass-rearing, sterilization, field assessment techniques, and pilot trials in different regions of the cotton belt, are necessary before the launching of a large-scale project can be contemplated (8). Geographical variation seems likely to be particularly significant in view of the wide range of ecological conditions over which the pest is found, and the fact that mating incompatibilities are already known to exist between races (10). A major problem with the use of SIRM for boll weevils is the extreme debilitating effect of sterilizing doses of irradiation or chemicals (10). Thus the search goes on for chemosterilants that will sterilize both sexes, but not severely affect male competitiveness. While progress has been made in this area (22), the situation is far from ideal. One measure which might improve matters here would be to develop a sex-sorting or female-killing system. Besides possible savings on rearing and distribution costs, a sexing system would reopen consideration of a whole range of chemosterilants, i.e. those that will sterilize males but are not as effective against females.

Regardless of the outcome of the boll weevil controversy, if SIRM is to find a place among the major systems of pest control, it must become less dependent on cataclysmic overkill approach advocated by many of its proponents. It needs to be replaced by strategies and tactics that permit a gradual evolution of a SIRM program demonstrating feasibility and paying for itself as it unfolds without dislocating the whole financial machinery of research or industry.

GENETIC AND OTHER BIOLOGICAL CONSIDERATIONS TO IMPROVE SIRM

Undoubtedly one of the biggest problems faced by entomologists in genetic control programs has been that of rearing, sterilizing, and distributing economically, large numbers of insects which are competitive under field conditions.

Although the general fitness of released insects is important for success in both genetic and other forms of biological control, there does exist some opportunity for a better adapted strain to be selected during the generations following the release of predators and parasites in classical biological control (40). The delayed effectiveness of the controlling agent in a number of cases of biological control may be thus explained. However, with SIRM, as well as with other autocidal programs to be described below, the opportunity for improvement in the field does not exist. If the sterilized males are not competitive at the time of release, then the program can be jeopardized.

This difficulty is in part offset by the fact that SIRM does not demand competitiveness in all spheres as does biological control. So long as sterilized males are competitive in inseminating females and evoking normal responses such as eliciting monocoitic behavior or terminating pheromone release, other major deficiencies, especially in the released females, tend to become less relevant. However, in alternative strategies of control (see below) where the release of fertile insects of both sexes is proposed, it becomes essential that released females as well as males are competitive in all areas that influence survival. Although there exists some opportunity for selection to occur in the field to improve performance of these strains, it will generally be too late. The outcome of genetic competition, unlike interspecific competition during biological control, is determined by the relative frequencies of the field insect and the released insect. The release ratio, in turn, is determined by the relative fitnesses of the released and field strains (24). Thus it becomes particularly important for any genetic control project to preserve fitness during the period of colonization. Three possible causes for inadequate fitness of insects released in genetic control programs are: 1. Some aspects of the rearing, sterilization, or distribution operation lead to production of less competitive insects. 2. Genetic deterioration occurs during the period of colonization or the strain colonized at the outset is genetically unsuitable for the overall conditions it will experience following release. 3. The insect's sojourn under laboratory regimes has resulted in short-term acclimatization to these artificial conditions. We envisage this acclimatization to be essentially nongenetic, but able to persist for at least one generation after the insect's transfer to a new set of environmental conditions before the reverse process is completed.

Significant advances have been made during recent years in mass-rearing and handling procedures for many important pests (72). There is every justification for optimism in supposing that entomologists will continue to find cheaper ways of rearing sufficiently large numbers to remove this particular obstacle to genetic control. Loss of competitiveness following induced sterilization is also becoming less of a problem for a variety of reasons. A better understanding of arthropod radiation biology (21), determination of the optimal stage and environmental conditions for irradiation treatment (72), and the availability of better chemosterilants (38) have all contributed to this progress. The release of insects that are not fully sterile, made possible both by a switch from eradication to suppression strategies (32, 79) and by the removal of females prior to release, would also lessen the problems of reduced competitiveness. Mosquitoes, fruit flies, and the boll weevil, once thought to be unsuited to SIRM because of radiosensitivity, can now be viewed with greater optimism.

It is well documented that laboratory colonies are capable of rapid evolution through selection for survival in a changed environment (40, 44, 59), and since economic constraints generally demand that mass rearing be conducted under quite artificial conditions, it becomes necessary to anticipate any genetic change which will eventuate in an insect becoming ill-suited for genetic control. Since artificial conditions, such as rearing in darkness or favoring inactive or flightless behavior, could provide a basis for adverse selection, each should be carefully examined to justify its retention. Frequent replacement of laboratory colonies with field material from projected release sites should minimize the problem of undesirable genetic change during mass rearing and, at the same time, eliminate the risk of colonizing unsuitable genetic material. Supplementing laboratory colonies with field material is unsatisfactory because of the uncertainty that successful introduction has occurred. Complete replacement of cultures is therefore recommended. Furthermore, there is no sound genetic basis for the pooling of genetic material from different origins, whether they be different geographic regions or from laboratory and the field. This admixture may facilitate selection of types better suited to colonial conditions or, by disrupting co-adapted gene complexes, it may simply generate genomes unsuited to any real set of field conditions. If a single mass rearing operation is to provide insects for different habitats then a case exists for the simultaneous rearing of strains derived from each habitat.

A general compromise to the problem of producing insects with the capabilities required for SIRM programs may be through the mating of individuals from separate strains in the last generation, relying on hybrid vigor to compensate for other deficiencies generated during mass rearing. Since a general statement cannot be made on the suitability of hybrid vigor, research on the performance of hybrids for individual species does seem warranted.

Positive results have been obtained with laboratory selection for certain desirable characteristics (2, 40), and there exists the possibility of choosing between field populations on the basis of life table parameters, such as high innate capacity to increase (11), but these measures are not without risk (40). The whole question of competitive ability, its association with other fitness characters, and its heritability, is reviewed by Sammeta & Levins (60). They argue that characters which impinge on fitness should normally exhibit little genetic variation, suggesting that loss of competitiveness under laboratory conditions may have a nongenetic basis.

The remaining aspect of reduced competitiveness that must be considered, acclimitization, could prove to be the most debilitating of the processes mentioned and it is the least well documented. We are not concerned with the mechanisms underlying acclimitization, whether it be genetic, paragenetic, or physiological, but rather with the relationship of this phenomenon to the success of genetic control and how it can be effectively countered.

Entomologists are familiar with the performance of field strains newly introduced to laboratory conditions or of laboratory cultures following radical change in diet. Usually there is a period of indifference, succeeded by rapid adaptation as evidenced by increased fecundity and survival (2). Some of this change is undoubtedly genetic but the repeatability of the process and the rapidity of the adaptation tend to suggest that the organism is adjusting to the new environment by an attunement of its own internal environment. It suffices for us to assume that, in addition to transmitting a complete genome to offspring, information is inherited via the cytoplasm which helps produce phenotypes better adapted to the prevailing conditions. For example, the maternal induction of larval diapause (67) is one of many instances of maternal influences on progeny. The transfer of insecticide from topically treated adults to their offspring (73) and the persisting effects of chlorinated hydrocarbons on characters such as fecundity and rate of development (4) provide some insight into possible mechanisms. One would also predict an apparently inherited state of enzyme induction where there is a carry-over of insecticides such as dieldrin (73) which is known to induce enzyme activity (69). If the adult's choice of food sources and oviposition sites is influenced by larval conditions then it becomes plausible to suggest that a released insect may not recognize its traditional heritage. The failure of chromosomally altered strains of *Drosophila melanogaster* to establish in tomato fields (7) could be due in part to an inability of the released insect to accept tomatoes as suitable oviposition media. The setting of biological clocks in a previous generation is also a possible factor in reducing competitiveness of released insects. Also closer attention should be paid to the possible role that larval diet plays in determining behavior in the adult phase. Some multivoltine species track cyclic changes in the environment by keying in on some component of the environment. For example, the degree of dark pigmentation in the adult alfalfa butterfly, *Colias eurytheme,* which influences flight time, and therefore affects mating, feeding, and oviposition, is determined by larval photoperiod (27). With more obscure characters it is conceivable that we are unwittingly releasing an insect quite unsuited to the circumstances it will encounter.

Thus it becomes very important to recognize any component of the mass rearing conditions which could reinforce inappropriate acclimitization. Frequent replacement of the colony, while remedying any unsuitability due to adverse genetic change, is unlikely to ameliorate poor acclimitization. During the final generations before release, then, it may prove necessary to provide a closer simulation of field conditions for any component of an insect's environment which influences acclimitization.

SEX-KILLING SYSTEMS

A major operational deficiency inherent in current SIRM programs is the need to rear, sterilize, and release both sexes. The role of the female in these programs ranges from being a definite liability to being a useful asset. On the debit side females may be disease vectors, consume some useful resource, cause ovipuncture damage (16), or lessen dispersal or mating competitiveness of released males (20, 51). They may simply add to the cost of mass rearing without providing any benefit, or the need to ensure their sterilization could demand harsher treatment of males leading to reduced competitiveness. For example, a dose of 2500 rad sterilizes male screwworm, but 7500 rad is routinely used to ensure that females are sterile and incapable of oviposition (5). Considerations such as these are a powerful stimulus to the development of sex-sorting and killing systems. While mechanical sorting systems based on pupal size (46) or color (76) have been reported, these do not permit savings in rearing costs, are often impractical for handling large numbers of insects, and are restricted to only a few insect species. The scope for developing genetic sexing systems appears to be much broader, and systems can be developed for mature or immature stages, depending on the requirement of the program. There is no general rule for devising genetic systems, other than that the genetic sex-determining mechanism of the species concerned must be exploited. Thus the haploid/diploid sex-determining mechanisms of *Tetranychus urticae* (45), the autosomal sex-determiner systems in *Musca domestica* (68), and the sex-linked translocations in *Bombyx mori* (66) and *Lucilia cuprina* (76), have been used to devise sexing systems for these species. If attention is restricted to insects with positive sex-determining genes or chromosomes (as opposed to chromosomal balance mechanisms), a practical method of developing sex-killing systems is to link sex-determining genes or chromosomes with conditional lethal mutations by suitable chromosome rearrangements. Temperature-sensitive lethal systems have been devised for *M. domestica* (68) and proposed for *L. cuprina* (23), but the major deterrent to the use of this sort of conditional lethal mutation is the prior need to develop the formal genetics of a species before such mutants can be isolated. On the other hand, a class of conditional lethal mutations whose possibilities have been overlooked until recently, already exists in most insect pest species, i.e. the susceptible alleles of insecticide resistance genes. Without extensive formal genetics, it should be possible to link insecticide resistance to sex chromosomes in a large number of species, in such a way that discriminating doses of insecticide applied during immature stages would kill all members of one or the other sex. The relative ease with which sex-killing systems based on insecticide resistance can be constructed, and the benefits which would accrue from their use, must provide sufficient incentive virtually to compel any SIRM program to include the development of such systems as a top priority.

NONINDUCED STERILITY FOR SIRM

So far, attention in this article has been focused on SIRM programs involving induced sterilization of insects, and some of the ways by which genetic techniques may be used to improve the efficiency or feasibility of such programs. The remainder of this review will be concerned principally with inherited sterility, genetic tricks which may allow the exploitation of certain naturally occurring forms of sterility, and various modifications of and departures from SIRM.

Inherited Sterility

Genetic methods of sterilization fall into two categories: zygotic inviability resulting from chromosomal imbalance, and true sterility caused by mutations which interfere with some stage of gametogenesis. The development of practical ways of producing large numbers of genetically sterile insects would provide an obvious boost to SIRM programs involving species which are greatly weakened by conventional sterilization treatments, and, for certain pests, may allow departures from SIRM which would considerably broaden the horizons of autocidal control.

Zygote inviability can be generated by chromosome segregation, either in strains carrying certain kinds of chromosome rearrangements, or in triploids. The fertility of insects heterozygous for reciprocal translocations approaches zero as the number and complexity of rearrangements increases (61, 68, 77). Multiple translocation heterozygotes with high sterility could be mass reared for release by crossing two strains differing for a number of homozygous interchanges (56, 72) provided that suitable sexing methods were available to make such crosses feasible. However, in the few pest species whose genetics are sufficiently known to make this approach possible, equivalent levels of sterility (approximately 90–99%) can be obtained far more easily by irradiation, with little or no reduction of fitness.

Compound autosomes may be a useful mechanism for sterilizing pests which have a small number of relatively large chromosomes, such as Diptera, since individuals bearing this type of rearrangement should be completely sterile when mated to normal individuals (24). Although the synthesis of compound autosomes depends on an extensive knowledge of genetics (23), serious thought should be given to undertaking the necessary research in such important pests as the Mediterranean fruit fly, *Ceratitis capitata,* which is notoriously prone to damage by conventional sterilization treatments (72). Unlike the isolation of translocations or compound autosomes, the breeding of triploid insects for release may not require such an extensive knowledge of genetics. Tetraploid *Bombyx mori* have been produced experimentally by several workers, and these have been shown to yield sterile triploid moths when crossed to diploids (66). Although it has not been possible to maintain tetraploid stocks of both sexes because of the low fertility of tetraploid males (66), Astaurov has reported a method of maintaining clones of tetraploid females through thermally induced parthenogenesis (1). If this could be repeated in a pest species of Lepidoptera, an ideal system would exist for obtaining virgin females for the production of sterile triploids. However, we could risk providing a pest species with the opportunity to resort to parthenogenesis, the one escape that genetic control is powerless to pursue.

The use of male-sterilizing mutations for SIRM programs has recently been suggested (18). The type of mutant required depends on the mating behavior and physiology of the species concerned. In insects which normally mate only once, sterile males must elicit the monocoitic response. In many dipterans this response appears to be caused by male accessory gland secretions (42, 58), so mutations which prevent spermiogenesis but do not interfere with synthesis or transfer of the secretions would presumably be acceptable. In polyandrous insects, sterile males must produce competitive sperm, or otherwise prevent entry of sperm from subsequent matings. Recently a class of male-sterile mutations has been reported in *Drosophila melanogaster* which produces sperm which are fully motile, and are transferred and stored normally, but do not enter the egg (18). However, there are several important problems to be solved before this source of sterility can be applied in SIRM programs (18).

Naturally Occurring Sterility

Hybrid sterility and cytoplasmic incompatibility are natural phenomena which have been investigated for possible use in SIRM programs, without much success to date (17, 48). A major bar to using hybrid sterility is that the reproductive system of hybrid males is often poorly developed (19), preventing them from producing competitive sperm and/or eliciting a monocoitic response (48). Even in cases where this is not an obstacle, the effective use of natural postmating barriers like hybrid sterility and cytoplasmic incompatibility may be prevented by the existence of premating barriers (17). Moreover, if allopatric forms are used either for direct release or for the production of hybrids, the released insects may be generally unfit, and unable to compete with the target species in its home range. This problem may be near solution in *Culex fatigans,* since recent studies on the mechanism of cytoplasmic incompatibility have suggested a simple way of transporting whole genomes between incompatible strains (79, 80). In cases where the major premating difference is not complex, or can be identified, such as the difference in stridulatory pattern between the crickets *Teleogryllus commodus* and *Teleogryllus oceanicus,* it may become possible to transfer the relevant portion of the genome from the target strain, without interfering with hybrid sterility (28). In general, however, the outlook is not favorable for the development of antidotes to premating and ecological isolating mechanisms.

ALTERNATIVES TO SIRM

Genetic Death Mechanisms

Delayed sterility, involving the release of fertile males whose progeny are sterile, would have no immediate effect on biotic potential, but might cause a far greater impact on a population than the release of the same number of fully sterile males. If such males could be liberated in the ratio of 9 released: one native, the genetic death ensuing in the following generation would be 99%, compared with 90% immediately, if the same ratio of fully sterile males were released (72). The use of delayed sterility was originally suggested for Lepidoptera, because of their poor competitiveness when completely sterilized by irradiation or chemicals, and because partially sterile males yield progeny whose sterility is far more pronounced than that of their parents (33). If such sterile progeny are competitive with native insects and suitable sexing systems become available, delayed sterility may provide a general solution to the low fitness of fully sterilized Lepidoptera. In addition, several ways of engineering delayed sterility for other pest groups may be considered: (*a*) the release of multiple homozygous translocation males, yielding sterile heterozygous progeny; (*b*) release of males of closely related species or races, to produce sterile hybrids; and (*c*) release of tetraploid males to give sterile triploid progeny.

Instead of maintaining high levels of genetic death in a population through the repeated release of sterile males, it may become possible to establish systems of recurring genetic death by releasing fertile insects of both sexes for a relatively short period, say at the beginning of a season (15, 23, 61, 77). This sort of approach may mean that we increase the biotic potential initially, but if this is done at a time of minimum population size and it leads to greater control than would be possible with SIRM or conventional methods, the ultimate gains would outstrip any losses.

Genetic death should occur repeatedly in any population in which two or more freely breeding races, with mutual postmating isolating mechanisms, coexist. Homozygous translocations (15, 61, 77), cytoplasmic incompatibility (80), and compound chromosomes (23) all lead to sterility or inviability in heterozygotes and therefore are possible sources of postmating isolation which could be used in this manner. Recurring genetic death systems would tend to be temporary, since they require the coexistence of several races of insects in a state of unstable equilibrium; however, they could theoretically be extended indefinitely with supplementary releases to maintain the strains in equilibrium (13).

Genetic Manipulation Methods

Manipulation of the genetic composition of a pest population may be aimed either at reducing the size of the population, or at removing the noxious characteristics of the pest. The major requirements of this type of control are: 1. a genetic mutation or strain with one of the above properties, and 2. a transporting mechanism to infiltrate the pest population with the desired genotype. Suitable genotypes for population suppression include conditional lethal factors such as insecticide susceptibility (77), temperature-sensitive mutations (77), and inability to survive on a particular host (25) or to enter diapause (31). If population suppression is not an attractive goal because of overriding density-dependent regulation or because an empty niche is created by the removal of one pest, it may be appropriate to consider displacing the pest with a less noxious form, e.g. one incapable of disease transmission.

The simplest sort of transporting device would be to flood a pest population with insects carrying dominant conditional lethal traits (25, 31). This device would require little knowledge of the genetics of a species. Meiotic drive has been suggested as a transporting mechanism but the biological causes of meiotic drive are only poorly understood, and no systems suitable for pest control are known to exist (23). A third type of mechanism would utilize negative heterosis. Since freely breeding races with postmating isolation cannot coexist indefinitely, they could be used to replace pest populations with strains fixed for desirable genotypes (14, 24, 77). Homozygous translocations, compound autosomes, and cytoplasmic incompatibility are possible sources of negative heterosis which could be used in this manner. Major obstacles to this sort of transport mechanism are the need for extensive formal genetic manipulation in the case of chromosome races, and the paucity of naturally occurring systems of cytoplasmic incompatibility.

CONCLUDING REMARKS

It must be stressed that theory currently far outstrips practice in the implementation of genetic approaches to insect control other than SIRM. Preliminary trials with released males have shown the successful introduction of genetic material into field populations (55, 68, 78), but, where full functioning of the released females was necessary for effective incorporation of new genetic material, the experiments were inconclusive (78). Competitiveness of the released female is essential, otherwise it may be impractical or undesirable to produce the numbers of insects required for these forms of genetic control.

Adequate information exists for some of the suggested conditional lethal conditions, such as temperature sensitivity (65) and insecticide susceptibility (4), but more data are needed on the genetic basis for innocuous traits like vector refractoriness and the relationship between its frequency in vector populations and disease incidence, before an evalution can be made.

Research on genetic control need not be viewed in a purely pragmatic light. If ecologists can construct an operational model to explain abundance and distribution of a pest species, opportunity exists to test these models. For example, Weidhaas and colleagues (74) have used SIRM to study the response of a field population to the imposition of a genetic load. If the ecological model can be used to predict population sensitivity to a particular control method, we become better informed when deciding whether to concentrate on genetic load, genetic manipulation, or whether to pursue nongenetic means of control.

Extensive formal genetics have already been developed for a number of important pest species (79). However, there are still many major pests such as *Culex fatigans, Anopheles* spp., *Cochliomyia hominivorax,* tephritid fruit flies, and nearly all lepidopterous, coleopterous, and hemipterous pests where the absence of formal genetics and a dearth of ecological information on population dynamics prevent consideration of much of the theory of genetic control for their management. It is to be hoped that more purposeful dialogue between population geneticists, chromosome mechanics, and ecologists will redress this situation.

Literature Cited

1. Astaurov, B. L. 1966. Experimental alterations of the developmental cytogenetic mechanism in mulberry silkworms: artificial parthenogenesis, polyploidy, gynogenesis, and androgenesis. *Advan. Morphol.* 6:199–257
2. Baumhover, A. H., Husman, C. N., Graham, A. J. 1966. Screwworms. *Insect Colonization and Mass Production,* ed. C. N. Smith, 533–54. New York: Academic. 618 pp.
3. Bogyo, T. P., Berryman, A. A., Sweeney, T. A. 1971. Computer simulation of population reduction by release of sterile insects. *Application of Induced Sterility for Control of Lepidopterous Populations,* 19–25. *Panel Proc., IAEA, Vienna, 1970,* 169 pp.
4. Brown, A. W. A., Pal, R. 1971. *Insecticide Resistance in Arthropods,* p. 49. WHO Monogr. Ser., No. 38. Geneva. 491 pp.
5. Bushland, R. C. 1971. Sterility principle for insect control. *Sterility Principle for Insect Control or Eradication,* 3–14. *Proceedings, IAEA, Vienna, 1970,* 542 pp.
6. Bushland, R. C. Personal communication
7. Cantelo, W. W., Childress, D. 1973. Laboratory and field studies with a compound chromosome strain of *Drosophila melanogaster. Theor. Appl. Genet.* In press
8. Carter, L. J. 1974. Eradicating the boll weevil: would it be a no-win war? *Science* 183:494–99
9. Chambers, D. L., Spencer, N. R., Tanaka, N., Cunningham, R. T. 1970. Sterile-insect technique for eradication or control of the melon fly and oriental fruit fly: review of current status. *Sterile-male Technique for Control of Fruit Flies,* 99–102. *Panel Proc., IAEA, Vienna, 1969,* 175 pp.
10. Cross, W. H. 1973. Biology, control and eradication of the boll weevil. *Ann. Rev. Entomol.* 18:17–46
11. Crovello, T. J., Hacker, C. S. 1972. Evolutionary strategies in life table characteristics among feral and urban strains of *Aedes aegypti. Evolution* 26:185–96
12. Cuellar, C. B. 1969. The critical level of interference in species eradication of mosquitoes. *Bull. WHO* 40:213–19
13. Curtis, C. F. 1968. A possible genetic method for the control of insect pests, with special reference to tsetse flies. *Bull. Entomol. Res.* 57:509–23
14. Curtis, C. F. 1968. Possible use of translocations to fix desirable genes in insect pest populations. *Nature* 218:368–69
15. Curtis, C. F., Hill, W. G. 1971. Theoretical studies on the use of translocations for the control of tsetse flies and other disease vectors. *Theor. Pop. Biol.* 2:71–90
16. de Murtas, I. D., Cirio, V., Guerrieri, G., Enkerlin, D. 1970. An experiment to control the medfly on the island of Procida by the sterile-insect technique. See Ref. 9, 59–70
17. Davidson, G., Odetoyinbo, J. A., Colussa, B., Coz, J. 1970. A field attempt to assess the mating competitiveness of sterile males produced by crossing 2 member species of the *Anopheles gambiae* complex. *Bull. WHO* 42: 55–67
18. Denell, R. E. 1973. Use of male sterilization mutations for insect control programmes. *Nature* 242:274–75
19. Dobzhansky, Th. 1951. *Genetics and the Origin of Species.* New York: Columbia Univ. Press. 3rd ed. 364 pp.
20. Donnelly, J. 1965. Possible causes of failure in a field test of the "sterile males" method of control. *Proc. Int. Congr. Entomol., 10th, London, 1964,* pp. 253–54
21. Ducoff, H. S. 1972. Causes of death in irradiated adult insects. *Biol. Rev.* 47:211–40
22. Flint, H. M., Earle, N., Eaton, J., Klassen, W. 1973. Chemosterilization of the female boll weevil. *J. Econ. Entomol.* 66:47–53
23. Foster, G. G., Whitten, M. J. 1974. The development of genetic methods of controlling the Australian sheep blowfly, *Lucilia cuprina. The Use of Genetics in Insect Control,* ed. R. Pal, M. J. Whitten, 19–43. Amsterdam: Elsevier/-North-Holland. 241 pp.

24. Foster, G. G., Whitten, M. J., Prout, T., Gill, R. 1972. Chromosome rearrangements for the control of insect pests. *Science* 176:875–80
25. Foster, J. E., Gallun, R. L. 1973. Controlling race B Hessian flies in field cages with the great plains race. *Ann. Entomol. Soc. Am.* 66:567–70
26. Hightower, B., Graham, O. H. 1968. Current status of screwworm eradication in the southwestern United States of America and the supporting research programme. *Control of Livestock Insect Pests by the Sterile-male Technique,* 51–54. *Panel Proc., IAEA, Vienna, 1967,* 102 pp.
27. Hoffman, R. J. 1973. Environmental control of seasonal variation in the butterfly *Colias eurytheme. Evolution* 27:387–97
28. Hogan, T. W. 1974. A genetic approcah to the population suppression of the common field cricket *Teleogryllus commodus.* See Ref. 23, 57–70
29. IAEA, 1969. *Insect Ecology and the Sterile-male Technique Panel Proc., IAEA, Vienna, 1967,* 102 pp.
30. James, S. H. 1970. Complex hybridity in *Isotoma petraea. Heredity* 25:53–77
31. Klassen, W., Knipling, E. F., McGuire, J. U. Jr. 1970. The potential for insect-population suppression by dominant conditional lethal traits. *Ann. Entomol. Soc. Am.* 63:238–55
32. Knipling, E. F. 1969. Concept and value of eradication or continuous suppression of insect populations. *Sterile-male Technique for Eradication or Control of Harmful Insects,* 19–32. *Panel Proc., IAEA, Vienna, 1968,* 142 pp.
33. Knipling, E. F. 1970. Suppression of pest Lepidoptera by releasing partially sterile males. A theoretical appraisal. *Bioscience* 20:465–70
34. La Brecque, G. C., Keller, J. C., eds. 1965. Early developments in the sterile-male technique. *Advances in Insect Population Control by the Sterile Male Technique,* 1–3. *IAEA Tech. Rep. Ser. No. 44., Vienna.* 79 pp.
35. Lewontin, R. C. 1967. Population genetics. *Ann. Rev. Genet.* 1:37–70
36. Lindquist, A. W. 1969. Biological information needed in the sterile-male method of insect control. See Ref. 32, 33–37
37. Lofgren, C. S. et al 1973. Release of chemosterilized males for the control of *Anopheles albimanus* in El Salvador: III. Field methods and population control. *Am. J. Trop. Med. Hyg.* In press
38. McDonald, F. J. 1974. The future of chemosterilants, their use against populations in the field, mammalian toxicity and hazards. See Ref. 23, 225–38
39. Mellado, L. 1971. La Tecnica de machos esteriles en el control do la mosca del mediterraneo. See Ref. 5, 49–53
40. Messenger, P. S., Wilson, F., Whitten M. J. 1974. Fitness and adaptability of natural enemies. *Theory and Practice of Biological Control,* ed. C. Huffaker, P. DeBach, P. S. Messenger. New York: Academic. In press
41. Monro, J. 1973. Some applications of computer modelling in population suppression by sterile males. *Computer Models and Application of the Sterile Males Technique,* 81–94. *Panel Proc., IAEA, Vienna, 1973*
42. Nelson, D. R., Adams, T. S., Pomonis, J. G. 1969. Initial studies on the extraction of the active substance inducing monocoitic behavior in house flies, black blow flies and screwworm flies. *J. Econ. Entomol.* 62:634–39
43. Newton, W. H., Ferguson, J. L. 1973. *Southwest Screwworm Eradication Program: Progress Reports.* Texas A & M Univ. (1962–1973)
44. Nicholson, A. J. 1957. The self-adjustment of populations to change. *Cold Spring Harbor Symp. Quant. Biol.* 22:153–72
45. Overmeer, W. P. 1974. Genetic control of spider mites. See Ref. 23, 45–56
46. Pal. R. 1974. WHO/ICMR programme of genetic control of mosquitoes in India. See Ref. 23, 73–95
47. Patterson, R. S., Weidhaas, D. E., Ford, H. R., Lofgren, C. S. 1970. Suppression and elimination of an island population of *Culex pipiens quinquefasciatus* with sterile males. *Science* 168:1368–70
48. Proshold, F. I., LaChance, L. E. Analysis of sterility in hybrids from interspecific crosses between *Heliothis virescens* and *H. subflexa.* Unpublished
49. Proverbs, M. D. 1969. Induced sterilization and control of insects. *Ann. Rev. Entomol.* 17:81–102
50. Proverbs, M. D. 1970. Procedures and experiments in population suppression of the codling moth *Laspeyresia pomonella* (L.) in British Columbia orchards by release of radiation sterilized moths. *Manitoba Entomol.* 4:46–52
51. Proverbs, M. D. Codling moth control by the sterility principle in British Columbia: estimated cost and some biological observations related to cost. *IAEA Panel Proc.* In press

52. Proverbs, M. D. 1974. Ecology and sterile release program, the measurement of relevant population processes before and during release, the assessment of results. See Ref. 23, 201–23
53. Proverbs, M. D. Personal communication
54. Rai, K. S. 1969. Status of the sterile-male technique for mosquito control. See Ref. 32, 107–14
55. Rai, K. S., Grover, K. K., Suguna, S. G. 1973. Genetic manipulation of *Aedes aegypti*—incorporation and maintenance of a genetic marker and a chromosomal translocation in natural populations. *Bull. WHO* 48:49–56
56. Rai, K. S., Lorimer, N., Hallinan, E. 1974. The current status of genetic methods for controlling *Aedes aegypti.* See Ref. 23, 119–32
57. Rhode, R. H. 1970. Application of the sterile-male technique in mediterranean fruit fly suppression. See Ref. 9, 43–50
58. Riemann, J. G., Thorson, B. J. 1969. Effect of male accessory material on oviposition and mating by female house flies. *Ann. Entomol. Soc. Am.* 62:828–34
59. Robertson, F. W. 1965. The analysis and interpretation of population differences. *The Genetics of Colonizing Species,* ed. H. G. Baker, G. L. Stebbins, 95–113. New York: Academic. 588 pp.
60. Sammeta, K. P. V., Levins, R. 1970. Genetics and ecology. *Ann. Rev. Genet.* 4:469–88
61. Serebrovsky, A. S. 1940. On the possibility of a new method for the control of insect pests. See Ref. 32, 123–37
62. Shaw, J. G., Lopez-D, F., Chambers, D. L. 1970. A review of research done with the Mexican fruit fly and the citrus blackfly in Mexico by the Entomology Research Division. *Bull. Entomol. Soc. Am.* 16:186–92
63. Smith, R. H. 1973. Letters. *Science* 182:775
64. Smith, R. H., von Borstel, R. C. 1972. Genetic control of insect populations. *Science* 178:1164–74
65. Suzuki, D. T. 1970. Temperature sensitive mutations in *Drosophila melanogaster. Science* 170:695–706
66. Tazima, Y. 1964. *The Genetics of the Silkworm.* London: Logos. 253 pp.
67. Vinogradova, E. B., Zinovjeva, K. B. 1972. Maternal inductions of larval diapause in the blowfly, *Calliphora vicina. J. Insect Physiol.* 18:2401–9
68. Wagoner, D. E., McDonald, I. C., Childress, D. 1974. The present status of genetic control mechanisms in the housefly, *Musca domestica* L. See Ref. 23, 183–97
69. Walker, C. R., Terriere, L. C. 1970. Induction of microsomal oxidases by dieldrin in *Musca domestica. Entomol. Exp. Appl.* 13:260–74
70. Wallace, B. 1970. *Genetic Load: Its Biological and Conceptual Aspects.* Englewood Cliffs, NJ: Prentice-Hall. 116 pp.
71. Wallace, B., Dobzhansky, Th. 1959. *Radiation, Genes and Man.* New York: Holt. 205 pp.
72. Waterhouse, D. F., LaChance, L., Whitten, M. J. 1974. Use of autocidal methods. See Ref. 40
73. Watts, W. S. 1969. Transmission of dieldrin from insects to their progeny. *Nature* 221:762–63
74. Weidhaas, D. E., Patterson, R. S., Lofgren, C. S., Ford, H. R. 1970. Bionomics of a population of *Culex pipiens fatigans.* WHO/VBC/70.190
75. White, M. J. D. 1970. Cytogenetics of speciation. *J. Aust. Entomol. Soc.* 9:1–6
76. Whitten, M. J. 1969. Automated sexing of pupae and its usefulness in control by sterile insects. *J. Econ. Entomol.* 62:272–73
77. Whitten, M. J. 1971. Insect control by genetic manipulation of natural populations. *Science* 173:682–84
78. Whitten, M. J., Foster, G. G., Kitching, R. L. 1973. The incorporation of laboratory-reared genetic material into a field population of the Australian sheep blowfly, *Lucilia cuprina. Can. Entomol.* 105:893–901
79. Whitten, M. J., Pal, R. 1974. Introduction. See Ref. 23, 1–16
80. Yen, J. H., Barr, A. R. 1974. Incompatibility in *Culex pipiens.* See Ref. 23, 97–118

AUTHOR INDEX

A

Abdallah, M. D., 431
Abdel-Salam, F., 318
Abushama, F. T., 265-67, 271
Adam, L., 368
Adams, C. H., 291, 305, 307, 308, 315
Adams, J. B., 303
Adams, J. R., 101, 229, 384
Adams, L., 251
Adams, M. E., 137
Adams, R. P., 88
Adams, T. S., 427, 471
Adashkevich, B. P., 300, 305
Adiyodi, K. G., 143
Adkins, H. G., 18
Adkisson, P. L., 299
Adlung, K. G., 83
Agarwal, R. A., 48
Agee, H. R., 226
Ahearn, G. A., 264, 268, 271-75
Ahlstrom, K. R., 307, 309
Ahmed, M. K., 287, 299, 308
Ahmed, M. S. H., 169
Ahmed, W., 318
Aidley, D. J., 151
Aizawa, K., 100, 107
Ajami, A. M., 447
Akert, K., 365
Akutsu, K., 102
Alagoda, A., 338
Alcaide, A., 208
Alemany, A., 339
Alexander, N., 222
Alexeyev, A. N., 245
Al-Hakkak, Z., 169
Allais, J. P., 208
Allan, R. E., 345
Allen, G. E., 3-6, 23, 24
Alley, E. G., 20
Allsopp, W. H. L., 34
Aloe, L., 135, 136
Al-Saqur, A., 169
Altner, H., 388, 390
Alverdes, F., 370
Amargier, A., 99
Anand, M., 339
Anderson, D. B., 82
Anderson, G. R., 139
Anderson, J. E., 447
Anderson, J. M., 81
Anderson, N. H., 82, 308
Anderson, R. F., 101
Anderson, S. M., 160
Anderson, S. O., 242
Anderson, W. H., 31, 37
Andres, F., 338
ANDRES, L. A., 31-46; 31, 36-40
Andrews, E. K., 21
Andrews, G. L., 101, 107, 109
Andrews, P., 384
Ankersmit, G. W., 291
Anstee, J. H., 136
Anthon, E. W., 288, 289
Anthony, D. W., 99, 107
Anwyl, R., 137
Aranda, L. C., 359
Arant, F. S., 12, 20, 21
Arif, B. M., 103, 104
Armanious, N. I., 419, 423, 437, 438
Arnott, H. J., 99, 100
Arnsman, J. C., 339
Arrand, J. C., 324
Aruga, H., 101, 107, 109
Asai, J., 108, 109
Ashburner, M., 427
Ashraf, M., 82
Ashrafi, S. H., 170, 171, 173, 176
Askew, R. R., 253
Asquith, D., 287, 321, 322
Astaurov, B. L., 471
Aston, R. J., 139, 140, 143
Atallah, Y. H., 289, 296, 300, 303, 307, 308, 312
Atger, P., 102
Athwal, A. S., 338
Athwal, D. S., 344
Atwood, C. E., 83, 84
Audy, J. R., 244
Austin, L., 82
Avault, J. W., 34
Awadallah, K. T., 299, 304
Axelsson, J., 156
Axelsson, S., 367, 368
Axtell, R. C., 290, 293, 383-85, 387, 388, 390
Aylward, J. B., 159
Azab, A. K., 299, 304

B

Babu, T. H., 441
Bacalangco, E. H., 344
Baccetti, B., 242
Bacetti, B., 54
Bae, S. H., 338
Baerends, G. P., 360
Baetcke, K. P., 21
Bagley, R. W., 422-24, 429, 430, 432, 434, 435, 438, 439, 442, 446
Bai, M. K., 252
Bailey, L., 110, 111
Bain, N. S., 57
Bains, S. S., 338
Baker, E. R., 125
Baker, J. E., 82, 84
Bakeyev, N. N., 247, 251
Bal, A. K., 383, 384
Balachowsky, A., 47, 48, 54
Balashov, Yu. S., 247
Ball, H. J., 373
Baloch, G. M., 35, 36, 39, 41, 42
Bancroft, T. A., 226
BANKS, W. A., 1-30; 16, 17, 20-22, 25
Banthorpe, D. V., 87
Baradat, P., 88
Baranowski, A., 32
Baranyovits, F., 54, 67
Barbier, M., 208
Barbieri, M. R., 402, 403
Barlin, M. R., 14
Barnes, A. M., 243
Barnes, D. F., 50, 53
Barnes, M. M., 324, 325
Barr, A. R., 471, 472
Barrera, A., 249
Barrett, G. W., 318
Barrett, J. R., 225
Barrington, E. J. W., 264
Barry, B. D., 176
Bartell, J. A., 169, 171-73
Bartels, E., 161
Barth, F. G., 386, 387, 389, 390
Barth, R. H., 133
Bartkowski, J., 103
Bartlett, A. C., 170, 171
Bartlett, B. R., 287, 288, 292, 298, 299, 301, 302, 304, 308
Bartlett, F. J., 12, 20, 22
Barton-Browne, L. B., 263, 271, 273
Basedow, T., 339, 341
Bastock, M., 360
Bates, J. K., 245
Batista, J. S., 406
Baud, L., 106
Baudry, N., 135
Bauer, H., 168, 169, 175, 176
Bauernfeind, J. C., 422-24, 429, 430, 432, 434, 435, 438, 439, 442, 446
Baumhover, A. H., 167, 212,

468
Bavaresco, F. A., 400, 410
Bayreuther, K., 243
Beament, J. W. L., 263, 270, 271, 273
Beams, H. W., 391
BEARDSLEY, J. W. JR., 47-73; 53, 60
Beattie, T. M., 135
Beavers, J. B., 57
Beck, A. J., 251
Beck, J., 367
Beck, S. D., 75, 76, 85, 209, 222, 339, 349
Beetsma, J., 445
Beford, G. O., 102
Beig, D., 408
Bell, M. R., 105
Bell, R. A., 135, 173
Bellett, A. J. D., 99, 104
Bellinger, F., 20
Belozerov, V. N., 274
Belton, P., 226
Benassy, C., 56, 63, 64
Benjamini, E., 248
Bennet-Clark, H. C., 241
Bennett, A. F., 207
BENNETT, F. D., 31-46; 31, 35, 36, 38, 39, 41, 42, 58, 59, 62, 64, 308
Bennett, W. H., 79
Benskin, J., 429, 437, 438, 447, 448
Benton, A. H., 245
Benton, C. V., 100
Benz, G., 420, 421, 441
Beranek, R., 156
Berg, C. O., 34, 36, 38, 39
Berg, M. A., 342
Bergoin, M., 97, 103-5, 110
Beri, R. M., 86
Berlese, A., 54, 385
Bern, H. A., 133, 134, 138, 139, 143
Bernard-Dagan, C., 88
Beroza, M., 81, 229, 427, 447
Berridge, M. J., 137, 141, 262
Berry, E. C., 225, 228-30, 431
Berryman, A. A., 82, 84, 462, 464
Bertkau, P., 386
Bess, H. A., 300, 308
Besse, N., 135
Bethe, A., 370
Bethell, R. S., 102
Betts, E., 276
Bhaskaran, G., 427
Bhatkar, A. P., 18, 24, 25
Bhatnagar-Thomas, P. L., 434, 435
Bhattacharya, A., 303
Bibikova, V. A., 247-50, 252
Bickley, W. E., 306
Bielenin, I., 54
Biesele, J. J., 242
Biever, K. D., 289-91, 298, 305, 308
Bigelow, R. S., 3
Binns, E. S., 299
Bird, F. T., 107, 108
Bird, R. G., 107
Bishop, P. M., 7, 10, 14, 17, 22, 25
Bjöklund, A., 367
Blackburn, R. D., 31-34, 36-41
Blais, J. R., 80
Bland, K. P., 139
Blaney, W. M., 384, 385
Blantran de Rozari, M., 223
Blest, A. D., 360, 368
Blickenstaff, C. C., 306, 421, 435
Bliss, C. L., 61-63
Blotta, L. F., 402, 407-10
Blum, A., 344
Blum, K., 160
Blum, M. S., 3, 10, 11, 14, 15, 25
Boch, R., 410
Bode, C., 214
Bodenheimer, F. S., 48, 55, 57, 58, 61-63, 65
Boemare, N., 107
Boening, O. P., 101
Boersma, A., 299
Boggino, P. A., 404
Bogush, P. P., 276
Bogyo, T. P., 462, 464
Bohm, M. K., 306, 445
Boistel, J., 151
Bonami, J. R., 104
Bonnemaison, L., 299, 301
Bonner, F. L., 300
Bookhout, C. G., 21
Boone, C. O., 306
Booth, G. M., 155
Boratynski, K. L., 54
Borchsenius, N. S., 47
Borden, J. H., 421
Borg, T. K., 82, 135, 139
Born, D., 299
Börner, C., 183, 190, 193-97
Borthwick, P. W., 21
Bosc, S., 139
Bostanian, N. J., 383, 385
Boswell, A. L., 307, 309, 319, 320, 440
Both, A., 412
Bourgogne, J., 183, 193, 197
Bouvarel, P., 78
Bowers, B., 136, 137
Bowers, W. S., 417-19, 421-23, 433-35, 442, 447
Bowlus, S. B., 447
Bowman, M. C., 340, 424
Boyce, H. R., 308, 318
Boyd, C. E., 32, 33
Boyd, J. C., 231
Boyden, B. L., 60
Bracken, G. K., 305, 420
Brady, J., 135, 373
Brakier-Gingras, L., 106
Brand, J. M., 14, 15
Bransby-Williams, W. R., 423, 425, 426, 435, 441
Bratkowski, T. A., 161
Brattsten, L. B., 296, 297
Braun, A. F., 191, 194
Bräuning, S., 243
Bravenboer, L., 289, 294
Bray, D. F., 231
Brecheen, K. G., 82
Bregliano, J. C., 112
Breillatt, J. P., 99
Breland, O. P., 242
Brès, N., 107
Brinck-Lindroth, G., 252
BRINDLEY, T. A., 221-39; 221-32, 349
Broadbent, B. M., 61-63
Brock, J. P., 189-92, 197, 198, 200
Brooks, G. T., 447
Brooks, L. D., 229
Brooks, M. A., 97, 100
Broome, J. R., 23, 101, 107, 109
Brower, J. H., 170
BROWN, A. W. A., 285-335; 286, 315, 426, 447, 468, 473
Brown, B. E., 137, 138
Brown, B. W., 222
Brown, C. E., 62, 64, 65
Brown, F., 110
Brown, G. L., 153
Brown, J. L., 38-40
Brown, L. L., 19
Brown, S. W., 47, 56, 58, 59, 62, 64, 66, 69
Brown, W. L. Jr., 19, 25
Browning, T. O., 273
Bruhn, J. C., 24
Brun, G., 112
Brunet, P. C. J., 140
Brunson, M. H., 308
Bryukhanova, L. V., 251
Brzostowski, H. W., 110
Buckner, C. H., 104
Bueno, D., 402, 404
Bugany, H., 213, 216
Bull, D. L., 292, 296, 298, 436, 442, 447
Bulla, L. A., 97
Bullock, H. R., 109, 110
Bullock, T. H., 360, 363, 370, 373, 381, 383
Burbutis, P. P., 228, 231
Buren, W. F., 2-6, 23, 24
Burges, H. D., 97, 100
Burke, H. R., 289, 290
Burke, J. M., 103, 104

Burkhardt, C. C., 228
Burkholder, W. E., 168-71
Burks, M. L., 207
Burnham, C. R., 175
Burns, E. C., 18
Burov, V. N., 442
Bursell, E., 268, 271, 273, 276
Burton, R. L., 340
Busck, A., 183, 191, 195, 197
Busgen, M., 51
Bushland, R. C., 167, 176, 462-64, 469
Byzova, J. B., 272

C

Cade, S. C., 83
Cain, J. D., 21
Caldas Filho, C. F., 404
Calderon, M., 169, 172, 173
Caldwell, R. M., 339, 346, 347
Callahan, P. S., 15
Calos, A., 139
Caltagirone, L. E., 324, 326
Cameron, M. D., 80
Camhi, J. M., 391
Camin, J. H., 267
Campbell, J. B., 424, 427
Campbell, R. L., 79, 84
Campbell, W. R., 104, 105
Campion, D. G., 441
Cantelo, W. W., 167, 468
Carayon, J., 440, 441
Carbonell, J., 17, 24
Carde, R. T., 224
Carlisle, D. B., 133, 268, 269, 276
Carlson, A. D., 141, 153
Carlson, D. A., 16
Carlson, R. E., 228
Carman, G. E., 57
Carnegie, A. M. J., 63
Carolin, V. M., 318
Carson, R., 19
Carter, J. B., 105
Carter, L. J., 462, 465
Carter, S. W., 227, 233, 234, 339-41
Carter, W., 53
Carthy, J. D., 386, 387
Cartwright, W. B., 339, 340, 347, 348
Casey, H., 32
Casida, J., 303
Casida, J. E., 424
Cassier, P., 134, 418
Catchside, D. G., 172
Cate, J. R., 288, 291, 296
Cazal, M., 134, 135, 142
Cazal, M. L., 139
Cerf, D. C., 447
Cerný, V., 440
Chada, H. L., 339, 340, 342, 345, 347, 351
Chadwick, M. J., 268, 270
Chakraborty, J., 176
Chalaye, D., 135
Chamberlain, W. F., 446
Chambers, D. L., 57, 80, 462
Chambers, H. W., 307, 312, 314
Chandrika, S., 291
Chang, C. C., 162
Changuex, J. P., 162
Chanussot, B., 135, 136, 139
Chapman, H. C., 105
Chapman, K. M., 385-90
Chapman, R. F., 276, 384, 385
Chapman, T. A., 184, 187, 193
Charlwood, B. V., 87
Charton, M., 159, 161
Charudattan, R., 39
Chatterji, S. M., 231, 233, 338, 339
Chaudhari, S., 339
Chaudhury, M. F. B., 225, 373
Chaud Netto, J., 399-412
Chen, W. Y., 359
Cheng, C. H., 338
Cheng, W. Y., 173
Cherbas, L., 213
Cherbas, P., 219
Cherry, E. T., 299
Chesnokov, P. G., 338, 341
Chesnut, T. L., 315
Chevalier, R. L., 385, 387, 388, 390, 391
Chew, R. M., 276
Chiang, H. C., 222, 223, 227-29, 339
Childress, D., 468-70, 473
Childs, L., 54
Chino, H., 216
Chippendale, G. M., 233, 421
Chiu, Y. J., 22
Chiyoda, Y., 208
Chou, W. D., 338
Christenson, D. M., 425
Christian, U., 384, 389
Chu, H. M., 209
Chu, I-Wu, 387
Chung-Der Huang, 427
Chu-Wang, I-Wu, 383-85, 387-90
Cirio, V., 462, 469
Clancy, D. W., 319, 320
Clarke, K. U., 136
Clay, T., 253
Clayton, R. B., 87, 209
Cleere, J. S., 438
Cleland, W. W., 161
Clench, H. K., 190, 191
Clever, I., 212
Clever, U., 212, 382
Clothier, S. E., 101
CLOUDSLEY-THOMPSON, J. L., 261-83; 261, 262, 264-76
Coaker, T. H., 290, 304
Cobb, D. L., 339
Coburn, G., 18
Cochrane, D. G., 156
Cockayne, E. A., 172
Cogburn, R. R., 168-71
Cohen, C. F., 207-9
Colburn, R., 287, 322
Colhoun, E. H., 155, 367
Collett, T., 368
Collier, S. J., 100
Collins, H. L., 5, 7-10, 17, 21, 22
Collins, W. J., 443
Colquhoun, W., 137
Colussa, B., 471
Colvin, I. B., 222
COMMON, I. F. B., 183-203; 184, 188, 189, 191, 195, 198
Compton, L. E., 339, 346, 347
Comstock, J. H., 183, 191, 193, 195
Cone, W. W., 309, 318, 324
Connin, R. V., 339, 421, 434
Connolly, K., 140
Conolly, D., 251
Cook, B. J., 138, 139, 156, 158
Cook, D. J., 142, 143
Cook, I. F., 208
Cooke, I. M., 143
Cooke, J. D., 160
Coombes, L. E., 426
Coon, D. W., 19
Cooper, K. W., 243
Corbiere-Tichane, G., 385, 387
Corey, E. J., 447
Corley, C., 229, 230
Cornejo, L. G., 402, 404, 413
Cornford, M. E., 225
Cosenza, G. W., 404, 406, 409-11, 413
Costa, M., 253
Costa, N., 17
Couch, J. A., 100
Coulson, J. R., 36-41
Coulter, W. K., 318
Coutts, M. P., 82
Cowan, C. B., 299
Cowan, S., 156
Cox, H. C., 340
Coz, J., 471
Craig, C. H., 10, 14, 17, 22, 25
Crance, J. H., 18
Crawford, C. S., 265, 267-71, 273, 274

Creech, J. L., 343
Creighton, W. S., 2
Cress, D. C., 339, 340
Cressman, A. W., 55, 61-63
Crisci, C., 17, 24
Critchley, B. R., 288, 290, 296, 304, 318, 441
CROFT, B. A., 285-335; 286, 288, 290, 297, 306-9, 311, 314, 317, 319-26
Croizier, G., 99, 100, 103, 111
Croker, S. G., 140
Cross, D. G., 33
Cross, W. H., 291, 307, 308, 315, 465
Crossan, D. F., 231
Crossley, A. C., 140
Crovello, T. J., 468
Crovetti, A., 276
Crowder, L. A., 139
Cruickshank, P. A., 443
Cuellar, C. B., 464
Cull-Candy, S. G., 161
Culpepper, G. H., 3
Cumming, J. E., 428
Cunningham, J. C., 103-5, 108
Cunningham, R. T., 462
Cupp, E. W., 2, 23, 444, 445
Curtis, B. C., 345, 351
Curtis, C. F., 472, 473
Cuzin-Roudy, J., 441
Cymborowski, B., 366, 367, 372, 373

D

da Cunha, A. B., 100
Dagan, D., 137
Dahl, F., 384
Dahm, K. H., 423, 424
Dahms, R. G., 338, 341, 346, 347
Dale, H. H., 153
Dale, J. L., 53
Dales, S., 103, 105
Dando, J., 136, 139
Dang, K., 339
Daniels, N. E., 339, 340, 345
Danilova, L. V., 176
Danzig, E. M., 50, 56
Darchen, R., 405
Darskaya, N. F., 247, 248, 250, 251
da Silva, R. M. B., 404
Dauterman, W. C., 290, 306, 307, 309, 314
Davey, K. G., 133, 141
DAVID, W. A. L., 97-117; 100, 101
Davidson, A., 102
Davidson, G., 471
Davies, R. G., 184, 198
Davis, C. J., 39
Davis, D. R., 8-10
Davis, D. W., 299, 300, 302, 317, 324
Davis, F. M., 339
Davis, J. R., 21
Day, H. O., 225
Dean, J. L., 34
Deay, H. O., 348
DeBach, P., 56, 62, 63, 302
Debolt, J. W., 169-72
de Buren, F. P., 248
DECK, E., 119-31
Decker, G. C., 223
Deckle, G. W., 66
De La Rose, H. H., 169
De Loof, A., 428
DeLotto, G., 58
Delphin, F., 135
Délye, G., 275
de Mello, R. P., 413
de Menezes, M., 3
Dempster, J. P., 304, 308, 318
de Murtas, I. D., 462, 469
Denell, R. E., 471
Denmark, H. A., 18
Deonier, D. L., 36
Deoras, P. J., 248
de Portugal-Araújo, V., 400, 406, 409, 410
DeSantis, L., 402, 404, 413
Deseö, K. V., 171, 420
de Souza, D. J., 413
Dethier, V. G., 185, 381, 384, 391
Detinova, T. S., 252
Devauchelle, G., 103, 104
Devine, T., 274
Dewhurst, S. A., 140
De Wilde, J., 421
Diatta, F., 112
Diaz, M., 288, 291, 298, 300, 301
Dicke, F. F., 221, 226-28, 231-33, 338-40, 344
Dickson, R. C., 66-68
Diekman, J., 418, 432-35
Dillier, J. H., 5, 7-11, 13, 25
Dishner, G. H., 342
Disselkamp, C., 54, 66-68
Ditman, L. P., 231
Djerassi, C., 418
Dobzhansky, T., 461, 471
Doering, G. W., 339
Doggett, H., 344, 351
Dogra, G. S., 136
Dolejš, L., 440
Dolezal, J. E., 82
Dollinger, E. J., 176, 225
Dominick, C. B., 308
do Nascimento, R. V., 413
Donnelly, J., 469
Dopp, H., 213, 216
dos Santos, M. W., 413
Doutt, R. L., 317, 324
Downes, J. A., 174
Downing, R. S., 317, 324
Dowson, R. J., 158
Drabkina, A. A., 442
Draper, C. C., 107
Draslar, K., 384
Drecktrah, H. G., 226
Drescher, W., 371
Dresco-Derouet, L., 267
Ducoff, H. S., 467
Dudnikova, A. F., 245, 252
Duelli, P., 266
Dugdale, J. S., 192, 193, 195-97
Dujardin, F., 361, 362
Duke, T. W., 21
Dumbleton, L. J., 194
Dumpert, K., 384
Dunbar, D. M., 300, 301, 305
Dunn, R. L., 426
Dunne, H. W., 110, 111
Dunnet, G. M., 243, 249, 252
Dupras, E. F. Jr., 426
Durant, J. A., 222
Durden, W. C., 37-39
Durzan, D. J., 108
Dustan, G. G., 308, 318
Duthoit, J. L., 99
Dutkowski, A., 366, 372
Dutky, R. C., 207
Dutky, S. R., 211, 212, 214
Dyar, H. G., 183, 186
Dyar, R. C., 229, 230
Dyer, R. E., 20
Dyte, C. E., 297, 447
Dzuik, L., 23

E

Eads, J. H., 12
Earle, N., 466
Earle, N. W., 172, 207
Eaton, J., 466
Eberhart, S. A., 344, 351
Eberson, L. E., 159
Ebert, B., 159
Ebling, W., 55, 263, 271
Ebstein, R. P., 68, 69
Echols, R. M., 88, 90
Eden, W. G., 21, 222
Edney, E. B., 262, 265, 268, 270-73, 275, 276
Edwards, A. M., 209
Edwards, C. A., 304
Edwards, C. R., 230
Edwards, L. H., 351
Egawa, K., 100
Eggers, F., 190
Eglinton, G., 86
Eidt, D. C., 80
Eikenbary, R. D., 298, 300

Eldefrawi, A. T., 161
Eldefrawi, M. E., 156, 161
Elgee, D. E., 101
Ellaby, S., 100
Eller, L. L., 21
Elliott, W. M., 290, 297, 299
Ellis, D. S., 107
Ellis, P. E., 276
Ellis, R. A., 385, 387-90
Elofsson, R., 135
El Rayah, E. A., 266, 268
El Sayed, E. I., 170, 171, 300
El-Ziady, S., 273, 274
Emmel, T. C., 176
Emschermann, P., 103
Engelmann, F., 133, 418
Engler, R. A., 102
Enkerlin, D., 462, 469
Enns, W. R., 309, 317
Enser, K., 53
Ensminger, A., 33
Eraker, J., 139
Erber, J., 372, 373
Ercelik, T. M., 170, 171
Eriksen, C. H., 269
Ernst, K. D., 388, 390
Esterline, A. L., 228
Etessami, S., 160
Evans, T. M., 249
Evenhuis, H. H., 298, 301
Everett, D. H., 274
Everett, T. R., 222
Everly, R. T., 342, 346
Everson, E. H., 342, 346
Eyer, J. R., 192
Ezzat, Y. M., 58

F

Fabre, J. H., 359
Facto, L. A., 226
Faeder, I. R., 151, 155, 156, 158, 162
Fahmy, H. S. M., 299, 304
Fahn, A., 87
Fain-Maurel, M. A., 134
Fairchild, M. L., 222, 223, 228, 338
Falcon, L. A., 102, 103
Fales, H. M., 14, 421
Fan, S. H., 298
Farley, R. D., 155-57, 160, 162, 265, 385, 387-89
Faulkner, P., 102
Fava, J. F. de M., 408
Federici, B. A., 99, 107
Fedorova, V. I., 248
Feeny, P. P., 85, 86, 297
Feingold, B. F., 248
Feldberg, W., 153
Fenemore, P. G., 86
Fenner, F., 251
Fenton, F. A., 308
Ferguson, D. E., 19, 286
Ferguson, J. L., 463
Ferkovich, S. M., 82
Fernando, E. F. W., 111
Ferrell, G. T., 82
Ferris, G. F., 54, 56, 66, 69
Fieldhouse, D. J., 231
Fields, E. S., 440
Fillon, C., 88
Finlayson, L. H., 133-35, 137, 140, 142, 143, 381, 387-89
Finley, M. T., 21
Finney, G. L., 61
Fiori, G., 276
Fischer, R. G., 252
Fisher, K. C., 81
Fisher, T. W., 34, 56, 62
Fisk, F. W., 443
Fitch, A., 337
Flaherty, D. W., 309, 324
Flake, R. H., 86
Flanders, S. E., 50, 303
Flatt, I., 266
Flattum, R. F., 139, 153
Fleet, R. R., 19
Fleschner, C. A., 303
Fleuriet, A., 112
Flint, H. M., 466
Flögel, J. H. L., 361-63
Fockler, C. E., 421
Foelix, R. F., 383-85, 388, 390, 391
Fogle, M. V., 20
Forbes, W. T. M., 183, 192, 197, 198
Ford, B., 244-49
Ford, H. R., 462, 473
Forel, A., 361
Forgash, A. J., 295, 384
Fort, S. W., 337, 340
Fossati, A., 177
FOSTER, G. G., 461-76; 462, 466, 469, 470, 472, 473
Foster, J. E., 349
Fowler, D. P., 80
Fox, P. M., 297, 300, 301, 315
Fracker, S. B., 183, 186
Fraenkel, G., 244, 252, 360, 385
Francis, M. J. O., 87
Franklin, R. T., 227
Frazer, E. D., 427, 428
Freeman, T. E., 34
Freyvogel, T. A., 249
Friedländer, M., 176
Friedman, K. J., 153
Friedman, S., 153
Friese, G., 193, 196, 197
Froggatt, W. W., 53, 300
Frontali, N., 139, 365-68
Frye, R. D., 228
Fuhr, A., 404
Fujioka, S., 215
Fukuda, T., 105
Fulcraud, J., 360
Furniss, M. M., 82, 88
Furuichi, Y., 108

G

Gabe, M., 133, 134
Gaffal, K. P., 382, 383, 388
Gaines, T. B., 21
Galacgac, N., 338
Galbraith, M. N., 209, 212, 215
Gallo, D., 24
GALLUN, R. L., 337-57; 231, 342, 345-50, 472
Galun, R., 249, 388, 389, 392
Gangstad, E. O., 40, 41
Gara, R. I., 83
Garcia, R. D., 169, 430
Gardenhire, J. H., 340, 345, 351
Gardiner, B. O. C., 101
Gargav, V. P., 289, 292, 297
Garofalo, C. A., 402
Gary, G. W., 110, 111
Garzon, S., 104
Gassner, G., 176
Gassner, G. III, 242
Gaudet, J. J., 32
Gay, F. J., 110
Gee, J. D., 140, 143
Geier, P. W., 62, 64
Geigy, R., 247
Gelbič, R., 419, 431
Geller, I., 160
Gentile, A. G., 53, 55, 58, 61-63
George, B. W., 339, 340
George, R. S., 252
Georghiou, G. P., 286, 306, 307, 311, 312, 314-16, 447
Gerasimova, N. G., 250-52
Gerhold, H. D., 78-80
Gerneck, R., 66
Gersch, M., 133, 142, 143
Gerson, U., 48, 68, 69
Gervet, J., 360
Gettrup, E., 385, 391
Gewecke, M., 391
Geyer, A., 448
Ghani, M. A., 35, 39, 41
Ghauri, M. S. K., 56
Ghent, A. W., 83
Ghilarov, M. S., 276
Gibbs, A. J., 97, 110
Gifford, J. R., 338
Gilbert, B. L., 81
Gilbert, L. I., 211, 213, 418, 419, 421, 427, 447, 448
Gilbert, M., 297
Gill, R., 466, 470, 473
Gill, S. S., 424

Gillott, C., 134
Gilmore, J. U., 300, 315
Gingrich, A. R., 446
Ginsborg, B. L., 139
Gitay, H., 103
GLANCEY, B. M., 1-30; 7, 10, 14, 17, 22, 25
Glasser, R. F., 306, 307, 315
Gnatzy, W., 382, 384, 385, 388, 390, 391
Godfrey, G. L., 301, 318
Godwin, P. A., 168, 170
Goeden, R. D., 39, 82
Goidanich, A., 48, 53, 69
Goldschmidt, R., 172
Goldsworthy, G. J., 133-36, 139-42
Goll, W., 363, 365, 366
Gomulinski, M. S., 339
Gonçalves, L. S., 399-412
Goncharov, A. I., 247, 252
Gonen, M., 169, 172, 173
GONZALEZ, R. H., 47-73
Good, E. E., 21
Goodnight, K. C., 209
Goodwin, T. W., 87, 208, 215
Goossen, M., 362, 364
Gordon, H. T., 297
Gordon, R. D., 40, 41
Gorell, T. A., 213
Görner, P., 384
Gosbee, J. L., 140, 143
Gotz, P., 103
Gough, D., 109, 110
Gould, H. J., 299, 318
Gouranton, J., 99, 111
Grace, T. D. C., 110
Graham, A. J., 468
Graham, H. M., 169, 173
Graham, O. H., 462-64
Graham, S. A., 76
Granados, R. R., 103
Granges, J., 177
Grant, M. N., 342
Grassmann, A., 448
Graves, J. B., 170, 171, 300
Graves, T. M., 426
Gray, E. G., 388
Greany, P. D., 385, 387-89
Greathead, D. J., 59, 61-63, 65
Green, A. J., 88
Green, E. E., 68, 300
Green, H. B., 7, 9, 10, 12, 18
Greene, G. L., 18
Greengaard, P., 140, 141
Greenslade, R. M., 299, 300
Griffiths, J. E., 262, 266
Grillot, J. P., 135
Grindeland, R. L., 232, 233
Grosch, D. S., 296, 298, 303, 304
Groth, U., 362, 364
Grover, K. K., 473
Grunberg, A., 60, 62
Gudz-Gorban, A. P., 106
Guerra, A. A., 109, 430
Guerrieri, G., 462, 469
Guirgis, S. S., 274
Gunn, D. L., 360
Günther, K. K., 246
Gupta, B. L., 137
Guthrie, D. M., 391
GUTHRIE, W. D., 221-39; 337-57; 176, 222, 225, 231-34, 347, 349, 350
Guy, R. D., 408, 412

H

Haas, G. E., 251, 252
Habib-ur-Rehman, 35, 41
Hacker, C. S., 468
Hackerott, H. L., 345
Hackett, J. T., 160
Hackman, R. H., 262, 270, 271
Haddow, J., 252
Hadley, N. F., 262, 264, 267, 271, 272, 274, 276
Haeselbarth, E., 252
Hafez, M., 265, 268, 273, 274
Hagen, K. S., 298, 299, 339
Häggendal, J., 139, 367
Hahn, S. K., 346, 348
Haile, D. G., 167
Haldane, J. B. S., 174
Hall, D. W., 99, 105
Hall, I. M., 101, 103, 110-12, 305
Hallauer, A. R., 232, 349, 350
Hallinan, E., 470
Hama, H., 105
Hambric, R. N., 37, 38
Hamilton, A. G., 272
Hamilton, E. W., 289, 290, 293, 298
Hamilton, R. J., 86
Hamilton, W. J. III, 262, 266, 275
Hamlen, R. A., 97, 101
Hamm, J. J., 339, 344
Hammock, B. D., 424
Hammond, A. M., 306
Hamori, J., 155
Hampson, G. F., 183
Hanegan, J. L., 268
Hangarter, W., 437, 438, 444
HANOVER, J. W., 75-95; 82, 86-88
Hansch, C., 160
Hansen, K., 382-84, 387, 388
Hanström, B., 362, 364
Haragsim, O., 445
Hardie, R. L., 159
Harding, J. A., 225, 229, 230
Hare, W. W., 340
Harker, J. E., 359, 373
Harlow, P. A., 155
Harmsen, R., 134, 135
Harper, J. D., 103
Harpez, A., 55
Harrap, K. A., 97, 99, 100, 109
Harrell, E. A., 340
Harris, D. J., 384, 385
Harris, P., 33, 35, 36, 43, 82
Harris, R. L., 427, 428
Harris, W. G., 18
Harrison, A. K., 110, 111
Hart, D. E., 139
Hartley, G. S., 299
Hartsock, J. G., 225
Harvey, T. L., 345
Haslam, S. M., 32
Hassanein, M. H., 288-90
Hatchett, J. H., 348
Hathaway, D. O., 171
Haupt, J., 383, 384
Hawke, S. D., 265, 385, 387-89
Hawkes, R. B., 31, 37
Hayashi, Y., 105, 107, 108
Hayes, D. K., 222
Haynes, J. M., 274
Hays, K. L., 12
Hays, S. B., 12, 20, 367
Hazard, E. I., 99, 107
Headlee, T. J., 192
Heath, J. E., 266, 268
Heath, M. S., 268
Hecker, H., 249
Hedrick, G. W., 88
Heed, W. B., 209
Heeg, J., 265, 273, 275
Hefnawy, T., 263, 264, 273, 274
Heikkenen, H. J., 81, 88
Heimpel, A. M., 101, 102
Heinrich, B., 142, 268
Heinrich, C., 197
Heinrich, G., 214
Hejno, K., 423, 444
Helle, W., 313, 314
Henderson, C. A., 339
Henderson, D. K., 22
Henderson, J. A., 22
Henderson, J. F., 102
Henneberry, T. J., 177, 307, 309, 319, 320
Hennessey, R. D., 36, 38, 40
Henrick, C. A., 423-25
Henry, J. E., 110, 111
Hensley, S. D., 18
Herald, F., 428
Herbert, H. J., 311

Herbin, G. A., 86
Herda, G., 448
Herforth, R. S., 112
Heriot, A. D., 51, 52
Herman, E., 302
Hermann, H. R., 3, 10, 11, 25
Herne, D. H. C., 306-8, 311, 314, 320
Heron, R. J., 80, 84
Herreid, C. F., 273
Hertel, W., 142
Hess, A., 366, 367
Hessel, J. H., 199
Hester, P., 424
Heuman, H. G., 384, 387, 388
Heymer, A., 360
Heyne, E. G., 345
Hibbs, E. T., 222
Hicks, D. M., 21
Hightower, B., 462-64
Hill, C. C., 351
Hill, R. B., 155, 156
Hill, R. D., 267
Hill, R. E., 222, 223, 227-29
Hill, S. O., 3, 10, 11, 20, 25
Hill, W. G., 472
Hille Ris Lambers, D., 436
Hills, T. M., 230
Himeno, M., 99, 100
Hink, W. F., 97, 102
Hinton, H. E., 183, 185-89, 194-97, 241, 245, 264, 268, 270, 272
Hirao, T., 392
Hiripi, L., 139, 140
Ho, K. K., 298
Hockenyos, G. L., 68
Hocks, P., 209
Hodek, I., 421, 436
Hodges, J. D., 82, 90
Hodgson, E. S., 392
Hodson, A. C., 222, 229
Hoekstra, P. E., 88, 90
Hoffman, A. C., 296, 298, 304
Hoffman, C., 381, 384, 387
Hoffman, J. A., 213, 214
Hoffman, R. J., 468
Hoffmann, D. F., 100
Hoffmeister, H., 209, 214
Hofmaster, R. N., 231
Hogan, T. W., 110, 471
Holbert, P. E., 384
Holdaway, F. G., 227, 339
Holdsworth, R. P., 319-22, 324
Holland, G. P., 241, 252
Holloway, J. K., 31
Holm, E., 275
Holm, J. G., 32, 33, 41
Holman, G. M., 138, 139, 156, 158
Holmes, N. D., 339, 342
Holmgren, N., 360
Holt, G. G., 167-78
Homberger, E., 441
Hoogers, B. J., 32
Hopkins, A. D., 81
Hopkins, D. E., 446
Hopkins, G. H. E., 243, 247
Hopla, C. E., 252
Hoppe, W., 209
Horber, E., 338
Horn, D. H. S., 205, 209, 211, 212, 215
Horridge, G. A., 359, 360, 363, 368, 370, 373, 381, 383
Horsburg, R. L., 321
Hoshino, M., 392
Hossain, M. M., 170
Hostetter, D. L., 101, 103
Hotchkiss, D. K., 225
Hough, W. S., 171
House, C. R., 138, 139
House, H. L., 83
Hower, A. A. Jr., 435
Howitt, A. J., 308, 309, 319-22, 324
HOWSE, P. E., 359-79; 362-65, 372-75, 381
Hoy, M., 309, 324
Hoyer, G. A., 213, 216
Hoyle, G., 137, 151, 162
Hoyt, S. C., 397, 308, 311, 314, 321, 324, 326
Hrdý, I., 438, 442, 443, 445
Hrutfiord, B. F., 81, 83, 88
Hsiao, C., 42, 421, 445
Hsiao, T. H., 42, 421, 445
Hsu, S. J., 338, 342
Huber, F., 359, 370, 371, 373
Huber, R., 209
Huddart, H., 151, 153
Hudon, M., 227, 228
Huffaker, C. B., 31, 33, 61, 308, 313, 316, 322, 324
Huger, A. M., 102, 103, 111
Huggans, J. L., 231, 233, 234, 340, 341
Hughes, M., 244-47, 249
Hughes-Schraeder, S., 58
Hukuhara, T., 101, 110
Hull, L. A., 322
Hull, R., 110
Hulley, P., 63-65
Hummel, H., 209
Hummeler, K., 111
Humphries, D. A., 245, 246, 249, 250, 252
Hunt, L. M., 420
Hunter, D. K., 100, 101
Hunter, P. E., 17
Hunter-Jones, P., 273
Hůrka, K., 251
Hurpin, B., 97, 104
Husman, C. N., 468
Husseiny, M., 169, 171, 172
Hussey, N. W., 97, 299, 318
Hutchins, R. F. N., 208
Hutt, R. B., 170

I

Ibrahim, M. M., 265
Idris, B. E. M., 275
Ignoffo, C. M., 97, 99-102, 105, 109, 111, 289-91, 298, 305, 308
Iijima, T., 170
Ikeda, K., 140
Ikekawa, N., 207, 208, 213
Ikeshoji, T., 428
Il'inskaya, V. L., 250, 251
Imai, H. T., 174, 176
Imai, K., 424
Imanishi, K., 100
Imms, A. D., 183
Injac, M., 99
Inserra, S., 302
Ioff, I. G., 253
Iqbal, Q. J., 246
Irving, S. N., 303
Ishaaya, I., 53
Ishibashi, M., 213
Israel, P., 338
Issi, I. V., 305
Iszard, R. E., 33
Ito, Y., 102
Iwashita, Y., 108

J

Jabball, I., 251
Jackson, B. R., 351
Jackson, R. D., 20, 225, 228-30
Jacobson, M., 81, 418, 424, 430, 442
Jacques, R. P., 101, 102
Jakob, W. L., 425-27
James, H. C., 59
James, S. H., 461
Jantz, O. K., 82, 421, 434
Janzen, D. H., 275
Jarolim, V., 423, 444
Jarvis, J. L., 223, 227
Jay, D. L., 101, 102
Jenkins, J. N., 75, 338, 340
Jennings, C. W., 232-34, 342
Jensen, F., 309, 324
Jensen, J. H., 121
Jenser, G., 48
Jeppson, L. R., 290, 307-9, 314, 326
Jobin, L., 433
Johnson, B., 135-37

Johnson, P. C., 82
Johnson, W. S., 424
Johnston, C. O., 345
Jolivet, P., 275
Joly, L., 134
Joly, P., 213, 214
Jones, A. L., 308, 309, 319-22, 324
Jones, E. P., 58, 61, 65
Jones, E. T., 345
Jones, J. A., 226
Jonescu, C., 361
Jordan, K., 188, 248, 250
Jorgensen, C. D., 317
Joseph, S. A., 244, 250
Jotwani, M. G., 289, 292, 339
Jousset, F. X., 111
Jouvenaz, D. P., 15-17, 20, 21, 25
Juckes, I. R. M., 110
Judy, K. J., 417, 419, 423
Juliano, J. C., 405
Jupin, N., 112
Juvonen, S., 82

K

Kabuta, H., 384
Kadatskaya, K. P., 247, 251
Kaissling, K. E., 383, 385
Kaluzhenova, Z. P., 250, 251
Kamm, J. A., 421, 439
Kane, W. R., 102
Kanjanasoon, P., 338
Kanta, S., 289, 293
Kanungo, K., 274
Kanwisher, J. W., 272
Kapadia, G. G., 209
KAPLANIS, J. N., 205-20; 205-7, 209, 211-14
Karlin, A., 161
Karlson, P., 209, 213, 214, 216
Karnavar, G. K., 434
Karpenko, C. P., 175, 176
Kartman, L., 248
Karuhize, G., 388, 390
Kasai, M., 162
Kasting, A., 339, 340
Kaston, B. J., 386
Katzenellenbogen, J. A., 447
Katznelson, M., 403, 404
Kaufman, T., 266
Kaur, B., 338
Kawahara, S., 294, 299, 300
Kawai, S., 50, 68
Kawamoto, F., 108, 109
Kawanishi, C. Y., 100
Kawarabata, T., 107
Kawase, S., 108, 109
Kaya, H. K., 300, 301, 305
Kearney, G., 2
Kearns, C. W., 296
Keaster, A. J., 222, 223, 227, 229
Kehat, M., 287, 289
Keiser, I., 80
Keith, D. L., 223
Keller, J. C., 462
Keller, S., 102
Kelly, D. C., 104, 105
Kelsey, J. M., 305
Kelsey, L. P., 231
Kempff Mercado, N., 404, 412, 413
Kempster, R. H., 226
Kenchington, W., 252
Keng, H., 87
Kennedy, J. S., 360, 444
Kennel, J., 190
Kennett, C. E., 61, 307, 308, 311
Kenyon, C. F., 365
Kerkut, G. A., 156
Kerr, S. H., 303
Kerr, W. E., 399-412
Kessel, E. L., 252
Khalil, F., 288-90
Khan, A. G., 39, 41
Khattar, N., 138
Khosaka, T., 99, 100
Kieckhefer, R. W., 289, 290, 293, 298
Killinen, R. G., 102, 305
Kim, K. C., 222
Kimbrough, R. D., 21
King, D. S., 211, 216, 418
King, R., 20
Kinn, D. N., 317, 324
Kirby, A. M. A., 325
Kircher, H. W., 209
Kiriakoff, S. G., 190
Kiritani, K., 294, 299, 300
Kir'yakova, A. N., 245, 251, 252
Kishaba, A. N., 110, 170, 171, 177
Kitching, R. L., 473
Klassen, W., 177, 466, 472, 473
Klauske, J., 433
Klemetson, D. J., 176
Klemm, M., 55
Klemm, N., 135, 136, 139, 366-68
Kloft, W., 48
Klots, A. B., 193
Klun, J. A., 224, 225, 349, 350
Klunsuwan, S., 139
Knapp, F. W., 428
Knerer, G., 83, 84
Knight, F. B., 76
Knight, K. L., 226
Knipling, E. F., 168, 177, 462, 467, 472, 473
Knisley, C. B., 317
Knowles, F. G. W., 133, 134, 137, 139, 144
Knudson, D. L., 111, 112
Knülle, W., 273, 274
Knyazeva, N. I., 391
Kobayashi, M., 99, 108, 109, 207
Koch, C., 262-64
Kok, L. T., 209
Kondo, S., 168
Kondrashkina, K. I., 245, 252
Kontev, C. A., 441, 442
Koolman, J., 213, 214
Koreeda, M., 215
Kostowski, W., 367
Kot, J., 291, 296, 301, 315
Kowalska, T., 301
Kozhanchikov, I. V., 184
Kramer, W. L., 428
Krampitz, H. E., 244
Kraus, J. F., 90
Křeček, J., 443
Krieg, A., 97, 103
Krieger, R. I., 297
Krishanan, G., 264
Krishnakumaran, A., 421, 448
Kristensen, N. P., 188, 189, 195, 197
Krupauer, V., 33
Krywienczyk, J., 108
Kühnelt, G., 266
Kuhr, R. J., 438
Kulakova, Z. G., 250
Kulinčevič, J. M., 409
Kunitskaya, N. T., 247
Kuraev, I. I., 252
Kurihara, M., 170
Kurstak, E., 97, 101, 104, 106, 107, 305
Kurtti, T. J., 97
Kuwabara, M., 368, 369, 384, 385, 388, 389
Kuwatsuka, S., 83
Kuznetsova, M. A., 106

L

Lábler, L., 440
La Brecque, G. C., 462
LaChance, L., 467, 470-72
LaChance, L. E., 167, 173-77
Ladle, M., 32
LaHue, D. W., 340
Lake, C. R., 140
Lal, R., 289, 293
Lamar, C., 158
Lamb, K. P., 438
Lambert, J. G., 273
Lambremont, E. N., 207
Lamparter, H. E., 365
Landa, V., 420, 434-36, 441
Landano, R. L., 170

Landolt, A. M., 365, 366
Lane, J. M., 90
Langley, P. A., 136
Larrivee, D. H., 248
Larsen, J. R., 153, 384
Larson, H., 325
Larson, R. I., 345
Lassota, A., 171
Laster, M. L., 174
Laszlo, I., 139
Laurentiaux, D., 184
Laverack, M. S., 136, 139
Lavigne, G. L., 412
Lavroushin, A., 102
Law, J. H., 16, 423
Lawrence, P. A., 420
Lawrence, P. O., 303
Lawson, G., 339
Lea, D. E., 168, 172
Lea, T. J., 157
Leader, J. P., 270
Leaf, G., 156
Leake, L. D., 156
Lebedeva, O. P., 106
Lebsock, K. L., 350
Lecomte, J., 410
Lee, A. H., 155
Lee, C. Y., 162
Lee, M. S., 317
Lee, P. E., 105
Lee, S. Y., 245
Lees, A. D., 274, 385, 436, 437
Legner, E. F., 309
Lekic, M., 35, 36, 41
Lemoine, G., 88
Lemoine, M., 78
Lemon, H. W., 86
Lennartz, F. E., 3-6, 24
Leroux, E. J., 55, 67, 311
Leslie, R. A., 138
Lester, J. J., 133
Letchworth, P. E., 423
Leuck, D. B., 338, 340
Leuthenegger, R., 100
Levi-Montalcini, R., 135, 136
Levins, R., 468
Levinson, A. S., 88
Levy, R., 22
Lewallen, L. L., 425
Lewandowski, L. J., 108
Lewis, C. T., 389
Lewis, J. G. E., 264, 275
Lewis, L. C., 222, 225, 226, 228, 233, 340, 431
Lewis, L. F., 425
Lewis, R. E., 242, 243, 252, 253
Lewontin, R. C., 461
Leydig, F., 383
L'Heritier, P., 112
Libera, W., 386
Lickerish, L. A., 299, 320
Lightly, P. M., 168
Lindauer, M., 360, 392
Lindquist, A. W., 464
Lingappa, S. S., 298, 300
Lingren, P. D., 177, 288-91, 296, 298-301, 308
Linn, J. D., 318
Lipa, J. J., 103
Little, E. C. S., 31-33
Little, E. L. Jr., 87
Livingston, J. M., 102
Lloyd-Jones, J. G., 208
Locher, J. T., 291
Loeb, J., 370
Loew, P., 424
LOFGREN, C. S., 1-30; 7, 12, 16, 17, 20-22, 25, 462, 464, 473
Lofty, J. R., 304
Loiselle, J. M., 106
Long, W. H., 340
Longworth, J. F., 99, 110
Lopez-D, F., 462
Lorimer, N., 470
Lorio, P. L. Jr., 90
Loughner, G. E., 224
Louloudes, S. J., 209, 212
Louw, G. N., 275
Lovely, W. G., 229, 230
Loveridge, J. P., 271, 272
Lowe, J. I., 21
Lowe, R. E., 105
Lucey, E. C. A., 241
Lucht-Bertram, E., 361, 362
Luckey, T. D., 303
Luco, J. V., 359
Lüdicke, M., 48, 62
Ludke, J. L., 21
Luginbill, P. Jr., 339, 342, 351
Lühl, R., 100, 101
Luk'yanova, A. D., 252
Lundquist, P. G., 382, 388, 390
Lunt, G. G., 157, 162
Lupo, V., 50, 54
Lüscher, M., 442, 443
Lusk, L., 21
Lwoff, A., 97
Lynch, R. E., 222, 225, 233, 340, 431
Lynn, C., 309, 324

M

Maa, T., 400, 407
Mabry, T. J., 87
MacConnell, J. G., 14, 15
Macdonald, D. R., 302, 308
MacDonald, M. D., 345
MacFarlane, J., 172
MacGillivray, H. G., 80
Machado-Allison, C. E., 252
Machili, P., 156-58
Maciel, H. B., 402, 407-10
MacKay, M. R., 184
Mackinnon, E. A., 102
Macleod, R., 110
MacPhee, A. W., 322
Maddox, D. M., 35-38, 40
Maddrell, S. H. P., 133, 135, 137, 139, 140, 143
Madhavan, K., 427
Madsen, H. F., 169, 171, 172, 309, 321
Major, R. T., 86
Majumdar, N., 303
Makino, S., 175
Maleki-Milani, H., 110
Mancini, G., 139, 365-68
Mandel'shtam, Y. E., 155
Mangum, C., 23
Mangum, C. L., 109
Manley, S. A. M., 80
Manlik, S., 68
Manser, P. D., 308
Manton, S. M., 261
Maramorosch, K., 97
March, R. B., 141, 155-57, 160, 162
Mardon, D. K., 252
Mardzhanyan, G. M., 308
Marek, J., 60
Margalit, J., 252, 392
Markin, G. P., 2, 3, 5, 7-11, 13, 14, 17, 20-22, 25
Markl, H., 391
Marks, E. P., 139
Marlatt, C. L., 59
Marpeau, A., 88
Marschall, K. J., 102
Marshall, G. A., 23
Martignoni, M. E., 97, 99, 107
Martin, H., 392
Martin, J. W., 423, 439
Martinez, E., 110
Marumo, S., 424
Maslennikova, V. A., 305
Masner, P., 420, 441, 444
Mason, C. A., 144
Mason, R. R., 82
Masuhr, T., 372, 373
Mathis, J., 62
Matolin, S., 419, 427, 441
Matsuda, M., 67
Matteson, J. W., 223
Maxwell, F. G., 75, 338-40
Maynard, D. M., 369
Mayo, Z. B. Jr., 340
McAfee, D. A., 141
McAlister, H. J., 319, 320
McCaman, R. E., 140
McCann, F. V., 153
McCarthy, W. J., 103, 104
McClanahan, R. J., 299
McCoy, C. E., 53
McDermott, R. E., 78
McDonald, F. J., 467
McDonald, I. C., 469, 470,

473
MCDONALD, T. J., 151-66; 155-62
McElroy, P. J., 274
McEwen, F. L., 102
McGaha, Y. J., 34, 39
McGinnis, A. J., 339, 340
McGovern, T. P., 427, 447
McGuire, J. U., 222
McGuire, J. U. Jr., 177, 472, 473
McIndoo, N. E., 386
MCIVER, S. B., 381-97; 383, 390
McKague, B., 428
McKenzie, H., 342
McKenzie, H. L., 47, 53, 60, 66, 69
McLaren, I. W., 50, 58, 63
McLaughlin, R. E., 105, 305
McMillian, W. W., 340, 344, 351
McMurtry, J. A., 313, 316, 318, 322
McNeal, F. H., 338, 342, 345, 350
McReynolds, R. D., 90
McWhorter, G. M., 228, 230
Mead-Briggs, A. R., 244-46
Meier, W., 437, 438
Meisch, M. V., 426
Melamed, J., 365
Melis, A., 55, 60, 62
Mellado, L., 462
Mellini, E., 276
Mello, M. L. S., 410
Méndez, E., 252
Menn, J. J., 418, 423
Menzel, R., 372, 373
Menzies, W., 251
Mergen, F., 88, 90
Messenger, P. S., 291, 295, 466-68
Mestriner, M. A., 407
Meszaros, J., 367
Metcalf, C. L., 68
Metcalf, R. L., 295-97, 299, 447
Metcalfe, M. E., 348
Metwally, M. M., 420, 434-36
Meyer, A. S., 419, 423, 424
Meyer, D., 442, 443
Meyer, H. J., 82
Meyer, R. H., 307, 309, 311, 319-21, 324, 325
Meynadier, G., 99, 100, 111
Meyrick, E., 183, 193, 198
Miaullis, J. B., 423
Michaeli, D., 248
Micheli, L., 371
MICHENER, C. D., 399-416
Michewicz, J. E., 33
Middleton, E. J., 209, 212, 215
Mihajlovic, L., 35
Mikulin, M. A., 253
Miledi, R., 162
Mill, P. J., 384, 385
Miller, D. M., 153
Miller, M. C., 227
Miller, P. L., 156, 272
Miller, R. W., 428
MILLER, T. A., 133-49; 135, 137, 139, 140, 143, 293, 318
Mills, G. D. Jr., 447
Mills, H. B., 21
Mills, R. B., 170, 171
Mills, R. R., 140
Millward, S., 108
Milne, R. G., 111
Mimura, K., 368, 369
Miner, R. C., 248
Minton, S. A. Jr., 276
Mirau, K., 108
Miskemen, G. W., 340
Mitsui, T., 424
Mittwoch, U., 172
Miura, T., 306, 426
Miya, K., 170
Miyazaki, H., 213
Moberly, B., 137
Mobley, G. S. Jr., 40, 41
Modder, W. W. D., 338
Moeck, H. A., 385, 388, 390
Moeur, J. E., 269
Moffitt, C., 57
Moffitt, H. R., 288, 289
Mohr, C. O., 251
Moilliet, T. K., 317
Molinoff, P., 162
Molyneux, D. H., 245
Monaco, R., 301
Monro, J., 464
Monroe, R. E., 207, 209
Monsarrat, P., 99, 111
Moorhead, M., 367
Moran, D. T., 385, 387-91
Mordue, W., 133-36, 139-42
Moreno, D. S., 57
Morgan, C. V. G., 308
Mori, C., 213
Moriarty, F., 302
Morisaki, M., 208
Morita, H., 368, 369
Moriyama, H., 213, 215
Morozova, I. V., 248, 250, 251
Morrill, W. L., 11, 13, 25
Morrison, F. O., 383, 385
Morrison, R. D., 345
Mosbacker, G. C., 172
Mosher, E., 187, 188, 196, 199
Mote, M. I., 143
Motoyama, N., 290, 306, 307, 309, 314
Moulins, M., 136, 139
Mountain, P., 408
Mowat, D. J., 290
Mrak, E. M., 121
Mucha, S., 82
Mulla, M. S., 428
Mullen, M. A., 190
Muller, V. A., 402, 413
Mulligan, H. F., 32
Mulligan, J. V., 140, 142, 143
Munakata, K., 338
Munshi, D. M., 248, 249
Munson, R. E., 230
Murakami, A., 168, 171, 174, 176
Murphy, F. A., 110, 111
Murty, B. R., 344
Murzakhemetova, K. M., 247, 249
Mustaque, M., 35, 41
Mutuura, A., 193, 195, 197
Myalkovskaya, S. A., 251
Myllymäki, A., 250

N

Naber, E. C., 21
Nagai, T., 137, 138
Nair, K. K., 305, 420
Nair, K. S. S., 434
Nair, M. R. G. K., 291
Nakanishi, K., 215
Nakazawa, H., 102
Naor, D., 252
Nascimento, C. B., 413
Nassar, S. G., 419, 423, 437-39
Natalizi, G. M., 139
Nathanson, J. A., 140, 141
Naveiro, J., 404
Nayar, K. K., 134, 137-39
Nazarova, I. V., 250
Neal, H., 161
Neal, J. W., 306
Neal, J. W. Jr., 435
Neese, V., 391
Nef, L., 305
Neilson, M. M., 101
Neiswander, C. R., 231, 338
Nel, R. G., 54, 55, 57, 58
Nelson, B. C., 252
Nelson, D. R., 471
Nelson, E. E., 288, 308, 309, 319-22
Nelson, L. L., 293, 318
Nemec, V., 444
Nettles, W. C., 296, 300
Neville, C., 241, 242, 245, 252
Newell, I. M., 267
Newman, J. F. E., 110

Newsom, L. D., 286, 289, 296, 300, 303, 306-8, 312, 314, 316-18
Newton, J. R., 167, 168
Newton, W. H., 463
Nicholas, H. J., 87
Nicholson, A. J., 285, 467
Nickerson, J. C., 24, 25
Nicklaus, R., 382, 384, 388, 390
Nielsen, R. A., 169-72
Nilles, G. P., 421, 434
Nishiitsutsuji-Uwo, J., 372
Noble-Nesbitt, J., 273
Nogueira-Neto, P., 400, 404
Noirot, C., 137, 140
Nordin, G. L., 100
Norland, R. L., 428
Normann, T. C., 133, 134, 136, 138-40, 142, 143
Norment, B. R., 23
Norris, D. M., 81, 82, 84, 209
Norris, M. J., 276
NORTH, D. T., 167-82; 167-78
Nosky, J. B., 288, 291, 298, 300, 301
Novak, A. F., 15
Novák, K., 432
Novák, V., 306, 430
Novák, V. J. A., 418, 419
Núñez, J. A., 408, 409
Nutting, W. L., 135

O

Oatman, E. R., 309, 318
Oberheu, J. C., 21
O'Brien, R. D., 155-58, 161, 162
Obrtel, R., 301
O'Conner, A., 155
Oda, T., 61
Odetoyinbo, J. A., 471
Odier, F., 107
Oho, N., 101, 102
Ohtaka, H., 208
Okamoto, D., 338
Okasha, A. Y. K., 273, 274
Okauchi, T., 215
Oksman, M., 404
Okubayashi, M., 208
Oliver, A. D., 12, 18, 40, 41
Oliver, B. F., 338
O'Loughlin, G. T., 110
Oloumi-Sadeghi, H., 224, 225
Oma, E. A., 110, 111
Omer, S. M., 276
O'Neal, J., 2, 7, 14, 23, 24, 444, 445
Onji, P. A., 104
Onodera, K., 99
Opler, P. A., 184
Orlab, K. B., 342
Ortman, E. E., 339
Osborne, M. P., 133-35, 137, 140, 142, 143, 152, 153, 381
O'Shea, M., 369
Oshima, Y., 83
Oswalt, R. M., 351
Otto, D., 359, 370, 374
Ou, S. H., 338
Outram, I., 430, 446
Ouye, M. T., 169
Overmeer, W. P., 469

P

Packard, A. S., 193, 195
Padovani, F., 170, 171, 173, 177, 339
Paget-Clarke, C. D., 22
Pailhé, L. A., 406
Painter, H., 338
Painter, R. H., 75, 76, 79, 84, 337, 339, 344, 348
Pal, R., 462-64, 467-69, 471, 473
Pallos, F. M., 418, 423
Pan, M. L., 421
Pan, Y. P., 340
Panaia, J. R., 24
Pandazis, G., 361
Panwar, V. P. S., 339
Paradis, E. R., 311
Paradis, S., 102
Parent, B., 311
Park, K. E., 135
Parker, A. H., 266
Parker, F. D., 101
Parker, J. H., 348
Parker, K., 241, 242, 245, 250, 252
Parr, T., 53
Parr, W. J., 299, 318
Parrish, P. R., 21
Parrott, W. L., 75, 338, 340
Pascal, A., 106
Paschke, J. D., 104, 105
Pass, H., 351
Pathak, M. D., 338, 339, 344, 346, 348
Patterson, F. L., 346
Patterson, J. W., 426, 442, 448
Patterson, R. S., 462, 473
Patton, R. F., 79
Pauly, G., 87, 88
Pausch, R. D., 244, 252
Pavan, C., 100
Pavlova, A. E., 252
Pawson, B. A., 449
Payne, C. C., 99
Pearson, L., 364-66
Peck, R. A., 351
Peckham, G. W., 360
Pener, M. P., 273
Penny, L. H., 232, 233, 344, 347
Penny, N. D., 222, 223
Penzlin, H., 136
Perkins, B. D., 35, 36, 38, 39, 41
Perron, J. M., 437, 438
Persson, K., 159
Pesho, G. R., 232, 233, 339, 340, 344
Pesson, P., 51, 52, 54
Peters, D. C., 230, 339
Peters, L. L., 338
Peters, W., 384
Peterson, L. K., 339, 342
Peus, F., 250-53
Peyer, B., 437, 438
Phillips, J. E., 273
Phillis, J. W., 151
Philogène, B. J. R., 75
Philpott, A., 189, 195
Piazzi, R., 366
Picard, D. J., 135
Pickard, L. S., 82
Picken, J. C., 225
Pickett, A. D., 285
Piek, T., 162
Pielou, D. P., 304, 306, 307, 315
Pierce, N. W., 20
Pieri, L., 367
Pinckney, J. S., 351
Pinnell, R. E., 101
Pinnock, D. E., 102
Pipa, R. L., 419
Pirie, N. W., 33
Pitman, G. B., 81
Pitman, R. M., 133, 140, 151
Pittendrigh, C. S., 372
Pizani, J., 408
Plank, G. H., 80
Plapp, F. W. Jr., 303, 447
Platt, R. B., 20
Pless, C. D., 299
Plewska, T., 296, 301
Plumley, J. K., 22
Plus, N., 111, 112
Poduška, K., 422, 441
Poe, S. L., 309, 317
Poe, W. E., 21
Poenicke, H-W., 242
Pogo, B. G. T., 103, 105
Polson, A., 103
Polunina, O. A., 247
Pomonis, J. G., 471
Popolizio, E. R., 406
Popov, G., 276
Porte, A., 134
Porter, K. B., 339, 340, 345
Porterfield, W. A., 34
Potter, L. T., 162
Pradhan, S., 289
Prasad, R. S., 244, 245, 248, 251, 252
Prestage, J. J., 391

Price, P. W., 308, 318
Pridham, J. B., 87
Priesner, H., 62, 64
Prince, W. T., 141
Pringle, J. W. S., 381, 385, 386
Prokopiyev, V. N., 247, 252
Proshold, F. I., 169, 171-74, 176, 177, 471
Prout, T., 466, 470, 473
Provansol, A., 135
Proverbs, M. D., 167-72, 462, 464, 465, 469
Pruszynski, S., 301, 309, 324
Pumphrey, R. J., 388
Pura, C. D., 344
Putman, W. L., 111, 112, 306-8, 311, 314, 320
Putnam, A. R., 309, 319-21, 324

Q

Quastel, D. M. J., 160
Quayle, H. J., 62, 65-67
Quennedey, A., 134, 137, 140
Quintana, V., 171, 177, 339
Quintana-Munez, V., 167-71, 173, 177
Quiot, J. M., 102
Quistad, G. B., 424
Qureshi, Z. A., 170, 171

R

Raabe, M., 135
Rabl-Rückhard, T., 361
Radovsky, F. J., 243, 244
Rafes, P. M., 275
Raheja, A. K., 100
Rahim, A., 39, 41
Rai, B. K., 289, 293
Rai, K. S., 462, 470, 473
Rainey, D. P., 84
Rainey, R. C., 276
Rajulu, G. S., 264
Ramaseshiah, G., 36, 41
Ramsay, J. A., 274
Rana, B. S., 344
Randall, P. J., 208
Randall, W. K., 79
Randolph, N. M., 339
Rao, C. B. J., 108
Rao, V. P., 36, 41, 42
Raros, R., 338
Ratcliffe, F. N., 251
Raun, E. S., 20, 225, 227-29, 233, 240
Rauston, J. R., 173
Rautapaa, J., 339
Raynes, J. J., 40, 41
Reagan, T. E., 18
Redfern, R. E., 447
Reece, R. W., 153
Reed, D. K., 110-12
Reed, E. M., 102
Reed, G. L., 222-24, 226, 227, 229, 233
Rees, D., 135, 137, 138, 143, 153, 156
Rees, H. H., 205, 208, 211, 215
Reh, L., 54
Řeháček, J., 252
Reichelderfer, C. F., 100, 101
Reicosky, D. A., 86
Reid, R. W., 82
Reineke, D. E., 97
Reinganum, C., 110
Reinhart, C., 249
Reissig, W. H., 439
Reitz, L. P., 343, 345, 348, 350
Remington, C. L., 183
Renwick, J. A. A., 83, 88
Retnakaran, A., 108, 430, 433, 435, 447
Revelo, M. A., 228
Reynolds, H. T., 299
Reynolds, P. S., 222
Rhoades, W. C., 8-10
Rhode, R. H., 462
Rice, M. J., 137, 388, 389, 392
Rice, R. E., 57, 306
Richard, R. D., 173
Richards, A. G., 190, 243, 249
Richards, A. M., 48
Richards, O. W., 184, 198
Richards, P. A., 243, 249
Richards, W. C., 108
Richardson, C. S., 337, 340
Richmond, C. E., 430
Richter, S., 384, 388
Rick, J. T., 140
Ricker, D. W., 39
Ricks, B. L., 12
Riddiford, L. M., 23, 82, 373, 418-20, 437, 438, 441, 445, 447
Ridgway, R. L., 288-92, 296, 298, 299, 308
Ridgway, W. O., 171
Riek, E. F., 184
Riemann, J. G., 171, 173, 175, 176, 471
Rinaldi, A. J. M., 406
Ringlund, K., 346
Ripper, W. E., 285, 298, 299, 320, 321
Ristich, S. S., 319, 320
Ritcher, P. O., 264
Ritter, F. J., 207
Rivers, C. F., 106, 107
Rivnay, E., 275
ROBBINS, W. E., 205-20; 205-9, 211-14
Robeau, R., 23
Robert, P. H., 104
Roberts, D. R., 88, 293, 318
Roberts, D. W., 103-5
Roberts, D. W. A., 342
Roberts, L. W., 293, 318
Roberts, S. K., 372, 373
Roberts, S. K. de F., 372
Robertson, F. W., 467
Robertson, H. A., 138, 140, 141
Robertson, J. G., 307, 315
Robertson, J. L., 295, 300, 302
Robertson, J. S., 99, 104, 105
Robinson, A. G., 339, 342
Robinson, J. F., 224
Robinson, R., 172, 174
Robson, T. O., 33
Rock, G. C., 290, 294, 306, 307, 309, 314, 319, 320, 324
Roe, R. A. II, 12, 18, 25
Roeder, K. D., 155, 359, 368, 371
Roelofs, W. L., 224
Roffey, J., 276
Rogoff, M. H., 101
Rohdendorf, E. B., 419, 420
Rojakovich, A. S., 141
Röller, H., 423, 424
Romañuk, M., 418, 423, 433, 441, 442
Root, R. B., 301, 318
Roppel, R. M., 137, 171, 173, 176
Rose, M., 433
Rosen, D., 291, 303
Rosenberry, T. L., 161
Roshdy, M. A., 263, 264, 384
Rosin, R., 267
Ross, H. H., 15
Ross, W. M., 345
Rossetto, C. J., 338
Roth, L. E., 176
Roth, L. M., 387
Rothenbuhler, W. C., 409, 410
ROTHSCHILD, M., 241-59; 241-53
Rottink, B. A., 88
Roussel, J. P., 142
Rowell, C. H. F., 369
Rowley, W. A., 225
Rozental, J., 82
Ruckelshaus, W. D., 21
Rudall, K. M., 252
Rudinsky, J. A., 81-83, 88, 90, 306
Rudolf, P. O., 79
Rule, H. D., 168, 170
Runeckles, V. C., 87

Ruppel, R., 342, 346
Rupprecht, R., 382
Russell, G. B., 86
Russell, W. A., 232-34, 339, 342, 349, 350
Ruttner, F., 400
Ruzicka, Z., 421, 436
Ryckman, R. E., 251
Ryder, J. C., 231

S

Sabilayev, A. S., 250
Sacher, R. M., 425
Sado, T., 170
St. George, R. A., 82
St. Onge, E., 425
Sakaguti, K., 252
Sakharov, D. A., 133
Salanki-Rozsa, K., 139, 140
Salmon, S. C., 348
Salpeter, M. M., 151, 155, 156, 158, 162, 386, 391
Samarasinghe, S., 55, 67
Samarina, G. P., 245
Sammeta, K. P. V., 468
Sana-Ullah, 35, 36, 42
Sandifer, J. B., 135
Sandow, A., 153
Sandri, C., 365, 366
Sanford, K. H., 319, 320, 322
Sanger, A. M. H., 306
Sankaran, T., 36, 41, 42
San Martin, Y. P., 24
Santamour, F., 90
Saradamma, K., 291
Sarmiento, R., 427
Sarup, P., 289, 292
Sasaba, T., 294
Satir, P., 385, 388, 390
Sato, Y., 306
Sauer, J. R., 140
Saxena, K. N., 441
Saxton, P. N., 347
Sazonov, A. P., 442
Scalla, R., 103
Schaefer, C. H., 207, 425, 426
Schafer, J. F., 346
Scharrer, B., 133
Schechter, M. S., 222
Scheidl, F., 449
Schein, Y., 135
Scherer, W. F., 110, 111
Schilling, P. E., 306, 425
Schillinger, J. A., 342, 346
Schlegel, P., 391
Schlehuber, A. M., 345, 351
Schlein, Y., 241, 242, 245, 250, 252
Schmialek, P., 417, 433, 448
Schmidt, C. H., 167, 176
Schmidt, K., 382, 384, 385, 388-91
Schmutterer, H., 48
Schneider, D., 87, 383, 385
Schneider, H., 288, 296
Schneiderman, H. A., 418, 419, 421, 423, 448
Schoener, T. W., 275
Schönherr, J., 102
Schoof, H. F., 425, 426
Schooley, D., 424
Schooneveld, H., 421, 434
Schoonhoven, A. V., 338
Schreiner, E. J., 78
Schuetz, R. D., 421, 434
Schultze, L., 263
Schulz, U., 249
Schürmann, F. W., 364-68, 373, 375
Schvester, D., 48
Schwartz, J. L., 419
Schwartzkopff, J., 381
Schwarz, M., 446
Schweig, C., 60, 62
Schwieter-Peyer, B., 424
Scopelliti, R., 366
Scott, G. E., 232, 233, 339, 344, 347
Scott, H. A., 105, 111
Scriven, G. T., 303
Sculthorpe, C. D., 32, 40
Seaman, D. E., 34
Searle, C. M. S., 288, 291
Seay, R. S., 102, 305
Seecof, R., 101, 112
Sehnal, F., 418-21, 423, 428-32, 434-36, 441, 442, 448
Sekhon, S. S., 384
Senior, L. J., 426
Seravin, L. N., 274
Serebrovskii, A. S., 167, 175, 462, 470, 472
Seymour, R. S., 267, 269, 273
Shaaya, E., 214
Shade, R. E., 342
Shane, J. L., 222
Shankland, D. L., 139
Shapira, A., 156
Shappirio, D. G., 420
Sharif, M., 242, 246
Sharma, D. P., 48
Sharma, G. C., 339
Sharma, K., 86
Sharp, D., 183, 193
Sharplin, J., 191, 197
Shaw, J., 271-74
Shaw, J. G., 57, 462
Sherman, M., 291, 295, 301, 303, 308
Sheta, I. B., 48
Shiga, M., 102
Shih-Coleman, C., 418
Shiranovich, P. I., 245
Shmaraev, G. E., 338
Shmuter, M. F., 250
Shorey, H. H., 301, 305
Shortino, T. J., 207-9
SHOWERS, W. B., 221-39; 222-24, 226, 227, 229, 233
Shulov, A. S., 252, 267, 273
Siddall, J. B., 423-25, 447
Sidiqui, K. H., 339
Sifuentes, J. A., 344
Sihler, H., 382, 385
Sikorowski, P. P., 23, 101, 107, 109
Silinsky, E. M., 139
Silva, T., 413
Silveira-Guido, A., 17, 24, 35, 38, 41
Silverstein, R. M., 36
Simons, J. N., 36
Singh, D. S., 289, 292
Singh, P., 86
Singh, S. P., 339
Singh, S. R., 348
Singh, Y. N., 419
Sinoir, Y., 381
Sisson, D. V., 226
Skaff, V., 425
Skalon, O. I., 253
Skangiel-Kramska, J., 366
Skuhravý, V., 428, 432
Skuratowicz, W., 253
Slade, M., 447
Slama, K., 418, 420, 422, 423, 427, 433, 440-42
Slatten, B. H., 207
Slifer, E. H., 381, 382, 384, 387, 388, 391
Smallman, B. N., 140, 143
Smart, E. C., 88
Smirnoff, W. A., 106
Smissman, E. E., 339
Smit, F. G. A. M., 247, 250, 252, 253
Smith, C. E., 97
Smith, D. S., 137, 242, 385, 388-90
Smith, E. L., 351
Smith, F. F., 307, 309, 319, 320, 440
Smith, F. G., 400, 405, 408, 410, 412
Smith, K. M., 97, 99-101, 104, 105, 107, 109, 110
Smith, L. J., 101
Smith, L. O., 288, 289
Smith, O. D., 345, 351
Smith, R. F., 298, 299
Smith, R. H., 81, 87, 463
Smola, U., 391
Snelling, R. O., 76
Snodgrass, R. E., 51, 242, 243, 247, 381, 385
Snow, J. W., 167
Sobey, W. R., 251
Socha, R., 419
Sohi, S. S., 102
Solina, L. T., 247-49

Solomon, G. B., 244
Solomon, K. R., 447
Solomon, M. E., 273, 274
Sommer, P. G., 404, 409, 412
Sonnet, P. E., 447
Soo Hoo, C. F., 102, 305
Sorm, F., 418, 422, 423, 433, 440, 441, 444
Sosunova, A. N., 248
Spadafora, R. R., 274
SPARKS, A. N., 221-39; 221-24, 226-28
Spates, G. E., 427, 445
Specht, H. B., 317
Spencer, N. R., 36, 38-40, 462
Spielman, A., 425
Spieth, H. T., 359
Spiller, D., 306, 315
Spiteller, G., 209
Splittstoesser, C. M., 102
Sprague, G. F., 338
Springett, B. P., 102
Squillace, A. E., 88, 90
Srivastava, U. S., 419, 427
STAAL, G. B., 417-60; 418, 419, 423-25, 437-39, 441, 444
Stafford, C. J., 304
Stafford, E., 50, 53
Stahl, J., 177
Staiger, L. E., 424
Stairs, G. R., 99, 102
Stanaard, L. J., 50
Stark, H. E., 253
Stark, R. W., 75, 90
STARKS, K. J., 337-57; 231, 298, 300, 339, 340, 344, 345, 351
Staten, R. T., 171
Steel, C. G. H., 134, 135
Steele, J. E., 139-41
Steelman, C. D., 306, 425
Steiger, U., 365, 366
Steiner, A., 360
Steiner, F. A., 367
Steinhaus, E. A., 23
Steiniger, F., 371
Stekol'nikov, A. A., 192, 193, 196
Stern, V. M., 298, 299, 301
Sternberg, S., 241, 242, 245, 250, 252
Sternburg, J., 296
Stetson, J. F., 176, 225
Stewart, P. G., 290, 309, 320
Stimmann, M. W., 169
Stobbart, R. H., 271-74
Stoltz, D. B., 100, 103-5
Storbeck, I., 212
Stort, A. C., 399-412
Stössel, W., 153
Stott, B., 33
Stower, W. J., 262, 266
Strausfeld, N. J., 363
Strickberger, M. W., 172
Strickland, A. H., 65
Strickland, E. H., 285
Strickler, J. R., 383, 384
Stringer, C. E., 7, 10, 12, 14, 17, 20, 22, 25
Stroh, H. H., 159
Stroh, R. C., 79
Strong, F. E., 426
Strong, F. M., 84
Strong, R. G., 432-35
Struble, G. R., 308
Stuart, A. M., 385, 388, 390
Stuermer, C. W., 110
Subak-Sharpe, J. H., 97
Suciu, M., 253
Sugai, E., 170, 173
Suguna, S. G., 473
Summerlin, J. W., 20, 21
Summers, F. M., 53, 55, 58, 61-63
Summers, M. D., 99, 100, 103-5
Suomalainen, E., 176
Surtees, G., 97
Suter, P. R., 245-47
Sutter, G. R., 103, 228
Sutton, D. L., 31, 33, 34
Suzuki, D. T., 473
Suzuki, M., 207
Suzuki, S., 170, 173
Suzuki, Y., 424
Svihra, P., 306
Sviridov, G. G., 251
SVOBODA, J. A., 205-20; 205-9, 211, 214
Swaine, G., 101
Sweeley, C. C., 423
Sweeney, T. A., 462, 464
Swenson, K. G., 421, 439
Swift, F. C., 309, 314, 317, 319, 320, 324
Swirski, E., 53, 287, 289
Szabó, I., 250, 253
Szczepanska, K., 301

T

Taba, H., 110
Takagi, S., 47, 50, 54, 67, 68
Takahashi, R., 50
Takahashi, R. M., 306, 426
Takezagua, H., 56
Tamaki, G., 437
Tamaki, Y., 68
Tamashiro, M., 291, 295, 301, 303, 304, 308
Tanada, Y., 97, 100, 101, 107, 305
Tanaka, N., 462
Tanaka, S., 109
Tang, P. M., 420
Tanimura, I., 170
Tarasevich, L. M., 97
Tarchalska, B., 367
Tash, J., 213
Tashiro, H., 57
Tatchell, R. J., 274
Tateda, H., 368, 369
Tawfik, M. F. S., 299, 304
Taxima, Y., 168, 170, 172, 175, 176
Taylor, E. A., 297, 302
Taylor, G., 100
Taylor, T. M., 31, 34
Tazima, Y., 168, 469, 471
Telford, A. D., 308
Ten Cate, J., 370
Teninges, D., 111, 112
Teodoro, G., 51, 52
Terbush, L. E., 102
Terent'eva, L. I., 249
Terriere, L. C., 468
Teucher, G., 78
Tevis, L. Jr., 267
Thebault, M., 425, 426
Theodor, O., 253
Thesleff, S., 156
Thielges, B. A., 79, 84
Thiem, H., 56, 66
Thobbi, V. V., 339
Thomas, B. R., 87
Thomas, E. D., 101
Thomas, J. P., 434, 435
Thomas, M., 111
Thommen, H., 441
Thompson, C. B., 361
Thompson, J. L., 12
THOMPSON, M. J., 205-20; 205-7, 209, 211-14, 417, 421, 423
Thompson, W. W., 309, 319-21, 324
Thomsen, E., 136, 138
Thomsen, M., 138
Thomson, J. A., 209
Thomson, R. H., 159
Thomson, W. W., 135, 137
Thorson, B. J., 471
Thouvenin, M., 440, 441
Thurm, U., 382, 385, 388-90
Thurston, R., 297, 300-2, 315
Tierney, P. G., 224
Tikhvinskaya, M. V., 250
Tillyard, R. J., 183, 185, 188, 191, 192, 194, 195, 198, 241
Tilton, E. W., 168-71
Timlin, J. S., 65
Timmons, F. L., 32
Tinbergen, N., 360
Tindale, N. B., 184, 192
Tinsley, T. W., 97, 104-6
Tippin, H. H., 54, 67, 68
Tipton, C. L., 349

Tipton, V. J., 252
Titze, J. F., 82
Toba, H. H., 170, 171, 177
Tobolski, J. J., 88
Tombes, A. S., 135, 137
Tomich, P. Q., 252
Tominaga, Y., 384
Tomlin, A. D., 295
Torres, J., 171, 177
Tournier, P., 97
Townsend, G. F., 405
Toye, S. A., 265, 271
Trahan, G. B., 338
Traub, R., 241-43, 245, 247-53
Traut, W., 172
Trautmann, H., 433
Treagan, L., 102
Treece, R. E., 427
Tremblay, E., 58
Triosi, S. J., 23
Triplehorn, C. A., 222
Triplett, R. F., 15, 19
Troisi, S. J., 444, 445
Trost, B. M., 423, 424
Trujillo-Cenoz, O., 365
Truman, J. W., 359, 373, 418, 420
Tsao, C. H., 190
Tschinkel, W. R., 17
Tseng, H. C., 420
Tsherny skev, W. B., 276
Tsizin, Y. S., 442
Tsuda, K., 207
Tunnicliff, G., 140
Turner, A. J., 191, 197
Tyrer, N. M., 135
Tyshchenko, V. P., 372, 375

U

Uchida, M., 56
Udine, U. J., 351
Uebel, E. C., 417, 421, 423, 428
Uga, S., 385, 388, 389
Ulmanen, I., 250
Unnithan, G. C., 134, 138, 139
Upholt, W. M., 102
Usherwood, P. N. R., 143, 151, 153, 155-58, 161
Usinger, R. L., 35
Ust'yan, A. K., 308

V

Vago, C., 97, 99, 102, 103, 110, 111
Vail, P. V., 101-3, 109, 110, 305
Valcovic, L. R., 303, 304
van Buijtenen, J. P., 90
Vance, D. E., 105
Vandel, A., 263
Van den Bosch, R., 298, 299
VanDenburg, R. S., 228, 231
Van der Kloot, W. G., 133, 135, 142, 366, 371
Van der Weij, H. G., 32
Vanderzant, E. S., 337, 340
Van de Veire, M., 428
Van de Vrie, M., 299, 313, 316, 322
Van Dinther, J. B. M., 51, 54, 290
Vane, F., 449
van Emden, H. F., 75
van Gelder, N. M., 157
Van Halteren, P., 300
Van Valin, C. C., 21
Van Zon, J. C. J., 33, 34, 43
Varela, F. G., 389
Varjas, L., 431
Vartell, R. J., 224
Vashchyonok, V. S., 247-49, 252
Vasseur, R., 48
Vaughan, J. A., 244-46
Vaughn, J. L., 97
Vavilov, N. I., 337
Vayssiere, P., 48
Velthuis, H. H. W., 291
Vercammen-Grandjean, P. H., 244
Vernoux, J. P., 106
Veyrunes, J. C., 103
Vickers, L., 19
Vidal Sarmiento, J. A., 402, 413
Vincent, J. F. V., 135
Vinegar, A., 267, 269, 273
Vinogradova, E. B., 468
Vinson, J. W., 446
Vinson, S. B., 12, 14, 23, 170, 429, 447, 448
Virkki, N., 170, 176
Vité, J. P., 81-83, 88, 90
Vogel, E., 40, 41
Vogt, G. B., 35, 40
Volyanskiy, Yu. Ye., 250
von Alten, H., 361
von Borstel, R. C., 463
Von Dehn, M., 437, 438
von Frisch, K., 360
von Rudloff, E., 86, 88
von Schantz, M., 82
von Schönborn, A., 80
Voth, V., 318
Vowles, D. M., 363, 369, 373, 374
Vysochkaya, S. O., 252

W

Wachmann, E., 243
Wada, H., 110
Wagoner, D. E., 469, 470, 473
Waheed, A., 443
Wahrman, J., 176
Walcott, C., 386, 391
Walker, C. R., 468
Walker, D. W., 167-73, 177, 339
Walker, J. R., 15
Walker, R. J., 156
Walker, W. F., 419, 433, 435
Wall, M. L., 228
Wallace, B., 461
Wallace, J. E., 160
Wallace, L. E., 338, 345, 350
Wallach, D. P., 158
Waloff, Z., 266, 276
Walsh, C. T., 16
Walsh, J. P., 17
Walter, E. V., 339, 344
Walther, C., 157
Wanyonyi, K., 442, 443
Warburg, M. R., 263-65, 270
Ware, G. W., 21
Warner, R. E., 41
Washino, R. K., 318
Wasserburger, H. J., 247
Watanabe, H., 99, 100, 109, 110
Waterfield, R. L., 231
Waterhouse, D. F., 140, 175, 467, 470, 472
Waters, R. M., 440
Waters, W. E., 168, 170
Waterson, J. M., 53
Watkinson, I. A., 139
Watson, J. A., 82
Watson, J. A. L., 273
Watts, W. S., 468
Way, M. J., 299
Webb, F. E., 302, 308
Webb, R. E., 440
Webb, S. R., 104, 105
Weber, H., 51
Weekman, G. T., 228
Weglarska, B., 54
Wehner, R., 266
Weiant, E. A., 155
Weibel, D. E., 345
Weidhaas, D. E., 7, 21, 462, 473
Weirich, G., 447
Weis-Fogh, T., 242, 271, 273
Weiss, C. M., 300
Weissmann, G., 87
Weitzman, M., 133
Weldon, L. W., 32, 33, 37-41
Wellenstein, G., 100-2
Wellington, W. G., 432
Wells, F. E., 102
Welsh, J. H., 367
Wenk, P., 247

Werker, E., 87
Wersäll, J., 382, 388, 390
Westermann, J., 433
Westfall, R. D., 88, 90
Westigard, P. H., 324, 440
Wetherly, A. H., 110
Wheeler, C. M., 425, 426
Whitaker, T., 110
Whitcomb, W. H., 3-6, 18, 24, 25, 228, 303
White, A. F., 139, 140, 143
White, D. F., 436, 438
White, L. D., 170
White, M. J. D., 172, 176, 177, 461
Whitehead, A. T., 138, 139
Whitesell, K. G., 318
Whitmore, D., 447
Whitmore, E., 447
Whitney, H. S., 82
Whittaker, R. H., 85, 86
Whitten, J., 418
WHITTEN, M. J., 461-76; 175, 462-64, 466-73
Wiackowski, S. K., 302
Widstrom, N. W., 339, 344, 351
Wientjens, W. H. J. M., 207
Wiese, H., 404, 407, 409, 412
Wigglesworth, V. B., 133-35, 155, 382, 391, 418, 422, 448
Wilbur, D. A., 170, 171, 388
Wilcox, T. A., 229
Wilder, W. H., 425, 426
Wildy, P., 97-99, 103, 104, 106, 107, 110, 111
Wilkens, J. L., 143
Wilkes, A., 306, 315
Wilkes, F. G., 300
Wilkin, P. J., 266, 268
Wilkinson, C. F., 215, 297, 447
Wilkinson, J. D., 289-91, 298, 305, 308
Wilkinson, R. C., 80, 82, 86
Williams, C. M., 82, 371, 419, 423, 425, 440, 441, 446
Williams, J. L. D., 362, 364-66, 369
Williams, J. R., 51, 60
Williams, P., 97
Williams, R. N., 3-6, 15, 24
Williams, R. T., 251
Williams, S. C., 263, 267
Williamson, D. L., 177
Willig, A., 215
Willis, J. H., 418, 420
Wilson, A. J. Jr., 21
Wilson, D. M., 88
Wilson, E. O., 2, 3, 9, 12, 13, 15, 16, 88, 90
Wilson, F., 31, 466-68
Wilson, H. J., 176
Wilson, L. F., 79
Wilson, M. C., 342
Wilson, N., 252
Wilson, N. L., 12, 13
Wilson, P. D., 21
Wilson, R. L., 339
Wing, M. W., 3
Winieski, J. A., 78
Winston, P. W., 271
Winteringham, F. P. W., 306
Wirtz, P., 445
Wiseman, B. R., 340, 344, 351
Wisseman, C. L., 251
Wojcik, D. P., 20-22
Wolf, W. W., 102, 305
Wolfenbarger, D. A., 288, 291, 298, 300, 301, 430
Womack, M., 207
Wood, D. L., 84
Wood, E. A., 345, 351
Wood, E. A. Jr., 342, 345, 347, 351
Wood, H. A., 107, 108
Woodard, D. B., 105
Woodard, G., 101
Woodrow, D. F., 276
Woodruff, R. E., 17
Wooten, R. C. Jr., 273
Woyke, J., 406
Wren, J., 447
Wressel, H. B., 227
Wright, J. E., 424, 427, 445, 446
Wright, J. W., 79, 86
Wright, R. D., 140
Wright, R. H., 80
Wyatt, G. R., 418, 421
Wyatt, I. J., 303

Y

Yadava, R. L., 100
Yamada, H., 101, 102
Yamamoto, H., 447
Yamamoto, I., 424
Yamamoto, R. T., 212, 213, 424
Yamamoto, T., 153
Yamaoka, R., 392
Yang, R. S. H., 215
Yasunaga, K., 83
Yeargan, D. R., 290, 294, 307, 319, 320
Yen, J. H., 471, 472
Yendol, W. G., 97, 101
Yeo, R. R., 34
Yin, C. M., 421
Yin, G. M., 134
Yinon, U., 252
Young, J. R., 221, 222
Young, N. L., 212-15
Young, S. Y., 102
Young, W. R., 339
Young, W. W., 293, 318
Younger, R. L., 427, 428
Yuan, C., 423
Yuill, J. S., 82
Yule, B. G., 105
Yunker, C. E., 274

Z

Zabik, M. J., 421, 434
Zacharuk, R. Y., 387, 390, 391
Ząoral, M., 422, 441
Ždárek, J., 420, 427, 441, 442, 445
Zebe, E., 153
Zeiger, C. F., 31, 39-41
Zelanzy, B., 102
Zelenko, A. P., 106
Zelený, J., 289-92, 295, 299, 438
Zettler, F. W., 34
Zeuner, F. E., 184
Zhovtyi, I. F., 251
Zielske, A. G., 36
Zimmack, H. L., 228
Zinovjeva, K. B., 468
Ziv, M., 275
Zuber, M. S., 338, 339
Zwart, K. W. R., 291
Zwick, R. W., 317, 324
Zwölfer, H., 35, 36, 38, 41

SUBJECT INDEX

A

Abscisic acid, 86
Acarapis woodi, 413
Acarina, 98
  and growth regulators, 447
Acclimatization, 275
  in mass rearing, 468-69
Acentropus, 185
Acetogenins, 85-86
Acetylcholine, 138, 153-55, 366-67
Acheta, 364
  domesticus, 153, 155, 303, 364-67
Acigona infusella, 38, 41
Acrididae
  and growth regulators, 444
Action potentials, 143
Adaptations
  to arid environment, 261-83
  to cold, 269-70
  to drought, 270-74
  to heat, 267-68
Adelges abietis, 79, 84
Adesmia antiqua, 266, 268-69
Adonia variegata, 289, 292
Adoxophyes orana, 102, 431
Adrenaline, 138, 141, 155
Aedes, 99
  aegypti, 106, 426
  nigromaculus, 426
  taeniorhynchus, 105
Aeshna, 361
African honey bee, 399-416
Agasicles, 37-40
  hygrophila, 31, 35-36, 40
Agathiphaga, 187, 194-95, 200
Agistemus fleschneri, 321
Aglais urticae, 107
Aiolopus thalassinus, 265-68
Alarm pheromones
  of Brazilian bee, 410, 412
Alary muscles, 137, 156-57
Alcohols
  and muscles, 159-60
Aldicarb, 291, 295-96
Aldrin, 290, 301, 303-4
  epoxidase, 297
Aleyrodidae
  and growth regulators, 440
Alfalfa butterfly, 468
Alfalfa looper, 305
Alfalfa weevil, 306
Aliesterase, 297
Alkaloids, 85-86, 297
  of fire ant venom, 14-15
Allatectomy, 447
Alligatorweed
  biological control of, 31, 33, 35-37, 39-40, 43
Alloxysta curvicornis, 302
Almond moth, 431
Altosid, 306
Amalgamation
  of fire ant colonies, 8
Amaranth, 36
Amber, 184
Amblyomma
  americanum, 18
  maculatum, 18, 463
Amblyseius
  fallacis, 290, 294, 307-9, 313-14, 316-18, 321-23
  hibisci, 307, 311
  potentillae, 299
American cockroach, 207
Amines
  and muscles, 156
Amino acids
  and muscles, 156-58
Aminobutyric acid, 138
Ammophila, 360
Amphetamine, 141, 156
Amphipsylla sibirica, 252
Amynothrips andersoni, 36, 40
Anacridium melanorhodon, 271
Anagasta kuhniella, 169
Anastrepha ludens, 462
Anatomy
  of the brain, 360-66
  of fleas, 247-48
Ancepaspis, 66, 69
Ancistropsylla, 245
Angoumois grain moth, 168
Anopheles, 473
  albimanus, 462
  gambia, 276
  stephensi, 107
Anoplura, 446
Antennae, 185, 188, 265
  sensilla of, 391
Antherea, 110
  pernyi, 373
  polyphemus, 83, 213
Anthidium, 361
Anthocorids, 299
Anthonomus grandis, 207, 462, 465
Anthororis confusus, 290
Antibiosis, 79, 342, 347
Antibiotics, 340
Anticoagulant, 248
Antidiuretic hormones, 137
Antigens, 104-6
Ants
  and arid environment, 275
  brain of, 361, 363, 367
  and growth regulators, 444
  sensilla of, 384, 391
Aonidiella
  aurantii, 49-50, 439
  citrina, 49-50
Apanteles, 302
  congregatus, 300-1, 315
  fumiferanae, 302
  glomeratus, 305
  marginiventris, 291
  melanoscelus, 295
  militaris, 305
  solotarius, 305
Aphelinus
  mali, 301, 305
  melinus, 302
Aphids
  behavior of, 360
  dispersal of, 65
  and fire ants, 18
  and growth regulators, 436
  and insecticides, 297, 304, 314
  and resistant cereals, 342, 347
  and tree resistance, 84
Aphis
  craccivora, 292
  fabae, 292, 438
  gossypii, 292
Apholate, 173, 304
Aphytis, 304
  holoxanthus, 291, 303-4
Apicotermes, 364
Apidae
  and growth regulators, 445
Apis cerana, 408
Apis mellifera, 361-62, 367, 400, 407
  adansonii, 400-1
  ligustica, 407
Apodicrania, 24
Apolysis, 391
Apple maggot, 325
Aquatic weeds
  biological control of, 31-46
Arachnids
  and arid environment, 266-67, 269-70, 274, 276
Aramite, 319

Archytas marmoratus, 227
Arctiidae
  and growth regulators, 431
Arenivaga, 265, 273
  investigata, 268
Argentine ant, 3, 302
Arid environment, 261-83
Armored scales, 47-73, 439
  aggregations in, 64
  chitin of, 68
  crawlers of, 60-64
  digestive system of, 54
  diapause of, 55-56
  dispersal of, 59-66
  ecology of, 47-73
  eradication of, 60
  gall and, 53
  haploids and, 58
  host specificity of, 48-50
  life history of, 54
  males of, 55-57
  mating behavior of, 57-58
  polyphenol polymers of, 68-69
Armyworm, 305
Arrowhead scale, 49, 56
Artificial diets, 339-41, 464
  for European corn borer, 233
  for fleas, 252
Artificial feeding, 249
Ascorbic acid, 159-60
Aspartate, 138
Aspidiotus
  ancylus, 50
  cornstocki, 50
  destructor, 49, 53, 60
  howardi, 50
  nerii, 49, 56
  simulans, 58
Asterolecanium variolosum, 53
Atherigona varia, 339
ATP
  and fleas, 249
Atropine, 367
Atta texana, 23
Attractants
  and tree resistance, 77, 81-83
Auditory center, 368
Aulacaspis
  greeni, 67
  rosae, 49
  tegalensis, 60, 62-63, 65
Autographa californica, 101-2
Autosomes, 473
  and control, 470
Axon, 133-36, 138-40, 143, 151-52, 387
Azinphosmethyl, 289-91, 294, 307-9, 311, 314, 319-20, 322, 325

B

Bacillus thuringiensis, 228-29, 305
Bacteria
  and European corn borer, 228, 233
  of fire ants, 23
  and parasites, 304
Baculovirus, 98-103
Bagous, 39, 42
Baits, 445
  for European corn borer, 229
  for fire ants, 12, 20-21, 25
Baltic amber, 184, 252
Bark beetles, 81, 84
Bat fleas, 251
Beekeeping
  and Brazilian bee, 412
Bees
  brain of, 363
  sensilla of, 391
  and viruses, 110-11
Beetles
  and fire ants, 17
  sensilla of, 386
Behavior
  and brain structure, 359-79
  and growth regulators, 421, 445
  and tree resistance, 77
Behavioral adaptations, 264-67
Bellura densa, 41
Bembidion lampros, 304
Benomyl, 319
BHC, 299, 301
Binapacryl, 289, 291, 319
Biogenic amines, 155-56
Biological control, 81-83, 101, 104
  of aquatic weeds, 31-46
  of European corn borer, 226
  of fire ants, 1, 23-25
Biology
  of fire ants, 5-17
Biotrol, 305
Biotypes, 337, 347-49
  of European corn borer, 222
Birds, 228
Blabera craniifer, 136, 139
Blaberus, 363
Black flies
  and growth regulators, 428
Black parlatoria scale, 49
Black thread scale, 49
Black widow spider, 161
Blaps sulcata, 266
Blastokinesis, 420
Blatta, 361, 363
Blattella germanica, 207, 443-44
Blattellidae
  and growth regulators, 443
Blissus leucopterus, 338
Blister-beetle, 266
Blowflies
  sensilla of, 391-92
Body temperature, 266, 270
Boettcherisia peregrina, 368
Boisduval scale, 49
Boll weevil, 207, 305, 462
  control of, 465
Bollworms, 109, 299
  and insecticides, 314
Bombyx, 170, 175
  larvae, 208, 215-16
  mori, 99, 107-9, 168, 207, 213, 225, 469, 471
Boophilus microplus, 274
Boreus, 242-43
Bovicola limbatus, 446
Brachymeria, 305
  intermedia, 291, 295
Bracon
  hebetor, 303-4, 315
  mellitor, 291, 305, 307-8, 315
Brain
  electrophysiology of, 368-70
  function, 373-75
  lesions, 370-73
  and neurosecretion, 134-35, 139
  structure and behavior, 359-79
Brassicasterol, 207
Brazilian bee, 399-416
  activity of, 408
  alarm pheromones of, 410, 412
  beekeeping and, 412
  color of, 406
  development of, 407
  diseases of, 413
  fernal colonies of, 409
  foraging by, 408
  males of, 400, 402
  migrating colonies of, 410
  morphology of, 406
  spread of, 400-6
  stinging of, 410-12
  variation in, 403-5
Breeding
  for resistance, 344-47
Brevicoryne brassicae, 438
Brood pheromone, 17
Bruchidae
  and growth regulators, 435
Bucculatrix thurberiella, 102
Budworm, 102
Bursa copulatrix, 193, 196, 246

Bursicon, 135

C

Cabbage aphid, 438
Cabbage looper, 109, 169-70, 172, 175, 177, 305
Cabbage worm, 305
Cadra cautella, 431
Calcium
and muscles, 153, 158
and neuroscretion, 143
California floodwater mosquito, 426
California red scales, 49, 53-54, 57-59, 61-63, 68
Calliphora
erythrocephala, 136-38, 140, 142-43, 153
stygia, 209, 212, 215
test, 214, 216
vicina, 212-16
Campaniform sensilla, 385-388, 390-91
Campesterol, 206-7, 209-10
Camphene, 88
Camphor scale, 49, 55, 61, 63
Campoletis, 227, 305
perdistinctus, 291, 296
sonorensis, 291
Camponotus
ligniperda, 365-66
pennsylvanicus, 17
Cancer, 120
Carabids
and insecticides, 300, 304
Carabus, 364
Carausius morosus, 134, 137, 140, 152
Carbaryl, 229, 289-92, 295, 297, 307-9, 319-22, 331
Carbofuran, 230-31, 320
Carbophenothion, 289, 294, 320
Carcinogen, 121
Cardioaccelerators, 141-42
Cardiosis, 266
Carene, 88, 90
Carpocapsa pomonella, 465
Carulaspis
carueli, 48-49
juniperi, 48-49, 52-53
visci, 53
Carzol, 319
Casebearers
and growth regulators, 432
Castes
and brain size, 361-62
determination of, 444
differentiation of, 443
of fire ants, 8, 12
and growth regulators, 443, 445
Cataglyphis bicolor, 266
Catecholamines, 366-68, 372
Cat flea, 245, 248
Cediopsylla, 244
Cell cultures, 252
Centipdes
and arid environment, 264-65
Centrioptera muricata, 268, 271
Cephus cinctus, 339
Ceratitis capitate, 462, 470
Ceratophyllus, 242, 246
gallinae, 246
styx, 245
Cerci, 384-86
Cereal leaf beetle, 339, 346
Cereal pests, 337-57
Certification
of applicators, 121
Cetonia, 364
Chaetosemata, 188
Chaetotaxy, 186, 194
Chaff scale, 49
Chalcidoma, 361
Chaoborus, 142
punctipennis, 293
Chelonus, 305
blackburni, 291
munakatae, 306
Chemoreceptors, 382, 384, 387-88, 390
Chemosterilants, 25, 169, 225, 304, 466-67
Chemotaxis, 63
Chewing lice
and growth regulators, 446
Chilecomadia, 191
Chilo, 338
suppressalis, 177, 306, 339, 346
zonellus, 339, 344
Chimaeropsylla potis, 245
Chinch bugs, 338
Chionaspis, 68
nyssae, 50
salicis, 54, 56
sylvatica, 50
Chironomidae
and growth regulators, 428
Chironomus
attenatus, 103-4
tentans, 212-13, 215
Chitin
of armored scales, 68
Chlordane, 291, 301
Chlordimeform, 291
Chlorfenson, 319-20
Chlorobenzilate, 289, 291, 319
Cholesterol, 205-10
Choline acetylase, 155
Cholinergic drugs, 155
Cholinesterase, 155, 297, 309, 366
Chordotonal sensilla, 381
Choristoneura, 104, 430
conflicata, 104
fumiferana, 80, 446
murinana, 102
Chorizagrotis, 430
Choromsome substitutions, 177
Chrysomelidae
and growth regulators, 434
Chrysomphalus
dictyospermi, 49
ficus, 49
Chrysopa, 305
californica, 303
carnea, 290, 292, 306
rufilabris, 303
Chrysopids
and European corn borer, 228
Chrysops, 363
Cicadellidae, 439
Circadian rhythms, 266-67, 275, 359, 366-67, 372-73, 375
Citrus mites, 112
Citrus red mite, 111
Citrus snow scale, 49
CL 47470, 230-31
Coccids
ecology of, 47
and growth regulators, 439
Coccinella
5-punctata, 292
7-punctata, 292, 297
11-punctata, 289
Coccinellida, 317
Coccinellids, 299, 303-4
and European corn borer, 228
and growth regulators, 435-36
and insecticides, 312, 315
Coccus hesperidum, 439
Cochliomyia hominivorax, 167
Cockroaches, 155, 158, 273, 359, 363-64, 367, 369, 371-73
and growth regulators, 443-44
sensilla of, 385-86
Coconut scale, 49, 60
Cocoon, 187-88, 252, 359
spinning, 371
Codling moth, 102, 167-68, 170-72, 308, 317, 325, 431, 462
control of, 464-65
Coleomegilla, 303, 315
maculata, 289, 300, 303, 307, 311, 314
Coleophora laricella, 432
Coleophoridae
and growth regulators, 432

Coleoptera
and growth regulators, 433-36
Colias eurytheme, 468
Colony
founding by fire ants, 5-8
growth of fire ants, 8
Color
of Brazilian bee, 406
Colorado potato beetle, 231
Comb cells, 406
Communication
by fire ants, 15
by honeybees, 360
Competitive displacement, 48
Confused flour beetle, 207
Contarinia sorghicola, 339
Control
of European corn borer, 226, 229-31
genetical, 461-76
of imported fire ants, 1-30
Copper
and muscles, 159
Copulation
by fleas, 245-47
Corn earworm, 177, 207-8, 339-40
Cornleaf aphids, 342, 346-48
Cornops
aquaticum, 35, 38, 41
longicorne, 38, 41
Corpora
allata, 134, 136, 138, 418
cardiaca, 135-43
pedunculata, 360-62
Corticosteroids
and fleas, 244
Cotton strainers, 441
Cranial sulci, 188
Crawlers
of armored scales, 60-64
Cremaster, 187
Crickets
behavior of, 359, 367, 370, 372, 374
and insecticides, 303
sensilla of, 384-85, 391
songs of, 370
and viruses, 110
Critical temperature, 270
Crochets, 186, 198, 200
Cross resistance, 309, 428
Crumena, 51-52
Cryptobiosis, 268, 270, 272
Cryptoglossa verrucosa, 271
Cryptolaemus, 304
Ctenocephalides, 246
canis, 247
felis, 244-45, 248, 252
Ctenolepisma
longicaudata, 265, 273
terebrans, 273
Culex
fatigans, 462, 471, 473
pipiens, 426
pipiens quinquefasciatus, 462
quinquefasciatus, 293
Culicidae
and growth regulators, 424-28
Curare, 153, 155
Curculionidae
and growth regulators, 435
Cuticular mechanoreceptors, 381-97
Cyanides, 297
Cyclic-AMP, 141
Cyclocephala pasadenae, 137
Cyclodiene insecticides, 301-2
Cycloneda limbifer, 289
Cydia pomonella, 465
Cymene, 88
Cyrtobagous singularis, 36, 39, 42
Cyrtopus fastuosus, 272
Cytoplasmic
incompatibility, 471-73
polyhedroses, 98, 107-10

D

2,4-D, 40, 303
Dacnonypha, 195-96
Dacus
cucurbitae, 462
dorsalis, 295, 301, 462
Danaus, 192
Dance language
of honeybees, 360
Dasineura, 428
Dasypsyllus, 243
Date scale, 60
DDT, 229, 231, 289-91, 300-4, 307-8, 312, 314, 319-20
Decamethonium, 155
Decapitation
and behavior, 359
Dehydrocholesterol, 206
Dehydrosolenopsin, 14
Delayed sterility, 472
Delphacidae
and growth regulators, 440
Demeton, 289-90, 292, 294, 320, 322
Demetonmethyl, 289, 292
Dendrite, 382-85, 387-90
Dendroctonus
brevicomis, 82-83
frontalis, 83
ponderosae, 82-83
pseudotsugae, 81-82
Densonucleosis viruses, 98, 106-7
Department of Agriculture and Pesticides, 122
Department of Transporation and Pesticides, 122
Dermacentor variabilis, 274
Dermestes maculatus, 433
Desert grasshopper, 265
Desert locust, 266, 276
Desert roaches, 268
Desmosterol, 206-9
Development
of Brazilian bee, 407
of European corn borer, 223-24
Diabrotica, 434
virgifera, 230, 339
Diapause
and acetylcholine, 366
and arid environments, 261, 267, 272, 275-76
of armored scales, 55-56
of European corn borer, 222-26
and genetics, 174
and growth regulators, 420-21, 434, 439
and neuroscretion, 135
Diaretiella rapae, 302
Diaspididae
ecology of, 47-73
and growth of regulators, 439
Diaspis
boisduvalii, 49
bromeliae, 49
Diatraea saccharalis, 18, 168-69, 339
Diazinon, 229, 289-90, 294, 309, 320, 322
Dibrachys cavus, 295, 302
Diceroprocta apache, 265
Dichlorvos, 292, 298
Dicofol, 289-91, 319-20
Dicrotophos, 289-90
Dictyoptera
brain of, 363
and growth regulators, 443-44
Dictyospermum scale, 49
Dieldrin, 20, 289-90, 300-1, 303-4, 307, 319
Digestion
by fleas, 248-49
Digestive system
of armored scales, 54
Digger wasps, 162
Dikar, 294, 319, 321
Dimethoate, 289-92, 297, 303, 309, 319-22
Dimetilan, 290
Dinobuton, 320
Dinocap, 294, 319, 321
Dinothrombium
pandorae, 267
tinctorium, 267
Dioxacarb, 230
Dioxathion, 291, 303, 320
Dipel, 305

Diptera
  and growth regulators, 424
  and insecticides, 288
Diseases
  of Brazilian bee, 413
Dispersal
  of armored scales, 59-66
  behavior, 312
  of Brazilian bee, 400
  of fire ants, 3
  of screwworm, 463
  of sterile males, 465
Disulfoton, 291, 296
Diterpenes, 87
Dithianon, 319
Ditrysia, 197-99
Diuretic hormone, 135, 139-40
DNA, 98-99, 103-6, 108, 222
Dodine, 319
Dopamine, 140-41, 367-68
Draeculacephala crassicornis, 439
Dragonflies
  behavior of, 360
Drosophila, 111-12, 175, 300, 359
  melanogaster, 172, 468, 471
  pachea, 209-10
Drugs
  and muscles, 161
Dudgeonea, 190, 199
Dufour gland, 16
Dynaspidiotus abietis, 69
Dysdercus, 424, 441
Dytiscus, 361, 364

E

Ecdysial
  canal, 382-85, 391
  glands, 215
Ecdysis, 187-89, 200, 391, 420
Ecdysones, 133, 205, 209
Echidnophaga
  gallinacea, 245-46, 248
  oschanini, 252
Eclosion hormone, 373
Ecological adaptations
  to arid environment, 274-76
Ectoparasites, 243
EDTA
  and muscles, 160
Eggs
  of Lepidoptera, 184-85
Electrophysiology
  of brain, 368-70
Eleodes, 268-69
  armata, 263, 271
Elm bark beetle, 81
Elm scale, 52
Embryonic development
  and growth regulators, 419-20
Encarsia formosa, 303
Endogenous rhythms, 265, 267
Endoparasites, 243
Endosulfan, 231, 289, 291-92, 301, 303, 319-20, 322
Endrin, 229, 289-91, 301, 303
Engraver bark beetle, 82
Enterobacterin, 305
Enteroviruses, 98, 110-11
Entomopox virus, 98
Environmental protection agency, 120-21
Enzymes
  inhibitors of, 158
  in saliva, 53
Ephestia cautella, 169
Epicauta aethiops, 266
Epicuticle, 262, 264, 270
Epidiaspis leperii, 49, 64
Epilachna varivestis, 209, 435
Epinephrine, 156
Epipagis albiguttalis, 36, 38, 41
Epizootiology
  of viruses, 97
EPN, 229, 320
Eradication
  of armored scales, 60
  of fire ants, 21, 26
  of screwworms, 463-64
Eriborus terebrons, 227
Erisoma lanigerum, 301
Estigmene acrea, 102-4, 172
Estrogens
  and fleas, 244
Ethion, 320
Ethyl methane sulfonate, 304
Eublemma amabilis, 303
Eugenol, 276
Euproctis chrysorrhoea, 432
Euonymus scale, 49
European bees, 400-1, 403-10, 413
European corn borer, 221-39, 339-41, 344, 347, 351-52, 431
  artificial diets for, 233
  bacteria and, 228, 233
  baits for, 229
  biological control of, 226
  control of, 226, 229-31
  development of, 223-24
  diapause of, 222-26
  flight of, 226
  fungi and, 229, 233
  mating behavior of, 224
  varietal resistance to, 231-33
  viruses and, 229
European fruit scale, 49
European pine sawfly, 79
European red mite, 111
Eurycotis floridana, 207
Eurygaster integriceps, 442
Eutetranychus banksi, 293
Evaluation
  of plant resistance, 341-43
Evolution
  of fleas, 249-50
  of Lepidoptera, 183-203
  of Siphonaptera, 241
Excretion
  and arid environment, 262
Exorista rossa, 295
Exoristes comstocki, 305
Exoteleia pinifoliella, 79

F

Face fly, 427
Fall armyworm, 207, 339, 344
Fall webworm, 431
False San Jose scale, 49
Farnesal, 417
Farnesol, 417, 424
Farnesyl methyl ether, 420, 422
Fat body, 104-6, 140, 142, 214-17, 247, 298-99
Fecundity
  and growth regulators, 420
Federal pesticide legislation, 119
Feeding
  behavior, 80
  deterrents, 84
  dispersal and tree resistance, 77
  effects of armored scales, 52-53
  by fleas, 248-49
  habits of European corn borer, 226
  stimulant, 36
Fenthion, 290-91
Fenitrothion, 294
Fensulfothion, 231
Feral colonies
  of Brazilian bee, 409
Ferban, 319-20
Fern scale, 49
Feronia melanaria, 290
FIFRA, 120-21
Fig scale, 49
Filter chamber, 54
Fiorinia, 69
  fioriniae, 49
  theae, 49
Fire ants, 1-30, 444
  bacteria of, 23
  baits for, 12, 20-21, 25
  biological control of, 1, 23-25

biology of, 5-17
foraging by, 13
fungi of, 23
life cycle of, 8
lipase of, 12
lures for, 25
mating behavior of, 10
spread of, 3-5
stinging by, 15
wing muscles of, 7
zoogeography of, 1, 3-5
Firebrat, 207, 273
Firefly, 141
Fish
and aquatic weeds, 33
Flea beetle, 36
Fleas, 241-59
artificial diets for, 252
ATP and, 249
digestion by, 248-49
estrogens and, 244
evolution of, 249-50
geographical distribution of, 251-52
life cycle of, 243-47
mating behavior of, 245-47
ovaries of, 247
zoogeography and, 252
Flight, 391
of European corn borer, 226
muscles, 242
Florida red scale, 49, 53, 60, 62, 68
Folbet, 319
Food
chain and insecticides, 299-300
and drug administration and pesticides, 120-22
exchange by fire ants, 13
of fire ants, 12
Foraging
by Brazilian bees, 408
by fire ants, 13
Forbes scale, 49
Forficula, 361
Form determination
and growth regulators, 438
Formica
lugubris, 365-66
pratensis, 365
rufa, 365-67
Formicidae
and growth regulators, 444
Fossil Lepidoptera, 184
Frit fly, 341
Frontal ganglia, 135-36
Fruitfly, 473
Fruit moth, 307
Fucosterol, 206-8
Fundal, 290, 320
Fungal associations, 84
Fungi
and European corn borer, 229, 233
of fire ants, 23

G

Galecron, 320
Gall
aphid, 79, 84
and armored scales, 53
midge, 338, 428
Galleria, 448
mellonella, 104-7, 167-70, 432
Gardona, 290, 307-9, 320, 322
Gene flow, 401-2
Genes
for resistance, 348-49
Genetic
control, 175, 451-76
death, 472-73
improvement for trees, 76
loads, 462
manipulation, 25, 174-77
Genetics
of resistance, 344-47
Genital rotation, 420, 425
Genitalia
of Lepidoptera, 183, 192-93
Geocoris punctipes, 290
Geographical distribution
of fleas, 251-52
Geometridae
and growth regulators, 433
Geotaxis, 63, 266
Geotrupes, 364, 367
German cockroach, 207-8, 443-44
Germ plasm, 343-44
Gibberellins, 86, 276
Gintraps, 433
Glaciopsyllus, 245
Glial cells, 151
Glischrochilus quadrisignatus, 228
Glover scale, 49
Glucosides, 114-16, 349
Glucuronides, 214
Glutamic acid, 156-58, 161-62
Glycogen, 106
Glycoside, 80, 85
Glyodin, 319
Golgi complex, 140
Gomphocerus rufus, 371
Gonometa podocarpi, 110
Grain aphid, 339, 341
Granuloses virus, 98-100, 305
Grasshoppers
and aquatic weeds, 41-42
and arid environment, 265, 268, 276
brain of, 371
and growth regulators, 444
and muscles, 158
sensilla of, 384
Gravity
detectors, 387
receptors, 391
Greedy scale, 49
Greenbug, 339-41, 345, 347, 351
Green peach aphid, 437
Gregarization, 276
Growth
factor, 84
regulators, 417-60
Gryllus, 269
bimaculatus, 275
campestris, 359
Gulf Coast tick, 18, 463
Gymnolaelops shealsi, 17
Gypsy moth, 168, 177
Gyrinus, 99
natator, 111

H

Habrobarcon juglandis, 298, 304
Haemaphysalis, 263
Haematobia irritans, 427
Hair sensilla, 382-83
Halictidae
and growth regulators, 445
Halimococcus, 47
Halteres, 385
Halticoptera aenea, 305
Haploids
and armored scales, 58
Harmine, 155
Harpalus
aeneus, 304
rufipes, 304
Heart
muscles, 156-57
and neurosecretion, 136-41
Heartbeat rate, 141-42
Heliothis, 102
subflexa, 174-75, 177
virescens, 169, 174-75, 177, 292, 297, 429-30, 432
zea, 100-2, 109, 169, 207, 339
Hemerocampa leucostigma, 108
Hemiberlesia
lataniae, 49, 56, 439
rapax, 49
Hemiptera
and growth regulators, 422, 436, 440
Hemocytes, 104, 157
Hempa, 435
Heptachlor, 29, 289, 320

Herbicides
and aquatic weeds, 33, 40
Herpetogramma bipunctalis, 39
Hessian fly, 337, 339-40, 345-52
HEW, 121
Hippodamia, 305
convergans, 289, 292-93
5-signata, 289, 292
Holcocera pulverea, 303
Homoptera
and growth regulators, 436
Honeybees
brain of, 361, 365, 367, 369, 371-72
Brazilian, 399-416
dance languages of, 360
and growth regulators, 445
and insecticides, 297
learning in the, 372-74
sensilla of, 388, 391
Honeycombs, 392
Honeydew, 54
and insecticides, 288, 302
and mites, 313
Hoplopsyllus, 243, 245
Hormoligosis, 303
Hornfly, 427-28
Hornworm, 315
Horogenes punctorius, 227
Host range, 35
Host relationships
of Siphonaptera, 241
Host specificity
and aquatic weeds, 35
of armored scales, 48-50
Houseflies
and growth regulators, 427
sensilla of, 387
and sterility, 167
Howardia biclavus, 49, 66
Hyalomma dromedarii, 263, 274
Hyalophora cecropia, 213, 359, 366, 371, 373
Hybrid sterility, 471
Hybridization
and sterility, 174-75
Hydrazines
and muscles, 158-59
Hydrellia, 36, 42
Hydroprene, 23, 419, 432, 440, 445-46
Hydroxylamine, 158
Hydroxytryptamine, 139, 155-56, 367
Hygrotaxis, 266
Hylemyia brassicae, 428
Hylobrius radicis, 79
Hymenia recurvalis, 39
Hymenoptera
brain of, 363
and growth regulators, 444
and viruses, 98
Hyperapostica, 435
Hyperparasites
and insecticides, 288, 302
Hyphantria cunea, 99, 431
Hystrichopsylla, 246-47

I

Imaginal disks, 105
Imidan, 290
Importation
of fire ants, 3-5
Incubation chamber, 66
Indian meal moth, 100, 143, 168, 170, 173
Indirect effect
of insecticides, 298-306
Indolamines, 366
Ingluvial ganglia, 135-36, 139
Inherited sterility, 167-82
induced, 168-74
noninduced, 470
Inhibitors
of sterol metabolism, 207
Inquilines
of fire ants, 17
Insecticides
and aquatic weeds, 39
and natural enemies, 285-335
resistance to, 325, 469
and synergism, 422
Insects
and aquatic weeds, 34-46
Integrated control, 76, 285, 298-99, 318
Intercastes, 442
Ion transport, 390
Ips
avulsus, 82
confusus, 82
grandicollis, 82
Iridomyrmex humilis, 3, 5
Iridoviruses, 98, 104-5
Irradiation
of European corn borer, 225
Ischnaspis longirostris, 49
Isopods
in arid environment, 263-65, 270
Isopropyl-parathion, 288, 295
Isoptera, 98
brain of, 363
and growth regulators, 442-43
Isozymes, 407
Italian pear scale, 49
Itoplectis behrensii, 295, 302

J

Juglone, 81, 86
Juniper scale, 49
Junonia coenia, 106-7
Juvabione, 421, 440
Juvenile hormone
of European corn borer, 225
and fire ants, 23
and fleas, 244, 249
as growth regulators, 417-60
and insecticides, 304
and neurosecretion, 133-34
and parasites, 305-6

K

Kairomone, 244
Kepone, 20, 231
Kuwanaspis, 48

L

Labauchena daguerri, 24
Labeling
of pesticides, 122
Labella, 384
Labium, 189
Labrum, 188, 194
Laccophilus, 306
fasciatus, 293
Lac insect, 303
Lady beetles
and growth regulators, 435-36
Lambdina fiscellaria, 433
Larch casebearer, 432
Larvae
of Lepidoptera, 185-87
Lasioderma, 435
Laspeyresia pomonella, 167, 169, 431, 462
Latania scale, 49, 439
Leafhoppers
and growth regulators, 439
Leaf miners, 186
Learning
in the honeybee, 372-74
Lepidochora argentogrisea, 266, 275
Lepidoptera
brains of, 361, 364
classification of, 183-203
eggs of, 184-85
evolution of, 183-203
and growth regulators, 422, 429-33
larvae of, 185-87
pupae of, 187-88
sterility of, 167-82
Lepidosaphes
beckii, 49
ficus, 49-50
gloverii, 49
newsteadi, 53
ulmi, 49, 51, 56, 68

Leptinotarsa
decemlineata, 231, 433
texana, 434
Leptophos, 309, 320
Leptopsylla segnis, 244
Lesions
and viruses, 106
Lesser snow scale, 49
Lethality
genetic, 461
Leucophea maderae, 138, 140, 158
Life cycle
of fire ants, 8
of fleas, 243-47
Light
and growth regulators, 448
traps, 225-26, 276
Lime-sulfur, 320
Limonene, 88
Lindane, 289, 295, 301-2, 304, 320
Lindorus, 304
Lipase
of fire ants, 12
Lipoprotein, 216
Liris nigra, 360
Litodactylis leucogaster, 41
Lixophaga, 227
Locusta migratoria, 135, 139, 142, 155, 208, 213-14, 269, 271-72
Locusts
brain of, 367-69, 372
and flight, 242
and growth regulators, 444
and neurosecretion, 136, 141
pharmacology of, 155, 158, 161
sensilla of, 388, 391
Lone star ticks, 18
Loopers
and growth regulators, 433
Lovozal, 320
Lucilia cuprina, 469
Luminescence, 141
Lures
for fire ants, 25
Lycosa pseudoannulata, 294, 299
Lydella thompsoni, 227
Lygaeidae
and growth regulators, 441-42
Lygaeonematus abietinus, 78
Lygus hesperus, 292, 442
Lymantria dispar, 99, 104,
Lymantriidae
and growth regulators, 432
Lyriform organs, 386
Lysergic acid, 155
Lysiphlebus fabarum, 291, 295
Lysosomes, 390

M

Machiloides delanyi, 265
Macrocentrus ancylivorous, 306-7, 315
Macrocheles muscaedomesticae, 290, 283
Macrosiphum avenae, 293, 339
Macrotermes, 364
Malacosoma
californicum, 432
disstria, 100, 108
Malaria
and viruses, 107
Malathion, 229, 289-92, 294-95, 298, 320, 440
Males
of armored scales, 55-57
of Brazilian bee, 400, 402
Mallophaga, 446
Malpighian tubules, 54, 111, 139, 142, 216
Mamestra brassicae, 107
Mandibles, 187-89, 199-200
Manduca, 216
sexta, 138, 142, 167, 205, 268, 297, 315, 432
Maneb, 319
Mantids, 300, 359
Mantis, 363
Maskellia globosa, 53
Mass-rearing, 467-69
Mating behavior
of armored scales, 57-58
of European corn borer, 224
of fire ants, 10
of fleas, 245-47
Mating flights
of bees, 402
of fire ants, 3, 5, 9-11
Maxillae, 185, 188-89, 200
Mayetiola destructor, 337, 339
Meal moth, 305
Mealybug
and fire ants, 18
sex ratio of, 59, 65
Mec, 381-97
Mediterranean flour moths, 173
Mediterranean fruit fly, 462, 470
Megachile, 361
Megachilidae
and growth regulators, 445
Melanaspis
bromeliae, 53
glomerata, 48
Melanin, 68
Melanization
and growth regulators, 440
Melolontha melolontha, 103-4, 107
Melon fly, 462
Melonontha, 364
Memory, 359, 372-73, 375
Menazon, 290, 319
Mercaptothion, 291
Meridic diet, 233-34, 340
Merope tuber, 243
Mesurol, 320
Metabolic water, 261, 273
Metabolism
of ecdysones, 211-17
of growth regulators, 447
of sterols, 205-20
Metamorphosis, 212, 419, 437
Metaphycus helvolus, 303
Meteorus, 305
leviventris, 291
Methoprene, 23, 306, 426-28, 430, 433, 440, 445, 447
Methoxychlor, 307, 320
Mevinphos, 289, 292
Mexican bean beetle, 209-10, 435
Mexican fruit fly, 462
Microbracon brevicornis, 291
Microctonus aethiops, 306
Micropalpus retroflexa, 305
Microplitis croceipes, 306
Micropterix, 185, 192-93
Micropyle, 184-85
Microsomal oxidase, 303
Microsporidians
and aquatic weeds, 42
and European corn borer, 228
and fire ants, 23
and parasites, 305-6
and weed controlling insects, 38
Microtubules, 387-90
Microvilli, 390
Migrating colonies
of Brazilian bee, 410
Milbex, 320
Millipedes
and arid environment, 264-65, 271, 274
Minim workers
of fire ants, 7
Mining scale, 49
Mirex, 1, 18, 20-21, 26, 445
Miridae
and growth regulators, 442
Mites
and aquatic weeds, 41
and arid environment, 267, 274
and fire ants, 17
and insecticides, 286, 303, 313

sensilla of, 385
and viruses, 97-117
Mitochondria, 387, 390
Mnesarchaea, 195
Molting hormones, 212-13, 215-16, 218, 421
Monoamines, 367
Monocrotophos, 291, 296, 320
Monoterpenes, 81, 83, 86-88
Monotrysia, 196-97
Mormoniella vitripennis, 291
Morphogenesis
and growth regulators, 418-19, 424
Morphological adaptations
to arid environment, 262-64
Morphology
of Brazilian bee, 406
of Lepidoptera, 184-93
Mosquitoes
and growth regulators, 306, 424-28
Mounds
of fire ants, 10-13
Mouthparts, 188, 198
Multiple resistance, 313
Musca
autumnalis, 427
domestica, 167, 266, 293, 427, 469
Muscidae
and growth regulators, 427
Muscidivorax raptor, 446
Muscles, 151
contraction of, 151, 153
in fleas, 242
ultrastructure of, 153
Mushroom bodies, 360-61, 363-75
Mutagens
and viruses, 100
Mutations
and insecticides, 304
Myrcene, 88
Myriapods
and arid environment, 264-65, 269, 271, 275
Myrmecaphodius excavaticollis, 17
Myrmecocystus, 14
Myrmecosaurus, 17
Myxomatosis, 251
Myzus persicae, 437-38

N

Nabis
americoferus, 290, 293
roseipennis, 306
Nagilactone, 86
Naled, 290
Namangana pectinicornis, 42
Nanophyes, 42
Naphthoquinones, 81
Natural enemies
and insecticides, 285-335
Natural insecticides, 297
Nauphoeta, 139
cinerea, 138, 139
Nectar
and insecticides, 288, 296
Neivamyrmex opacithorax, 24
Nematodes
and fire ants, 24
and weed controlling insects, 38
Nemeritis canescens, 305
Neochetina
bruchi, 36, 38, 41
eichhorniae, 35-36, 38-39, 41, 43
Neodiprion
abietes, 83
pratti, 83, 207
sertifer, 79
Nephotettix, 338
cincticeps, 294, 299
impicticeps, 344, 346
Neophylax, 99
Neopseustis, 195
Neopsylla setosa, 245
Neurilemma, 389
Neuroanatomy, 365-66
Neuromuscular
junction, 134
morphology, 151-53
pharmacology, 151-66
transmission, 151
Neurons, 381, 383-84, 387
Neurosecretion, 133-49
Neurosecretomotor innervation, 136-41
Neurosecretory
cells, 142-44, 372-73
fibers, 134
granules, 134-40
neurons, 134
Neurotransmitters, 134, 138-39, 141, 366-68
Nicotine, 289, 291-92, 300-1, 315
Nilaparavata lugens, 338, 344, 346
Noctuids
and growth regulators, 429
and insecticides, 314
Nodamura virus, 110-11
Nomada, 361
Nomadacris septemfasciata, 276
Noninduced sterility, 470
Nonpreference, 79, 344, 346-47, 351
Noradrenaline, 138, 140-41, 155, 367
Norepinephrine, 156
Northern corn rootworm, 339
Nosema, 228-29, 233, 240
polyvora, 305
Nosopsyllus fasciatus, 244-45
Notonecta undulata, 293
Nuclear polyhedrosis, 98-99, 305
Nutrients
and aquatic weeds, 32
balance of, 83-84
Nutrition
and resistance, 77, 83-84, 349
Nutritional
deficiency, 84
requirements, 83, 340
Nymphula diminutalis, 42

O

Oakworm, 302
Ocelli, 185, 198, 200, 369
Ocnera hispida, 266, 268-69
Octopamine, 140-41
Odonaspis ruthae, 52, 56
Oedothorax insecticeps, 294
Oleander scale, 48-49, 56, 62
Oleoresin, 81, 86-91
Olethreutidae
and growth regulators, 431
Olive scale, 49, 53, 58-60
Omite, 290, 294, 308, 319, 321
Oncopeltus fasciatus, 138-39, 442
Ontogenesis
and growth regulators, 420
Onymacris
plana, 268
rugatipennis, 275
Oogenesis, 170
Operophtera brumata, 86
Ophyra leucostoma, 307, 312
Opius
longicaudatus, 295
oophilus, 291, 295, 301-2
persulcatus, 295
Optic lobes, 135, 361-62, 365, 367-68, 371-73
Orasema, 24
Orgyia leucostigma, 108
Oriental fruit fly, 301, 462
Oriental fruit moth, 314
Orius insidiosus, 227-28, 290
Ornithodoros savignyi, 263, 274
Orthogalumna terebrantis, 41
Orthoptera
brain of, 364
and growth regulators, 422

Oryctes rhinoceros, 99, 102, 111
Oryzaephylus, 435
Oscinella frit, 341
Osmia, 361
Osmoreceptors, 249
Osmoregulation, 272
Ostrinia nubilalis, 221, 431
Oulema melanopus, 339, 434
Ovaries
of fleas, 247
Ovipositor, 197, 387
Oxythioquinox, 308, 319-20
Oystershell scale, 49, 55, 67

P

Pacemaker mechanisms, 143
Pachydiplosis oryzae, 338
Paederus alfierii, 290
Panonychus
citri, 111
ulmi, 111, 294, 314
Panorpa, 242
Panstronglyus, 442
Paper factor, 421
Parabiosis, 373
Paralysis, 153
Paramyelois transitella, 169
Parapoynx stratiotata, 35-36, 41
Parasite-host relationships, 300-2
Parasites
of European corn borer, 226-27
of fire ants, 24
and growth regulators, 431-32, 438
and insecticides, 286, 298, 300, 302
resistance of, 306, 308
of weed controlling insects, 38
Parathion, 289-92, 294-95, 298, 301, 307-8, 319-20
Parathionmethyl, 289-91, 307-8, 311, 315
Paratrechina melanderia arenivaga, 24
Parlatoria
blanchardi, 49
cinerea, 50
oleae, 49
pergandii, 49-50
pittospori, 65
proteus, 49
theae, 49
Parlatoria ziziphi, 49
Parthenogenesis, 56, 58, 417, 439
Parvoviruses, 98, 106-7
Pathogens
and aquatic weeds, 34, 39, 42-43
and European corn borer, 228
and fire ants, 23
and insecticides, 304
and weed controlling insects, 38
Paulinia, 43
acuminata, 35-36, 38, 42
peregrina, 291
Pectines, 386-87
Pectinophora gossypiella, 109, 169, 173
Pediculus humanus humanus, 446
Pelopaeus
behavior of, 359
Pentatomidae
and growth regulators, 442
Peregrinus maidis, 440
Pericardial cells, 140
Periplaneta americana, 135-39, 141-43, 155, 207, 272, 363, 365, 367, 371
Perisympathetic organs, 142
Peritrophic membrane, 105, 109
Persistance
of growth regulators, 448
Pest
control, 97, 461-76
management, 285-86, 326, 417
Pesticides
application of, 122
disposal of, 122
distribution of, 122
labeling of, 122
licensing dealers of, 123
regulation and legislation, 119-31
residues, 120, 122
storage, 122
Phagostimulants
for fire ants, 12
Phanerotoma flavotestacea, 305
Pharmacology
neuromuscular, 151-66
Pharoscymnus numidicus, 289
Phase change, 276
Phellandrene, 88
Phenacaspis
nyssae, 50
pinifoliae, 49
Phenolics, 84-86
Phenological adaptations
to arid environment, 274-76
Phenylhydrazine, 158-59, 161
Pheromone traps, 225
Pheromones
of armored scales, 57-58, 64
of European corn borer, 224-25
and fire ants, 15-16, 24-25
Philanthus, 360
triangulum, 162
Philosamia, 107
Phoenicococcus, 47
Phorate, 230, 289-90, 292
Phormia terrae-novae, 137
Phosalone, 309, 320, 322
Phosmet, 290, 294, 307-9, 319, 322, 325
Phosphamidon, 231, 289-91, 304, 309, 319-20, 322
Photodiapsis, 69
Photoperiod, 171, 222-24
Phototaxis, 63, 266, 367
Photuris, 141
Phryganidea californica, 295
Phthorimaea operculella, 102, 231
Physiological adaptations
to arid environment, 267-74
Physiology
of tree resistance, 75-95
Phytoecdysones, 306
Phytoseiid mites, 308, 311, 311, 313-14, 316, 318
Phytosterol metabolism, 206
Pieris, 363
brassicae, 99, 107, 169, 175
rapae, 304
Pigmentation, 444
Pimelia grandis, 266, 268
Pine needle miner, 79
Pine needle scale, 49, 65
Pine sawfly, 207
Pine weevils, 79, 435
Pineapple scale, 49
Pinene, 81, 88, 90
Pink bollworm, 173
Pink stem borer, 339
Pinnaspis
aspidistrae, 49
strachani, 49
Piperonyl butoxide, 297, 422, 447
Pirimicarb, 303
Pissodes strobi, 79, 435
Pit scale, 53
Plant
resistance, 337-57
toxins, 297
Plictran, 290, 294, 319-21
Plodia interpunctella, 100, 168-69, 431
Plum curculio, 325
Plutella xylostella, 102
Poecilocerus hierglyphicus, 265-66, 271
Polistes, 306
Pollen
and European corn borer, 228, 231

and insecticides, 288
and viruses, 111
Pollution
and aquatic weeds, 39
control, 120
Polyhedra, 99, 102, 108-10
Polymorphism, 421
Polypedilum, 272
vanderplanki, 268, 270
Polyphenol polymers
of armored scales, 68-69
Polyploidy
and sterility, 175-76
Pompilus, 360
Population
dynamics, 463
genetic, 461
Porthetria dispar, 99, 168, 295
Postmentum, 185
Poxvirus, 98, 103-4
Pracocaspis diversa, 54
Praon abjectum, 291, 295
Predation
by fire ants, 18
Predator-prey relationships, 298-300
Predators
of European corn borer, 227
of fire ants, 24
and insecticides, 286, 298-300, 302-3
resistance of, 306, 308
of weed controlling insects, 38
Preference, 79, 344
Prenolepis, 14
Proctodone, 222
Prolegs, 186, 194, 199-200
Propoxur, 230
Proprioceptors, 381, 385
Prostigmine, 155
Protein synthesis, 372
Proteolytic enzymes, 100
Proteus scale, 49
Prothoracic glands, 215-16, 421
Protodiaspis agrifoliae, 66
Protozea
and parasites, 304
Pseudacteon, 24
Pseudaletia unipuncta, 100
Pseudaonidia
claviger, 66
duplex, 49
Pseudaulacaspis pentagona, 49, 54
Pseudococcids
and growth regulators, 439
Pseudococcus citri, 439
Pseudocossus, 191
Pseudophonus rufipes, 290
Pseudotracheae, 263-64
Psyllidae
and growth regulators, 440
Psythirus, 361
Pterostichus vulgaris, 290
Public health
and fire ants, 19
Pulex irritans, 252
Pungenin, 80, 84
Pupae
of Lepidoptera, 187-88
Puparia, 427
Pupillarial armored scales, 69
Purple scale, 49, 55, 62-
Pygidium, 54-55, 58, 67
Pyralidae
and growth regulators, 431
Pyrausta nubilalis, 339
Pyrethrins, 291-92, 440
Pyridoxal phosphate, 158
Pyrrhocoridae
and growth regulators, 441
Pyrrhocoris apterus, 420, 441

Q

Quadraspidiotus
forbesi, 49
juglansregiae, 49
ostreaeformis, 49
perniciosus, 49
pyri, 49
Queen excluders, 413
Queen-tending pheromone, 17
Queens
of bumble bees, 400, 402-3, 409-10
of fire ants, 5, 7, 10-11, 14-15, 24
Quinones, 86

R

Rabbit flea, 245-46, 249
Radiation
and sterility, 167-68, 170
Rat fleas, 246
Rearing
and control, 466
of European corn borer, 233
Receptor
membrane, 389
potential, 390
Recessive lethals, 168, 172
Rectal pads, 247
Red-banded leafroller, 325
Red scale, 50, 62, 302, 439
Red mites
and predators, 299
Reduviidae, 442
Repellants
and tree resistance, 77, 80-81
Repellency
of insecticides, 296
Repletes
of fire ants, 14
Reproduction
in armored scale, 56-59
of fire ants, 10
of fleas, 243-47
and growth regulators, 420
Reproductive isolation, 402-3
Reserpine, 367
Residue tolerances, 120, 123
Resilin, 241-42, 252, 385
Resin, 81, 86-87
ducts, 79-80
midge, 82
Resistance
development of, 312-16
to disease, 346
of European corn borer, 230
to feeding, 338
genetics of, 344-47
to growth regulators, 447
inheritance of, 337-38
to insecticides, 286, 297, 309, 311, 313-15
mechanism, 79
in natural enemies, 306-16
nature of, 349-50
to oviposition, 338
of plants, 337-57
of trees to insects, 75-95
and viruses, 100
Resistant natural enemies
in pest management, 323-25
Resistant varieties
in plants, 350-52
Reticuliteromes, 443
flavipes, 361
Retinodiplosis, 82
Rhabdoviruses, 98, 111-12
Rhinoceros beetle, 111
Rhodnius, 139-40
prolixus, 137, 155, 442
Rhopalosiphum
maidis, 342
padi, 339
Rhyacionia buoliana, 82
Rhythms, 261, 266
Rhyzopertha, 435
Rice borer, 177, 339
Rice leafhoppers, 299
Rice weevil, 347
Ring glands, 215
RNA, 107-8, 110-11, 222
viruses, 98
Romalea microptera, 154, 158-61
Rose scale, 49
Rotenoids, 297
Rotenone, 291
Rufous scale, 49

S

Sabatinca, 185, 194
Sacbrood virus, 110
Saliva, 53, 248
Salivary glands, 138-39, 247-48
Salt march caterpillars, 302
Salvinia, 42
Samea multiplicalis, 36, 42
Samia cynthia, 419
San Jose scale, 48-49, 53, 58-60, 62, 67-68
Sarcolemma, 153-54
Sarcophaga
 bullata, 143, 153
 falculata, 135
Satin moth, 305
Sawflies, 83-84, 363
Scales, 194, 383
 covering of, 66-69
Scapanes australis grossepunctatus, 102
Sceliphron, 367
Schistocerca gregaria, 135, 137, 139, 142, 156-59, 269, 271, 276, 364, 367, 372
Schizaphis gramium, 339, 439
Schizodactylus monstrosus, 138
Schradan, 289, 295, 320
Scolopidia, 390
Scolopidial sensilla, 381
Scolytus
 multistriatus, 81
 ventralis, 82
Scorpions
 and arid enviornment, 263, 267, 269-70
 sensilla of, 386-87
Screwworm, 167
 eradication, 463-64
Selectivity
 of insecticides, 318-21
Selenaspidus articulatus, 49
Semicarbazide, 158
Sensilla, 381-97
 chaetica, 382
 squamiformia, 383
 trichodea, 225, 382
Sensillar structure, 382
Sensory
 neurons, 386
 transduction, 382
Septicemia, 228
Sesamia inferens, 339
Sesamix, 447
Sesoxane, 422
Sesquiterpenes, 17, 87
Sex
 attraction, 224-25
 chromosomes, 243
 determination, 58-59, 172
 -killing systems, 469-70
 pheromones, 57
 ratio, 58-59, 172, 312
Sexual maturation, 276
Sigma virus, 112
Silk
 of fleas, 252
Silkworm, 168, 207-8, 216
 sensilla of, 392
 viruses, 99
Simpiesis viridula, 227
Simuliidae
 and growth regulators, 428
Siphonaptera, 241-59
 evolution of, 241
 host relationships of, 241
Sirex noctilio, 82
Sitophilus, 435
 granarius, 137
 zeamais, 338
Sitoplosis mosellana, 339
Sitosterol, 206-10
Sitotroga, 301
 cerealella, 168-69
Snails
 and aquatic weeds, 34
Soft scales, 439
Solenopsin, 14
Solenopsis, 444
 daquerri, 24
 geminata, 5, 16-17
 invicta, 1-30
 molesta, 24
 richteri, 1-30
 saevissima, 2, 23
 xyloni, 5, 16, 18
Solenotus intermedius, 305
Song patterns, 359
Songs
 of crickets, 370
Sorghum
 midge, 339
 shoot fly, 339
 stem borer, 339
Southern armyworm, 215
Spalangia endius, 441
Sperm
 of bees, 403
 of fleas, 242-46
 transfer, 173-74
Spermatheca, 137, 173-74, 193, 246, 248, 251
Spermatogenesis, 170, 173-74, 225
Spermatophores, 174
Sphingidae, 432
Sphingonotus carinatus, 265, 268
Sphinx, 365
Spider
 mites, 316-18, 321, 323
 sensilla of, 386, 391
 venom, 161
Spilopsyllus, 244-45
 cuniculi, 243, 245-48
Spinneret, 185, 194, 199
Spiracles, 185, 263-64, 271-72
Spiracular control, 271-72
Spodoptera, 430
 eridania, 215
 exigua, 102, 169
 frugiperda, 100, 207, 339-40
Spotted alfalfa aphid, 351-52
Spread
 of Brazilian bee, 400-6
 of fire ants, 3-5
Spruce budworm, 80, 84, 302, 430
Spruce gall aphid, 79
Stablefly, 427-28, 445-46
Starvation
 and neurosecretion, 134, 136
Stegobium, 435
Stem borer, 338
Stenoponia, 247
 tripectinata, 248
Sterile
 heterozygotes, 472
 hybrids, 472
 release, 462-69
 males, 25, 167
  dispersal of, 465
 triploids, 472
Sterility
 in Lepidoptera, 167-82
Sterilization and control, 466
 of European corn borer, 225
 and growth regulators, 420
Steroids
 metabolism of, 205-20
Sterols, 84-86
 and fleas, 244
 inhibitors of, 207
Stethorus
 pauperculus, 289, 293
 punctillum, 289, 294
Stigmaeid mites, 323
Stigmasterol, 206-10
Stilbenes, 86
Sting appatatus
 of fire ants, 15
Stinging
 of Brazilian bee, 410-12
 by fire ants, 15
Stingless bees, 412
Stinkbug
 and growth regulators, 442
Stivalius robinsoni, 253
Stomoxys calcitrans, 427
Stretch receptors, 381
Stylet penetration, 50-52
Stylet renewal
 by armored scales, 50
Subesophageal ganglia, 135, 139
Sublethal levels
 of insecticides, 302-4
Suckers
 of fleas, 245
Sugarcane borer, 168, 171, 177, 339-40

Sugarcane scale, 49, 60, 65
Sulfur, 303, 320
Sunlight
 and growth regulators, 424
Supercooling, 269-70
Supernumeraries, 437-38, 446
Supracide, 309
Suseptibility
 to insecticides, 298
 of plants, 344
Swarms
 of Brazilian bees, 400, 409, 413
 of fire ants, 11
Symbiotes, 205
Synapses, 151
Synaptic vesicles, 137-38, 151-53, 161, 366
Synephrine, 140-41
Synergism
 of growth regulators, 447
Synthetic diets, 340
Syrphids
 and insecticides, 304
Systemic
 insecticides, 299
 translocation, 441

T

Tabanus, 361
Tachinids, 302
Tactile setae, 186
TD-5032, 230
Tea scale, 49
Telenomus alsophilae, 305
Teleogryllus
 commodus, 471
 oceanicus, 471
Tenebrio molitor, 208, 217, 433
Tenebrionidae
 and growth regulators, 433-34
Tent caterpillar, 108
Tenthredo, 363
Tepa, 169, 435
Tepp, 289-91, 295, 309
Termites, 267
 brain of, 361, 364, 374
 and growth regulators, 442
 and viruses, 110
Terpenes, 83, 85-86, 88, 276
Terpinolene, 88
Tetradifon, 289, 291, 319-20
Tetranychid mites, 314, 316-18
Tetranychus
 mcdanieli, 314, 324
 pacificus, 314
 urticae, 294, 314, 321, 469
Thaumetopoea pityocampa, 109-10
Therioaphis maculata, 292
Thermobia domestica, 207, 273
Thermonectes, 306
Thermoperiod, 224
Thermoregulation, 270, 405
Thigmotaxis, 63-64
Thiometon, 289, 291-92, 295
Thonazin, 290, 304
Thorax
 of fleas, 241-43
 of Lepidoptera, 189
Thrassis, 253
Thuricide HPC, 305
Ticks
 and arid environment, 263, 273-74
 sensilla of, 390
Timarcha, 275
Tipula, 105
Tischeria complanellae, 197
Tissue culture, 97, 102
Tobacco budworm, 173, 306, 315
Tobacco hornworm, 167, 171, 205, 207-8, 211-13, 217, 300-1, 432
Tolerance
 to insecticides, 308
 and plant resistance, 79
Tormogen cell, 389-90
Tortricidae
 and growth regulators, 430
Tortrix viridana, 430
Toxaphene, 289-91, 303, 308
Toxicology
 of fire ant venum, 15
Tracheal system, 185, 192
Trail pheromone
 of fire ants, 13, 15-16
Translocations
 and sterility, 175
Transpiration
 and arid environment, 270-71
Transportation
 of pesticides, 122
Tree resistance, 75-95
Triatoma, 442
Tribolium, 207-10
 castaneum, 433
 confusum, 207, 433
Trichlorfon, 289-92, 298
Trichobothria, 382, 384-86
Trichogen cell, 389-90
Trichogramma
 evanescens, 291, 301, 315
 nubilalum, 227
Trichoplusia ni, 101-2, 110, 169, 297
Trichospilus pupivora, 291
Triparanol, 207
Triploid insects, 470
Trogoderma, 436
 granarium, 433-35
 inclusum, 434
Tryptamine, 155-56
Tsetse flies, 262, 392
Tuberworm, 231
Tunga
 monositus, 243-44, 246-48
 penetrans, 247
Tussock moth, 108, 432
Tympanal organs, 190-91, 226, 368
Typhlodromus
 caudiglans, 307, 311, 314, 317
 longipilus, 294
 occidentalis, 290, 307-10, 313-14, 316-18, 321, 323-26
 pyri, 307, 311, 314, 317
 reticulatus, 308
Tyramine, 141

U

Unaspis
 citri, 49, 439
 euonymi, 49
 yanonensis, 49, 56
Uric acid, 106, 262
Uropsylla tasmanica, 243

V

Vamidothion, 289, 292, 297
Variation
 in Brazilian bees, 403-5
Varietal resistance
 to European corn borer, 231-33
Veins, 194
Venation, 183, 191, 197-98, 200
Venoms, 276
 of fire ants, 14-15
 and muscles, 161-62
Vespidae
 and growth regulators, 445
Virions, 97, 99-100, 102-4, 106-11
Viruses
 and European corn borer, 229
 and parasites, 304
 pathogenic
  for insects and mites, 97-117
Visceral organs
 and neurosecretion, 133-49
Vitamins, 84, 244
Vogtia, 40
 malloi, 36, 38-40
Voria, 305
 ruralis, 291

W

Waggle dances, 408
Walnut scale, 49
Wasps
 behavior of, 359-60, 363
 and growth regulators, 445

sensilla of, 387
Water
balance, 276
loss, 262-63, 271
regulation, 271
uptake, 273-74
Waterhyacinth
insect control of, 35-36, 39-41, 43
Watermilfoil
insect control of, 41
Wax
and arid environment, 270, 273
of armored scales, 68
Waxes
and tree resistance, 85-86
Webworm
and aquatic weeds, 39
Weevils
and aquatic weeds, 39
Western corn borer, 339
Western corn rootworm, 230, 339-40
Wheat midge, 339
Wheat stem sawfly, 339-40, 342, 345, 350-52
Whiteflies, 303
and growth regulators, 440
White peach scale, 49, 58-59, 61, 63-64, 68
Wind dispersal, 64
Wing coupling of, 183, 197
Wing muscles
of fire ants, 7
Wings
of Lepidoptera, 191-92
sensilla of, 391
Winter moth, 86
Woodlice
and arid environment, 263-65, 268, 270, 274
Woolly apple aphid, 52, 315

X

Xanthogramma aegyptium, 304
Xenopsylla
astia, 244, 249
cheopis, 243-44, 247-49, 251-52, 273
conformis, 251
Xyleborus ferrugineus, 209-10
Xyleutes, 191

Y

Yellow scale, 49-50, 57
Yponomeuta malinella, 432
Yponomeutidae
and growth regulators, 432

Z

Zeadiatrea grandissella, 339
Zectran, 295, 302
Zelotypia, 192
Zetzellia mali, 321
Zeugloptera, 193-95
Zineb, 319-20
Zoogeography
of fire ants, 1, 3-5
and fleas, 252
Zootermopsis, 364, 442

CUMULATIVE INDEXES

CONTRIBUTING AUTHORS VOLUMES 11-20

A

Alexander, C. P., 14:1
Alexander, R. D., 12:495
Alloway, T. M., 17:43
Anderson, D. T., 11:23
Anderson, L. D., 13:213
Andres, L. A., 20:31
Ashhurst, D. E., 13:45
Atkins, E. L. Jr., 13:213

B

Bailey, L., 13:191
Baker, H. G., 13:385
Banks, W. A., 20:1
Barth, R. H., 18:445
Bateman, M. A., 17:493
Bay, E. C., 19:441
Beardsley, J. W. Jr., 20:47
Benjamini, E., 13:137
Bennett, F. D., 20:31
Bergerard, J., 17:57
Blum, M. S., 14:57
Bohart, G. E., 17:287
Brindley, T. A., 20:221
Brooks, M. A., 16:27
Brown, A. W. A., 20:285
Brundin, L., 12:149
Buckner, C. H., 11:449
Burgdorfer, W., 12:347
Burts, E. C., 19:231
Butcher, J. W., 16:249
Butts, W. L., 11:515

C

Caltagirone, L. E., 18:421
Chapman, H. C., 19:33
Chapman, P. J., 18:73
Chiang, H. C., 18:47
Cloudsley-Thompson, J. L., 20:261
Common, I. F. B., 20:183
Coope, G. R., 15:97
Cope, O. B., 16:325
Cranham, J. E., 11:491
Croft, B. A., 20:285
Cross, W. H., 18:17
Cummins, K. W., 18:183

D

Dadd, R. H., 18:381
Danilevsky, A. S., 15:201
David, W. A. L., 20:97
Davis, R. E., 15:405
DeBach, P., 11:183
Deck, E., 20:119
DeLong, D. M., 16:179
Detinova, T. S., 13:427
Downes, J. A., 14:271
Dupuis, C., 19:1
Durham, W. F., 17:123

E

Eastop, V. F., 14:197
Ebeling, W., 16:123
Edmunds, G. F. Jr., 17:21
Engelmann, F., 13:1
Evans, H. E., 11:123
Ewing, A. W., 12:471

F

Feingold, B. F., 13:137
Feir, D., 19:81
Foster, G. G., 20:461
Fritz, R. F., 17:75
Fuzeau-Braesch, S., 17:403

G

Gallun, R. L., 20:337
Geier, P. W., 11:471
Gelperin, A., 16:365
Gillies, M. T., 19:345
Gilula, N. B., 18:143
Glancey, B. M., 20:1
Glasgow, J. P., 12:421
Gonzalez, R. H., 20:47
Goryshin, N. I., 15:201
Gradwell, G. R., 15:1
Graham, K., 12:105
Gray, B., 17:313
Gressitt, J. L., 19:293
Guthrie, W. D., 20:221; 337

H

Hagen, K. S., 13:325
Hanover, J. W., 20:75
Harcourt, D. G., 14:175
Harris, C. R., 17:177
Harris, P., 16:159
Harshbarger, J. C., 13:159
Haworth, J., 17:75
Haydak, M. H., 15:143
Headley, J. C., 17:273
Heimpel, A. M., 12:287
Helle, W., 18:97
Hendrickson, J. A. Jr., 18:227
Hensley, S. D., 17:149
Hille Ris Lambers, D., 11:47
Hinton, H. E., 14:343
Hocking, B., 16:1
Hodek, I., 12:79
Hoogstraal, H., 11:261; 12:377
Howden, H. F., 14:39
Howe, R. W., 12:15
Howse, P. E., 20:359
Hoyt, S. C., 19:231
Huffaker, C. B., 14:125
Hughes, R. D., 14:197
Hurd, P. D. Jr., 13:385
Hynes, H. B. N., 15:25

I

Ilan, Joseph, 18:167
Ilan, Judith, 18:167

J

Jacobson, M., 11:403
Jamnback, H., 18:281
Johnson, C. G., 11:233
Jones, H. L., 17:453

K

Kaplanis, J. N., 16:53; 20:205
Kenchington, W., 16:73
Kerr, W. E., 19:253
Khan, M. A., 14:369
Knight, F. B., 12:207
Kring, J. B., 17:461
Kroeger, H., 11:1
Kulman, H. M., 16:289
Kurtii, T. J., 16:27

L

LaChance, L. E., 19:269
Leonard, D. E., 19:197
Le Pelley, R. H., 18:121
Lester, L. J., 18:445
Leston, D., 15:273

Lewis, D. J., 19:363
Lezzi, M., 11:1
Lindauer, M., 12:439
Ling, L., 19:177
Lloyd, J. E., 16:97
Lofgren, C. S., 15:321; 20:1
Long, W. H., 17:149
Lubischew, A. A., 14:19

M

Madelin, M. F., 11:423
Madsen, H. F., 15:295
Manning, A., 12:471
Mansingh, A., 14:387
Martin, E. C., 18:207
Matthews, R. W., 19:15
McCann, F. V., 15:173
McDonald, T. J., 20:151
McGregor, S. E., 18:207
McIver, S. B., 20:381
McMurtry, J. A., 14:125
Menzie, C. M., 17:199
Metcalf, R. L., 12:229
Michaeli, D., 13:137
Michener, C. D., 14:299; 20:399
Mickel, C. E., 18:1
Miller, T. A., 20:133
Morgan, C. V. G., 15:295
Morgan, F. D., 13:239
Moss, W. W., 18:227
Mulkern, G. B., 12:59

N

Newsom, L. D., 12:257
Noirot, C., 19:61
Nørgaard Holm, S., 11:155
Norris, D. M., 12:127
North, D. T., 20:167
Nüesch, H., 13:27

O

O'Brien, R. D., 11:369
Oldfield, G. N., 15:343
Oliver, D. R., 16:211
Osmun, J. V., 11:515
Ossiannilsson, F., 11:213
Ostmark, H. E., 19:161
Overmeer, W. P. J., 18:97

P

Pal, R., 19:269
Pathak, M. D., 13:257
Plumb, R. T., 17:425
Poinar, G. O. Jr., 17:103
Proverbs, M. D., 14:81

Q

Quennedey, A., 19:61

R

Rainey, R. C., 19:407
Remington, C. L., 13:415
Rettenmeyer, C. W., 15:43
Robbins, W. E., 16:53; 20:205
Ross, E. S., 15:157
Ross, H. H., 12:169
Roth, L. M., 15:75
Rothschild, M., 20:241
Roulton, W. J., 15:381
Rudall, K. M., 16:73
Ryckman, R. E., 11:309

S

Salkeld, E. H., 11:331
Satir, P., 18:143
Schaller, F., 16:407
Schlinger, E. I., 19:323
Schneider, F., 14:103
Schoonhoven, L. M., 13:115
Scudder, G. G. E., 16:379
Selander, R. K., 19:117
Shorey, H. H., 18:349
Showers, W. B., 20:221
Shulman, S., 12:323
Slifer, E. H., 15:121
Smallman, B. N., 14:387
Smith, E. H., 11:331
Snider, R., 16:249
Snider, R. J., 16:249
Sparks, A. N., 20:221
Spielman, A., 16:231
Spieth, H. T., 19:385
Staal, G. B., 20:417
Stairs, G. R., 17:355
Starks, K. J., 20:337
Stern, V. M., 18:259
Svoboda, J. A., 16:53; 20:205

T

Taylor, R. L., 13:159
Terriere, L. C., 13:75
Thompson, M. J., 16:53; 20:205
Throckmorton, L. H., 13:99
Torii, T., 13:295
Treherne, J. E., 12:43
Tremblay, E., 18:421
Turnbull, A. L., 18:305
Tuxen, S. L., 12:1
Tyshchenko, V. P., 15:201

U

Ulrich, W., 17:1
Usinger, R. L., 11:309

V

van den Bosch, R., 13:325
Vanderzant, E. S., 19:139
van de Vrie, M., 14:125
van Emden, H. F., 14:197; 19:455
Varley, G. C., 15:1
Varma, M. G. R., 12:347

W

Wagner, R. P., 19:117
Waters, T. F., 17:253
Watson, M. A., 17:425
Way, M. J., 14:197
Weaver, N., 11:79
Weiser, J., 15:245
Wharton, R. H., 15:381
Whitcomb, R. F., 15:405
Whitten, J. M., 17:373
Whitten, M. J., 20:461
Wilhm, J., 17:223
Williams, C. H., 17:123
Williams, G. F., 19:455
Willis, J. H., 19:97
Wilson, D. M., 11:103
Winteringham, F. P. W., 14:409
Wright, J. W., 17:75
Wygodzinsky, P., 11:309

Y

Yamamoto, I., 15:257
Yasumatsu, K., 13:295

Z

Zwölfer, H., 16:159

CHAPTER TITLES VOLUMES 11-20

ACARICIDES
see Insecticides and Toxicology
AGRICULTURAL ENTOMOLOGY
Management of Insect Pests | P. W. Geier | 11:471-90
Tea Pests and Their Control | J. E. Cranham | 11:491-514
Food Selection by Grasshoppers | G. B. Mulkern | 12:59-78
Consequences of Insecticide Use on Nontarget Organisms | L. D. Newsom | 12:257-86
Ecology of Common Insect Pests on Rice | M. D. Pathak | 13:257-94
Impact of Parasites, Predators, and Diseases on Rice Pests | K. Yasumatsu, T. Torii | 13:295-324
Impact of Pathogens, Parasites, and Predators on Aphids | K. S. Hagen, R. van den Bosch | 13:325-84
Entomology of the Cocoa Farm | D. Leston | 15:273-94
Pome Fruit Pests and Their Control | H. F. Madsen, C. V. G. Morgan | 15:295-320
Insect Pests of Sugar Cane | W. H. Long, S. D. Hensley | 17:149-76
Economics of Agricultural Pest Control | J. C. Headley | 17:273-86
A Critique of the Status of Plant Regulatory and Quarantine Activities in the United States | H. L. Jones | 17:453-60
The Ecology of Fruit Flies | M. A. Bateman | 17:493-518
Coffee Insects | R. H. Le Pelley | 18:121-42
Economic Thresholds | V. M. Stern | 18:259-80
Economic Insect Pests of Bananas | H. E. Ostmark | 19:161-76
Plant Pest Control on the International Front | L. Ling | 19:177-96
Recent Developments in Ecology and Control of the Gypsy Moth | D. E. Leonard | 19:197-229
Integrated Control of Fruit Pests | S. C. Hoyt, E. C. Burts | 19:231-52
Biometerology and Insect Flight: Some Aspects of Energy Exchange | R. C. Rainey | 19:407-39
Physiology of Tree Resistance to Insects | J. W. Hanover | 20:75-95
Recent Research Advances on the European Corn Borer in North America | T. A. Brindley, A. N. Sparks, W. B. Showers, W. D. Guthrie | 20:221-39
Plant Resistance to Insects Attacking Cereals | R. L. Gallum, K. J. Starks, W. D. Guthrie | 20:337-57
Insect Growth Regulators with Juvenile Hormone Activity | G. B. Staal | 20:417-60
APICULTURE AND POLLINATION
The Utilization and Management of Bumble Bees for Red Clover and Alfalfa Seed Production | S. Nørgaard Holm | 11:155-82
Recent Advances in Bee Communication and Orientation | M. Lindauer | 12:439-70
Honey Bee Pathology | L. Bailey | 13:191-212
Pesticide Usage in Relation to Beekeeping | L. D. Anderson, E. L. Atkins Jr. | 13:213-38
Honey Bee Nutrition | M. H. Hayak | 15:143-56
Management of Wild Bees for the Pollination of Crops | G. E. Bohart | 17:287-312
Changing Trends in Insect Pollination of Commerical Crops | E. C. Martin, S. E. McGregor | 18:207-26
Advances in Cytology and Genetics of Bees | W. E. Kerr | 19:253-68
The Brazilian Bee Problem | C. D. Michener | 20:399-416
APPLICATION OF INSECTICIDES
Pest Control | J. V. Osmun, W. L. Butts | 11:515-48
Pesticide Usage in Relation to Beekeeping | L. D. Anderson, E. L. Atkins Jr. | 13:213-38
Ultralow Volume Applications of Concentrated Insecticides in Medical and Veterinary Entomology | C. S. Logren | 15:321-42

Title	Author(s)	Vol:Pages
BEHAVIOR		
Insect Walking	D. M. Wilson	11:103-22
The Behavior Patterns of Solitary Wasps	H. E. Evans	11:123-54
Recent Advances in Bee Communication and Orientation	M. Lindauer	12:439-70
The Evolution and Genetics of Insect Behaviour	A. W. Ewing, A. Manning	12:471-94
Acoustical Communication in Arthropods	R. D. Alexander	12:495-526
Alarm Pheromones	M. S. Blum	14:57-80
The Swarming and Mating Flight of Diptera	J. A. Downes	14:271-98
Comparative Social Behavior of Bees	C. D. Michener	14:299-342
Insect Mimicry	C. W. Rettenmeyer	15:43-74
Blood-Sucking Behavior of Terrestrial Arthropods	B. Hocking	16:1-26
Bioluminescent Communication in Insects	J. E. Lloyd	16:97-122
Indirect Sperm Transfer by Soil Arthropods	F. Schaller	16:407-46
Learning and Memory in Insects	T. M. Alloway	17:43-56
Flight Behavior of Aphids	J. B. Kring	17:461-92
Behavioral Responses to Insect Pheromones	H. H. Shorey	18:349-80
Neuro-Hormonal Control of Sexual Behavior in Insects	R. H. Barth, L. J. Lester	18:445-72
Courtship Behavior in Drosophila	H. T. Spieth	19:385-405
Brain Structure and Behavior in Insects	P. E. Howse	20:359-79
BIOLOGICAL CONTROL		
A Critical Review of Bacillus thuringiensis var. thuringiensis Berliner and Other Crystalliferous Bacteria	A. M. Heimpel	12:287-322
Bionomics and Physiology of Aphidophagous Syrphidae	F. Schneider	14:103-24
The Ecology of Tetranychid Mites and Their Natural Control	C. B. Huffaker, M. van de Vrie, J. A. McMurtry	14:125-74
Host Specificity Determination of Insects for Biological Control of Weeds	H. Zwölfer, P. Harris	16:159-78
Biology of Braconidae	R. W. Matthews	19:15-32
Biological Control of Mosquito Larvae	H. C. Chapman	19:33-59
Biological Control of Aquatic Weeds	L. A. Andres, F. D. Bennett	20:31-46
BIOGEOGRAPHY		
see Systematics, Evolution, and Biogeography		
BIONOMICS		
see also Ecology		
Bionomics and Ecology of Predaceous Coccinellidae	I. Hodek	12:79-104
Bionomics of Siricidae	F. D. Morgan	13:239-56
Bionomics and Physiology of Aphidophagous Syrphidae	F. Schneider	14:103-24
The Bionomics of Leafhoppers	D. M. DeLong	16:179-210
Life History of the Chironomidae	D. R. Oliver	16:211-30
Bionomics of Autogenous Mosquitoes	A. Spielman	16:231-48
Biology, Control, and Eradication of the Boll Weevil	W. H. Cross	18:17-46
Bionomics of the Northern and Western Corn Rootworms	H. C. Chiang	18:47-72
Bionomics of the Apple-Feeding Tortricidae	P. J. Chapman	18:73-96
Recent Developments in Control of Blackflies	H. Jamnback	18:281-304
Biology of Braconidae	R. W. Matthews	19:15-32
The Biology and Ecology of Armored Scales	J. W. Beardsley Jr. R. H. Gonzalez	20:47-73
ECOLOGY		
see also Bionomics, Population Ecology, and Behavior		
A Functional System of Adaptive Dispersal by Flight	C. G. Johnson	11:233-60
Food Selection by Grasshoppers	G. B. Mulkern	12:59-78
Bionomics and Ecology of Predaceous Coccinellidae	I. Hodek	12:79-104
Chemosensory Bases of Host Plant Selection	L. M. Schoonhoven	13:115-36
Intrafloral Ecology	H. G. Baker, P. D. Hurd Jr.	13:385-414
The Ecology of Tetranychid Mites and Their		

Title	Authors	Volume:Pages
Natural Control	C. B. Huffaker, M. van de Vrie, J. A. McMurtry	14:125-74
The Development and Use of Life Tables in the Study of Natural Insect Populations	D. G. Harcourt	14:175-96
The Ecology of Myzus persicae	H. F. van Emden, V. F. Eastop, R. D. Hughes, M. J. Way	14:197-270
The Swarming and Mating Flight of Diptera	J. A. Downes	14:271-98
The Ecology of Stream Insects	H. B. N. Hynes	15:25-42
Bioecology of Edaphic Collembola and Acarina	J. W. Butcher, R. Snider, R. J. Snider	16:249-88
Interactions Between Pesticides and Wildlife	O. B. Cope	16:325-64
Graphic and Mathematical Analyses of Biotic Communities in Polluted Streams	J. Wilhm	17:223-52
The Drift of Stream Insects	T. F. Waters	17:253-72
The Ecology of Fruit Flies	M. A. Bateman	17:493-518
Trophic Relations of Aquatic Insects	K. W. Cummins	18:183-206
Ecology of the True Spiders (Araneomorphae)	A. L. Turnbull	18:305-48
Recent Developments in Ecology and Control of the Gypsy Moth	D. E. Leonard	19:197-229
Predator-Prey Relationships Among Aquatic Insects	E. C. Bay	19:441-53
Insect Stability and Diversity in AgroEcosystems	H. F. van Emden, G. F. Williams	19:455-75
The Biology and Ecology of Armored Scales	J. W. Beardsley Jr., R. H. Gonzalez	20:47-73
Adaptations of Arthropoda to Arid Environments	J. L. Cloudsley-Thompson	20:261-83
Responses of Arthropod Natural Enemies to Insecticides	B. A. Croft, A. W. A. Brown	20:285-335
EVOLUTION		
see Systematics, Evolution, and Biogeography		
FOREST ENTOMOLOGY		
The Role of Vertebrate Predators in the Biological Control of Forest Insects	C. H. Buckner	11:449-70
Fungal-Insect Mutualism in Trees and Timber	K. Graham	12:105-26
Systemic Insecticides in Trees	D. M. Norris	12:127-48
Evaluation of Forest Insect Infestations	F. B. Knight	12:207-28
Bionomics of Siricidae	F. D. Morgan	13:239-56
Effects of Insect Defoliation on Growth and Mortality of Trees	H. M. Kulman	16:289-324
Economic Tropical Forest Entomology	B. Gray	17:313-54
Pathogenic Microorganisms in the Regulation of Forest Insect Populations	G. R. Stairs	17:355-72
GENETICS		
Regulation of Gene Action in Insect Development	H. Kroeger, M. Lezzi	11:1-22
The Evolution and Genetics of Insect Behaviour	A. W. Ewing, A. Manning	12:471-94
The Population Genetics of Insect Introduction	C. L. Remington	13:415-26
Variability in Tetranychid Mites	W. Helle, W. P. J. Overmeer	18:97-120
Advances in Cytology and Genetics of Bees	W. E. Kerr	19:253-68
The Operational Feasibility of Genetic Methods for Control of Insects of Medical and Veterinary Importance	R. Pal, L. E. LaChance	19:269-91
Inherited Sterility in Lepidoptera	D. T. North	20:167-82
Genetical Methods of Pest Control	M. J. Whitten, G. G. Foster	20:461-76
HISTORICAL		
The Entomologist, J. C. Fabricius	S. L. Tuxen	12:1-14
Baron Osten Sacken and His Influence on American Dipterology	C. P. Alexander	14:1-18
Hermann Burmeister, 1807 to 1892	W. Ulrich	17:1-20
John Ray: Indefatigable Student of Nature	C. E. Mickel	18:1-16
Pierre André Latreille (1762-1833): The Foremost Entomologist of His Time	C. Dupuis	19:1-13
INSECTICIDES AND TOXICOLOGY		
The Use and Action of Ovicides	E. H. Smith, E. H. Salkeld	11:331-68
Mode of Action of Insecticides	R. D. O'Brien	11:369-402

Title	Author(s)	Volume:Pages
Systemic Insecticides in Trees	D. M. Norris	12:127-48
Mode of Action of Insecticide Synergists	R. L. Metcalf	12:229-56
Consequences of Insecticide Use on Nontarget Organisms	L. D. Newsom	12:257-86
Insecticide-Cytoplasmic Interactions in Insects and Vertebrates	L. C. Terriere	13:75-98
The Cholinergic System in Insect Development	B. N. Smallman, A. Mansingh	14:387-408
Mechanisms of Selective Insecticidal Action	F. P. W. Winteringham	14:409-42
Mode of Action of Pyrethroids, Nicotinoids, and Rotenoids	I. Yamamoto	15:257-72
Resistance of Ticks to Chemicals	R. H. Wharton, W. J. Roulston	15:381-404
Sorptive Dusts for Pest Control	W. Ebeling	16:123-58
Interactions Between Pesticides and Wildlife	O. B. Cope	16:325-64
Mutagenic, Teratogenic, and Carcinogenic Properties of Pesticides	W. F. Durham, C. H. Williams	17:123-48
Factors Influencing the Effectiveness of Soil Insecticides	C. R. Harris	17:177-98
Fate of Pesticides in the Environment	C. M. Menzie	17:199-222
Federal and State Pesticide Regulations and Legislation	E. Deck	20:119-31
MEDICAL AND VETERINARY ENTOMOLOGY		
Ticks in Relation to Human Diseases Caused by Viruses	H. Hoogstraal	11:261-308
Allergic Responses to Insects	S. Shulman	12:323-46
Trans-Stadial and Transovarial Development of Disease Agents in Arthropods	W. Burgdorfer, M. G. R. Varma	12:347-76
Ticks in Relation to Human Diseases Caused by Rickettsia Species	H. Hoogstraal	12:377-420
Recent Fundamental Work on Tsetse Flies	J. P. Glasgow	12:421-38
The Allergic Responses to Insect Bites	B. F. Feingold, E. Benjamini, D. Michaeli	13:137-58
Age Structure of Insect Populations in Medical Importance	T. S. Detinova	13:427-50
Systemic Pesticides for Use on Animals	M. A. Khan	14:369-86
Resistance of Ticks to Chemicals	R. H. Wharton, W. J. Roulston	15:381-404
Blood-Sucking Behavior of Terrestrial Arthropods	B. Hocking	16:1-26
Changing Concepts of Vector Control in Malaria Eradication	J. W. Wright, R. F. Fritz, J. Haworth	17:75-102
Mutagenic, Teratogenic, and Carcinogenic Properties of Pesticides	W. F. Durham, C. H. Williams	17:123-48
The Operational Feasibility of Genetic Methods for Control of Insects of Medical and Veterinary Importance	R. Pal, L. E. LaChance	19:269-91
Methods for Assessing the Density and Survival of Blood-Sucking Diptera	M. T. Gillies	19:345-62
The Biology of Phlebotomidae in Relation to Leishmaniasis	D. J. Lewis	19:363-84
Biology and Control of Imported Fire Ants	C. S. Lofgren, W. A. Banks, B. M. Glancey	20:1-30
Recent Advances in Our Knowledge of the Order Siphonaptera	M. Rothschild	20:241-59
MORPHOLOGY		
Regulation of Gene Action in Insect Development	H. Kroeger, M. Lezzi	11:1-22
The Comparative Embryology of the Diptera	D. T. Anderson	11:23-46
Polymorphism in Aphididae	D. Hille Ris Lambers	11:47-78
Temperature Effects on Embryonic Development in Insects	R. W. Howe	12:15-42
The Role of the Nervous System in Insect Morphogenesis and Regeneration	H. Nüesch	13:27-44
The Connective Tissues of Insects	D. E. Ashhurst	13:45-74
Respiratory Systems of Insect Egg Shells	H. E. Hinton	14:343-68
The Structure of Arthropod Chemoreceptors	E. H. Slifer	15:121-42
Comparative Morphology of Insect Genitalia	G. G. E. Scudder	16:379-406
Comparative Anatomy of the Tracheal System	J. M. Whitten	17:373-402

Title	Author(s)	Vol:Pages
Pigments and Color Changes	S. Fuzeau-Braesch	17:403-24
The Fine Structure of Membranes and Intercellular Communication in Insects	P. Satir, N. B. Gilula	18:143-66
Fate of Polar Bodies in Insects	E. Tremblay, L. E. Caltagirone	18:421-44
Fine Structure of Insect Epidermal Glands	C. Noirot, A. Quennedey	19:61-80
Morphogenetic Action of Insect Hormones	J. H. Willis	19:97-115
Structure of Cuticular Mechanoreceptors of Arthropods	S. B. McIver	20:381-97
NUTRITION		
see Physiology		
PATHOLOGY		
Fungal Parasites of Insects	M. F. Madelin	11:423-48
A Critical Review of Bacillus thuringiensis var. thuringiensis Berliner and Other Crystalliferous Bacteria	A. M. Heimpel	12:287-322
Neoplasms of Insects	J. C. Harshbarger, R. L. Taylor	13:159-90
Honey Bee Pathology	L. Bailey	13:191-212
Recent Advances in Insect Pathology	J. Weiser	15:245-56
Nematodes as Facultative Parasities of Insects	G. O. Poinar Jr.	17:103-22
Pathogenic Microorganisms in the Regulation of Forest Insect Populations	G. R. Stairs	17:355-72
The Status of Viruses Pathogenic for Insects and Mites	W. A. L. David	20:97-117
PHYSIOLOGY		
Polymorphism in Aphididae	D. Hille Ris Lambers	11:47-78
Physiology of Caste Determination	N. Weaver	11:79-102
Insect Walking	D. M. Wilson	11:103-22
Chemical Insect Attractants and Repellants	M. Jacobson	11:403-22
Temperature Effects on Embryonic Development in Insects	R. W. Howe	12:15-42
Gut Absorption	J. E. Treherne	12:43-58
Endocrine Control of Reproduction in Insects	F. Engelmann	13:1-26
The Role of the Nervous System in Insect Morphogenesis and Regeneration	H. Nüesch	13:27-44
Chemosensory Bases of Host Plant Selection	L. M. Schoonhoven	13:115-36
Alarm Pheromones	M. S. Blum	14:57-80
Induced Sterilization and Control of Insects	M. D. Proverbs	14:81-102
Respiratory Systems of Insect Egg Shells	H. E. Hinton	14:343-68
The Cholinergic System in Insect Development	B. N. Smallman, A. Mansingh	14:387-408
Honey Bee Nutrition	M. H. Haydak	15:143-56
Physiology of Insect Hearts	F. V. McCann	15:173-200
Biological Rhythms in Terrestrial Arthropods	A. S. Danilevsky, N. I. Goryshin, V. P. Tyshchenko	15:201-44
Insect Cell and Tissue Culture	M. A. Brooks, T. J. Kurtii	16:27-52
Steroid Metabolism in Insects	W. E. Robbins, J. N. Kaplanis, J. A. Svoboda, M. J. Thompson	16:53-72
Arthropod Silks: The Problem of Fibrous Proteins in Animal Tissues	K. M. Rudall, W. Kenchington	16:73-96
Regulation of Feeding	A. Gelperin	16:365-78
Environmental and Physiological Control of Sex Determination and Differentiation	J. Bergerard	17:57-74
Pigments and Color Changes	S. Fuzeau-Braesch	17:403-24
Protein Synthesis and Insect Morphogenesis	J. Ilan, J. Ilan	18:167-82
Insect Nutrition: Current Developments and Metabolic Implications	R. H. Dadd	18:381-420
Oncopeltus fasciatus: A Research Animal	D. Feir	19:81-96
Morphogenetic Action of Insect Hormones	J. H. Willis	19:97-115
Isozymes in Insects and Their Significance	R. P. Wagner, R. K. Selander	19:117-38
Development, Significance, and Application of Artificial Diets for Insects	E. S. Vanderzant	19:139-60
Biometerology and Insect Flight: Some Aspects of Energy Exchange	R. C. Rainey	19:407-39
Neurosecretion and the Control of Visceral Organs in Insects	T. A. Miller	20:133-49
Neuromuscular Pharmacology of Insects	T. J. McDonald	20:151-66
Recent Developments in Insect Steroid Metabolism	J. A. Svoboda, J. N. Kaplanis,	

	W. E. Robbins, M. J. Thompson	20:205-20
POLLINATION		
see Apiculture and Pollination		
POPULATION ECOLOGY		
The Competitive Displacement and Coexistence Principles	P. DeBach	11:183-212
Insects and the Problem of Austral Disjunctive Distribution	L. Brundin	12:149-68
The Population Genetics of Insect Introduction	C. L. Remington	13:415-26
The Development and Use of Life Tables in the Study of Natural Insect Populations	D. G. Harcourt	14:175-96
Recent Advances in Insect Population Dynamics	G. C. Varley, G. R. Gradwell	15:1-24
Graphic and Mathematical Analyses of Biotic Communities in Polluted Streams	J. Wilhm	17:223-52
SOCIAL INSECTS		
Physiology of Caste Determination	N. Weaver	11:79-102
The Utilization and Management of Bumble Bees for Red Clover and Alfalfa Seed Production	S. Nørgaard Holm	11:155-82
Comparative Social Behavior of Bees	C. D. Michener	14:299-342
SYSTEMATICS, EVOLUTION, AND BIOGEOGRAPHY		
The Biosystematics of Triatominae	R. L. Usinger, P. Wygodzinsky, R. E. Ryckman	11:309-30
The Evolution and Past Dispersal of the Trichoptera	H. H. Ross	12:169-206
Biochemistry and Taxonomy	L. H. Throckmorton	13:99-114
Philosophical Aspects of Taxonomy	A. A. Lubischew	14:19-38
Effects of the Pleistocene on North American Insects	H. F. Howden	14:39-56
Evolution and Taxonomic Significance of Reproduction in Blattaria	L. M. Roth	15:75-96
Interpretations of Quaternary Insect Fossils	G. R. Coope	15:97-120
Biosystematics of the Embioptera	E. S. Ross	15:157-72
Biogeography and Evolution of Ephemeroptera	G. F. Edmunds Jr.	17:21-42
Numerical Taxonomy	W. W. Moss, J. A. Hendrickson Jr.	18:227-58
Insect Biogeography	J. L. Gressitt	19:293-321
Continental Drift, Nothofagus, and Some Ecologically Associated Insects	E. I. Schlinger	19:323-43
Evolution and Classification of the Lepidoptera	I. F. B. Common	20:183-203
VECTORS OF PLANT PATHOGENS		
Insects in the Epidemiology of Plant Viruses	F. Ossiannilsson	11:213-32
Mite Transmission of Plant Viruses	G. N. Oldfield	15:343-80
Mycoplasma and Phytarboviruses as Plant Pathogens Persistently Transmitted by Insects	R. F. Whitcomb, R. E. Davis	15:405-64
Transmission of Plant-Pathogenic Viruses by Aphids	M. A. Watson, R. T. Plumb	17:425-52